Learning Analytics Methods and Tutorials

Mohammed Saqr • Sonsoles López-Pernas
Editors

Learning Analytics Methods and Tutorials

A Practical Guide Using R

Springer

Editors
Mohammed Saqr
School of Computing
University of Eastern Finland
Joensuu, Finland

Sonsoles López-Pernas
School of Computing
University of Eastern Finland
Joensuu, Finland

https://github.com/sn-code-inside/labook-code

ISBN 978-3-031-54463-7 ISBN 978-3-031-54464-4 (eBook)
https://doi.org/10.1007/978-3-031-54464-4

This work was supported by University of Eastern Finland.

This Springer imprint is published by the registered company Springer Nature Switzerland AG
The registered company address is: Gewerbestrasse 11, 6330 Cham, Switzerland

To our families, friends, and loved ones.

A special dedication from Mohammed to his daughters, the exquisite roses, Carmen and Layla.

Foreword

Learning analytics and its sister community of educational data mining have, over the years, matured into a field where an increasing community of scholars and practitioners work together to understand the best methods for exploring the high-volume data available about learners and their learning. This has led to an increasing number of handbooks and online courses. However, none of these prior resources had focused on providing step-by-step tutorials for how to carry out the various types of analyses conducted in learning analytics. This book, by Saqr and López-Pernas, and their colleagues, fills this important gap, offering clear and concise step-by-step tutorials combined with high level explanations of why each step is necessary. As such, it is an important contribution to the field and represents a manuscript that I am sure many learners will find extremely valuable—both newcomers and relatively experienced practitioners looking to learn a new method.

Professor of education and computer science — Ryan Baker
University of Pennsylvania

Foreword

My introduction to learning analytics was unexpected. A brief conversation with George Siemens in early June 2010 sparked a transformative journey for me, with lasting impact on many. George proposed co-organizing a conference on "learning analytics." It was the first time I'd heard of the phrase. The phrase resonated deeply, connecting with my previous collaborations with Jelena Jovanovic, Chris Brooks, Marek Hatala, Griff Richards, Gord McCalla, Colin Knight, and other colleagues in Canada's LORNET research network (mid 2000s).

Excited to collaborate, I was initially a bit skeptical about George's ambition to organize the conference by September 2010. However, I learned to appreciate his eagerness to implement ideas swiftly. We eventually reconnected and agreed on a late February 2011 date, giving us extra six months. Despite remaining nervous, I underestimated the potential impact. The experience proved me wrong in the most positive way, and I'm forever grateful to George, a dear friend, for his unwavering vision and belief.

The conference we organized, the 1st International Conference on Learning Analytics and Knowledge (LAK), was held in Banff, Canada, in late February and early March 2011. The frigid temperatures (–35 °C) were matched by the conference's energy. It attracted a large audience and fostered an open, vibrant, productive, and transformative community of researchers and practitioners. This conference is often considered the birth of learning analytics, and the definition we used in the chairs' message in the conference proceedings remains widely used in the field.

Learning analytics is now a well-established field. Since our early conversations, we've emphasized the importance of methods. My dear friend and esteemed colleague, Shane Dawson, a co-founder of the Society for Learning Analytics Research and field pioneer, first referred to learning analytics as a "bricolage field," one that borrows methods and theories from data science, artificial intelligence, network science, psychology, psychometrics, sociology, and natural language processing.

As founding editors-in-chief of the *Journal of Learning Analytics*, Shane and I recognized the need to introduce and promote diverse learning analytics methods. This led us to collaborate with Mykola Pechenizkiy, then-president of the International Educational Data Mining Society, to edit a special section on learning analytics tutorials featuring five papers on different methods. These papers, based on tutorials and workshops from the first two Learning Analytics Summer Institutes, became some of the journal's most impactful publications. It demonstrated the need for learning analytics methods tutorials. However, 8 years have passed, a significant timeframe considering the rapid advancements in artificial intelligence and the methods they've driven.

Therefore, I'm delighted to see *Learning Analytics Methods and Tutorials* edited by Mohammed Saqr and Sonsoles López-Pernas. This timely book addresses a critical need in learning analytics.

The book offers a comprehensive guide to data analysis methods for researchers and practitioners of all experience levels. It starts with the basics, equipping beginners with R programming and data analysis skills through chapters on data cleaning and exploration. These skills are fundamental for understanding student data and preparing it for further analysis. Even for experts, the book offers advanced methods while emphasizing the broader applicability of these techniques beyond education.

The core of the book explores various analytical approaches. Machine learning methods receive well-deserved attention, including introductions to commonly used methods—specifically, predictive modeling and clustering. Predictive modeling is frequently used in learning analytics to identify at-risk students or classify online discussions by analyzing past data patterns. This allows for early intervention to support students at different progress levels. Cluster analysis, another machine learning technique, groups students based on similar characteristics, behaviors, or learning outcomes. It's commonly used to analyze learning strategies in different learning environments and can provide educators with valuable insights to tailor teaching support to diverse student needs.

We've long recognized the dynamic nature of learning, with many learning processes unfolding over time. This highlights the need for temporal analytic methods in learning analytics. I'm pleased to see the book provides a rich guide to various temporal methods that analyze the order and timing of events in data about learning and learners. Techniques like sequence analysis and process mining leverage student activity traces to understand how learning unfolds over time. This is crucial for studying longitudinal processes like student engagement throughout a program or explaining connections between cognitive and metacognitive processes in solo and group learning activities.

Network analysis has been a prominent methodological approach since the early days of learning analytics. The book offers excellent coverage of network analytic approaches that explore the relational aspects of data about learners and learning environments. By examining interactions between learners, teachers, and topics, researchers and practitioners can understand collaboration patterns and identify student communities. These methods can be further enriched by combining them with temporal analysis to explore how these relationships evolve over time. The book also covers recent advancements in quantitative ethnography, introducing epistemic network and ordered network analysis, techniques I've extensively used with collaborators in recent years.

I have recently been advocating for the need to establish stronger collaborative ties between learning analytics and psychometrics. Psychometrics is a well-established discipline that can offer many relevant methods to learning analytics. Specifically, psychometrics can help us address issues that have received insufficient attention in learning analytics—reliability and validity. Without addressing these issues, learning analytics can be jeopardized. Therefore, I am delighted that the final section of the book delves into psychometrics, a field that investigates relationships between psychological constructs (like metacognition) and observable data (like test scores). The book explores techniques like factor analysis and structural equation modeling, which help researchers test hypotheses and theories about these relationships. By mastering these methods, learning analytics researchers and practitioners can gain valuable insights from student data to improve learning experiences and outcomes.

The book editors and chapter authors must be commended for their valuable contributions to learning analytics. What is outstanding about this book is that it provides source code illustrating all the methods introduced. Readers can directly use the code to build hands-on skills on the many high-importance methods for learning analytics that are so thoughtfully introduced in this book. The book can be directly used in any graduate or postgraduate course on learning analytics. I wish this book had been available 3 years ago when we developed the graduate certificate in learning analytics at Monash University. Many of its chapters would have been directly used as part of the graduate certificate program. And, I can easily see that many similar programs around the world will greatly benefit from this outstanding book. The book is equally suitable for practitioners and researchers in education institutions, schools, or industry who either want to develop basic data analytic skills or advance their skills with methods they haven't used before.

The overall learning analytics community is significantly richer because of this book. For that, we all owe immense thanks to Mohammed Saqr and Sonsoles López-Pernas, the book editors, for their incredible leadership and vision!

Distinguished Professor of Learning Analytics, Dragan Gašević
Director of Research in the Department of Human
Centred Computing,
Director of the Centre for Learning Analytics,
Monash University
Clayton, VIC, Australia

Preface

Learning analytics is a fast-paced discipline at the cross-section of cutting-edge methodological advances and theoretical innovations. Most of the methods have been born long before the field of learning analytics and continue to be developed and advanced across diverse fields. Therefore, researchers and students have to navigate through fragmented literature from several fields and learn to apply such methods. Notwithstanding the difficulties of identifying which literature or guides one can use, several methodological questions remain unanswered in these books. This is of course because such literature was prepared for other disciplines and offers case studies that are oftentimes not relevant.

Being learning analytics researchers ourselves, we have been through these difficulties. Learning a new methodology took several trials and searching for answers that were rather hard to get. Collaboration with other researchers from learning analytics and other fields—e.g., statistics and methodology—has helped us navigate these difficulties. In many situations, we would contact statisticians or package developers to help with solving an issue or collaborate on a paper. With the publication of our papers, we got several emails from researchers asking about technical details. It has become clear to us that there is a need for a well-documented resource for learning analytics methods. In fact, this is how we—the editors of this book—met. Sonsoles was doing mobility in Sweden, where Mohammed was working, and she wanted to do a learning analytics study for a dataset she collected from a programming course. Lacking any resource that could help her kick-start her exploration of the emerging field, she sought collaboration. Ever since, we had an interesting journey of working together on so many projects.

The lack of resources and methodological guidance was a problem then and continues to be a problem today. We thought that the arduous journey in learning analytics should not be endured by everyone, and we decided to make that resource with the help of the community as well as our collaborators. More than a year of intensive work is within your hands now. We hope that we have contributed to LA by providing an informative resource that students, researchers, and practitioners can use.

In this book, we tried to include all the basics of R as a programming language as well as the basics of data cleaning, statistics, and data manipulation. In doing so, we wanted the newcomers to find an easy entry to the field. We also tried to be as comprehensive as we could and included almost all major methodologies. For every method, we started with the basics, explaining the main concepts, the essential techniques, and basic functions. In subsequent chapters, we went deeper into advanced methods that are at the forefront of novel methodological innovations. We also used real-life learning analytics data and made it readily available to researchers. To do so, we collaborated with world-renowned researchers, package developers, and methodological experts from other fields to offer an unprecedented resource on novel topics that are hard to find any resource for.

We hope the readers find this book useful as a guide through learning analytics methods, highlighting the ways in which data-driven insights can benefit educators, learners, and researchers.

Joensuu, Finland

Mohammed Saqr
Sonsoles López-Pernas

Competing Interests

The authors have no conflicts of interest to declare that are relevant to the content of this book.

Acknowledgments

We are thankful for the great work everyone has done to make this book happen: authors have spent valuable time compiling these chapters and bringing their expertise to the wider audience of learning analytics. Authors from other fields did a great job and took the effort to navigate concepts they had little contact with to present their methods to a completely different audience than what they are used to.

We are also grateful to the reviewers who offered valuable feedback to refine and improve the chapters.

A special thanks goes to the associate editors who worked with us all over the last year to process this large number of chapters and handle the peer review process.

We would like to acknowledge the generous funding by the Academy of Finland (Research Council of Finland) for the project TOPEILA, Decision Number 350560, which was received by Mohammed Saqr.

Lastly, we would like to thank the generous funding from the University of Eastern Finland and, in particular, Markku Tukianen and Kari Lehtinen, who made this book open access for everyone.

Contents

List of Contributors

Editors

Mohammed Saqr University of Eastern Finland, Joensuu, Finland

Sonsoles López-Pernas University of Eastern Finland, Joensuu, Finland

Associate Editors

Miguel Ángel Conde-González University of León, León, Spain

Rogers Kaliisa University of Oslo, Oslo, Norway

Kamila Misiejuk University of Bergen, Bergen, Norway

Authors

Emorie Beck UC Davis, Davis, CA, USA

Javier Conde Universidad Politécnica de Madrid, Madrid, Spain

Miguel Ángel Conde-González University of León, León, Spain

Carlos Cuenca-Enrique Universidad Politécnica de Madrid, Madrid, Spain

Laura Del-Río-Carazo Universidad Politécnica de Madrid, Madrid, Spain

Marion Durand University of Eastern Finland, Joensuu, Finland

Merja Heinäniemi University of Eastern Finland, Kuopio, Finland

Jouni Helske University of Jyväskylä, Jyväskylän yliopisto, Finland

Satu Helske University of Turku, Turku, Finland

Ángel Hernández-García Universidad Politécnica de Madrid, Madrid, Spain

Rogers Kaliisa University of Oslo, Oslo, Norway

Jelena Jovanovic University of Belgrade, Belgrade, Serbia

Joran Jongerling Tilburg University, Tilburg, Netherlands

Juho Kopra University of Eastern Finland, Joensuu, Finland

Sonsoles López-Pernas University of Eastern Finland, Joensuu, Finland

Kamila Misiejuk University of Bergen, Bergen, Norway

Keefe Murphy Maynooth University, Maynooth, Ireland

Gilbert Ritschard University of Geneva, Geneva, Switzerland

A. R. Ruis University of Wisconsin, Madison, WI, USA

Mohammed Saqr University of Eastern Finland, Joensuu, Finland

Marieke J. Schreuder KU Leuven, Leuven, Belgium

Luca Scrucca Università degli Studi di Perugia, Perugia, Italy

Matthias Studer University of Geneva, Geneva, Switzerland

David Shaffer University of Wisconsin, Madison, WI, USA

Zachari Swiecki Monash University, Clayton, Australia

Yuanru Tan University of Wisconsin, Madison, WI, USA

Santtu Tikka University of Jyväskylä, Jyväskylä, Finland

Adrienne Traxler University of Copenhagen, København, Denmark

Leonie V. D. E. Vogelsmeier Tilburg University, Tilburg, Netherlands

Reviewers

Gökhan Akçapınar Hacettepe University, Ankara, Turkey

Daniel Amo-Filva Universitat Ramon Llull, Barcelona, Spain

Enrique Barra Universidad Politécnica de Madrid, Madrid, Spain

Umar Bin Qushem University of Turku, Turku, Finland

Manuel Castejón-Limas Universidad de León, León, Spain

Javier Conde Universidad Politécnica de Madrid, Madrid, Spain

Tudor Cristea TU Eindhoven, Eindhoven, Netherlands

Laura Del-Río-Carazo Universidad Politécnica de Madrid, Madrid, Spain

Ramy Elmoazen University of Eastern Finland, Joensuu, Finland

Dilrukshi Gamage Tokyo Tech, Tokyo, Japan

Brian P. Godor Erasmus University Rotterdam, Rotterdam, Netherlands

Francisco José García Peñalvo Universidad de Salamanca, Salamanca, Spain

Sami Heikkinen LAB University of Applied Sciences, Lahti, Finland

Rogers Kaliisa University of Oslo, Oslo, Norway

Qinyi Liu University of International Business and Economics, Beijing, China

Sonsoles López-Pernas University of Eastern Finland, Joensuu, Finland

Kamila Misiejuk University of Bergen, Bergen, Norway

Andrés Munoz-Arcentales Universidad Politécnica de Madrid, Madrid, Spain

Nicolae Nistor Ludwig-Maximilians-Universität München, Munich, Germany

Merlijn Olthof Radboud University, Nijmegen, Netherlands

Dijana Oreški Sveučilište u Zagrebu, Zagreb, Croatia

Sambit Praharaj Ruhr University Bochum, Bochum, Germany

Miroslava Raspopović Milić Belgrade Metropolitan University, Belgrade, Serbia

Mohammed Saqr University of Eastern Finland, Joensuu, Finland

Yuanru Tan University of Wisconsin-Madison, Madison, WI, USA

Santtu Tikka University of Jyväskylä, Jyväskylä, Finland

Tiina Törmänen University of Oulu, Oulu, Finland

Adrienne Traxler University of Copenhagen, Copenhagen, Denmark

Daphne van de Bongardt Erasmus University Rotterdam, Rotterdam, Netherlands

Thijs Vroegh Netherlands eScience Center, Amsterdam, Netherlands

Jin Zhou Central China Normal University, Wuhan, Hubei, China

List of Abbreviations

AGNES	Agglomerative Nesting
AHC	Agglomerative Hierarchical Clustering
AIC	Akaike Information Criterion
ANOVA	Analysis of Variance
ASW	Average Silhouette Width
AUC	Area Under the Curve
BIC	Bayesian Information Criterion
DFG	Directly Follows Graph
DIANA	Divisive Analysis
DNA	Deoxyribonucleic Acid
ECTS	European Credit Transfer System
EEG	Electroencephalogram
EFA	Exploratory Factor Analysis
EM	Expectation-Maximization
ENA	Epistemic Network Analysis
FMM	Finite Mixture Model
FN	False Negative
FP	False Positive
GBTM	Growth-Based Trajectory Model
GMM	Gaussian Mixture Modeling
GPA	Grade Point Average
HMM	Hidden Markov Model
ICL	Integrated Complete Likelihood
IDE	Integrated Development Environment
IP	Internet Protocol
IQR	Interquartile Range
JPEG	Joint Photographic Experts Group
KMO	Kaiser-Meyer-Olkin
KMS	Knowledge Management System
LA	Learning Analytics
LATUX	Learning Awareness Tools–User eXperience

LCA	Latent Class Analysis
LCP	Longest Common Prefix
LCS	Longest Common Subsequence
LMS	Learning Management System
LPA	Latent Profile Analysis
MAE	Mean Absolute Error
MAP	Maximum A Posteriori
MAR	Missing at Random
MBC	Model-Based Clustering
MDS	Multidimensional Scaling
MHMM	Mixture Hidden Markov Model
MI	Multiple Imputation
ML	Maximum Likelihood
MLE	Maximum Likelihood Estimation
MLR	Robust Maximum Likelihood
MM	Markov Model
MMM	Mixture Markov Model
MOOC	Massive Open Online Course
MPQ	Multicultural Personality Questionnaire
MULAS	Model of User-Centered Learning Analytics Systems
OM	Optimal Matching
ONA	Ordered Network Analysis
PAM	Partitioning Around Medoids
RF	Random Forest
RMSE	Root Mean Square Error
RMSEA	Root Mean Square Error of Approximation
ROC	Receiver Operating Characteristic
RSS	Really Simple Syndication
SD	Standard Deviation
SEM	Structural Equation Modeling
SNA	Social Network Analysis
SRL	Self-Regulated Learning
SRMR	Standardized Root Mean Square Residual
TAM	Technology Acceptance Model
TWCSS	Total Within-Cluster Sum-of-Squares
TN	True Negative
TP	True Positive
TSS	Total Sum of Squares
VIF	Variance Inflation Factor
WCTD	Within-Cluster Total Distance

Capturing the Wealth and Diversity of Learning Processes with Learning Analytics Methods

Sonsoles López-Pernas, Kamila Misiejuk, Rogers Kaliisa, Miguel Ángel Conde-González, and Mohammed Saqr

1 Introduction

The official birth of the field of learning analytics is often ascribed to the first Learning Analytics & Knowledge Conference in 2011, where the widely used definition was coined as "the measurement, collection, analysis, and reporting of data about learners and their contexts, for purposes of understanding and optimizing learning and the environments in which it occurs" [1]. Over the years, learning analytics has grown in scope, scale, and diversity and has since attracted researchers from diverse backgrounds bringing several disciplines together under one umbrella. Such increasing interest has resulted in a large number of publications, research groups, specialized units, and funding of large projects. Although other data-intensive fields started before learning analytics (e.g., academic analytics in 2007 and educational data mining in 2008), interest in such methods—and in closely related fields at large—has surged only after the rapid adoption of learning analytics. The unique position of learning analytics at the intersection of education and computer science while reaching out to several other disciplines such as statistics, psychometrics, econometrics, mathematics, and linguistics has accelerated the

S. López-Pernas (✉) · M. Saqr
School of Computing, University of Eastern Finland, Joensuu, Finland
e-mail: sonsoles.lopez@uef.fi

K. Misiejuk
Centre for the Science of Learning & Technology (SLATE), University of Bergen, Bergen, Norway

R. Kaliisa
Department of Education, University of Oslo, Oslo, Norway

M. Á. Conde-González
Robotics Research Group, Engineering School, University of León, León, Spain

M. Saqr, S. López-Pernas (eds.), *Learning Analytics Methods and Tutorials*,
https://doi.org/10.1007/978-3-031-54464-4_1

growth and expansion of the field. Therefore, it is a crucial endeavor for learning analytics researchers to stay abreast of the latest methodological and computational advances to drive their research forward. The diversity and complexity of the existing methods can make this task overwhelming both for newcomers to the learning analytics field and for experienced researchers. When conducting learning analytics research, researchers need to decide which data can and must be collected from learners to give answers to their research questions. They need to understand and decide which analytical methods may apply to such data and their potential limitations. They also need to interpret the findings and contextualize them in light of the existing literature and learning theories. With the motivation to accompany researchers in this challenging journey, this book aims to provide a methodological guide for researchers to study, consult, and take the first steps toward innovation.

Every method described in this book was developed before learning analytics, for instance, predicting students' performance can be traced back to almost a century ago [2], and using social network analysis to understand students' networks dates back to over five decades ago [3]. Similarly, process mining, sequence analysis, and Markov models can all be traced to times before the birth of the field of learning analytics (e.g., [4–6]). Not only were these methods born outside learning analytics, but they also continue to be developed, refined and advanced in their relevant fields. Nevertheless, learning analytics has succeeded in taking advantage of these methodological developments as well as the increasingly available digital data, computational resources, and data science, and in popularizing data-intensive research in education to bring this field together [7].

The diversity of methods and fields that have given rise to the field of analytics is evident in the list of authors of the chapters of this book, which include authors of R packages, and experts in methods and applications to drive a state-of-the-art blend. In the first category, we have the world renowned sequence analysis experts Gilbert Ritschard and Matthias Studer (University of Geneva), who are the creators of `TraMineR`, which is the most central R package for sequence analysis. Closely related, we have Satu and Jouni Helske (University of Turku), developers of `seqHMM`, an innovative R package for Mixture Hidden Markov Models of sequential data with tons of visualizations and statistical analysis potentials. Also, our list of authors includes Luca Scrucca (University of Perugia), author of `mclust`, one of the most widely used clustering packages for classification, and density estimation using Gaussian finite mixture models, and Keefe Murphy (Maynooth University), creator of several R packages such as `MoEClust` for Gaussian parsimonious clustering models with covariates. Our list of authors also includes David Shaffer (University of Wisconsin-Madison), the creator of `rENA` (an R package for Epistemic Network Analysis) along with other members of the `rENA` and the Ordered Network Analysis groups: Yuanru Tan and Andrew Ruis (University of Wisconsin-Madison), and Zachari Swiecki (Monash University). We also have Leonie V.D.E. Vogelsmeier (Tilburg University), author of `lmfa`, an innovative R package for latent Markov factor analysis which offers transition, and mixture factor analysis. We have Santtu Tikka (University of Jyväskylä), author of the `dynamite` R package for dynamic multivariate panel models. In addition, among our authors, we have prominent methodology experts in several domains: Joran Jongerling (Tilburg University); Emorie Beck (UC Davis); Merja Heinäniemi and Juho Kopra (University of Eastern Finland), and Marieke Schreuder (KU Leuven). Lastly, we count on several senior and emerging learning analytics researchers: Jelena Jovanović (University of Belgrade); Ángel Hernández-García, Javier Conde, Laura Del-Río-Carazo, and Carlos Cuenca-Enrique (Universidad Politécnica de Madrid); Adrienne Traxler (University of Copenhagen); Kamila Misiejuk (University of Bergen); Miguel Ángel Conde (University of Leon), and Marion Durand (University of Eastern Finland). Besides, the editors (Mohammed Saqr and Sonsoles López-Pernas) from the University of Eastern Finland have a long history of learning analytics research working with diverse methods and interdisciplinary research that spans most learning analytics methods.

Thanks to the breadth and diversity of authors' backgrounds and expertise, the book offers a comprehensive array of methods that are described thoroughly. A step-by-step tutorial using the R programming language with real-life datasets and case studies is presented for each method. The book starts with an introductory section for readers to get up-to-speed with the R programming language, in which we cover the basics of the language, data preprocessing, basic statistics, and data visualization. Then, we move to classic machine learning methods, such as prediction and clustering applied to educational data. These methods enable readers to predict achievement, dropout, and students at risk, as well as to group students into different groups or profiles according to certain characteristics such as motivation or engagement. This is followed by an extensive section devoted to temporal methods such as sequence analysis, Markovian modeling, and process mining, which allow taking advantage of the endless possibilities of trace log data. The book then moves on to discuss network analysis in its many forms including social network analysis, epistemic network analysis, ordered network analysis, and temporal networks. Such methods are crucial for understanding collaboration and relationships between individuals and concepts, which are key aspects of learning. The book concludes with a section on psychometrics such as psychological networks, factor analysis, and structural equation modeling, which are fundamental tools for the analysis of self-reported data among others. We hope the readers find this book useful as a guide through learning analytics methods, highlighting the ways in which data-driven insights can benefit educators, learners, and researchers alike.

The book targets learning analytics researchers (and education researchers at large) at all stages. It is suitable to teach newcomers to the field, even with no experience in R, since the introductory chapters are aimed at getting readers acquainted with the basics of R programming and data analysis. The book also covers advanced methods that may be of interest to experts in the field of learning analytics or data science in general. Moreover, the skills taught are transferable to other fields, i.e., can be applied in other contexts outside education.

We hope the readers find this book useful as a guide through learning analytics methods, highlighting the ways in which data-driven insights can benefit educators, learners, and researchers.

2 How the Book Is Structured

2.1 Introductory Chapters

The first section of the book provides the basis for getting up to speed with the R programming language and the data that will be analyzed throughout the book. This section covers the fundamental steps of the data analysis process, such as data preprocessing and exploratory analysis. During data preprocessing, educational data is cleaned and prepared for further analysis. Many crucial decisions about building and conceptualizing learning indicators from raw data are made in this essential step of the learning analytics process. Exploratory analysis enables an early detection of interesting phenomena that can be discovered in the data using visualizations or simple statistics. Using these techniques helps to guide the direction of the in-depth analysis and the selection of more advanced analytical methods.

Chapter 2: A Broad Collection of Datasets for Educational Research Training and Application [8]

Sonsoles López-Pernas, Mohammed Saqr, Javier Conde, Laura Del-Río-Carazo

Since the goal of this book is to provide a guide and tutorial on how to implement learning analytics methods, the use of relevant data is a key aspect of the contextualization of these methods within learning analytics research. Chapter 2 kicks off the book with an introduction to the most relevant types of data in learning analytics and provides a diverse collection of curated datasets that we will use throughout the book to illustrate the different methods. Understanding the data under examination is a crucial step for the interpretability of the analyses that we will learn to perform in this book, and therefore, readers should familiarize themselves with the datasets described in this chapter to facilitate following the tutorials presented in subsequent ones.

Chapter 3: Getting Started with R for Education Research [9]

Santtu Tikka, Juho Kopra, Merja Heinäniemi, Sonsoles López-Pernas, Mohammed Saqr

The first tutorial-like chapter of the book provides an introduction to the basics of R programming, with a focus on the Rstudio integrated development environment and the `tidyverse` programming paradigm. The R programming language has become a popular tool for conducting data analysis in the field of learning analytics. The chapter covers topics such as data types and structures, control structures, pipes, functions, loops, and input/output operations. By the end of the chapter, readers should have a solid understanding of the basics of R programming and have the necessary tools to learn more in-depth topics such as data wrangling and basic statistics using R.

Chapter 4: An R Approach to Data Cleaning and Wrangling for Education [10]

Juho Kopra, Santtu Tikka, Merja Heinäniemi, Sonsoles López-Pernas, Mohammed Saqr

After learning the basics of the R programming language, Chap. 4 goes one step further by introducing the reader to data wrangling, also known as data cleaning and preprocessing. Data wrangling is a critical step in the data analysis process, particularly in the context of learning analytics. This chapter provides an introduction to data wrangling using R and covers topics such as data importing, cleaning, manipulation, and reshaping with a focus on tidy data. Specifically, readers will learn how to read data from different file formats (e.g., CSV, Excel), how to manipulate data using the `dplyr` package, and how to reshape data using the `tidyr` package. Additionally, the chapter covers techniques for combining multiple data sources.

Chapter 5: Introductory Statistics with R for Educational Researchers [11]

Santtu Tikka, Juho Kopra, Merja Heinäniemi, Sonsoles López-Pernas, Mohammed Saqr

Statistics play a fundamental role in learning analytics, providing a means to analyze and make sense of the vast amounts of data generated by learning environments, visualize relationships, test hypotheses, and make comparisons. Chapter 5 in this book provides an introduction to basic statistical concepts using R and covers topics such as measures of central tendency, variability, correlation, and regression analysis. Specifically, readers will learn how to compute descriptive statistics, test hypotheses, and perform simple linear regression analysis. The chapter also includes practical examples using realistic data sets from the field of learning analytics. By the end of the chapter, readers should have a solid understanding of the basic statistical concepts and methods commonly used in learning analytics, as well as a practical understanding of how to use R to conduct statistical analysis of learning data.

Chapter 6: Visualizing and Reporting Educational Data with R [12]

Sonsoles López-Pernas, Kamila Misiejuk, Santtu Tikka, Mohammed Saqr, Juho Kopra, Merja Heinäniemi

Visualizing data is central in learning analytics research, underpins learning dashboards, and is a prime method for reporting results and insights to stakeholders. Chapter 6 guides the reader through the process of generating meaningful and aesthetically pleasing visualizations of different types of datasets using well-known R packages. The main visualization types will be demonstrated with an explanation of their usage and use cases. Furthermore, learning-related examples will be discussed in detail. For instance, readers will learn how to visualize learners' logs extracted from learning management systems to show how trace data can be used to track students' learning activities. Readers will also be able to generate professional-looking tables with summary statistics.

2.2 *Machine Learning Methods*

The next section follows with some of the classic machine learning methods of learning analytics, which date to the early beginnings of the field: predictive modelling and cluster analysis. Predictive modelling is a supervised learning method used widely in learning analytics research, where past data patterns are analysed to predict students' future outcomes. Clustering is an unsupervised learning method that detects similar patterns in the data and is typically used to group students based on their personal characteristics, observed behavior, or learning outcomes. Both methods are applied to address various challenges in education, such as preventing student drop-out, comparing strategies to improve academic performance, or identifying disengaged students. In addition, the results are often used to trigger specific interventions to help students succeed or to raise awareness about students' performance based on specific indicators.

Chapter 7: Predictive Modelling in Learning Analytics Using R [13]

Jelena Jovanovic, Sonsoles López-Pernas, Mohamed Saqr

Prediction of learners' course performance has been a central theme in learning analytics since the inception of the field. The main motivation behind it has been to identify learners who are at risk of low achievement so that they could be offered timely support based on intervention strategies derived from analysis of learners' data. To predict student success, numerous indicators, from varying data sources, have been examined and reported in the literature, as well as various predictive algorithms. Chapter 7 introduces the reader to predictive modeling, through a review of the main objectives, indicators, and algorithms that have been operationalized in previous works as well as a step-by-step tutorial on how to perform predictive modeling in learning analytics using R. The tutorial demonstrates how to predict student success using learning traces originating from a learning management system, guiding the reader through all the required steps from the data preparation to the evaluation of the built models.

Chapter 8: Dissimilarity-Based Clustering Educational Data Using R [14]

Keefe Murphy, Sonsoles López-Pernas, Mohammed Saqr

Chapter 8 presents another central method in learning analytics research: clustering. Clustering is a collective term that refers to techniques aimed at uncovering patterns and subgroups within the data. Finding patterns or differences among students enables teachers and researchers to improve their understanding of the diversity of students—and their learning processes—and tailor their support to different needs. This chapter introduces the theory underpinning the dissimilarity-based clustering methods. Then, the focus is placed on some of the most widely-used heuristic dissimilarity-based clustering algorithms; namely, K-means, K-medoids, and agglomerative hierarchical clustering. Methods for choosing the optimal number of clusters are provided, particularly the criteria that can guide the choice of clustering solution among multiple competing methodologies, and not only the choice of the number of clusters K for a given method. All of these are demonstrated in detail with a tutorial in R using a real-life educational dataset.

Chapter 9: An Introduction and R Tutorial to Model-Based Clustering in Education via Latent Profile Analysis [15]

Luca Scrucca, Mohammed Saqr, Sonsoles López-Pernas, Keefe Murphy

Chapter 9 presents an alternative approach for capturing different patterns or subgroups within students' behavior or functioning. Assuming that there is an average pattern that represents the entirety of student populations requires the measured construct to have the same causal mechanism, the same development pattern, and affect students in exactly the same way. Using a person-centered method (Finite Gaussian mixture model or latent profile analysis), this chapter offers an introduction to model-based clustering that includes the principles of the methods, a guide to the choice of number of clusters, an evaluation of clustering results and a detailed guide with code and a real-life dataset. The tutorial part shows how to uncover the heterogeneity within engagement data by identifying latent or unobserved clusters. The discussion elaborates on the interpretation of the results, the advantages of model-based clustering as well as how this method compares with others.

2.3 *Temporal Methods*

We continue our journey with an introduction to temporal methods in learning analytics. Unlike the methods based on mere counts of events or activities, temporal methods acknowledge the order and temporality of events, as well as the transitions thereof, which are key aspects of learning. Temporal methods have garnered increasing attention since they allow researchers to take advantage of the trace log data that students leave behind when using educational technology and also to study longitudinal processes (e.g., a whole study program). Such methods originate in social sciences and have been imported and adapted into the learning analytics field. We provide three chapters focused on sequence analysis, and two on transition analysis through Markovian modeling and process mining.

Chapter 10: Sequence Analysis in Education: Principles, Technique, and Tutorial with R [16]

Mohammed Saqr, Sonsoles López-Pernas, Satu Helske, Marion Durand, Keefe Murphy, Matthias Studer, Gilbert Ritschard

Sequence analysis is a data mining technique that is increasingly gaining ground in learning analytics. Sequence analysis enables researchers to extract meaningful insights from sequential data, i.e., to summarize the sequential patterns of learning data and classify those patterns into homogeneous groups. Chapter 10 introduces readers to sequence analysis techniques and tools through real-life step-by-step examples of sequential trace log data of students' online activities. Readers are guided on how to visualize the common sequence plots and interpret such visualizations. An essential part of sequence analysis is the discovery of patterns within sequences through clustering techniques. Therefore, this chapter demonstrates the various sequence clustering methods, calculation of cluster indices, and evaluation of clustering results.

Chapter 11: Modeling the Dynamics of Longitudinal Processes in Education. A Tutorial with R for The VaSSTra Method [17]

Sonsoles López-Pernas, Mohammed Saqr

Building upon the knowledge acquired in the previous chapter, Chap. 11 covers VaSSTra, a method for analyzing multiple variables across multiple time points. The idea behind this method is to summarize multiple variables at each time point into a single state using person-based methods. Then, sequence analysis can be used to analyze the sequences of such states for each person, and clustering techniques can be implemented to detect similar trajectories of the evolution of such states. The method is illustrated in a case study about engagement. Several engagement-related variables are derived from students' online activities (frequency of each activity, regularity, etc.). These variables are used for clustering students into three states (active, moderate, and disengaged) at each course. Then, sequence analysis is used to map the sequence of engagement states across a whole program. Lastly, clustering mechanisms are used to detect distinct trajectories of engagement.

Chapter 12: A Modern Approach to Transition Analysis and Process Mining with Markov Models in Education [18]

Jouni Helske, Satu Helske, Mohammed Saqr, Sonsoles López-Pernas, Keefe Murphy

Chapter 12 presents Markov models, a widely used technique to model temporal processes. Contrary to the deterministic approach seen in the previous sequence analysis chapters, Markovian models are probabilistic models, focusing on the transitions between states instead of studying sequences as a whole. The chapter provides an introduction to Markov models and differentiates between its most common variations: first-order Markov models, hidden Markov models, mixture Markov models, and hidden mixture Markov models. All implementations are illustrated with a step-by-step tutorial using the R package seqHMM using students' longitudinal data. The chapter also provides a complete guide to performing stochastic process mining with Markovian models as well as plotting, comparing, and clustering different process models

Chapter 13: Multichannel Sequence Analysis in Educational Research Using R [19]

Sonsoles López-Pernas, Satu Helske, Mohammed Saqr, Keefe Murphy

When dealing with learners' data, sometimes one single source of information is not enough to capture all of the dimensions of the learning process. Fortunately, sequence analysis as a method supports the examination of multiple sequences (termed channels) at the same time as long as they follow the same time scheme. Chapter 13 covers multi-channel sequence analysis, allowing the reader to study and visualize synchronized sequences together, and cluster them into distinct trajectories based on the values of the various channels. We present two methods for clustering: one distance-based (see Chap. 10) and one based on Markovian models (see Chap. 12). We illustrate the method by studying the longitudinal association between student engagement and achievement across a study program.

Chapter 14: The Why, the How, and the When of Educational Process Mining in R [20]

Sonsoles López-Pernas, Mohammed Saqr

Process mining is a recent analytical method that enables the extraction of meaningful insights from time-ordered event logs. The goal of process mining is to discover processes from the data, evaluate process efficiency, and help or enhance processes. Since its introduction in education, process mining has been used to map students' learning processes, visualize learners' strategies, as well as demonstrate differences in approach to learning across different learning groups. Chapter 14 illustrates how to prepare learners' data for process mining and how to visualize the process data using the `bupaverse` framework. Moreover, readers will learn how to examine the transitions between phases or activities within a learning process.

2.4 Network Analysis

The next section of the book deals with the relational aspects of analyzing educational data such as relationships between students, teachers, and topics. Network analysis is the underlying method used to study such relational aspects. Social network analysis allows researchers to study collaboration and discussion between peers and understand the role each student occupies in the network. Moreover, community finding allows the detection of distinct groups of peers in the network that interact with each other more than with the rest. We can even combine network analysis with the temporal methods presented in the last section through temporal networks or use epistemic or ordered network analysis to explore topic or construct co-occurrence.

Chapter 15: Social Network Analysis: A Primer, a Guide and a Tutorial in R [21]

Mohammed Saqr, Sonsoles López-Pernas, Miguel Ángel Conde, Ángel Hernández-García

For five decades, learning networks have been used to map collaboration networks among students, study the influence of peers, and capture the relational dimension of collaborative learning. Additionally, networks have been used to study the semantics of discourse, relations between behaviors, and patterns of relations among teachers. Networks offer a powerful framework with vast potential for data analysis. Chapter 15 introduces the concept and methods of social network analysis and a detailed guide on how researchers can use network analysis using real-world data. The chapter demonstrates network analysis and visualisation with an emphasis on learners' roles and relevance to the educational context. The chapter further provides a mathematical analysis and interpretation of the different social network metrics such as centrality and betweenness measures with several examples of how they can be used in practice.

Chapter 16: Community Detection in Learning Networks Using R [22]

Ángel Hernández-García, Carlos Cuenca-Enrique, Adrienne Traxler, Sonsoles López-Pernas, Miguel Ángel Conde, Mohammed Saqr

In learning situations, communities can be groups of students within a whole cohort who collaborate with each to a larger extent than with other students in a learning situation. Finding these communities is integral to understanding the interaction process, the structure and behaviour of the formed groups and how they contribute to the overall learning process. Chapter 16 builds on the principles of social networks from Chap. 15 and introduces the topic of community detection. The main aim of community detection is to identify different groups or clusters of nodes within the network that share some similar characteristics. One way of understanding communities in social networks is as subnetworks where the number of internal connections is larger than the number of external connections, and therefore members of a community have a higher probability of being connected to each other than to members of other communities. The chapter focuses on detecting communities (groups of highly connected nodes) within a wider network and shows how to visualize them using R.

Chapter 17: Temporal Network Analysis: Introduction, Methods, and Analysis with R [23]

Mohammed Saqr

Learning can be viewed as relational, interdependent, and temporal and therefore, methods that account for such multifaceted dynamic processes that unfold overtime are required. Chapter 17 combines the temporal and relational aspects in a single analytics framework: temporal networks. Temporal networks allow modeling of the temporal learning processes i.e., the emergence and flow of activities, communities, and social processes through fine-grained dynamic analysis. This can provide insights into phenomena like knowledge co-construction, information flow, and relationship building. This chapter introduces the basic concepts of temporal networks, their types (i.e, contact and interval), and techniques. The chapter further provides a detailed guide to temporal network analysis, which involves network building, visualization, and statistical analysis at the graph and node level.

Chapter 18: Epistemic Network Analysis and Ordered Network Analysis in Learning Analytics [24]

Yuanru Tan, Zachari Swiecki, Andrew Ruis, David Shaffer

The increasing use of technology in many areas of society and life has led to an increasing amount of Big Data about human behavior and interaction. However, this volume of data is usually too large and strains the capabilities of human interpretation and the traditional social science research approaches. Chapter 18 presents two quantitative ethnographic approaches that link the power of statistics and in-depth ethnographic approaches to understand learning behaviour through large-scale qualitative data. Epistemic Network Analysis (ENA) and Ordered Network Analysis (ONA), are two methods for quantifying, visualizing, and interpreting network data. Taking coded data as input, ENA and ONA represent associations between codes (e.g., topics or categories) in undirected or directed weighted network models, respectively. Both techniques measure the strength of association among codes and illustrate the structure of connections in network graphs, quantify changes in the composition and strength of those connections over time, and enable comparison of networks. The chapter presents a thorough description of the methods and a step-by-step guide on how to implement them with R.

2.5 Psychometrics

We finalize the book with a section on psychometrics. In the field of educational psychology, psychometrics aims to study how psychological constructs (e.g., intelligence or aptitude) are related to observable variables (e.g., test scores). Traditionally, psychometric methods in educational psychology have relied on self-reported data from validated questionnaire-like instruments, although nowadays researchers have begun to make use of digital data. We present several techniques to investigate the relationship between measured variables and to test hypotheses and theories: psychological networks, factor analysis, and structural equation modeling (SEM).

Chapter 19: Psychological Networks: A Modern Approach to Analysis of Learning and Complex Learning Processes [25]

Mohammed Saqr, Emorie Beck, Sonsoles López-Pernas

When analyzing psychological phenomena that take place in educational settings, a multitude of variables are at play that may interact with, trigger, and influence each other. To understand such dependency between variables, it is not enough to analyze the linear relationships between each pair of variables, but rather such complexity calls for using more sophisticated methods that capture the full breadth of the interplay between variables: psychological networks. As opposed to social networks where nodes often represent people and edges represent the interactions or relations between them, the nodes in psychological networks represent observed psychological variables and edges represent a statistical relationship between them. Chapter 19 opens the section on psychometric methods by presenting the concept of psychological networks as well as a tutorial for their estimation, visualization, and interpretation with R.

Chapter 20: Factor Analysis in Education Research Using R [26]

Leonie V. D. E. Vogelsmeier, Mohammed Saqr, Sonsoles López-Pernas, Joran Jongerling

Chapter 20 presents factor analysis, a method employed to reduce a large number of variables into fewer numbers of factors. The method is commonly used to identify which observable indicators are representative of latent, not directly-observed constructs. This is a key step in developing valid instruments to assess latent constructs such as student engagement in educational research. The chapter describes the two main approaches for conducting factor analysis in detail and provides a tutorial on how to implement both techniques. The first is confirmatory factor analysis (CFA), a more theory-driven approach, in which a researcher actively specifies the number of underlying constructs as well as the pattern of relations between these dimensions and observed variables. The second is exploratory factor analysis (EFA), a more data-driven approach, in which the number of underlying constructs is inferred from the data, and all underlying constructs are assumed to influence all observed variables (at least to some degree).

Chapter 21: Structural Equation Modeling with R for Education Scientists [27]

Joran Jongerling, Sonsoles López-Pernas, Mohammed Saqr, Leonie V. D .E. Vogelsmeier

Chapter 21 presents the last method in our book: Structural Equation Modeling (SEM). SEM is a suitable and useful method for modeling the multitude of relationships between latent variables and the observable indicators, as well as the relationship between the latent variables themselves to test theories. In its most common form, SEM combines CFA (covered in Chap. 20) with another method named path analysis. Just like CFA, SEM relates observed variables to latent variables that are measured by those observed variables and, as path analysis does, SEM allows for a wide range of regression-type relations between sets of variables (both latent and observed). This chapter presents an introduction to SEM, an integrated strategy for conducting SEM analysis that is well-suited for educational sciences, and a tutorial on how to carry out an SEM analysis in R.

3 The Companion Code and Data

To enhance your learning experience and practical understanding of the concepts discussed in this book, we have developed a companion code repository that accompanies each chapter. The repository contains all the code included in the step-by-step tutorials that illustrate how to implement the different learning analytics methods covered in the book chapters. The code also contains custom functions that automate complex operations, making it easier for anyone to apply the techniques to their own datasets. Moreover, the code will guide the reader on how to generate visualizations, graphs, and plots aimed at helping to interpret and communicate findings effectively. The companion code repository can be accessed at:

https://github.com/lamethods/code

Along with the code, we provide a collection of datasets carefully curated to represent educational scenarios, allowing readers to experiment with the techniques discussed in the book and beyond. Each dataset is described in detail in Chap. 2.

https://github.com/lamethods/data

The combination of theoretical knowledge and practical application through the companion code and datasets will prepare the reader to begin their journey into the multidisciplinary field of learning analytics.

References

1. Conole G, Gašević D, Long P, Siemens G (2011) Message from the LAK 2011 general & program chairs. In: Conole G, Gašević D (eds) Proceedings of the 1st international conference on learning analytics and knowledge. Association for Computing Machinery (ACM), Banff
2. Cornog J, Stoddard GD (1925) Predicting performance in chemistry. J Chem Educ 2:701. https://doi.org/10.1021/ed002p701

3. Platts CV, Wyant G (1969) Network analysis and the possibility of its use in education. Educ Rev 21:109–119. https://doi.org/10.1080/0013191690210203
4. Pechenizkiy M, Trcka N, Vasilyeva E, Van der Aalst W, De Bra P (2009) Process mining online assessment data. ERIC Clearinghouse
5. Soller AL, Lesgold A (2003) A computational approach to analyzing online knowledge sharing interaction. In: Proceedings of artificial intelligence in education
6. Wang F-H (2002) On analysis and modeling of student browsing behavior in web-based asynchronous learning environments. In: Advances in web-based learning. Springer, Berlin, pp 69–80
7. Adam K, Bakar NAA, Fakhreldin MAI, Majid MA (2018) Big data and learning analytics: a big potential to improve e-learning. Adv Sci Lett 24:7838–7843. https://doi.org/10.1166/asl.2018.13028
8. López-Pernas S, Saqr M, Conde J, Del-Río-Carazo L (2024) A broad collection of datasets for educational research training and application. In: Saqr M, López-Pernas S (eds) Learning analytics methods and tutorials: a practical guide using R. Springer, this volume, 2024
9. Tikka S, Kopra J, Heinäniemi M, López-Pernas S, Saqr M (2024) Getting started with r for education research. In: Saqr M, López-Pernas S (eds) Learning analytics methods and tutorials: a practical guide using R. Springer, this volume, 2024
10. Kopra J, Tikka S, Heinäniemi M, López-Pernas S, Saqr M (2024) An r approach to data cleaning and wrangling for education research. In: Saqr M, López-Pernas S (eds) Learning analytics methods and tutorials: a practical guide using R. Springer, this volume, 2024
11. Tikka S, Kopra J, Heinäniemi M, López-Pernas S, Saqr M (2024) Introductory statistics with r for educational researchers. In: Saqr M, López-Pernas S (eds) Learning analytics methods and tutorials: a practical guide using R. Springer, this volume, 2024
12. López-Pernas S, Misiejuk K, Tikka S, Saqr M, Kopra J, Heinäniemi M (2024) Visualizing and reporting educational data with r. In: Saqr M, López-Pernas S (eds) Learning analytics methods and tutorials: a practical guide using R. Springer, this volume, 2024
13. Jovanovic J, López-Pernas S, Saqr M (2024) Predictive modelling in learning analytics using R. In: Saqr M, López-Pernas S (eds) Learning analytics methods and tutorials: a practical guide using R. Springer, this volume, 2024
14. Murphy K, López-Pernas S, Saqr M (2024) Dissimilarity-based cluster analysis of educational data: a comparative tutorial using R. In: Saqr M, López-Pernas S (eds) Learning analytics methods and tutorials: a practical guide using R. Springer, this volume, 2024
15. Scrucca L, Saqr M, López-Pernas S, Murphy K (2024) An introduction and r tutorial to model-based clustering in education via latent profile analysis. In: Saqr M, López-Pernas S (eds) Learning analytics methods and tutorials: a practical guide using R. Springer, this volume, 2024
16. Saqr M, López-Pernas S, Helske S, Durand M, Murphy K, Studer M, Ritschard G (2024) Sequence analysis in education: principles, technique, and tutorial with r. In: Saqr M, López-Pernas S (eds) Learning analytics methods and tutorials: a practical guide using R. Springer, this volume, 2024
17. López-Pernas S, Saqr M (2024) Modeling the dynamics of longitudinal processes in education. A tutorial with r for the VaSSTra method. In: Saqr M, López-Pernas S (eds) Learning analytics methods and tutorials: a practical guide using R. Springer, this volume, 2024
18. Helske J, Helske S, Saqr M, López-Pernas S, Murphy K (2024) A modern approach to transition analysis and process mining with Markov models in education. In: Saqr M, López-Pernas S (eds) Learning analytics methods and tutorials: a practical guide using R. Springer, this volume, 2024
19. López-Pernas S, Saqr M, Helske S, Murphy K (2024) Multichannel sequence analysis in educational research using R. In: Saqr M, López-Pernas S (eds) Learning analytics methods and tutorials: a practical guide using R. Springer, this volume, 2024
20. López-Pernas S, Saqr M (2024) The why, the how, and the when of educational process mining in R. In: Saqr M, López-Pernas S (eds) Learning analytics methods and tutorials: a practical guide using R. Springer, this volume, 2024

21. Saqr M, López-Pernas S, Conde MÁ, Hernández-García Á (2024) Social network analysis: a primer, a guide and a tutorial in r. In: Saqr M, López-Pernas S (eds) Learning analytics methods and tutorials: a practical guide using R. Springer, this volume, 2024
22. Hernández-García Á, Cuenca-Enrique C, Traxler A, López-Pernas S, Conde MÁ, Saqr M (2024) Community detection in learning networks using R. In: Saqr M, López-Pernas S (eds) Learning analytics methods and tutorials: a practical guide using R. Springer, this volume, 2024
23. Saqr M (2024) Temporal network analysis: introduction, methods, and analysis with r. In: Saqr M, López-Pernas S (eds) Learning analytics methods and tutorials: a practical guide using R. Springer, this volume, 2024
24. Tan Y, Swiecki Z, Ruis A, Shaffer D (2024) Epistemic network analysis and ordered network analysis in learning analytics. In: Saqr M, López-Pernas S (eds) Learning analytics methods and tutorials: a practical guide using R. Springer, this volume, 2024
25. Saqr M, Beck E, López-Pernas S (2024) Psychological networks: a modern approach to analysis of learning and complex learning processes. In: Saqr M, López-Pernas S (eds) Learning analytics methods and tutorials: a practical guide using R. Springer, this volume, 2024
26. Vogelsmeier LVDE, Saqr M, López-Pernas S, Jongerling J (2024) Factor analysis in education research using R. In: Saqr M, López-Pernas S (eds) Learning analytics methods and tutorials: a practical guide using R. Springer, this volume, 2024
27. Jongerling J, López-Pernas S, Saqr M, Vogelsmeier LV (2024) Structural equation modeling with r for education scientists. In: Saqr M, López-Pernas S (eds) Learning analytics methods and tutorials: a practical guide using R. Springer, this volume, 2024

Part I
Getting Started

A Broad Collection of Datasets for Educational Research Training and Application

Sonsoles López-Pernas, Mohammed Saqr, Javier Conde, and Laura Del-Río-Carazo

1 Introduction

Learning analytics involves the combination of different types of data such as behavioral data, contextual data, performance data, and self-reported data to gain a comprehensive understanding of the learning process [1, 2]. Each type of data provides a unique perspective of the learning process and, when analyzed together, can provide a more complete picture of the learner and the learning environment. Throughout the book, we will work with different types of learning analytics data to illustrate the analysis methods covered. This chapter explores the most common types of data that are used in learning analytics and that we will work with in the subsequent book chapters. Such data include demographic and other contextual data about students, performance data, online activity, interactions with other students and teachers, and self-reported data.

This chapter also describes a set of datasets that will be used throughout the book, as well as additional datasets that may be useful for readers to put the newly learned methods into practice. We will discuss the characteristics, structure, and contents of each dataset, as well as the context in which they have been used within the book. The goal of this chapter is to give readers a solid foundation for working with the datasets used in the book, as well as to provide a starting point for those interested in exploring additional data sources.

S. López-Pernas (✉) · M. Saqr
School of Computing, University of Eastern Finland, Joensuu, Finland
e-mail: sonsoles.lopez@uef.fi

J. Conde · L. Del-Río-Carazo
Universidad Politécnica de Madrid, Madrid, Spain

M. Saqr, S. López-Pernas (eds.), *Learning Analytics Methods and Tutorials*,
https://doi.org/10.1007/978-3-031-54464-4_2

2 Types of Data

2.1 Contextual Data

Contextual data refer to data that provide information about the environment in which learning takes place, such as demographic information, socioeconomic status, and prior academic achievement. This type of data can be used to understand how external factors may impact learning, to identify students with similar profiles, and to develop targeted interventions. Demographic data can be used to understand the characteristics of the learners, such as age, gender, race, and ethnicity [3]. Socioeconomic data can be used to examine the impact of the socio-economic status of learners, such as income, employment status, and education level [4]. Prior academic achievement data can be used to understand how the academic background of the learners, such as their previous grades and test scores, may influence their learning at present [5]. The data about the learning context is also relevant to better understand and support students; for example, the level and difficulty of the educational program and courses, format (online vs. face-to-face), or pedagogical approach (e.g., flipped classroom, laboratory course, etc.) [6].

Descriptive statistics can be used to summarize and describe the main characteristics of the contextual data, such as the mean, median, and standard deviation. In Chapters 3 [7] and 4 [8], we will learn how to clean and manipulate data and how to summarize it using descriptive statistics. In Chapter 6 [9], we will learn how to create different types of plots that will allow us to better understand the data. Cluster analysis can also be used to group similar students together. This can be used to identify patterns in the data and, for example, to understand which different groups of students exist in a course or degree and whether such groups differ in terms of, e.g., performance [10]. We cover clustering in Chapters 8 and 9 [11, 12].

It is important to bear in mind that contextual data are essential to understand learners' and the learning process, but they should be used in combination with other types of data to obtain a comprehensive understanding [13]. It is also crucial to comply with data protection laws and regulations and to consider the ethical implications of collecting and operationalizing this type of data, especially when it comes to the existence of bias when making decisions based on demographic data [14].

2.2 Self-reported Data

Self-reported data refers to data provided by students themselves (or other relevant stakeholders), such as data collected through surveys or questionnaires. This type of data can provide valuable insight into learners' and teachers' attitudes, motivation, and perspectives on their learning experiences, and can be used to inform the design of educational programs [15]. It is important to keep in mind that the data should be cleaned and pre-processed before applying any analytical techniques, especially when dealing with qualitative data (e.g., free text, video, or recordings), and the results should be interpreted with caution, keeping in mind the limitations of self-reported data [16].

Regarding the techniques employed to analyze self-reported data, descriptive statistics and data visualization are commonly used to understand the distribution of responses and to identify patterns in the data (see Chapters 4 [8] and 6 [9]). Moreover, inferential statistics can be used to make inferences about a population based on a sample of data. This can include techniques such as t-tests and analysis of variance to identify significant differences between groups or chi-squared tests to identify associations in the data. Chapter 5 will help us better understand the most common statistical tests and how to implement them with R [17]. Depending on the research question, the type of data, and the level of detail required, a more sophisticated choice of analytical techniques might be needed. For instance, Factor Analysis is a statistical technique that can be used to identify underlying factors or dimensions that explain the relationships between multiple variables [18]. We will learn about it in Chapter 20 [19]. Similarly, Structural Equation Modeling (SEM) can be used to test complex models that involve multiple observed and latent variables that depend on one another. We cover this method in Chapter 21 [20]. Moreover, self-reported data can be analyzed using psychological networks, a relatively new approach in the field of psychology that focuses on understanding psychological phenomena as interconnected networks of individual components. We cover this method in Chapter 19 [21]. Lastly, text mining can be used to analyze unstructured data, such as open-ended responses to surveys or interviews. It can be used to identify key concepts and themes, perform sentiment analysis, and summarize text [22]. This type of analysis is beyond the scope of this book.

2.3 Activity Data

Activity data in learning analytics refers to the data that is collected about a student's interactions with educational technology. Activity data can include information such as the learning resources a student visits, the time spent on a resource, the buttons clicked, and the messages posted [23]. Data can be collected automatically by the learning management system (LMS) or other educational technology (e.g., a game, an intelligent tutoring system, eBooks, or coding environments). Log activity data can be used to track student progress, identify areas where students may be struggling, and personalize instruction [24]. For example, if a student is spending an excessive amount of time on a particular concept, it may indicate that they are having difficulty understanding that concept. In this case, the teacher can provide additional support or re-teach the concept to help the student improve. Log activity data can also be used to measure student engagement with the course content and to identify students who are not engaging with the material [25]. Log activity data have been used to detect students' online tactics and strategies [26] paying attention not only to the frequency but to the order and timing of students' events.

Besides basic analysis using descriptive and inferential statistics, activity logs have been operationalized in many ways in learning analytics, especially using temporal methods that allow to take advantage of the availability of large amounts of timestamped data. For example, process mining—which we cover in Chapter 14 [27]—has been used to investigate how students navigate between different online activities [28]. Sequence analysis has been used to detect and interpret students' online tactics and strategies based on the order of learning activities within learning sessions [29]. We dedicate several chapters to this technique [30–33]. Such analyses have been complemented with cluster analysis, which allows to detect distinct patterns of students with different online behavior [34] (see Chapters 8 and 9 [11, 12]).

2.4 Social Interaction Data

Social interaction data in learning analytics refers to the data collected about students' interactions with each other (and sometimes teachers too) in a learning environment, social media, or messaging platforms. This can include data such as the frequency and nature of interactions, the content of discussions, and the participation of students in group work or collaborative activities. Social interaction data can be used to understand how students are engaging with each other and to identify patterns or roles that students assume [35]. For example, if a student is not participating in group discussions, it may indicate that they are feeling disengaged or are having difficulty understanding the material. Furthermore, social interaction data can be used to study how students' depth of contributions to the discussion influences performance [36]. For example, an analysis of social interaction data may reveal that students who receive more replies from other students perform better in the course than students whose contributions do not spark a lot of interest.

Social Network Analysis (SNA) is the most common method to study social interaction data. SNA comprises a wealth of quantitative metrics that summarize relationships in a network. In most cases in learning analytics, this network is formed based on students' interactions. These metrics, named centrality measures, pave the path to apply other analytical methods such as cluster analysis to detect collaboration roles [37], or predictive analytics to determine whether performance can be predicted from students' centrality measures [38]. We cover the basics of SNA in Chapter 15 of the book [39], community detection in Chapter 16 [40], and temporal network analysis in Chapter 17 [41]. Moreover, the nature and content of students' interactions can be analyzed with Epistemic Network Analysis (ENA), a method for detecting and quantifying connections between elements in coded data and representing them in dynamic network models [42]. We cover this method in Chapter 18 [43].

2.5 Performance Data

Performance data refers to data that measures how well learners are able to apply what they have learned. This type of data can be used to evaluate the effectiveness of a particular educational activity, to identify areas where additional support may be needed, or to detect students at risk. Performance includes assessment data from tests, quizzes, projects, essays, exams, and other forms of evaluation used to track students' progress. Assessment can be performed by different entities, such as teachers, peers or automated assessment tools. Moreover, assessment data can have different levels of granularity: it can be the grade for a specific task, a midterm or final exam, or a project; it can be the final grade for a course, or even a whole program GPA [44]. Performance data used for learning analytics may not necessarily be assessment data. For instance, pre-test and post-test data are used to evaluate the effectiveness of a particular educational intervention [45]. Another example is the data captured by audience response systems (ARSs) [46], which are often used to evaluate learners' knowledge retention during lectures.

Performance data are rarely analyzed on its own, but rather used in combination with other sources of data. For example, a common use case in learning analytics is to correlate or predict grades with indicators from several sources [13, 47], such as demographic data, activity data or interaction data. In the book, we cover predictive modelling in Chapter 7 [48]. Moreover, grades are often compared among groups or clusters of students, for example, to evaluate the performance of students that use different online learning strategies [29] or to establish whether students' assuming different levels of collaboration also show differences in performance [49]. Clustering is covered in Chapters 8 and 9 [11, 12].

2.6 Other Types of Data

In recent years, the landscape of data used for learning analytics has undergone a remarkable expansion beyond demographics, grades, surveys and digital logs [50]. This evolution has led to the incorporation of novel methodologies designed to capture a more holistic understanding of students' learning experiences, including their physiological responses [51]. This progression encompasses a diverse range of data acquisition techniques, such as eye-tracking data that traces the gaze patterns of students, electrodermal activity which measures skin conductance and emotional arousal, EEG (electroencephalogram) recordings that capture brain activity patterns, heartbeat analysis reflecting physiological responses to learning stimuli, and motion detection capturing physical movements during learning activities [52]. These physiological datasets are often combined with other forms of information, such as video recordings (e.g., [50]). Through the combination of various data modalities, researchers and educators can better understand how students engage with educational content and respond to different teaching methodologies [51]. This analysis goes beyond the capabilities of conventional digital learning tools, offering insights into the emotional, cognitive, and physical aspects of learning that might otherwise remain concealed [53, 54]. This synergistic analysis of multiple data sources is often referred to as "multimodal learning analytics". In Chapter 13, we will cover multi-channel sequence analysis, a method suitable for analyzing several modalities of data at the same time [33].

3 Dataset Selection

The present section describes a set of curated datasets that will be used throughout the book. In the introductory chapters, the reader will learn how to import datasets in different formats [7], clean and transform data [8], conduct basic statistics [17], and create visualizations [9]. Each of the remaining chapters covers a specific method, which is illustrated in a tutorial-like way using one or more of the datasets described below. All the datasets are available on Github.[1]

3.1 LMS Data from a Blended Course on Learning Analytics

Link to the dataset

The first dataset in our book is a synthetic dataset generated from a real blended course on Learning Analytics offered at the University of Eastern Finland. The course has been previously described in a published article [55] which used the original (non-synthetic) dataset. The lectures in the course provided the bases for understanding the field of learning analytics: the recent advances in the literature, the types of data collected, the methods used, etc. Moreover, the course covered learning theories as well as ethical and privacy concerns related to collecting and using learners' data. The course had multiple practical sessions which allowed students to become skilled in learning analytics methods such as process mining and social network analysis using real-life datasets and point-and-click software.

Students in the course were required to submit multiple assignments; most of them were practical, in which they had to apply the methods learned in the course, but others were focused on discussing learning theories, ethics, and even conducting a small review of the literature. The course had a final project that accounted for 30% of the course final grade in which students had to analyze several datasets in multiple ways and comment and discuss their findings. Moreover, there was a group project in which students had to present an implementation of learning analytics application in an institutional setting, discussing the sources of data collection, the analyses that could be conducted, and how to present and make use of the data and analyses. The

[1] Github repository: https://github.com/lamethods/data.

course was implemented in a blended format: instruction was face-to-face while the learning materials and assignments were available online, in the Moodle LMS. Discussions among students in the group project also took place online in the LMS forum.

The dataset contains four files: a file containing students' online activities in Moodle, a file containing their grades, a file containing their demographic data, and a file that aggregates all the information. It is shared with a CC BY 4.0 license, which means that anyone is free to share, adapt, and distribute the data as long as appropriate credit is given. The dataset has been used in the introductory chapters of the book to learn the basics of R [7], data cleaning [8], basic statistics [17] and data visualization [9]. Moreover, it has been used in two additional chapters to illustrate to well-known learning analytics methods: sequence analysis [30] and process mining [27]. Below, we provide further details on each of the files of the dataset.

3.1.1 Events

The `Events.xlsx` file contains 95,580 timestamped Moodle logs for 130 distinct students. The activities include viewing the lectures, discussing on forums, and working on individual assignments, as well as discussion in small groups, among other events. The logs were re-coded to balance granularity with meaningfulness, i.e., grouping together logs that essentially represent the same action. For example, the activities related to the group project were all coded as `Group_work`, log activities related to feedback were coded as `Feedback`, logs of students' access to practical resources or assignments were coded as `Practicals`, social interactions that are unrelated to learning were coded as `Social`, etc. Below we describe the columns of the dataset. A preview of the dataset can be seen in Table 1. In Fig. 1, we show the distribution of events per student.

- **Event.context**: Resource of the LMS where the event takes place, for example "Assignment: Literature review".
- **user**: User name in the LMS.
- **timecreated**: Timestamp in which each event took place, ranging from September 9th 2019 to October 27th 2019.
- **Component**: Type of resource involved in the event. There are 13 distinct entries, such as Forum (39.11%); System (34.33%); Assignment (15.50%) and 10 others.
- **Event.name**: Name of the event in Moodle. There are 27 distinct entries, such as Course module viewed (35.89%); Course viewed (26.28%); Discussion viewed (13.77%) and 24 others.
- **Action**: Column coded based on the combination of the event name and context. There are 12 distinct entries, such as Group_work (34.25%); Course_view (26.45%); Practicals (10.48%) and 9 others.

Table 1 Preview of the LA course Moodle Events dataset

	Event.context	User	Timecreated	Component	Event.name	Action
1	Assignment: Final Project	9d744e5bf	1572082632	Assignment	Course module viewed	Assignment
2	Assignment: Final Project	91489f7a9	1572080974	Assignment	The status of the submission has been viewed.	Assignment
3	Assignment: Final Project	278a75edf	1571400328	Assignment	Course module viewed	Assignment
4	Assignment: Final Project	53d6ab60c	1571491717	Assignment	The status of the submission has been viewed.	Assignment
5	Assignment: Final Project	aab7ad69f	1571182693	Assignment	Course module viewed	Assignment
6..95625						
95626	Zoom meeting: Group presentation 8 AM	215b886a6	1571119806	Zoom meeting	Clicked join meeting button	Group_work

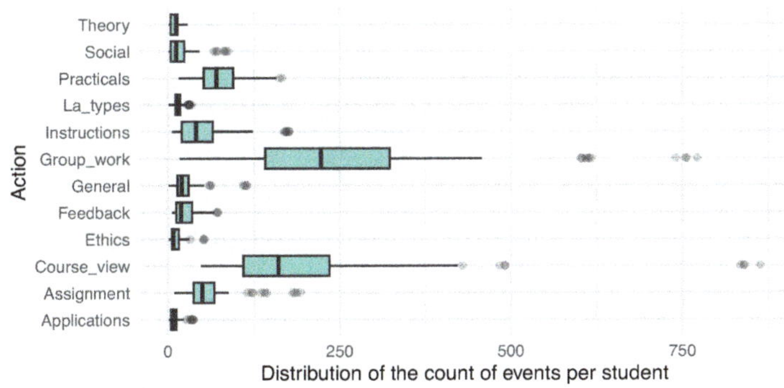

Fig. 1 Number of actions per student per type

Table 2 Preview of the LA course Demographic dataset

	User	Name	Surname	Origin	Gender	Birthdate	Location	Employment
1	6eba3ff82	Amanda	Mora	Costa Rica	F	28.2.1998	On campus	None
2	05b604102	Lian	Abdullah	Yemen	F	19.11.1996	On campus	None
3	111422ee7	Bekim	Krasniqi	Kosovo	M	30.1.1999	On campus	None
4	b4658c3a9	Yusuf	Kaya	Turkey	M	16.6.1998	On campus	None
5	e6ec47f29	Zoran	Babić	Serbia	M	29.10.1998	On campus	Part-time
6..129								
130	278a75edf	Marwa	Singh	Qatar	F	29.3.1996	Remote	None

3.1.2 Demographics

The `Demographics.xlsx` file contains simulated demographic data on the students, including name, date of birth, gender, location (on-campus vs. remote student), and employment status. Table 2 shows a preview of the data.

- **user**: User identifier in the learning management system, with 130 distinct entries.
- **Name**: User's first name.
- **Surname**: User's last name.
- **Origin**: Country of origin.
- **Gender**: User's gender: F (Female, 50%) or M (Male, 50%).
- **Birthdate**: Date of birth.
- **Location**: Whether the student is on campus or studying remotely, with 2 distinct entries, such as On campus (81.54%); Remote (18.46%).
- **Employment**: Whether the student is working part-time, full-time or not at all, with 3 distinct entries, such as None (70.77%); Part-time (25.38%); Full-time (3.85%).

3.1.3 Results

Performance data is provided in the `Results.xlsx` file, including grades per assignment and the overall course grade. Table 3 shows a preview of the data (the column names have been abbreviated in the preview). Figure 2 shows the distribution of grades per graded item.

- **user**: User name in the learning management system (it matches the previous files).
- **Grade.SNA_1**: Grade of the first SNA assignment (0–10).
- **Grade.SNA_2**: Grade of the second SNA assignment (0–10).
- **Grade.Review**: Grade of the studies' review assignment (0–10).
- **Grade.Group_self**: Individual grade of the group project (0–10).
- **Grade.Group_All**: Group grade of the group project (0–10).
- **Grade.Excercises**: Grade of the practical exercises (0–10).
- **Grade.Project**: Final project grade (0–10).
- **Grade.Literature**: Grade of the literature review assignment (0–10).
- **Grade.Data**: Grade of the data analysis assignment (0–5).
- **Grade.Introduction**: Grade of the introductory assignment (0–10).
- **Grade.Theory**: Grade of the theory assignment (0–10).
- **Grade.Ethics**: Grade of the ethics assignment (0–10).
- **Grade.Critique**: Grade of the critique assignment (0–10).
- **Final_grade**: Final course grade (0–10).

3.1.4 AllCombined

This file `AllCombined.xlsx` contains the students' demographics, grades, and frequency of each action in the LMS. The columns from students' demographic data and grades remain the same, while the event data has been grouped per student for each type of event (`Action` column in the `Events` dataset), i.e., there is a new column for each type of event that contains the number of events of that type that each student had. Moreover, a new column `AchievingGroup` separates the high achievers from the low achievers. Table 4 shows a preview of the new columns of the data (the column names have been abbreviated) that were not shown in the previous previews (Fig. 3).

- **Frequency.Applications**: Number of events related to the "Applications" resource.
- **Frequency.Assignment**: Number of events related to the assignments' submission.
- **Frequency.Course_view**: Number of visits to the course main page.
- **Frequency.Feedback**: Number of views of the assignment feedback.
- **Frequency.General**: Number of events related to general learning resources.
- **Frequency.Group_work**: Number of events related to the group work.

Table 3 Preview of the LA course results (grades) dataset

	User	SNA_1	SNA_2	Review	Group_self	Group_All	Excercises	Project	Literature	Data	Introduction	Theor	Ethics	Critique	Final_grade
1	6eba3ff82	0	0	6.67	5	4.00	10	0.00	6.67	4	6	2	2	4	2.626970
2	05b604102	8	10	6.67	1	3.00	10	7.00	6.67	3	6	2	8	4	4.670169
3	111422ee7	10	10	10.00	10	9.11	10	9.33	10.00	5	10	10	10	10	9.244600
4	b4658c3a9	5	5	0.00	1	4.00	10	6.00	4.33	3	4	2	2	6	0.000000
5	e6ec47f29	10	10	10.00	10	9.18	10	5.33	10.00	5	10	10	10	10	8.238179
6..129															
130	278a75edf	9	10	7.33	10	8.56	10	5.33	7.33	4	6	2	6	6	7.026270

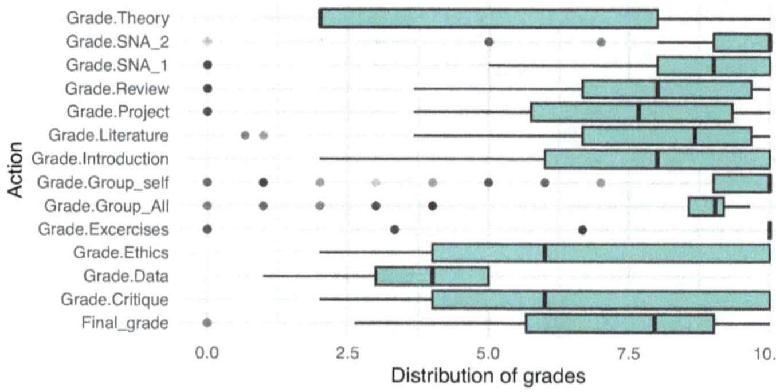

Fig. 2 Grade per student and graded item

- **Frequency.Instructions**: Number of events related to assignment instructions.
- **Frequency.La_types**: Number of events related to the "LA types" resource.
- **Frequency.Practicals**: Number of events related to the practicals.
- **Frequency.Social**: Number of events related to the forum discussion.
- **Frequency.Ethics**: Number of events related to the "Ethics" resource.
- **Frequency.Theory**: Number of events related to the "Theory" resource.
- **Frequency.Total**: Number of events overall.
- **AchievingGroup**: Categorization as high achievers (top 50% grades) and lows achievers (bottom 50% grades).

3.2 *LMS Data from a Higher Education Institution in Oman*

Link to the dataset

The next dataset includes the log data of students enrolled in a computing specialization in a higher education institution in Oman. The dataset contains data from students enrolled in the sixth semester or beyond. The data was recorded across five modules (Spring 2017–2021) and includes the records of 326 students with 40 features in total, including the students' academic information (24 features), logs of students' activities performed on the Moodle LMS (10 features), and students' video interactions on the eDify mobile application (6 features). The academic data includes demographic data, academic data, study plan, and academic violations. The Moodle activity data includes students' timestamped activities in Moodle. The eDify data contains video interactions in the eDify mobile application.

The dataset has been described in an article by Hasan et al. [56] and was originally made available in the Zenodo repository [57] with a CC BY 4.0 license, which means anyone can share and adapt the dataset but must give credit to the author and cannot apply any further restrictions to the dataset. Besides the raw data,

Table 4 Preview of the LA course combined dataset

	F.Applications	F.Assignment	F.Course_view	F.Feedback	F.General	F.Group_work	F.Instructions	F.La_types	F.Practicals	F.Social	F.Ethics	F.Theory	F.Total	AchievingGroup
1	2	121	103	7	10	55	17	1	89	12	0	0	417	Low achiever
2	0	62	294	22	19	323	68	12	69	32	14	3	918	Low achiever
3	0	41	53	0	11	19	6	7	47	0	0	15	199	Low achiever
4	0	44	49	0	10	19	7	8	48	0	0	14	199	Low achiever
5	0	9	119	9	9	218	7	1	61	0	0	3	436	Low achiever
6..129														
130	7	84	491	56	49	457	176	25	145	67	32	28	1617	High achiever

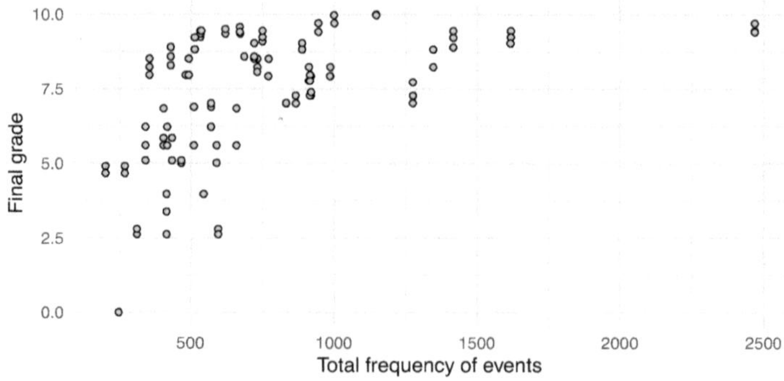

Fig. 3 Relationship between total frequency of events and final grade

the dataset includes a processed file that contains a series of indicators that can be used for different purposes such as predictive modeling or clustering of students' profiles. It has been used in several publications with the purpose of predicting student performance using different algorithms and approaches [58, 59]. The main files of the dataset are described below.

3.2.1 Student Academic Information

Students' academic information downloaded from the institutional information system. The data are spread in 15 files (starting with `KMS`), but have been combined in a single file (`KMSmodule.RDS`) for convenience. Table 5 shows a preview of the data. The resulting dataset has the following structure:

- **file**: Original file name that contains the module offering identifier (e.g., "KMS Module 1 F19.csv"). There are 15 distinct entries (one for each file).
- **ModuleCode**: Module identifier (Module 1–5).
- **ModuleTitle**: Name of the module (Course 1–5).
- **SessionName**: Class section in which the student has been enrolled (Session-A or Session-B).
- **RollNumber**: Student identifier (e.g., "Student 83"). There are 306 distinct students.
- **CGPA**: Students' cumulative GPA (1–4).
- **AttemptCount**: Number of attempts per student and module.
- **Advisor**: Students' advisor identifier (e.g., 'Advisor 16'). There are 50 distinct advisors.
- **RemoteStudent**: Whether the student is studying remotely. There are 2 possible values: Yes (0.31%) or No (99.69%).
- **Probation**: Whether the student has previous incomplete modules. There are 2 possible values: Yes (3.36%) or No (96.64%).

Table 5 Preview of the Higher Education Institution Academic data

	File	Module-Code	Module-Title	Session-Name	RollNumber	CGPA	Attempt-Count	Advisor	Remote-Student	Probation	High-Risk	Term-Exceeded	AtRisk	AtRisk-SSC	Other-Modules	Prerequisite-Modules	Plagiarism-History	MalPractice-History
1	KMS Module 1 F19.csv	Module 1	Course 1	Session-A	Student 221	2.50	1	Advisor 16	No	No	No	No	No	No	2	No	Module 3	No History
2	KMS Module 1 F19.csv	Module 1	Course 1	Session-A	Student 208	3.25	1	Advisor 25	No	No	No	No	No	No	3	No	No History	No History
3	KMS Module 1 F19.csv	Module 1	Course 1	Session-A	Student 209	2.50	1	Advisor 3	No	No	No	No	No	No	2	No	No History	Module 2
4	KMS Module 1 F19.csv	Module 1	Course 1	Session-A	Student 210	2.50	1	Advisor 8	No	No	No	No	No	No	3	No	No History	No History
5	KMS Module 1 F19.csv	Module 1	Course 1	Session-A	Student 211	2.50	1	Advisor 14	No	No	No	No	No	No	3	No	No History	No History
6..326																		
327	KMS Module 5 SP18.csv	Module 5	Course 5	Session-A	Student 89	2.00	1	Advisor 1	No	No	No	Yes	No	No	2	Yes	Module 20	Module 18

- **HighRisk**: Whether the module has a high risk of failure. There are 2 possible values: Yes (6.73%) or No (93.27%).
- **TermExceeded**: Whether the student is progressing in their degree plan. There are 2 possible values: Yes (1.83%) or No (98.17%).
- **AtRisk**: Whether the student has failed two or more modules in the past. There are 2 possible values: Yes (23.55%) or No (76.45%).
- **AtRiskSSC**: Whether the student has been registered for having any educational deficiencies. There are 2 possible values: Yes (4.59%) or No (95.41%).
- **OtherModules**: Number of other modules the student is enrolled in on the same semester.
- **PrerequisiteModules**: Whether the student has enrolled in the prerequisite module.
- **PlagiarismHistory**: Modules for which the student has a history of plagiarism.
- **MalPracticeHistory**: Modules for which the student has a history of academic malpractice.

3.2.2 Moodle

The Moodle data is spread around 15 files (starting with `Moodle`), which contain all the clicks that students performed in the Moodle LMS in each module. The 15 files have been combined in a single file (`moodleEdify.RDS`) for convenience. Table 6 shows a preview of the data. The resulting dataset has the following structure:

- **csv**: Module offering identifier. There are 15 distinct entries (one for each file).
- **Time**: Timestamp in which the event occurred, ranging between January 1st 2018 to December 9th 2020.
- **Event context**: Resource of the LMS where the event takes place, for example "File: Lecture 4".
- **Component**: Type of resource involved in the event. There are 16 distinct entries, such as System (51.71%); File (22.30%); Turnitin Assignment 2 (15.04%) and 13 others.
- **Event name**: Name of the event in Moodle. There are 70 distinct entries, such as Course viewed (36.59%); Course module viewed (25.98%); List Submissions (11.08%) and 67 others.

3.2.3 Activity

Aggregated information per student based on the Moodle data, including the activity on campus and at home. The data are also spread in 15 files (starting with `Activity`), but has been combined in a single file (`ActivityModule.RDS`) for convenience. Table 7 shows a preview of the data. The resulting dataset has the following structure:

Table 6 Preview of the Higher Education Institution Moodle data

	csv	Time	Event context	Component	Event name
1	Moodle Module 1 F19	17/10/21, 12:39	Course: Object Oriented Programming(Fall 2019 - F19 COMP 0328.1)	Logs	Log report viewed
2	Moodle Module 1 F19	17/10/21, 12:38	Course: Object Oriented Programming(Fall 2019 - F19 COMP 0328.1)	System	Course viewed
3	Moodle Module 1 F19	19/08/20, 12:40	Course: Object Oriented Programming(Fall 2019 - F19 COMP 0328.1)	System	Course viewed
4	Moodle Module 1 F19	6/08/20, 21:00	Course: Object Oriented Programming(Fall 2019 - F19 COMP 0328.1)	System	Course viewed
5	Moodle Module 1 F19	4/08/20, 12:51	File: MIG	File	Course module viewed
6..98696					
98697	Moodle Module 5 SP18	29/02/18, 18:59	Course: Business Technology Management(Spring 2018 - S18 COMP 30009)	System	Course viewed

Table 7 Preview of the Higher Education Institution Activity data

	File	RollNumber	Online C	Online O
1	Activity Module 1 F19.csv	Student 208	35	108
2	Activity Module 1 F19.csv	Student 209	332	256
3	Activity Module 1 F19.csv	Student 210	158	20
4	Activity Module 1 F19.csv	Student 211	345	90
5	Activity Module 1 F19.csv	Student 212	352	459
6..325				
326	Activity Module 5 SP18.csv	Student 89	216	43

Table 8 Preview of the Higher Education Institution grade data

	File	RollNumber	SessionName	CW1	CW2	ESE
1	Result Module 1 F19.csv	Student 208	Session-A	29	34	63
2	Result Module 1 F19.csv	Student 209	Session-A	18	36	54
3	Result Module 1 F19.csv	Student 210	Session-A	16	18	34
4	Result Module 1 F19.csv	Student 211	Session-A	15	28	43
5	Result Module 1 F19.csv	Student 212	Session-A	20	34	54
6..325						
326	Result Module 5 SP18.csv	Student 89	Session-A	79	80	30

- **file**: Original file name that contains the module offering identifier (e.g., "Activity Module 1 F19.csv"). There are 15 distinct entries (one for each file).
- **RollNumber**: Student identifier (e.g., "Student 208").
- **Online C**: Duration of the activity (in minutes) within campus.
- **Online O**: Duration of the activity (in minutes) off campus.

3.2.4 Results

Student performance data in each module. The data are also spread in 10 files (starting with `Result`), but has been combined in a single file (`ResultModule.RDS`) for convenience. Table 8 shows a preview of the data. The resulting dataset has the following structure:

- **file**: Original file name that contains the module offering identifier (e.g., "Result Module 1 F19.csv"). There are 10 distinct entries (one for each file).
- **RollNumber**: Student identifier (e.g., "Student 208"). There are 326 distinct students.
- **SessionName**: Class section, Session-A (84.66%), or Session-B (9.82%).
- **CW1**: Grade of students' first assignment (0–100).
- **CW2**: Grade of students' second assignment (0–100).
- **ESE**: Grade of students' end-of-semester examination (0–100).

Table 9 Preview of the Higher Education Institution eDify data

	File	RollNumber	Played	Paused	Likes	Segment
1	VL Module 1 F19.csv	Student 208	5	0	1	3
2	VL Module 1 F19.csv	Student 209	1	4	3	3
3	VL Module 1 F19.csv	Student 210	5	0	2	5
4	VL Module 1 F19.csv	Student 211	5	0	1	3
5	VL Module 1 F19.csv	Student 212	4	0	2	3
6..325						
326	VL Module 5 SP18.csv	Student 89	5	7	1	4

3.2.5 eDify

Student aggregated activity data in the eDify mobile application. The data are also spread in 15 files (starting with `VL`), but has been combined in a single file (`VLModule.RDS`) for convenience. Table 9 shows a preview of the data. The resulting dataset has the following structure:

- **file**: Original file name that contains the module offering identifier (e.g., "Result Module 1 F19.csv"). There are 15 distinct entries (one for each file).
- **RollNumber**: Student identifier (e.g., "Student 208"). There are 326 distinct students.
- **Played**: Number of times the student has played a video.
- **Paused**: Number of times the student has paused a video.
- **Likes**: Number of times the student has liked a video.
- **Segment**: Number of times a student has scrolled to a specific part of the video.

3.3 *School Engagement, Academic Achievement, and Self-regulated Learning*

Link to the dataset

This dataset includes measures of school engagement, self-regulation and academic performance of a group of primary school students in northern Spain [60]. The data contains responses to the questionnaire from 717 primary education students. The subjects were recruited using convenience sampling from 15 schools (5 privately funded and 10 publicly funded) in northern Spain. The objective of collecting these data was to characterize school engagement and to explore to which extent different engagement profiles are associated with academic performance and self-regulation. In the questionnaire, engagement was assessed with the school engagement measure [61], which allows to differentiate between behavioral, cognitive and emotional engagement. Self-regulation was assessed using the self-regulation strategy inventory [62], which allows measuring students' approaches

Fig. 4 Self-reported grades of students in Mathematics and Spanish

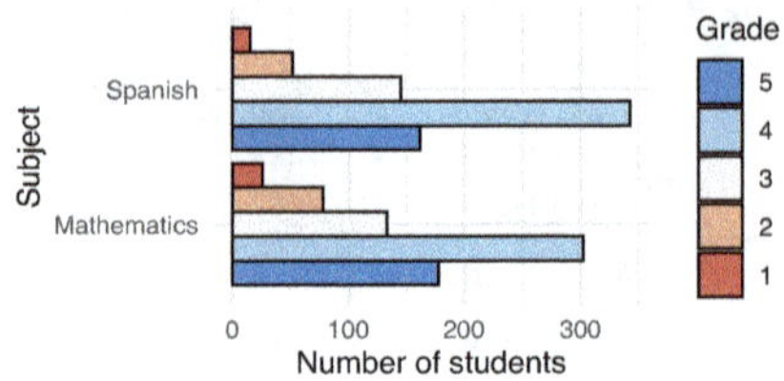

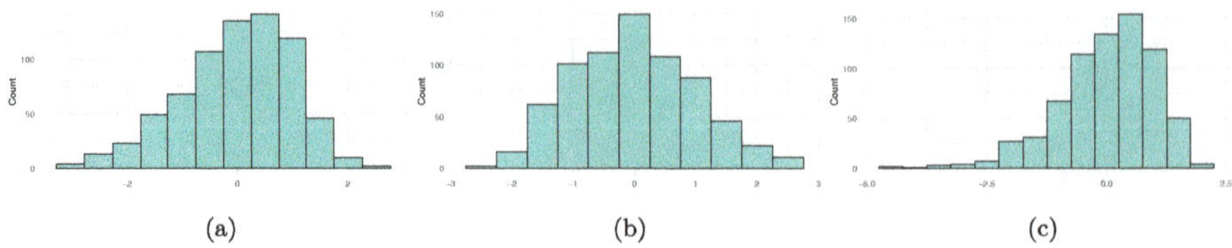

Fig. 5 School engagement results of the survey. (**a**) Emotion engagement z-score. (**b**) Cognitive engagement z-score. (**c**) Behavior engagement z-score

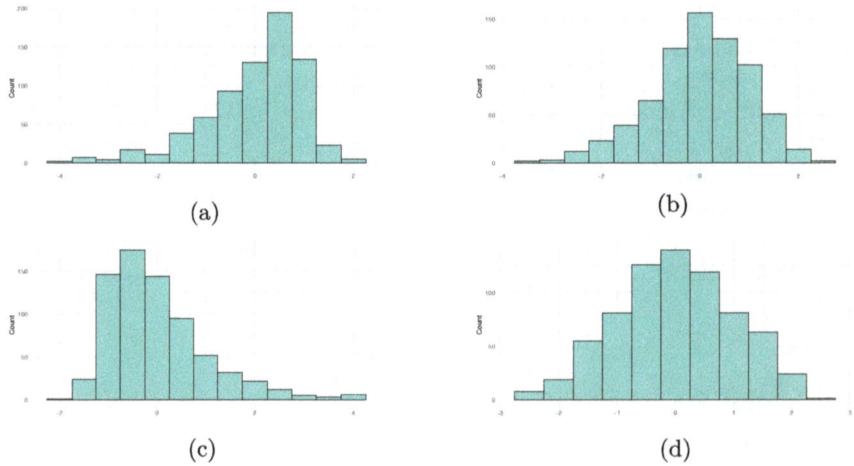

Fig. 6 Self-regulation results of the survey. (**a**) Environment management z-score. (**b**) Information help management z-score. (**c**) Maladaptative behavior z-score. (**d**) Time management z-score

to seeking and learning information, maladaptive regulatory behavior, environment management, and time management. The measure used for achievement was students' self-reported information about their grades in Spanish and mathematics on a scale of 1 (fail) to 5 (outstanding).

Figure 4 summarizes the responses of the students in both subjects, Fig. 5 presents a set of histograms with the engagement measures, and Fig. 6 includes a set of histograms with the self-regulation measures. The dataset was analyzed in a previous article [63], where the authors carried out cluster analysis using LPA (Latent Profile Analysis) to identify different groups of students according to their engagement and self-regulation and to compare the performance between these groups. The dataset is used in Chapter 9 [11] of this book, which covers model-based clustering. The dataset is published with a CC BY 4.0 license, which means that you can share, copy and modify this dataset so long as appropriate credit is given. Table 10 shows a preview of the dataset. Below, we describe the dataset's main variables.

- **alumno**: Student identifier in the school.
- **sexo**: Student's gender (1 = Male, 48.40%; 2 = Female, 51.46%).
- **coleg**: School identifier (1–15).
- **curso**: Grade that students were in 5th grade (62.48%) or 6th grade (37.52%).
- **grup**: Class section (1–3).
- **ren.mat**: Mathematics self-reported academic achievement (1–5).
- **ren.leng**: Spanish self-reported academic achievement (1–5).
- **Emotion_Engage**: Emotional engagement (z-score).
- **Cognitive_Engage**: Cognitive engagement (z-score).
- **Behavior_Engage**: Behavioural engagement (z-score).
- **Enviroment_Manage**: Environment management (z-score).
- **Information_help_Manage**: Information and help management (z-score).
- **Maladapative_Behavior**: Maladaptative self-regulation (z-score).
- **Time_Manage**: Time management self-regulation (z-score).

3.4 *Teacher Burnout Survey Data*

Link to the dataset

The next dataset presents the responses collected from a survey about teacher burnout in Indonesia [64]. The survey questionnaire contains 18 items in five different categories. The first category contains five items to assess the teacher self-concept (TSC), from the TSC Evaluation Scale [65]. The second category is teacher efficacy (TE), with 5 items adapted from [66]. The remaining categories are Emotional Exhaustion (EE, 5 items), Depersonalization (DP, 3 items), and Reduced Personal Accomplishment (RPA, 5 items), adapted from the Maslach burnout inventory [67]. The survey items were measured using a 5-point Likert scale, where 1 represents "never", and 5 represents "always".

Table 11 shows a preview of the dataset with the questions and answers, and Fig. 7 represents it in a Likert scale chart. The dataset has been analyzed using several statistical methods [68]: Content Validity Index (CVI), Exploratory Factor Analysis (EFA), Confirmatory Factor Analysis (CFA), Covariance-Based SEM (CB-SEM). The main aim was to determine the factors that may be predictors of teacher burnout. In this book, we also make use of this dataset to illustrate Factor Analysis [19] and SEM [20]. The files associated with this dataset are licensed under a CC BY 4.0 license, which means you can share, copy and modify this dataset so

Table 10 Preview of the School Engagement, Academic Achievement, and SRL data

	alumno	sexo	coleg	curso	grup	ren.mat	ren.leng	Emotion_ Engage	Cognitive_ Engage	Behavior_ Engage	Enviroment_ Manage	Information_ help_Manage	Maladapative_ Behavior
1	1	2	1	6	1	4	4	0.8989251	0.25057461	−0.66817340	−0.1483424	0.2489923	0.08695216
2	2	1	1	6	1	4	3	−2.4712139	1.50596375	0.55572643	0.6627681	0.3436559	−0.79488941
3	3	2	1	6	1	4	4	0.2614537	1.43763063	−0.49791451	0.3090213	0.1281156	−0.13740548
4	4	2	1	6	1	3	3	−0.5421392	−0.19963989	−0.48553522	−0.0706820	−0.7501752	0.31906640
5	5	1	1	6	1	5	3	0.5337485	−0.05256182	−0.04035667	0.1241936	0.2688936	−0.77730497
6..716													
717	25	2	15	6	2	5	5	0.1609796	−0.88981754	1.28726373	0.3358852	1.4474958	0.05243749

Table 11 Preview of the burnout survey data answers

	TSC1	TSC2	TSC3	TSC4	TSC5	TE1	TE2	TE3	TE4	TE5	EE1	EE2	EE3	EE4	EE5	DE1	DE2	DE3	RPA1	RPA2	RPA3	RPA4	RPA5
1	4	4	4	4	4	4	4	4	4	4	4	4	4	4	4	4	4	4	4	4	4	4	4
2	4	4	4	4	4	4	4	4	4	4	4	4	4	4	4	4	4	4	4	4	4	4	4
3	4	4	5	4	5	5	4	4	5	4	4	5	5	4	5	5	5	5	5	4	5	4	5
4	4	4	5	4	5	5	4	4	5	4	4	5	5	4	5	5	5	5	5	4	5	4	5
5	4	5	3	4	4	4	4	4	4	4	4	4	4	4	4	4	4	4	4	4	4	4	4
6..875																							
876	5	5	5	5	5	5	5	5	5	5	5	5	5	5	5	5	4	5	5	5	5	5	5

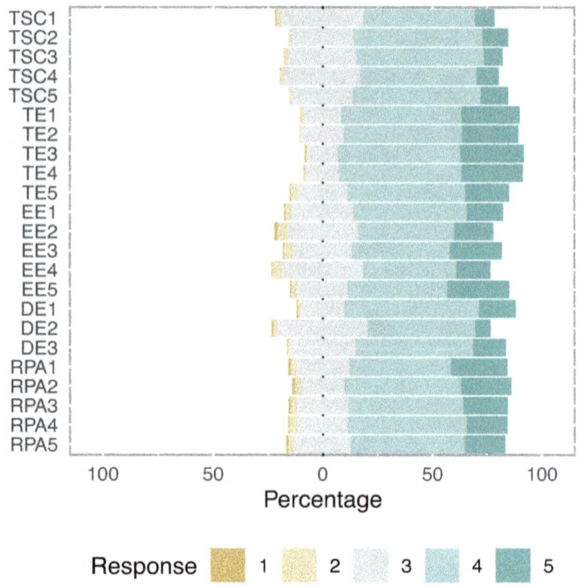

Fig. 7 Results of the burnout survey

long as appropriate credit is given to the authors. The variables of the dataset are described below.

- **Teacher self-concept**
  - **TSC1**: Response to the item "I think I have good teaching skills and ability".
  - **TSC2**: Response to the item "I have a reputation for being an efficient teacher".
  - **TSC3**: Response to the item "My colleagues regard me as a competent teacher".
  - **TSC4**: Response to the item "I feel I am a valuable person".
  - **TSC5**: Response to the item "Generally speaking, I am a good teacher".
- **Teacher efficacy**
  - **TE1**: Response to the item "I help my students value learning".
  - **TE2**: Response to the item "I motivate students who show low interest in schoolwork".
  - **TE3**: Response to the item "I improve understanding of students who are failing".
  - **TE4**: Response to the item "I provide appropriate challenges for very capable students".
  - **TE5**: Response to the item "I get students the students to follow classroom rules".
- **Emotional exhaustion**
  - **EE1**: Response to the item "I feel emotionally drained from my work".

  - **EE2**: Response to the item "I feel used up at the end of the workday".
  - **EE3**: Response to the item "I feel fatigued when I get up in the morning and have to face another day on the job".
  - **EE4**: Response to the item "I feel burnt out from my work".
  - **EE5**: Response to the item "Working with people all day is really a strain on me".

- **Depersonalization**

  - **DE1**: Response to the item "I've become more callous toward people since I took this job".
  - **DE2**: Response to the item "I worry that this job is hardening me emotionally".
  - **DE3**: Response to the item "I feel frustrated by my job".

- **Reduced Personal Accomplishment**

  - **RPA1**: Response to the item "I cannot easily create a relaxed atmosphere with my students".
  - **RPA2**: Response to the item "I do not feel exhilarated after working closely with my students".
  - **RPA3**: Response to the item "I have not accomplished many worthwhile things in this job".
  - **RPA4**: Response to the item "I do not feel like I'm at the end of my rope".
  - **RPA5**: Response to question "In my work, I do not deal with emotional problems very calmly".

3.5 Interdisciplinary Academic Writing Self-efficacy

Link to the dataset

This dataset contains the result of nursing students' responses to the Situated Academic Writing Self-Efficacy Scale (SAWSES) [69] questionnaire for interdisciplinary students (543 undergraduate and 264 graduate). Participating students were recruited from three higher education institutions and from the general public using social media. The survey contains 16 items based on Bandura's self-efficacy theory [70] and the model proposed by [71].

The questionnaire items are related to three dimensions. The first dimension is Writing Essentials, with three items related to synthesis, emotional control, and language. The second dimension is Relational Reflective Writing, with seven items related to relationship building with writing facilitators and the self through reflection. The third dimension is Creative Identity, with six items related to gaps in student achievement of transformative writing. Demographic data for age, gender, years in post-secondary, English language status, English writing status, and writing attitude are also included.

Table 12 Preview of the SAWSES demographic data

	Age	Gender	Writing-Attitude	TypeStudent	Ugyears	Gryears	WriteEnglish	SpeakEnglish
1	2	2	3	1	5	NA	1	1
2	2	2	1	1	5	NA	1	1
3	4	2	1	1	1	NA	3	4
4	2	2	3	1	3	NA	1	1
5	2	2	3	1	4	NA	1	1
6..806								
807	3	3	1	2	NA	3	1	1

The survey has been validated in a published article [72]. We make use of this dataset in Chapter 8 [12], devoted to clustering algorithms. The dataset is published under the CC0 1.0, which means that anyone can copy, modify, distribute it, even for commercial purposes, without asking permission from the authors. The dataset variables are described below.

First, we describe the **demographic** variables, which can be previewed in Table 12.

- **Age**: Age.
- **Gender**: Student's gender, 1 = male (24.91%), 2 = female (72.12%), 3 = non-binary (2.97%).
- **WritingAttitude**: Writing attitude, 1 = dislikes writing (44.73%), 2 = somewhere in between (14.37%), or 3 = likes writing (44.73%).
- **TypeStudent**: Academic level, 1 = undergraduate (67.29%), 2 = graduate (32.71%).
- **Ugyears**: Undergraduate years in the school for undergraduate students.
- **Gryears**: Graduate years in the school for graduate students.
- **WriteEnglish**: English writing status (1–5).
- **SpeakEnglish**: English language status (1–5).

Next, we present the questionnaire responses for each of the dimensions. The responses are on a scale of 0–100. Table 13 shows a preview of the data, and Fig. 8 presents a box plot with some the response distribution of the questionnaire.

- **Writing Essentials**
  - **overcome**: Response to the item "Even when the writing is hard, I can find ways to overcome my writing difficulties".
  - **words**: Response to the item "I can successfully use scholarly academic words and phrases when writing in my courses".
  - **synthesize**: Response to the item "I can combine or synthesize multiple sources I've read to create an original product or text".

Table 13 Preview of the SAWSES questionnaire responses

	Overcome	Words	Synthesize	Creativity	Meaning	Improve	Reflect	Spark	Ideas	Overall	Voice	Original	Discipline	Wander	Adapt	Feedback
1	100	100	100	100	100	100	100	100	80	100	80	100	97	82	100	86
2	100	100	100	61	70	100	100	64	92	100	100	100	100	93	100	100
3	100	100	100	100	100	100	100	65	81	100	100	100	100	95	100	71
4	51	82	49	20	42	76	93	84	91	83	46	18	75	25	21	53
5	16	100	37	75	100	100	80	67	100	59	59	47	100	27	90	100
6..806																
807	71	93	95	94	93	100	100	100	100	74	76	100	91	94	100	100

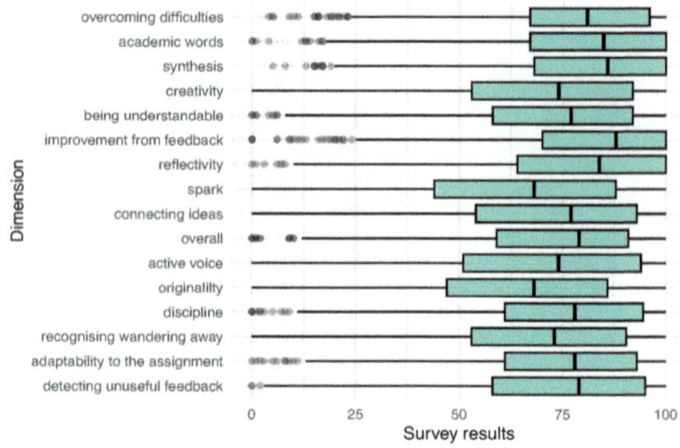

Fig. 8 Results of the SAWSES survey

- **Relational Reflective Writing**
  - **meaning**: Response to the item "When I write, I can think about my audience and write so they clearly understand my meaning".
  - **improve**: Response to the item "When I receive feedback on my writing, no matter how it makes me feel, I can use that feedback to improve my writing in the future".
  - **reflect**: Response to the item "When I reflect on what I am writing I can make my writing better".
  - **ideas**: Response to the item "When I read articles about my topic, the connections I feel with the ideas of other authors can inspire me to express my own ideas in writing".
  - **overall**: Response to the item "When I look at the overall picture I've presented in my writing, I can assess how all the pieces tell the complete story of my topic or argument".
  - **wander**: Response to the item "I can recognize when I've wandered away from writing what my audience needs to know and have begun writing about interesting, but unrelated, ideas".
  - **adapt**: Response to the item "With each new writing assignment, I can adapt my writing to meet the needs of that assignment".
  - **feedback**: Response to the item "When I seek feedback on my writing, I can decide when that feedback should be ignored or incorporated into a revision in my writing".
- **Creative Identity**
  - **creativity**: Response to the item "I can use creativity when writing an academic paper".
  - **spark**: Response to the item "I feel I can give my writing a creative spark and still sound professional".
  - **voice**: Response to the item "I feel I can develop my own writing voice (ways of speaking in my writing that are uniquely me)".

  - **original**: Response to the item "Even with very specific assignment guidelines, I can find ways of writing my assignment to make it original or unique".
  - **discipline**: Response to the item "I can comfortably express the concepts, language, and values of my discipline or major in my writing assignments".

3.6 Educators' Discussions in a MOOC (SNA)

Link to the dataset

This dataset belongs to two offerings of the MOOC "The Digital Learning Transition in K-12 Schools" [73]. The course was aimed at helping school and district leaders implement digital learning initiatives in K-12 education. The objectives of the course were for participants to understand the advantages of digital learning in schools, to assess the specific goals for their own school, and to devise a plan to achieve such goals. The course consisted of five learning units dealing with the schools of the future, teaching and learning culture, successful digital transition, leading the transition, and crowd sourcing. The MOOCs were offered to American as well as international teachers. There were two offerings of the MOOC, with minor variations regarding duration and groups. The dataset contains the interactions of the MOOC discussion forums and concerns teachers' communications throughout the courses. The dataset also contains the characteristics of the teachers, e.g., their professional roles, and experience.

The dataset is extensively described in a dedicated publication where the authors give details about the context and the courses, the files, and the fields contained in each file [74]. In this book, we use this dataset to illustrate dissimilarity-based clustering [39], Social Network Analysis [39] and Temporal Network Analysis [41]. The dataset is available with a CC0 1.0 license. Therefore, permission to copy, modify, and distribute, even for commercial purposes, is granted. As a standard Social Network Analysis dataset, it comes in two files for each of the courses (four in total), which we describe in detail below.

- **Edges file**: This file defines who interacted (the source or the sender of the communication) with whom (the target or the receiver of the communication). The edges file comes with other metadata, such as time, discussion topic and group. Table 14 presents a preview of one of the edge files, and Fig. 9 shows the network of collaboration between forum contributors.
  - **Sender**: Source of the communication identifier (1–445).
  - **Receiver**: Target of the communication identifier (1–445).
  - **Timestamp**: Timestamp of the intervention in "m/d/Y H:M" format, ranging from April 10th 2013 to June 8th 2013.
  - **Discussion Title**: Title of the discussion.
  - **Discussion Category**: Category of the discussion.

Table 14 Preview of the MOOC for educators discussion data (edges)

	Sender	Receiver	Timestamp	Discussion title	Discussion category
1	360	444	4/4/13 16:32	Most important change for your school or district?	Group N
2	356	444	4/4/13 18:45	Most important change for your school or district?	Group D-L
3	356	444	4/4/13 18:47	DLT Resources—Comments and Suggestions	Group D-L
4	344	444	4/4/13 18:55	Most important change for your school or district?	Group O-T
5	392	444	4/4/13 19:13	Most important change for your school or district?	Group U-Z
6..2528					
2529	11	11	6/16/13 17:12	How to make changes in teacher's understanding of importance of innovative teaching?	Curriculum & Instruction

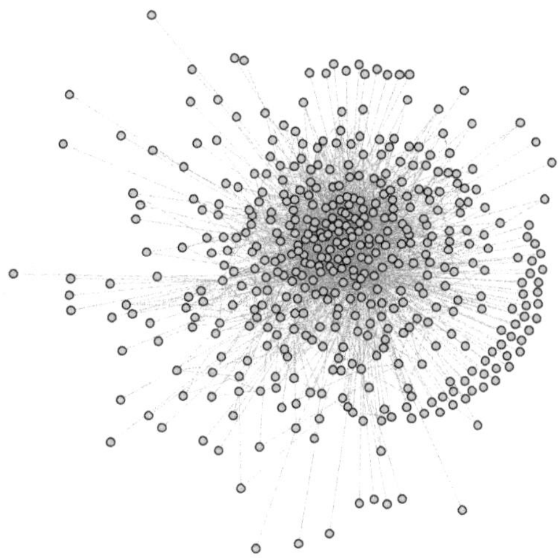

Fig. 9 MOOC participants' network

- **Nodes file**: The file defines the characteristics of the interacting teachers, their IDs, their professional status and expertise level. Table 15 shows a preview of one of the nodes file data.
  - **UID**: Teacher identifier (1–445).
  - **role1**: Role of the teacher.
  - **experience**: Level of experience (1–3).
  - **experience2**: Years of experience.
  - **country**: Country of origin.
  - **gender**: Teachers' gender, female (68.09%); male (31.69%).
  - **expert**: Level of expertise (0–1).

- **Centralities file**: The file contains the centrality measures of the participants which indicate their number of contributions (`OutDegree`), replies (`InDegree`), position in the network (`Closeness_total`), worth of their connections (`Eigen`), spread of their ideas (`Diffusion_degree`), and more. For more information on how to calculate centralities from interaction data, refer to Chapter 15 [39]. Table 16 shows a preview.
  - **name**: Teacher identifier (1–445).
  - **InDegree**: In-degree centrality. Number of responses received.
  - **OutDegree**: Out-degree centrality. Number of messages sent.
  - **Closeness_total**: Closeness centrality. Position in the network.
  - **Betweenness**: Betweenness centrality. Influential position.
  - **Eigen**: Eigen centrality. Worth of connections.
  - **Diffusion.degree**: Diffusion degree centrality [75]. Spread of ideas.
  - **Coreness**: Coreness centrality. Spreading capability.
  - **Cross_clique_connectivity**: Cross clique connectivity. Facilitation of information propagation.

Table 15 Preview of the MOOC for educators discussion data (nodes)

	UID	Facilitator	role1	Experience	experience2	Grades	Location	Region	Country	Group	Gender	Expert	Connect
1	1	0	libmedia	1	6 to 10	secondary	VA	South	US	UZ	female	0	1
2	2	0	classteaching	1	6 to 10	secondary	FL	South	US	DL	female	0	0
3	3	0	districtadmin	2	11 to 20	generalist	PA	Northeast	US	OT	female	0	1
4	4	0	classteaching	2	11 to 20	middle	NC	South	US	N	female	0	0
5	5	0	otheredprof	3	20+	generalist	AL	South	US	AC	female	0	0
6..444													
445	445	1	otheredprof	3	20+	generalist	NC	South	US	N	female	0	0

Table 16 Preview of the MOOC for educators discussion data (nodes)

	Name	InDegree	OutDegree	Closeness_total	Betweenness	Eigen	Diffusion.degree	Coreness	Cross_clique_connectivity
1	1	20	33	0.0010952903	1258.14319	0.205523262	1865	18	305
2	2	2	5	0.0008084074	26.52429	0.010717789	218	6	13
3	3	2	4	0.0007987220	30.60112	0.008623924	191	6	11
4	4	2	14	0.0010193680	72.52345	0.080264834	965	13	37
5	5	16	17	0.0010604454	309.03274	0.161503654	1508	18	154
6..444									
445	445	276	56	0.0013175231	16690.42611	0.699247801	3574	31	2218

3.7 *High School Learners' Interactions (SNA)*

Link to the dataset

The next dataset [76] concerns a course of interactions among 30 students in a high school in Kenitra, Morocco. The course under examination had a duration of two months and covered topics related to computer science: the computer information system, algorithms and programming. The course was implemented in the Moodle LMS, using the forum as a discussion space. Students' interactions were aimed at communicating, discussing and exchanging knowledge among them.

The dataset has been analyzed using social network analysis, is briefly described in an article [77], and is shared under Creative Commons license CC BY 4.0, which means that anyone can share, copy and modify this dataset so long as appropriate credit is given. The dataset includes two files described below.

- **Edges file**: The file contains the interactions source, target and weights. Table 17 shows a preview of the edge files. Figure 10 presents the graph of all the interactions on the network.
  - **source**: Source node identifier (1–21).
  - **Target**: Target node identifier (1–21).
  - **W**: Weight of the link (the value is always 1).

Table 17 Preview of the High school learners' interactions data (edges)

	Source	Target	W
1	17	6	1
2	17	5	1
3	17	11	1
4	17	17	1
5	17	6	1
6..221			
222	21	8	1

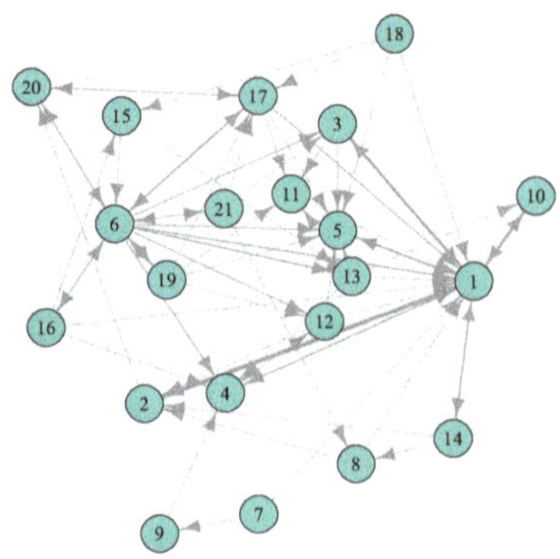

Fig. 10 High school learners' network

Table 18 Preview of the High school learners' interactions data (nodes)

	ID	Username	Name	Genre	Date de naissance
1	1	L1	Aya	F	22/11/2002
2	2	L2	Nouhaila Mahsana	F	25/10/2002
3	3	L3	Said Rhioui	M	26/02/2001
4	4	L4	Mehdi ouchne	M	04/06/2000
5	5	L5	ilyass liagoubi	M	03/04/2001
6..29					
30	30	L30	Kaoula Elyamani	F	06/05/2002

- **Nodes file**: contains the characteristics of the interacting students, e.g., gender and age. Table 18 presents a preview of the dataset.
  - **ID**: Student identifier (1–30).
  - **Username**: Username of the student.
  - **name**: Name of the student.
  - **genre**: Gender of the student F (n = 23); M (n = 7).
  - **Date de naissance**: Birthdate of the student in format "D/M/Y".

3.8 Interactions in an LMS Forum from a Programming Course (SNA)

Link to the dataset

This dataset includes message board data collected from a programming undergraduate course in a higher education institution in Spain using the Moodle LMS. The most particular characteristic of the course is that it follows the CTMTC (Comprehensive Training Model of the Teamwork Competence) methodology [78], which allows individualized training and assessment of teamwork across all stages of teamwork-based learning: storming, norming, performing, delivery and documentation [79].

This is a mandatory course with a workload of 6 ECTS. The data dates back to the first semester of the 2014–2015 academic year. The course offers foundational knowledge on programming and numeric calculus, has a strong focus on the development of algorithms, and a hands-on approach to teaching. The course starts with a two-hour long introductory session; after this session, students work in class and out of class during the whole semester on a team project, following the CTMTC methodology. Individual evidence of teamwork is collected from forum activity, where the work phases are presented, and group evidence of teamwork is collected from Dropbox and wikis.

The dataset refers to individual evidence of teamwork; in other words, it contains information about forum interactions during the course. The original dataset, obtained from Moodle logs, was processed with the help of GraphFES [80] to provide condensed information. The output of GraphFES consists of three different datasets: views, or the number of times user `a` read a message posted by user `b`; replies, which informs about the number of replies from user `a` to user `b`; and messages, which provides a network with the hierarchical structure of messages in the forum. This dataset has been used previously in [81] and is now being publicly released under a CC 4.0 BY-NC-SA license, which means that anyone is free to share, adapt, and distribute the data as long as appropriate credit is given, it is not used for commercial purposes, and the original license is kept. The dataset is used in Chapter 16 [40] of this book, about community detection. This dataset presents the replies network, a directed graph, and consists of an edges file and a nodes file.

- **Edges file**: this file includes information about who (attribute `source`) interacted with (replied to) whom (attribute `target`); in other words, the sender and the receiver of the informational exchange, and how many times that exchange happened during the course (attribute `weight`), considering all messages exchanged. The dataset includes a total of 662 weighed edges. Table 19 presents a preview of the edges file data, and Fig. 11 represents the complete network of interactions among the Moodle users.
  - **source**: Source node identifier (108 distinct values).
  - **target**: Target node identifier (108 distinct values).
  - **weight**: Weight of the link.
- **Nodes file**: this file contains information about all users with access to the course in the Moodle space, including students, instructors and administrators. The file includes a total of 124 nodes (users); of these, 110 users are students, distributed in 19 groups of between 5 and 7 members each. The file includes an identifier for each user (attribute `id`), the username (attribute `user`; after anonymization, all usernames have the format `user_id`), the number of initial posts, which refers to the number of first posts in a thread (attribute `initPosts`), the number of replies, or posts that were a reply to another post (attribute `replyPosts`) and the total number of posts by that user in the forum (attribute `totalPosts`), which is the sum of `initPosts` and `replyPosts`. It is worth noting that `user_55`, a central node of the network, corresponds to the main instructor of the course. Table 20 shows a preview of the nodes file.
  - **User**: User identifier. There are 124 distinct users.

Table 19 Preview of the programming course interaction data (edges)

	Source	Target	Weight
1	192	164	1
2	164	55	2
3	139	142	1
4	142	55	2
5	175	55	5
6..661			
662	194	153	1

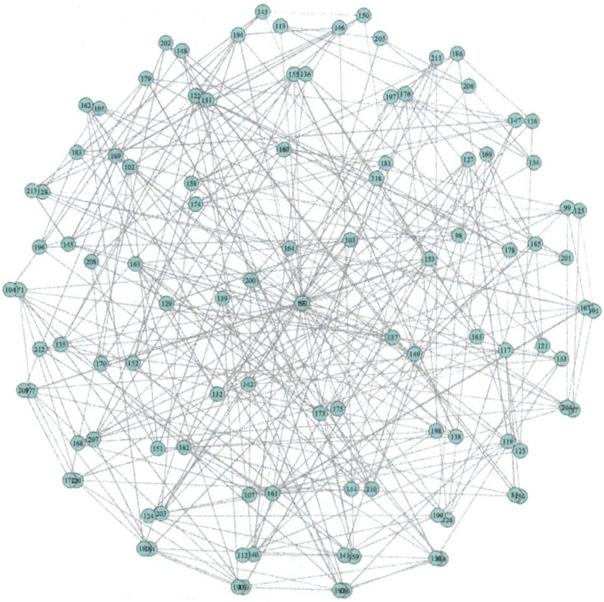

Fig. 11 Interactions of the users in the Moodle discussion forum

Table 20 Preview of the programming course interaction data (nodes)

	Id	User	initPosts	replyPosts	totalPosts
1	54	user_54	0	0	0
2	55	user_55	23	2	25
3	56	user_56	0	0	0
4	57	user_57	0	0	0
5	58	user_58	0	0	0
6..123					
124	214	user_214	1	0	1

- **initPosts**: Number of first posts in a thread.
- **replyPosts**: Number of replies to posts in a thread.
- **totalPosts**: Total number of posts by a user in the forum.

3.9 *Engagement and Achievement Throughout a Study Program*

Link to the dataset

This dataset contains simulated data of students' online engagement and academic achievement throughout a study program. The dataset has been simulated based on the results of a published article [82]. The article used students' logs extracted from a university's Moodle LMS for all the subjects and for all the students who attended the years: 2015, 2016, 2017 and 2018. The logs were used to derive indicators of engagement, such as frequency of performing learning activities (course browsing, forum consumption, forum contribution, and lecture viewing), session count, total session time, active days and regularity of each activity. Regularity of viewing the course main page, for example, was calculated by dividing main page browse actions for the given student over the total main page browse actions over the course duration; the resulting probabilities are used within a Shannon entropy formula to calculate the entropy.

Then, Latent Class Analysis was used to cluster students into engagement states for each course: Active, Average or Disengaged. Achievement was measured through course final grades, which were divided into tertiles: Achiever, Intermediate and Low. Hence, for each course, students had an engagement state, and an achievement state. The motivation and process of deriving these states from students' engagement indicators is explained in Chapter 11 [31].

The simulated dataset contains the data for 142 students for 8 sequential courses, including their engagement and achievement states, as well as covariate data. It is shared with a CC BY 4.0 license, which means that anyone is free to share, adapt, and distribute the data, as long as appropriate credit is given. The dataset is used in Chapter 13 [33] of this book, to illustrate multi-channel sequence analysis. A preview of the dataset can be seen on Table 22. A visual representation of the evolution of engagement and achievement throughout the program is depicted in Fig. 13. The dataset contains two files:

3.9.1 Longitudinal Engagement Indicators and Grades

The file `LongitudinalEngagement.csv` contains all of the engagement indicators per student and course (frequency, duration and regularity of learning activities) as well as the final grade. Each column is described below and a preview can be seen in Table 21 and in Fig. 12.

- **UserID**: User identifier. There are 142 distinct users.
- **CourseID**: Course identifier. There are 38 distinct courses.
- **Sequence**: Course sequence for the student (1–8).
- **Freq_Course_View**: Number of views of the course main page.
- **Freq_Forum_Consume**: Number of views of the forum posts.
- **Freq_Forum_Contribute**: Number of forum posts created.
- **Freq_Lecture_View**: Number of lectures viewed.
- **Regularity_Course_View**: Regularity of visiting the course main page.
- **Regularity_Lecture_View**: Regularity of visiting the lectures.
- **Regularity_Forum_Consume**: Regularity of reading forum posts.
- **Regularity_Forum_Contribute**: Regularity of writing forum posts.
- **Session_Count**: Number of online learning sessions.
- **Total_Duration**: Total activity time online.

Table 21 Preview of engagement indicators and final grades throughout the eight courses

	UserID	CourseID	Sequence	FCV	FFCs	FFCt	FLV	RCV	RLV	RFCs	RFCt	SC	TD	AD	FG
1	D2C5F64E	C6107FC4	1	150	251	79	132	0.42	0.38	0.18	0.83	121	53330	12	50.04
2	D2C5F64E	4C3F37F0	2	98	84	17	108	0.26	0.49	0.11	0.03	53	14465	5	38.86
3	D2C5F64E	E54A52A3	3	254	354	34	284	0.29	0.56	0.14	0.07	159	64324	12	44.73
4	D2C5F64E	AB7EC624	4	332	825	185	256	0.63	0.84	0.53	0.53	287	122821	19	48.99
5	D2C5F64E	B0E95213	5	386	960	233	356	0.74	0.80	0.75	0.56	321	148792	23	47.84
6..1135															
1136	B235D544	4B912491	8	152	546	203	56	0.32	0.23	0.28	0.48	125	88432	10	69.12

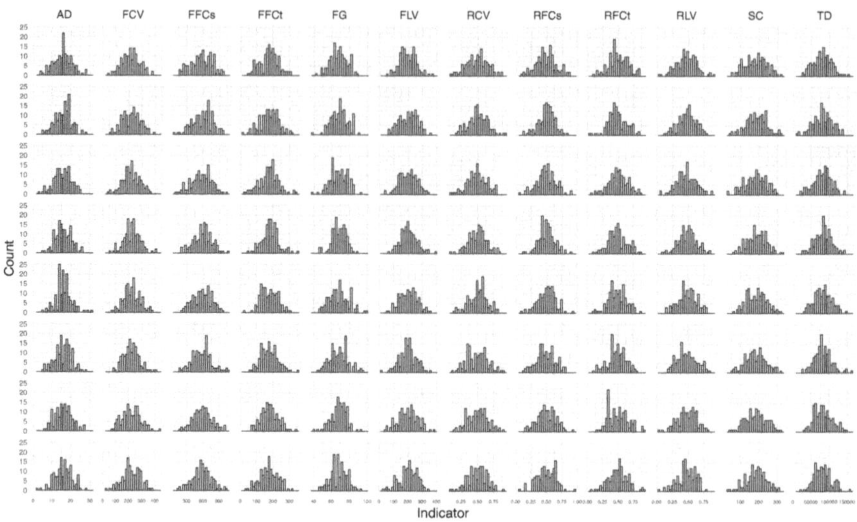

Fig. 12 Histogram of engagement indicators and final grade throughout the eight courses

- **Active_Days**: Number of active days (with online activity).
- **Final_Grade**: Final grade (0–100).

3.9.2 Longitudinal Engagement and Achievement States

The file `SequenceEngagementAchievement.xlsx` contains students' engagement and achievement states for each course as well as covariates (their previous grade, their attitude towards learning, and their gender). Engagement states were obtained by applying model-based clustering techniques to the engagement indicators in the previous file. Achievement states were obtained in a similar way for the final grade. The dataset columns are described below and a preview of the data can be seen at Table 22 and a graphical representation is shown in Fig. 13.

- **UserID**: User identifier. There are 142 distinct users.
- **CourseID**: Course identifier. There are 38 distinct courses.
- **Sequence**: Course sequence for the student (1–8).
- **Engagement**: Engagement state. There are 3 distinct states: Active (29.93%), Average (47.98%), amd Disengaged (22.10%).
- **Final_Grade**: Final grade of each student for each course (1–100).
- **Achievement**: Achievement state calculated using model-based clustering. There are 3 distinct states: Achiever (39.26%), Intermediate (23.33%), and Low (37.41%).

Table 22 Preview of the dataset about engagement and achievement throughout the eight courses

	UserID	CourseID	Sequence	Engagement	Final_Grade	Achievement	AchievementNtile	Prev_grade	Attitude	Gender
1	00050F0E	4C3F37F0	1	Average	72.27	Achiever	Intermediate	6.826613	12.92833	Male
2	00050F0E	E54A52A3	2	Disengaged	72.56	Achiever	Achiever	6.826613	12.92833	Male
3	00050F0E	AB7EC624	3	Average	78.78	Achiever	Achiever	6.826613	12.92833	Male
4	00050F0E	B0E95213	4	Average	74.19	Achiever	Achiever	6.826613	12.92833	Male
5	00050F0E	0B301F55	5	Average	87.35	Achiever	Achiever	6.826613	12.92833	Male
6..1135										
1136	FE2E4C85	79ED6873	8	Active	87.69	Achiever	Achiever	4.189420	20.00000	Male

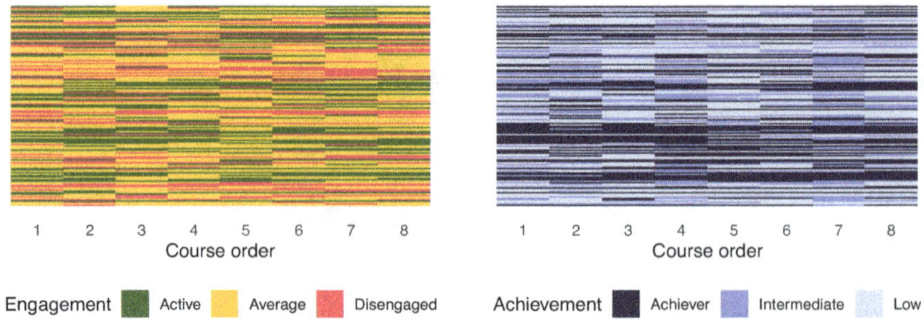

Fig. 13 Sequence and engagement for each student across eight courses

- **AchievementNtile**: Achievement state calculated using tertiles. There are 3 distinct states: Achiever (33.27%), Intermediate (33.36%), and Low (33.36%).
- **Prev_grade**: GPA with which the student applied to the program (1–10).
- **Attitude**: Attitude towards learning (0–20).
- **Gender**: Gender (male/female). There are 44% females and 56% males.

3.10 University Students' Basic Need Satisfaction, Self-regulated Learning and Well-Being During COVID-19

Link to the dataset

This dataset contains the results of a survey investigating students' psychological characteristics related to their well-being during the COVID-19 pandemic. The variables under study are related to the satisfaction of basic psychological needs (relatedness, autonomy, and experienced competence), self-regulated learning, positive emotion and intrinsic learning motivation. Moreover, the dataset contains demographic variables, such as country, gender, and age. The data were collected from 6071 students from Austria and Finland. There are, however, 564 records with missing responses to at least one item.

This dataset has been used in a published study to examine the relationships between the different variables using SEM [83]. The dataset has been used in Chapter 19 [21] of this book, to illustrate the implementation of psychological networks. The dataset has been published under a CC BY 4.0 license, which means that you are free to use and adapt the data but you must give appropriate credit, provide a link to the license, and indicate if changes were made. A preview of the dataset can be found on Table 23. A summary of the responses is in Fig. 14. Below, we describe each of the dataset columns.

- **Demographic data**
  - **country**: Country of the student: 0 = Austria (78.60%), 1 = Finland (21.40%).

Table 23 Preview of the COVID-19 well-being survey data

	Country	Gender	Age	sr1	sr2	sr3	comp1	comp2	comp3	auto1	auto2	auto3	pa1	pa2	pa3	lm1	lm2	lm3	gp1	gp2	gp3
1	0	1	36	5	4	2	2	3	3	3	4	NA	4	3	2	5	5	2	NA	NA	NA
2	0	2	36	4	4	1	4	4	3	4	4	3	4	4	4	5	5	4	3	3	4
3	0	2	36	1	1	1	3	4	2	5	4	4	4	4	3	4	4	4	2	1	2
4	0	2	36	1	2	1	2	2	2	4	3	3	2	2	2	3	3	3	2	1	2
5	0	1	36	1	1	3	3	2	3	4	4	1	3	3	2	4	3	3	2	5	3
6..7723																					
7724	1	1	69	2	2	1	2	2	2	2	2	2	3	3	3	1	1	1	2	2	2

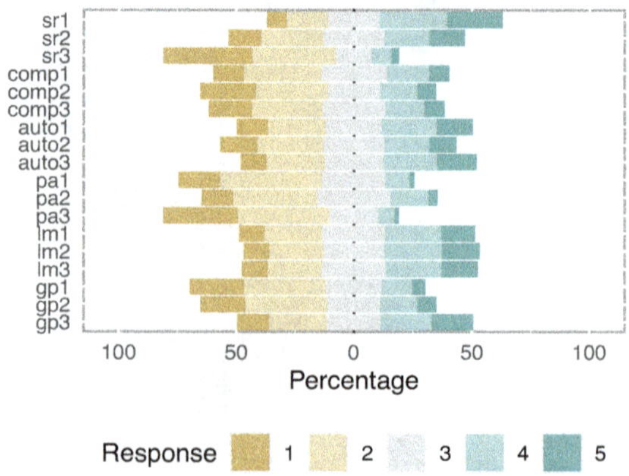

Fig. 14 Results of COVID-19 well-being survey

- **gender**: Gender of the student: 1 = Female (70.74%), 2 = Male (28.25%), 3 = Other (0.70%).
- **age**: Age of the student.

Below we describe the items of the questionnaire for each construct. The possible responses are: 1 = strongly agree, 2 = agree, 3 = somewhat agree, 4 = disagree, 5 = strongly disagree.

- **Basic Psychological Needs: Relatedness**
  - **sr1**: Response to the item "Currently, I feel connected with my fellow students".
  - **sr2**: Response to the item "Currently, I feel supported by my fellow students".
  - **sr3**: Response to the item "Currently, I feel connected with the people who are important to me (family, friends)".
- **Basic Psychological Needs: Competence**
  - **comp1**: Response to the item "Currently, I am dealing well with the demands of my studies".
  - **comp2**: Response to the item "Currently, I have no doubts about whether I am capable of doing well in my studies".
  - **comp3**: Response to the item "Currently, I am managing to make progress in studying for university".
- **Basic Psychological Needs: Autonomy**
  - **auto1**: Response to the item "Currently, I can define my own areas of focus in my studies".
  - **auto2**: Response to the item "Currently, I can perform tasks in the way that best suits me".
  - **auto3**: Response to the item "In the current home-learning situation, I seek out feedback when I need it".

- **Positive Emotion**
  - **pa1**: Response to the item "I feel good".
  - **pa2**: Response to the item "I feel confident".
  - **pa3**: Response to the item "Even if things are difficult right now, I believe that everything will turn out all right".
- **Intrinsic learning motivation**
  - **lm1**: Response to the item "Currently, doing work for university is really fun".
  - **lm2**: Response to the item "Currently, I am really enjoying studying and doing work for university".
  - **lm3**: Response to the item "Currently, I find studying for university really exciting".
- **Self-regulated learning**
  - **gp1**: Response to the item "In the current home-learning situation, I plan my course of action".
  - **gp2**: Response to the item "In the current home-learning situation, I think about how I want to study before I start".
  - **gp3**: Response to the item "In the current home-learning situation, I formulate learning goals that I use to orient my studying".

4 Discussion

In this chapter, we have provided an overview of the types of data operationalized in learning analytics research. We covered a wide spectrum of data types, ranging from foundational demographic information to the footprints left by the interactions of students with online learning technologies, including clicks, activities, social interactions, and assessment data. We have pointed to some of the most commonly employed analytical techniques for each type of data and we referred the reader to the chapters of the book that have covered each type of analysis. Thereafter, we presented a curation of illustrative datasets. We have described each dataset in detail, describing and representing the relevant variables. We also acknowledged the ways each dataset have been analyzed throughout the remaining chapters of the book.

We must disclose that collecting learners' data is not an easy endeavor. First and foremost, it is crucial to consider the ethical implications of collecting and using different types of data, and to comply with data protection laws and regulations [14]. Moreover, it is important to ensure the data quality to draw relevant conclusions from the data, especially in scenarios where data come from heterogeneous sources and are provided in large quantities [84], such as in the educational field [85]. These requirements make finding good-quality open datasets online extremely challenging. In this regard, we hope that the selection offered in this chapter is useful for the reader beyond the scope of the book. A few articles have offered other dataset collections suitable for learning analytics [86] or educational data mining [87]. Moreover, the reader is encouraged to consult open data repositories where datasets are continuously published in multiple fields: Zenodo (https://zenodo.org), US Department of Education Open Data Platform (https://data.ed.gov), Harvard Dataverse (https://dataverse.harvard.edu), European Data Portal (https://data.europa.eu), Mendeley Data (https://data.mendeley.com), openICPSR (https://www.openicpsr.org), Google Dataset Search (https://datasetsearch.research.google.com), figshare (https://figshare.com), Open Science Framework (https://osf.io), or data.world (https://data.world).

References

1. Sclater N (2017) Data. In: Learning analytics explained. Routledge, New York, pp 78–87
2. Nistor N, Hernández-García Á (2018) What types of data are used in learning analytics? An overview of six cases. Comput Human Behav 89:335–338. https://doi.org/10.1016/j.chb.2018.07.038
3. Li W, Sun K, Schaub F, Brooks C (2021) Disparities in Students' propensity to consent to learning analytics. Int J Artif Intell Educ 32:564–608. https://doi.org/10.1007/s40593-021-00254-2
4. Rodríguez-Hernández CF, Cascallar E, Kyndt E (2020) Socio-economic status and academic performance in higher education: a systematic review. Educ Res Rev 29:100305. https://doi.org/10.1016/j.edurev.2019.100305
5. Mengash HA (2020) Using data mining techniques to predict student performance to support decision making in university admission systems. IEEE Access 8:55462–55470. https://doi.org/10.1109/access.2020.2981905
6. Mullen CA (2019) Does modality matter? A comparison of aspiring leaders' learning online and face-to-face. J Further Higher Educ 44:670–688. https://doi.org/10.1080/0309877x.2019.1576859
7. Tikka S, Kopra J, Heinäniemi M, López-Pernas S, Saqr M (2024, this volume) Basics of R. In: Saqr M, López-Pernas S (eds) Learning analytics methods and tutorials: a practical guide using R. Springer
8. Kopra J, Tikka S, Heinäniemi M, López-Pernas S, Saqr M (2024, this volume) Data cleaning and wrangling. In: Saqr M, López-Pernas S (eds) Learning analytics methods and tutorials: a practical guide using R. Springer
9. López-Pernas S, Misiejuk K, Tikka S, Saqr M, Kopra J, Heinäniemi M (2024, this volume) Visualizing and reporting educational data with r. In: Saqr M, López-Pernas S (eds) Learning analytics methods and tutorials: a practical guide using R. Springer
10. Meaney M, Fikes T (2022) Adding a demographic lens to cluster analysis of participants in entry-level massive open online courses (MOOCs). In: Proceedings of the Ninth ACM conference on learning @ scale. https://doi.org/10.1145/3491140.3528306
11. Scrucca L, Saqr M, López-Pernas S, Murphy K (2024, this volume) An introduction and r tutorial to model-based clustering in education via latent profile analysis. In: Saqr M, López-Pernas S (eds) Learning analytics methods and tutorials: a practical guide using R. Springer
12. Murphy K, López-Pernas S, Saqr M (2024, this volume) Dissimilarity-based cluster analysis of educational data: a comparative tutorial using r. In: Saqr M, López-Pernas S (eds) Learning analytics methods and tutorials: a practical guide using R. Springer
13. Du X, Yang J, Shelton BE, Hung J-L, Zhang M (2019) A systematic meta-review and analysis of learning analytics research. Behav Inf Technol 40:49–62. https://doi.org/10.1080/0144929x.2019.1669712
14. Slade S, Prinsloo P (2013) Learning analytics. Am Behav Sci 57:1510–1529. https://doi.org/10.1177/0002764213479366

15. Tempelaar D, Rienties B, Nguyen Q (2021) The contribution of dispositional learning analytics to precision education. Educ Technol Soc 24:109–122. https://www.jstor.org/stable/26977861
16. Brenner PS, DeLamater J (2016) Lies, damned lies, and survey self-reports? Identity as a cause of measurement bias. Soc Psychol Quart 79:333–354. https://doi.org/10.1177/0190272516628298
17. Tikka S, Kopra J, Heinäniemi M, López-Pernas S, Saqr M (2024, this volume) Basic statistics with R. In: Saqr M, López-Pernas S (eds) Learning analytics methods and tutorials: a practical guide using R. Springer
18. Oster M, Lonn S, Pistilli MD, Brown MG (2016) The learning analytics readiness instrument. In: Proceedings of the sixth international conference on learning analytics & knowledge - LAK '16. https://doi.org/10.1145/2883851.2883925
19. Vogelsmeier LVDE, Saqr M, López-Pernas S, Jongerling J (2024, this volume) Factor analysis in education research using R. In: Saqr M, López-Pernas S (eds) Learning analytics methods and tutorials: a practical guide using R. Springer
20. Jongerling J, López-Pernas S, Saqr M, Vogelsmeier L (2024, this volume) Structural equation modeling with R for education scientists. In: Saqr M, López-Pernas S (eds) Learning analytics methods and tutorials: a practical guide using R. Springer
21. Saqr M, Beck E, López-Pernas S (2024, this volume) Psychological networks. In: Saqr M, López-Pernas S (eds) Learning analytics methods and tutorials: a practical guide using R. Springer
22. Ullmann T, Rienties B (2021) Using text analytics to understand open-ended student comments at scale: insights from four case studies. Springer International Publishing, Berlin, pp 211–233
23. Henrie CR, Bodily R, Larsen R, Graham CR (2017) Exploring the potential of LMS log data as a proxy measure of student engagement. J Comput Higher Educ 30:344–362. https://doi.org/10.1007/s12528-017-9161-1
24. Alvarez P, Fabra J, Hernandez S, Ezpeleta J (2016) Alignment of teacher's plan and students' use of LMS resources. Analysis of moodle logs. In: 2016 15th international conference on information technology based higher education and training (ITHET). https://doi.org/10.1109/ithet.2016.7760720
25. Saqr M, López-Pernas S (2021) The longitudinal trajectories of online engagement over a full program. Comput Educ 175:104325. https://doi.org/10.1016/j.compedu.2021.104325
26. Jovanović J, Gašević D, Dawson S, Pardo A, Mirriahi N (2017) Learning analytics to unveil learning strategies in a flipped classroom. Internet Higher Educ 33:74–85. https://doi.org/10.1016/j.iheduc.2017.02.001
27. López-Pernas S, Saqr M (2024, this volume) Process mining. In: Saqr M, López-Pernas S (eds) Learning analytics methods and tutorials: a practical guide using R. Springer
28. Ahmad Uzir N, Gašević D, Matcha W, Jovanović J, Pardo A, Lim L-A, Gentili S (2019) Discovering time management strategies in learning processes using process mining techniques. Springer International Publishing, Berlin, pp 555–569
29. Saqr M, López-Pernas S, Jovanović J, Gašević D (2023) Intense, turbulent, or wallowing in the mire: a longitudinal study of cross-course online tactics, strategies, and trajectories. Internet Higher Educ 57:100902. https://doi.org/10.1016/j.iheduc.2022.100902
30. Saqr M, López-Pernas S, Helske S, Durand M, Murphy K, Studer M, Ritschard G (2024) Sequence analysis in education: principles, technique, and tutorial with r. In: Saqr M, López-Pernas S (eds) Learning analytics methods and tutorials: a practical guide using R. Springer, Berlin
31. López-Pernas S, Saqr M (2024, this volume) Modelling the dynamics of longitudinal processes in education. A tutorial with R for the VaSSTra method. In: Saqr M, López-Pernas S (eds) Learning analytics methods and tutorials: a practical guide using R. Springer
32. Helske J, Helske S, Saqr M, López-Pernas S, Murphy K (2024, this volume) A modern approach to transition analysis and process mining with Markov models: a tutorial with R. In: Saqr M, López-Pernas S (eds) Learning analytics methods and tutorials: a practical guide using R. Springer
33. López-Pernas S, Saqr M, Helske S, Murphy K (2024, this volume) Multichannel sequence analysis in educational research using r. In: Saqr M, López-Pernas S (eds) Learning analytics methods and tutorials: a practical guide using R. Springer

34. Matcha W, Gašević D, Ahmad Uzir N, Jovanović J, Pardo A, Maldonado-Mahauad J, Pérez-Sanagustín M (2019) Detection of learning strategies: a comparison of process, sequence and network analytic approaches. Springer International Publishing, Berlin, pp 525–540
35. Saqr M, López-Pernas S (2022) How CSCL roles emerge, persist, transition, and evolve over time: a four-year longitudinal study. Comput Educ 189:104581. https://doi.org/10.1016/j.compedu.2022.104581
36. Saqr M, López-Pernas S (2021) Modelling diffusion in computer-supported collaborative learning: a large scale learning analytics study. Int J Comput-Support Collab Learn 16:441–483. https://doi.org/10.1007/s11412-021-09356-4
37. Dowell NMM, Nixon TM, Graesser AC (2018) Group communication analysis: a computational linguistics approach for detecting sociocognitive roles in multiparty interactions. Behav Res Methods 51:1007–1041. https://doi.org/10.3758/s13428-018-1102-z
38. Saqr M, Elmoazen R, Tedre M, López-Pernas S, Hirsto L (2022) How well centrality measures capture student achievement in computer-supported collaborative learning? – A systematic review and meta-analysis. Educ Res Rev 35:100437. https://doi.org/10.1016/j.edurev.2022.100437
39. Saqr M, López-Pernas S, Conde MÁ, Hernández-García Á (2024, this volume) Social network analysis: a primer, a guide and a tutorial in R. In: Saqr M, López-Pernas S (eds) Learning analytics methods and tutorials: a practical guide using R. Springer
40. Hernández-García Á, Cuenca-Enrique C, Traxler A, López-Pernas S, Conde MÁ, Saqr M (2024, this volume) Community detection in learning networks using R. In: Saqr M, López-Pernas S (eds) Learning analytics methods and tutorials: a practical guide using R. Springer
41. Saqr M (2024, this volume) Temporal network analysis: introduction and methods and analysis with R. In: Saqr M, López-Pernas S (eds) Learning analytics methods and tutorials: a practical guide using R. Springer
42. Shaffer DW, Collier W, Ruis AR (2016) A tutorial on epistemic network analysis: analyzing the structure of connections in cognitive, social, and interaction data. J Learn Anal 3:9–45. https://doi.org/10.18608/jla.2016.33.3
43. Tan Y, Swiecki Z, Ruis A, Shaffer D (2024, this volume) Epistemic network analysis and ordered network analysis in learning analytics. In: Saqr M, López-Pernas S (eds) Learning analytics methods and tutorials: a practical guide using R. Springer
44. Teasley SD (2019) Learning analytics: where information science and the learning sciences meet. Inf Learn Sci 120:59–73. https://doi.org/10.1108/ils-06-2018-0045
45. Gordillo A, Lopez-Fernandez D, López-Pernas S, Quemada J (2020) Evaluating an educational escape room conducted remotely for teaching software engineering. IEEE Access 8:225032–225051. https://doi.org/10.1109/access.2020.3044380
46. Li KC, Wong BTM (2020) The use of student response systems with learning analytics: a review of case studies (2008–2017). Int J Mob Learn Organ 14:63. https://doi.org/10.1504/ijmlo.2020.103901
47. Namoun A, Alshanqiti A (2020) Predicting student performance using data mining and learning analytics techniques: a systematic literature review. Appl Sci 11:237. https://doi.org/10.3390/app11010237
48. Jovanovic J, López-Pernas S, Saqr M (2024, this volume) Predictive modelling in learning analytics using R. In: Saqr M, López-Pernas S (eds) Learning analytics methods and tutorials: a practical guide using R. Springer
49. Saqr M, López-Pernas S (2022) The curious case of centrality measures: a large-scale empirical investigation. J Learn Anal 9:13–31. https://doi.org/10.18608/jla.2022.7415
50. Blikstein P (2013) Multimodal learning analytics. In: Proceedings of the third international conference on learning analytics and knowledge. Association for Computing Machinery, New York, pp 102–106
51. Mu S, Cui M, Huang X (2020) Multimodal data fusion in learning analytics: a systematic review. Sensors 20. https://doi.org/10.3390/s20236856
52. Sharma K, Giannakos M (2020) Multimodal data capabilities for learning: What can multimodal data tell us about learning? Br J Educ Technol 51:1450–1484. https://doi.org/10.1111/bjet.12993

53. Kubsch M, Caballero D, Uribe P (2022) Once more with feeling: emotions in multimodal learning analytics. In: Giannakos M, Spikol D, Di Mitri D, Sharma K, Ochoa X, Hammad R (eds) The multimodal learning analytics handbook. Springer International Publishing, Cham, pp 261–285
54. Bleck M, Le N-T (2022) A physiology-aware learning analytics framework. In: Giannakos M, Spikol D, Di Mitri D, Sharma K, Ochoa X, Hammad R (eds) The multimodal learning analytics handbook. Springer International Publishing, Cham, pp 231–257
55. Saqr M, López-Pernas S (2024, this volume) Why learning and teaching learning analytics is hard: an experience from a real-life LA course using LA methods. In: Proceedings of the eleventh international conference on technological ecosystems for enhancing multiculturality (TEEM'23). Springer, Bragança
56. Hasan R, Palaniappan S, Mahmood S, Abbas A, Sarker KU (2021) Dataset of students' performance using student information system, moodle and the mobile application 'eDify'. Data 6: https://doi.org/10.3390/data6110110
57. Hasan R (2021) Dataset of Student's Performance using Student Information System, Moodle and Mobile Application 'eDify'
58. Hasan R, Palaniappan S, Raziff ARA, Mahmood S, Sarker KU (2018) Student academic performance prediction by using decision tree algorithm. In: 2018 4th international conference on computer and information sciences (ICCOINS). IEEE, pp 1–5
59. Hasan R, Palaniappan S, Mahmood S, Sarker KU, Abbas A (2020) Modelling and predicting student's academic performance using classification data mining techniques. Int J Bus Inf Syst 34:403–422
60. Rodríguez S, Valle A, Piñeiro I, Vieites T, González-Suárez R, Rodríguez-Llorente C (2020) School engagement, SRL and academic achievement
61. Fredricks JA, Blumenfeld P, Friedel J, Paris A (2005) School engagement. What do children need to flourish? Conceptualizing and measuring indicators of positive development, pp 305–321
62. Cleary TJ (2006) The development and validation of the self-regulation strategy inventory—self-report. J School Psychol 44:307–322
63. Estévez I, Rodríguez-Llorente C, Piñeiro I, González-Suárez R, Valle A (2021) School engagement, academic achievement, and self-regulated learning. Sustainability 13: https://doi.org/10.3390/su13063011
64. Prasojo LD, Habibi A, Yaakob MFM, Pratama R, Yusof MR, Mukminin A, Suyanto, Hanum F (2020) Teachers' burnout: a SEM analysis in an Asian context. Heliyon 6:e03144. https://doi.org/10.1016/j.heliyon.2019.e03144
65. Villa A, Calvete E (2001) Development of the teacher self-concept evaluation scale and its relation to burnout. Stud Educ Eval 27:239–255. https://doi.org/10.1016/s0191-491x(01)00028-1
66. Yu G, Xin T, Shen J (1995) Teacher's sense of teaching efficacy: its structure and influencing factors. Acta Psychol Sin 27:159
67. Champion DF, Westbrook BW (1984) Maslach burnout inventory. Meas Eval Couns Dev 17:100–102. https://doi.org/10.1080/07481756.1984.12022754
68. Prasojo LD, Habibi A, Yaakob MFM, Pratama R, Yusof MR, Mukminin A, Suyanto, Hanum F (2020) Dataset relating to the relationship between teacher self-concept and teacher efficacy as the predictors of burnout: a survey in Indonesian education. Data Brief 30:105448. https://doi.org/10.1016/j.dib.2020.105448
69. Mitchell K (2020) Interdisciplinary undergraduate and graduate student data. https://doi.org/10.7910/DVN/M07HQ7. Harvard Dataverse
70. Bandura A, Freeman WH, Lightsey R (1999) Self-efficacy: the exercise of control. J Cogn Psychother 13:158–166. https://doi.org/10.1891/0889-8391.13.2.158
71. Pajares F, Valiante G (2006) Self-efficacy beliefs and motivation in writing development. In: Handbook of writing research. The Guilford Press, New York, pp 158–170
72. Mitchell KM, McMillan DE, Lobchuk MM, Nickel NC, Rabbani R, Li J (2021) Development and validation of the situated academic writing self-efficacy scale (SAWSES). Assess Writ 48:100524. https://doi.org/10.1016/j.asw.2021.100524

73. Kellogg S, Edelmann A (2015) Massively Open Online Course for Educators (MOOC-Ed) network dataset. https://doi.org/10.7910/DVN/ZZH3UB. Harvard Dataverse
74. Kellogg S, Edelmann A (2015) Massively Open Online Course for Educators (MOOC-Ed) network dataset. Br J Educ Technol 46:977–983. https://doi.org/10.1111/bjet.12312
75. Saqr M, López-Pernas S (2021) Modelling diffusion in computer-supported collaborative learning: a large scale learning analytics study. Int J Comput-Support Collab Learn 16:441–483. https://doi.org/10.1007/s11412-021-09356-4
76. Adraoui M, Akachar E, Retbi A, Idrissi MK, Bennani S (2022) Dataset of learners' interactions in forum discussions [dataset]. Mendeley. https://doi.org/10.17632/CKNF9FVYBR.1
77. Adraoui M, Retbi A, Idrissi MK, Bennani S (2017) Social learning analytics to describe the learners' interaction in online discussion forum in moodle. In: 2017 16th international conference on information technology based higher education and training (ITHET). IEEE
78. Lerís D, Fidalgo Á, Sein Echaluce ML (2014) A comprehensive training model of the teamwork competence. Int J Learn Intellect Cap 11:1. https://doi.org/10.1504/ijlic.2014.059216
79. Fidalgo-Blanco Á, Lerís D, Sein-Echaluce ML, García-Peñalvo FJ, et al. (2015) Monitoring indicators for CTMTC: comprehensive training model of the teamwork competence in engineering domain. Int J Eng Educ 31(Extra 3):829–838
80. Chaparro-Peláez J, Acquila-Natale E, Iglesias-Pradas S, Suárez-Navas I (2015) A web services-based application for LMS data extraction and processing for social network analysis. In: New information and communication technologies for knowledge management in organizations. Springer International Publishing, Berlin, pp 110–121
81. Hernández-García Á, Suárez-Navas I (2016) GraphFES: a web service and application for moodle message board social graph extraction. In: Big data and learning analytics in higher education. Springer International Publishing, Berlin, pp 167–194
82. Saqr M, López-Pernas S (2021) The longitudinal trajectories of online engagement over a full program. Comput Educ 175:104325. https://doi.org/10.1016/j.compedu.2021.104325
83. Holzer J, Lüftenegger M, Korlat S, Pelikan E, Salmela-Aro K, Spiel C, Schober B (2021) Higher education in times of COVID-19: University students' basic need satisfaction, self-regulated learning, and well-being. AERA Open 7:233285842110031. https://doi.org/10.1177/23328584211003164
84. Becker D, King TD, McMullen B (2015) Big data, big data quality problem. In: 2015 IEEE international conference on big data (big data), pp 2644–2653
85. Klašnja-Milićević A, Ivanović M, Budimac Z (2017) Data science in education: big data and learning analytics. Comput Appl Eng Educ 25:1066–1078. https://doi.org/10.1002/cae.21844
86. Dietze S, Siemens G, Taibi D, Drachsler H (2016) Editorial: datasets for learning analytics. J Learn Anal 3:307–311. https://doi.org/10.18608/jla.2016.32.15
87. Mihaescu MC, Popescu PS (2021) Review on publicly available datasets for educational data mining. WIREs Data Min Knowl Discov 11. https://doi.org/10.1002/widm.1403

Getting Started with R for Education Research

Santtu Tikka, Juho Kopra, Merja Heinäniemi, Sonsoles López-Pernas, and Mohammed Saqr

1 Introduction

R is a free, versatile, and open source programming language and software environment specifically designed for statistical computing and data analysis. R has a vast library of packages that enable data manipulation, visualization, modeling and machine learning [1]. Such a wealth of packages enable a wide range of functionalities that saves the users time, effort or the need to write complex code. The fact that R is freely available makes all these functionalities accessible to all. R has a large community of developers, users and researchers who support the development of the platforms as well as provide support and shared knowledge on popular sites such as StackExchange. Thereupon, R is becoming an increasingly popular choice for students, researchers and data scientists [2].

Being open source and accessible to researchers, several packages are added continuously to expand the possibilities and functions that R offers. Some of the R packages included in this book have been added by researchers during the last few years to address contemporary scientific problems and state-of-the-art innovations [3]. For example, R software packages for the analysis of psychological networks were developed in the past five years and ever since have grown tremendously due to contributions from a large base of researchers [4].

S. Tikka (✉)
Department of Mathematics and Statistics, University of Jyväskylä, Jyväskylä, Finland
e-mail: santtu.tikka@jyu.fi

J. Kopra · S. López-Pernas · M. Saqr
School of Computing, University of Eastern Finland, Joensuu, Finland

M. Heinäniemi
Institute of Biomedicine, University of Eastern Finland, Kuopio, Finland

M. Saqr, S. López-Pernas (eds.), *Learning Analytics Methods and Tutorials*,
https://doi.org/10.1007/978-3-031-54464-4_3

Although many of the methods described in this book can be implemented with other software tools, it is hard to find a comprehensive platform that can be used to perform practically all the existing learning analytics methods with such maturity, performance and range of possibilities. For instance, Social Network Analysis (SNA) can be performed with several programming languages (e.g., Python) and desktop software applications (e.g., Gephi) [5]. However, both options provide limited capabilities compared to what R provides for the analysis of SNA. This includes a wider range of SNA centrality measures, mathematical models, community finding algorithms and generative models. Sequence analysis is another example in which the possibilities offered by R are hard to match with other software solutions.

This book does not make any assumptions about the superiority of R over any other platform. Other languages and software platforms are indeed very helpful and have vast capabilities for researchers. For instance, Python has remarkable tools for machine learning and Gephi offers beautiful graphics for the visualization of networks. Oftentimes, readers may need to learn or use other tools to do specific tasks. Put another way, where R offers a rich toolset for researchers, there is a space for other tools that researchers can use to accomplish certain tasks. In summary, investing time in learning R is a worthwhile endeavor that helps interested researchers to perform and expand their research skills and toolset. Since R is a large platform, it represents a doorway to the vast capabilities of its ever expanding repertoire of functions and packages.

2 Learning R

The goal of programming is to write, i.e. code, a program that performs a desired task. A program consists of several commands, each of which does something very simple. In the statistics and data analysis context, R is typically used to write short programs called scripts. R is therefore not intended for developing games or other complicated programs. R is also not a language originally intended for web programming, although with the right packages you can also make web applications with R.

R is a high-level programming language. This means that there are many ready-made commands in R, which have much more code "underneath" that the R programmer does not have to touch. For example, a statistical t-test requires several mathematical intermediate steps, but an R programmer can perform the test with a single command (`t.test`) that provides all the necessary computations and information about the test.

The best way to learn how to use and program R code is by doing. This text has R code embedded between the text in gray boxes, as in the example below. Lines starting with two hashes, i.e. ##, are not code, but output generated by running the code. Let's first take the classic "Hello, world!" command as an example:

```
print("Hello, world!")
```

```
[1] "Hello, world!"
```

The print function prints the given text to the console. It is convenient, for example, for testing the operation of a program and monitoring the progress of a longer program. R can also be used as a calculator. In the example below, we calculate the price of a product after a 35% discount that was originally priced at 80 euros.

```
80 * (1 - 0.35)
```

```
[1] 52
```

However, running individual commands is usually not useful unless the results can be saved somewhere. In programming languages, data is stored in variables, which you will become familiar with later.

3 RStudio

R has a large array of tools and integrated development environments (IDEs) that make writing and managing code easier and more accessible. The most widely used R IDE is RStudio, which is a free open source software that—similarly to R—runs on all major operating systems [6]. RStudio provides a comprehensive and user-friendly interface for writing, running, and debugging R code, which makes it easier for users to get started and become more productive. Together, R and RStudio allow for the creation of reproducible research. The code and results can be easily shared and replicated, making R and RStudio great tools for collaboration and transparency.

R and RStudio can be very useful in analyzing and visualizing different types of data. This can help researchers, educators and administrators make data-driven decisions and improve the learning experience for students. Whether you are analyzing student performance, demographic data, or tracking the effectiveness of instructional interventions, R and RStudio provide a flexible and efficient platform for achieving your goal.

First install R (step 1) and then RStudio Desktop for your operating system. R and RStudio are available for many operating systems. The interface of RStudio shown in Fig. 1 has the following default components:

(1) *Editor*: The editor is used to write files containing R code, i.e., R scripts. Scripts will be introduced later, but they are simply a collection of R commands that carry out a specific task when placed together, for example analyze the data of

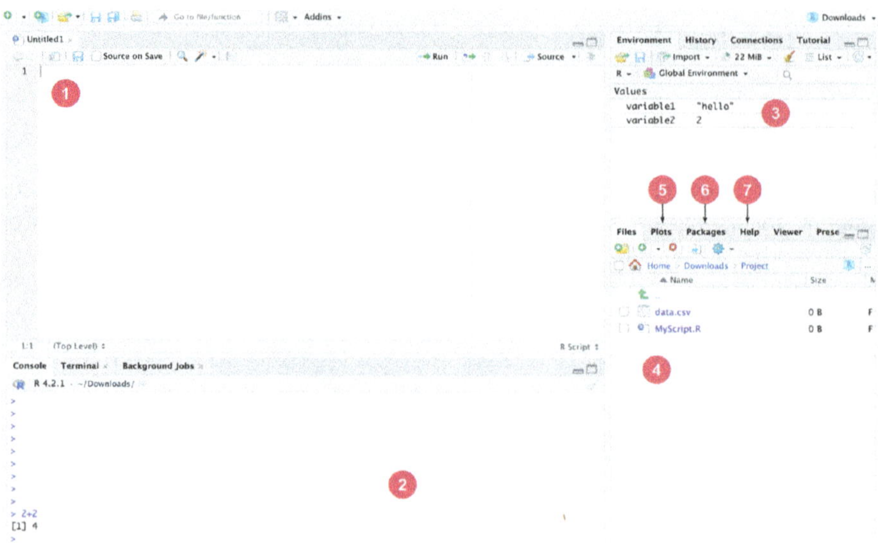

Fig. 1 RStudio user interface

a research project or draw figures of the finished results. A new script can be opened from the "File" menu by selecting "New File" and then "R script". The same can be accomplished via a keyboard shortcut by pressing Ctrl + Shift + N on Windows and Linux. On macOS, users can use Cmd + Shift + N. Also, see the "Keyboard Shortcuts Help" under the "Tools" menu for all the shortcuts available. The code written in the editor can be run line by line by pressing Ctrl + Enter at the line. Several lines can also be selected and run at once. The Source button at the top runs all the code in the current file. R scripts can be saved just like other files and their file extension is .R. All the code you use to analyze your data should be written in scripts. When you save your code in this way, the next time you can simply run the script to perform the desired task instead of rewriting the code from scratch.

(2) *Console*: R commands are executed in the console. If the code written in the editor is run, RStudio automatically executes the commands in the console. In the console, just pressing Enter is enough to execute a line of code. You can try to write a calculation in the console, such as 2 * 3 and press Enter, and the result will be printed in the console. You can also write code in the editor and press Ctrl + Enter to accomplish the same result. Possible messages, warnings and errors are also printed into the console. The main difference between the console and the editor is that the commands written in the console are not saved in any file. So if you want to keep your code, it should be written in the editor and saved in a script file. Commands made during the same session can be scrolled in the console with the up and down arrows. In addition, the command history can be viewed in the History tab in RStudio.

(3) *Workspace*: Displays the variables in the current working environment of the R session.
(4) *Files*: Shows the directory structure of the operating system, by default the working directory.
(5) *Plots*: Graphics drawn with R appear here.
(6) *Packages*: Here you can manage the installed packages (instructions for installing the packages are below).
(7) *Help*: Here you can browse the R manual with instructions for every R command. You can try running the `?print` command in the editor or console, which opens the help page for the `print` function.

4 Best Practices in Programming

As the word "script" already suggests, data analysis often requires a "script" of various steps, such as reading measurements from a file, organizing a table, or describing the results of an analysis, for example by calculating the averages of different sample groups and drawing figures. Writing the R commands that perform these steps in a script provides a documentation of how the analysis was done, and it's also easy to come back to and possibly add a few additional steps, such as performing a t-test. The script file is also easy to share with others who work with similar data analysis tasks, because with small modifications (file names, table structure) the same code also works for different datasets and workflows. Comments are often used inside the code. Comments are text written along the R commands, which are not written in a programming language, and which are ignored when running the code. The purpose of comments is to describe the behavior and purpose of the code. It is good to practice to comment your own code from the beginning, even if the code for the first tasks is very simple. In R, comments are marked with the # symbol.

```
# Assign arbitrary numbers to two variables
x <- 3
y <- 5
# Sum of two variables
z <- x + y
# Print the results
z
```

```
[1] 8
```

4.1 R Markdown

While scripts can only store code and comments, a more comprehensive format called R Markdown is also available in RStudio. R Markdown is an extension of the Markdown markup language that allows users to create dynamic reports and interactive notebooks that can integrate text, code, and visualizations. R Markdown documents are created in a plain text format and can be rendered into various output formats, such as HTML, PDF, Word, or even presentations. To create a new R Markdown document in RStudio, go to the "File" menu, select "New File" and finally "R Markdown". In the dialog that opens, choose the output format you want to use, and give your document a title. Figure 2 shows the RStudio editor panel for a default R Markdown document of R Studio.

The YAML metadata wrapped between the `---` delimiters at the top of the document can be used to customize the output and contents in various ways. For example, you can change the title, author, or add a table of contents. In this example we have defined the title, the output format (HTML) and the date, but many more options are available besides these. Use Markdown syntax to format your text, and enclose your R code in code chunks with the ```` ```{r} ```` and ```` ``` ```` tags which can also be provided other options. For example, our first code chunk has been given the label `setup` and the following option `include = FALSE` means that this chunk will be executed but its output will not be printed into the final document. Within the chunk, a global `knitr` option is set so that all code chunks will print (echo) their

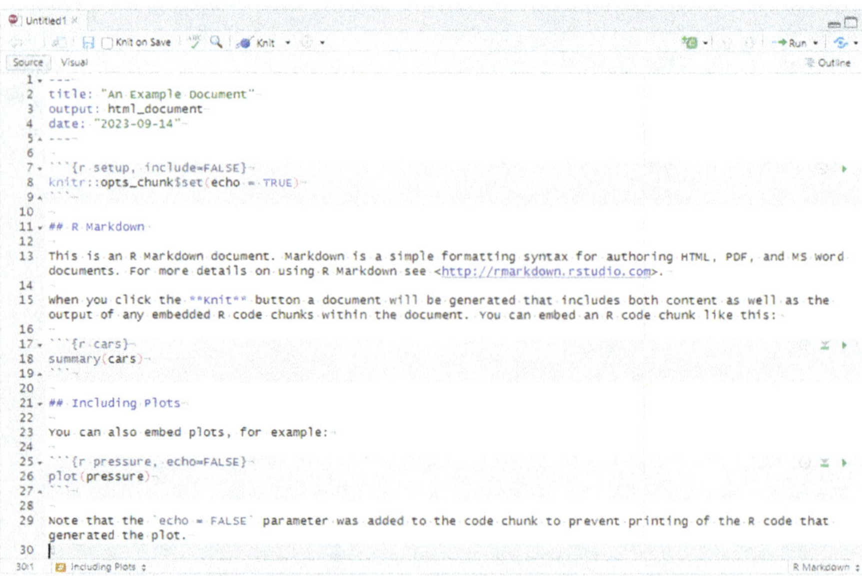

Fig. 2 An example R Markdown document in R Studio

output by default. You can also use inline chunks to quickly reference the results of computations or other R objects. Some examples of standard markdown syntax in the document are second level headers marked with ## (note that # is not used for comments in an R Markdown document) and bold text denoted by wrapping the text with **. Once you are satisfied with your document, you can render it into the chosen output format by clicking the "Knit" button in RStudio, or use the keyboard shortcut "Ctrl + Shift + K". Figure 3 shows the corresponding rendered HTML document.

This simple example shows only a fraction of the full features of R Markdown documents and the Markdown syntax. R Markdown is a powerful tool for creating reproducible research reports, teaching materials, or even websites. It allows users to integrate code and output seamlessly into their written work, making it easier to share and reproduce analyses. As an alternative to R Markdown documents, R Studio also supports the creation of R Notebooks, which are in essence interactive R Markdown documents. R Notebooks can be useful for example when the goal is not to produce one comprehensive analysis report but instead to keep track of the code and try out various approaches to a problem interactively. For an in-depth guide to R Markdown, see [7] which is also freely available online at https://bookdown.org/yihui/rmarkdown/.

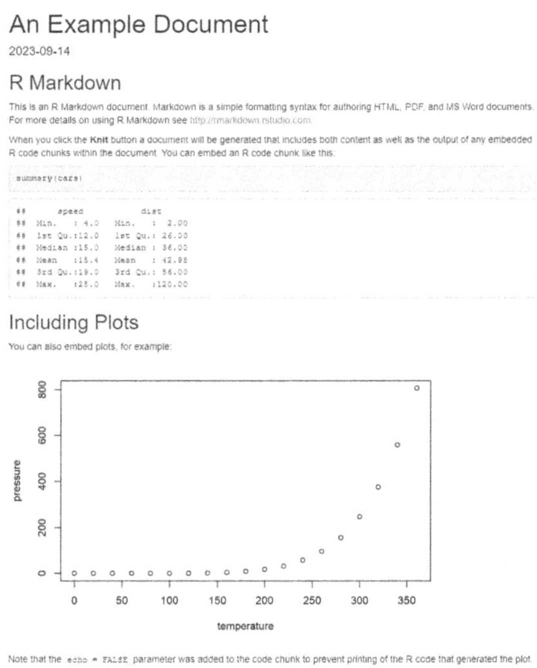

An Example Document

2023-09-14

R Markdown

This is an R Markdown document. Markdown is a simple formatting syntax for authoring HTML, PDF, and MS Word documents. For more details on using R Markdown see http://rmarkdown.rstudio.com.

When you click the **Knit** button a document will be generated that includes both content as well as the output of any embedded R code chunks within the document. You can embed an R code chunk like this:

```
summary(cars)
```

```
##      speed           dist       
##  Min.   : 4.0   Min.   :  2.00  
##  1st Qu.:12.0   1st Qu.: 26.00  
##  Median :15.0   Median : 36.00  
##  Mean   :15.4   Mean   : 42.98  
##  3rd Qu.:19.0   3rd Qu.: 56.00  
##  Max.   :25.0   Max.   :120.00  
```

Including Plots

You can also embed plots, for example:

Note that the `echo = FALSE` parameter was added to the code chunk to prevent printing of the R code that generated the plot.

Fig. 3 The example R Markdown document rendered into HTML

4.2 How Is Code Developed?

Code development typically follows similar steps:

1. Design parts of the code.
2. Start by writing a small piece of code.
3. Test whether the code you wrote works. If it doesn't, find out why and fix it.
4. Go to the next piece of code and continue accordingly, always testing piece by piece whether your code works.

Along with this material, many packages include a "Cheat Sheet" as a summary of basic tasks and functions related to the package. Cheat sheets provide a quick and easy reference for checking how something is done in R if you don't remember it by heart. There are cheat sheets for various R packages and other entities on the Internet e.g., the base R cheat sheet, or the tidyr [8] cheat sheet.

In addition to the basic commands presented in this chapter of the book, practical R programming relies to a great extent on the use of various packages developed by the scientific community. Packages are collections of code that contain new functions, classes and data, i.e., they extend R. Most R packages are available from the Comprehensive R Archive Network (CRAN). They can be installed with the `install.packages()` function, or via RStudio's installation window which in practice calls the `install.packages()` function. You can also install several packages at once. The command below installs the `dplyr` [9] and `tidyr` packages:

```
install.packages(c("dplyr", "tidyr"))
```

In order to use the commands contained in an R package, the package must be installed and attached to the R workspace. This is done with the `library()` command:

```
library("tidyr")
```

Now that the `tidyr` package is loaded, we can use the commands it provides, for example to manage the learning analytics data in data frame format that we will address later. If you don't want to attach the entire package, you can use individual commands from packages with the format `name_of_the_package::name_of_the_command()`.

5 Basic Operations

Basic operations in R consist of arithmetic operations, logical operations, and assignment. In addition, there are several commands that are often helpful when starting a new project or managing the working directory. For example, the current working directory can be obtained with the following command:

```
getwd()
```

```
[1] "/home/sonsoles/labook/chapters/ch03-intro-r"
```

5.1 Arithmetic Operators

R can be used to compute basic arithmetic operations such as addition (+), subtraction (-), multiplication (*), division (/), and exponentiation (^). These operations follow standard precedence rules, and additional brackets can be added to control the evaluation order if needed.

```
1 + 1 # Addition
```

```
[1] 2
```

```
2 - 1 # Subtraction
```

```
[1] 1
```

```
2 * 4 # Multiplication
```

```
[1] 8
```

```
5 / 2 # Division
```

```
[1] 2.5
```

```
2 ^ 4 # Exponentiation
```

```
[1] 16
```

5.2 Relational Operators

Relational operators compare objects or values to other objects or values. These operators are often required for conditional data filtering, for example when selecting a subset of individuals that satisfy some criterion. In R, there are six such operators: smaller than, greater than, smaller or equal to, greater or equal to, equal to, and not equal to. There operators have the following syntax in R:

```
1 < 2  # Smaller than
```

```
[1] TRUE
```

```
3 > 2  # Greater than
```

```
[1] TRUE
```

```
2 <= 2 # Smaller or equal to
```

```
[1] TRUE
```

```
3 >= 3 # Greater or equal to
```

```
[1] TRUE
```

```
5 == 5 # Equal to
```

```
[1] TRUE
```

```
1 != 2 # Equal to
```

```
[1] TRUE
```

In the previous example we used these operators to compare integers, but we may also use them to compare other types of values, such as characters:

```
"a" == "b"
```

```
[1] FALSE
```

5.3 *Logical Operators*

Similar to relational operators, logical operators are used to evaluate the logical value of a conjunction of two logical values. In R, there are five logical operators: negation, AND, OR, elementwise AND, and elementwise OR. These operators have the following syntax:

```
!TRUE         # Negation
```

```
[1] FALSE
```

```
TRUE && TRUE  # Logical AND
```

```
[1] TRUE
```

```
TRUE || FALSE # Logical OR
```

```
[1] TRUE
```

```
TRUE & TRUE   # Elementwise AND
```

```
[1] TRUE
```

```
TRUE | FALSE  # Elementwise OR
```

```
[1] TRUE
```

The elementwise operators `&` and `|` can be used to compare multiple pairs of logical values simultaneously, for example

```
c(TRUE, FALSE, TRUE) | c(TRUE, FALSE, FALSE)
```

```
[1]  TRUE FALSE  TRUE
```

whereas the operators `&&` and `||` only accept single values. In the previous example, we also used one of the most important operations of R: the `c()` function (the letter ‘c’ is short for “combine”) which we used to combine the logical values into a vector, i.e., an ordered sequence of values of the same type. Vectors and other important data types will be discussed in greater details in the next section.

5.4 Special Operators

Another core functionality in R is the assignment operator <-. Assignment can be used to store the results of computations into variables which can then be used again in other computations without having to redo the original computations. For instance

```
x <- 5 # Assign value 5 into variable named x
y <- 7 # Assign value 7 into variable named y
x      # Access value of x (value is printed into the
                                console in RStudio)
```

```
[1] 5
```

```
y      # Access value of y
```

```
[1] 7
```

```
x + y  # Compute the sum of x and y
```

```
[1] 12
```

```
x > y  # Is x greater than y?
```

```
[1] FALSE
```

Here we chose the names `x` and `y` for our variables, but the names are arbitrary with the caveat that one should avoid assigning values to objects with names that R already uses internally, such as names of common functions like `c()`, `exp()`, or `lm()` to name a few. Variables currently assigned in the working environment can be displayed with the following command:

```
ls()
```

```
[1] "x" "y" "z"
```

It is also possible to use the equal sign = as the assignment operator, but this is often not recommended, because the equal sign also has other purposes in the R language and may cause confusion if used for assignment. There are also some special instances, where the equals sign does not function identically to <-. Therefore, we recommend always using the standard assignment operator <-.

When constructing vectors, the function `c()` can be cumbersome in some scenarios. For instance, say we wanted to create a vector that contains all integers from 1 to 100. With `c()`, we would have to write each value individually. Fortunately, such sequences can be constructed effortlessly using the `:` (colon) operator:

```
1:100
```

```
  [1]   1   2   3   4   5   6   7   8   9  10  11  12  13  14  15  16  17  18  19
 [20]  20  21  22  23  24  25  26  27  28  29  30  31  32  33  34  35  36  37  38
 [39]  39  40  41  42  43  44  45  46  47  48  49  50  51  52  53  54  55  56  57
 [58]  58  59  60  61  62  63  64  65  66  67  68  69  70  71  72  73  74  75  76
 [77]  77  78  79  80  81  82  83  84  85  86  87  88  89  90  91  92  93  94  95
 [96]  96  97  98  99 100
```

Often we may wish to select only specific values from a vector. Values in a vector can be accessed by their index, starting from 1. This is accomplished by using the subset operator which uses the following square bracket syntax. For example, we could select the first value of a vector `x` by writing `x [1]`. For a more involved example, we could simultaneously select the first, 50th and 100th value of the vector `1:100` by writing:

```
x <- 1:100
x[c(1, 50, 100)]
```

```
[1]   1  50 100
```

Essentially, we write the positions of the values that we wish to select inside the square brackets as a vector. Alternatively, we can create a vector of logical values that has the same length as the vector we are selecting from, and the value `TRUE` for the values we wish to select, and `FALSE` otherwise. In the next section, we also show how to select values based on a condition.

6 Basic Data Types and Variables

In the examples of the previous section, we already used several of the most common data types that users are most likely to encounter in practical data analyses. Each object in R has a type, which can be determined with the function `typeof()`. Types are used to describe what kind of data our variables of interest contain and what kind of operations can be carried out on them. Perhaps the most common type is the `numeric` type, which describes values that can be interpreted as numbers. Two special instances of the `numeric` type are `integer` and `double`, which correspond to integer values and decimal values, respectively.

```
typeof(1.0)
```

```
[1] "double"
```

```
typeof(1L)
```

```
[1] "integer"
```

The capital `L` on the second line denotes that we mean the integer 1, and not the decimal number 1.0. Text data is often represented by the `character` type. Unlike in some other programming languages, in R the `character` type does not necessarily describe individual characters as the name would suggest, but character strings, for instance:

```
typeof("a")
```

```
[1] "character"
```

```
typeof("hello world!")
```

```
[1] "character"
```

As we can see, both have the same type.

The `logical` type is used to represent the boolean values `TRUE` and `FALSE`. Logical values are typically not as common in actual data where such values may often be represented by the integers 1 and 0 instead. However, their importance lies in forming conditions for data filtering and manipulations that we may wish to carry out based on a criterion that is true for only a subset of subjects in our data. For example, suppose we have the values 1, 2, 3, 4, 5, and we wish to programmatically select only those values that are greater than 2. We can accomplish this as follows:

```
some_numbers <- 1:5
some_numbers[some_numbers > 2]
```

```
[1] 3 4 5
```

Let's walk through what the code above does step by step. On the first line, we create a vector of numeric values 1 through 5 by using the `:` operator and assign the result to the variable named `some_numbers`. On the second line, we use the square brackets (the subset operator) to select only those values of `some_numbers` for which the condition `some_numbers > 2` is true. Here we introduce the vectorization feature of R which applies to a vast majority of arithmetic and relational operators. By writing `some_numbers > 2`, we actually evaluate the condition for every value in the vector `some_numbers`:

```
some_numbers > 2
```

```
[1] FALSE FALSE  TRUE  TRUE  TRUE
```

All of the aforementioned types are called atomic types, meaning that vectors of such values can only contain values of one specific atomic type. For example, a vector cannot contain values of both character and integer types:

```
c(0L, "a")
```

```
[1] "0" "a"
```

We see that the result was automatically converted to an atomic character vector.

Alongside type, objects in R also have a class, which can be viewed with the `class()` function. For atomic types, their class is the same as their type, but for more complicated types of variables, a class essentially describes a special instance of a type. Objects within a specific class typically have their own set of functions or methods that can only be applied for that specific class. A common example of a class is `factor`. Factors are a special class of `integer` vectors designed to represent categorical and ordinal data. Variables that are of the `factor` class have levels, which differentiates `factor` variables from ordinary `integer` variables. For example, a factor could represent group membership in a randomized experiment indicating inclusion in the control group or the treatment group for each individual. The levels of this factor could be called `"control"` and `"treatment"` for example. Supposing that 0 indicates that an individual belongs in the control group and 1 indicates that an individual belongs in the treatment group, we could create such a factor by writing

```
group <- c(0, 0, 1, 0, 1, 0, 1, 1)
factor(group, labels = c("control", "treatment"))
```

```
[1] control   control   treatment control   treatment control   treatment treatment
Levels: control treatment
```

Factors are important when fitting statistical models or when performing statistical tests, because if we would simply use the corresponding integer values, they may be erroneously interpreted as continuous. The levels of the factor also often help to provide more informative output. In the next section we will discuss more complicated data types including data frames, which can contain data of various types simultaneously.

7 Basic R Objects

In this section we will discuss the concepts of data frame and tibble. Let's assume that your data can be stored in a two-dimensional array where the columns represent variables and each row represents a case of measurement. In R, a data frame is a concept for storing such a two-dimensional array of data. Let's first study how data frames work in R and we will then move on to see how another concept called `tibble` extends the capabilities of a data frame.

A data frame which has been loaded into R under the name `grades` and printed in the R console will look as follows.

```
grades
```

```
  group grade
1     2  4.67
2     2  4.90
3     3  2.63
4     4  3.39
5     4  6.89
```

In the above printout we can see that R prints the data frame just as we would expect the data to look like. One issue with a standard data frame is that if there are very many columns or rows, then the printout may be difficult to read. In RStudio, a neater (and more flexible) way of inspecting the data is by using a command called `View()`:

```
View(grades)
```

To go further with the data, we need tools for data manipulation. We begin with the very basic tools which are commonplace in any data analysis workflow. More comprehensive knowledge about data transforming, cleaning etc. can be found in Chap. 2.

A data frame can be constructed directly in the R by using the function called `data.frame()` and then listing variable names and their values. The `grades` data above has two columns, `group` and `grade`, and each row represents a student. The variable `group` is indicates the number of a group in which each person has studied. The variable `grade` stores the final grade that the student has received from a course.

8 Working with Dataframes

To extract a column from a data frame, one needs to start with the name of the data frame object and connect the column name with the name of the data frame object by using the dollar symbol ($). Thus, extracting the column `group` from `grades` data can be accomplished by writing

```
grades$group
```

```
[1] 2 2 3 4 4
```

Next, let's use a couple of basic functions which are often needed when developing R code. First of all, to get a summary of every variable of a data frame, we can call

```
summary(grades)
```

```
     group          grade      
 Min.   :2   Min.   :2.630  
 1st Qu.:2   1st Qu.:3.390  
 Median :3   Median :4.670  
 Mean   :3   Mean   :4.496  
 3rd Qu.:4   3rd Qu.:4.900  
 Max.   :4   Max.   :6.890  
```

The result show us the minimum, maximum, median, mean and quartiles of both variables. You may note that as `group` was intended to be a categorical variable, so computing its mean value in the data does not make sense. We can change that behavior by converting the `group` column into a factor.

```
grades$group <- as.factor(grades$group)
summary(grades)
```

```
 group      grade      
 2:2   Min.   :2.630  
 3:1   1st Qu.:3.390  
 4:2   Median :4.670  
       Mean   :4.496  
       3rd Qu.:4.900  
       Max.   :6.890  
```

On the other hand, we might want to calculate the mean or the sample standard deviation of the variable `grade`. This can be done with functions called `mean()` and `sd()`, respectively.

```
mean(grades$grade)
```

```
[1] 4.496
```

```
sd(grades$grade)
```

```
[1] 1.630178
```

The functions above can only take numeric vectors as input. If we tried using another type of argument, we would encounter an error message.

```
sd(grades)
```

```
Error in is.data.frame(x): 'list' object cannot be coerced to type 'double'
```

The above message can actually teach us a couple of things. First of all, the error comes from the function `is.data.frame()`, which is called somewhere in the definition of `sd()`. Second, the actual error message tells us that the object which we gave to the function as its argument is a `list` object. List is an object type of R on top of which `data.frame` objects have been built. We will describe lists in greater detail later. Further, the message tells us that R has tried to coerce the argument object into the `double` type. This means that the object we supplied to the function is not of the right type and cannot easily be converted into the proper format.

8.1 *tibble*

A `tibble` is an expansion of `data.frame` objects which is used in the tidyverse programming paradigm [10]. To use tidyverse, we advise to load the `tidyverse` [11] (meta)package which loads all key tidyverse packages.

```
# load tidyverse to use as_tibble
library("tidyverse")
# convert a data frame as tibble
grades2 <- as_tibble(grades)
```

Next, let's see what a `tibble` looks like when printed in the console

```
grades2
```

```
# A tibble: 5 x 2
  group grade
  <fct> <dbl>
1 2      4.67
2 2      4.9
3 3      2.63
4 4      3.39
5 4      6.89
```

Tibbles behave similarly compared to data frames when printed, but they also describe the dimensions of the data and the types of the columns right under the column names. For instance, `<fct>` refers to a `factor` column and `<dbl>` refers to a `double` column. In order to discover the column types when using data frames, one would need to apply the `class()` or `typeof()` function to the columns, or write `str(grades)` to see the types of the columns and the structure of the data. Another useful property of `tibble` tables is that if a tibble has a large number of observations or variables, then only the rows or the columns which can fit on to the screen are printed.

Tidyverse and tibbles also support so called lazy evaluation, which is useful when your data is stored in a database, for instance. With lazy evaluation, the commands that you use on your data would be evaluated directly in the database (if possible). Without lazy evaluation, the entire data would be downloaded onto your computer only after which the commands would be evaluated. Lazy evaluation can perform many tasks faster and it can also alleviate memory usage of the computer.

9 Pipes

Piping is a fairly recent concept in R and was very rarely used in R code just a few years ago. The concept of a pipe originates from a package called `magrittr`, and pipes are commonly used under the tidyverse programming paradigm, but the pipe was also later added to base R. The notations for a tidyverse pipe and a native R pipe are `%>%` and `|>`, respectively. The idea of a pipe is that you can connect multiple function calls sequentially while keeping the code more readable. Pipes also serve to unnest standard R code which often involves using many nested parentheses, and can quickly become hard to read as one has to read the code based on the order of the operations instead of reading it linearly. For example, consider the following code where we apply a sequence of operations on a numeric vector `x`.

```
x <- 1:10
round(mean(diff(log(x))), digits = 2)
```

```
[1] 0.26
```

This code computes the rounded mean differences of the logarithms of the vector x, however this description does not match the order of operations, where the logarithm is computed first. To accomplish the same result using pipes we would write

```
x <- 1:10
x |> log() |> diff() |> mean() |> round(digits = 2)
```

```
[1] 0.26
```

Here, the order of operations can be easily read from left to right. Next, we will discuss the use of pipes in more detail.

9.1 magrittr pipe %>%

Let's have a look at an example, where we call the `summary()` function for the `grades2` data using the magrittr pipe %>%, and we also define that results should be printed with two digits.

```
grades2 %>%
  summary(digits = 2)
```

```
 group      grade
 2:2   Min.   :2.6
 3:1   1st Qu.:3.4
 4:2   Median :4.7
       Mean   :4.5
       3rd Qu.:4.9
       Max.   :6.9
```

In the code above, the object `grades2` is taken by the pipe operator %>% and forwarded to the first argument of the `summary()` function. The `summary()` function also has a second argument, which is defined by `digits = 2`. Thus, the pipe only takes the object mentioned before the pipe operator and forwards it to the function after the pipe as the first free argument. It is very common and recommended to structure R code so that there is only one pipe per row and that a new line is started after each pipe.

Although the above example is easy to understand as we already know the `summary()` function, there is also a more general way to compute summarized information following the tidyverse style. The function `summarise()` can be used to compute arbitrary statistics from the data, for example the number of observations (via the function `n()`) and the mean and the sample standard deviation of the variable `grade`.

```
grades2 %>%
  summarise(
    n = n(),
    mean = mean(grade),
    sd = sd(grade)
  )
```

```
# A tibble: 1 x 3
      n  mean    sd
  <int> <dbl> <dbl>
1     5  4.50  1.63
```

The `summarise()` function also produces a `tibble` enabling further operations via piping, if desired.

9.2 *Native pipe* |>

In R version 4.1, the native pipe `|>` was introduced to the R language, which does not require any external packages to use. In most scenarios, it does not matter whether the native or the `magrittr` pipe is used. However, there are two technical differences between the `magrittr` pipe and the native pipe. First, the `magrittr` pipe is actually a function

```
class(`%>%`)
```

```
[1] "function"
```

while the native pipe is not, and simply converts the written code into a non-piped version, i.e., into a form that one would write without using the pipe:

```
x <- 1:5
quote(x |> sum())
```

```
sum(x)
```

We see that providing x to the sum function via the native pipe is exactly the same as writing sum(x) directly. What this means in practice is that the native pipe may have better performance for example when passing a large dataset through a large number of pipes. The reason for this is that the magrittr pipe incurs an additional computational function call overhead each time it is called. The second difference between the pipes is that parentheses have to be provided for function calls when using the native pipe, but they can be omitted when using the magrittr pipe

```
x %>% sum
```

```
[1] 15
```

```
x %>% sum()
```

```
[1] 15
```

```
x |> sum()
```

```
[1] 15
```

```
x |> sum # produces an error
```

```
Error: The pipe operator requires a function call
         as RHS (<text>:1:6)
```

10 Lists

Earlier, we already briefly mentioned lists in the context of data frames. Lists are one the most common types of data in R and they resemble basic vectors in many aspects. Like vectors, a list is an ordered sequence of elements, but unlike vectors, lists can contain elements of different types simultaneously and may even contain other lists. For example, we could construct a list that contains a logical value, a numeric value and a character value using the list() function as follows.

```r
y <- list(TRUE, 7.2, "this is a list")
```

Subsetting a `list` object works slightly differently compared to vectors. When single brackets are used, a sublist is selected, i.e., a `list` object that contains the elements at the supplied indices, for example:

```r
y[1:2]
```

```
[[1]]
[1] TRUE

[[2]]
[1] 7.2
```

```r
typeof(y[1:2])
```

```
[1] "list"
```

To extract an actual element of a list, double brackets should be used:

```r
y[[2]]
```

```
[1] 7.2
```

The elements of a list may also be named, which enables subsetting via the dollar sign operator similar to data frames, or by giving the element name in double brackets instead of the index:

```r
z <- list(bool = TRUE, num = 7.2,
                            description = "this is another list")
z$bool
```

```
[1] TRUE
```

```r
z[["description"]]
```

```
[1] "this is another list"
```

One benefit of using the dollar sign is that it is not necessary to provide the full element name, unlike when using the double brackets. It is sufficient to provide a prefix of the element name so that the full name can be uniquely determined from the prefix. Because all the names of our elements in the previous list z start with a different letter, the first letter of the name suffices as the prefix. The same functionality also applies when using the dollar sign to select columns of data frames.

```
z$b
```

```
[1] TRUE
```

```
z$n
```

```
[1] 7.2
```

```
z$d
```

```
[1] "this is another list"
```

11 Functions

A function is a set of statements that when organized together perform a specific task. Each function in R has a name, a set of arguments, a body and a return object. The name of the function usually describes the purpose of the function. For example, the base R function `mean()` computes the arithmetic mean of the argument vector. The result is returned as a numeric value.

```
x <- 1:5
mean(x)
```

```
[1] 3
```

Functions are often much more complicated, which is why it is often helpful to view the documentation of a function before using it in practice. To view the documentation pages of a function, one can simply write the name of the function prefixed by a question mark.

```
?mean
```

In RStudio, the documentation will open in the "Help" tab in the bottom right pane by default. Functions will only be executed when they are called, i.e., when arguments are supplied to them. Simply writing the function name without parentheses will instead print the body of the function to the console, meaning the code that the function consists of and which is executed if the function is called.

In the previous sections, we've already familiarized ourselves with some commonly used basic functions such as `c()`, `sd()`, and `summary()`. Base R has a wide range of function to accomplish common tasks needed in data analysis, which is further extended by `tidyverse` and other R packages. This means that one does not typically have to write their own functions when programming in R. We will explore several of the functions provided by the `tidyverse` in later chapters.

12 Conditional Statements

Sometimes we may only wish to execute a piece of code when a certain condition is met. Conditional statements in R can be defined via the `if` and `else` clauses. The `if` clause evaluates a condition, which is an R expression that evaluates to a single `logical` value, and if this condition evaluates to `TRUE`, the expression following the clause if executed. Further, if an `else` clause is also provided, the expression following `else` will be executed instead if the condition evaluates to `FALSE`. As R code, the syntax for these clauses is

```
if (cond) expr
if (cond) expr else alt_expr
```

where `cond` is the condition being evaluated, `expr` is the expression that will be evaluated if `cond == TRUE`, and `alt_expr` will be evaluated if `cond == FALSE`.

Note that `if` will only evaluate a single condition. If `cond` is a vector, an error will be produced:

```
cond <- c(TRUE, FALSE)
if (cond) {
  print("This will not be printed")
}
```

```
Error in if (cond) {: the condition has length > 1
```

As expected, the error message tells us that the condition contained more than one element when it was evaluated. However, there are often scenarios where we may wish to conditionally select or define values based on a vector of conditions. For such instances, the function `ifelse()` can be used. This function has three arguments: `test`, `yes`, and `no`. When called, the function will pick those elements of the vector `yes` for which the logical vector `test` evaluates to `TRUE`, and those elements of the vector `no` for which `test` evaluates to `FALSE`:

```
cond <- c(TRUE, FALSE, FALSE, TRUE)
x <- 1:4
y <- -(1:4)
ifelse(cond, x, y)
```

```
[1]  1 -2 -3  4
```

13 Looping Constructs

In some cases, we may wish to execute the same piece of code multiple times under varying conditions. Instead of writing the same code multiple times for each condition, we can use a looping construct. There are two types of loops in R: the `for` loop and the `while` loop. The main difference between the two loops is that `for` always executes the code associated with the loop a fixed number of times whereas `while` will continue executing the code as long as a specific condition remains satisfied. The syntax of these loops is

```
for (var in seq) expr
while (cond) expr
```

In other words, `for` will execute the expression `expr` for every element `var` in some object `seq` that can be indexed. For-loops are very general, and can be used to loop over most ordered structures such as vectors and lists. Similarly, `while` will execute the expression `expr` as long as the condition `cond` evaluates to `TRUE`. Care must be taken when using while-loops to ensure that the condition will eventually evaluate to `FALSE`, otherwise the loop will simply run indefinitely and the program will be stuck. As an example, we will print the number 1 through 5 to the console using both `for` and `while` loops:

```
x <- 1:5
for (i in x) {
  print(x[i])
}
```

```
[1] 1
[1] 2
[1] 3
[1] 4
[1] 5
```

```
i <- 0
while (i < length(x)) {
  i <- i + 1
  print(x[i])
}
```

```
[1] 1
[1] 2
[1] 3
[1] 4
[1] 5
```

In contrast to explicit `for` and `while` loops, the so-called `apply` family of functions can often be a simpler alternative (see `?apply`). As the name suggests, these functions apply an operation to each element of a list or a vector (and other more general data structures). For example, the above loop example could also be accomplished with the `lapply()` function as follows:

```
y <- lapply(x, print)
```

```
[1] 1
[1] 2
[1] 3
[1] 4
[1] 5
```

14 Discussion and Other Resources for Learning R

The main aim behind this chapter was to introduce R to new users. This chapter is, of course, an initial step and can hardly cover all the basics. Interested users are advised to consult other resources e.g., open access books, tutorials, cheat sheets, and package manuals for more information. An introductory book "An Introduction to R" packaged with each R installation can be accessed from the RStudio Help panel by first selecting "Show R Help" and then selecting "An Introduction to R" under "Manuals". This book covers a wide range of topics on base R programming in great detail. The book "R for Data Science" by Hadley Wicham and Garret Grolemund provides a comprehensive tutorial on using R for data science under the tidyverse paradigm. The book is free to use and readily available online at https://r4ds.had.co.nz/. As we go further, several questions will emerge and the reader will learn by doing and by consulting the literature and help files. In doing so, the reader will build knowledge and experience that helps advance their skills.

References

1. R Core Team (2022) R: a language and environment for statistical computing. R Foundation for Statistical Computing, Vienna
2. Wickham H, Grolemund G (2016) R for data science: import, tidy, transform, visualize, and model data. O'Reilly Media, Inc., Sebastopol
3. Wickham H (2015) R packages: organize, test, document, and share your code. O'Reilly Media, Inc., Sebastopol
4. Epskamp S, Fried EI (2018) A tutorial on regularized partial correlation networks. Psychol Methods 23:617–634. https://doi.org/10.1037/met0000167
5. Bastian M, Heymann S, Jacomy M (2009) Gephi: an open source software for exploring and manipulating networks. In: Third international AAAI conference on weblogs and social media, pp 361–362
6. RStudio Team (2020) RStudio: integrated development environment for R. RStudio, PBC, Boston
7. Xie Y, Allaire JJ, Grolemund G (2019) R markdown: the definitive guide, 1st edn. Chapman & Hall/CRC, Boca Raton
8. Wickham H, Vaughan D, Girlich M (2023) tidyr: tidy messy data. https://CRAN.R-project.org/package=tidyr
9. Wickham H, François R, Henry L, Müller K, Vaughan D (2023) dplyr: a grammar of data manipulation. https://CRAN.R-project.org/package=dplyr
10. Müller K, Wickham H (2023) tibble: simple data frames. https://CRAN.R-project.org/package=tibble
11. Wickham H, Averick M, Bryan J, Chang W, McGowan LD, François R, Grolemund G, Hayes A, Henry L, Hester J, Kuhn M, Pedersen TL, Miller E, Bache SM, Müller K, Ooms J, Robinson D, Seidel DP, Spinu V, Takahashi K, Vaughan D, Wilke C, Woo K, Yutani H (2019) Welcome to the tidyverse. J Open Source Softw 4:1686. https://doi.org/10.21105/joss.01686

An R Approach to Data Cleaning and Wrangling for Education Research

Juho Kopra, Santtu Tikka, Merja Heinäniemi, Sonsoles López-Pernas, and Mohammed Saqr

1 Introduction

When analyzing data, it is crucial that the data is in a suitable format for the tools you will be using. This makes data wrangling essential. Data preparation and cleaning, such as extracting information from raw data or removing erroneous measurements, must also be done before data is ready for analysis. Data wrangling often takes up the majority of the time spent on analysis, sometimes up to 80%. To reduce the amount of work required, it is beneficial to use tools that follow the same design paradigm to minimize the time spent on data wrangling. The `tidyverse` [1] programming paradigm is currently the most popular approach for this in R.

The `tidyverse` has several advantages that make it preferable over other alternatives such as simply using base R for your data wrangling needs. All packages in the `tidyverse` follow a consistent syntax, making it intuitive to learn and use new `tidyverse` packages. This consistency also makes the code more easier to read, and maintain, and reduces the risk of errors. The `tidyverse` also has a vast range of readily available packages that are actively maintained, reducing the need for customized code for each new data wrangling task. Further, these packages integrate seamlessly with one another, facilitating a complete data analysis pipeline.

J. Kopra (✉) · S. López-Pernas · M. Saqr
School of Computing, University of Eastern Finland, Kuopio, Finland
e-mail: juho.kopra@uef.fi

S. Tikka
Department of Mathematics and Statistics, University of Jyväskylä, Finland

M. Heinäniemi
Institute of Biomedicine, University of Eastern Finland, Kuopio, Finland

M. Saqr, S. López-Pernas (eds.), *Learning Analytics Methods and Tutorials*,
https://doi.org/10.1007/978-3-031-54464-4_4

To fully realize the benefits of the `tidyverse` programming paradigm, one must first understand the key concepts of tidy data and pivoting. Tidy data follows three rules:

1. Each variable must have its own column.
2. Each observation must have its own row.
3. Each value must have its own cell.

Let's consider examples of tidy data. For instance, if you have data from a Moodle course where two attempts of an exam for each student are located in a single column. This example data violates the first rule, because there are two variables in a single column instead of separate columns for each variable. What is needed to make this data tidy is to pivot it to a longer format. The pivoted data would have two rows for each student, both of which are different observations (exam attempts 1 and 2). Thus the pivoted data would not conflict with second rule.

Data can also be too long, but in practice, this is much more rare. This can occur if two or more variables are stored on a single column across multiple rows. A key indicators of this is if the different rows of the same column have different measurements units (e.g. lb vs. kg). It may also occur that your raw data has multiple values in a single cell. In these cases, it is necessary to split the cells to extract the necessary information. In a simple case, where you have two values systematically in a one cell, the values can be easily separated into their own columns.

Overall, using `tidyverse` and understanding the key concepts of tidy data and pivoting can streamline the data analysis process and make code easier to work with and maintain. The rest of this chapter will guide readers through the process of data cleaning and wrangling with R in the field of learning analytics. We demonstrate how data can be grouped and summarized, how to select and transform variables of interest, and how data can be rearranged, reshaped and joined with other datasets. We will strongly rely on the `tidyverse` programming paradigm for a consistent and coherent approach to data manipulation, with a focus on tidy data.

2 Reading Data into R

Data files come in many formats, and getting your data ready for analysis can often be a daunting task. The `tidyverse` offers much better alternatives to base R functions for reading data, especially in terms of simplicity and speed when reading large files. Additionally, most of the file input functions in the `tidyverse` follow a similar syntax, meaning that the user does not have to master every function for reading every type of data individually.

Often, just before the data can be read into R, user must specify the location of data files by setting a working directory. Perhaps most useful way to do that is to create a project in RStudio and then create a folder called "data" within the project folder. Data files can be put into that folder and user can refer to those files just by telling R-functions relative path of data file (e.g. "data/Final%20Dataset.csv") while the project takes care of the rest of the path. A more traditional way of setting this, which also works without RStudio, is by using a command such as `setwd("/home/Projects/LAproject/data/Final%20Dataset.csv")`. Here, a function called `setwd()` is used to set up a folder into location mentioned in a character string given as its argument. A `getwd()` lists current working directory, which can also be seen in RStudio just above the Console output.

Some of the most common text data formats are comma-separated files or semicolon-separated files, both of which typically have the file extension .csv. These files can be read into R using the `readr` [2] package and the functions `read_csv()` and `read_csv2()`, respectively. For instance, we can read a comma-separated file R as follows

```
library("readr")
url <- "https://github.com/lamethods/data/raw/main/2_moodleEdify/"
lms <- read_csv(paste0(url, "Final%20Dataset.csv"))
```

Functions in `readr` provide useful information about how the file was read into R, which can be used to assess if the input was successful and what assumptions about the data were made during the process. In the printout above, the `read_csv()` function tells us the number of rows and columns in the data and the column specification, i.e., what type of data is contained within each column. In this case, we have 17 columns with `character` type of data, and 4 columns of `double` type of data. Functions in `readr` try to guess the column specification automatically, but it can also be specified manually when using the function. For more information about this dataset, please refer to Chapter 2 in this book [3].

Data from Excel worksheets can be read into R using the `import()` function from the `rio` [4] package. We will use synthetic data generated based on a real blended course of learning analytics for the remainder of this chapter. These data consist of three Excel files which we will first read into R.

```
library("rio")
url <- "https://github.com/lamethods/data/raw/main/1_moodleLAcourse/"
events <- import(paste0(url, "Events.xlsx"), setclass = "tibble")
results <- import(paste0(url, "Results.xlsx"), setclass = "tibble")
demographics <- import(paste0(url, "Demographics.xlsx"), setclass = "tibble")
```

The data files contain information on students' Moodle events, background information such as their name, study location and employment status, and various grades they've obtained during the course. For more information about the dataset, please refer to Chapter 2 in this book [3]. These data are read in the `tibble` [5] format, a special type of `data.frame` commonly used by `tidyverse` packages. We also load the `dplyr` [6] package which we will use for various tasks throughout this chapter.

```
library("dplyr")
```

3 Grouping and Summarizing Data

Instead of individual-level data metrics, we may be interested in specific groups as specified by some combination of data values. For example, we could compute the number of students studying in each location by gender. To accomplish this, we need to start by creating a grouped dataset with the function `group_by()`. To compute the number of students, we can use the `summarise()` function which we already used previously in Chapter 1 and the function `count()`, which simply returns the number of observations in each category of its argument.

```
demographics |>
  group_by(Gender) |>
  count(Location)
```

```
# A tibble: 4 x 3
# Groups:   Gender [2]
  Gender Location       n
  <chr>  <chr>      <int>
1 F      On campus     55
2 F      Remote        10
3 M      On campus     51
4 M      Remote        14
```

The column `n` lists the number of students in each group. When a `tibble` that contains grouped data is printed into the console, the grouping variable and the number of groups will be displayed in the console below the dimensions. Next, we will compute the total number of Moodle events of each student, which we will also use in the subsequent sections.

```
events_summary <- events |>
  group_by(user) |>
  tally() |>
  rename(Frequency.Total = n)
events_summary
```

```
# A tibble: 130 x 2
  user      Frequency.Total
  <chr>               <int>
1 00a05cc62             417
2 042b07ba1             918
3 046c35846             199
4 05b604102             199
5 0604ff3d3             436
# i 125 more rows
```

Here, the function `tally()` simply counts the number of number rows in the data related to each student, reported in the column `n` which we rename to `Frequency.Total` with the `rename()` function. We could also count the number of events by event type for each student

```
events |>
  group_by(user, Action) |>
  count(Action)
```

```
# A tibble: 1,439 x 3
# Groups:   user, Action [1,439]
  user      Action           n
  <chr>     <chr>        <int>
1 00a05cc62 Applications     2
2 00a05cc62 Assignment     121
3 00a05cc62 Course_view    103
4 00a05cc62 Feedback         7
5 00a05cc62 General         10
# i 1,434 more rows
```

4 Selecting Variables

In the `tidyverse` paradigm, selecting columns, i.e., variables from data is done using the `select()` function. The `select()` function is very versatile, allowing the user to carry out selections ranging from simple selection of a single variable to highly complicated selections based on multiple criteria. The most basic selection selects only a single variable in the data based on its name. For example, we can select the employment statuses of students as follows

```
demographics |>
  select(Employment)
```

```
# A tibble: 130 x 1
  Employment
  <chr>
1 None
2 None
3 None
4 None
5 Part-time
# i 125 more rows
```

Note that using `select()` with a single variable is not the same as using the `$` symbol to select a variable, as the result is still a `tibble`, `select()` simply produces a subset of the data, where only the selected columns are present. Select is more similar to `subset()` in base R, which can accomplish similar tasks as `select()` and `filter()` in the `tidyverse`. However, we do not recommend using `subset()`, as it may not work correctly when the working environment has variables that have the same name as columns in the data, which can lead to undesired outcomes.

To extract the values of the selected column as a vector, we can use the function `pull()`. We use the `head()` function here to limit the console output to just the first few values of the vector (default is 6 values).

```
demographics |>
  pull(Employment) |>
  head()
```

```
[1] "None"      "None"      "None"      "None"      "Part-time" "Part-time"
```

The `select()` function syntax supports several operations that are similar to base R. We can select ranges of consecutive variables using `:`, complements using `!`, and combine selections using `c()`. The following selections illustrate some of these features:

```
demographics |>
  select(user:Origin)
```

```
# A tibble: 130 x 4
  user      Name   Surname  Origin
  <chr>     <chr>  <chr>    <chr>
1 6eba3ff82 Amanda Mora     Costa Rica
2 05b604102 Lian   Abdullah Yemen
3 111422ee7 Bekim  Krasniqi Kosovo
4 b4658c3a9 Yusuf  Kaya     Turkey
5 e6ec47f29 Zoran  Babić    Serbia
# i 125 more rows
```

```
demographics |>
  select(!Gender)
```

```
# A tibble: 130 x 7
  user      Name   Surname  Origin     Birthdate Location  Employment
  <chr>     <chr>  <chr>    <chr>      <chr>     <chr>     <chr>
```

```
1 6eba3ff82 Amanda Mora     Costa Rica 28.2.1998  On campus None
2 05b604102 Lian   Abdullah Yemen      19.11.1996 On campus None
3 111422ee7 Bekim  Krasniqi Kosovo     30.1.1999  On campus None
4 b4658c3a9 Yusuf  Kaya     Turkey     16.6.1998  On campus None
5 e6ec47f29 Zoran  Babić    Serbia     29.10.1998 On campus Part-time
# i 125 more rows
```

```
demographics |>
  select(c(user, Surname))
```

```
# A tibble: 130 x 2
  user      Surname
  <chr>     <chr>
1 6eba3ff82 Mora
2 05b604102 Abdullah
3 111422ee7 Krasniqi
4 b4658c3a9 Kaya
5 e6ec47f29 Babić
# i 125 more rows
```

In the first selection, we select all variables starting from `user` on the left to `Origin` on the right. In the second, we select all variables except `Gender`. In the third, we select both `user` and `Surname` variables.

Sometimes, our selection might not be based directly on the variable names themselves as in the examples above but instead on vectors that contain the names of columns we wish to select. In such cases, we can use the function `all_of()`. We can consider the intersections or unions of such selections using `&` and `|`, respectively.

```
cols_a <- c("user", "Name", "Surname")
cols_b <- c("Surname", "Origin")
demographics |>
  select(all_of(cols_a))
```

```
# A tibble: 130 x 3
  user      Name   Surname
  <chr>     <chr>  <chr>
1 6eba3ff82 Amanda Mora
2 05b604102 Lian   Abdullah
3 111422ee7 Bekim  Krasniqi
4 b4658c3a9 Yusuf  Kaya
5 e6ec47f29 Zoran  Babić
# i 125 more rows
```

```
demographics |>
  select(all_of(cols_a) & all_of(cols_b))
```

```
# A tibble: 130 x 1
  Surname
  <chr>
1 Mora
2 Abdullah
3 Krasniqi
4 Kaya
5 Babić
# i 125 more rows
```

```
demographics |>
  select(all_of(cols_a) | all_of(cols_b))
```

```
# A tibble: 130 x 4
  user      Name   Surname  Origin
  <chr>     <chr>  <chr>    <chr>
1 6eba3ff82 Amanda Mora     Costa Rica
2 05b604102 Lian   Abdullah Yemen
3 111422ee7 Bekim  Krasniqi Kosovo
4 b4658c3a9 Yusuf  Kaya     Turkey
5 e6ec47f29 Zoran  Babić    Serbia
# i 125 more rows
```

Often the names of data variables follow a similar pattern, and these patterns can be used to construct selections. Selections based on a prefix or a suffix in the variable name can be carried out with the functions `starts_with()` and `ends_with()`, respectively. The function `contains()` is used to look for a specific substring in the names of the variables, and more complicated search patterns can be defined with the function `matches()` that uses regular expressions (see `?tidyselect::matches` for further information).

```
results |>
  select(starts_with("Grade"))
```

```
# A tibble: 130 x 13
  Grade.SNA_1 Grade.SNA_2 Grade.Review Grade.Group_self Grade.Group_All
        <dbl>       <dbl>        <dbl>            <dbl>           <dbl>
1           0           0         6.67                5               4
2           8          10         6.67                1               3
```

```
3           10          10          10                    10          9.11
4            5           5           0                     1          4
5           10          10          10                    10          9.18
# i 125 more rows
# i 8 more variables: Grade.Excercises <dbl>, Grade.Project <dbl>,
#   Grade.Literature <dbl>, Grade.Data <dbl>, Grade.Introduction <dbl>,
#   Grade.Theory <dbl>, Grade.Ethics <dbl>, Grade.Critique <dbl>
```

```
results |>
  select(contains("Data"))
```

```
# A tibble: 130 x 1
  Grade.Data
       <dbl>
1          4
2          3
3          5
4          3
5          5
# i 125 more rows
```

So far, our selections have been based on variable names, but other conditions for selection are also feasible. The general-purpose helper function `where()` is used to select those variables for which a function provided to it returns `TRUE`. For example, we could select only those columns that contain `character` type data or `double` type data.

```
results |>
  select(where(is.character))
```

```
# A tibble: 130 x 1
  user
  <chr>
1 6eba3ff82
2 05b604102
3 111422ee7
4 b4658c3a9
5 e6ec47f29
# i 125 more rows
```

```
results |>
  select(where(is.double))
```

```
# A tibble: 130 x 14
  Grade.SNA_1 Grade.SNA_2 Grade.Review Grade.Group_self Grade.Group_All
        <dbl>       <dbl>        <dbl>            <dbl>           <dbl>
1           0           0         6.67                5            4
2           8          10         6.67                1            3
3          10          10        10                  10            9.11
4           5           5         0                   1            4
5          10          10        10                  10            9.18
# i 125 more rows
# i 9 more variables: Grade.Excercises <dbl>, Grade.Project <dbl>,
#   Grade.Literature <dbl>, Grade.Data <dbl>, Grade.Introduction <dbl>,
#   Grade.Theory <dbl>, Grade.Ethics <dbl>, Grade.Critique <dbl>, Final_grade
<dbl>
```

5 Filtering Observations

In contrast to selection which relates to obtaining a subset of the columns of the data, filtering refers to obtaining a subset of the rows. In the `tidyverse`, data filtering is carried out with the `dplyr` package function `filter()`, which should not be confused with the base R `filter()` function in the `stats` package. As we have attached the `dplyr` package, the base R `filter()` function is masked, meaning that when we write code that uses `filter()`, the `dplyr` version of the function will automatically be called.

Filtering is often a much simpler operation than selecting variables, as the filtering conditions are based solely on the values of the data variables. Using `filter()` is analogous to the base R subset operator `[`, but the filtering condition is given as an argument to the `filter()` function instead. It is good to remind that in R a single equal sign (=) is merely for arguments of function calls, while double equal sign (==) is needed for comparison of two values. And example of filter:

```
demographics |>
  filter(Origin == "Bosnia") |>
  select(Name, Surname)
```

```
# A tibble: 2 x 2
  Name  Surname
  <chr> <chr>
1 Hamza Hodžić
2 Davud Delić
```

The code above first filters our student demographics data to only those students whose country of origin is Bosnia. Then, we select their first and last names.

Multiple filtering conditions can be refined and combined using base R logical operators, such as `&` and `|`.

```
demographics |>
  filter(Gender == "F" & Location == "Remote")
```

```
# A tibble: 10 x 8
  user       Name    Surname      Origin      Gender Birthdate  Location Employment
  <chr>      <chr>   <chr>        <chr>       <chr>  <chr>      <chr>    <chr>
1 d93f7f0d3 Zahra   Gul          Afghanistan F      22.11.1999 Remote   None
2 93d1f2e82 Louise  Bernard      France      F      5.9.1998   Remote   Part-time
3 417892918 Miora   Rakotomalala Madagascar  F      9.12.1995  Remote   None
4 f98e6e2b8 Linda   Mensah       Ghana       F      7.2.1991   Remote   None
5 590846fe3 Lucija  Horvat       Croatia     F      22.7.1998  Remote   None
# i 5 more rows
```

Here, we filtered our data to female students who are studying remotely. The same result could also be obtained by using the `filter()` function two times

```
demographics |>
  filter(Gender == "F") |>
  filter(Location == "Remote")
```

```
# A tibble: 10 x 8
  user       Name    Surname      Origin      Gender Birthdate  Location Employment
  <chr>      <chr>   <chr>        <chr>       <chr>  <chr>      <chr>    <chr>
1 d93f7f0d3 Zahra   Gul          Afghanistan F      22.11.1999 Remote   None
2 93d1f2e82 Louise  Bernard      France      F      5.9.1998   Remote   Part-time
3 417892918 Miora   Rakotomalala Madagascar  F      9.12.1995  Remote   None
4 f98e6e2b8 Linda   Mensah       Ghana       F      7.2.1991   Remote   None
5 590846fe3 Lucija  Horvat       Croatia     F      22.7.1998  Remote   None
# i 5 more rows
```

This type of approach may improve the readability of your code especially when there are several independent filtering conditions to be applied simultaneously.

Filters can naturally be based on numeric values as well. For example, we could select those students whose final grade is higher than 8.

```
results |>
  filter(Final_grade > 8)
```

```
# A tibble: 58 x 15
  user      Grade.SNA_1 Grade.SNA_2 Grade.Review Grade.Group_self Grade.Group_All
  <chr>           <dbl>       <dbl>        <dbl>            <dbl>           <dbl>
1 111422ee7          10          10        10                  10            9.11
2 e6ec47f29          10          10        10                  10            9.18
3 4951e7386          10          10         7                   9            8.56
4 9d744e5bf          10          10        10                  10            9.29
5 0ef305578          10          10         9.33               10            8.56
# i 53 more rows
```

```
# i 9 more variables: Grade.Excercises <dbl>, Grade.Project <dbl>,
#   Grade.Literature <dbl>, Grade.Data <dbl>, Grade.Introduction <dbl>,
#   Grade.Theory <dbl>, Grade.Ethics <dbl>, Grade.Critique <dbl>, Final_grade <dbl>
```

Similarly, we could select students based on their total number of Moodle events.

```
events_summary |>
  filter(Frequency.Total > 100 & Frequency.Total < 500)
```

```
# A tibble: 44 x 2
  user      Frequency.Total
  <chr>               <int>
1 00a05cc62             417
2 046c35846             199
3 05b604102             199
4 0604ff3d3             436
5 0af619e4b             268
# i 39 more rows
```

6 Transforming Variables

In the best-case scenario, our data is already in the desired format after it has been read into R, but this is rarely the case with real datasets. We may need to compute new variables that were not present in the original data, convert measurements to different units, or transform text data into a numeric form. In the `tidyverse`, data transformations are carried out by the `mutate()` function of the `dplyr` package. This function can be used to transform multiple variables at the same time or to construct entirely new variables. The syntax of the function is the same in both cases: first, the name of the variable should be provided followed by an R expression that defines the variable. The transformed data is not automatically assigned to any variable, enabling transformations to be used as temporary variables within a chain of piped operations.

As a simple example, we could convert the students' locations into a factor variable.

```
demographics |>
  mutate(Location = factor(Location))
```

```
# A tibble: 130 x 8
  user      Name   Surname Origin     Gender Birthdate Location  Employment
  <chr>     <chr>  <chr>   <chr>      <chr>  <chr>     <fct>     <chr>
1 6eba3ff82 Amanda Mora    Costa Rica F      28.2.1998 On campus None
```

```
2 05b604102 Lian    Abdullah Yemen      F       19.11.1996 On campus None
3 111422ee7 Bekim   Krasniqi Kosovo     M       30.1.1999  On campus None
4 b4658c3a9 Yusuf   Kaya     Turkey     M       16.6.1998  On campus None
5 e6ec47f29 Zoran   Babič    Serbia     M       29.10.1998 On campus Part-time
# i 125 more rows
```

As we see from the `tibble` printout, the `Location` variable is a factor in the transformed data as indicated by the `<fct>` heading under the variable name. Note that the original `demographics` data was not changed, as we did not assign the result of the computation.

The gender and employment status of the students could also be used as factors, which we could do in a single `mutate()` call

```
demographics |>
  mutate(
    Gender = factor(Gender),
    Location = factor(Location),
    Employment = factor(Employment)
  )
```

```
# A tibble: 130 x 8
  user      Name    Surname  Origin     Gender Birthdate  Location  Employment
  <chr>     <chr>   <chr>    <chr>      <fct>  <chr>      <fct>     <fct>
1 6eba3ff82 Amanda  Mora     Costa Rica F      28.2.1998  On campus None
2 05b604102 Lian    Abdullah Yemen      F      19.11.1996 On campus None
3 111422ee7 Bekim   Krasniqi Kosovo     M      30.1.1999  On campus None
4 b4658c3a9 Yusuf   Kaya     Turkey     M      16.6.1998  On campus None
5 e6ec47f29 Zoran   Babič    Serbia     M      29.10.1998 On campus Part-time
# i 125 more rows
```

However, writing out individual identical transformations manually is cumbersome when the number of variables is large. For such cases, the `across()` function can be leveraged, which applies a function across multiple columns. This function uses the same selection syntax that we already familiarized ourselves with earlier to define the columns that will be transformed. To accomplish the same three transformations into a factor format, we could write

```
demographics |>
  mutate(across(c(Gender, Location, Employment), factor))
```

```
# A tibble: 130 x 8
  user      Name   Surname  Origin     Gender Birthdate  Location  Employment
  <chr>     <chr>  <chr>    <chr>      <fct>  <chr>      <fct>     <fct>
1 6eba3ff82 Amanda Mora     Costa Rica F      28.2.1998  On campus None
2 05b604102 Lian   Abdullah Yemen      F      19.11.1996 On campus None
3 111422ee7 Bekim  Krasniqi Kosovo     M      30.1.1999  On campus None
4 b4658c3a9 Yusuf  Kaya     Turkey     M      16.6.1998  On campus None
5 e6ec47f29 Zoran  Babić    Serbia     M      29.10.1998 On campus Part-time
# i 125 more rows
```

The first argument to the `across()` function is the selection that defines the variables to be transformed. The second argument defines the transformation, in this case, a function, to be used.

Working with dates can often be challenging. When we read the student demographic data into R, the variable `Birthdate` was assumed to be a `character` type variable. If we would like to use this variable to e.g., compute the ages of the students, we need to first convert it into a proper format using the `as.Date` function. Since the dates in the data are not in any standard format, we must provide the format manually. Afterwards, we can use the `lubridate` [7] package to easily compute the ages of the students, which we will save into a new variable called `Age`. We will also construct another variable called `FullName` which formats the first and last names of the students as `"Last, First"`.

```
library("lubridate")
demographics |>
  mutate(
    Birthdate = as.Date(Birthdate, format = "%d.%m.%Y"),
    Age = year(as.period(interval(start = Birthdate, end = date("2023-03-12")))),
    FullName = paste0(Surname, ", ", Name)
  ) |>
  select(Age, FullName)
```

```
# A tibble: 130 x 2
    Age FullName
  <dbl> <chr>
1    25 Mora, Amanda
2    26 Abdullah, Lian
3    24 Krasniqi, Bekim
4    24 Kaya, Yusuf
5    24 Babić, Zoran
# i 125 more rows
```

The computation of the ages involves several steps. First, we construct a time interval object with the `interval()` function from the birthdate to the date for which we wish to compute the ages. Next, the `as.period()` function converts this interval into a time duration, from which we lastly get the number of years with the `year()` function.

Suppose that we would like to construct a new variable `AchievingGroup` that categorizes the students into top 50% achievers and bottom 50% achievers based on their final grade on the course. We leverage two functions from the `dplyr` package to construct this new variable: `case_when()` and `ntile()`. The function `case_when()` is used to transform variables based on multiple sequential conditions. The function `ntile()` has two arguments, a vector `x` and an integer `n`, and it splits `x` into `n` equal-sized groups based on the ranks of the values in `x`.

```
results <- results |>
  mutate(
    AchievingGroup = factor(
      case_when(
        ntile(Final_grade, 2) == 1 ~ "Low achiever",
        ntile(Final_grade, 2) == 2 ~ "High achiever"
      )
    )
  )
```

The syntax of `case_when()` is very simple: we describe the condition for each case followed by ~ after which we define the value that the case should correspond to. We assign the result of the computation to the `results` data, as we will be using the `AchievingGroup` variable in later chapters.

We would also like to categorize the students based on their activity level, i.e., the number of total Moodle events. Our goal is to create three groups of equal size consisting of low activity, moderate activity and high activity students. The approach we applied to categorizing the achievement level of the students is also applicable for this purpose. We name our new variable as `ActivityGroup`, and we assign the result of the computation, as we will also be using this variable in later chapters.

```
events_summary <- events_summary |>
  mutate(
    ActivityGroup = factor(
      case_when(
        ntile(Frequency.Total, 3) == 1 ~ "Low activity",
        ntile(Frequency.Total, 3) == 2 ~ "Moderate activity",
        ntile(Frequency.Total, 3) == 3 ~ "High activity"
      )
    )
  )
```

7 Rearranging Data

Sometimes we may want to reorder the rows or columns of our data, for example in alphabetical order based on the names of students on a course. The `arrange()` function from the `dplyr` package orders the rows of the by the values of columns selected by the user. The values are sorted in ascending order by default, but the order can be inverted by using the `desc()` function if desired. The variable order in the selection defines how ties should be broken when duplicate values are encountered in the previous variables of the selection. For instance, the following code would arrange the rows of our `demographics` data by first comparing the surnames of the students, and then the given names for those students with the same surname. Missing values are placed last in the reordered data.

```
demographics |>
  arrange(Surname, Name)
```

```
# A tibble: 130 x 8
  user      Name   Surname  Origin     Gender Birthdate  Location  Employment
  <chr>     <chr>  <chr>    <chr>      <chr>  <chr>      <chr>     <chr>
1 ba76ebfab Bismah Abbasi   Pakistan   F      2.4.1996   Remote    Full-time
2 05b604102 Lian   Abdullah Yemen      F      19.11.1996 On campus None
3 d2c3ce9a4 Amir   Ait      Morocco    M      19.6.1997  On campus None
4 68a377c82 Saliha Akmatova Kyrgyzstan F      19.5.1999  Remote    None
5 7e2726f3c Kazi   Akter    Bangladesh M      22.12.1992 On campus None
# i 125 more rows
```

A descending order based on both names can be obtained by applying the `desc()` function.

```
demographics |>
  arrange(desc(Surname), desc(Name))
```

```
# A tibble: 130 x 8
  user      Name    Surname  Origin    Gender Birthdate Location  Employment
  <chr>     <chr>   <chr>    <chr>     <chr>  <chr>     <chr>     <chr>
1 a48165ad5 Liam    Zambrano Ecuador   M      4.4.1998  On campus None
2 1115dae61 Poema   Wong     Tahiti    F      22.1.1999 Remote    Part-time
3 0ef305578 Jack    White    Australia M      22.4.1995 Remote    None
4 f753ce9bf Dechen  Wangmo   Bhutan    F      29.4.1999 On campus None
5 f87eaa00c Prasert Wang     Thailand  M       9.4.1997 On campus None
# i 125 more rows
```

Column positions can be changed with the `relocate()` function of the `dplyr` package. Like `arrange()`, we first select the column or columns we wish to move into a different position in the data. Afterwards, we specify the position where the columns should be moved to in relation to positions of the other columns. In our `demographics` data, the user ID column `user` is the first column. The following code moves this column after the `Employment` column so that the `user` column becomes the last column in the data.

```
demographics |>
  relocate(user, .after = Employment)
```

```
# A tibble: 130 x 8
  Name   Surname  Origin     Gender Birthdate  Location  Employment user
  <chr>  <chr>    <chr>      <chr>  <chr>      <chr>     <chr>      <chr>
1 Amanda Mora     Costa Rica F      28.2.1998  On campus None       6eba3ff82
2 Lian   Abdullah Yemen      F      19.11.1996 On campus None       05b604102
3 Bekim  Krasniqi Kosovo     M      30.1.1999  On campus None       111422ee7
4 Yusuf  Kaya     Turkey     M      16.6.1998  On campus None       b4658c3a9
5 Zoran  Babić    Serbia     M      29.10.1998 On campus Part-time  e6ec47f29
# i 125 more rows
```

The mutually exclusive arguments `.before` and `.after` of `relocate()` specify the new column position in relation to columns that were not selected. These arguments also support the `select()` function syntax for more general selections.

8 Reshaping Data

Broadly speaking, tabular data typically take one of two formats: wide or long. In the wide format, there is one row per subject, where the subjects are identified by an identifier variable, such as the `user` variable in our Moodle data, and multiple columns for each measurement. In the long format, there are multiple rows per subject, and the columns describe the type of measurement and its value. For example, the `events` data is in a long format containing multiple Moodle events per student, but the `results` and `demographics` data are in a wide format with one row per student.

In the previous section, we constructed a summary of the users' Moodle events in total and of different types. The latter data is also in a long format with multiple rows per subject, but we would instead like to have a column for each event type with one row per user, which means that we need to convert this data into a wide format. Conversion between the two tabular formats is often called *pivoting*, and the corresponding functions `pivot_wider()` and `pivot_longer()` from the `tidyr` [8] package are also named according to this convention. We will create a wide format data of the counts of different event types using the `pivot_wider()` function as follows

```
library("tidyr")
events_types <- events |>
  group_by(user, Action) |>
  count(Action) |>
  pivot_wider(
    names_from = "Action",
    names_prefix = "Frequency.",
    values_from = "n",
    values_fill = 0
  )
events_types
```

```
# A tibble: 130 x 13
# Groups:   user [130]
  user      Frequency.Applications Frequency.Assignment Frequency.Course_view
  <chr>                      <int>                <int>                 <int>
1 00a05cc62                      2                  121                   103
2 042b07ba1                      0                   62                   294
3 046c35846                      0                   41                    53
4 05b604102                      0                   44                    49
5 0604ff3d3                      0                    9                   119
# i 125 more rows
# i 9 more variables: Frequency.Feedback <int>, Frequency.General <int>,
#   Frequency.Group_work <int>, Frequency.Instructions <int>,
#   Frequency.La_types <int>, Frequency.Practicals <int>, Frequency.Social <int>,
#   Frequency.Ethics <int>, Frequency.Theory <int>
```

Here, we first specify the column name that the names of the wide format data should be taken from in the long format data with `names_from`. In addition, we specify a prefix for the new column names using `names_prefix` that helps to distinguish what these new columns will contain, but in general, the prefix is optional. Next, we specify the column that contains the values for the new columns with `values_from`. Because not every student necessarily has events of every type, we also need to specify what the value should be in cases where there are no events of a particular type by using `values_fill`. As we are considering the frequencies of the events, it is sensible to select 0 to be this value. We save the results to `events_types` as we will use the event type data in later sections and chapters.

9 Joining Data

Now that we have computed the total number of events for each student and converted the event type data into a wide format, we still need to merge these new data with the demographics and results data. Data merges are also called *joins*, and the dplyr package provides several functions for different kinds of joins. Here, we will use the left_join() function that will preserve all observations of the first argument.

```
left_join(demographics, events_summary, by = "user")
```

```
# A tibble: 130 x 10
  user     Name  Surname Origin Gender Birthdate Location Employment Frequency.Total
  <chr>    <chr> <chr>   <chr>  <chr>  <chr>     <chr>    <chr>                <int>
1 6eba3ff~ Aman~ Mora    Costa~ F      28.2.1998 On camp~ None                   312
2 05b6041~ Lian  Abdull~ Yemen  F      19.11.19~ On camp~ None                   199
3 111422e~ Bekim Krasni~ Kosovo M      30.1.1999 On camp~ None                   532
4 b4658c3~ Yusuf Kaya    Turkey M      16.6.1998 On camp~ None                   246
5 e6ec47f~ Zoran Babić   Serbia M      29.10.19~ On camp~ Part-time              356
# i 125 more rows
# i 1 more variable: ActivityGroup <fct>
```

In essence, the above left join adds all columns from events_summary to demographics whenever there is a matching value in the by column. To continue, we can use additional left joins to add the remaining variables from the results data, and the Moodle event counts of different types from events_types to have all the student data in a single object.

```
all_combined <- demographics |>
  left_join(events_types, by = "user") |>
  left_join(events_summary, by = "user") |>
  left_join(results, by = "user")
all_combined
```

```
# A tibble: 130 x 37
  user      Name   Surname  Origin     Gender Birthdate  Location  Employment
  <chr>     <chr>  <chr>    <chr>      <chr>  <chr>      <chr>     <chr>
1 6eba3ff82 Amanda Mora     Costa Rica F      28.2.1998  On campus None
2 05b604102 Lian   Abdullah Yemen      F      19.11.1996 On campus None
3 111422ee7 Bekim  Krasniqi Kosovo     M      30.1.1999  On campus None
4 b4658c3a9 Yusuf  Kaya     Turkey     M      16.6.1998  On campus None
5 e6ec47f29 Zoran  Babić    Serbia     M      29.10.1998 On campus Part-time
# i 125 more rows
# i 29 more variables: Frequency.Applications <int>, Frequency.Assignment <int>,
#   Frequency.Course_view <int>, Frequency.Feedback <int>, Frequency.General <int>,
#   Frequency.Group_work <int>, Frequency.Instructions <int>,
#   Frequency.La_types <int>, Frequency.Practicals <int>, Frequency.Social <int>,
#   Frequency.Ethics <int>, Frequency.Theory <int>, Frequency.Total <int>,
#   ActivityGroup <fct>, Grade.SNA_1 <dbl>, Grade.SNA_2 <dbl>, ...
```

We will use this combined dataset in the following chapters as well.

10 Missing Data

Sometimes it occurs that learning analytics data has cells for which the values are missing for some reason. The Moodle event data which we have utilized in this chapter does not naturally contain missing data. Thus, to have an example, we need to create a data which does. Second, handling of missing data is a vast topic of which we can only discuss some of the key points very briefly from a practical perspective. For a more comprehensive overview, we recommend reading [9] and [10] for a hands on approach. A short overview of missingness can be found in [11].

The code below will create missing values randomly to each column of `events_types` data (`user` column is an exception). To do that, we use the `mice` [12] package which also has methods for the handling of missing data. Unfortunately, `mice` is not part of the `tidyverse`. For more information about `mice`, a good source is `miceVignettes` at https://www.gerkovink.com/miceVignettes/. Now, let's create some missing data.

```
library("mice")
set.seed(44)
events_types <- events_types |>
  rename(
    "Ethics" = "Frequency.Ethics",
    "Social" = "Frequency.Social",
    "Practicals" = "Frequency.Practicals"
  )
ampute_list <- events_types |>
  ungroup(user) |>
  select(Ethics:Practicals) |>
  as.data.frame() |>
  ampute(prop = 0.3)
events_types_mis <- ampute_list$amp |>
  as_tibble()
events_types_mis[2, "Practicals"] <- NA
```

Above, we also rename the variables that contain the frequencies of Moodle events related to ethics, social and practicals into `Ethics`, `Social` and `Practicals`, respectively. Let's now see some of the values of `events_types_mis`

```
events_types_mis
```

```
# A tibble: 130 x 3
  Ethics Social Practicals
   <int>  <int>      <int>
1     NA     12         89
```

```
2      14      NA          NA
3       0       0          47
4       0       0          48
5       0       0          61
# i 125 more rows
```

We can see that now the data contains `NA` values in some of the cells. These are the cells in which a missing value occurs, meaning that a value for those measurements has not been recorded. A missing data pattern, that is how missing of one variable affects missingness of other variables, can be show as:

```
md.pattern(events_types_mis, rotate.names = TRUE)
```

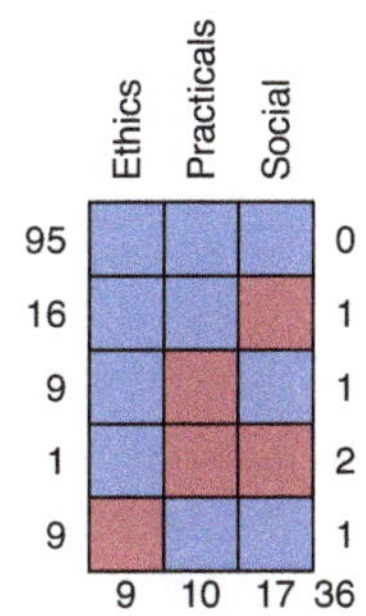

Above, each red square indicates a missing value while blue squares stand for observed ones. We can see that there are 95 complete rows, 10 for which `Practicals` are missing, 17 have missingness on `Social` and 9 are missing on `Ethics`. Also, one row has two red squares indicating a missing value on both `Social` and `Practicals`.

Let's now discuss options of handling missing data briefly. There are four classes of statistical methods for analyzing data with missing values: complete case (CC) methods, weighting methods, imputation methods, and model-based methods. The simplest of these is complete case analysis, which leaves missing values out of the analysis and only uses observations with all variables recorded. This can be done with the `tidyr` [8] package function `drop_na()`:

```
events_types_mis |>
  drop_na()
```

```
# A tibble: 95 x 3
  Ethics Social Practicals
   <int>  <int>      <int>
1      0      0         47
```

```
2        0        0           48
3        0        0           61
4        0       24          102
5        4       18           71
# i 90 more rows
```

We can see that after using this method, our data has only 95 rows as those were the rows without any columns having missing values. This made our data much smaller! If there are a lot of missing values, the data may become too small to use for practical purposes.

A more novel group of methods are imputation methods. One of the options is using single imputation (SI) where the mean of each variable will determine the imputed value. The single mean imputation can be done as follows:

```
imp <- mice(events_types_mis, method = "mean",
m = 1, maxit = 1 , print = FALSE)
complete(imp) |>
  head()
```

```
     Ethics   Social Practicals
1  7.553719 12.00000   89.00000
2 14.000000 15.64602   74.80833
3  0.000000  0.00000   47.00000
4  0.000000  0.00000   48.00000
5  0.000000  0.00000   61.00000
6  0.000000 24.00000  102.00000
```

We can see from above that the imputed values are not integers anymore. However, if we aim to estimate means or regression coefficients (see Chapter 5 [13] for details) that is not a problem. One of the problems with mean imputation is that the variance and standard error estimates will become downward biased. A mean of `Ethics` for mean imputation is:

```
fit <- with(imp, lm(Ethics ~ 1))
summary(fit)
```

```
# A tibble: 1 x 6
  term        estimate std.error statistic  p.value  nobs
  <chr>          <dbl>     <dbl>     <dbl>    <dbl> <int>
1 (Intercept)     7.55     0.811      9.32 4.20e-16   130
```

Next, let's briefly have a look at how we can utilize multiple imputation (MI) which is an improvement over single imputation. The multiple imputation approach generates more than one imputation thus creating many complete data sets for us.

For each of these datasets, we can perform any analysis that we are interested in. After the analysis, one must pool the results from the impured datasets to get the final result. Here, we utilize a method called predictive mean matching (`method = "pmm"` in the code below), which uses the neighbour values of data as imputations.

```
imp2 <- mice(events_types_mis, method = "pmm",
m= 10, maxit = 100, print = FALSE)
fit2 <- with(imp2, lm(Ethics ~ Practicals))
pool_fit <- pool(fit2)
# Multiple imputation
summary(pool_fit)
```

```
         term   estimate  std.error statistic       df     p.value
1 (Intercept) 1.62080372 2.18285149 0.7425167 101.7304 0.459485402
2  Practicals 0.08049328 0.02616765 3.0760606 106.0802 0.002668431
```

```
# Complete cases
summary(lm(Ethics ~ Practicals, events_types_mis))["coefficients"]
```

```
$coefficients
              Estimate Std. Error  t value   Pr(>|t|)
(Intercept) 2.15459162 2.12235051 1.015191 0.31226283
Practicals  0.06220884 0.02605001 2.388054 0.01865793
```

```
# Without missingness
summary(lm(Ethics ~ Practicals, events_types))["coefficients"]
```

```
$coefficients
              Estimate Std. Error   t value     Pr(>|t|)
(Intercept) 1.05409529 2.17821479 0.4839262 0.6292651258
Practicals  0.08891892 0.02590313 3.4327482 0.0008053447
```

From the results above, we can see that in this particular case the multiple imputation performs well in comparison to CC approach. The regression coefficient for full data without any missing values is 0.089, and it is 0.080 for multiple imputation, while complete case analysis gives 0.062. As all of them have very similar standard errors, this yields that MI and full data give statistically significant p-values for significance level 0.01, while CC does not.

11 Correcting Erroneous Data

Let's imagine that our data has an error on the surname variable `Surname` and that all the names ending with "sen" should end with "ssen". What we can do is that we can use regular expressions to detect the erroneous rows and we can also use them to replace the values. Let's first figure out which last names contain a name ending with "sen". We can use a function `str_detect()` to return TRUE/FALSE for each row from `stringr` [14] package within a `filter()` function call. We define `pattern = "sen$"` where $ indicates the end of the string.

```
library("stringr")
demographics |>
  filter(str_detect(string = Surname, pattern = "sen$")) |>
  pull(Surname)
```

```
[1] "Nielsen"   "Johansen" "Joensen"   "Jansen"    "Olsen"
```

After pulling the filtered surnames, there seems to be five surnames ending with "sen". Next, let's try to replace "sen" with "ssen". On the next row we filter just as previously to limit output.

```
demographics |>
  mutate(Surname = str_replace(
    string = Surname, pattern = "sen$", replacement = "ssen")
  ) |>
  filter(str_detect(string = Surname, pattern = "sen$")) |>
  pull(Surname)
```

```
[1] "Nielssen"   "Johanssen" "Joenssen"   "Janssen"    "Olssen"
```

Thus, the following code updates the data so that all the surnames ending with "sen" now end with "ssen" instead.

```
demographics <- demographics |>
  mutate(Surname = str_replace(
    string = Surname, pattern = "sen$", replacement = "ssen")
  )
```

12 Conclusion and Further Reading

Data wrangling is one of the most important steps in any data analysis pipeline. This chapter introduced the `tidyverse`, tidy data, and several commonly used R packages for data manipulation and their use in basic scenarios in the context of learning analytics. However, the `tidyverse` is vast and can hardly be fully covered in a single chapter. We refer the reader to additional resources such as those found on the tidyverse website at https://www.tidyverse.org/learn/ and the book "R for Data Science" by Hadley Wicham and Garret Grolemund. The book is free to use and readily available online at https://r4ds.had.co.nz/.

References

1. Wickham H, Averick M, Bryan J, Chang W, McGowan LD, François R, Grolemund G, Hayes A, Henry L, Hester J, Kuhn M, Pedersen TL, Miller E, Bache SM, Müller K, Ooms J, Robinson D, Seidel DP, Spinu V, Takahashi K, Vaughan D, Wilke C, Woo K, Yutani H (2019) Welcome to the tidyverse. J Open Source Softw 4:1686. https://doi.org/10.21105/joss.01686
2. Wickham H, Hester J, Bryan J (2024) readr: read rectangular text data. R package version 2.1.5. https://CRAN.R-project.org/package=readr
3. López-Pernas S, Saqr M, Conde J, Del-Río-Carazo L (2024) A broad collection of datasets for educational research training and application. In: Saqr M, López-Pernas S (eds) Learning analytics methods and tutorials: a practical guide using R. Springer, Berlin
4. Chan C, Leeper T, Becker J, Schoch D (2023) rio: a swiss-army knife for data file I/O. https://cran.r-project.org/package=rio
5. Müller K, Wickham H (2023) tibble: simple data frames. R package version 3.2.1. https://CRAN.R-project.org/package=tibble
6. Wickham H, François R, Henry L, Müller K, Vaughan D (2023) dplyr: a grammar of data manipulation. R package version 1.1.4. https://CRAN.R-project.org/package=dplyr
7. Grolemund G, Wickham H (2011) Dates and times made easy with lubridate. J Stat Softw 40(3):1–25. https://www.jstatsoft.org/v40/i03/
8. Wickham H, Vaughan D, Girlich M (2023) tidyr: tidy messy data. R package version 1.3.0. https://CRAN.R-project.org/package=tidyr
9. Little RJA, Rubin DB (2020) Statistical analysis with missing data, 3rd edn. Wiley, Hoboken
10. van Buuren S (2018) Flexible imputation of missing data. Chapman et Hall/CRC Press, Boca Raton. https://doi.org/10.1201/9780429492259
11. Kopra J (2018) Statistical modelling of selective non-participation in health examination surveys. Report/University of Jyväskylä, Department of Mathematics and Statistics
12. van Buuren S, Groothuis-Oudshoorn K (2011) mice: multivariate imputation by chained equations in R. J Stat Softw 45:1–67. https://doi.org/10.18637/jss.v045.i03
13. Tikka S, Kopra J, Heinäniemi M, López-Pernas S, Saqr M (2024) Introductory statistics with R for educational researchers. In: Saqr M, López-Pernas S (eds). Springer, Berlin
14. Wickham H (2023) stringr: simple, consistent wrappers for common string operations. R package version 1.5.1. https://CRAN.R-project.org/package=stringr

Introductory Statistics with R for Educational Researchers

Santtu Tikka, Juho Kopra, Merja Heinäniemi, Sonsoles López-Pernas, and Mohammed Saqr

1 Introduction

Learning analytics involves the practical application of statistical methods to quantitative data, which can represent various aspects of the learning process such as student engagement, progress, and outcomes. Thus, knowledge about basic statistical methods is essential. Let's start with how statistics connects with the research process and how it is also linked with philosophy of science. According to Niiniluoto [1], the research process can be described with the following eight steps:

1. Setting up a problem.
2. Disambiguation of the problem. Building a research strategy.
3. Collecting data.
4. Describing the data.
5. Analysis of data.
6. Interpreting the analyses.
7. Writing the report.
8. Publishing the results.

Steps 1 and 2 require knowledge and skills related to the applied field, but also general scientific aptitude. Knowledge about statistics is central in steps 3, 4, 5, and

S. Tikka (✉)
Department of Mathematics and Statistics, University of Jyväskylä, Jyväskylä, Finland
e-mail: santtu.tikka@jyu.fi

J. Kopra · S. López-Pernas · M. Saqr
School of Computing, University of Eastern Finland, Joensuu, Finland

M. Heinäniemi
Institute of Biomedicine, University of Eastern Finland, Kuopio, Finland

M. Saqr, S. López-Pernas (eds.), *Learning Analytics Methods and Tutorials*,
https://doi.org/10.1007/978-3-031-54464-4_5

6. Finally, steps 7 and 8 mostly require skills in writing and communication. Overall, it can be argued that a solid understanding of statistics and statistical methods is crucial for anyone conducting research with quantitative data.

This chapter of the book concentrates on steps 4, 5, and 6 of the research process. We start with descriptive statistics, which are statistics that describe the overall features of the data. In contrast, inferential statistics are used to draw conclusions and make inferences about the population under study. Afterwards, we explain the basics of statistical hypothesis testing, which is the most common—although not the only—way to analyze data. The most common statistical tests, such as Student's t-test, Chi-squared test, Analysis of variance, Levene's test, and Shapiro-Wilk test are covered in this chapter. We also explain how to interpret the results of each test. We also present the linear regression model, which is not a statistical test but one of the most powerful statistical tools. Basic understanding of linear regression is essential for anyone interested in more advanced regression techniques, including logistic regression which is covered in the final section of this chapter. For a more in-depth view on the statistical tests covered in this chapter, we refer the reader to works such as [2, 3].

2 Descriptive Statistics

Descriptive statistics are used to provide a meaningful quantitative overview of data, and to summarize potentially vast amounts of information into more easily comprehensible and manageable quantities. In general, descriptive statistics are used as a first step in a data analysis workflow. In this section we will focus on numeric descriptors while visualizations are the topic of Chapter 6 [4]. For this chapter, we will use the combined Moodle data with students' demographics, results, and summarized Moodle event activity. For more information about the dataset, please refer to Chapter 2 in this book [5]. We begin by installing all R packages that we will use in this chapter.

```
install.packages(
  c("car", "rio", "see", "dplyr", "tidyr",
    "broom", "report", "correlation", "performance")
)
```

We use the `rio` [6] package to read the data files into R via the `import()` function. The `Events` dataset contains log data on the student's Moodle activity such as Moodle event types and names. The `Demographics` dataset contains background information on the students such as their gender and location of study (categorical variables). Finally, the `Results` data contains grades on various aspects of the Moodle course including the final course grade (numeric variables). We also create a new variable called `AchievingGroup` that categorizes the students into bottom and top 50% of achievers in terms of the final grade. We will leverage the `dplyr` [7] and `tidyr` [8] packages to wrangle the data into a single combined dataset. We begin by reading the data files and by constructing the `AchievingGroup` variable.

```
library("rio")
library("dplyr")
library("tidyr")
url <- "https://github.com/lamethods/data/raw/main/1_moodleLAcourse/"
events <- import(paste0(url, "Events.xlsx"), setclass = "tibble")
demographics <- import(paste0(url, "Demographics.xlsx"), setclass = "tibble")
results <- import(paste0(url, "Results.xlsx"), setclass = "tibble") |>
  mutate(
    AchievingGroup = factor(
      case_when(
        ntile(Final_grade, 2) == 1 ~ "Low achiever",
        ntile(Final_grade, 2) == 2 ~ "High achiever"
      )
    )
  )
```

Next, we summarize the student's engagement based on their Moodle activity into three groups: Low activity, Moderate activity and High Activity.

```
events_summary <- events |>
  group_by(user) |>
  tally() |>
  rename(Frequency.Total = n) |>
  mutate(
    ActivityGroup = factor(
      case_when(
        ntile(Frequency.Total, 3) == 1 ~ "Low activity",
        ntile(Frequency.Total, 3) == 2 ~ "Moderate activity",
        ntile(Frequency.Total, 3) == 3 ~ "High activity"
      )
    )
  )
```

We also count the different types of Moodle events.

```
events_types <- events |>
  group_by(user, Action) |>
  count(Action) |>
  pivot_wider(
    names_from = "Action",
```

```
    names_prefix = "Frequency.",
    values_from = "n",
    values_fill = 0
  )
```

Finally, we combine the data.

```
all_combined <- demographics |>
  left_join(events_types, by = "user") |>
  left_join(events_summary, by = "user") |>
  left_join(results, by = "user")
```

The various steps of the combined data construction are discussed in greater detail in Chapter 4 [9].

2.1 Measures of Central Tendency

A typical way to summarize a univariate data sample is to describe its "middle point" using an appropriate statistic depending on the measurement scale of the data. The most common statistics to describe such a value are the *mean*, the *median*, and the *mode*.

For data on the interval or ratio scales (and sometimes also on the ordinal scale), the most common option is to use the arithmetic mean, which is available via the base R function `mean()`. This function takes a vector of values as its input.

```
all_combined |>
  summarise(
    mean_grade = mean(Final_grade),
    mean_total = mean(Frequency.Total)
  )
```

```
# A tibble: 1 x 2
  mean_grade mean_total
       <dbl>      <dbl>
1       7.25       736.
```

The means are reported as a `tibble` with a single column for both variables.

For data on the ordinal scale (or interval or ratio scales), the median can be used which is defined as the value that separates the lower half from the bottom half of the data sample, i.e., the 50% quantile. The median can be computed in R using the built-in `median()` function. Similarly to `mean()`, this function also takes a vector of values as its input.

```
all_combined |>
  summarise(
    median_grade = median(Final_grade),
    median_total = median(Frequency.Total)
  )
```

```
# A tibble: 1 x 2
  median_grade median_total
         <dbl>        <dbl>
1         7.95          606
```

Just like before, the medians are reported as a `tibble` with each value in its own column.

For data on the nominal or ordinal scale, the mode is a suitable choice as it describes the category with the highest number of observations. Unfortunately, there is no readily available function in R to compute the mode, and the reader should take care not to mistakenly use the `mode()` function, which is used to determine the internal storage mode of a variable (similar to the `typeof()` function). However, we can easily write our own function to compute the statistical mode as follows:

```
stat_mode <- function(x) {
  u <- unique(x)
  u[which.max(tabulate(match(x, u)))]
}
```

Functions in R are written using the following syntax. First, we define the name of the function, just like we would define the name of a variable when assigning data into it, in this case the name is `stat_mode`. Next, we assign the function definition, which starts with the keyword `function`. Next, we describe the function arguments within the parentheses, in this case we call our argument `x`, which we assume contains the data vector we wish to compute the mode for. Next, we define the body of the function within the braces. The body determines what the function does and what its output should be. Within the body, we first determine the unique values in the data vector `x`, and assign the result to a variable `u`. Next, we need to count the number of occurrences of each unique value. To start, we first match each observed value in `x` to the unique values in `u` to get the corresponding indices, which we will then count using tabulate. We obtain the index of the value with the highest number of occurrences with the function `which.max()`, and finally the corresponding unique value by selecting it from `u` using the subset operator, i.e., the brackets. Our function will now work on all types of data.

```
all_combined |>
  summarise(
    mode_gender = stat_mode(Gender),
```

```
    mode_location = stat_mode(Location)
  )

# A tibble: 1 x 2
  mode_gender mode_location
  <chr>       <chr>
1 F           On campus
```

The output is now similar to the mean and median functions that we used earlier, showing the modes of Gender and Location as a two-column tibble. For nominal variables, it is common to also compute the frequencies of each category. This can easily be done with the base R function table()

```
table(all_combined$Gender)


 F  M
65 65

table(all_combined$Location)


On campus    Remote
      106        24
```

The function outputs the names of the categories and the frequency of each category as an integer vector.

2.2 Measures of Dispersion

For data on the interval and ratio scales (and sometimes also on the ordinal scale), it is also meaningful to describe how clustered or scattered the values in the sample are, i.e., how far apart the values are from one another. Commonly used measures of statistical dispersion include the *variance*, *standard deviation*, and the *interquartile range*. Typically, such measures have the value 0 when all values in the sample are identical, and the value increases as the dispersion in the data grows.

All three measures can be readily computed with built-in R functions var(), sd(), and IQR() respectively. Like mean() and median(), these functions accept a vector or numeric values as input.

```
all_combined |>
  summarise(
    var_grade = var(Final_grade),
```

```
    sd_grade = sd(Final_grade),
    iqr_grade = IQR(Final_grade)
  )

# A tibble: 1 x 3
  var_grade sd_grade iqr_grade
      <dbl>    <dbl>     <dbl>
1      4.81     2.19      3.34
```

The variance, standard deviation and interquartile range of the final grade are returned as a `tibble` with three columns.

2.3 Covariance and Correlation

Covariance and correlation measure the linear dependence between two variables. Correlation is a unitless measure between -1 and 1, whereas covariance is not, and its scale depends on the scale of the variables. The sign of both measures indicates the tendency of the relationship. Positive sign means that as the value of one variable increases, the value of the other variable tends to increase as well. Conversely, a negative sign indicates that the value of the second variable tends to decrease as the value of the first variable increases.

Both covariance and correlation and be computed directly in R using the functions `cov()` and `cor()`, respectively.

```
all_combined |>
  summarise(
    cov_grade_total = cov(Final_grade, Frequency.Total),
    cor_grade_total = cor(Final_grade, Frequency.Total),
  )

# A tibble: 1 x 2
  cov_grade_total cor_grade_total
            <dbl>           <dbl>
1            504.           0.513
```

We obtain the covariance and correlation between the final grade and total number of Moodle events. We will familiarize ourselves with correlations in greater depth in Sect. 4.

2.4 Other Common Statistics

The extreme values of a data sample can be found using the function `range()`, which computes the minimum and maximum values of the sample as a vector. These values can also be computed individually with the corresponding functions `min()` and `max()`. Because `summarise()` only allows a single value as output per row, we use the `reframe()` function instead when computing the range.

```
all_combined |>
  reframe(
    range_grade = range(Final_grade)
  )
```

```
# A tibble: 2 x 1
  range_grade
        <dbl>
1           0
2          10
```

```
all_combined |>
  summarise(
    min = min(Final_grade),
    max = max(Final_grade)
  )
```

```
# A tibble: 1 x 2
    min   max
  <dbl> <dbl>
1     0    10
```

With `reframe()`, we obtain a `tibble` with two rows, the first containing the minimum value and the second the maximum value of the final grade. If we instead use `summarise()` like before, we can only obtain one value per computed variable. The `summary()` function can also be used to quickly compute several of the most common descriptive statistics for all variables of a dataset.

```
results |>
  select(Grade.SNA_1:Grade.Group_self) |>
  summary()
```

```
  Grade.SNA_1       Grade.SNA_2       Grade.Review     Grade.Group_self
 Min.   : 0.000   Min.   : 0.000   Min.   : 0.000   Min.   : 0.000
 1st Qu.: 8.000   1st Qu.: 9.000   1st Qu.: 6.670   1st Qu.: 9.000
 Median : 9.000   Median :10.000   Median : 8.000   Median :10.000
 Mean   : 8.346   Mean   : 9.262   Mean   : 7.724   Mean   : 8.085
 3rd Qu.:10.000   3rd Qu.:10.000   3rd Qu.: 9.670   3rd Qu.:10.000
 Max.   :10.000   Max.   :10.000   Max.   :10.000   Max.   :10.000
```

The output shows the minimum and maximum values, the quartiles, and the mean of each variable that we selected.

3 Statistical Hypothesis Testing

Statistical hypothesis testing aims to evaluate hypotheses about a population of interest using probabilistic inference. The starting point of any statistical test is a so-called *null hypothesis* (denoted by H_0), which typically corresponds to a scenario, where evidence supporting a specific hypothesis is a result of pure chance. For example, when evaluating whether a new drug is an efficient form of treatment via a randomized controlled trial, the null hypothesis could be that the drug has no effect on the response. A null hypothesis is always associated with an *alternative hypothesis* (denoted by H_1), which is typically the inverse of the null hypothesis and corresponds to the hypothesis of interest, e.g., the drug has an effect on the response.

Statistical tests operate by assuming that the null hypothesis is true, and highly unlikely events under this assumption are typically regarded as giving cause for rejecting the null hypothesis. A statistical test is associated with a *test statistic*, which is a measure of how much the observations deviate from the null hypothesis scenario. The distribution of the test statistic under the null hypothesis and the sample test statistic can be used to compute the probability of obtaining a test statistic as extreme or more extreme than the one observed, assuming that the null hypothesis is true. This probability is known as the *p-value*, which is often mischaracterized even in scientific literature. For instance, the p-value is not the probability that the null hypothesis is true or that the alternative hypothesis is false. The p-value also does not quantify the size of the observed effect, or its real-world importance.

Typically, a *confidence level* is decided before applying a statistical test (usually denoted by α), and the null hypothesis is rejected if the observed p-value is smaller than this confidence level. If the p-value is greater than the confidence level, the null hypothesis is not rejected. Traditionally, the confidence level is 0.05, but this convention varies by field, and should be understood as being arbitrary, i.e., there is nothing special about the value 0.05. If the p-value falls below the confidence level, the result is regarded as *statistically significant*.

Hypothesis testing is a powerful tool for drawing conclusions from data, but it is important to use it appropriately and to understand its limitations. Every statistical test is associated with a set of assumptions which are often related to the distribution of the data sample. If these assumptions are violated, then the results of the test may be unreliable. In the following sections, some of the most common statistical tests are introduced. We will take advantage of the `report` [10] package and the corresponding function `report()` to showcase the results of the various statistical tests. For more information on the concepts and principles related to statistical hypothesis testing, see e.g., [2, 3].

3.1 *Student's t-test*

Student's t-test [11] is one of the most well-known statistical tests. It compares the mean values of variables either between two populations or between a single population and a reference level and is thus applicable to continuous variables. The test assumes homogeneity of variance and that the data originates from a normal distribution. For nonhomogeneous data, the test can still be performed by using an approximation [12]. In R, all variants of the t-test test can be applied using the function `t.test()`.

Our goal is to compare the students' Moodle activity with respect to their final grade. For this purpose, we use the binary variable called `AchievingGroup` which categorizes the students into top and bottom 50% achievers in terms of the final grade.

3.1.1 One-Sample t-test

The one-sample t-test compares the mean of a data sample against a reference value, typically defined by the null hypothesis. Let us begin by testing the hypothesis that the average number of Moodle events (`Frequency.Total`) is 600 ($H_0 : \mu = 600$). The function `t.test()` can be used in various ways, but in this example we provide the function with a `formula` object `Frequency.Total ~ 1` as the first argument. The formula syntax is a standard method for defining statistical models and other dependency structures in R. The formula defines that the left-hand side of the `~` symbol is a response variable which is explained by the terms on the right-hand side. Because we're not conducting the test with respect to any other variable, the right-hand side of the formula is simply `1`, which means that it is a constant in the R formula syntax. This does not mean for example, that our null hypothesis would be that the number of Moodle events is 1. The expected value that the test is applied against (i.e., the value we assume μ to have under the null hypothesis) is defined via the argument `mu`, which by default has the value 0 for a one-sample t-test. Argument `data` defines in which environment the formula should be evaluated. By providing the `all_combined` data, we do not have to explicitly extract the `FrequencyTotal` variable from the data in the formula by writing `all_combined$Frequency.Total` or by using `pull()`. This is especially useful when the formula contains several variables from the same data.

```
ttest_one <- t.test(Frequency.Total ~ 1, data = all_combined, mu = 600)
ttest_one
```

```

	One Sample t-test

data:  Frequency.Total
t = 3.4511, df = 129, p-value = 0.0007553
```

```
alternative hypothesis: true mean is not equal to 600
95 percent confidence interval:
 657.8530 813.3163
sample estimates:
mean of x
 735.5846
```

As a result, we obtain the value of the test statistic (`t = 3.4511`), the degrees of freedom (`df = 129`), and the p-value of the test (`p-value = 0.0007553`). Because the p-value is very small (much smaller than the standard 0.05 confidence level), we reject the null hypothesis, which means that the average number of Moodle events is significantly different from 600. The output of the test result object also describes the alternative hypothesis H_1 under `alternative hypothesis`, and the confidence interval of the test statistic.

We produce a summarized report of the test results with the `report()` function.

```
report(ttest_one)
```

```
Warning: Unable to retrieve data from htest object.
  Returning an approximate effect size using t_to_d().
```

```
Effect sizes were labelled following Cohen's (1988) recommendations.

The One Sample t-test testing the difference between Frequency.Total (mean =
735.58) and mu = 600 suggests that the effect is positive, statistically
significant, and small (difference = 135.58, 95% CI [657.85, 813.32], t(129) =
3.45, p < .001; Cohen's d = 0.30, 95% CI [0.13, 0.48])
```

This produces a description of the results of the test that is easier to read and interpret than the direct output of the test result object. We note that a warning is also produced which we can safely ignore in this case. The warning occurs because the test result object `ttest_one` does not retain the original data `all_combined` which we used to carry out the test. If non-approximate effect sizes are desired, the test should be carried out by supplying the variables being compared directly without using the formula interface of the `t.test()` function. For more information on the effect size, see e.g., [13, 14].

3.1.2 Two-Sample t-test

In contrast to the one-sample t-test, the two-sample t-test compares the means of two data samples against one another. For example, suppose that we're interested in the hypothesis that the average number of Moodle events is the same for the top and bottom 50% achievers ($H_0 : \mu_1 = \mu_2$). We can once again leverage the formula syntax, but instead of the constant `1` on the right-hand side of the formula, we will now replace it with the variable `Achievement` which defines the achievement level.

```r
ttest_two <- t.test(Frequency.Total ~ AchievingGroup, data = all_combined)
ttest_two
```

```
	Welch Two Sample t-test

data:  Frequency.Total by AchievingGroup
t = 4.4749, df = 95.988, p-value = 2.102e-05
alternative hypothesis:
true difference in means between group 1 and group 2 is not equal to 0
95 percent confidence interval:
 182.6427 473.8496
sample estimates:
mean in group High achiever  mean in group Low achiever
                   899.7077                    571.4615
```

The contents of the result object are mostly the same as in the case of the one-sample t-test. The result is again statistically significant (using the 0.05 confidence level) meaning that according to the test, the average number of Moodle events is higher for the top 50% achievers. The `report()` function can be used to produce a similar summary as in the case of the one-sample t-test.

```r
report(ttest_two)
```

```
Warning: Unable to retrieve data from htest object.
  Returning an approximate effect size using t_to_d().

Effect sizes were labelled following Cohen's (1988) recommendations.

The Welch Two Sample t-test testing the difference of Frequency.Total by
AchievingGroup (mean in group High achiever = 899.71, mean in group Low achiever =
571.46) suggests that the effect is positive, statistically significant, and large
(difference = 328.25, 95% CI [182.64, 473.85], t(95.99) = 4.47, p < .001; Cohen's d
= 0.91, 95% CI [0.49, 1.33])
```

This produces the same warning as before in the one-sample case, but we can safely ignore it again.

3.1.3 Paired Two-Sample t-test

Instead of directly comparing the means of two groups, it may sometimes be of interest to compare differences between pairs of measurements. Such a scenario typically arises in an experiment, where subjects are paired, or two sets of measurements are taken from the same subjects. While our Moodle event data does not contain such measurement pairs, we could still imagine that our data was organized such that each student in the bottom 50% achievers was paired with a student in the top 50% achievers and that there is a one-to-one correspondence between the two achievement groups. This dependency between the two groups is the key difference between the paired test and the two-sample test.

A more suitable approach for paired data is to test the differences between the pairs, e.g., the differences between the number of Moodle events in our scenario. We supply the `t.test()` function with the argument `paired = TRUE` so that it will take the measurement pairs into account. In this test, the null hypothesis is that the mean difference between the student pairs is zero ($H_0 : \mu_d = 0$).

```
ttest_paired <- t.test(
  Frequency.Total ~ AchievingGroup, data = all_combined, paired = TRUE
)
ttest_paired
```

```
	Paired t-test

data:  Frequency.Total by AchievingGroup
t = 4.3733, df = 64, p-value = 4.599e-05
alternative hypothesis: true mean difference is not equal to 0
95 percent confidence interval:
 178.3014 478.1910
sample estimates:
mean difference
       328.2462
```

The result is once again statistically significant, and we reject the null hypothesis. Because the mean difference between the pairs is positive, this means average number of Moodle events is higher for the top 50% achievers of the pairs.

Paired two-sample t-test is also supported by `report()`.

```
report(ttest_paired)
```

```
Warning: Unable to retrieve data from htest object.
  Returning an approximate effect size using t_to_d().
```

```
Effect sizes were labelled following Cohen's (1988) recommendations.

The Paired t-test testing the difference of Frequency.Total by AchievingGroup (mean
difference = 328.25) suggests that the effect is positive, statistically
significant, and medium (difference = 328.25, 95% CI [178.30, 478.19], t(64) =
4.37, p < .001; Cohen's d = 0.55, 95% CI [0.28, 0.81])
```

We can once again safely ignore the produced warning message.

3.2 Chi-Squared Test

The chi-squared test is used for the analysis of contingency tables; it tests whether two categorical variables are independent or not [15]. A typical use case for this test is to investigate differences between groups such as student attendance by location or gender. The basic idea of the test is to compare the observed contingency table to a table under the null hypothesis where the variables are independent. The chi-squared test is based on the cell-specific differences between these two tables. As a general rule, the test assumes that the expected value is at least 5 in at least 80% of the cells, and that no expected values are below 1. If these assumptions are violated, the results of the test may not be reliable. In such cases, Fisher's exact test [16] can be used instead via the function `fisher.test()`, but it may be computationally slow for large contingency tables. Both the chi-squared test and Fisher's exact test assume that the data is a random sample from the population.

We will use the combined Moodle data to investigate whether the achievement level and the activity level of the students are independent. First, we must create the contingency table from the individual-level data. We use the `table()` function for this purpose.

```
tab <- table(all_combined$ActivityGroup, all_combined$AchievingGroup)
tab
```

```
                  High achiever Low achiever
High activity                27           16
Low activity                 14           30
Moderate activity            24           19
```

The table shows the observed frequencies in each cell, i.e., for each combination of activity and achievement. Next, we apply the chi-squared test using the `chisq.test()` function.

```
Xsq_test <- chisq.test(tab)
Xsq_test
```

```

    Pearson's Chi-squared test

data:  tab
X-squared = 9.2135, df = 2, p-value = 0.009984
```

Printing the test result object shows the test statistic (`X-squared`), the associated degrees of freedom (`df`) and the p-value (`p-value`). The p-value is very small, meaning that we reject the null hypothesis. In other words, the achievement and activity levels of the students are not independent. This is not a surprising result, as more active students are more likely to engage with the course content and perform better in terms of the learning outcomes. We can confirm that the assumptions of the test related to the expected values of the cells were not violated by using the test result object, which contains the expected values of the cells in the element `expected`.

```
all(Xsq_test$expected >= 1)
```

```
[1] TRUE
```

```
mean(Xsq_test$expected >= 5) >= 0.80
```

```
[1] TRUE
```

All expected values were greater than one, and over 80% of the expected values were greater than 5. This means that the assumptions are satisfied for our data and thus the results are reliable. Here, we used the function `all()` which takes a `logical` vector as input and returns `TRUE` if all elements of the vector were `TRUE`. Otherwise, the function returns `FALSE`. Unfortunately, the `report()` function does not support the chi-squared test.

3.3 Analysis of Variance

Analysis of variance (ANOVA) [17] can be viewed as the generalization of Student's t-test, where instead of one or two groups, the means of a variable are compared across multiple groups simultaneously. The name of the method comes from its test statistic, which is based on a decomposition of the total variance of the variable into variance within the groups and between the groups. ANOVA makes several assumptions: the observations are independent, the residuals of the underlying linear model follow a normal distribution, and that the variance of the variable is the same across groups (homoscedasticity). If these assumptions are violated, the results of the test may not be reliable. One alternative in such instances is to use the non-parametric Kruskal-Wallis test [18] instead, which is available in R via the function `kruskal.test()`. This test uses the ranks of the observations, and the null hypothesis is that the medians are the same for each group.

We use our combined Moodle data to demonstrate ANOVA. Instead of comparing the total number of Moodle events between top and bottom 50% of achievers, this time we will compare the final grade of the students across three activity groups: low activity, moderate activity, and high activity, described by the variable `ActivityGroup`. Thus the null and alternative hypotheses are in this case:

- H_0: The expected values of the final grade are the same across the three activity groups ($\mu_1 = \mu_2 = \mu_3$),

- H_1: At least one activity group has a different expected final grade ($\mu_i \neq \mu_j$ for at least one pair $i \neq j$).

To carry out the analysis, we apply the aov function, which uses the same formula syntax to define the response variable and the groups as the `t.test()` function does. Next, we apply the `summary()` function to the `aov()` function return object `fit`, as the default output of `aov()` is not very informative.

```
fit <- aov(Final_grade ~ ActivityGroup, data = all_combined)
summary(fit)
```

```
               Df Sum Sq Mean Sq F value   Pr(>F)    
ActivityGroup   2  175.7   87.87   25.11 6.47e-10 ***
Residuals     127  444.4    3.50                     
---
Signif. codes: 0 '***' 0.001 '**' 0.01 '*' 0.05 '.' 0.1 ' ' 1
```

The summary contains the following columns: `Df` describes the degrees of freedom of the F-distribution associated with the test, `Sum Sq` reports the sum of squares related to the groups and the residuals, `Mean Sq` reports the corresponding mean sum of squares, `F value` is the value of the test statistic, and finally `Pr(>F)` is the p-value of the test. For this example, the p-value is very small, which means that the null hypothesis is rejected, and there are statistically significant differences in the final grade between the groups according to the test. In the following sections we will learn how to test for the assumptions related to normality and homoscedasticity. The `report()` function can be used for the output of `aov()` as well.

```
report(fit)
```

```
The ANOVA (formula: Final_grade ~ ActivityGroup) suggests that:

  - The main effect of ActivityGroup is statistically significant and large (F(2,
127) = 25.11, p < .001; Eta2 = 0.28, 95% CI [0.17, 1.00])

Effect sizes were labelled following Field's (2013) recommendations.
```

This output also reports the degrees of freedom, the test statistic value and the p-value but in a more easily readable format.

Note that ANOVA simply measures if there are differences between the groups but does not provide information on how these differences emerge. For example, there could be a single group that is different from all the rest, or two subgroups where the means are similar within each group, but different between the subgroups. Visualizations can be a helpful tool for gaining more insight into the differences, and post-hoc pairwise tests can be carried out to compare the pairs of groups.

3.4 Levene's Test

Levene's test is used to investigate whether the variance of a variable is the same across two or more groups [19]. Compared to alternatives such as Bartlett's test [20], Levene's test is less sensitive to non-normal observations. The test is not available in base R, but it can be found in the car package as the function `leveneTest()`. The function uses the same formula syntax as `t.test()` and `aov()`. We will investigate the homogeneity of the variance of the final grade between the activity groups.

```
library("car")
leveneTest(Final_grade ~ ActivityGroup, data = all_combined)
```

```
Levene's Test for Homogeneity of Variance (center = median)
       Df F value   Pr(>F)
group   2  5.7204 0.004181 **
      127
---
Signif. codes: 0 '***' 0.001 '**' 0.01 '*' 0.05 '.' 0.1 ' ' 1
```

The output of `leveneTest()` is analogous to the output of the ANOVA summary, and it contains the degrees of freedom (`Df`), the value of the test statistic (`F value`) and the p-value of the test (`Pr(>F)`). The p-value is very small, so we reject the null hypothesis meaning that the variance of the final grade is not the same across the groups according to the test. This means that the assumption of homoscedasticity is violated for the analysis of variance of the final grade, and thus the results may not be reliable. The `report()` function is not supported for `leveneTest()`.

3.5 Shapiro-Wilk Test

Shapiro-Wilk test tests the null hypothesis that a data sample originated from a normal distribution [21]. The test is available in base R as the function `shapiro.test()`. Unfortunately, this function does not support the formula syntax unlike the other test functions we have used thus far. The function only accepts a single `numeric` vector as its argument. Therefore, to test the normality of multiple groups simultaneously, the data must first be split into the groups to be tested. We apply the test to the final grade in each achievement group. With the help of the `broom` package [22], we wrap the test results into a tidy format.

```
library("broom")
all_combined |>
  # Performs the computations in each activity group
  group_by(ActivityGroup) |>
```

```
  # Apply a function in each group
  group_modify(~{
    # Apply the Shapiro test in each group and create tidy output
    shapiro.test(.$Final_grade) |>
      tidy()
  }) |>
  # Selection of variables to keep in the output
  select(ActivityGroup, statistic, p.value)
```

```
# A tibble: 3 x 3
# Groups:   ActivityGroup [3]
  ActivityGroup     statistic  p.value
  <fct>                 <dbl>    <dbl>
1 High activity         0.918 0.00448
2 Low activity          0.909 0.00215
3 Moderate activity     0.862 0.000103
```

This is also a great example of the tidyverse paradigm. First, we group the data by `ActivityGroup` using `group_by()`, and then apply a function in each group using `group_modify()`. We apply the `shapiro.test()` function to the `Final_grade` variable, and then we convert the test results into a tidy tibble using the `tidy()` function from the `broom` package. We also use the special dot notation `.` to select the final grade variable from the data in each group. Finally, we select the grouping variable (`ActivityGroup`), the test statistic (`statistic`) and the p-value (`p.value`) of each test using `select()` and print the results. The resulting object is a tibble with three columns: `ActivityGroup`, `statistic` and `p.value`, the last two of which give the test statistic and p-value of the test for the activity group of the first column.

We can see that according to the test, the `Final_grade` variable is not normally distributed in any of the activity groups, as the p-values are very small. As a consequence, the results of the analysis of variance carried out earlier may not be reliable.

4 Correlation

In Sect. 2.3, we briefly described covariance and correlation, and showcased the base R functions to compute them. However, there are several powerful and user-friendly packages for the analysis and reporting of correlations, such as the `correlation` [23] package which we will demonstrate in this section.

```
library("correlation")
```

For example, the package can easily compute all pairwise correlations between the numeric variables of the data with the function `correlation()`. The argument `select` can be used to compute the correlations only for a subset of the variables.

```
corrs <- correlation(
  all_combined,
  select = c("Frequency.Total", "Grade.Theory", "Final_grade")
)
corrs
```

```
# Correlation Matrix (pearson-method)

Parameter1      |   Parameter2 |    r |        95% CI | t(128) |         p
-------------------------------------------------------------------------
Frequency.Total | Grade.Theory | 0.31 | [0.15, 0.46] |   3.75 | < .001***
Frequency.Total |  Final_grade | 0.51 | [0.37, 0.63] |   6.76 | < .001***
Grade.Theory    |  Final_grade | 0.45 | [0.30, 0.58] |   5.69 | < .001***

p-value adjustment method: Holm (1979)
Observations: 130
```

The columns `Parameter1` and `Parameter2` describe the variables that the correlation was computed for, `r` is the value of the sample correlation, and the remaining columns report the 95% confidence interval, the value of the test statistic, and the p-value of the test (a t-test for correlations) along with the statistical significance. By default, Pearson's correlation coefficient is calculated, but the package also supports many alternative correlation measures. The correlation coefficient to be computed can be selected with the argument `method` that has the value `"pearson"` by default. Selecting for example `method = "spearman"` would compute the Spearman correlation coefficient instead. We can also obtain a correlation matrix by using `summary()`

```
summary(corrs)
```

```
# Correlation Matrix (pearson-method)

Parameter       | Final_grade | Grade.Theory
--------------------------------------------
Frequency.Total |     0.51*** |      0.31***
Grade.Theory    |     0.45*** |

p-value adjustment method: Holm (1979)
```

By default, redundant correlations are omitted, but they can be obtained by setting `redundant = TRUE` in the call to `summary()`. A plot of the correlation matrix can be produced with the help of the package `see` [24].

```
library("see")
corrs |>
  # Also include redundant correlations
  summary(redundant = TRUE) |>
  plot()
```

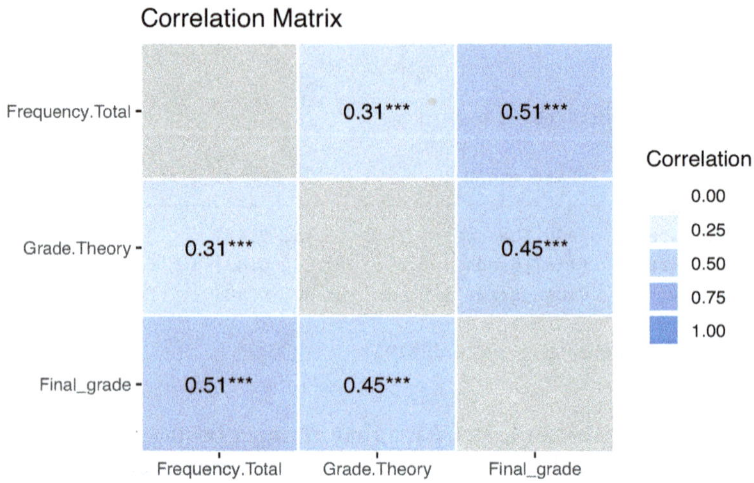

The plot shows the strength of the correlations where darker colors imply stronger correlations. Visualizations will be covered at greater length in Chapter 6 [4].

5 Linear Regression

Linear regression is a statistical tool where one continuous variable is explained by the values of other variables. The variable of interest is said to be a dependent, while the other variables are called predictors. Predictors may also be called explanatory variables, independent variables or covariates depending on the context, applied field, and perspective.

Consider a very simple case, where we only have one predictor, which happens to be a continuous variable. In this case, fitting a linear regression model is merely the same as fitting a straight line to a scatterplot. It is assumed that deviations from this line are simply a result of random variation.

Now, let's go through the formal definition of a linear regression model. Let Y be a dependent variable with measurements $y_1 \ldots, y_n$, and let $X_1, X_2, \ldots, X_k$ be predictor variables with measurements $x_{1i}, \ldots, x_{ki}$ for all $i = 1, \ldots, n$ where i refers to an individual measurement. Then, the regression equation is

$$y_i = \beta_0 + \beta_1 x_{1i} + \beta_2 x_{2i} + \cdots + \beta_k x_{ki} + \varepsilon_i, \quad \varepsilon_i \sim N(0, \sigma^2), \quad i = 1, \ldots, n$$

where we have the regression coefficients $\beta_0, \beta_1, \ldots \beta_k$ and the error variance σ^2. The first parameter β_0 is called the intercept that models the conditional expectation of Y when all the predictors have the value 0. From the regression equation, several assumptions become apparent. First, as the name of the model suggests, a linear relationship is assumed between the response and the predictors. Next, the variance σ^2 of the errors ε_i is constant, and does not depend on the values of the predictors (homoscedasticity). The errors are also assumed independent. The predictor variables are assumed fixed and their values perfectly measured without error.

Let's fit a linear regression model that predicts the final grade with the number of Moodle events of different types. To simplify the exposition, we will only use three types of Moodle events as predictors. We use the `lm()` function which has the same formula interface that we are already familiar with. First, we must define the dependent variable on the left-hand side of the formula, followed by the predictors on the right-hand side separated by a + sign. We must also supply the `data` argument, which tells the function where the actual values of the variables can be accessed.

```
fit <- lm(
  Final_grade ~ Frequency.Applications + Frequency.Assignment +
    Frequency.La_types,
  data = all_combined
)
summary(fit)
```

```
Call:
lm(formula = Final_grade ~ Frequency.Applications +
    Frequency.Assignment + Frequency.La_types, data = all_combined)

Residuals:
    Min      1Q  Median      3Q     Max
-7.0382 -0.8872  0.3665  1.2372  3.4422

Coefficients:
                        Estimate Std. Error t value Pr(>|t|)
(Intercept)             5.800211   0.405963  14.288  < 2e-16 ***
Frequency.Applications  0.076516   0.022294   3.432 0.000811 ***
Frequency.Assignment   -0.005049   0.005734  -0.881 0.380225
Frequency.La_types      0.088252   0.027314   3.231 0.001574 **
---
Signif. codes:  0 '***' 0.001 '**' 0.01 '*' 0.05 '.' 0.1 ' ' 1

Residual standard error: 1.914 on 126 degrees of freedom
```

```
Multiple R-squared:  0.2559,    Adjusted R-squared:  0.2382
F-statistic: 14.45 on 3 and 126 DF,  p-value: 3.798e-08
```

The `summary()` function provides a compact overview of the model fit for `lm` objects. First, a summary of the residuals (i.e., the differences between the observed and predicted values) is provided under `Residuals`. Next, a summary of the regression coefficients β is provided under `Coefficients`, including their estimates (`Estimate`), standard errors (`Std. Error`), test statistics for t-tests that test whether the coefficients are significantly different from zero (`t value`), p-values of the tests (`Pr(>|t|)`) and statistical significance (indicated by the asterisks). For instance, we see that the number of group work events is statistically significant. The notation used for the significance levels of the tests is described following `Signif. codes`. Estimate of the square root of the error variance σ^2 is reported following `Residual standard error`. The two `R-squared` values are estimates of the proportion of variance of the data that is explained by the model. Finally, the `F-statistic` reports the results of ANOVA when applied with the same model formula that was used for the linear regression model.

The `report()` function provides a more comprehensive summary of the model fit and the regression coefficients:

```
report(fit)
```

```
We fitted a linear model (estimated using OLS) to predict Final_grade with
Frequency.Applications, Frequency.Assignment and Frequency.La_types (formula:
Final_grade ~ Frequency.Applications + Frequency.Assignment + Frequency.La_types).
The model explains a statistically significant and moderate proportion of variance
(R2 = 0.26, F(3, 126) = 14.45, p < .001, adj. R2 = 0.24). The model's intercept,
corresponding to Frequency.Applications = 0, Frequency.Assignment = 0 and
Frequency.La_types = 0, is at 5.80 (95% CI [5.00, 6.60], t(126) = 14.29, p < .001).
Within this model:

  - The effect of Frequency Applications is statistically significant and positive
(beta = 0.08, 95% CI [0.03, 0.12], t(126) = 3.43, p < .001; Std. beta = 0.32, 95%
CI [0.13, 0.50])
  - The effect of Frequency Assignment is statistically non-significant and negative
(beta = -5.05e-03, 95% CI [-0.02, 6.30e-03], t(126) = -0.88, p = 0.380; Std. beta =
-0.08, 95% CI [-0.26, 0.10])
  - The effect of Frequency La types is statistically significant and positive (beta
= 0.09, 95% CI [0.03, 0.14], t(126) = 3.23, p = 0.002; Std. beta = 0.31, 95% CI
[0.12, 0.49])

Standardized parameters were obtained by fitting the model on a standardized
version of the dataset. 95% Confidence Intervals (CIs) and p-values were computed
using a Wald t-distribution approximation.
```

Again, the report output has condensed the information about the model fit into a format that can be read in a straightforward manner.

The assumption of normality of the residuals can be assessed with a quantile-quantile plot, or q-q plot for short. The residuals of the model fit can be accessed with the function `resid()`. The function `qqnorm()` draws the quantiles of the residuals against the quantiles of the normal distribution. The function `qqline()` adds a straight line through the plot that passes through the second and third quantiles, by default. Ideally, the residuals should fall on this line, and large deviations indicate that the normality assumption may not hold.

```
# Draw the quantiles of the residuals and the theoretical quantiles
qqnorm(resid(fit))
# Add a line through the theoretical quantiles
qqline(resid(fit))
```

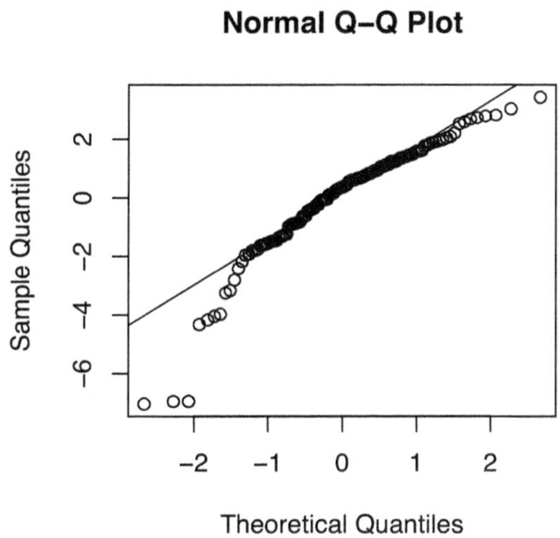

The vast majority of residuals fall nicely onto the line for our model. Besides the q-q plot, we can obtain more model diagnostics with the help of the `performance` [25] package. This package provides a wide array of tools to assess how well models fit to the data. The general-purpose function `check_model(` provides a visual overview of the model fit using several metrics.

```
library("performance")
check_model(fit, theme = see::theme_lucid(base_size = 10))
```

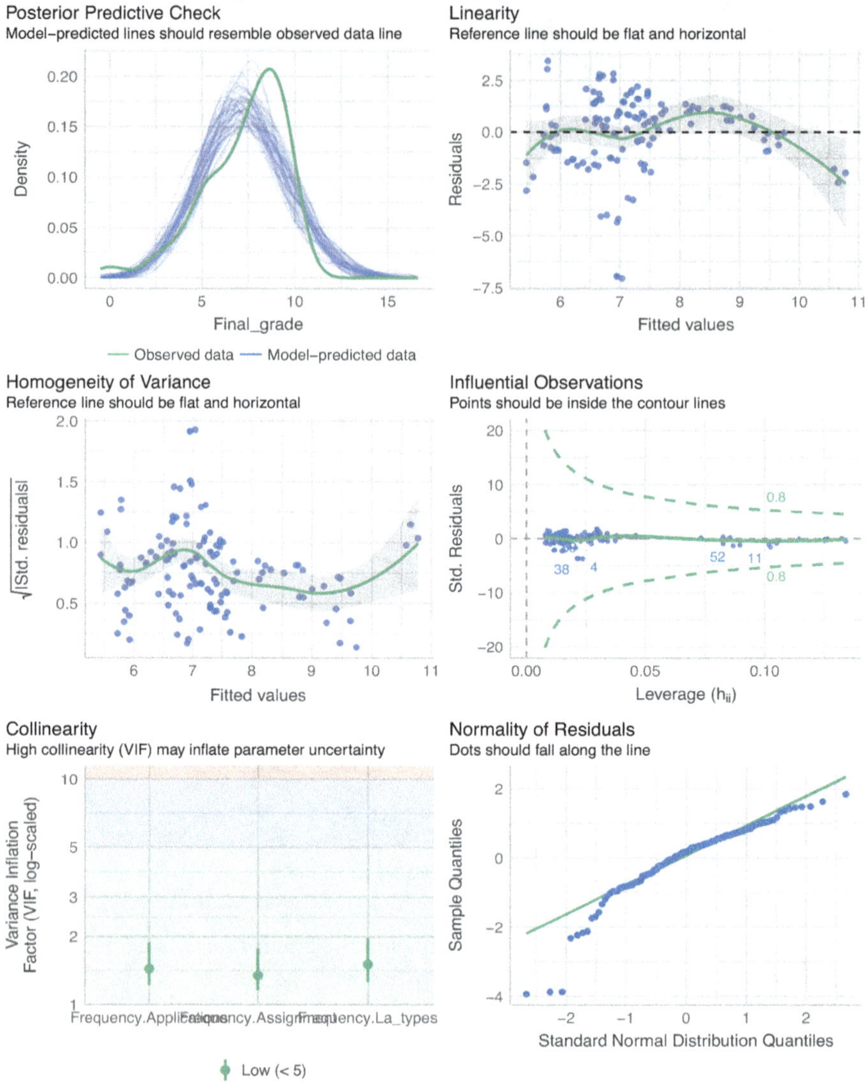

The functions performs various tests related to the assumptions of the linear regression model. For example, the bottom right panel contains the same q-q plot that we previously constructed using the `qqnorm()` and `qqline()` functions. We refer the reader to the documentation of the `performance` package for more information on the remaining tests.

6 Logistic Regression

Logistic regression is a similar tool to linear regression, but with a binary outcome instead of a continuous one. Instead of modeling the outcome variable directly, a linear model is constructed for the logarithmic odds of the probability of "success" for the binary outcome, e.g., obtaining a passing grade. There is also no explicit error term ε in the model, as the uncertainty in the outcome is already captured by the success probability. Formally, the model is

$$\text{logit}\,(P(y_i = 1)) = \beta_0 + \beta_1 x_{1i} + \beta_2 x_{2i} + \cdots + \beta_k x_{ki}, \quad i = 1, \ldots, n,$$

where the logit-function is defined as $\text{logit}(x) = \log(x/(1-x))$. Here, the logit-function serves as the so-called *link function* that connects the expected value of the response to the predictors.

We fit a logistic regression model where the outcome variable is the level of achievement (AchievingGroup) and the predictors are the Moodle event counts of each type. The logistic regression model is a *generalized linear model*: a class of models that extend the linear regression model and that can be fitted in R with the function glm(). The syntax of glm() is analogous to lm(), but we must also specify the distribution of the outcome and the link function via the family argument. We use the function binomial() and supply the argument link = "logit" to define that the model should be a logistic regression model (in this case the link argument is optional, as "logit" is the default value). Because AchievingGroup is a factor, we must first convert it into a binary response that attains values 1 and 0 (or TRUE and FALSE). We can do this within the formula via the I() function, so that we do not have to modify our data. When used in a formula, this function will first compute its argument expression when evaluated, so that the expression is not mistaken for a variable name in the data (that does not exist). We select the high achievers as the "success" category for the outcome.

```
fit_logistic <- glm(
  # Use the I() function to construct a binary response in the formula
  I(AchievingGroup == "High achiever") ~ Frequency.Applications +
    Frequency.Assignment + Frequency.La_types,
  data = all_combined,
  # Our response is binary, so we use the binomial family with logit link
  family = binomial(link = "logit")
)
summary(fit_logistic)
```

```
Call:
glm(formula = I(AchievingGroup == "High achiever") ~ Frequency.Applications +
    Frequency.Assignment + Frequency.La_types, family = binomial(link = "logit"),
    data = all_combined)

Coefficients:
```

```
                        Estimate Std. Error z value Pr(>|z|)
(Intercept)             -0.66272    0.62914  -1.053  0.29217
Frequency.Applications   0.30443    0.07778   3.914 9.07e-05 ***
Frequency.Assignment    -0.04477    0.01402  -3.193  0.00141 **
Frequency.La_types       0.12245    0.04710   2.600  0.00933 **
---
Signif. codes:  0 '***' 0.001 '**' 0.01 '*' 0.05 '.' 0.1 ' ' 1

(Dispersion parameter for binomial family taken to be 1)

    Null deviance: 180.22  on 129  degrees of freedom
Residual deviance: 120.21  on 126  degrees of freedom
AIC: 128.21

Number of Fisher Scoring iterations: 6
```

The summary of a `glm()` function output is very similar to the output of a `lm()` summary. First, the `Call` is reported, which simply restates how the model was fitted. Next, `Coefficients` reports the estimates of the regression coefficients β, and their standard errors and statistical significance. Lastly, two deviance measures and their degrees of freedom are reported. The null deviance is twice the difference between the log-likelihood of the saturated model and the null model, and residual deviance is twice the difference between the saturated model and the model that was fitted. In simpler terms, the saturated model is a perfect model in a sense that there is a parameter for each observation. Conversely, the null model only has the intercept term. The deviance serves as a generalization of the residual sum of squares of the linear regression model, and it can be used to assess the quality of the model fit [26].

The `report()` function is applicable to models fitted with `glm()`.

```
report(fit_logistic)
```

```
We fitted a logistic model (estimated using ML) to predict AchievingGroup with
Frequency.Applications, Frequency.Assignment and Frequency.La_types (formula:
I(AchievingGroup == "High achiever") ~ Frequency.Applications +
Frequency.Assignment + Frequency.La_types). The model's explanatory power is
substantial (Tjur's R2 = 0.38). The model's intercept, corresponding to
Frequency.Applications = 0, Frequency.Assignment = 0 and Frequency.La_types = 0, is
at -0.66 (95% CI [-1.94, 0.56], p = 0.292). Within this model:

  - The effect of Frequency Applications is statistically significant and positive
(beta = 0.30, 95% CI [0.17, 0.48], p < .001; Std. beta = 1.62e-14, 95% CI
[-73784.14, 73784.14])
  - The effect of Frequency Assignment is statistically significant and negative
(beta = -0.04, 95% CI [-0.07, -0.02], p = 0.001; Std. beta = -9.05e-16, 95% CI
[-71374.92, 71374.92])
  - The effect of Frequency La types is statistically significant and positive (beta
= 0.12, 95% CI [0.04, 0.22], p = 0.009; Std. beta = -2.52e-17, 95% CI [-75547.24,
75547.24])
```

```
Standardized parameters were obtained by fitting the model on a standardized
version of the dataset. 95% Confidence Intervals (CIs) and p-values were computed
using a Wald z-distribution approximation.
```

The output describes succinctly the model that was fitted, and the effects of the predictors on the response. The `performance` package is also applicable to models fitted with the `glm()` function.

```
check_model(fit_logistic)
```

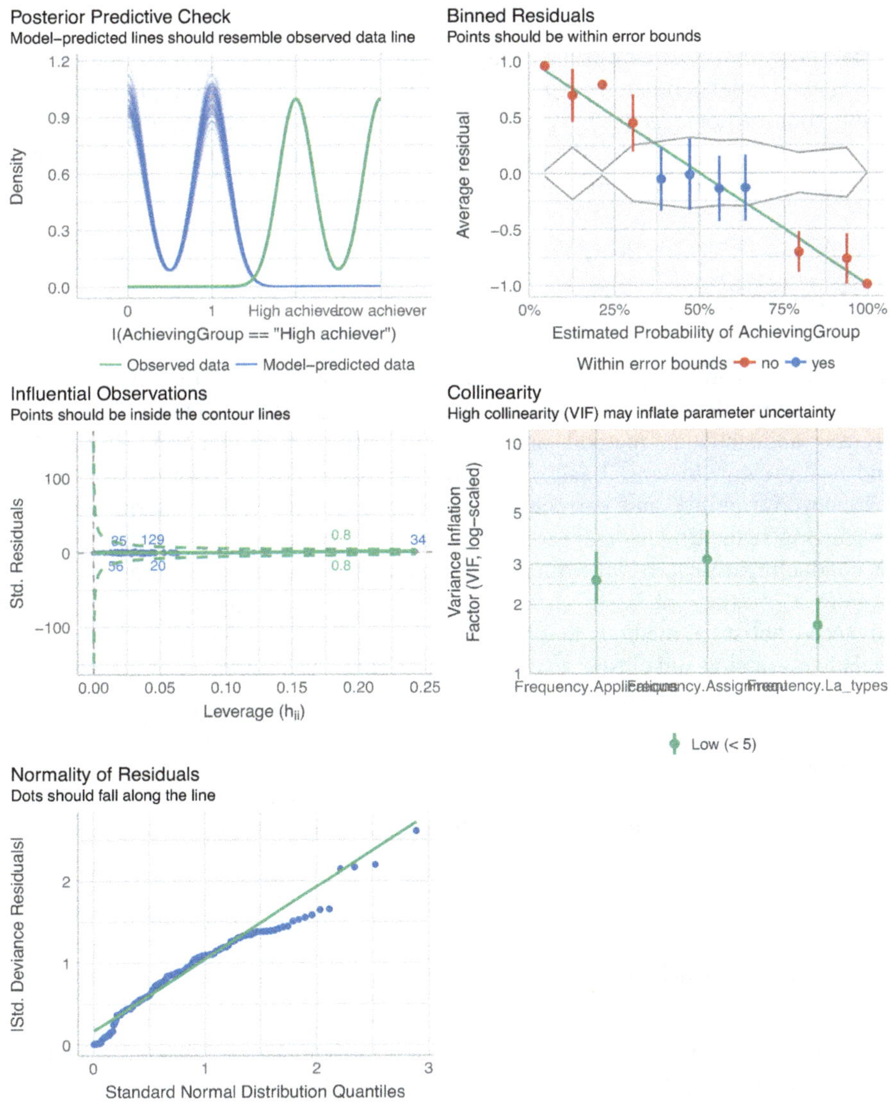

Table 1 Summary of the statistical tests, their null hypotheses, and the number of groups they compare simultaneously

Test	Null Hypothesis	Groups	R function
Student's t-test	Equal means	One, two or paired	`t.test()`
Chi-squared test	Independence	One	`chisq.test()`
Fisher's test	Independence	One	`fisher.test()`
ANOVA	Equal means	Two or more	`aov()`
Kruskal-Wallis test	Equal medians	Two or more	`kruskal.test()`
Levene's test	Homoscedasticity	Two or more	`leveneTest()`
Shapiro-Wilk test	Normality	One	`shapiro.test()`

We note that a different set of diagnostic checks is carried out for the logistic regression model compared to the linear regression model. For example, there is no assumption of homoscedasticity of variance as there is no explicit error term ε in the model. Again, we refer the reader to the documentation of the `performance` package for more details on these checks.

7 Conclusion

Basic statistics are an essential component of learning analytics. Learning analytics involves the collection, analysis, and interpretation of data related to the learning process, and statistical methods are used to identify patterns and trends in this data and to draw conclusions. Basic descriptive statistics such as measures of central tendency, variability and correlation are crucial for analyzing, interpreting, and visualizing data. Understanding these concepts is important for anyone involved in conducting research with quantitative data in the field of learning analytics. Moreover, mastery of basic statistics can facilitate the comprehension of more advanced statistical methods that are commonly used in learning analytics, such as logistic regression and cluster analysis. Table 1 contains a summary of the statistical tests that were introduced in this chapter.

We emphasize that when using any statistical test or a statistical model, it is important to keep the various assumptions related to the chosen method in mind, and to assess them beforehand whenever possible. If the assumptions are violated, the results of the method may not be reliable, and thus suitable alternatives should be considered.

8 Further Reading

This chapter scratched the surface of the full features of packages such as `correlation`, `report` and `performance` that can streamline the statistical analysis and reporting process. We refer the reader to the documentation of these packages to gain a more thorough understanding of their features. These packages are part of a package collection called *easystats* [27] (https://github.com/easystats/easystats). There are several other packages in this collection that were not discussed in this chapter that can be useful for R users working with learning analytics. The book "Learning Statistics with R" by Danielle Navarro is freely available online and provides a comprehensive introduction to statistics using R (https://learningstatisticswithr.com/). For a general introductory text to statistical methods and inference, see e.g., [2].

References

1. Niiniluoto I (1980) Johdatus tieteenfilosofiaan: Käsitteen- ja teorianmuodostus. Otava
2. Ross SM (2010) Introductory statistics, 4th edn. Elsevier, Amsterdam
3. Lehmann EL, Romano JP (2022) Testing statistical hypotheses, 4th edn. Springer, Cham
4. López-Pernas S, Misiejuk K, Tikka S, Saqr M, Kopra J, Heinäniemi M (2024) Visualizing and reporting educational data with R. In: Saqr M, López-Pernas S (eds) Learning analytics methods and tutorials: a practical guide using R. Springer, Berlin
5. López-Pernas S, Saqr M, Conde J, Del-Río-Carazo L (2024) A broad collection of datasets for educational research training and application. In: Saqr M, López-Pernas S (eds) Learning analytics methods and tutorials: a practical guide using R. Springer, Berlin
6. Chan C, Chan GC, Leeper TJ, Becker J (2021) rio: a swiss-army knife for data file I/O. https://cran.r-project.org/package=rio
7. Wickham H, François R, Henry L, Müller K, Vaughan D (2023) dplyr: a grammar of data manipulation
8. Wickham H, Vaughan D, Girlich M (2023) tidyr: tidy messy data
9. Kopra J, Tikka S, Heinäniemi M, López-Pernas S, Saqr M (2024) An R approach to data cleaning and wrangling for education research. In: Saqr M, López-Pernas S (eds) Learning analytics methods and tutorials: a practical guide using R. Springer, Berlin
10. Makowski D, Lüdecke D, Patil I, Thériault R, Ben-Shachar MS, Wiernik BM (2023) Automated results reporting as a practical tool to improve reproducibility and methodological best practices adoption. CRAN
11. Gosset W"Student"S (1908) The probable error of a mean. Biometrika 6:1–25. https://doi.org/10.1093/biomet/6.1.1
12. Welch BL (1947) The generalization of "student's" problem when several different population variances are involved. Biometrika 34:28–35. https://doi.org/10.1093/biomet/34.1-2.28
13. Ellis PD (2010) The essential guide to effect sizes: statistical power, meta-analysis, and the interpretation of research results. Cambridge University Press, Cambridge
14. Cohen J (1988) Statistical power analysis for the behavioral sciences. Routledge, New York
15. Pearson K (1900) On the criterion that a given system of deviations from the probable in the case of a correlated system of variables is such that it can be reasonably supposed to have arisen from random sampling. Philos Mag Ser 5 50:157–175. https://doi.org/10.1080/14786440009463897
16. Fisher RA (1922) On the interpretation of χ^2 from contingency tables, and the calculation of P. J R Stat Soc 85:87–94
17. Fisher RA (1921) On the "probable error" of a coefficient of correlation deduced from a small sample. Metron 1:3–32
18. Kruskal WH, Wallis WA (1952) Use of ranks in one-criterion variance analysis. J Am Stat Assoc 47:583–621. https://doi.org/10.1080/01621459.1952.10483441

19. Levene H (1960) Robust tests for equality of variances. In: Olkin I, Hotelling H, et al (eds) Contributions to probability and statistics: essays in honor of harold hotelling. Stanford University Press, Redwood City, pp 278–292
20. Bartlett MS (1937) Properties of sufficiency and statistical tests. Pro R Stat Soc Ser A 160:268–282
21. Shapiro SS, Wilk MB (1965) An analysis of variance test for normality (complete samples). Biometrika 52:591–611. https://doi.org/10.1093/biomet/52.3-4.591
22. Robinson D, Hayes A, Couch S (2023) broom: convert statistical objects into tidy tibbles
23. Makowski D, Ben-Shachar MS, Patil I, Lüdecke D (2020) Methods and algorithms for correlation analysis in R. J Open Source Softw 5:2306. https://doi.org/10.21105/joss.02306
24. Lüdecke D, Patil I, Ben-Shachar MS, Wiernik BM, Waggoner P, Makowski D (2021) See: an R package for visualizing statistical models. J Open Source Softw 6:3393. https://doi.org/10.21105/joss.03393
25. Lüdecke D, Ben-Shachar MS, Patil I, Waggoner P, Makowski D (2021) Performance: an R package for assessment, comparison and testing of statistical models. J Open Source Softw 6:3139. https://doi.org/10.21105/joss.03139
26. Dobson AJ (1990) An introduction to generalized linear models. Chapman Hall, Boca Raton
27. Lüdecke D, Ben-Shachar MS, Patil I, Wiernik BM, Bacher E, Thériault R, Makowski D (2022) easystats: framework for easy statistical modeling, visualization, and reporting. CRAN

Visualizing and Reporting Educational Data with R

Sonsoles López-Pernas, Kamila Misiejuk, Santtu Tikka, Juho Kopra, Merja Heinäniemi, and Mohammed Saqr

1 Introduction

Data visualization can be defined as "the representation and presentation of data that exploits our visual perception abilities in order to amplify cognition" [1]. It has the power to transform complex information into stories that inform and inspire action. Data visualization is an effective tool for learning analytics, as it helps to present learners' data in a way that is easily understandable and intuitive for students, teachers, researchers, and other stakeholders. Through the use of graphs, charts, and other visual aids, it is possible to quickly identify patterns, trends, and relationships within data that may not be immediately apparent through purely numerical data analysis methods.

Visualization in learning analytics has two distinct applications. On the one hand, the use of visual dashboards has become the main vehicle for putting learning analytics into practice. Presenting data in visually appealing and intuitive ways can help promote data literacy among students and other stakeholders, encouraging

S. López-Pernas (✉) · J. Kopra
School of Computing, University of Eastern Finland, Joensuu, Finland
e-mail: sonsoles.lopez@uef.fi

K. Misiejuk
Centre for the Science of Learning & Technology (SLATE), University of Bergen, Bergen, Norway

S. Tikka
Department of Mathematics and Statistics, University of Jyväskylä, Jyväskylä, Finland

M. Heinäniemi
Institute of Biomedicine, University of Eastern Finland, Kuopio, Finland

M. Saqr
School of Computing, University of Eastern Finland, Joensuu, Finland

M. Saqr, S. López-Pernas (eds.), *Learning Analytics Methods and Tutorials*,
https://doi.org/10.1007/978-3-031-54464-4_6

greater engagement with data and fostering a culture of continuous improvement. On the other hand, learning analytics scientific production heavily relies on data visualization to present research findings in a clear and accessible manner, making it easier for readers from different scholarly backgrounds to understand and act upon research insights. Regardless of the context, the power of visualization in learning analytics lies in its ability to take complex data and turn it into meaningful insights that support better decision-making and drive improvement.

In this chapter, the reader will be guided through the process of generating meaningful and aesthetically pleasing visualizations of different types of datasets using well-known R packages. Relevant plots and plot types will be demonstrated with an explanation of their usage and usage cases. Furthermore, learning-related examples will be discussed in detail. For instance, readers will learn how to visualize learners' logs extracted from learning management systems (LMSs) to show how trace data can be used to track students' learning activities. Other examples of common research scenarios in which learners' data are visualized will be illustrated throughout the chapter. In addition to creating compelling plots, readers will also be able to generate professional-looking tables with summary statistics to report descriptive statistics.

2 Visualization in Learning Analytics

Developing visualizations is a challenging task of balancing the cognitive load of users while not compromising on conveying specific insights from data [2]. Visualizations for practice in learning analytics are mostly developed for two main stakeholders: learners and instructors. Depending on the target group, a visualization or a dashboard (i.e., a collection of visualizations depicting multiple indicators) have different goals.

Learner-facing visualizations are meant to make learners aware of their own learning and to provide them with actionable feedback on their learning. Visualizations display learners' performance on a specific metric and compare it with a reference frame: other peers, desirable learning achievement, or their own progress over time [3]. Sense-making questions triggering reflection can be added to a visualization [4, 5], or some elements of the visualizations can be highlighted and described in words using layered storytelling [6, 7]. Another option is to gamify a dashboard, for example, by using badges [8]. To provide feedback to learners, visualizations can be augmented with links to recommended resources [9], information about specific topics to review to close the achievement gap [6], or explanations of the meaning of visualizations and their implications for the learner [10]. Current learner-facing dashboards mostly show resource use and assessment data [11], compare learners to their peers [12], display descriptive analytics rather than predictive or prescriptive analytics [10], and use self-regulated learning theory as their framework [12, 13]. Some reviews found a positive effect on student outcomes [10], while others reported mixed results [11, 14]. Showing visualizations to learners can change their behavior. For example, social network analysis visualizations have resulted in fewer cross-group commenting [15], while a visualization comparing individual submission patterns with the top 25% of students in a class led to earlier homework submissions [16].

In comparison, the goal of instructor-facing visualizations is to support teachers and their decision-making process by tracking student progress. Two main types can be distinguished. Mirroring or descriptive visualizations provide insights about the learners on an aggregated or an individual level using either descriptive or comparative data. Advising or prescriptive visualizations show not only information about the learners but also alert the instructor to undertake a pedagogical action [17, 18]. Current instructor-facing visualizations mostly display course-wide information about the learners or track group work [14]. These visualizations can support teachers in facilitating student collaboration [19], planning and collecting student feedback on learning activities [20], or obtaining insights into student interactions within an online environment, such as simulations, virtual labs or online games [21, 22]. However, interpreting dashboard information is a challenging task for instructors. Although some teachers use dashboards as complementary sources of information, others act based only on the dashboard information without further investigation [23].

A common point of criticism of learning analytics dashboards is that most of them are not grounded in learning theories [13, 14]. Data-driven evaluations of dashboards focused on dashboard acceptance, usefulness, or usability are more prevalent than pedagogically-focused evaluations [24]. Some approaches were developed to mitigate these issues. The model of user-centered learning analytics systems (MULAS) presents a set of recommendations on four interconnected dimensions: theory, design, evaluation, and feedback, and can be used to guide dashboard development [14]. Another approach is an iterative five-stage Learning Awareness Tools—User eXperience (LATUX) workflow, including problem identification, low-fidelity prototyping, high-fidelity prototyping, pilot studies, and classroom use, that can be used to develop visual analytics [25]. Finally, open learner model research could be used as a source of insights while developing learning analytics visualizations, such as dashboards [9].

3 Generating Plots with `ggplot2`

In the previous section, we have seen how central visualization is to learning analytics. In the remainder of the chapter, we will learn how to create different types of visualizations that are relevant to different types of data related to teaching and learning. We will mostly rely on `ggplot2`, a popular data visualization package in R that was developed by Hadley Wickham [26]. It is based on the grammar of graphics [27], which is a systematic way of thinking about and constructing visualizations. The `ggplot2` library provides a flexible and intuitive framework for creating a wide range of graphics, from basic scatter plots to complex visualizations with multiple layers. It is known for its ability to produce visually appealing and informative graphics with relatively few lines of code. It enables users to define aesthetics, such as color and size, and add layers, such as points and lines, to create customized and interactive plots. In addition, `ggplot2` allows for easy customization of plot features, such as titles, axis labels, and legends.

Overall, `ggplot2` is a powerful and versatile tool for data visualization in R, and is widely used by data scientists, statisticians, and researchers in a variety of fields. In this chapter, we will cover the fundamental concepts and techniques of `ggplot2`, including how to create basic plots, and customize their appearance. We will start by introducing the building blocks of a `ggplot2` plot, including aesthetics, layers, and scales. Then, we will create a plot from scratch step by step, showing how to customize its appearance, including how to change theme, colors, and scales. We will then explore the different types of plots that can be created with `ggplot2`, such as scatter plots, bar charts, and histograms.

Throughout this section, we will use datasets of students' learning data to demonstrate how to create effective visualizations for learning analytics with `ggplot2`. Please, refer to Chapter 2 of this book [28] to learn more about the datasets used. By the end of this section, you will have a solid foundation in `ggplot2` and be able to create basic, yet compelling visualizations to explore your data.

3.1 The `ggplot2` Grammar

The `ggplot2` library is based on Wilkinson's grammar of graphics [27]. The main idea is that every plot can be broken down into a set of components, each of which can be customized and combined in a flexible way. These components are:

- **Data**: This is the data we want to visualize. It can be in the form of a dataframe, tibble or any other structured data format.
- **Aesthetic mapping** (`aes`): It defines how variables in the data are mapped to visual properties of the plot, such as position, color, shape, size, and transparency.
- **Geometric object** (`geom`): It represents the actual visual elements of the plot, such as points, lines, bars, and polygons.
- **Statistical transformation** (`stat`): It summarizes or transforms the data in some way, such as by computing means, medians, or proportions, or by smoothing or summarizing data, or grouping them into bins.
- **Scale** (`scale`): It maps values in the data to visual properties of the plot, such as color, size, or position.
- **Coordinate system** (`coord`): It defines the spatial or geographic context in which the plot is displayed, such as Cartesian coordinates, polar coordinates, or maps.
- **Facet** (`facet`): It allows to split the data into subsets and display each subset in a separate panel. It often useful for visualizing data with multiple categories or groups.

Through the combination and customization of these components, we can create a wide variety of complex and informative visualizations in `ggplot2`. The idea behind the graphics grammar is to provide a consistent framework for constructing plots, allowing users to focus on the data and the message they want to convey, rather than on the technical details of the visualization. In the following section, we will create a plot from scratch step by step to become familiar with the most relevant components.

3.2 Creating Your First Plot

We will now create our first plot using `ggplot2`. Our example deals with a widely studied matter in learning analytics, which is the relationship between online activity and achievement. We will use a bar chart to represent the number of students that have low, moderate and high activity levels in each achievement group (high achievers vs. low achievers). In order to become familiar with the syntax of `ggplot2`, we will recreate the plot step by step, explaining each of the elements in the plot. Below is the final result we aim at accomplishing (Fig. 1):

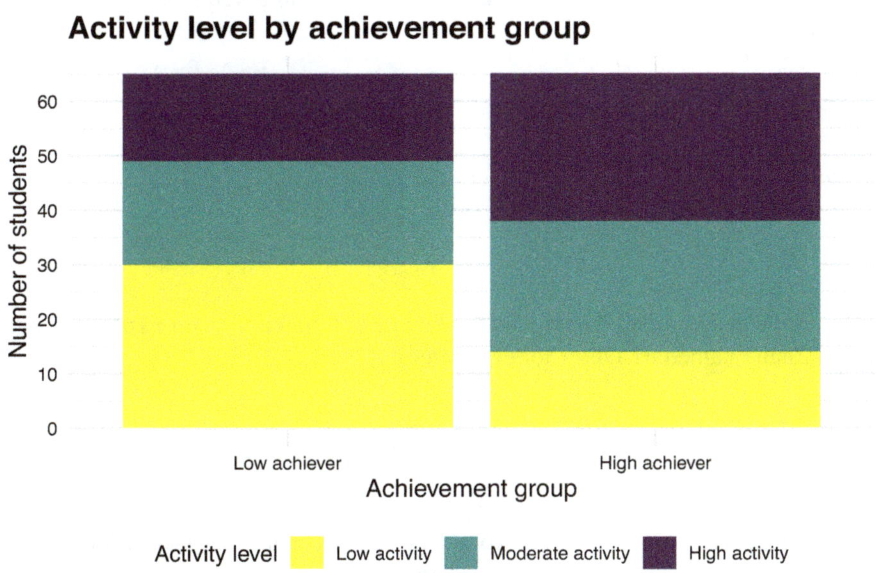

Fig. 1 First plot with ggplot2

3.2.1 Installing ggplot2

Our first step is installing the ggplot2 library. This is usually the first step in any R script that makes use of external libraries.

```
install.packages("ggplot2")
```

To import ggplot2 we just need to use the library command and specify the ggplot2 library:

```
library(ggplot2)
```

3.2.2 Downloading the Data

Next, we need to import the data that we are going to plot. For this chapter, we are using synthetic data from a blended course on learning analytics. For more details about this dataset, refer to Chap. 2 in this book. The data is in Excel format. We can use the library rio since it makes it easy to read data in several formats. We first install the library:

```
install.packages("rio")
```

And import it so we can use its functions:

```
library(rio)
```

Now we can download the data using the import function from rio and assign it to a variable named df (short for dataframe).

```
demo_url =
"https://github.com/lamethods/data/raw/main/1_moodleLAcourse/AllCombined.xlsx"
df <- import(demo_url)
```

We can use the head command to get an idea of what the dataset looks like. To recreate the plot above we will need the AchievingGroup column —which indicates whether students' are high achievers (to 50%) or low achievers (bottom 50%), according to their final grade— and the ActivityGroup column —which indicates whether students have a high level of activity (top 33%), moderate activity (middle 33%), or low activity (bottom 33%), according to their total number of events in the LMS.

```
head(df)
```

```
# A tibble: 130 x 37
   User      Name   Gender ActivityGroup AchievingGroup Surname Origin Birthdate
   <chr>     <chr>  <chr>  <chr>         <chr>          <chr>   <chr>  <chr>
 1 00a05cc62 Wan    M      Low activity  Low achiever   Tan     Malay~ 12.12.19~
 2 042b07ba1 Daniel M      High activity Low achiever   Tromp   Aruba  28.5.1999
 3 046c35846 Sarah  F      Low activity  Low achiever   Schmit  Luxem~ 25.4.1997
 4 05b604102 Lian   F      Low activity  Low achiever   Abdull~ Yemen  19.11.19~
 5 0604ff3d3 Nina   F      Low activity  Low achiever   Borg    Malta  13.6.1994
 6 077584d71 Moham~ M      High activity High achiever  Gamal   Egypt  13.7.1998
 7 081b100cf Maxim~ M      Moderate act~ High achiever  Gruber  Austr~ 20.12.19~
 8 0857b3d8e Hugo   M      High activity High achiever  Pérez   Spain  22.12.19~
 9 0af619e4b Aylin  F      Low activity  Low achiever   Barat   Kazak~ 14.8.1995
10 0ec99ce96 Polina F      Moderate act~ Low achiever   Novik   Belar~ 9.10.1996
# i 120 more rows
# i 29 more variables: Location <chr>, Employment <chr>,
#   Frequency.Applications <dbl>, Frequency.Assignment <dbl>,
#   Frequency.Course_view <dbl>, Frequency.Feedback <dbl>,
#   Frequency.General <dbl>, Frequency.Group_work <dbl>,
#   Frequency.Instructions <dbl>, Frequency.La_types <dbl>,
#   Frequency.Practicals <dbl>, Frequency.Social <dbl>, ...
```

3.2.3 Creating the Aesthetic Mapping

Now that we have our data, we can pass it on to ggplot2 as follows:

```
ggplot(df)
```

Fig. 2 Empty plot

We still do not see anything because we have not selected the type of chart or the variables of the data that we want to plot (Fig. 2). First, let us specify that we want to plot the AchievingGroup column (high vs. low achievers) on the x-axis. Assigning columns of our dataset to different elements of the plot is called constructing an aesthetic mapping. We can do it by calling the aes function from ggplot2, specifying that we want to map the AchievingGroup column to the x-axis, and then passing this call to aes to our plot using the second argument of ggplot:

3.2.4 Add the Geometry Component

```
ggplot(df, aes(x = AchievingGroup))
```

Fig. 3 Empty plot with AchievingGroup in x-axis labels

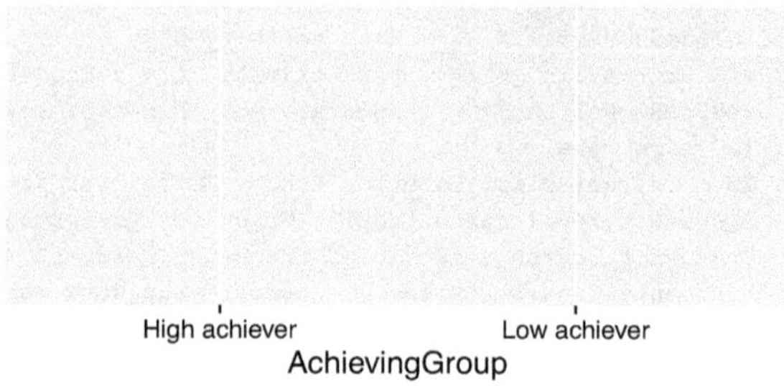

We now see that the x-axis has the two possible values of AchievingGroup: "High achiever" and "Low achiever" (Fig. 3). We still need to tell ggplot2 the type of chart we want to use to plot the number of students of each type. To do that we need to add a geometrical (geom) component to our plot in which we specify that we want a bar chart. We do it by adding a + sign after our call to ggplot and calling geom_bar() (the name of the geometry that represents a bar chart).

```
ggplot(df, aes(x = AchievingGroup)) + geom_bar()
```

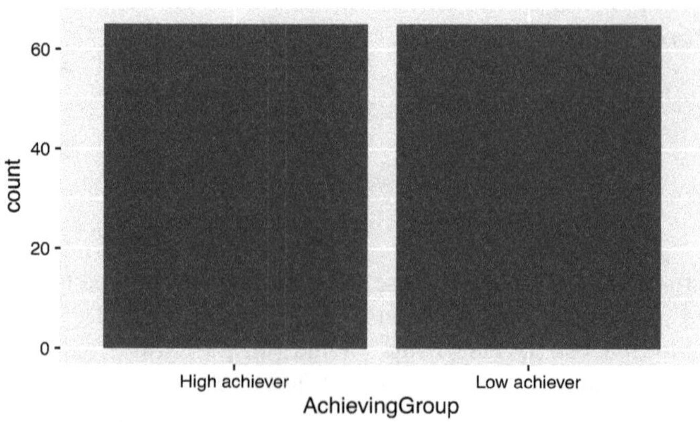

Fig. 4 Basic bar plot showing students by achievement group

Now the plot can actually be called a plot. Notice that we have not specified what we want to plot in the y-axis. When not specified, `ggplot2` assumes that we want to use the count of rows (Fig. 4).

We also notice that the bars are in the wrong order. By default, `ggplot2` orders the values in an ascending way (alphabetically in the case of text values). If we want to enforce our own order, we need to convert the `AchievingGroup` column of `df` into a factor and provide the ordered list of values to the `levels` argument.

```
df$AchievingGroup = factor(df$AchievingGroup,
                           levels = c("Low achiever", "High achiever"))
```

If we generate our plot again, we see that the bars are now in the order we want them to be (Fig. 5):

```
ggplot(df, aes(x = AchievingGroup)) + geom_bar()
```

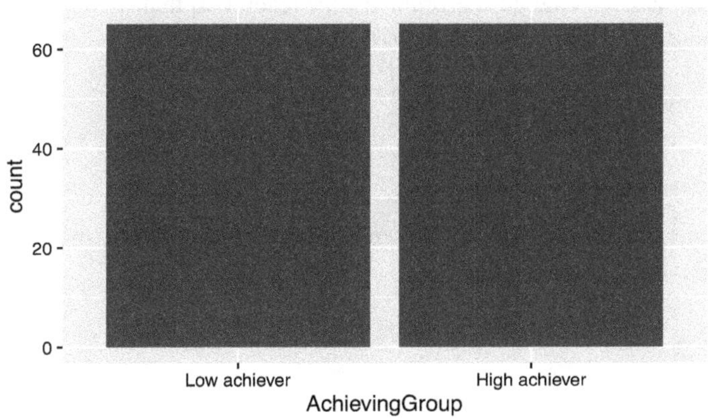

Fig. 5 Basic bar plot showing students by achievement group after transforming the x-axis variable into a factor

3.2.5 Adding the Color Scale

We still need to color our bar chart according to students' activity level. We do that by mapping the fill aesthetic to the `ActivityLevel` column inside the `aes`. When we provide the `fill` property, `ggplot` will automatically create the appropriate legend (Fig. 6).

```
ggplot(df, aes(x = AchievingGroup, fill = ActivityGroup)) + geom_bar()
```

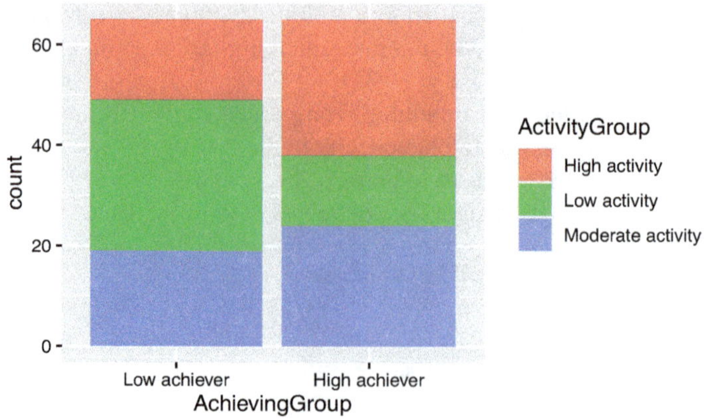

Fig. 6 Basic bar plot showing students' activity level by achievement group and colored by activity level

Again, we need to change the order of our legend so that it follows the logical semantic order for the activity levels (low-moderate-high):

```
df$ActivityGroup = factor(df$ActivityGroup,
                          levels = c("Low activity", "Moderate activity",
                          "High activity"))
```

If we generate the plot again, we see now that the legend is in the right order (Fig. 7):

```
ggplot(df, aes(x = AchievingGroup, fill = ActivityGroup)) + geom_bar()
```

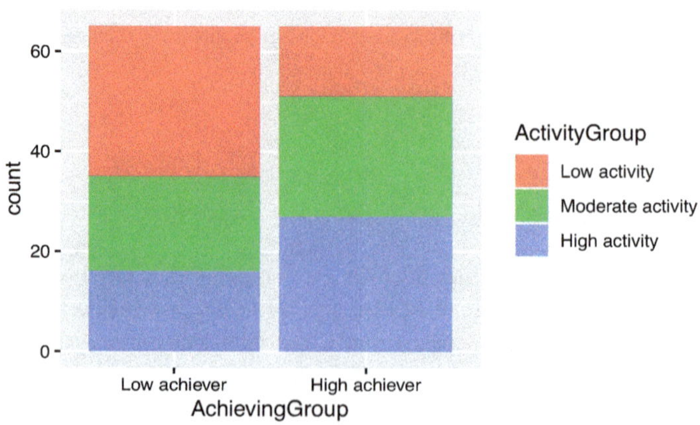

Fig. 7 Basic bar plot showing students' activity level by achievement group and colored by activity level after ordering the legend

However, the stacks are still not in the right order, being the low activity students at the top of the bar, and the high activity students at the bottom, which might be counter-intuitive. To change this, we need to reverse the position of the bar using `position = position_stack(reverse = TRUE)` inside `geom_bar` (Fig. 8):

```
ggplot(df, aes(x = AchievingGroup, fill = ActivityGroup)) +
  geom_bar(position = position_stack(reverse = TRUE))
```

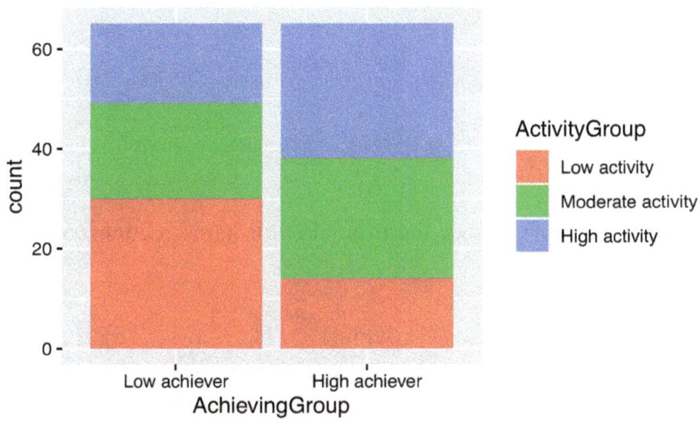

Fig. 8 Basic bar plot showing students' activity level by achievement group and colored by activity level after ordering the stacks

We are getting closer but the color scheme does not quite match our intended result. To add a color scheme to our plot we need to add a `scale` layer. In this case, the scale is for the `fill` property, which is the color of the bars in our chart. There are many ways to specify the color scheme. One option is to use sequential colors from the same palette. For that we add a new layer to our plot named `scale_fill_brewer` and we pass the palette that we want as an argument. For example, palette number 15 would look like this (Fig, 9):

```
ggplot(df, aes(x = AchievingGroup, fill = ActivityGroup)) +
  geom_bar(position = position_stack(reverse = TRUE)) +
  scale_fill_brewer(palette = 15)
```

Another option is to provide a manual scale with the colors of our choice. For that we use `scale_fill_manual` and specify a `values` vector as an argument. We need to specify as many colors as unique elements in your scale. In this case we have three activity groups (for low, moderate or high activity), so we must provide three colors. There are tons of resources online where you can find or create your own palettes (e.g., Coolors, Adobe Color or Lospec). You have to provide the

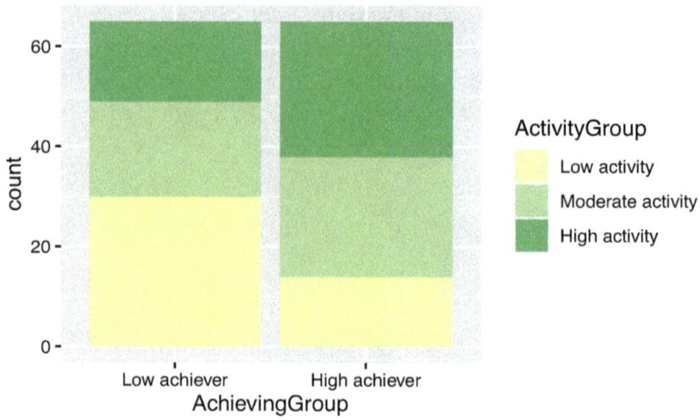

Fig. 9 Bar plot showing students' activity level by achievement group with sequential color scale

hexadecimal code of each color or the official color name recognized by R. Below is an example (Fig. 10):

```
ggplot(df, aes(x = AchievingGroup, fill = ActivityGroup)) +
  geom_bar(position = position_stack(reverse = TRUE)) +
  scale_fill_manual(values = c("#ef6461", "#7AE7C7", "#8E518D"))
```

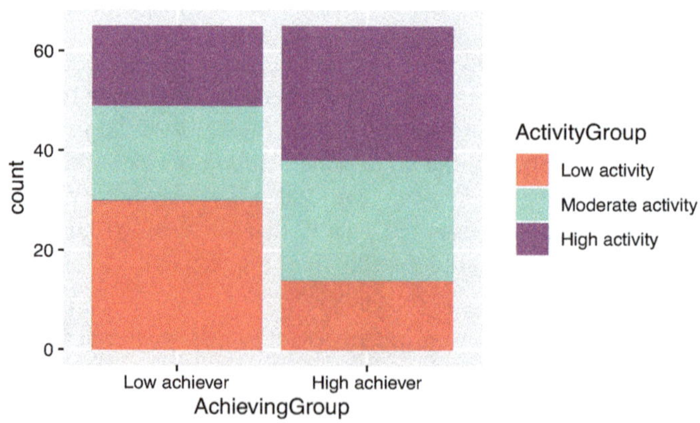

Fig. 10 Bar plot showing students' activity level by achievement group with manual color scale

Lastly, a very common color scale used is *Viridis*. It is designed to be perceived by viewers with common forms of color blindness. To use it in our plot we just add `scale_fill_viridis_d()` (Fig. 11).

```
ggplot(df, aes(x = AchievingGroup, fill = ActivityGroup)) +
  geom_bar(position = position_stack(reverse = TRUE)) +
  scale_fill_viridis_d()
```

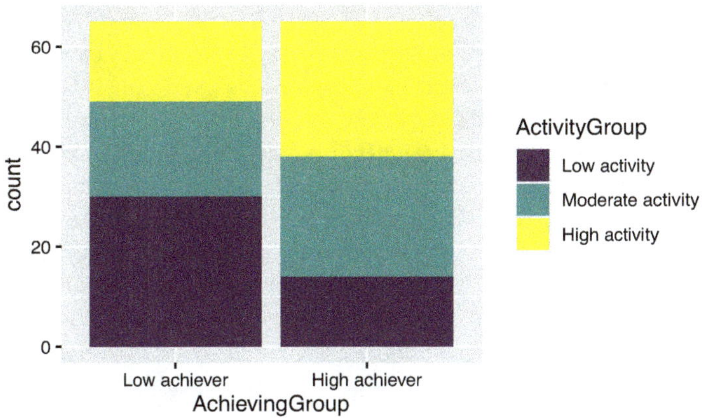

Fig. 11 Bar plot showing students' activity level by achievement group with viridis color scale

Viridis is the palette we need to replicate our target plot. However, the order of the color needs to be reversed so the most dense color represents the higher activity level. We do this by reversing the direction of the palette as follows (Fig. 12):

```
ggplot(df, aes(x = AchievingGroup, fill = ActivityGroup)) +
  geom_bar(position = position_stack(reverse = TRUE)) +
  scale_fill_viridis_d(direction = -1)
```

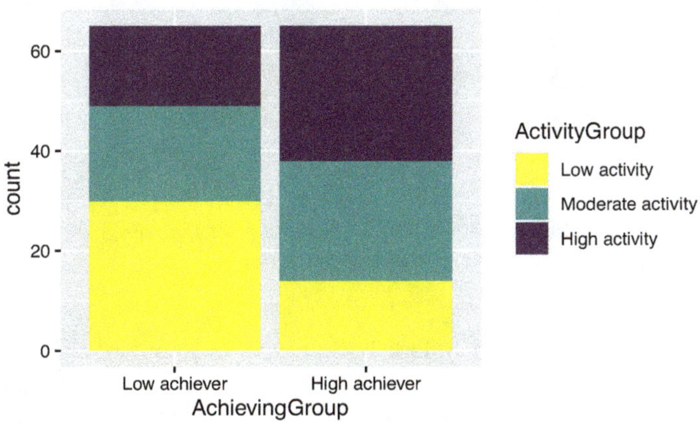

Fig. 12 Bar plot showing students' activity level by achievement group with viridis color scale

3.2.6 Working with Themes

Now that the geometry and color scheme of the bars looks like our initial plot, we notice that there are still some differences. An important one is the grey background of the plot. To change the general appearance of our plot, we may use the ggplot2 themes. Below are some examples (Fig. 13):

```
ggplot(df, aes(x = AchievingGroup, fill = ActivityGroup)) +
  geom_bar(position = position_stack(reverse = TRUE)) +
  scale_fill_viridis_d(direction = -1) + theme_dark()
ggplot(df, aes(x = AchievingGroup, fill = ActivityGroup)) +
  geom_bar(position = position_stack(reverse = TRUE)) +
  scale_fill_viridis_d(direction = -1) + theme_classic()
ggplot(df, aes(x = AchievingGroup, fill = ActivityGroup)) +
  geom_bar(position = position_stack(reverse = TRUE)) +
  scale_fill_viridis_d(direction = -1) + theme_void()
ggplot(df, aes(x = AchievingGroup, fill = ActivityGroup)) +
  geom_bar(position = position_stack(reverse = TRUE)) +
  scale_fill_viridis_d(direction = -1) + theme_minimal()
```

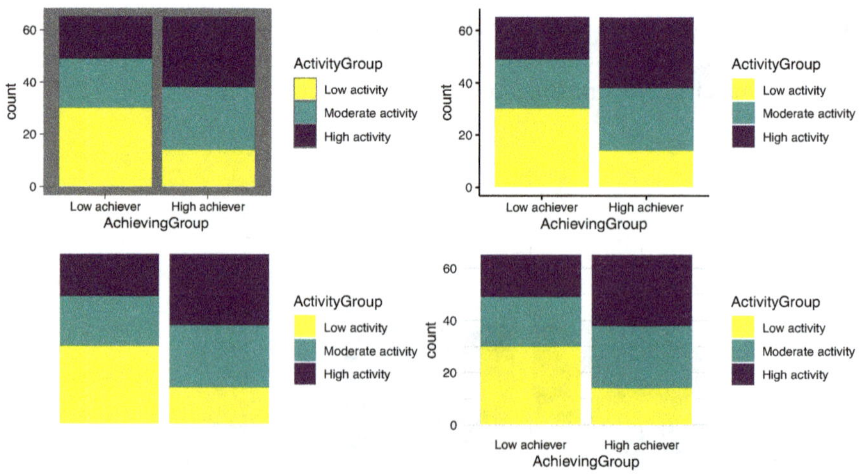

Fig. 13 Bar plot using different themes: `theme_dark` (top left), `theme_classic` (top right), `theme_void` (bottom left), and `theme_minimal` (bottom right)

We have `theme_dark` with a dark background and border, `theme_classic` with thick axes and no grid lines, `theme_void` which is completely empty, and `theme_minimal` with a minimalistic look. There are more available in the ggplot2 documentation and even more third-party implementations. To recreate our goal plot, we select the `theme_minimal`. To avoid having to add the theme to all of our plots from now on, we can set a default theme for our whole project by using `theme_set`:

```
theme_set(theme_minimal())
```

Notice how now we get `theme_minimal` even when we do not specify it in our code (Fig. 14):

```
ggplot(df, aes(x = AchievingGroup, fill = ActivityGroup)) +
  geom_bar(position = position_stack(reverse = TRUE)) +
  scale_fill_viridis_d(direction = -1)
```

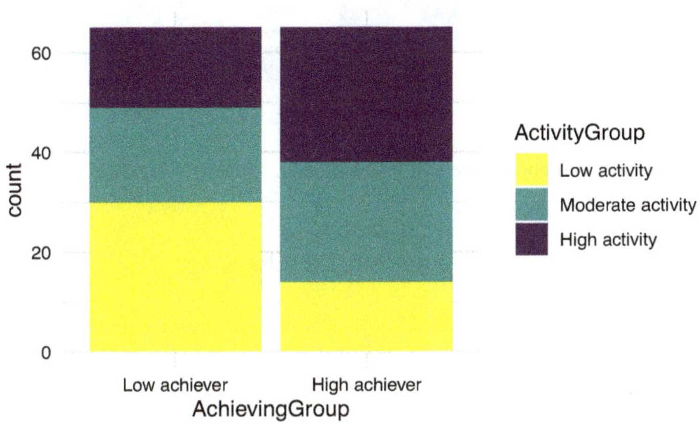

Fig. 14 Bar plot with theme `minimal` by default

3.2.7 Changing the Axis Ticks

You may have not noticed that another difference with our goal plot is the ticks in our y-axis. In the goal plot we count 10 by 10, whereas in our last plot we do so 20 by 20. Just like we modified the scale of the `fill` aesthetic when we changed the color of our bars, we can also modify the `y` aesthetic to adjust to our needs. We use the `scale_y_continuous` layer and we try different number of breaks (`n.breaks`), until we find what we like best (Fig. 15):

```
ggplot(df, aes(x = AchievingGroup, fill = ActivityGroup)) +
  geom_bar(position = position_stack(reverse = TRUE)) +
  scale_fill_viridis_d(direction = -1) +
  scale_y_continuous(n.breaks = 15)
ggplot(df, aes(x = AchievingGroup, fill = ActivityGroup)) +
  geom_bar(position = position_stack(reverse = TRUE)) +
  scale_fill_viridis_d(direction = -1) +
  scale_y_continuous(n.breaks = 3)
ggplot(df, aes(x = AchievingGroup, fill = ActivityGroup)) +
  geom_bar(position = position_stack(reverse = TRUE)) +
  scale_fill_viridis_d(direction = -1) +
  scale_y_continuous(n.breaks = 7)
```

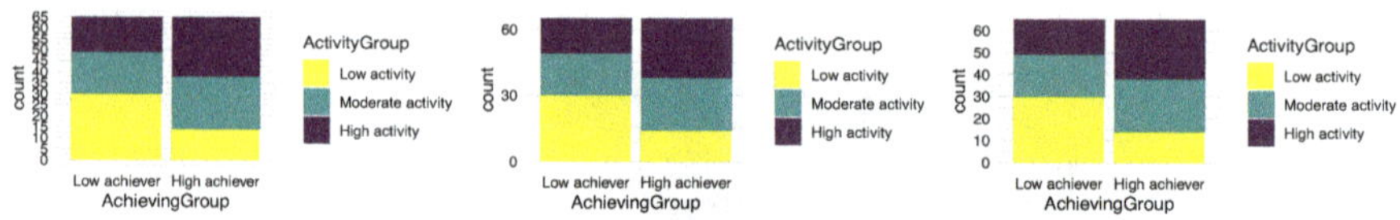

Fig. 15 Bar plot with different numbers of y.axis breaks: 15 (left), 3 (middle), and 7 (right)

We choose 7 breaks to obtain our desired result.

3.2.8 Titles and Labels

Our plot is still missing some slight modifications to be 100% equal to the original one. For instance, the axes' titles are not the same. To specify the y-axis label, we add a new layer to our plot named `ylab` and we pass a string with our desired label "Number of students" (Fig. 16):

```
ggplot(df, aes(x = AchievingGroup, fill = ActivityGroup)) +
  geom_bar(position = position_stack(reverse = TRUE)) +
  scale_fill_viridis_d(direction = -1)  +
  scale_y_continuous(n.breaks = 7) +
  ylab("Number of students")
```

Fig. 16 Bar plot with y-axis label

We do the same for the x-axis using `xlab`, and for the legend using `labs` (Fig. 17):

```
ggplot(df, aes(x = AchievingGroup, fill = ActivityGroup)) +
  geom_bar(position = position_stack(reverse = TRUE)) +
  scale_fill_viridis_d(direction = -1)  +
  scale_y_continuous(n.breaks = 7) +
  ylab("Number of students") +
  xlab("Achievement group") +
  labs(fill = "Activity level")
```

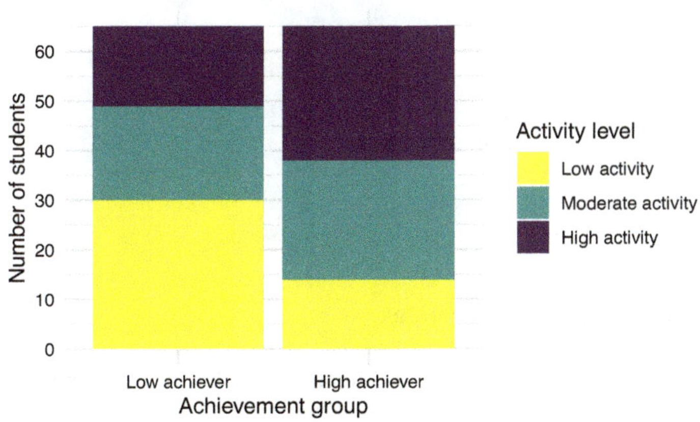

Fig. 17 Bar plot with all labels

More importantly, we are missing the overall title of the plot. To add it we use `ggtitle` and we pass our intended plot title "Activity level by achievement group". Keep in mind that, whenever possible, it is better to add a caption to the image rather than a title on the plot. A caption is more accessible for visually impaired users since it is compatible with screen readers. In scientific papers, it is also more common to have a Figure caption than a title within the plot. In social media, it is frequent to see the title on the plot as images are often shared without context. However, many social media platforms allow to provide an *alternative text* which is what screen readers will read as a substitute for the image, and that is also the case in learning analytics dashboards (Fig. 18).

```
ggplot(df, aes(x = AchievingGroup, fill = ActivityGroup)) +
  geom_bar(position = position_stack(reverse = TRUE)) +
  scale_fill_viridis_d(direction = -1)  +
  scale_y_continuous(n.breaks = 7) +
  ylab("Number of students") +
  xlab("Achievement group") +
  labs(fill = "Activity level") +
  ggtitle("Activity level by achievement group")
```

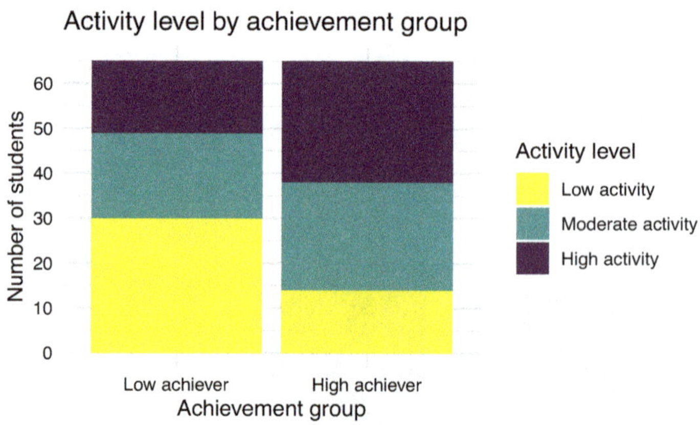

Fig. 18 Bar plot with title

3.2.9 Other Cosmetic Modifications

Lastly, we need to do some slight modifications to the overall appearance of the plot. We do this through the generic `theme` function of `ggplot2`. We first modify the position of the legend by setting `legend.position` to "bottom". We then increase the size of the axes titles, by setting `axis.title` to `element_text(size = 12)`. Finally, we make the plot title bigger as well and put it in bold by setting `plot.title` to `element_text(size = 15, face = "bold"))`. With these last changes, we have an exact replica of our original plot (Fig. 19).

```
ggplot(df, aes(x = AchievingGroup, fill = ActivityGroup)) +
  geom_bar(position = position_stack(reverse = TRUE)) +
  scale_fill_viridis_d(direction = -1) +
  scale_y_continuous(n.breaks = 7) +
  ylab("Number of students") +
  xlab("Achievement group") +
  labs(fill = "Activity level") +
  ggtitle("Activity level by achievement group") +
  theme(legend.position = "bottom",
        axis.title = element_text(size = 12),
        plot.title = element_text(size = 15, face = "bold"))
```

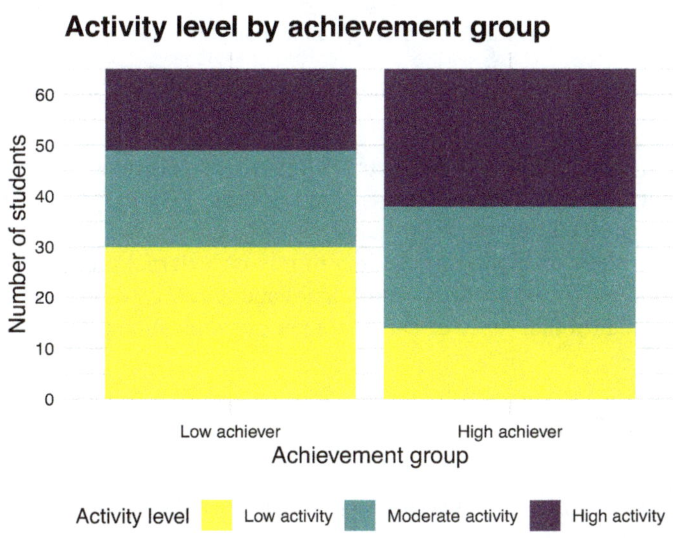

Fig. 19 Bar plot with theme modifications

3.2.10 Saving the Plot

Since we have obtained the desired result, we may now save it as an image to be able to use it elsewhere. For that, we first need to assign the plot to a variable (e.g., `myplot`).

```r
myplot <- ggplot(df, aes(x = AchievingGroup, fill = ActivityGroup)) +
  geom_bar(position = position_stack(reverse = TRUE)) +
  scale_fill_viridis_d(direction = -1) +
  scale_y_continuous(n.breaks = 7) +
  ylab("Number of students") +
  xlab("Achievement group") +
  labs(fill = "Activity level") +
  ggtitle("Activity level by achievement group") +
  theme(legend.position = "bottom", axis.title = element_text(size = 12),
        plot.title = element_text(size = 15, face = "bold"))
```

We then use `ggsave` to save the plot to our filesystem. We need to specify the file path (including the extension, such as PNG, JPEG, etc.) where we want to save the plot (e.g., "bar.png") as the first argument and pass the variable where we saved our plot (`myplot`) as a second argument. If we do not do this, `ggplot2` assumes we want to save the latest plot that we created. Lastly, we may specify the width, height and resolution (dpi) of our plots. If we are submitting our figure to a scientific journal, we probably need a high resolution image. If we are using the figure in social media, we do not want the resolution to be so high as it would take a long time to load.

```r
ggsave("bar.png", myplot, width = 10000, height = 5000, units = "px", dpi = 900)
```

Throughout this section, we have learned how we can create a plot from scratch using only the `ggplot2` library and a simple dataset. We have seen the many customization possibilities (theme, scales, titles) that we can achieve using the different plot components without needing to rely on external tools for retouching our final graph. In the next section we will learn about new types of plots that might be more suitable for other types of data and their customization possibilities.

3.3 *Types of Plots*

The `ggplot2` library offers many types of plots (or `geoms`) that you can choose from to visualize your data in several ways. In this section, we go over some of the most common types and present examples using students' learning data.

3.3.1 Bar Plot

We have seen how to construct a bar plot in the previous section as an example of how to use `ggplot2`. But when should we use a bar plot? Bar plots are useful when we want to represent counts or any numerical variable broken down by categories. The y-axis would represent the count (or other continuous numerical variable) and the x-axis would represent the categories. Keep in mind that if the categories follow a natural order, the x-axis should respect it (for example: "Morning", "Afternoon", "Evening"; or "Children", "Adults", "Elders"). Otherwise, you can just order the x-axis alphabetically or from highest to lowest value in the y-axis (Fig. 20).

```
ggplot(df, aes(x = AchievingGroup)) +
  geom_bar(position = position_stack(reverse = TRUE))
```

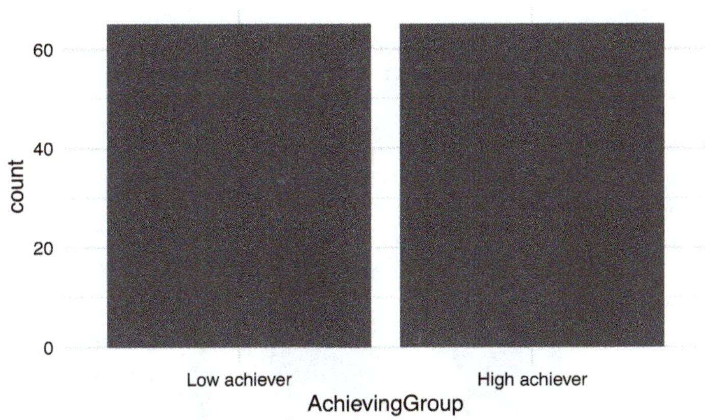

Fig. 20 Basic bar plot of students by achievement group

Remember that you can add a "third dimension" to the plot by using the `fill` property. This is known as a 'stacked' bar chart and helps highlight the proportion of, in this case, students' activity level (`ActivityGroup`) (Fig. 21).

```
ggplot(df, aes(x = AchievingGroup, fill = ActivityGroup)) +
  scale_fill_viridis_d(direction = -1) +
  geom_bar(position = position_stack(reverse = TRUE))
```

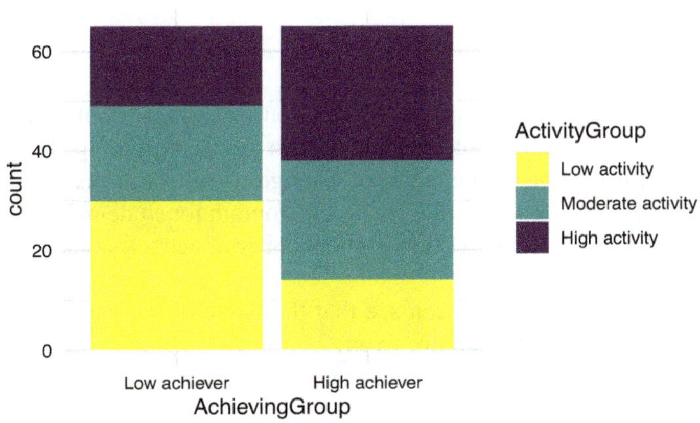

Fig. 21 Basic bar plot of students by achievement group filled by activity level

If we care more about the actual number rather than the proportion of students with each activity level, instead of a stacked bar chart we can keep each 'stack' as a whole bar of their own. This plot is very useful to compare values among categories. We accomplish this by passing the `position` argument with the value "dodge" to the `geom_bar` component (Fig. 22):

```
ggplot(df, aes(x = AchievingGroup, fill = ActivityGroup)) +
  scale_fill_viridis_d(direction = -1) + geom_bar(position = "dodge")
```

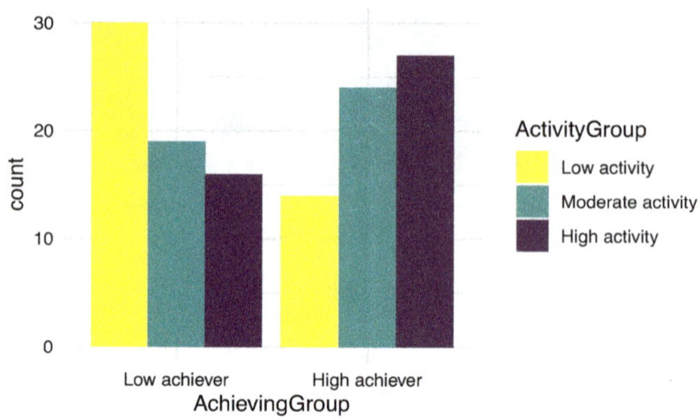

Fig. 22 Basic bar plot of students by achievement group filled by activity level with position dodge instead of stacked

We can now see that the highest group is represented by the low achievers with low activity, followed by the high achievers with high activity.

3.3.2 Histogram

Histograms allow us to represent the distribution of a single continuous variable. It is inherently a bar chart, but instead of each bar representing the count of a single category, it represents the count of a range of values in the x-axis (what is known as a bin). Let us, for example, create a histogram for students' online activity. Specifically, let us see the distribution of the number of accesses to the course main page online.

If we look at our dataset, we can see that the name of the variable that we are interested in is `Frequency.Course_view`:

```
head(df)
```

```
# A tibble: 130 x 37
   User       Name        Surname  Origin     Gender Birthdate Location Employment
   <chr>      <chr>       <chr>    <chr>      <chr>  <chr>     <chr>    <chr>
 1 00a05cc62 Wan          Tan      Malaysia   M      12.12.19~ Remote   None
 2 042b07ba1 Daniel       Tromp    Aruba      M      28.5.1999 Remote   None
 3 046c35846 Sarah        Schmit   Luxembourg F      25.4.1997 On camp~ None
 4 05b604102 Lian         Abdullah Yemen      F      19.11.19~ On camp~ None
 5 0604ff3d3 Nina         Borg     Malta      F      13.6.1994 On camp~ None
 6 077584d71 Mohamed      Gamal    Egypt      M      13.7.1998 On camp~ Part-time
 7 081b100cf Maximilian Gruber     Austria    M      20.12.19~ On camp~ None
 8 0857b3d8e Hugo         Pérez    Spain      M      22.12.19~ On camp~ None
 9 0af619e4b Aylin        Barat    Kazakhstan F      14.8.1995 On camp~ None
10 0ec99ce96 Polina       Novik    Belarus    F      9.10.1996 On camp~ None
# i 120 more rows
# i 29 more variables: Frequency.Applications <dbl>,
#   Frequency.Assignment <dbl>, Frequency.Course_view <dbl>,
#   Frequency.Feedback <dbl>, Frequency.General <dbl>,
#   Frequency.Group_work <dbl>, Frequency.Instructions <dbl>,
#   Frequency.La_types <dbl>, Frequency.Practicals <dbl>,
#   Frequency.Social <dbl>, Frequency.Ethics <dbl>, Frequency.Theory <dbl>, ...
```

To create a histogram for this variable we may use the `geom_histogram` feature of `ggplot2`. We just pass our dataset and map the `Frequency.Course_view` variable to the x axis, and we add the geometry `geom_histogram` (Fig. 23):

```
ggplot(df, mapping = aes(x = Frequency.Total)) + geom_histogram()
```

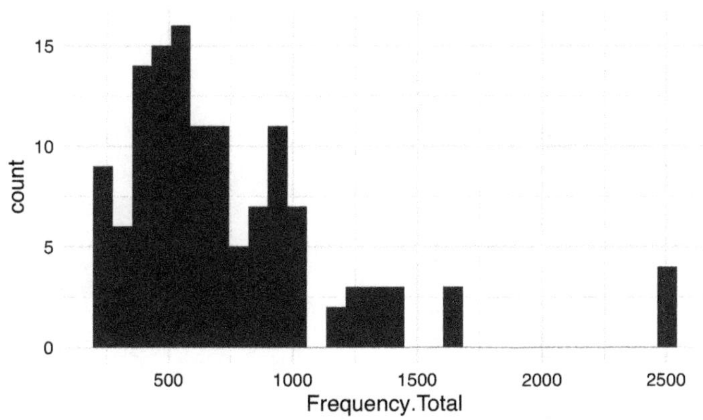

Fig. 23 Histogram of students' course page views

We can provide our own value to the `bins` argument in `geom_histogram` to personalize how many bins we want in our plot (Fig. 24):

```
ggplot(df, mapping = aes(x = Frequency.Total)) +
  geom_histogram(bins = 50)
```

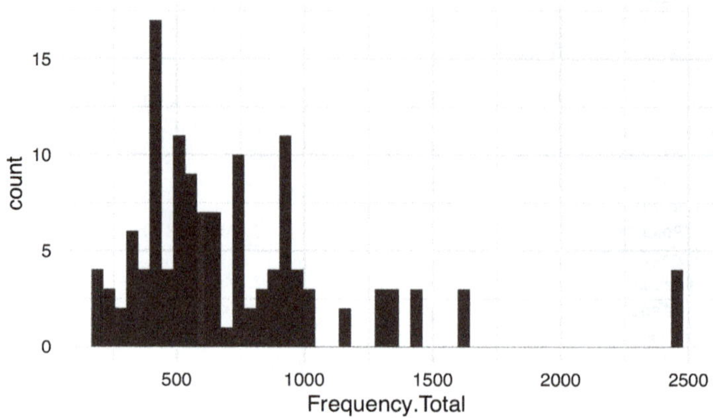

Fig. 24 Histogram of students' course page view with 50 bins

We can also personalize the color scheme using `fill` for the background of the bars (Fig. 25):

```
ggplot(df, mapping = aes(x = Frequency.Total)) +
  geom_histogram(bins = 20,  fill = "deeppink" ) +
  scale_x_continuous(n.breaks = 10)
```

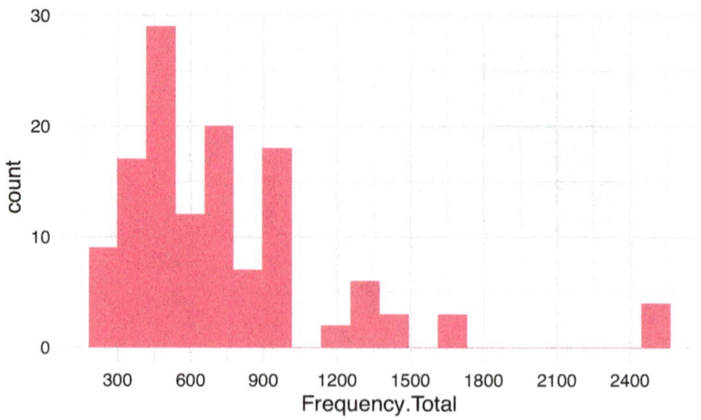

Fig. 25 Histogram of students' course page view with color, fill and linewidth

The histogram allows us to acknowledge that most students had around 400–500 events, with another peak around 900–1000. Students with more than 1000 events were rare.

3.3.3 Line Plot

Another very widely used type of plot is the line plot. Like the histogram, it is also appropriate when we have both a numerical continuous x-axis and y-axis but it gives us a bit more liberty of what we plot and it is suitable for when we want to plot several series of data together. A very common scenario for a line plot is when we deal with timelines and we wish to visualize the evolution of a certain variable over time. Let us, for instance, plot the students' daily events in the LMS throughout the course, a common plot in learning analytics dashboards. In the dataset that we have been using, we have the total count of events per user but not the timestamp of each event. We need to import the original event data from the dataset:

```
ev_url <- "https://github.com/lamethods/data/raw/main/1_moodleLAcourse/Events.xlsx"
events <- import(ev_url)
```

The `Events.xlsx` file contains all the actions that the students enrolled in this course performed in the LMS (`Action`) with their corresponding timestamp (`timecreated`): clicking on a lecture file, viewing the assignment instructions, etc.

```
head(events)
```

```
# A tibble: 95,626 x 7
   Event.context      user  timecreated          Component Event.name Log   Action
   <chr>              <chr> <dttm>               <chr>     <chr>      <chr> <chr>
 1 Assignment: Fina~ 9d74~ 2019-10-26 09:37:12 Assignme~ Course mo~ Assi~ Assig~
 2 Assignment: Fina~ 9148~ 2019-10-26 09:09:34 Assignme~ The statu~ Assi~ Assig~
 3 Assignment: Fina~ 278a~ 2019-10-18 12:05:28 Assignme~ Course mo~ Assi~ Assig~
 4 Assignment: Fina~ 53d6~ 2019-10-19 13:28:37 Assignme~ The statu~ Assi~ Assig~
 5 Assignment: Fina~ aab7~ 2019-10-15 23:38:13 Assignme~ Course mo~ Assi~ Assig~
 6 Assignment: Fina~ 82ed~ 2019-10-18 17:51:43 Assignme~ Course mo~ Assi~ Assig~
 7 Assignment: Fina~ 4178~ 2019-10-18 15:22:56 Assignme~ Course mo~ Assi~ Assig~
 8 Assignment: Fina~ 82ed~ 2019-10-22 13:46:51 Assignme~ The statu~ Assi~ Assig~
 9 Assignment: Fina~ f2e9~ 2019-10-15 14:58:17 Assignme~ Submissio~ Assi~ Assig~
10 Assignment: Fina~ 53d6~ 2019-10-19 13:28:38 Assignme~ Course mo~ Assi~ Assig~
# i 95,616 more rows
```

Instead of mapping `timecreated` directly to the `x` aesthetic, we can plot the timeline of the number of events per day by using `as.Date(timecreated)` and the `geom_line` geometry from `ggplot2`. Notice that, unlike `geom_bar`, if we do not provide a `y` aesthetic and want `ggplot2` to count the number of events per day for us, we need to make it explicit by passing the `stat` argument with value `"count"` to `geom_line` (Fig. 26).

```
ggplot(events, aes(x = as.Date(timecreated) )) + geom_line(stat = "count")
```

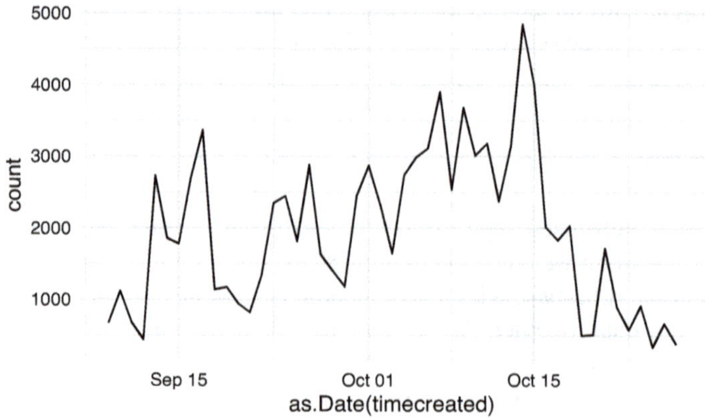

Fig. 26 Line plot of number of events per day

The line plot of students' events allows us to identify periods of increased activity. We can see that it was low at the very beginning of the course, with some peaks corresponding to the assignment deadlines and one last peak for the final project. When the course is over, activity begins to decrease.

To make our plot more aesthetically pleasing, we can customize the color and line width. We do so by tweaking the `color` and `linewidth` properties of the `geom_line`. We can also fix the axes' titles as we learned before (Fig. 27). For example:

```
ggplot(events, aes(x = as.Date(timecreated) )) +
  geom_line(stat = "count", color = "turquoise", linewidth = 2) +
  xlab ("Date") + ylab("Number of events")
```

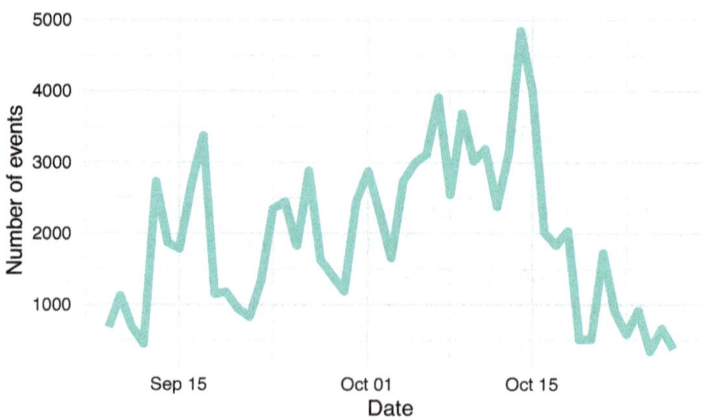

Fig. 27 Line plot of number of events per day with color, linewidth, and custom labels

We can also add a point to mark each date using `geom_point` (Fig. 28):

```
ggplot(events, aes(x = as.Date(timecreated) )) +
  geom_line(stat = "count", color = "turquoise", linewidth = 1.5)  +
  geom_point(stat = "count",  color = "purple", size = 2, stroke = 1)  +
  xlab ("Date")  +
  ylab("Number of events")
```

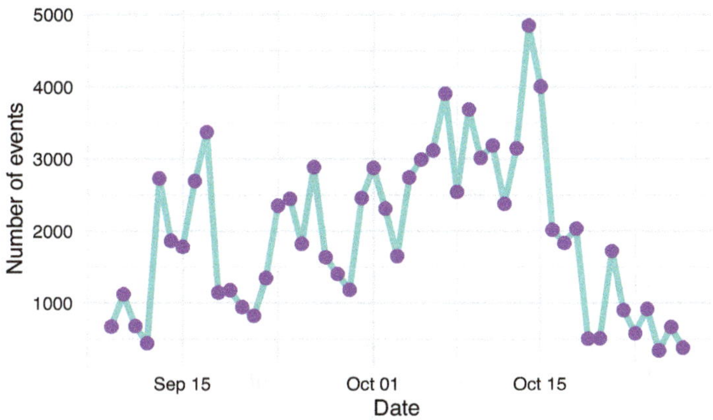

Fig. 28 Line plot of number of events per hour with points every hour

Besides visualizing the events for all the students of the course, we can pinpoint specific students to follow their progress and offer them personalized support. To do this, we would need to filter our data before handing it over to `ggplot2`. We can filter the data using the `filter` function from `dplyr`, as we learned in Chapter 4 [29]. We first install `dplyr` if we do not have it:

```
install.packages("dplyr")
```

Then, we import it as usual:

```
library(dplyr)
```

We can now filter the data and pass it on to `ggplot2` (Fig. 29):

```
events |> filter(user == "9d744e5bf") |> ggplot(aes(x = as.Date(timecreated) )) +
  geom_line(stat = "count", color = "turquoise", linewidth = 2)  +
  geom_point(stat = "count",  color = "purple", size = 2, stroke = 1)  +
  xlab ("Date")  +
  ylab("Number of events")
```

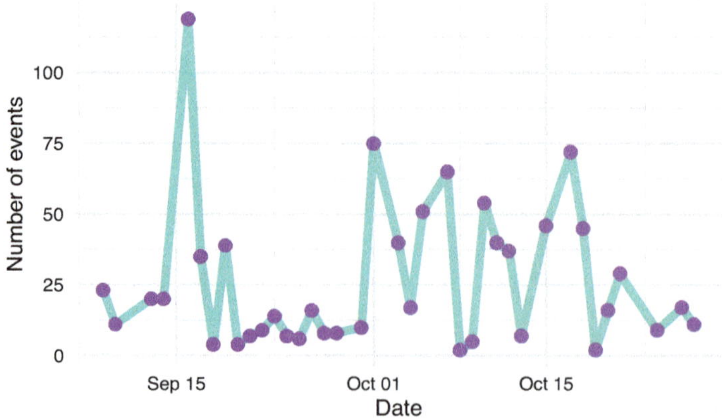

Fig. 29 Line plot of number of events per date for a single student

3.3.4 Jitter Plots

In the previous plots we have seen aggregated information for all the cohort of students as well as information for a single student. However, in some occasions, it is very useful to see the general picture while accounting for possible individual differences. For example, using our original `df` dataset, we can plot the number of events on the LMS, differentiating between high achievers and low achievers.

One option is to use `geom_point` to represent each students' count of events as a single point. To do this, we map the `Event` column to the `x` aesthetic, the `Frequency` column to the `y` aesthetic, and the `User` column to the `group` aesthetic (Fig. 30):

```
ggplot(df, aes(x = AchievingGroup, y = Frequency.Total)) +
  geom_point() +
  xlab("Achieving group") +
  ylab("Number of events") +
  theme(legend.position = "bottom",
        legend.text = element_text(size = 7),
        legend.title = element_blank())
```

However, there are many points that overlap. If we use `geom_jitter` instead, we take advantage of the horizontal gap between the event names to spread the points and avoid the overlap:

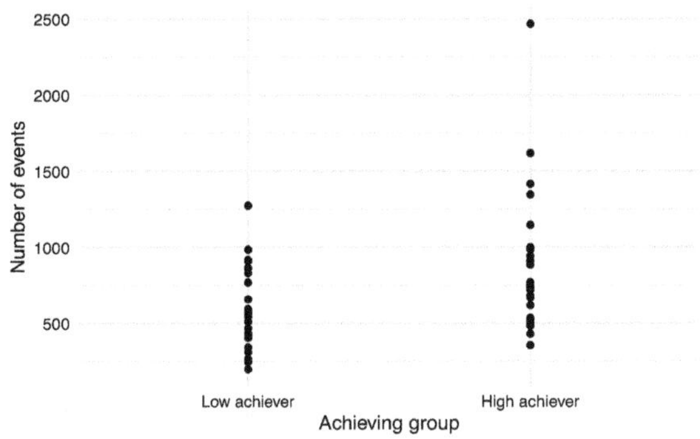

Fig. 30 Jitter plot of number of events per achievement group using `geom_point`

```
ggplot(df, aes(x = AchievingGroup, y = Frequency.Total)) +
  geom_jitter() +
  xlab("Achieving group") +
  ylab("Number of events") +
  theme(legend.position = "bottom",
        legend.text = element_text(size = 7),
        legend.title = element_blank())
```

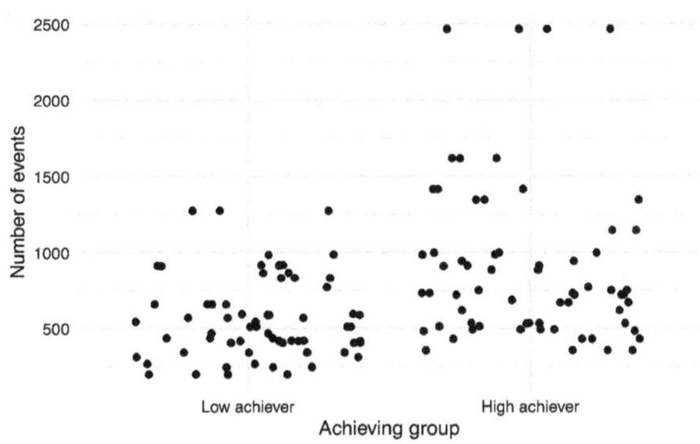

Fig. 31 Jitter plot of number of events per achievement group using `geom_jitter`

The plot shows that students that are high achievers generally have a higher number of events than low achievers (Fig. 31).

3.3.5 Box Plot

When we have too many data points, it is often more useful to visualize summary statistics instead of all the points. Box plots are very useful in summarizing data distributions. We can create a box plot for the number of events per achievement group using `geom_boxplot`:

```
ggplot(df, aes(x = AchievingGroup, y = Frequency.Total)) + geom_boxplot() +
  xlab("Achieving group") + ylab("Number of events")
```

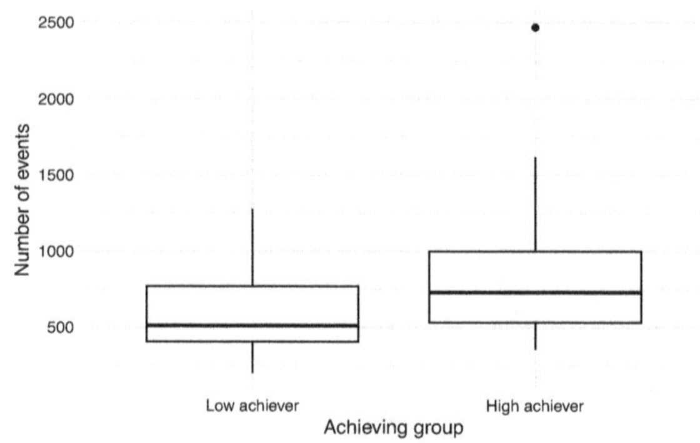

Fig. 32 Box plot of activity per achievement group

The lower hinge of each box indicates the 25% percentile, the thick middle line is the median, and the top hinge is the top 75% percentile. The upper whisker extends from the hinge up to the maximum value within 1.5 * IQR (inter-quantile range), whereas the lower whisker extends to the minimum value within 1.5 * IQR of the hinge. The points outside the whisker represent outliers in the distribution (i.e., values outside of the 1.5 * IQR range). As the jitter plot already hinted, the median number of events is higher in the high achieving group (Fig. 32).

3.3.6 Violin Plot

We can also visualize the distribution of the number of events for each group using violin plots (`geom_violin`), but these are recommended when we have a large amount of data (Fig. 33):

```
ggplot(df, aes(x = AchievingGroup, y = Frequency.Total)) + geom_violin() +
  xlab("Achieving group") + ylab("Number of events")
```

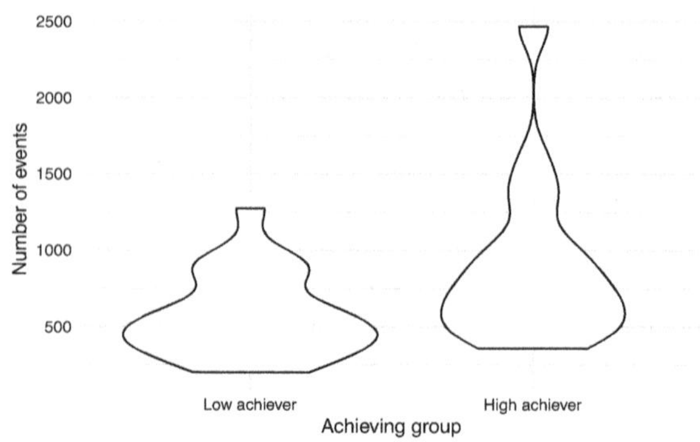

Fig. 33 Violin plot of total activity per achievement group

3.3.7 Scatter Plots

The examples we have seen so far have dealt with plotting a single variable alone or divided in categories. Another common scenario is to investigate the direct relationship between two or more variables. Scatter plots are used to visualize how two numerical variables relate to each other. For example, we can use them to see how LMS activity relates to grades (Fig. 34).

```
ggplot(df, aes(x = Frequency.Total, y = Final_grade)) +
  geom_point() +
  ylab("Final grade") + xlab("Number of events")
```

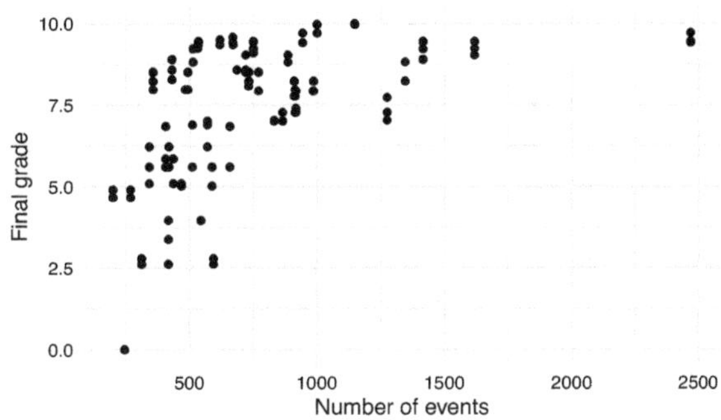

Fig. 34 Scatter plot of number of events vs. final grade

In the plot, each point represents a student. Students at the right side of the plot represent students with higher activity, while students closer to the left side of the plot, represent students with lower activity. At the same time, students with low grades are closer to the bottom of the plot, while students with high grades are closer to the top. Overall, se see an upward trend whereby students with higher activity indeed obtain better grades.

We can add another dimension by coloring points according to another variable. For example, we can color the points according to high vs. low achievers (Fig. 35), so we can now where the division between the two groups is:

```
ggplot(df, aes(x = Frequency.Total, y = Final_grade, color = AchievingGroup)) +
  geom_point() +
  ylab("Final grade") + xlab("Number of events") +
  labs(color = "Achievement")
```

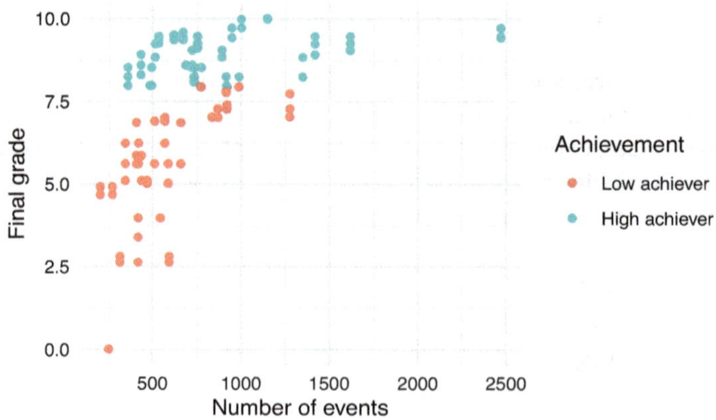

Fig. 35 Scatter plot of number of events vs. final grade colored by achievement group

We can add yet another dimension by mapping the `size` aesthetic to another variable, for example `Frequency.Group_work` which represents the number of events related to group work (Fig. 36).

```
ggplot(df, aes(x = Frequency.Total, y = Final_grade,
               fill = AchievingGroup, size = Frequency.Group_work)) +
  geom_point(color = "black", pch = 21) +
  scale_size_continuous(range = c(1, 7)) +
  ylab("Final grade") + xlab("Number of events") +
  labs(size = "Group work", fill = "Achievement")
```

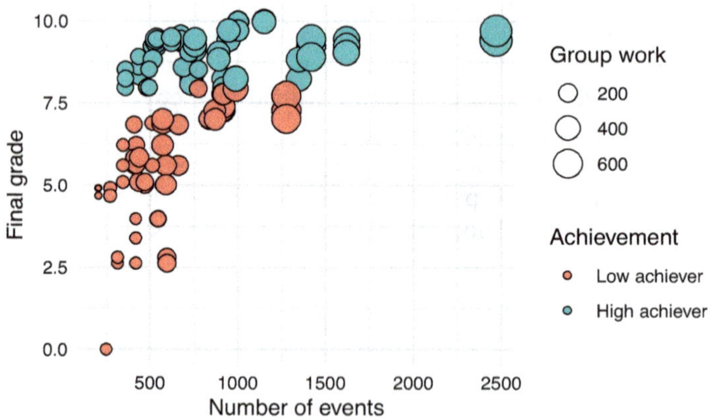

Fig. 36 Scatter plot of number of events vs. final grade colored by achievement group and sized by frequency of group work

3.4 *Advanced Features*

3.4.1 Plot Grids

Sometimes, adding all the information in a single plot can be overwhelming and hard to interpret. For example, take a look at the following line plot that shows the number of events per day for each of the course online components (Fig. 37):

```
ggplot(events, aes(x = as.Date(timecreated), color = Action )) +
  scale_fill_viridis_d() +
  geom_line(stat = "count") +
  xlab("Date") +
  ylab("Number of events")
```

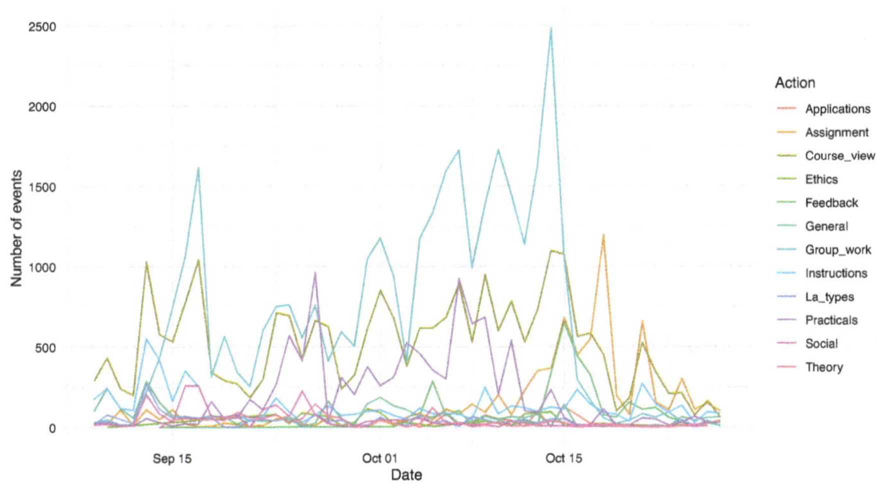

Fig. 37 Multiple series line plot

If we had only a few (2–5) lines, the plot would probably look good, but as the number of categories grow, the plot becomes unintelligible. Instead of showing all the lines together, the plot would be easier to understand if each component had their own plot. To do this, instead of mapping the `Action` column to the `color` aesthetic, we add a new component to our plot using `facet_wrap` and we pass the name of the column as a character string (`"Action"`). We can change the `geom_line` to a `geom_area` to enhance the visualization (Fig. 38).

```
ggplot(events, aes(x = as.Date(timecreated))) +
  geom_area(stat = "count", fill = "turquoise", color = "black") +
  facet_wrap("Action") +
  xlab("Date") +
  ylab("Number of events")
```

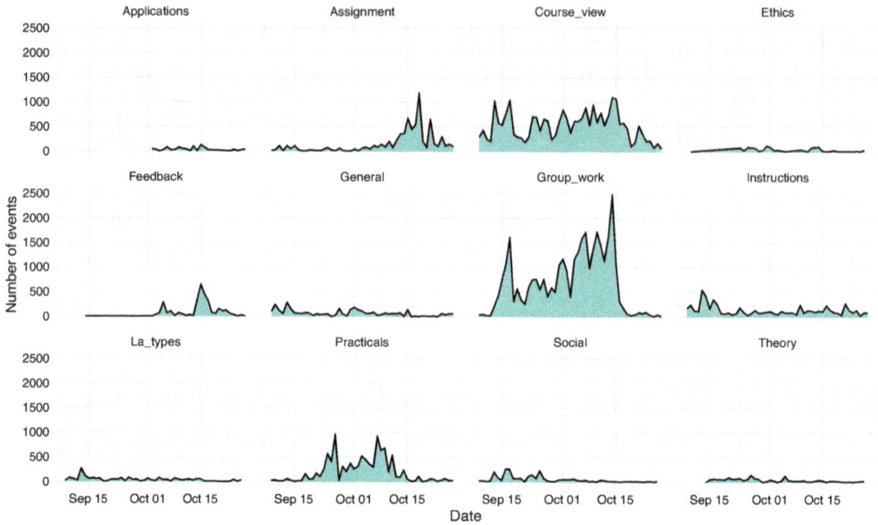

Fig. 38 Grid of multiple plots

3.4.2 Combining Multiple Plots

In the previous example, we saw how to split a plot into multiple plots. But what happens if we want to combine multiple independent plots? For that purpose, we can use the library `patchwork`. Install it first if you do not have it already:

```
install.packages("patchwork")
```

We import the `patchwork` library:

```
library(patchwork)
```

We have to create the plots that we want to combine and assign each of them to a different variable. We can use previous examples from this chapter and assign them to variables named p1, p2, and p3.

```
p1 <- ggplot(df, aes(x = Frequency.Total, y = Final_grade)) +
  geom_point() + ylab("Grade") +
  xlab("Total number of events")

p2 <- ggplot(df, aes(x = AchievingGroup, fill = ActivityGroup )) +
  geom_bar(position = position_fill(reverse = T)) +
  scale_fill_viridis_d(direction = -1) +
  xlab("Achievement group") +
  ylab("Number of events") +
  labs(fill = "Activity level")

p3 <- ggplot(events, aes(x = as.Date(timecreated) )) +
  geom_line(stat = "count", color = "turquoise", linewidth = 1.5)  +
  geom_point(stat = "count",  color = "purple", size = 2, stroke = 1)  +
  xlab ("Date")  +
  ylab("Number of events")
```

Now, if we add the three variables together separated by the + sign, the plots will be placed horizontally next to each other (Fig. 39):

```
p1 + p2 + p3
```

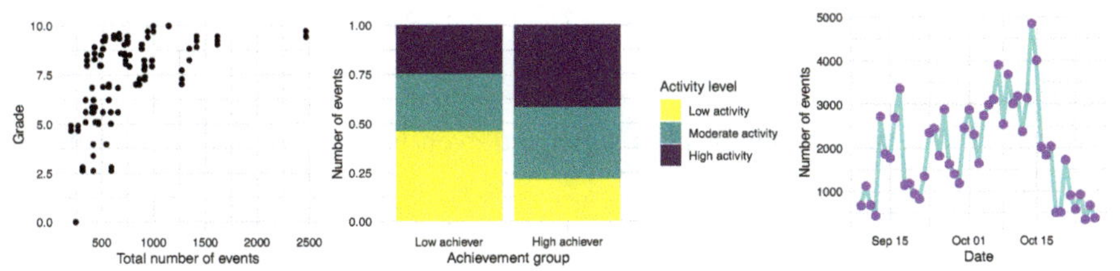

Fig. 39 Multiple plots stacked horizontally

If we use the / character side instead, we lay them out vertically (Fig. 40):

```
p1 / p2 / p3
```

Fig. 40 Multiple plots stacked vertically

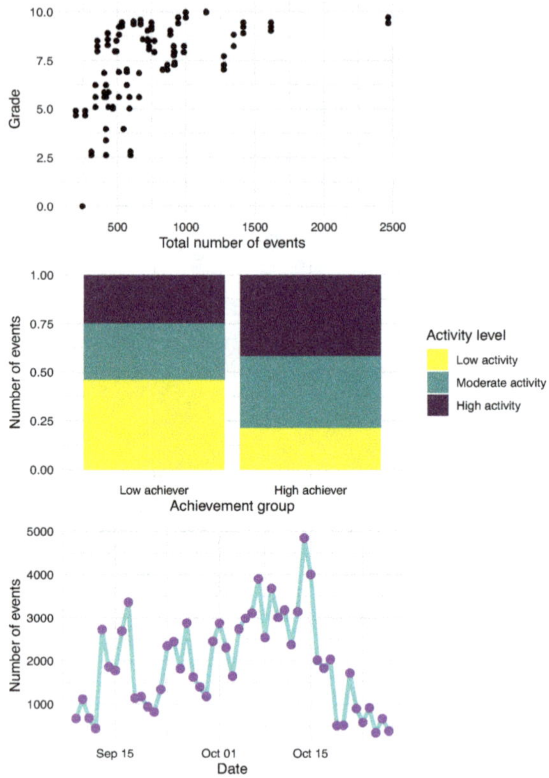

We can use combinations of both signs and even leave blank spaces as follows (Fig. 41):

```
(p1 + p2) / ( p3 + plot_spacer())
```

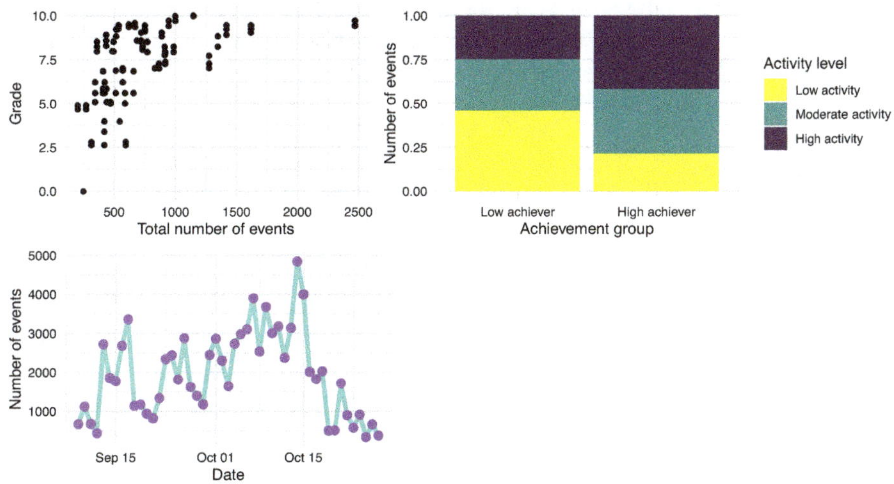

Fig. 41 Multiple plots in a grid

Putting plots side by side can be very useful to compare datasets and discuss the differences. Some publication venues limit the number of figures or pages of their articles, so combining several plots together can be very useful to overcome this limitation.

4 Creating Tables with gt

We have seen earlier in this chapter multiple types of visualizations that are suitable for diverse scenarios in learning analytics. However, we must not forget the other main way of reporting results or metrics, i.e., tables. When we display a data frame in Rstudio, it is by default presented as a table, but we need to be able to extract this table and display it in a dashboard, a report or a scientific article. The library gt can help us with this endeavor. First, install it if you do not have it yet:

```
install.packages("gt")
```

We then import it, as usual:

```
library(gt)
```

Let us create a table, for example, to display the descriptive statistics of students' events in the LMS. Using the events dataset, we first count the number of events of each type (Event.name) per student (user) using group_by and count from dplyr. We then group by Event.name only and use the summarize function, also from dplyr, to create the mean, and standard deviation of the number of events of each type per student, as we learned in Chapter 5 [30].

```
events |>
  group_by(user, Action) |>
  count() |>
  group_by(Action) |>
  summarize(Mean = mean(n), SD = sd(n))
```

```
# A tibble: 12 x 3
   Action        Mean    SD
   <chr>        <dbl> <dbl>
 1 Applications  11.1  9.83
 2 Assignment    56.7  34.1
 3 Course_view  195.  152.
 4 Ethics        11.7  10.7
 5 Feedback      24.7  16.2
 6 General       25.7  21.4
 7 Group_work   252.  163.
 8 Instructions  49.8  40.3
 9 La_types      14.5  7.58
10 Practicals    77.1  33.8
11 Social        18.1  19.0
12 Theory        11.1  6.92
```

Now that we have a data frame with the shape that we like, we can use gt to create the formatted table by simply adding gt to the pipeline of operations (Table 1):

```
events |>
  group_by(user, Action) |>
  count() |>
  group_by(Action) |>
  summarize(Mean = mean(n), SD = sd(n)) |>
  gt()
```

Table 1 Table created with gt

Action	Mean	SD
Applications	11.07143	9.825022
Assignment	56.68462	34.129492
Course_view	194.56154	151.656947
Ethics	11.68182	10.669050
Feedback	24.71429	16.243082
General	25.73846	21.390991
Group_work	251.90769	162.899810
Instructions	49.80000	40.272213
La_types	14.54615	7.583245
Practicals	77.07692	33.751627
Social	18.10744	19.034093
Theory	11.10484	6.922120

We might add some tweaks by forcing the numerical columns to have two decimals and the first column to be aligned left. You can also apply themes to the table using the library `gtExtras` (Table 2).

```
events |>
  group_by(user, Action) |>
  count() |>
  group_by(Action) |>
  summarize(Mean = mean(n), SD = sd(n)) |>
  gt() |>
  fmt_number(decimals = 2, columns = where(is.numeric)) |>
  cols_align(align = "left", columns = 1)
```

Table 2 Table created with gt with formatting

Action	Mean	SD
Applications	11.07	9.83
Assignment	56.68	34.13
Course_view	194.56	151.66
Ethics	11.68	10.67
Feedback	24.71	16.24
General	25.74	21.39
Group_work	251.91	162.90
Instructions	49.80	40.27
La_types	14.55	7.58
Practicals	77.08	33.75
Social	18.11	19.03
Theory	11.10	6.92

5 Discussion

The use of data visualization in the context of learning analytics has the potential to greatly enhance our understanding of student behavior and performance. Using tools such as ggplot2, instructors and researchers can create informative and visually appealing plots that highlight important patterns and trends in student activity, providing insights into factors that may be impacting student success and therefore inform instructional decisions and improve student outcomes.

As we have already seen throughout the chapter, we often use different plots when dealing with categorical variables or numerical variables; when plotting a single variable or two (or more), etc. Moreover, on some occasions when we need very detailed information, a table might be more informative compared to a figure. As a summary for the possible visualizations, Table 3 gathers the most commonly used visualization types that we have seen throughout this chapter according to the number of variables and the data type. It also points to the ggplot2 geometry that is used to create each visualization.

Table 3 Summary of the types of visualization for each data type and number of variables

Number of variables	Variable types	Type of visualization	ggplot2 geometry
One variable	Continuous	Histogram	geom_hist()
	Discrete	Bar chart	geom_bar()
Two or more variables	Both continuous	Scatter plot	geom_point()
	One discrete time and one continuous	Line chart	geom_line()
		Area chart	geom_area()
	One discrete and one continuous	Bar chart	geom_bar()
		Box plot	geom_boxplot()
		Jitter plot	geom_jitter()
		Violin plot	geom_violin()
	Both discrete	Stacked bar chart	geom_bar()

Another way to decide which visualization to use is to think what kind of story we want to tell or which aspect of our data we want to highlight. Figure 42 shows a flowchart that can help choose the most suitable visualization for our data. There are many other decision charts online made for this purpose. For example, "From Data to Viz"[1] leads you to the most appropriate graph for your data and also links to the code to build it and lists common caveats you should avoid.

Throughout the rest of the book, we will see other forms of data visualization that are inherent to specific learning analytics methods. For example, in Chapter 15 [31], we will learn how to represent students' discussions in the form of

[1] Data to Viz https://www.data-to-viz.com/.

Fig. 42 Flowchart to decide the most appropriate visualization for your data

social networks, and in Chapter 10 [32], we will represent students' sequences of activities using sequence analysis. The foundations learned in this chapter are key to understanding more complex visualizations in learning analytics and are, of course, transferable to other fields as well. We encourage readers to expand their knowledge of data visualization by referring to the recommended resources in the next section. Especially readers that would like to take their visualizations to the next step should consider using `shiny`,[2] a web framework for R that allows creating fully interactive web apps for data analyses such as dashboards.

6 Additional Material

- Wilke, Claus. 2019. *Fundamentals of Data Visualization*. O'Reilly. https://clauswilke.com/dataviz/.
- Rahlf, Thomas. 2019 *Data visualisation with R: 111 Examples*. Springer. https://doi.org/10.1007/978-3-030-28444-2.
- Wickham, Hadley, Danielle Navarro, and Thomas Lin Pedersen. 2019. *`ggplot2`: Elegant Graphics for Data Analysis (Use R)* https://ggplot2-book.org/index.html.
- Sahin, Muhittin and Dirk Ifenthaler. 2021. *Visualizations and Dashboards for Learning Analytics*. Springer. https://doi.org/10.1007/978-3-030-81222-5.

[2] Shiny https://mastering-shiny.org/.

- Dougherty, Jack and Ilya Ilyankou. 2021. *Hands-On Data Visualization: Interactive Storytelling from Spreadsheets to Code* https://handsondataviz.org/spreadsheet.html.
- *From Data to Viz*. https://www.data-to-viz.com/about.html
- Wickham, Hadley. 2021. *Mastering shiny*. O'Reilly. https://mastering-shiny.org/.

References

1. Kirk A (2012) Data visualization: a successful design process. Packt Publishing, Birmingham
2. Demmans Epp C, Bull S (2015) Uncertainty representation in visualizations of learning analytics for learners: current approaches and opportunities. IEEE Trans Learn Technol 8:242–260. https://doi.org/10.1109/tlt.2015.2411604
3. Jivet I, Scheffel M, Drachsler H, Specht M (2017) Awareness is not enough: pitfalls of learning analytics dashboards in the educational practice. Springer, Berlin, pp 82–96
4. Park Y, Jo I-H (2019) Factors that affect the success of learning analytics dashboards. Edu Technol Res Develop 67:1547–1571. https://doi.org/10.1007/s11423-019-09693-0
5. Jivet I, Wong J, Scheffel M, Valle Torre M, Specht M, Drachsler H (2021) Quantum of choice: how learners' feedback monitoring decisions, goals and self-regulated learning skills are related. In: LAK21: 11th international learning analytics and knowledge conference. https://doi.org/10.1145/3448139.3448179
6. Sedrakyan G, Malmberg J, Verbert K, Järvelä S, Kirschner PA (2020) Linking learning behavior analytics and learning science concepts: designing a learning analytics dashboard for feedback to support learning regulation. Comput Human Behav 107:105512. https://doi.org/10.1016/j.chb.2018.05.004
7. Martinez-Maldonado R, Echeverria V, Fernandez Nieto G, Buckingham Shum S (2020) From data to insights: a layered storytelling approach for multimodal learning analytics. In: Proceedings of the 2020 CHI conference on human factors in computing systems. https://doi.org/10.1145/3313831.3376148
8. de Freitas S, Gibson D, Alvarez V, Irving L, Star K, Charleer S, Verbert K (2017) How to use gamified dashboards and learning analytics for providing immediate student feedback and performance tracking in higher education. In: Proceedings of the 26th international conference on world wide web companion - WWW'17 companion. https://doi.org/10.1145/3041021.3054175
9. Bodily R, Kay J, Aleven V, Jivet I, Davis D, Xhakaj F, Verbert K (2018) Open learner models and learning analytics dashboards. In: Proceedings of the 8th international conference on learning analytics and knowledge. https://doi.org/10.1145/3170358.3170409
10. Susnjak T, Ramaswami GS, Mathrani A (2022) Learning analytics dashboard: a tool for providing actionable insights to learners. Int J Edu Technol Higher Edu 19:12. https://doi.org/10.1186/s41239-021-00313-7
11. Bodily R, Verbert K (2017) Review of research on student-facing learning analytics dashboards and educational recommender systems. IEEE Trans Learn Technol 10:405–418. https://doi.org/10.1109/tlt.2017.2740172
12. Valle N, Antonenko P, Dawson K, Huggins-Manley AC (2021) Staying on target: a systematic literature review on learner-facing learning analytics dashboards. British J Edu Technol https://doi.org/10.1111/bjet.13089
13. Perez-Alvarez R, Jivet I, Perez-Sanagustin M, Scheffel M, Verbert K (2022) Tools designed to support self-regulated learning in online learning environments: a systematic review. IEEE Trans Learn Technol 15:508–522. https://doi.org/10.1109/tlt.2022.3193271

14. Matcha W, Uzir NA, Gasevic D, Pardo A (2020) A systematic review of empirical studies on learning analytics dashboards: a self-regulated learning perspective. IEEE Trans Learn Technol 13:226–245. https://doi.org/10.1109/tlt.2019.2916802
15. Cheng J, Lei J (2020) A description of students' commenting behaviours in an online blogging activity. E-Learn Digit Media 18:209–225. https://doi.org/10.1177/2042753020954971
16. Duan X, Wang C, Rouamba G (2022) Designing a learning analytics dashboard to provide students with actionable feedback and evaluating its impacts. In: Proceedings of the 14th international conference on computer supported education. https://doi.org/10.5220/0011116400003182
17. van Leeuwen A, Rummel N (2020) Comparing teachers' use of mirroring and advising dashboards. In: Proceedings of the tenth international conference on learning analytics & knowledge. https://doi.org/10.1145/3375462.3375471
18. Isaias P, Backx Noronha Viana A (2020) On the design of a teachers' dashboard: requirements and insights. Springer, Berlin, pp 255–269
19. Verbert K, Govaerts S, Duval E, Santos JL, Van Assche F, Parra G, Klerkx J (2013) Learning dashboards: an overview and future research opportunities. Person Ubiq Comput 18:1499–1514. https://doi.org/10.1007/s00779-013-0751-2
20. Chavan P, Mitra R (2022) Tcherly. J Learn Anal 9:125–151. https://doi.org/10.18608/jla.2022.7555
21. López-Pernas S, Gordillo A, Barra E, Quemada J (2021) Escapp: a web platform for conducting educational escape rooms. IEEE Access 9:38062–38077. https://doi.org/10.1109/access.2021.3063711
22. López Tavares D, Perkins K, Kauzmann M, Aguirre Velez C (2019) Towards a teacher dashboard design for interactive simulations. J Phys Conf Ser 1287:012055. https://doi.org/10.1088/1742-6596/1287/1/012055
23. Li Y, Zhang M, Su Y, Bao H, Xing S (2022) Examining teachers' behavior patterns in and perceptions of using teacher dashboards for facilitating guidance in CSCL. Edu Technol Res Develop 70:1035–1058. https://doi.org/10.1007/s11423-022-10102-2
24. Jivet I, Scheffel M, Specht M, Drachsler H (2018) License to evaluate. In: Proceedings of the 8th international conference on learning analytics and knowledge. https://doi.org/10.1145/3170358.3170421
25. Martinez-Maldonado R, Pardo A, Mirriahi N, Yacef K, Kay J, Clayphan A (2015) The LATUX workflow. In: Proceedings of the fifth international conference on learning analytics and knowledge. https://doi.org/10.1145/2723576.2723583
26. Wickham H (2016) ggplot2: elegant graphics for data analysis. Springer, New York
27. Wilkinson L (1999) The grammar of graphics. Springer, New York
28. López-Pernas S, Saqr M, Conde J, Del-Río-Carazo L (2024) A broad collection of datasets for educational research training and application. In: Saqr M, López-Pernas S (eds) Learning analytics methods and tutorials: a practical guide using R. Springer, Berlin
29. Kopra J, Tikka S, Heinäniemi M, López-Pernas S, Saqr M (2024) Data cleaning and wrangling. In: Saqr M, López-Pernas S (eds) Learning analytics methods and tutorials: a practical guide using R. Springer, Berlin
30. Tikka S, Kopra J, Heinäniemi M, López-Pernas S, Saqr M (2024) Introductory statistics with R for educational researchers. In: Saqr M, López-Pernas S (eds) Learning analytics methods and tutorials: a practical guide using R. Springer, Berlin
31. Saqr M, López-Pernas S, Conde MÁ, Hernández-García Á (2024) Social network analysis: a primer, a guide and a tutorial in R. In: Saqr M, López-Pernas S (eds) Learning analytics methods and tutorials: a practical guide using R. Springer, Berlin
32. Saqr M, López-Pernas S, Helske S, Durand M, Murphy K, Studer M, Ritschard G (2024) Sequence analysis in education: principles, technique, and tutorial with R. In: Saqr M, López-Pernas S (eds) Learning analytics methods and tutorials: a practical guide using R. Springer, Berlin

Part II
Machine Learning

Predictive Modelling in Learning Analytics: A Machine Learning Approach in R

Jelena Jovanovic, Sonsoles López-Pernas, and Mohammed Saqr

1 Introduction

Prediction of students' performance has been a central theme within the field of learning analytics (LA) since the early days [1]. In fact, the initial conceptualization of the field has highlighted the use of digital data collected from learners to predict their success—among other usages. Such predictions hold the promise to help identify those who are at risk of low achievement, in order to proactively offer early support and appropriate intervention strategies based on insights derived from learners' data [1, 2]. Nevertheless, the prediction of students' performance is not unique to LA and was an important theme in related fields even before LA, e.g., academic analytics [3], educational data mining [4], and even far earlier in education research at large [5].

Such widespread, longstanding and continuous centrality of early and accurate prediction of students' performance lends itself to the premise that detection of early signs could allow a timely prevention of, e.g., dropout, low achievement, or undesired outcomes in general [6]. More importantly, identifying the predictors could help inform interventions, explain variations in outcomes and inform educators of why such outcomes happened—a form of predictive modelling that is often referred to as explanatory modelling [7]. Noticeable is the famous example of the *Signal* system at the University of Purdue, where predictions were based on digital data collected from an online learning platform [8]. *Signal* produced predictions and classified students into three categories according to "safety" and presented the students with traffic-light-inspired dashboard signs, where at-risk

J. Jovanovic (✉)
University of Belgrade, Belgrade, Serbia

S. López-Pernas · M. Saqr
School of Computing, University of Eastern Finland, Joensuu, Finland

M. Saqr, S. López-Pernas (eds.), *Learning Analytics Methods and Tutorials*,
https://doi.org/10.1007/978-3-031-54464-4_7

students get a red light. However, the influence of *Signal* on retention rates is unclear and often debated [9]. Several other systems were designed, built and applied in practice, e.g., *OU analyse* at the Open University, where the system offers informative dashboards to students and teachers as well as predictive models to forecast students' performance [10].

Successful prediction of students' performance has been demonstrated repeatedly in several LA studies across the years [11]. In general, the majority of the published studies used features extracted from logged learning trace data (i.e., data about students' interactions with online learning activities and resources) and achieved accurate predictions for a considerable number of students. Yet, most of such studies examined a single course or what is often referred to as a convenience sample (i.e., a course with a sufficiently large and accessible dataset) [11]. Studies that attempted to apply predictive modelling across several courses have not found similar success [12–15]. For instance, Finnegan et al. [15] examined 22 courses across three academic domains using student log-trace data recorded from the learning management system. The authors found considerable differences among predictive models developed for individual courses regarding their predictive power as well as the significance of features. Similar results were reported by Gašević et al. [12] who used data from nine undergraduate courses in different disciplines to examine how instructional variations affected the prediction of academic success. Gašević et al. [12] found that predictors were remarkably different across courses with no consistent pattern that would allow for having one model applicable across all courses. Similarly, Conijn et al. [13] examined 17 courses across several subjects and confirmed the considerable variability of the indicators and predictive models across courses.

Studies within the same domain have also found significant differences in predictors and predictive models. For instance, a recent study [14] examined 50 courses with a similar course design and homogeneous pedagogical underpinning. The authors found variations among different offerings of the same course, that is, the same predictor was statistically significantly correlated with performance in one course offering, but not in the same course offered to similar students in the next year. Furthermore, some predictors were more consistent than others e.g., the frequency of online sessions was more consistent than the frequency of lectures. In a similar vein, Jovanović et al. [16] applied mixed-effect linear modelling to data from fifty combined courses and developed several predictive models with different combinations of features. All predictive models in the work by Jovanović et al. [16] were able to explain only a limited proportion of variations in students' grades. The intraclass correlation coefficient (a measure of source of variability) of all models revealed that the main source of variability were students themselves, that is, students' specific features not captured in the logged data, pointing to the importance of taking students' international conditions into account.

The goal of this chapter is to introduce the reader to predictive LA. The next section is a review of the existing literature, including the main objectives, indicators and algorithms that have been operationalized in previous works. The remainder of the chapter is a step-by-step tutorial of how to perform predictive LA using R. The tutorial describes how to predict student success using students' online trace log data extracted from a learning management system. The reader is guided through all the required steps to perform prediction, including the data preparation and exploration, the selection of the relevant indicators (i.e., feature engineering) and the actual prediction of student success.

2 Predictive Modelling: Objectives, Features, and Algorithms

Extensive research in the LA field has been devoted to the prediction of different measures of student success, as proven by the existence of multiple reviews and meta-analyses on the topic [17–20]. Among the measures of student success that have been examined in the literature are student retention [21], grades [22], and course completion [23]. Predicting lack of success has also been a common target of predictive analytics, mostly in the form of dropout [24], with special interest in the early prediction of at-risk students [25, 26].

To predict student success, numerous indicators from varying data sources have been examined in the literature. Initially, indicators were derived from students' demographic data and/or academic records. Some examples of such indicators are age, gender, and previous grades [27]. More recent research has focused on indicators derived from students' online activity in the learning management system (LMS) [17, 20]. Many of such indicators are derived directly from the raw log data such as the number of total clicks, number of online sessions, number of clicks on the learning materials, number of views of the course main page, number of assignments completed, number of videos watched, number of forum posts [13, 14, 28–31]. Other indicators are related to time devoted to learning, rather than to the mere count of clicks, such as login time, login frequency, active days, time-on-task, average time per online session, late submissions, and periods of inactivity [13, 14, 32–35]. More complex indicators are often derived from the time, frequency, and order of online activities, such as regularity of online activities, e.g., regularity of accessing lecture materials [16, 36, 37], or regularity of active days [14, 16]. Network centrality measures derived from network analysis of interactions in collaborative learning settings were also considered, as they compute how interactions relate to each other and their importance [38]. Research has found that predictive models with generic indicators are only able to explain just a small portion of the overall variability in students' performance [36]. Moreover, it is important to take into account learning design as well as quality and not quantity of learning [17, 20].

The variety of predictive algorithms that have been operationalized in LA research is also worth discussing. Basic algorithms, such as linear and logistic regression, or decision trees, have been used for their explainability, which allows teachers to make informed decisions and interventions related to the students "at risk" [37]. Other machine learning algorithms have also been operationalized such as kNN or random forest [39, 40], although their interpretability is less straightforward. Lastly, the most cutting-edge techniques in the field of machine learning have also made their way to LA, such as XGBoost [41] or Neural Networks [42]. Despite the fact that the accuracy achieved by these complex algorithms is often high, their lack of interpretability is often pointed out as a reason for teachers to avoid making decisions based on their outcomes [7, 43].

It is beyond the scope of this review to offer a comprehensive coverage of the literature. Interested readers are encouraged to read the cited literature and the literature reviews on the topics [11, 14, 17–20]

3 Predicting Students' Course Success Early in the Course

3.1 Prediction Objectives and Methods

The overall objective of this section is to illustrate predictive modelling in LA through a typical LA task of making early-in-the-course predictions of the students' course outcomes based on the logged learning-related data (e.g., making predictions of the learners' course outcomes after log data has been gathered for the first 2–3 weeks). The course outcomes will be examined and predicted in two distinct ways: (1) as success categories (high vs. low achievement), meaning that the prediction task is approached with classification models; (2) as success score (final grades), in which case the development of regression models is required.

To meet the stated objectives, the following overall approach will be applied: create several predictive models, each one with progressively more learning trace data (i.e., logged data about the learners' interactions with course resources and activities), as they become available during the course. In particular, the first model will be built using the learning traces available at the end of the first week of the course; the second model will be built using the data available after the completion of the second week of the course (i.e., the data logged over the first 2 weeks); then, the next one will be built by further accumulating the data, so that we have learning traces for the first 3 weeks, and so on. In all these models, the outcome variable will be the final course outcome (high/low achievement for classification models, that is, the final grade for regression models). We will evaluate all the models on a small set of properly chosen evaluation metrics and examine when (that is, how early in the course) we can make reasonably good predictions of the course outcome. In addition, we will examine which learning-related indicators (i.e., features of the predictive models) had the highest predictive power.

3.2 Context

The context of the predictive modelling presented in this chapter is a postgraduate course on learning analytics (LA), taught at University of Eastern Finland. The course was 6 weeks long, though some assignments were due in the week after the official end of the course. The course covered several LA themes (e.g., Introductory topics, Learning theories, Applications, Ethics), and each theme was covered roughly in 1 week of the course. Each theme had a set of associated learning materials, mostly slides, and reading resources. The course reading resources included seminal articles, book chapters, and training materials for practical work. The course also contained collaborative project work (referred to as group projects). In the group project, students worked together in small groups to design an LA system. The group project was continuous all over the course and was designed to align with the course themes. For instance, when students learned about LA data collection, they were required to discuss the data collection of their own project. The group project has two grades, one for the group project as a whole and another for the individual contribution to the project. It is important to note here that the dataset is based on a synthetic anonymized version of the original dataset and was augmented to three times the size of the original dataset. For more details on the course and the dataset, please refer to the dataset chapter [44] of the book.

3.3 An Overview of the Required Tools (R Packages)

In addition to a set of tidyverse packages that facilitate general purpose data exploration, wrangling, and analysis tasks (e.g., `dplyr`, `tidyr`, `ggplot2`, `lubridate`), in this chapter, we will also need a few additional R packages relevant for the prediction modelling tasks:

- The `caret` (Classification And REgression Training) package [45] offers a wide range of functions that facilitate the overall process of development and evaluation of prediction models. In particular, it includes functions for data pre-processing, feature selection, model tuning through resampling, estimation of feature importance, and the like. Comprehensive documentation of the package, including tutorials, is available online.[1]

- The `randomForest` package [46] provides an implementation of the Random Forest prediction method [47] that can be used both for the classification and regression tasks.
- The `performance` package [48] offers utilities for computing indices of model quality and goodness of fit for a range of regression models. In this chapter, it

[1] https://topepo.github.io/caret/.

will be used for estimating the quality of linear regression models. The package documentation, including usage examples, is available online.[2]

- The **corrplot** package [49] allows for seamless visual exploration of correlation matrices and thus facilitates understanding of connections among variables in high dimensional datasets. A detailed introduction to the functionality the package offers is available online.[3]

3.4 Data Preparation and Exploration

The data that will be used for predictive modelling in this chapter originates from the LMS of a blended course on LA. The dataset is publicly available in a GitHub repository,[4] while its detailed description is given in the book's chapter on datasets [44]. In particular, we will make use of learning trace data (stored in the `Events.xlsx` file) and data about the students' final grades (available in the `Results.xlsx` file).

We will start by familiarising ourselves with the data through exploratory data analysis.

```
library(tidyverse)
library(lubridate)
library(rio)
```

After loading the required packages, we will load the data from the two aforementioned data files:

```
events = import(
  "https://github.com/lamethods/data/raw/main/1_moodleLAcourse/Events.xlsx")
results = import(
  "https://github.com/lamethods/data/raw/main/1_moodleLAcourse/Results.xlsx")
```

We will start by exploring the events data, and looking first into its structure:

```
glimpse(events)
```

```
Rows: 95,626
Columns: 7
$ Event.context <chr> "Assignment: Final Project", "Assignment: Final Project"~
$ user          <chr> "9d744e5bf", "91489f7a9", "278a75edf", "53d6ab60c", "aab~
```

[2] https://easystats.github.io/performance/.

[3] https://cran.r-project.org/web/packages/corrplot/vignettes/corrplot-intro.html.

[4] https://github.com/sonsoleslp/labook-data/tree/main/1_moodleLAcourse.

```
$ timecreated    <dttm> 2019-10-26 09:37:12, 2019-10-26 09:09:34, 2019-10-18 12~
$ Component      <chr> "Assignment", "Assignment", "Assignment", "Assignment", ~
$ Event.name     <chr> "Course module viewed", "The status of the submission ha~
$ Log            <chr> "Assignment: Final Project", "Assignment: Final Project"~
$ Action         <chr> "Assignment", "Assignment", "Assignment", "Assignment", ~
```

Since we intend to build separate predictive models for each week of the course, we need to be able to organise the events data into weeks. Therefore, we will extend the events data frame with additional variables that allow for examining temporal aspects of the course events from the weekly perspective. To that end, we first order the events data based on the events' timestamp (`timecreated`) and then add three auxiliary variables for creating the course_week variable: weekday of the current event (`wday`), weekday of the previous event (`prev_wday`), and indicator variable for the start of a new week (`new_week`). The assumption applied here is that each course week starts on Monday and the beginning of a new week (`new_week`) can be identified by the current event being on Monday (`wday=="Mon"`) while the previous one was on any day other than Monday (`prev_wday!="Mon"`) :

```
events |>
  arrange(timecreated) |>
  mutate(wday = wday(timecreated,
                     label = TRUE,
                     abbr = TRUE,
                     week_start = 1)) |>
  mutate(prev_wday = lag(wday)) |>
  mutate(new_week = ifelse((wday == "Mon") & (is.na(prev_wday) | prev_wday != "Mon"),
                           yes = TRUE, no = FALSE)) |>
  mutate(course_week = cumsum(new_week)) -> events
```

Having created the variable that denotes the week of the course (`course_week`), we can remove the three auxiliary variables, to keep our data frame tidy:

```
events |> select(-c(wday, prev_wday, new_week)) -> events
```

We can now explore the distribution of the events across the course weeks. The following code will give us the count and proportion of events per week (with proportions rounded to the fourth decimal):

```
events |>
  count(course_week) |>
  mutate(prop = round(n/nrow(events), 4))
```

The output of the above lines show that we have data for 7 weeks: 6 weeks of the course plus one more week, right after the course officially ended but students were still able to submit assignments. We can also observe that the level of students' interaction with course activities steadily increased up until week 5 and then started going down.

Let us now move to examining the factor variables that represent different types of actions and logged events. First, we can check how many distinct values each of these variables has:

```
events |>
  summarise(across(c(Event.context, Component:Action), n_distinct))
```

```
  Event.context Component Event.name Log Action
1            80        13         27  80     12
```

We can also examine unique values of each of the four variables, but it is better to examine them together, that will help us better understand how they relate to one another and get a better idea of the semantics of events they denote. For example, we can examine how often distinct *Component*, *Event*, and *Action* values co-occur (Table 1):

```
events |>
  count(Component,Event.name, Action) |>
  arrange(Component, desc(n))
```

Likewise, we can explore how Action and Log values are related (i.e., co-occur) (Table 2):

Table 1 Count of all combinations of Component, Event, and Action

	Component	Event.name	Action	n
1	Assignment	Course module viewed	Practicals	3463
2	Assignment	Course module viewed	Assignment	2926
3	Assignment	The status of the submission has been viewed	Practicals	2698
4	Assignment	The status of the submission has been viewed	Assignment	2427
5	Assignment	Course module viewed	Group_work	676
6...102				
103	Zoom meeting	Clicked join meeting button	Group_work	25

Table 2 Count of all combinations of Log and Action

	Action	Log	n
1	Applications	File: Case studies	141
2	Applications	File: Features students really expect from learning analytics	153
3	Applications	File: OU Analyse: Analysing at-risk students at The Open University	161
4	Applications	File: Whitelock-Wainwright et al-2019-Journal of Computer Assisted Learning	42
5	Applications	URL: E2Coach	90
6..80			
81	Theory	URL: Theory video lecture	207

```
events |>
  count(Action, Log) |>
  arrange(Action)
```

Having explored the four categorical variables that capture information about the students' interactions with course resources and activities, we will select the Action variable as the most suitable one for further analysis. The reason for choosing the Action variable is twofold: (1) it is not overly granular (it has 12 distinct values), and thus allows for the detection of patterns in the learning trace data; (2) it captures sufficient information about the distinct kinds of interaction the events refer to. In fact, the Action variable was manually coded by the course instructor to offer a more nuanced way of analysis. The coding was performed to group actions that essentially indicate the same activities under the same label. For instance, logs of viewing feedback from the teacher were grouped under the label *feedback*. Practical activities (Social network analysis or Process mining) were grouped under the label *practicals*. In the same way, accessing the group work forums designed for collaboration, browsing, reading others' comments, or writing were all grouped under the label *group_work* [50].

We will rename some of the `Action` values to make it clear that they refer to distinct topics of the course materials:

```
topical_action <- c("General", "Applications", "Theory",  "Ethics", "Feedback", "La_types")

events |>
  mutate(action = ifelse(test = Action %in% topical_action,
                         yes = str_glue("Materials_{Action}"),
                         no = Action),
         .keep = "unused") -> events
```

Let us now visually examine the distribution of events across different action types and course weeks:

```
# Compute event counts across action types and course weeks
events |>
  count(course_week, action) |>
  arrange(course_week, desc(n)) -> action_dist_across_weeks

# Visualise the event distribution
action_dist_across_weeks |>
  mutate(Action = as.factor(action)) |>
  ggplot(aes(x = course_week, y = n, fill = action)) +
  geom_col(position = position_fill()) +
  scale_fill_brewer(palette = 'Paired') +
  scale_x_continuous(breaks = seq(1,7)) +
  labs(x = "\nCourse week", y = "Proportion\n", fill ="Action") +
  theme_minimal()
```

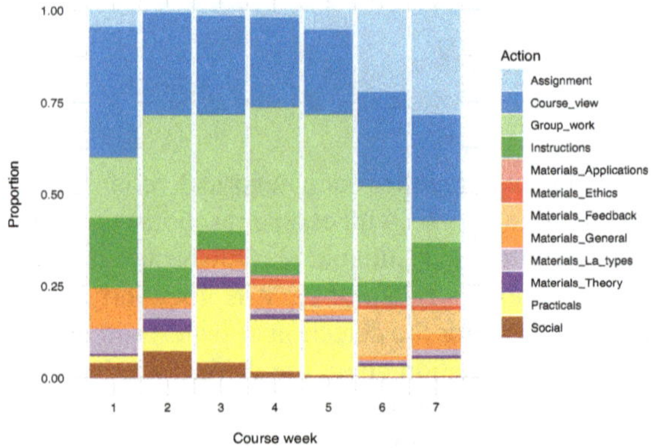

Fig. 1 Distribution of action types across the course weeks

From the plot produced by the above lines of code (Fig. 1), we can observe, for example, that group work (`Group_work`) was the most represented type of actions from week 2 till the end of the course (week 6). It is followed by browsing the main page of the course containing the course materials, announcements and updates (`Course_view`) and working on practical tasks (`Practicals`). We can also note that the assignment-related actions (`Assignment`) are present mostly towards the end of the course.

Now that we have familiarised ourselves with the events data and done some initial data preparation steps, we should do some final 'polishing' of the data and store it to have it ready for further analysis.

```
# Keep only the variables to be used for further analysis and
# Rename some of the remaining ones to keep naming consistent
events |>
  select(user, timecreated, course_week, action) |>
  rename(week = course_week, ts = timecreated) -> events

# Save the prepared data in the R native format
dir.create("preprocessed_data")
saveRDS(events, "preprocessed_data/events.RDS")
```

The next step is to explore the grades data that we previously loaded into the results data frame

```
glimpse(results)
```

```
Rows: 130
Columns: 15
$ user          <chr> "6eba3ff82", "05b604102", "111422ee7", "b4658c3a9",~
$ Grade.SNA_1   <dbl> 0, 8, 10, 5, 10, 7, 9, 10, 10, 10, 7, 10, 9, 9, 9, ~
$ Grade.SNA_2   <dbl> 0, 10, 10, 5, 10, 10, 9, 10, 10, 10, 8, 10, 10, 10,~
$ Grade.Review  <dbl> 6.67, 6.67, 10.00, 0.00, 10.00, 9.67, 6.67, 7.00, 1~
```

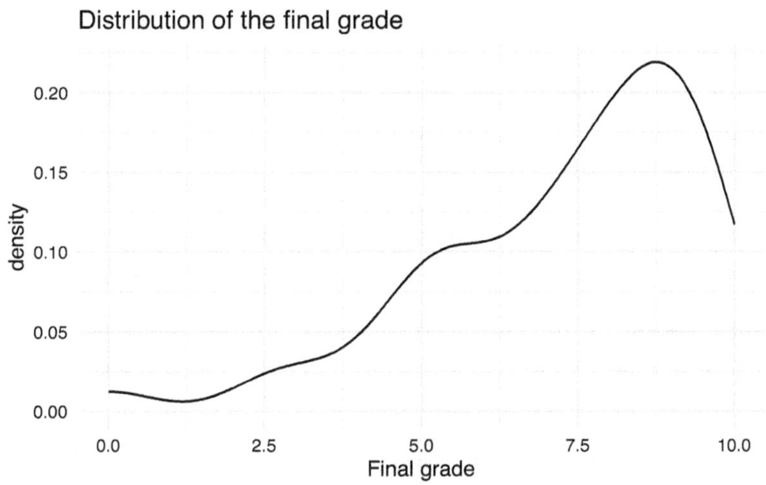

Fig. 2 Distribution of the final course grade

```
$ Grade.Group_self    <dbl> 5, 1, 10, 1, 10, 6, 10, 9, 10, 10, 6, 10, 10, 10, 9~
$ Grade.Group_All     <dbl> 4.00, 3.00, 9.11, 4.00, 9.18, 4.00, 8.56, 8.56, 9.2~
$ Grade.Excercises    <dbl> 10.00, 10.00, 10.00, 10.00, 10.00, 3.33, 10.00, 10.~
$ Grade.Project       <dbl> 0.00, 7.00, 9.33, 6.00, 5.33, 7.67, 0.00, 9.33, 10.~
$ Grade.Literature    <dbl> 6.67, 6.67, 10.00, 4.33, 10.00, 9.67, 5.00, 6.67, 1~
$ Grade.Data          <dbl> 4, 3, 5, 3, 5, 5, 1, 4, 5, 4, 5, 4, 4, 5, 3, 4, 5, ~
$ Grade.Introduction  <dbl> 6, 6, 10, 4, 10, 10, 4, 8, 10, 8, 10, 10, 6, 8, 6, ~
$ Grade.Theory        <dbl> 2, 2, 10, 2, 10, 10, 2, 8, 6, 2, 8, 2, 2, 8, 8, 6, ~
$ Grade.Ethics        <dbl> 2, 8, 10, 2, 10, 10, 4, 6, 10, 6, 10, 10, 6, 8, 4, ~
$ Grade.Critique      <dbl> 4, 4, 10, 6, 10, 10, 2, 2, 10, 6, 10, 10, 6, 6, 6, ~
$ Final_grade         <dbl> 2.626970, 4.670169, 9.244600, 0.000000, 8.238179, 5~
```

Even though the results dataset includes the students' grades on individual assignments, we will be able to use just the final grade (`Final_grade`) since we do not have information when during the course the individual assignment grades became available.

To get an overall understanding of the final grade distribution, we will compute the summary statistics and plot the density function for the `Final_grade` variable:

```
summary(results$Final_grade)
```

```
  Min. 1st Qu.  Median    Mean 3rd Qu.    Max.
 0.000   5.666   7.954   7.254   9.006  10.000
```

```
ggplot(results, aes(x = Final_grade)) +
  geom_density() +
  labs(x = "Final grade",
       title = "Distribution of the final grade") +
  theme_minimal()
```

We can clearly notice both in the summary statistics and the distribution plot (Fig. 2) that the final grade is not normally distributed, but skewed towards higher grade values.

As noted in Sect. 3.1, we will build two kinds of prediction models: models that predict the final grade (regression models) as well as models that predict whether a student belongs to the group of high or low achievers (classification models). For the latter group of models, we need to create a binary variable (e.g., Course_outcome) indicating if a student is in the high or low achievement group. Students whose final grade is above the 50th percentile (i.e., above the median) will be considered as being high achievers in this course (High), the rest will be considered as having low course achievement (Low):

```
results |>
  mutate(Course_outcome = ifelse(test = Final_grade > median(Final_grade),
                                 yes = "High", no = "Low")) |>
  mutate(Course_outcome = factor(Course_outcome)) -> results
```

Now that we have prepared the outcome variables both for regression and classification models (`Final_grade` and `Course_outcome`, respectively), we can save them for later use in model building:

```
results |>
  select(user, Final_grade, Course_outcome) |>
  saveRDS("preprocessed_data/final_grades.RDS")
```

3.5 Feature Engineering

After the data has been preprocessed, we can focus on feature engineering, that is, the creation of new variables (features) to be used for model development. This step needs to be informed by the course design and any learning theory that underpins the course design, so that the features we create and use for predictive modelling are able to capture relevant aspects of the learning process in the given learning settings. In addition, we should consult the literature on predictive modelling in LA (see Sect. 2), to inform ourselves about the kinds of features that were good predictors in similar learning settings. Following such an approach, we have identified the following event-based features as potentially relevant:

A. Features based on learning action counts
    1. Total number of each type of learning actions
    2. Average number of actions (of any type) per day
    3. Entropy of action counts per day
B. Features based on learning sessions:
    1. Total number of learning sessions
    2. Average (median) session length (time)
    3. Entropy of session length
C. Features based on number of active days (= days with at least one learning session)
    1. Number of active days
    2. Average time distance between two consecutive active days

In addition to the course specific features (A1), the feature set includes several course-design agnostic (i.e., not directly related to a specific course design) features (e.g., A2 and A3) that proved as good predictors in similar (blended) learning settings [14, 16, 36, 51]. Furthermore, the chosen features allow for capturing both the amount of engagement with the course activities (features A1, A2, B1, B2, C1) and regularity of engagement (features A3, B3, C2) at different levels of granularity (actions, sessions, days).

To compute features based on action counts per day (group A), we need to extend the `events` dataset with date as an auxiliary variable:

```
events |> mutate(date = as.Date(ts)) -> events
```

To compute features based on learning sessions, we need to add sessions to the events data. It is often the case that learning management systems and other digital learning platforms do not explicitly log beginning and end of learning sessions. Hence, LA researchers have used heuristics to detect learning sessions in learning events data. An often used approach to session detection consists of identifying overly long periods of time between two consecutive learning actions (of the same student) and considering them as the end of one session and beginning of the next one [14, 16, 36]. To determine such overly long time periods that could be used as "session delimiters", LA researchers would examine the distribution of time periods between consecutive events in a time-ordered dataset, and set the delimiter to the value corresponding to a high percentile (e.g., 85th or 90th percentile) of the time distance distribution. We will rely on this approach to add sessions to the event data.

First, we need to compute time distance between any two consecutive actions of each student:

```
events |>
  group_by(user) |>
  arrange(ts) |>
  mutate(ts_diff = ts - lag(ts)) |>
  ungroup() -> events
```

Next, we should examine the distribution of time differences between any two consecutive actions of each student, to set up a threshold for splitting action sequences into sessions:

```
events |> pull(ts_diff) -> ts_diff

ts_diff_hours = as.numeric(ts_diff, units = 'hours')

summary(ts_diff_hours)
```

```
    Min.   1st Qu.    Median      Mean   3rd Qu.      Max.      NA's
 0.00000   0.00028   0.00028   1.40238   0.01694 307.05028       130
```

As summary statistics is not sufficiently informative, we should examine the upper percentiles:

```
quantile(ts_diff_hours, probs = seq(0.8, 1, 0.01), na.rm = TRUE) |> round(3)
```

```
    80%     81%     82%     83%     84%     85%     86%     87%     88%     89%
  0.017   0.017   0.034   0.034   0.034   0.050   0.067   0.084   0.117   0.167
    90%     91%     92%     93%     94%     95%     96%     97%     98%     99%
  0.234   0.350   0.617   1.000   1.834   3.367   7.434  14.300  22.035  39.767
   100%
307.050
```

Considering the computed percentile values, on one hand, and the expected length of learning activities in the online part of the course (which included long forums discussions), we will set 1.5 hours (value between 93th and 94th percentile) as the threshold for splitting event sequences into learning sessions:

```
events |>
  mutate(ts_diff_hours = as.numeric(ts_diff, units = 'hours')) |>
  group_by(user) |>
  arrange(ts) |>
  mutate(new_session = (is.na(ts_diff_hours)) | (ts_diff_hours >= 1.5)) |>
  mutate(session_nr = cumsum(new_session))|>
  mutate(session_id = paste0(user,"_", "session_",session_nr)) |>
  ungroup() -> events_with_sessions
```

We will also add session length variable, which can be computed as the difference between the first and last action in each session, as it will be required for the computation of some of the features (B2 and B3):

```
events_with_sessions |>
  group_by(session_id) |>
  mutate(session_len = as.numeric(max(ts) - min(ts), units = "secs")) |>
  ungroup() -> events_with_sessions
```

After adding the necessary variables for feature computation, we can tidy up the dataset before proceeding to the feature computation. In particular, we will keep only the variables required for feature computation:

```
events_with_sessions <- events_with_sessions |>
  select(user, ts, date, week, action, session_nr, session_id, session_len)
```

All functions for computing the event-based features outlined above are given in the *feature_creation* R script. In the following, we give a quick overview of those functions, while the script with further explanations is available at the book's GitHub repository:

- the `total_counts_per_action_type` function computes the set of features labelled as A1, that is, counts of each type of learning action, up to the current week
- the `avg_action_cnt_per_day` function computed feature A2, that is, average (median) number of learning actions per day, up to the current week
- the `daily_cnt_entropy` function computes feature A3, namely entropy of action counts per day, up to the current week
- the `session_based_features` function computes all session-based features up to the current week: total number of sessions (B1), average (median) session length (B2), and entropy of session length (B3)
- the `active_days_count` function computes the number of active days (C1), up to the current week
- the `active_days_avg_time_dist` function computes avg. (median) time distance between two consecutive active days (C2), up to the current week
- finally, the `create_event_based_features` function makes use of the above functions to compute all event-based features, up to the given week

Having defined the feature set and functions for feature computation, cumulatively for each week of the course, we can proceed to the development of predictive models. In the next section (Sect. 3.6), we will present the creation and evaluation of models for predicting the overall course success (high/low), whereas Sect. 3.7 will present models for predicting the final course grade.

3.6 Predicting Success Category

To build predictive (classification) models, we will use Random Forest [47]. This decision was motivated by the general high performance of this algorithm on a variety of prediction tasks [52] as well as its high performance on prediction tasks specific to the educational domain [43].

Random forest (RF) is a very efficient machine learning method that can be used both for classification and regression tasks. It belongs to the group of ensemble methods, that is, machine learning methods that build and combine multiple individual models to do the prediction. In particular, RF builds and combines the output of several decision or regression trees, depending on the task at hand (classification or regression). The way it works can be briefly explained as follows: the method starts by creating a number of bootstrapped training samples to be used for building a number of decision trees (e.g. 100). When building each tree, each time a split in a tree is to be made, instead of considering all predictors, a random sample of predictors is chosen as split candidates from the full set of predictors (typically the size of the sample is set to be equal to the square root of the number of predictors). The reason for choosing a random sample of predictors is to make a diverse set of trees, which has proven to increase the performance. After all the trees have been built, each one is used to generate a prediction, and those predictions are then aggregated into the overall prediction of the RF model. In case of a classification task, the aggregation of predictions is done through majority vote, that is, the class voted (i.e., predicted) by the majority of the classification trees is the final prediction. In case of a regression task, the aggregation is done by averaging predictions of individual trees. For a thorough explanation of the RF method (with examples in R), an interested reader is referred to Chapter 8 of [53].

We will first load the additional required R packages as well as R scripts with functions for feature computation and model building and evaluation:

```
library(caret)
library(randomForest)

source("feature_creation.R")
source("model_develop_and_eval.R")
```

The following code snippet shows the overall process of model building and evaluation, one model for each week of the course, starting from week 1 to week 5. Note that week 5 is set as the last week for prediction purposes since it is the last point during the course when some pedagogical intervention, informed by the model's output, can be applied by the course instructors.

```
models <- list()
eval_measures <- list()

for(k in 1:5) {

  ds <- create_dataset_for_course_success_prediction(events_with_sessions,
                                                      k, results)
  set.seed(2023)
  train_indices <- createDataPartition(ds$Course_outcome,
```

```
                                    p = 0.8, list = FALSE)
    train_ds <- ds[train_indices,] |> select(-user)
    test_ds <- ds[-train_indices,] |> select(-user)

    rf <- build_RF_classification_model(train_ds)
    eval_rf <- get_classification_evaluation_measures(rf, test_ds)

    models[[k]] <- rf
    eval_measures[[k]] <- eval_rf
  }
```

The process consists of the following steps, each of which will be explained in more detail below:

(1) Creation of a dataset for prediction of the course outcomes, based on the logged events data (`events_with_sessions`) up to the given week (`k`) and the available course outcomes data (`results`)
(2) Splitting of the dataset intro the part for training the model (`train_ds`) and evaluating the model's performance (`test_ds`)
(3) Building a RF model based on the training portion of the dataset
(4) Evaluating the model based on the test portion of the dataset

All built models and their evaluation measures are stored (in `models` and `eval_measures` lists) so that they can later be compared.

Going now into details of each step, we start with the creation of a dataset to be used for predictive modelling in week *k*. This is done by first computing all features based on the logged events data (`events_data`) up to the week *k*, and then adding the course outcome variable (`Course_outcome`) from the dataset with course results (`grades`):

```
create_dataset_for_course_success_prediction <- function(events_data,
                                                         current_week,
                                                         grades) {
  features <- create_event_based_features(events_data, current_week)
  grades |>
    select(user, Course_outcome) |>
    inner_join(features, by = "user")
}
```

Next, to be able to properly evaluate the performance of the built model, we need to test its performance on a dataset that the model "has not seen". This requires the splitting of the overall feature set into two parts: one for training the model (training set) and the other for testing its performance (test set). This is done in a way that a larger portion of the dataset (typically 70–80%) is used for training the model, whereas the rest is used for testing. In our case, we use 80% of the feature set for training (`train_ds`) and 20% for evaluation purposes (`test_ds`). Since observations (in this case, students) are randomly selected for the training and test sets, to assure that we can replicate the obtained results, we initiate the random process with an (arbitrary) value (`set.seed`).

In the next step, we use the training portion of the dataset to build a RF model, as shown in the code snippet below. We train a model by tuning its `mtry` hyper-parameter and choose the model with optimal `mtry` value based on the Area under the ROC curve (AUC ROC) metric. The `mtry` hyper-parameter defines the number of features that are randomly chosen at each step of tree branching, and thus controls how much variability will be present among the trees that RF will build. It is one of the key hyper-parameters for tuning RF models and its default value (`default_mtry`) is equal to the square root of the number of features (`n_features`). Hence, we create a grid that includes the default value and a few values around it.

```
build_RF_classification_model <- function(dataset) {

  #defining the model hyperparameter (mtry) that we want to tune
  n_features <- ncol(dataset)-1
  default_mtry <- round(sqrt(n_features))
  grid <- expand.grid(.mtry = (default_mtry-1):(default_mtry+1))

  #setting that we want to train the model through 10-fold cross-validation
  ctrl <- trainControl(method = "CV",
                       number = 10,
                       classProbs = TRUE,
                       summaryFunction = twoClassSummary)

  # initiating the training process and setting the evaluation measure
  # (ROC) for choosing the best value of the tuned hyperparameter
  rf <- train(x = dataset |> select(-Course_outcome),
              y = dataset$Course_outcome,
              method = "rf",
              metric = "ROC",
              tuneGrid = grid,
              trControl = ctrl)

  rf$finalModel
}
```

The parameter tuning is done through 10-fold cross-validation (CV). K-fold CV is a widely used method for tuning parameters of machine learning models. It is an iterative process, consisting of *k* iterations, where the training dataset is randomly split into *k* folds of equal size, and in each iteration, k-1 folds are used for training the model whereas the k-th fold is used for evaluating the model on the chosen performance measure (e.g., ROC AUC, as in our case). In particular, in each iteration, a different fold is used for evaluation purposes, whereas the remaining k-1 folds are used for training. When this iterative process is finished, the models' performance, computed in each iteration, are averaged, thus giving a more stable estimate of the performance for a particular value of the parameter being tuned. CV is often done in 10 iterations, hence the name 10-fold CV.

The final step is to evaluate each model based on the test data. To that end, we compute four standard evaluation metrics for classification models—Accuracy, Precision, Recall, and F1—as shown in the code snippet below. These four metrics are based on the so-called confusion matrix, which is, in fact, a cross-tabulation of the actual and predicted counts for each value of the outcome variable (i.e., class).

```
get_classification_evaluation_measures <- function(model, test_data) {

  # use the model to make predictions on the test set
  predicted_vals <- predict(model,
                            test_data |> select(-Course_outcome))
  actual_vals <- test_data$Course_outcome

  # create the confusion matrix (see Fig. 3)
  cm <- table(actual_vals, predicted_vals)

  TP <- cm[2,2]
  TN <- cm[2,2]
  FP <- cm[1,2]
  FN <- cm[2,1]

  # compute evaluation measures based on the confusion matrix
  accuracy = sum(diag(cm)) / sum(cm)
  precision <- TP / (TP + FP)
  recall <- TP / (TP + FN)
  F1 <- (2 * precision * recall) / (precision + recall)

  c(Accuracy = accuracy,
    Precision = precision,
    Recall = recall,
    F1 = F1)
}
```

In our case, the confusion matrix has the structure as shown on Fig. 3. In rows, it has the counts of the *actual* number of students in the high and low achievement groups, whereas the columns give the *predicted* number of high and low achievers. We consider low course achievement as the positive class, since we are primarily interested in spotting those students who might benefit from a

Fig. 3 Confusion matrix for the prediction of the students' overall course success

		Predicted values	
		High	Low
Actual values	High	**TN**	**FP**
	Low	**FN**	**TP**

pedagogical intervention (to prevent a poor course outcome). Hence, TP (True Positive) is the count of students who had low course achievement and were predicted by the model as such. TN (True Negative) is the count of those who were high achieving in the course and the model predicted they would be high achievers. FP (False Positive) is the count of those who were high achievers in the course, but the model falsely predicted that they would have low achievement. Finally, FN (False Negative) is the count of students who were predicted to have high achievement in the course, but actually ended up in the low achievement group. These four count-based values forming the confusion matrix serve as the input for computing the aforementioned standard evaluation measures (Accuracy, Precision, Recall, and F1) based on the formuli given in the code snippet above.

After the predictive models for weeks 1–5 are built and evaluated, we combine and compare their performance measures:

```
eval_df <- bind_rows(eval_measures)
eval_df |>
  mutate(week = 1:5) |>
  mutate(across(Accuracy:F1, \(x) round(x, digits = 4))) |>
  select(week, Accuracy:F1)
```

Table 3 shows the resulting comparison of the built models. According to all measures, models 2 and 3, that is, models with the data from the first two and first 3 weeks of the course, are the best. In other words, the students' interactions with the course activities in the first 2–3 weeks are the most predictive of their overall course success. In particular, the accuracy of these models is 84%, meaning that for 84 out of 100 students, the models will correctly predict if the student would be a high or low achiever in this course. These models have precision of 75%, meaning that out of all the students for whom the models predict will be low achievers in the course, 75% will actually have low course achievement. In other words, the models will underestimate students' performance in 25% of predictions they make, by wrongly predicting that students would have low course achievement. The two best models have perfect recall (100%), meaning that the models would identify all the students who will actually have low course performance. These models outperform the other three models also in terms of the F1 measure, which was expected considering that this measure combines precision and recall giving them equal relevance. Interestingly, the studies exploring predictive models on weekly

Table 3 Comparison of prediction models for successive course weeks

Week	Accuracy	Precision	Recall	F1
1	0.7917	0.7500	0.8182	0.7826
2	0.8400	0.7500	1.0000	0.8571
3	0.8400	0.7500	1.0000	0.8571
4	0.7692	0.7333	0.8462	0.7857
5	0.7692	0.7333	0.8462	0.7857

basis have found similar high predictive power for models developed around the second week of the course [54].

RF allows for estimating the relevance of features used for model building. In a classification task, RF estimates feature relevance as the total decrease in the impurity (measured by the Gini index) of leaf nodes from splitting on a particular feature, averaged over all the trees that a RF model builds [46]. We will use this RF's functionality to compute and plot the importance of features in the best model. The function that does the computation and plotting is given below.

```
compute_and_plot_variable_importance <- function(rf_model) {
  importance(rf_model, type = 2) |>
    as.data.frame() |>
    (function(x) mutate(x, variable = rownames(x)))() -> var_imp_df

  row.names(var_imp_df) <- 1:nrow(var_imp_df)
  colnames(var_imp_df) <- c("importance","variable")

  ggplot(var_imp_df,
         aes(x = reorder(variable, importance), y = importance)) +
    geom_col(width = 0.35) +
    labs(x = "", y = "", title = "Feature importance") +
    coord_flip() +
    theme_minimal()
}
```

The plot produced by this function for one of the best models (Model 2) is given in Fig. 4.

```
compute_and_plot_variable_importance(models[[2]])
```

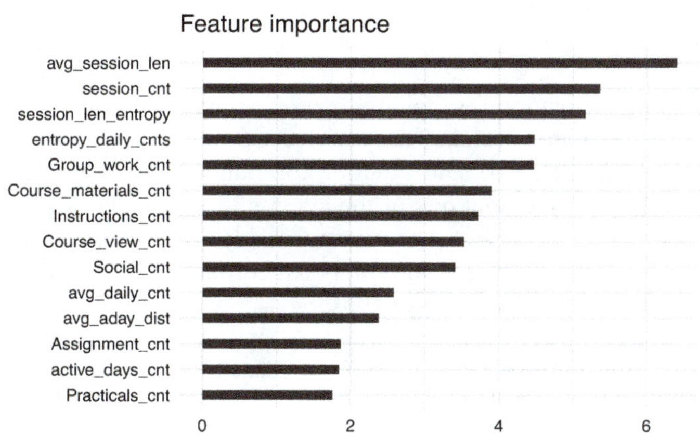

Fig. 4 The importance of features in the best course outcome prediction model, as estimated by the RF algorithm

As Fig. 4 shows, features denoting the overall level of activity (avg_session_len, session_cnt) are those with the highest predictive power. They are followed by entropy-based features, that is, features reflective of the regularity of study. These findings are in line with the LA literature (e.g., [14, 36, 37]. It should be also noted that the feature reflective of the level of engagement in the group work (Group_work_cnt) is among the top 5 predictors, which can be explained by the prominent role of group work in the course design.

3.7 Predicting Success Score

To predict the students' final grades, we will first try to build linear regression models, since linear regression is one of the most often used regression methods in LA [43]. To that end, we will first load a few additional R packages:

```
library(performance)
library(corrplot)
```

Considering that a linear regression model can be considered valid only if it satisfies a set of assumptions that linear regression, as a statistical method, is based upon (linearity, homogeneity of variance, normally distributed residuals, and absence of multicollinearity and influential points), we will first examine if our data satisfies these assumptions. In particular, we will compute the features based on the events data from the first week of the course, build a linear regression model using the computed features, and examine if the resulting model satisfies the assumptions. Note that we limit our initial exploration to the logged events data over the first week of the course since we aim to employ a regression method that can be applied to any number of course weeks; so, if the data from the first course week allow for building a valid linear regression model, we can explore the same method further; otherwise, we need to choose a more robust regression method, that is, method that is not so susceptible to imperfections in the input data.

Having created the dataset for final grade prediction based on the week 1 events data, we will split it into training and test sets (as done for the prediction of the course outcome, Sect. 3.6), and examine correlations among the features. The latter step is due to the fact that one of the assumptions of linear regression is the absence of high correlation among predictor variables. In the code below, we use the corrplot function to visualise the computed correlation values (Fig. 5), so that highly correlated variables can be easily observed.

```
ds <- create_dataset_for_grade_prediction(events_with_sessions, 1, results)

set.seed(2023)
train_indices <- createDataPartition(ds$Final_grade, p = 0.8, list = FALSE)
train_ds <- ds[train_indices,] |> select(-user)
```

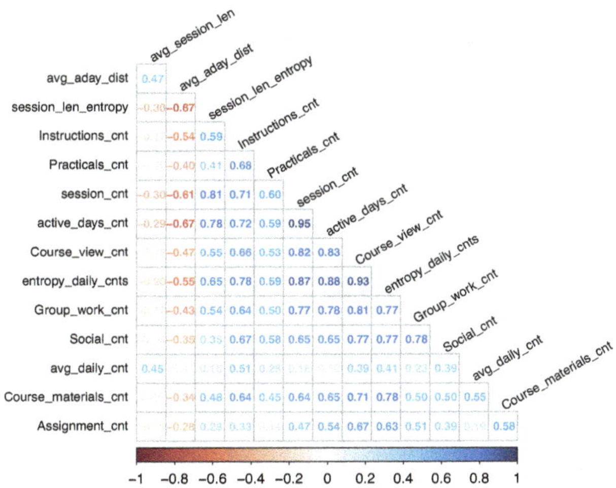

Fig. 5 Correlations among variables in the feature set

```
test_ds <- ds[-train_indices,] |> select(-user)

# examine correlations among the variables: for a linear regression model,
# they must not be highly mutually correlated
corrplot(train_ds |> select(-Final_grade) |> cor(),
         method = "number", type = "lower",
         diag = FALSE, order = 'hclust',
         tl.cex = 0.75, tl.col = 'black', tl.srt = 30, number.cex = 0.65)
```

Figure 5 indicates that there are a couple of features that are highly mutually correlated. These will be removed before proceeding with the model building. While there is no universal agreement on the correlation threshold above which features should be considered overly correlated, correlation coefficients of 0.75 and −0.75 are often used as the cut-off values [53].

```
train_ds |> select(-c(session_cnt, Course_view_cnt,
                      active_days_cnt, entropy_daily_cnts)) -> train_ds_sub
```

We can now build a model and check if it satisfies the assumptions of linear regression:

```
lr <- lm(Final_grade ~ ., data = train_ds_sub)
check_model(lr)
```

The check_model function from the *performance* R package [48] allows for seamless, visual verification of whether the assumptions are met. The output of this function when applied to our linear regression model (`lr`) is shown on Fig. 6.

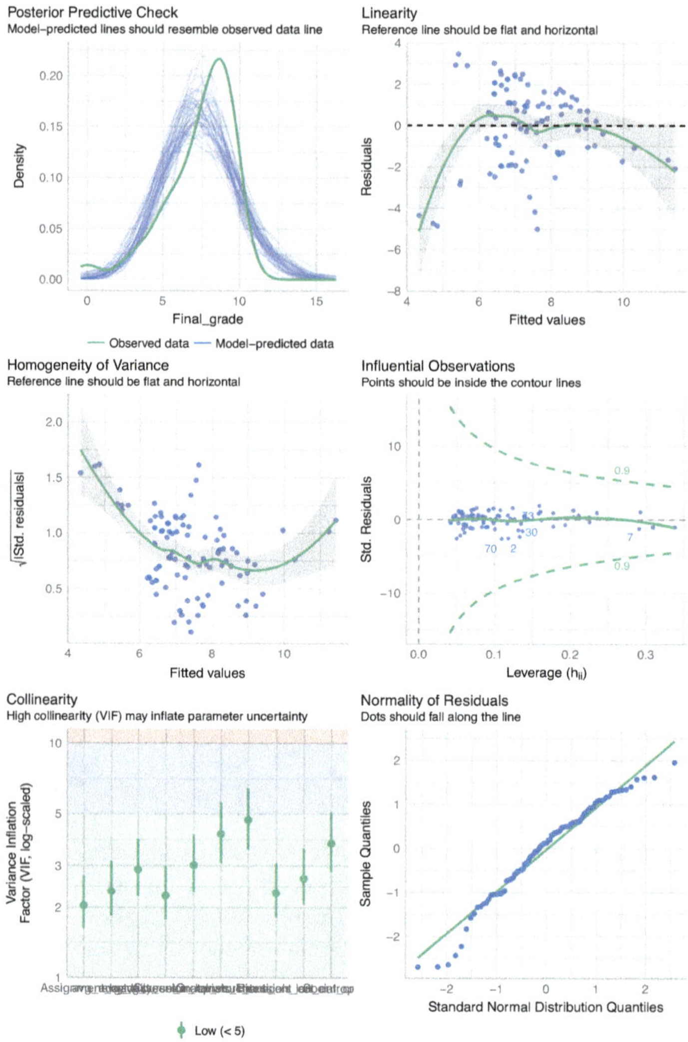

Fig. 6 The output of the check_model function enables visual verification of the assumptions that linear regression is based upon

As the figure shows, two important assumptions of linear models are not met, namely linearity and homoscedasticity (i.e. homogeneity of variance). Therefore, linear regression cannot be used with the given feature set. Instead, we have to use a regression method that does not impose such requirements on the data distribution. Since Random forest is such a method and it has already proven successful with our dataset on the classification task (Sect. 3.6), we will use it to build regression models that predict students' final grades.

Before moving to regression with Random forest, it is worth noting that, in addition to checking all model assumptions at once, using the `check_model` function, one can also check each assumption individually using appropriate functions from the `performance` R package. For example, from Fig. 6, one can not clearly see the X-axis of the collinearity plot and might want to explore this assumption more closely. That can be easily done using the `check_collinearity` function:

```
check_collinearity(lr)
```

```
# Check for Multicollinearity

Low Correlation

                Term  VIF   VIF 95% CI Increased SE Tolerance Tolerance 95% CI
 Course_materials_cnt 3.02 [2.32, 4.08]         1.74      0.33     [0.25, 0.43]
       Group_work_cnt 4.10 [3.09, 5.59]         2.02      0.24     [0.18, 0.32]
     Instructions_cnt 4.70 [3.52, 6.44]         2.17      0.21     [0.16, 0.28]
       Practicals_cnt 2.30 [1.81, 3.08]         1.52      0.43     [0.32, 0.55]
           Social_cnt 3.74 [2.83, 5.09]         1.93      0.27     [0.20, 0.35]
       Assignment_cnt 2.04 [1.63, 2.71]         1.43      0.49     [0.37, 0.61]
        avg_daily_cnt 2.89 [2.23, 3.90]         1.70      0.35     [0.26, 0.45]
      avg_session_len 2.24 [1.77, 3.00]         1.50      0.45     [0.33, 0.56]
  session_len_entropy 2.65 [2.06, 3.56]         1.63      0.38     [0.28, 0.49]
       avg_aday_dist 2.34 [1.84, 3.13]         1.53      0.43     [0.32, 0.54]
```

From the function's output, we can clearly see the VIF (Variance Inflation Factor) values for all the features and a confirmation that the assumption of the absence of multicollinearity is satisfied. The documentation of the `performance`[5] package provides the whole list of functions for different ways of checking regression models.

To build and compare regression models in each course week, we will follow a similar procedure to the one applied when building classification models (Sect. 3.6); the code that implements it is given below. The differences are in the way that the dataset for grade prediction is built (`create_dataset_for_grade_prediction`), the way that regression models are built (`build_RF_regression_model`) and evaluated (get_regression_evaluation_measures), and these will be explained in more detail below.

```
regression_models <- list()
regression_eval <- list()
for(k in 1:5) {
  print(str_glue("Starting computations for week {k} as the current week"))

  ds <- create_dataset_for_grade_prediction(events_with_sessions, k, results)

  set.seed(2023)
  train_indices <- createDataPartition(ds$Final_grade, p = 0.8, list = FALSE)
  train_ds <- ds[train_indices,] |> select(-user)
  test_ds <- ds[-train_indices,] |> select(-user)
```

[5] https://easystats.github.io/performance/reference/index.html.

```
  rf <- build_RF_regression_model(train_ds)
  eval_rf <- get_regression_evaluation_measures(rf, train_ds, test_ds)

  regression_models[[k]] <- rf
  regression_eval[[k]] <- eval_rf
}
```

To create a dataset for final grade prediction in week *k*, we first compute all features based on the logged events data (events_data) up to the week *k*, and then add the final grade variable (`Final_grade`) from the dataset with course results (`grades`):

```
create_dataset_for_grade_prediction <- function(events_data, current_week, grades) {
  features <- create_event_based_features(events_data, current_week)
  grades |>
    select(user, Final_grade) |>
    inner_join(features, by = "user")
}
```

As can be observed in the code snippet below, building a RF regression model is very similar to building a RF classification model. The main difference is in the evaluation measure that is used for selecting the optimal *mtry* value in the cross-validation process - here, we are using RMSE (Root Mean Squared Error), which is a standard evaluation measure for regression models [53]. As its name suggests, RMSE is the square root of the average squared differences between the actual and predicted values of the outcome variable on the test set.

```
build_RF_regression_model <- function(dataset) {

  n_features <- ncol(dataset)-1
  default_mtry <- round(sqrt(n_features))
  grid <- expand.grid(.mtry = (default_mtry-1):(default_mtry+1))

  ctrl <- trainControl(method = "CV",
                       number = 10)

  rf <- train(x = dataset |> select(-Final_grade),
              y = dataset$Final_grade,
              method = "rf",
              metric = "RMSE",
              tuneGrid = grid,
              trControl = ctrl)

  rf$finalModel
}
```

Finally, to evaluate each model on the test data, we compute three standard evaluation metrics for regression models, namely RMSE, MAE (Mean Absolute

Error), and R^2. MAE is the average value of the absolute differences between the actual and predicted values of the outcome variable (final grade) on the test set. Finally, R^2 (R-squared) is a measure of variability in the outcome variable that is explained by the given regression model. The computation of the three evaluation measures is shown in the code below.

```
get_regression_evaluation_measures <- function(model, train_ds, test_data) {

  predicted_vals <- predict(model,
                            test_data |> select(-Final_grade))
  actual_vals <- test_data$Final_grade

  # R2 = 1 - RSS/TSS
  # RSS - Residual Sum of Squares
  RSS <- sum((predicted_vals - actual_vals)^2)
  # TSS - Total Sum of Squares
  TSS <- sum((median(train_ds$Final_grade) - actual_vals)^2)
  R2 <- 1 - RSS/TSS

  # RMSE = sqrt(RSS/N)
  RMSE <- sqrt(RSS/nrow(test_ds))

  # MAE = avg(abs(predicted - actual))
  MAE <- mean(abs(predicted_vals - actual_vals))

  c(R2 = R2, RMSE = RMSE, MAE = MAE)
}
```

After the regression models for weeks 1–5 are built and evaluated, we combine and compare their performance measures, with the results reported in Table 4.

```
regression_eval_df <- bind_rows(regression_eval)
regression_eval_df |>
  mutate(WEEK = 1:5) |>
  mutate(across(R2:MAE, \(x) round(x, digits = 4))) |>
  select(WEEK, R2, RMSE, MAE)
```

As shown in Table 4, in this case, we do not have a clear situation as it was with the classification task (Table 3), since the three evaluation measures point to

Table 4 Comparison of grade prediction models for successive course weeks

WEEK	R2	RMSE	MAE
1	0.7315	0.8123	0.6614
2	0.8423	0.6215	0.4531
3	0.8387	0.6285	0.4474
4	0.9067	0.7625	0.5915
5	0.9179	0.7151	0.5777

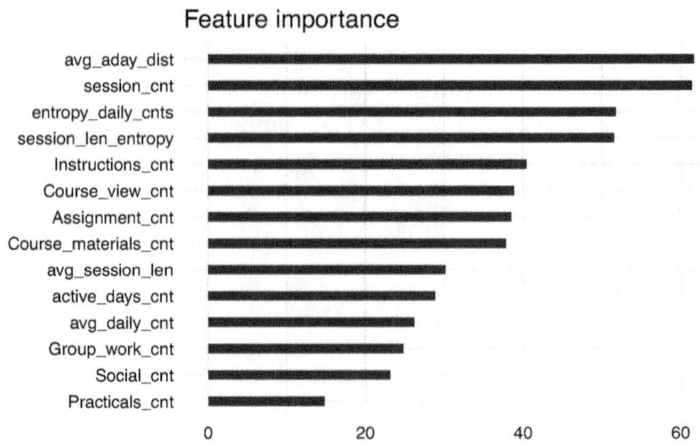

Fig. 7 The importance of features in the best final grade prediction model, as estimated by the RF algorithm

different models as potentially the best ones. In particular, according to R^2, the best model would be model 5 (i.e., the model based on the data from the first 5 weeks), whereas the other two measures point to the second or third model as the best. Considering that (1) RMSE and MAE measures are considered more important than R^2 when evaluating the predictive performance of regression models [13] and (2) RMSE and MAE values for models 2 and 3 are very close, while the second model is better in terms of R^2, we will conclude that the second model, that is, the model based on the logged event data from the first 2 weeks of the course is the best model. This model explains 84.23% of variability in the outcome variable (final grade), and predicts it with an average absolute error of 0.4531, which can be considered a small value with respect to the value range of the final grade [0–10].

To estimate the importance of features in the best regression model, we will again leverage the RF's ability. We will use the same function as before (`compute_and_plot_variable_importance`) to estimate and plot feature importance. The only difference will be that the importance function (from the *randomForest* package) will internally use residual sum of squares as the measure of node impurity when estimating the features importance. Figure 7 shows that, as in the case of predicting the overall course success (Fig. 4), regularity of study features (`avg_aday_dist`, `entropy_daily_cnts`, `session_len_entropy`) are among the most important ones. In addition, the number of learning sessions (`session_cnt`), as an indicator of overall activity in the course, is also among the top predictors.

```
compute_and_plot_variable_importance(regression_models[[2]])
```

4 Concluding Remarks

The results of predictive modelling presented in the previous section show that, in the examined postgraduate course on LA, we can make fairly accurate predictions of the students' course outcomes already in the second week of the course. In fact, both classification and regression models, that is, prediction of the students' overall course success and final grades, proved to be the most accurate when based on the logged learning events data from the first two or three course weeks. That students' learning behaviour in the first part of the course is highly predictive of their course performance, which is in line with related research on predictive modelling (e.g., [54–56]). It should be also noted that the high performance of the presented predictive models can be partially explained by the well chosen feature set and the used algorithm (Random forest) that generally performs well on prediction tasks. However, it may also be due to the relatively large dataset. As noted in Sect. 3.1, we used a synthetic anonymized version of the original dataset that is three times larger than the original dataset.

Considering the features that proved particularly relevant for predicting the students' course performance, we note that in both kinds of predictive tasks—course success and final grade prediction—features reflective of regularity of study stand out. In addition, features denoting the overall level of engagement with online learning activities and resources also have high predictive power. It is also worth noting that the highly predictive features are session level features, suggesting that learning session is the right level of granularity (better than actions or active days) for predictive modelling in the given course. In fact, this finding is in line with earlier studies that examined predictive power of a variety of features derived from learning traces [13, 14, 36]. Note that due to the purpose of this chapter to serve as introductory reading to predictive modelling in LA, we based the feature set on relatively simple features and used only one data source for feature creation. For more advanced and diverse feature creation options, interested readers are referred to, for example [57–59].

The algorithm used for building predictive models, namely Random forest, offers the advantage of flexibility in terms of the kinds of data it can work with (unlike, for example, linear regression which is based on several assumptions about data distribution) as well as fairly good prediction results it tends to produce. On the other hand, the algorithm is not as transparent as simpler algorithms are (e.g., linear regression or decision trees) and thus its use might raise issues of teachers' trust and willingness to rely on the models' output. On the positive side, the algorithm offers an estimate of feature importance thus shedding some light on the underlying "reasoning" process that led to its output (i.e., predictions).

To sum up, predictive modelling, as applied in LA, can bring about important benefits in the form of early in the course detection of students who might be struggling with the course and pointing out indicators of learning behaviour that are associated with poor course outcomes. With such insights available, teachers can make better informed decisions as to the students who need support and the kind of support they might benefit from. However, predictive modelling is also associated with challenges, especially practical challenges related to the development and use of such models, including availability and access to the data, interpretation of models and their results, and the associated issue of trust in the models' output.

5 Suggested Readings

- Max Kuhn & Julia Silge (2022). *Tidy Modeling with R: A Framework for Modeling in the Tidyverse*. O'Reilly. https://www.tmwr.org/
- Bradley Boehmke & Brandon Greenwell (2020). *Hands-On Machine Learning with R*. Taylor & Francis. https://bradleyboehmke.github.io/HOML/
- James, G., Witten, D., Hastie, T., & Tibshirani, R. (2021). *An Introduction to Statistical Learning: With Applications in R. 2nd Edition*. Springer US. https://doi.org/10.1007/978-1-0716-1418-1

References

1. Siemens G, Long P (2011) Penetrating the fog: Analytics in learning and education. EDUCAUSE Rev 46:30
2. Siemens G (2013) Learning analytics: The emergence of a discipline. Am Behav Sci 57:1380–1400. https://doi.org/10.1177/0002764213498851
3. Campbell JP, DeBlois PB, Oblinger DG (2007) Academic analytics. Educause Rev 42:40–57
4. Baker RS, Yacef K, et al (2009) The state of educational data mining in 2009: A review and future visions. J. Educ. Data Mining 1:3–17
5. Cornog J, Stoddard GD (1925) Predicting performance in chemistry. J. Chem. Educ. 2:701. https://doi.org/10.1021/ed002p701
6. Tomcsik D, Joksimovic J, Juhász J, Mihályi K (2014) Early warning systems in six European countries desk research report on study visit countries in the frame of CROCOOS-cross-sectoral cooperation focused solutions for the prevention of early school leaving project interim report. https://cpi.si/wp-content/uploads/2020/08/12-Sistemi-zgodnjega-opozarjanja-v-EU-EN.pdf
7. Brooks C, Thompson C (2017) Predictive modelling in teaching and learning. In: Handbook of learning analytics. Society for Learning Analytics Research (SoLAR), pp 61–68
8. Arnold KE, Pistilli MD (2012) Course signals at Purdue. In: Proceedings of the 2nd International Conference on Learning Analytics and Knowledge - LAK '12 267–267. https://doi.org/10.1145/2330601.2330666
9. Caulfield M (2013) What the course signals "kerfuffle" is about, and what it means to you. EDUCAUSE edu
10. Kuzilek J, Hlosta M, Herrmannova D, Zdrahal Z, Vaclavek J, Wolff A (2015) OU analyse: Analysing at-risk students at the open university. Learn Anal Rev LAK15-1:1–16
11. Ifenthaler D, Yau JYK (2020) Utilising learning analytics to support study success in higher education: A systematic review. Educ Technol Res Dev ETR & D. https://doi.org/10.1007/s11423-020-09788-z
12. Gašević D, Dawson S, Rogers T, Gasevic D (2016) Learning analytics should not promote one size fits all: The effects of instructional conditions in predicting academic success. Internet High Educ 28:68–84. https://doi.org/10.1016/j.iheduc.2015.10.002

13. Conijn R, Snijders C, Kleingeld A, Matzat U (2017) Predicting student performance from LMS data: A comparison of 17 blended courses using moodle LMS. IEEE Trans Learn Technol 10:17–29. https://doi.org/10.1109/TLT.2016.2616312
14. Saqr M, Jovanović J, Viberg O, Gašević D (2022) Is there order in the mess? A single paper meta-analysis approach to identification of predictors of success in learning analytics. Stud High Educ 47:2370–2391. https://doi.org/10.1080/03075079.2022.2061450
15. Finnegan C, Morris LV, Lee K (2008) Differences by course discipline on student behavior, persistence, and achievement in online courses of undergraduate general education. J College Student Retention Res Theory Pract 10:39–54. https://doi.org/10.2190/CS.10.1.d
16. Jovanović J, Saqr M, Joksimović S, Gašević D (2021) Students matter the most in learning analytics: The effects of internal and instructional conditions in predicting academic success. Comput Educ 172:104251. https://doi.org/10.1016/j.compedu.2021.104251
17. Ahmad A, Schneider J, Griffiths D, Biedermann D, Schiffner D, Greller W, Drachsler H (2022) Connecting the dots – a literature review on learning analytics indicators from a learning design perspective. J Comput Assisted Learn. https://doi.org/10.1111/jcal.12716
18. Albreiki B, Zaki N, Alashwal H (2021) A systematic literature review of student' performance prediction using machine learning techniques. Educ Sci 11:552. https://doi.org/10.3390/educsci11090552
19. Shafiq DA, Marjani M, Habeeb RAA, Asirvatham D (2022) Student retention using educational data mining and predictive analytics: A systematic literature review. IEEE Access 10:72480–72503. https://doi.org/10.1109/ACCESS.2022.3188767
20. Wang Q, Mousavi A (2023) Which log variables significantly predict academic achievement? A systematic review and meta-analysis. Br J Educ Technol J Council Educ Technol 54:142–191. https://doi.org/10.1111/bjet.13282
21. Gray CC, Perkins D (2019) Utilizing early engagement and machine learning to predict student outcomes. Comput Educ 131:22–32. https://doi.org/10.1016/j.compedu.2018.12.006
22. Hussain S, Khan MQ (2021) Student-performulator: Predicting students' academic performance at secondary and intermediate level using machine learning. Ann Data Sci. https://doi.org/10.1007/s40745-021-00341-0
23. Nouri J, Larsson K, Saqr M (2019) Identifying factors for master thesis completion and non-completion through learning analytics and machine learning. In: Lecture notes in computer science. Springer International Publishing, Cham, pp 28–39
24. Sani NS, Fikri A, Ali Z, Zakree M, Nadiyah K (2020) Drop-out prediction in higher education among B40 students. Int J Adv Comput Sci Appl. IJACSA 11. https://doi.org/10.14569/ijacsa.2020.0111169
25. Adnan M, Habib A, Ashraf J, Mussadiq S, Raza AA, Abid M, Bashir M, Khan SU (2021) Predicting at-risk students at different percentages of course length for early intervention using machine learning models. IEEE Access 9:7519–7539. https://doi.org/10.1109/ACCESS.2021.3049446
26. Bañeres D, Rodríguez ME, Guerrero-Roldán AE, Karadeniz A (2020) An early warning system to detect at-risk students in online higher education. NATO Adv Sci Inst Ser E Appl Sci 10:4427. https://doi.org/10.3390/app10134427
27. Jorgensen S, Ferraro V, Fichten C, Havel A (2009) Predicting college retention and dropout: Sex and disability. ERIC Clearinghouse
28. Joksimović S, Gašević D, Kovanović V, Riecke BE, Hatala M (2015) Social presence in online discussions as a process predictor of academic performance. J Comput Assisted Learn 31:638–654. https://doi.org/10.1111/jcal.12107
29. Ober TM, Hong MR, Rebouças-Ju DA, Carter MF, Liu C, Cheng Y (2021) Linking self-report and process data to performance as measured by different assessment types. Comput Educ 167:104188. https://doi.org/10.1016/j.compedu.2021.104188
30. Scheffel M, Drachsler H, Kraker J de, Kreijns K, Slootmaker A, Specht M (2017) Widget, widget on the wall, am I performing well at all? IEEE Trans Learn Technol 10:42–52. https://doi.org/10.1109/TLT.2016.2622268

31. Wu Z, Zhao B, Wang Y (2021) Analysis of students' learning behavior under network learning environment. In: 2021 IEEE 3rd international conference on computer science and educational informatization (CSEI), pp 46–50
32. Stadler M, Hofer S, Greiff S (2020) First among equals: Log data indicates ability differences despite equal scores. Comput Hum Behav 111:106442. https://doi.org/10.1016/j.chb.2020.106442
33. Tempelaar D, Rienties B, Nguyen Q (2020) Subjective data, objective data and the role of bias in predictive modelling: Lessons from a dispositional learning analytics application. PloS One 15:e0233977. https://doi.org/10.1371/journal.pone.0233977
34. You JW (2016) Identifying significant indicators using LMS data to predict course achievement in online learning. Internet High Educ 29:23–30. https://doi.org/10.1016/j.iheduc.2015.11.003
35. Zarrabi F, Bozorgian H (2020) EFL students' cognitive performance during argumentative essay writing: A log-file data analysis. Comput Compos 55:102546. https://doi.org/10.1016/j.compcom.2020.102546
36. Jovanovic J, Mirriahi N, Gašević D, Dawson S, Pardo A (2019) Predictive power of regularity of pre-class activities in a flipped classroom. Comput Educ 134:156–168. https://doi.org/10.1016/j.compedu.2019.02.011
37. Saqr M, Fors U, Tedre M (2017) How learning analytics can early predict under-achieving students in a blended medical education course. Medical Teacher 39:757–767. https://doi.org/10.1080/0142159X.2017.1309376
38. Agudo-Peregrina ÁF, Iglesias-Pradas S, Conde-González MÁ, Hernández-García Á (2014) Can we predict success from log data in VLEs? Classification of interactions for learning analytics and their relation with performance in VLE-supported F2F and online learning. Comput Hum Behav 31:542–550. https://doi.org/10.1016/j.chb.2013.05.031
39. Ho LC, Jin Shim K (2018) Data mining approach to the identification of at-risk students. In: 2018 IEEE international conference on big data (big data), pp 5333–5335
40. Jokhan A, Sharma B, Singh S (2019) Early warning system as a predictor for student performance in higher education blended courses. Stud High Educ 44:1900–1911. https://doi.org/10.1080/03075079.2018.1466872
41. Asselman A, Khaldi M, Aammou S (2021) Enhancing the prediction of student performance based on the machine learning XGBoost algorithm. Interactive Learn Environ 1–20. https://doi.org/10.1080/10494820.2021.1928235
42. Badal YT, Sungkur RK (2023) Predictive modelling and analytics of students' grades using machine learning algorithms. Educ Inf Technol 28:3027–3057. https://doi.org/10.1007/s10639-022-11299-8
43. Sghir N, Adadi A, Lahmer M (2022) Recent advances in predictive learning analytics: A decade systematic review (2012–2022). Educ Inf Technol 1–35. https://doi.org/10.1007/s10639-022-11536-0
44. López-Pernas S, Saqr M, Conde J, Del-Río-Carazo L (2024, this volume) A broad collection of datasets for educational research training and application. In: Saqr M, López-Pernas S (eds) Learning analytics methods and tutorials: A practical guide using R. Springer
45. Kuhn M (2008) Building predictive models in r using the caret package. J Stat Softw 28:1–26. https://doi.org/10.18637/jss.v028.i05
46. Liaw A, Wiener M (2002) Classification and regression by randomForest. R news 2:18–22
47. Breiman L (2001) Random forests. Mach Learn 45:5–32. https://doi.org/10.1023/A:1010933404324
48. Lüdecke D, Ben-Shachar M, Patil I, Waggoner P, Makowski D (2021) Performance: An r package for assessment, comparison and testing of statistical models. J Open Source Softw 6:3139. https://doi.org/10.21105/joss.03139
49. Wei T, Simko V (2021). R package corrplot: Visualization of a Correlation Matrix. (Version 0.92). https://github.com/taiyun/corrplot
50. Saqr M, López-Pernas S (2024) Why learning and teaching learning analytics is hard: An experience from a real-life LA course using LA methods. In: Proceedings of the eleventh international conference on technological ecosystems for enhancing multiculturality (TEEM'23). Springer, in press

51. Gitinabard N, Xu Y, Heckman S, Barnes T, Lynch CF (2019) How widely can prediction models be generalized? Performance prediction in blended courses. IEEE Trans Learn Technol 12:184–197. https://doi.org/10.1109/TLT.2019.2911832
52. Fernandez-Delgado M, Cernadas E, Barro S, Amorim D (2014) Do we need hundreds of classifiers to solve real world classification problems? J Mach Learn Res JMLR 15:3133–3181
53. James G, Witten D, Hastie T, Tibshirani R (2021) An introduction to statistical learning: With applications in r. Springer US
54. Saqr M, Nouri J (2020) High resolution temporal network analysis to understand and improve collaborative learning. In: Proceedings of the tenth international conference on learning analytics & knowledge. ACM, New York, NY, USA, pp 314–319
55. Chen W, Brinton CG, Cao D, Mason-Singh A, Lu C, Chiang M (2019) Early detection prediction of learning outcomes in online short-courses via learning behaviors. IEEE Trans Learn Technol 12:44–58. https://doi.org/10.1109/TLT.2018.2793193
56. Jovanović J, Dawson S, Joksimović S, Siemens G (2020) Supporting actionable intelligence: Reframing the analysis of observed study strategies. In: Proceedings of the tenth international conference on learning analytics & knowledge. Association for Computing Machinery, New York, NY, USA, pp 161–170
57. Bulut O, Gorgun G, Yildirim-Erbasli SN, Wongvorachan T, Daniels LM, Gao Y, Lai KW, Shin J (2023) Standing on the shoulders of giants: Online formative assessments as the foundation for predictive learning analytics models. Br J Educ Technol J Council Educ Technol 54:19–39. https://doi.org/10.1111/bjet.13276
58. Deeva G, De Smedt J, De Weerdt J (2022) Educational sequence mining for dropout prediction in MOOCs: Model building, evaluation, and benchmarking. IEEE Trans Learn Technol 15:720–735. https://doi.org/10.1109/TLT.2022.3215598
59. Marras M, Vignoud JTT, Kaser T (2021) Can feature predictive power generalize? Benchmarking early predictors of student success across flipped and online courses. In: Proceedings of the 14th international conference on educational data mining, pp 150–160

Dissimilarity-Based Cluster Analysis of Educational Data: A Comparative Tutorial Using R

Keefe Murphy, Sonsoles López-Pernas, and Mohammed Saqr

1 Introduction

Cluster analysis is a term used to describe a broad range of techniques which have the goal of uncovering groups of observations in a data set. The typical aim of cluster analysis is to categorise observations into groups in such a way that observations within the same group are in some sense more similar to each other, while being relatively dissimilar to those in other groups [1]. In other words, clustering methods uncover group structure in heterogeneous populations by identifying more homogeneous groupings of observations which may represent distinct, meaningful subpopulations. Using machine learning terminology, cluster analysis corresponds to unsupervised learning, whereby groups are identified by relying solely on the intrinsic characteristics (typically dissimilarities), without guidance from unavailable ground truth group labels. Indeed, foreknowledge of the fixed number of groups in a data set is characteristic of supervised learning and the distinct field of classification analysis.

It is important to note that there is no universally applicable definition of what constitutes a cluster [2, 3]. Indeed, in the absence of external information in the form of existing "true" group labels, different clustering methods can reveal different things about the same data. There are many different ways to cluster the same data set, and different methods may yield solutions with different assignments, or even differ in the number of groups they identify. Consequently, we present several cluster analysis algorithms in this chapter; namely, K-means

K. Murphy (✉)
Department of Mathematics and Statistics, Hamilton Institute, Maynooth University, Maynooth, Ireland
e-mail: keefe.murphy@mu.ie

S. López-Pernas · M. Saqr
School of Computing, University of Eastern Finland, Joensuu, Finland

M. Saqr, S. López-Pernas (eds.), *Learning Analytics Methods and Tutorials*,
https://doi.org/10.1007/978-3-031-54464-4_8

[4] in Sect. 3.1, a generalisation thereof, K-medoids [5], in Sect. 3.1.2.3, and agglomerative hierarchical clustering [6] in Sect. 3.2. We apply each method in a comparative study in our tutorial using R [7], with applications to a data set about the participants from discussion forum of a massive open online course (MOOC) for teachers.

The clustering methods we review in this chapter are designed to find mutually exclusive, non-overlapping groups in a data set, i.e., they recover a hard partition whereby each observation belongs to one group only. This is in contrast to soft clustering approaches under which each observation is assigned a probability of belonging to each group. One such example of soft clustering is the model-based clustering paradigm, which is discussed in the later Chapter 9 [8]. While model-based clustering methods, as the name suggests, are underpinned by the assumption of generative probabilistic models [9], the more traditional dissimilarity-based methods on which this chapter is focused are purely algorithmic in nature and rely on heuristic criteria regarding the pairwise dissimilarities between objects.

Heuristic clustering algorithms can be further divided into two categories; partitional clustering (e.g., K-means and K-medoids) and hierarchical (of which we focus on the agglomerative variant). Broadly speaking, partitional clustering methods start with an initial grouping of observations and iteratively update the clustering until the "best" clustering is found, according to some notion of what defines the "best" clustering. Hierarchical clustering methods, on the other hand, is more sequential in nature; a tree-like structure of nested clusterings is built up via successive mergings of similar observations, according to a defined similarity metric. In our theoretical expositions and our applications in the R tutorial, we provide some guidelines both on how to choose certain method-specific settings to yield the best performance and how to determine the optimal clustering among a set of competing methods.

The remainder of this chapter proceeds as follows. In Sect. 2, we review relevant literature in which dissimilarity-based clustering methods have been applied in the realm of educational research. In Sect. 3, we describe the theoretical underpinnings of each method in turn and discuss relevant practical guidelines which should be followed to secure satisfactory performance from each method throughout. Within Sect. 4, we introduce the data set of our case study in Sect. 4.1 and give an overview of some required pre-processing steps in Sect. 4.1.1, and then present a tutorial using R for each clustering method presented in this chapter in Sect. 4.2, with a specific focus on identifying the optimal clustering solution in Sect. 4.2.4. Finally, we conclude with a discussion and some recommendations for related further readings in Sect. 5, with a particular focus on some limitations of dissimilarity-based clustering which are addressed by other frameworks in the broader field of cluster analysis.

2 Clustering in Education: Review of the Literature

In education, clustering is among the oldest and most common analysis methods, predating the field of learning analytics and educational data mining by several decades. Such early adoption of clustering is due to the immense utility of cluster analysis in helping researchers to find patterns within data, which is a major pursuit of education research [10]. Interest was fueled by the increasing attention to heterogeneity and individual differences among students, their learning processes, and their approaches to learning [11, 12]. Finding such patterns or differences among students allows teachers and researchers to improve their understanding of the diversity of students and tailor their support to different needs [13]. Finding subgroups within cohorts of students is a hallmark of so-called "person-centered methods", to which clustering belongs [8].

A person-centered approach stands in contrast to the variable-centered methods which consider that most students belong to a coherent homogeneous group with little or negligible differences [14]. Variable-centered methods assume that there is a common average pattern that represents all students, that the studied phenomenon has a common causal mechanism, and that the phenomenon evolves in the same way and results in similar outcomes amongst all students. These assumptions are largely understood to be unrealistic, "demonstrably false, and invalidated by a substantial body of uncontested scientific evidence" [15]. In fact, several analytical problems necessitate clustering, e.g., where opposing patterns exist [16]. For instance, in examining attitudes towards an educational intervention using variable-centered methods, we get an average that is simply the sum of negative and positive attitudes. If the majority of students have a positive attitude towards the proposed intervention—combined with a minority against—the final result will imply that students favour such intervention. The conclusions of this approach are not only wrong but also dangerous as it risks generalising solutions to groups to whom it may cause harm.

Therefore, clustering has become an essential method in all subfields of education (e.g., education psychology, education technology, and learning analytics) having operationalised almost all quantitative data types and been integrated with most of the existing methods [10, 17]. For instance, clustering became an essential step in sequence analysis to discover subsequences of data that can be understood as distinct approaches or strategies of students' behavior [18]. Similar applications can be found in social network analysis data to identify collaborative roles, or in multimodal data analysis to identify moments of interest (e.g., synchrony). Similarly, clustering has been used with survey data to find attitudinal patterns or learning approaches to mention a few [17]. Furthermore, identifying patterns within students' data is a prime educational interest in its own right and therefore, has been used extensively as a standalone analysis technique to identify subgroups of students who share similar interests, attitudes, behaviors, or background.

Clustering has been used in education psychology for decades to find patterns within self-reported questionnaire data. Early examples include the work of Beder and Valentine [19] who used responses of a motivational questionnaire to discover subgroups of students with different motivational attitudes. Similarly, Clément et al. [20] used the responses of a questionnaire that assessed anxiety and motivation towards learning English as a second language to find clusters which differentiated students according to their attitudes and motivation. Other examples include the work by Fernandez-Rio et al. [21] who used hierarchical clustering and K-means to identify distinct student profiles according to their perception of the class climate. With the digitalisation of educational institutions, authors also sought to identify students profiles using admission data and learning records. For instance, Cahapin et al. [22] used several algorithms such as K-means and agglomerative hierarchical clustering to identify patterns in students' admission data.

The rise of online education opened many opportunities to investigate students' behavior in learning management systems based on the trace log data that students leave behind them when working on online activities. An example is the work by Saqr et al. [23], who used K-means to cluster in-service teachers' approaches to learning in a MOOC according to their frequency of clicks on the available learning activities. Recently, research has placed a focus on the temporality of students' activities, rather than the mere count. For this purpose, clustering has been integrated within sequence analysis to find subsequences which represent meaningful patterns of students behavior. For instance, Jovanovic et al. [24] used hierarchical clustering to identify distinct learning sequential patterns based on students' online activities in a learning management system within a flipped classroom, and found a significant association between the use of certain learning sequences and learning outcomes. Using a similar approach, López-Pernas et al. [25] used hierarchical clustering to identify distinctive learning sequences in students' use of the learning management system and an automated assessment tool for programming assignments. They also found an association between students' activity patterns and their performance in the course final exam. Several other examples exist for using clustering to find patterns within sequence data [16, 26, 27].

A growing application of clustering can be seen in the study of computer-supported collaborative learning. Saqr and López-Pernas [28] used cluster analysis to discover students with similar emergent roles based on their forum interaction patterns. Using the K-means algorithm and students' centrality measures in the collaboration network, they identified three roles: influencers, mediators, and isolates. Perera et al. [29] used K-means to find distinct groups of similar teams and similar individual participating students according to their contributions in an online learning environment for software engineering education. They found that several clusters which shared some distinct contribution patterns were associated with more positive outcomes. Saqr and López-Pernas [30] analysed the temporal unfolding of students' contributions to group discussions. They used hierarchical clustering to identify patterns of distinct students' sequences of interactions that have a similar start and found a relationship between such patterns and student achievement.

An interesting application of clustering in education concerns the use of multimodal data. For instance, Vieira et al. [31] used K-means clustering to find patterns in students' presentation styles according to their voice, position, and posture data. Other innovative uses involve students' use of educational games [32, 33], virtual reality [34], or artificial intelligence [35]. Though this chapter illustrates traditional dissimilarity-based clustering algorithms with applications using R to data on the centrality measures of the participants of a MOOC, along with related demographic characteristics, readers are also encouraged to read Chapter 9 [8] and Chapter 10 [18] in which further tutorials are presented and additional literature is reviewed, specifically in the contexts of clustering using the model-based clustering paradigm and clustering longitudinal sequence data, respectively. We now turn to an explication of the cluster analysis methodologies used in this chapter's tutorial.

3 Clustering Methodology

In this section, we describe some of the theory underpinning the clustering methods used in the later R tutorial in Sect. 4. We focus on some of the most widely-known heuristic dissimilarity-based clustering algorithms; namely, K-means, K-medoids, and agglomerative hierarchical clustering. In Sect. 3.1, we introduce K-means clustering by describing the algorithm, outline the arguments to the relevant R function `kmeans()`, and discuss some of the main limitations and practical concerns researchers should be aware of in order to obtain the best performance when running K-means. We also discuss the related K-medoids algorithm and the associated function `pam()` in the `cluster` library [36] in R, and situate this method in the context of an extension to K-means designed to overcome its reliance on squared Euclidean distances. In Sect. 3.2, we introduce agglomerative hierarchical clustering and the related R function `hclust()`, while outlining various choices available to practitioners and their implications. Though method-specific strategies of choosing the optimal number of clusters K are provided throughout Sects. 3.1 and 3.2, we offer a more detailed treatment of this issue in Sect. 3.3, particularly with regard to criteria that can guide the choice of clustering solution among multiple competing methodologies, and not only the choice of the number of clusters K for a given method.

3.1 K-Means

K-means is a widely-used clustering algorithm in learning analytics and indeed data analysis and machine learning more broadly. It is an unsupervised technique that seeks to divide a typically multivariate data set into some pre-specified number of clusters, based on the similarities between observations. More specifically, K-means aims to partition n objects $\mathbf{x}_1, \ldots, \mathbf{x}_n$, each having measurements on $j = 1, \ldots, d$ strictly continuous covariates, into a set of K groups $\mathcal{C} = \{C_1, \ldots, C_K\}$, where C_k is the set of n_k objects in cluster k and the number of groups $K \leq n$ is pre-specified by the practitioner and remains fixed. K-means constructs these partitions in such a way that the squared Euclidean distance between the row vector for observations in a given cluster and the centroid (i.e., mean vector) of the given cluster are smaller than the distances to the centroids of the remaining clusters. In other words, K-means aims to learn both the cluster centroids and the cluster assignments by minimising the within-cluster sums-of-squares (i.e., variances). Equivalently, this amounts to maximising the between-cluster sums-of-squares, thereby capturing heterogeneity in a data set by partitioning the observations into homogeneous groups. What follows is a brief technical description of the K-means algorithm in Sect. 3.1.1, after which we describe some limitations of K-means and discuss practical concerns to be aware of in order to optimise the performance of the method in Sect. 3.1.2. In particular, we present the widely-used K-medoids extension in Sect. 3.1.2.3.

3.1.1 K-Means Algorithm

The origins of K-means are not so straightforward to summarise. The name was initially used by James MacQueen in 1967 [4]. However, the standard algorithm was first proposed by Stuart Lloyd in a Bell Labs technical report in 1957, which was later published as a journal article in 1982 [37]. In order to understand the ideas involved, we must first define some relevant notation. Let $\mu_k^{(j)}$ denote the centroid value for the j-th variable in cluster C_k by

$$\mu_k^{(j)} = \frac{1}{n_k} \sum_{\mathbf{x}_i \in C_k} x_{ij}$$

and the complete centroid vector for cluster C_k by

$$\boldsymbol{\mu}_k = \left(\mu_k^{(1)}, \ldots, \mu_k^{(p)}\right)^\top .$$

These centroids therefore correspond to the arithmetic mean vector of the observations in cluster C_k. Finding both these centroids $\boldsymbol{\mu}_1, \ldots, \boldsymbol{\mu}_K$ and the clustering partition $\mathcal{C}$ is computationally challenging and typically proceeds by iteratively alternating between allocating observations to clusters and then updating the centroid vectors. Formally, the objective is to minimise the total within-cluster sum-of-squares

$$\sum_{i=1}^{n} \sum_{k=1}^{K} z_{ik} \|\mathbf{x}_i - \boldsymbol{\mu}_k\|_2^2, \tag{1}$$

where $\|\mathbf{x}_i - \boldsymbol{\mu}_k\|_2^2 = \sum_{j=1}^{p} \left(x_{ij} - \mu_k^{(j)}\right)^2$ denotes the squared Euclidean distance to the centroid $\boldsymbol{\mu}_k$—such that $\|\cdot\|_2$ denotes the ℓ^2 norm—and $\mathbf{z}_i = (z_{i1}, \ldots, z_{iK})^\top$ is a latent variable such that z_{ik} denotes the cluster membership of observation i;

$z_{ik} = 1$ if observation i belongs to cluster C_k and $z_{ik} = 0$ otherwise. This latent variable construction implies $\sum_{i=1}^{n} z_{ik} = n_k$ and $\sum_{k=1}^{K} z_{ik} = 1$, such that each observation belongs wholly to one cluster only. As the total variation in the data, which remains fixed, can be written as a sum of the total within-cluster sum-of-squares (TWCSS) from Eq. (1) and the between-cluster sum-of-squares as follows

$$\sum_{i=1}^{n} (\mathbf{x}_i - \bar{\mathbf{x}})^2 = \sum_{i=1}^{n} \sum_{k=1}^{K} z_{ik} \|\mathbf{x}_i - \boldsymbol{\mu}_k\|_2^2 + \sum_{k=1}^{K} n_k \|\boldsymbol{\mu}_k - \bar{\mathbf{x}}\|_2^2,$$

where $\bar{\mathbf{x}}$ denotes the overall sample mean vector, minimising the TWCSS endeavours to ensure that observations in the same cluster are maximally similar to observations in the same cluster and maximally dissimilar to those in other clusters.

Using the notation just introduced, a generic K-means algorithm would proceed as follows:

1. *Initialise*: Select the number of desired clusters K and define K initial centroid vectors $\boldsymbol{\mu}_1, \ldots, \boldsymbol{\mu}_K$.
2. *Allocate*: Find the optimal z_{ik} values that minimise the objective, holding the μ_k values fixed.

Calculate the squared Euclidean distance between each observation and each centroid vector and allocate each object to the cluster corresponding to the initial centroid to which it is closest in terms of squared Euclidean distance. Looking at the objective function in Eq. (1) closely and examining the contribution of observation i, we need to choose the value of $\mathbf{z}_i$ which minimises the expression $\sum_{k=1}^{K} z_{ik} \|\mathbf{x}_i - \boldsymbol{\mu}_k\|_2^2$. This is achieved by setting $z_{ik} = 1$ for the value of k that has smallest $\|\mathbf{x}_i - \boldsymbol{\mu}_k\|_2^2$ and setting $z_{ik'} = 0 \, \forall \, k' \neq k$ everywhere else.

3. *Update*: Find the optimal $\boldsymbol{\mu}_k$ values that minimise the objective, holding the z_{ik} values fixed.

If we re-write the objective function in Eq. (1) as

$$\sum_{i=1}^{n} \sum_{k=1}^{K} \sum_{j=1}^{p} z_{ik} \left(x_{ij} - \mu_k^{(j)} \right)^2,$$

we can use the fact that

$$\frac{\partial}{\partial \mu_k^{(j)}} \left(x_{ij} - \mu_k^{(j)} \right)^2 = -2 \left(x_{ij} - \mu_k^{(j)} \right)$$

to obtain

$$\frac{\partial}{\partial \mu_k^{(j)}} \sum_{i=1}^{n} \sum_{k=1}^{K} \sum_{j=1}^{p} z_{ik} \left(x_{ij} - \mu_k^{(j)} \right)^2 = -2 \sum_{i=1}^{n} z_{ik} \left(x_{ij} - \mu_k^{(j)} \right).$$

Solving this expression for $\mu_k^{(j)}$ yields

$$\mu_k^{(j)} = \frac{\sum_{i=1}^{n} z_{ik} x_{ij}}{\sum_{i=1}^{n} z_{ik}} = \frac{1}{n_k} \sum_{\mathbf{x}_i \in C_k} x_{ij}.$$

4. *Iterate*: One full iteration of the algorithm consists of an allocation step (Step 2) and an update step (Step 3). Steps 2 and 3 are repeated until no objects can be moved between clusters, at which point the algorithm has converged to at least a local minimum of Eq. (1).

Upon convergence, we obtain not only the estimated partition $\mathcal{C} = \{C_1, \ldots, C_K\}$, indicating the cluster membership of each observation, but also the estimated cluster centroids $\boldsymbol{\mu}_1, \ldots, \boldsymbol{\mu}_K$ which act as cluster prototypes, efficiently summarising the characteristics of each cluster. The algorithm just described is just one standard variant of K-means. there have been several algorithms proposed for the same objective which derive their names from the author who proposed them; namely MacQueen [4], Lloyd [37], Forgy [38], and Hartigan and Wong [39]. They differ in some subtle ways, particularly with regard to how initial centroids in Step 1 are chosen and whether Steps 3 and 4 are applied to all n observations simultaneously or whether allocations and centroids are updated for each observation one-by-one (i.e., some variants iterate over Step 2 and Step 3 in a loop over $i = 1, \ldots, n$). Without going into further details, we note that the default option for `kmeans()` in R uses the option `algorithm="Hartigan-Wong"` and this is what we will henceforth adopt throughout.

3.1.2 *K*-means Limitations and Practical Concerns

Though K-means is a useful tool in many application contexts due to its conceptual and computational simplicity—so ubiquitous, in fact, that the `kmeans()` function in R is available without loading any additional libraries—it suffers from numerous limitations and some care is required in order to obtain reasonable results. We now discuss some of the main limitations in turn, but note that each is addressed explicitly throughout the K-means application portion of the R tutorial in Sect. 4.2.1. The concerns relate to choosing K (Sect. 3.1.2.1), choosing initial centroids $\boldsymbol{\mu}_1, \ldots, \boldsymbol{\mu}_K$ (Sect. 3.1.2.2), and relaxing the reliance on squared Euclidean distances with the more general K-medoids method (Sect. 3.1.2.3).

3.1.2.1 Fixed *K* and the Elbow Method

The first major drawback is that the number of clusters K must be pre-specified. This is a key input parameter: if K is too low, dissimilar observations will be wrongly grouped together; if K is too large, observations will be partitioned into many small, similar clusters which may not be meaningfully different. Choosing the optimal K necessitates running the algorithm at various fixed values of K and finding the single value of K which best balances interpretability, parsimony, and fit quality. Fit quality is measured by the TWCSS, i.e., the objective function in Eq. (1). Increasing K indefinitely will cause the TWCSS to decrease indefinitely, but this is not what we want. Instead, we seek a K value beyond which the decrease in TWCSS is minimal, in order to yield a parsimonious solution with a reasonable number of clusters to interpret, without overfitting the data or merely subdividing the actual groups. Thus, a commonly used heuristic graphical method for determining the optimal K value is to plot a range of K values against the corresponding obtained TWCSS values and look for an "elbow" or kink in the resulting curve. Such a plot will guide the choice of K in the K-means portion of R tutorial which follows in Sect. 4.2.1.

3.1.2.2 Initialisation and K-means^{++}

An especially pertinent limitation which must be highlighted is that K-means is liable to converge to sub-optimal local minima, i.e., it is not guaranteed to converge to the global minimum of the objective function in Eq. (1). Many practitioners have observed that the performance of K-means is particularly sensitive to a good choice of initial cluster centroids in Step 1. Indeed, as different initial settings can lead to different clusterings of the same data, good starting values are vital to the success of the K-means algorithm. Typically, K random vectors are used to define the initial $\boldsymbol{\mu}_1, \ldots, \boldsymbol{\mu}_K$ centroids. One means of mitigating (but not completely remedying) the problem is to run K-means with a suitably large number of random starting values and choose the solution associated with the set of initial centroids which minimise the TWCSS criterion.

In order to contextualise this issue, it is prudent to first describe some of the main arguments to the `kmeans()` R function. The following list is a non-exhaustive list of the available arguments to `kmeans()`:

- `x`: a numeric matrix of data or a `data.frame` with all numeric columns.
- `centers`: either the number of clusters K or a set of K initial (distinct) cluster centroids.
- `nstart`: if `centers` is specified as a number, this represents the number of random sets of K initial centroids with which to run the algorithm.
- `iter.max`: the maximum number of allocation/update cycles allowed per set of initial centroids.

The arguments `nstart` and `iter.max` have default values of `1` and `10`, respectively. Thus, a user running `kmeans()` with `centers` specified as a number will, by default, only use one random set of initial centroids, to which the results are liable to be highly sensitive, and will have the algorithm terminate after just ten iterations, regardless of whether convergence was achieved. It would seem be an improvement, therefore, to increase the values of `nstart` and `iter.max` from these defaults.

Fortunately, the function automatically returns the single optimal solution according to the random initialisation which yields the lowest TWCSS when `nstart` exceeds `1`.

Though this will generally lead to better results, this approach can be computationally onerous if the number of observations n, number of features d, or size of the range of fixed K values under consideration is large. An alternative strategy, which greatly reduces the computational burden and the sensitivity of the final solution to the initial choices of $\boldsymbol{\mu}_1, \ldots, \boldsymbol{\mu}_k$ is to choose a suitable set of informed starting values in a data-driven fashion. To this end, the so-called K-means^{++} algorithm was proposed [40] in order to improve the performance of K-means by replacing Step 1 with an iterative distance-weighting scheme to select the initial cluster centroids. Though there is still randomness inherent in K-means^{++}, this initialisation technique ensures that the initial centroids are well spread out across the data space, which increases the likelihood of converging to the true global minimum. The K-means^{++} algorithm works as follows:

(A) Choose an initial centroid uniformly at random from the rows of the data set.
(B) For each observation not yet chosen as a centroid, compute $D^2(\mathbf{x}_i)$, which represents the squared Euclidean distance between $\mathbf{x}_i$ and the nearest centroid that has already been chosen.
(C) Randomly sample a new observation as a new centroid vector with probability proportional to $D^2(\mathbf{x}_i)$.
(D) Repeat Steps B and C until K centroids have been chosen. If any of the chosen initial centroids are not distinct, add a small amount of random jitter to distinguish the non-unique centroids.
(E) Proceed as per Steps 2–4 of the traditional K-means algorithm (or one of its variants).

Although these steps take extra time, K-means itself tends to converge very quickly thereafter and thus K-means^{++} actually reduces the computational burden. A manual implementation of the K-means^{++} is provided by the function `kmeans_pp()` below. Its output can be used as the `centers` argument when running `kmeans()`.

```
kmeans_pp <- function(X, # data
                      K  # number of centroids
                      ) {

  # sample initial centroid from distinct rows of X
  X         <- unique(as.matrix(X))
  new_center_index <- sample(nrow(X), 1)
  centers <- X[new_center_index,, drop=FALSE]

  # let x be all observations not yet chosen as a centroid
  X <- X[-new_center_index,, drop=FALSE]

  if(K >= 2) {
```

```
14      # loop over remaining centroids
15      for(kk in 2:K) {
16
17        # calculate distances from all observations to all already chosen
18          centroids
19        distances <- apply(X, 1, function(x) min(sum((x - centers)^2)))
20
21        # sample new centroid with probability proportional to squared
22          Euclidean distance
23        probabilities <- distances/sum(distances)
24        new_center_index <- sample(nrow(X), 1, prob=probabilities)
25
26        # record the new centroid and remove it for the next iteration
27        centers <- rbind(centers, X[new_center_index,, drop=FALSE])
28        X <- X[-new_center_index,, drop=FALSE]
29      }
30    }
31
32    # add random jitter to distinguish non-unique centroids and return
33    centers[duplicated(centers)] <- jitter(centers[duplicated(centers)])
34    return(centers)
35  }
```

However, it should be noted that there is still inherent randomness in K-means^{++}—note the use of `sample()` in lines 7 and 22—and the algorithm is liable to produce different initial centroids in different runs on the same data. In effect, K-means^{++} does not remove the burden of random initialisation; it is merely a way to have more informed random initialisations. Thus, it would be prudent to run K-means *with* K-means^{++} initialisation and select the solution which minimises the TWCSS, to transfer the burden of requiring multiple runs of K-means with random starting values to fewer runs of K-means^{++} followed by K-means with more informed starting values. We adopt this strategy in the later R tutorial in Sect. 4.2.1.

3.1.2.3 K-medoids and Other Distances

The K-means objective function in Eq. (1) explicitly relies on squared Euclidean distances and requires all features in the data set to be strictly continuous. An artefact of this distance measure is that it is generally recommended to standardise all features of have a mean of zero and unit variance prior to running K-means. In general, standardisation is advisable if the values are of incomparable units (e.g., height in inches and weight in kilogram). More specifically for K-means, it is desirable to ensure all features have comparable variances to avoid having variables with higher magnitudes and variances dominate the distance calculation and have an undue prominence on the clustering partition obtained. While we employ such normalisation to the data used in our R tutorial when applying some pre-processing steps in Sect. 4.1.1, we note that this is not sufficient to overcome all shortcomings of relying on squared Euclidean distances.

For these reasons and more, K-medoids—otherwise known as partitioning around medoids (PAM)—was proposed as an extension to K-means which allows using any alternative dissimilarity measure [5]. The K-medoids objective function is given by

$$\sum_{i=1}^{n} \sum_{k=1}^{K} z_{ik} d\left(\mathbf{x}_i, \boldsymbol{\psi}_k\right),$$

where $d\left(\mathbf{x}_i, \boldsymbol{\psi}_k\right)$ can be any distance measure rather than squared Euclidean and $\boldsymbol{\psi}_k$ is used in place of the mean vector $\boldsymbol{\mu}_k$. The PAM algorithm works in much the same fashion as K-means, alternating between an allocation step which assigns each observation to the cluster with the closest $\boldsymbol{\psi}_k$ (according to the specified distance measure) and an update step which minimises the within-cluster total distance (WCTD). Notably, when minimising with respect to $\boldsymbol{\psi}_k$, the notion of a cluster centroid $\boldsymbol{\mu}_k$ is redefined as a cluster medoid $\boldsymbol{\psi}_k$, which is selected among the rows of the observed data set, i.e., the medoid is the observation $\mathbf{x}_i$ from which the distance to all other observations currently allocated to the same cluster, according to the specified distance measure, is minimised. Similar to K-means, the medoids obtained at convergence again enable straightforward characterisation of a "typical" observation from each cluster and the elbow method from Sect. 3.1.2.1 can be adapted to guide the choice of K in K-medoids by plotting a range of candidate K values against the within-cluster total distance.

This reformulation has three main advantages. Firstly, the distance $d\left(\mathbf{x}_i, \boldsymbol{\psi}_k\right)$ is not squared, which diminishes the influence of outliers. As K-means relies on squared Euclidean distances, which inflates the distances of atypical observations, and defines centroids as means, it is not robust to outliers. Secondly, by defining the medoids as observed rows of the data, rather than finding the value of $\boldsymbol{\psi}_k$ that minimises in general, which could potentially be difficult to estimate for complex data types or particularly sophisticated dissimilarity measures, the algorithm can be much more computationally efficient. It requires only a pre-calculated pairwise dissimilarity matrix as input. Finally, the flexibility afforded by being able to modify the dissimilarity measure enables data which are not strictly continuous to be clustered. In other words, K-medoids is applicable in cases where the mean is undefined. As examples, one could use the Manhattan or general Minkowski distances as alternatives for clustering continuous data, the Hamming [41], Jaccard [42], or Sørensen-Dice [43, 44] distances for clustering binary data, or the Gower distance [45] for clustering mixed-type data with both continuous and categorical features. The closely related K-modes algorithm [46] has also been proposed specifically for purely categorical data applications, as well as the K-Prototypes algorithm [47] for mixed-type variables applications; neither will considered further in this chapter). As their names suggest, they again redefine the notion of a centroid but otherwise proceed much like K-means and K-medoids.

The function `pam()` in the `cluster` library in R provides an implementation of K-medoids, with options for implementing many recent additional speed improvements and improved initialisation strategies [48]. We will discuss these in the K-medoids portion of the later R tutorial in Sect. 4.2.2. Most dissimilarity measures we will use are implement in the base-R function `dist()`, with the exception of the Gower distance which is implemented in the `daisy()` function in the `cluster` library.

3.2 Agglomerative Hierarchical Clustering

Hierarchical clustering is another versatile and widely-used dissimilarity-based clustering paradigm. Though also dissimilarity-based, hierarchical clustering differs from partitional clustering methods like K-means and K-medoids in that it typically doesn't avail of the notion of computing distances to a central prototype, be that a centroid mean vector or a medoid, but instead greedily builds a hierarchy of clusters based on dissimilarities between observations themselves and sets of observations. Consequently, a hierarchical clustering solution provides a set of partitions, from a single cluster to as many clusters as observations, rather than the single partition obtained by K-means and K-medoids. The results of a hierarchical clustering are usually presented in the form of a dendrogram visualisation, which illustrates the arrangement of the set of partitions visited and can help guide the decision of the optimal single clustering partition to extract. However, hierarchical clustering shares some of the advantages K-medoids has over K-means. Firstly, any valid measure of distance can be used, so it is not restricted to squared Euclidean distances and not restricted to clustering purely continuous data. Secondly, hierarchical clustering algorithms do not require the data set itself as input; all that is required is a matrix of pairwise distances.

Broadly speaking, there are two categories of hierarchical clustering:

- *Agglomerative*: Starting from the bottom of the hierarchy, begin with each observation in a cluster of its own and successively merged pairs of clusters while moving up the hierarchy, until all observations are in one cluster. This approach is sometimes referred to as agglomerative nesting (AGNES; [6]).
- *Divisive*: Starting from the top of the hierarchy, with all observations in one cluster, recursively split clusters while moving down the hierarchy, until all observations are in a cluster of their own. This approach is sometimes referred to as divisive analysis (DIANA; [49]).

However, divisive clustering algorithms such as DIANA are much more computationally onerous for even moderately large data sets. Thus, we focus here on the agglomerative variant of hierarchical clustering, AGNES, which is implemented in both the `agnes()` function in the `cluster` library and the `hclust()` function in base R. We adopt the latter in the hierarchical clustering portion of the R tutorial in Sect. 4.2.3. That being said, even agglomerative hierarchical clustering has significant computation and memory burdens when n is large [50, 51].

There are three key decisions practitioners must make when employing agglomerative hierarchical clustering. The first of these, the distance measure, has already been discussed in the context of K-medoids. We now discuss the other two in turn; namely, the so-called linkage criterion for quantifying the distances between merged clusters as the algorithm moves up the hierarchy, and the criterion used for cutting the resulting dendrogram to produce a single partition.

3.2.1 Linkage Criteria

Agglomerative hierarchical clustering employs two different notions of dissimilarity. There is the distance measure, d, such as Euclidean, Manhattan, or Gower distance, which is used to quantify the distance between pairs of *single* observations in the data set. Different choices of distance measure can lead to markedly different clustering results and it is thus common to run the hierarchical clustering algorithm with different choices of distance measure and compare the results. However, in the agglomerative setting, individual observations are successively merged into clusters. In order to decide which clusters should be combined, it is necessary to quantify the dissimilarity of *sets* of observations as a function of the pairwise distances of observations in the sets. This gives rise to the notion of a linkage criterion. At each step, the two clusters separated by the shortest distance are combined; the linkage criteria is precisely the definition of 'shortest distance' which differentiates different agglomerative approaches. Again, the choice of linkage criterion can have a substantial impact on the result of the clustering so multiple solutions with different combinations of distance measure and linkage criterion should be evaluated.

There are a number of commonly-used linkage criteria which we now describe. A non-exhaustive list of such linkages follows—only those which we use in the later R tutorial in Sect. 4—in which we let $\mathcal{A}$ and $\mathcal{B}$ denote two sets of observations, $|\cdot|$ denote the cardinality of a set, and $d(a, b)$ denote the distance between observations in those corresponding sets according to the chosen distance measure. We note that ties for the maximum or minimum distances for complete linkage and single linkage, respectively, are broken at random.

- *Complete* linkage: Define the dissimilarity between two clusters as the distance between the two elements (one in each cluster) which are furthest away from each other according to the chosen distance measure d:

$$\max_{a \in \mathcal{A}, b \in \mathcal{B}} d(a, b).$$

- *Single* linkage: Define the dissimilarity between two clusters as the distance between the two elements (one in each cluster) which are closest to each other according to the chosen distance measure d:

$$\min_{a \in \mathcal{A}, b \in \mathcal{B}} d(a, b).$$

- *Average* linkage: Define the dissimilarity between two clusters as the average distance according to the chosen distance measure d between all pairs of elements (on in each cluster):

$$\frac{1}{|\mathcal{A}| \times |\mathcal{B}|} \sum_{a \in \mathcal{A}} \sum_{b \in \mathcal{B}} d(a, b).$$

- *Centroid* linkage: Define the dissimilarity between two clusters $\mathcal{A}$ and $\mathcal{B}$ as the distance, according to the chosen distance measure d, between their corresponding centroid vectors $\boldsymbol{\mu}_{\mathcal{A}}$ and $\boldsymbol{\mu}_{\mathcal{B}}$:

$$d(\boldsymbol{\mu}_{\mathcal{A}}, \boldsymbol{\mu}_{\mathcal{B}}).$$

- *Ward* linkage [52]: Instead of measuring the dissimilarity between clusters directly, define the dissimilarity as the cost of merging two clusters as the increase in total within-cluster variance after merging. In other words, minimise the total within-cluster sum-of-squares by finding the pair of clusters at each step which leads to minimum increase in total within-cluster variance after merging, where $\mathcal{A} \cup \mathcal{B}$ denotes the cluster obtained after merging, with corresponding centroid $\boldsymbol{\mu}_{\mathcal{A} \cup \mathcal{B}}$:

$$\frac{|\mathcal{A}| \times |\mathcal{B}}{|\mathcal{A} \cup \mathcal{B}|} \|\boldsymbol{\mu}_{\mathcal{A}} - \boldsymbol{\mu}_{\mathcal{B}}\|_2^2 = \sum_{x \in \mathcal{A} \cup \mathcal{B}} \|x - \boldsymbol{\mu}_{\mathcal{A} \cup \mathcal{B}}\|_2^2 - \sum_{x \in \mathcal{A}} \|x - \boldsymbol{\mu}_{\mathcal{A}}\|_2^2 - \sum_{x \in \mathcal{B}} \|x - \boldsymbol{\mu}_{\mathcal{B}}\|_2^2.$$

The Ward and centroid linkage criteria differ from the other linkage criteria in that they are typically meant to be used only when the initial pairwise distances between observations are squared Euclidean distances. All of the above linkage criteria are implemented in the function `hclust()`, which is available in R without requiring any add-on libraries to be loaded and specifically performs agglomerative hierarchical clustering. Its main arguments are `d`, a pre-computed pairwise dissimilarity matrix (as can be created from the function `dist()`), and `method`, which specifies the linkage criterion (e.g., `"complete"`, `"single"`, `"average"`, and `"centroid"`). Special care must be taken when employing Ward's linkage criterion as two options are available: `"ward.D"`, which assumes that the initial pairwise distance matrix already consists of *squared* Euclidean distances, and `"ward.D2"`, which assumes the distances are merely Euclidean distances and performs the squaring internally [53].

3.2.2 Cutting the Dendrogram

One might notice that when calling `hclust()`, the number of clusters K is not specified in advance, as it is when calling `kmeans()` or `pam()`. Instead, `hclust()` returns an object which describes the hierarchy of the tree produced by the clustering process. A visualisation of such a tree is referred to as a dendrogram, which can be thought of as a representation of a set of candidate partitions. In a dendrogram representation of an agglomerative hierarchical clustering solution, each observation is initially in a singleton cluster on its own, along the x-axis, according to their similarities. Thereafter, each observation, and subsequently each set of observations, are merged along the y-axis in a nested fashion. The scale along the y-axis is proportional to the distance, according to the chosen linkage criterion, at which two clusters are combined. In the end, the groups formed towards the bottom of the graph are close together, whereas those at the top of the graph are far apart.

Obtaining a single hard partition of objects into disjoint clusters is obtained by cutting the dendrogram horizontally at the corresponding height. In other words, observations are allocated to clusters by cutting the tree at an appropriate height. Generally, the lower this height, the greater the number of clusters (theoretically, there can be as many clusters as there are observations, n), while the greater the height, the lower the number of clusters (theoretically, there can be as few as only one cluster, corresponding to no group structure in the data). Thus, an advantage of hierarchical clustering is that the user need not know K in advance; the user can manually select K after the fact by examining the constructed tree and fine-tuning the output to find clusters with a desired level of granularity. There is no universally applicable rule for determining the optimal height at which to cut the tree, but it is common to select a height in the region where there is the largest gap between merges, i.e., where there is a relatively wide range of distances over which the number of clusters in the resulting partition does not change. This is, of course, very much guided by the visualisation itself.

In R, one can visualise the dendrogram associated with a particular choice of distance measure and linkage criterion by calling `plot()` on the output from `hclust()`. Thereafter, the function `cutree()` can be used to obtain a single partition. This function takes the arguments `tree`, which is the result of a call to `hclust()`, `h` which is the height at which the tree should be cut, and `k` which more directly allows the desired number of clusters to be produced. Specifying `k` finds the corresponding height which yields `k` clusters and overrides the specification of `h`.

However, it is often the case that certain combinations of dissimilarity measure and linkage criterion produce undesirable dendrograms. In particular, complete linkage is known to perform poorly in the presence of outliers, given its reliance on maximum distances, and single linkage is known to produce a "chaining" effect on the resulting dendrogram, whereby, due to its reliance on minimum distances, observations tend to continuously join increasingly larger, existing clusters rather than being merged with other observations to form new clusters. A negative consequence of this phenomenon is lack of cohesion: observations at opposite ends of the same cluster in a dendrogram could be quite dissimilar. These limitations can also be attributed to hierarchical clustering—regardless of the linkage employed—optimising a local criterion for each merge, unlike K-means and K-medoids which endeavour to optimise global objectives.

3.3 Choosing the Number of Clusters

Determining the number of clusters in a data set is a fraught task. Throughout Sects. 3.1 and 3.2, method-specific strategies for guiding the choice of K were presented. However, they are not without their limitations. For K-means and K-medoids, the elbow method is somewhat subjective and unreliable. Often, the presence of an elbow is not so clear at a single value of K. Likewise, for agglomerative hierarchical clustering, choosing a height at which to cut the dendrogram as the criterion for choosing K has also been criticised as an overly subjective method. Moreover, these strategies are only capable of identifying the best K value conditional on the chosen method and do not help to identify the overall best solution among multiple competing methods. Moreover, we are required to choose more than just the optimal K value when using K-medoids (for which different solutions with different dissimilarity measures and different K values can be obtained) and agglomerative hierarchical clustering (for which different solutions can be obtained using different dissimilarity measures and linkage criteria).

Overcoming these ambiguities and identifying a more general strategy for comparing the quality of clustering partitions is a difficult task for which many criteria have been proposed. Broadly speaking, cluster quality measures fall into two categories:

1. Comparing of the uncovered partition to a reference clustering (or known grouping labels).
2. Measuring of internal cluster consistency without reference to ground truth labels.

The first is typically conducted using the Rand index [54] or adjusted Rand index [55], which measure the agreement between two sets of partitions. However, as we will be exploring clustering in exploratory, unsupervised settings, using data for which there is no assumed "true" group structure, we will instead focus on a quality measure of the latter kind. As previously stated, a large number of such criteria have been proposed in the literature: several are summarised in Table 7 of Chapter 10 of this book [18], where they are used to guide the choice of K for agglomerative hierarchical clustering in the context of sequence analysis. Here, however, for the sake of brevity, we describe only one commonly used criterion which we later employ in the R tutorial in Sect. 4—which is itself a dissimilarity-based measure and is thus universally applicable to all clustering algorithms we employ—even if in most applications it would be wise to inform the choice of K with several such quantitative criteria. Moreover, the practitioner's own subject matter expertise and assessment of the interpretability of the obtained clusters should also be used to inform the choice of K.

The quantitative criterion we employ is referred to as the average silhouette width (ASW) criterion [56], which is routinely used to assess the cohesion and separation of the clusters uncovered by dissimilarity-based methods. Cohesion refers to the tendency to group similar objects together and separation refers to the tendency to group dissimilar objects apart in non-overlapping clusters. As the name implies, the ASW is computed as the average of observation-specific silhouette widths. Under the assumption that $K > 1$, silhouette widths and the ASW criterion are calculated as follows:

(A) Let $a(i)$ be the average dissimilarity from observation i to the other members of the same cluster to which observation i is assigned.
(B) Compute the average dissimilarity from observation i to the members of all $K - 1$ other clusters: let $b(i)$ be the minimum such distance computed.
(C) The silhouette for observation i is then defined to be

$$s(i) = \frac{b(i) - a(i)}{\max\{a(i), b(i)\}}$$
$$= \begin{cases} 1 - \frac{a(i)}{b(i)} & \text{if } a(i) < b(i) \\ 0 & \text{if } a(i) = b(i) \\ \frac{b(i)}{a(i)} - 1 & \text{if } a(i) > b(i), \end{cases}$$

unless observation i is assigned to a cluster of size 1, in which case $s(i) = 0$. Notably, $a(i)$ and $b(i)$ need not be calculated using the same dissimilarity measure with which the data were clustered; it is common to adopt the Euclidean distance.
(D) Define the ASW for a given partition $\mathcal{C}$ as: ASW $(\mathcal{C}) = \frac{1}{n}\sum_{i=1}^{n} s_i$.

Given that $-1 \leq s(i) \leq 1$, the interpretation of $s(i)$ is that a silhouette close to 1 indicates that the observation has been well-clustered, a silhouette close to -1 indicates that the observation would be more appropriately assigned to another cluster, and a silhouette close to zero indicates that the observation lies on the boundary of two natural clusters. If most values of $s(i)$ are high, then the clustering solution can be deemed appropriate. This occurs when $a(i) \ll b(i)$, meaning that observation i must be well-matched its own cluster and poorly-matched to all other clusters. Conversely, if most $s(i)$ values are low or even negative, this provides evidence that K may be too low or too high.

The values of $s(i)$ can be averaged over all observations assigned to the same cluster, as a measure of how tightly grouped the observations in the given cluster are. The ASW criterion itself is simply the mean silhouette width over all observations in the entire data set. Generally, clustering solutions with higher ASW are to be preferred, however it is also prudent to dismiss solutions with many negative silhouettes or particularly low cluster-specific mean silhouettes. A silhouette plot can help in this regard; such a plot depicts all values of $s(i)$, grouped according to the corresponding cluster and in decreasing order within a cluster. In the R tutorial which follows, ASW values and silhouette plots will guide the choice of an optimal clustering solution in a comparison of multiple methods in Sect. 4.2.4.

4 Tutorial with R

In this section, we will learn how to perform clustering using the R programming language [7], using all methods described throughout Sect. 3. We start by loading the necessary libraries. We will use `cluster` [36] chiefly for functions related to K-medoids and silhouettes. As per other chapters in this book, we use `tidyverse` [57] for data manipulation and `rio` [58] for downloading the data: see Sect. 4.1.1 for details the on the data pre-processing steps employed.

```
library(cluster)
library(rio)
library(tidyverse)
```

We note that hierarchical clustering and K-means are implemented in base R and thus no dedicated libraries need to be loaded.

4.1 *The Data Set*

Our case study will be to identify different groups of participants that have a similar role in the discussion forum of a massive open online course (MOOC) for teachers. For that purpose, we will rely on the centrality measures of the participants which indicate their number of contributions (`OutDegree`), replies (`InDegree`), position in the network (`Closeness_total`), worth of their connections (`Eigen`), spread of their ideas (`Diffusion_degree`), and more. For more details about the data set, please refer to the data chapter of the book (Chapter 2; [59]). To learn more about centrality measures and how to calculate them, refer to the social network analysis chapter (Chapter 15; [60]). We will henceforth refer to these data as "the MOOC centralities data set". We can download and preview (Table 1) the data with the following commands:

```
URL <-"https://github.com/lamethods/data/raw/main/6_snaMOOC/"
df <- import(paste0(URL, "Centralities.csv"))
df
```

Table 1 Preview of MOOC centralities data set

	name	InDegree	OutDegree	Closeness_total	Betweenness	Eigen	Diffusion.degree	Coreness	Cross_clique_connectivity
1	1	20	33	0.001	1258.143	0.206	1865	18	305
2	2	2	5	0.001	26.524	0.011	218	6	13
3	3	2	4	0.001	30.601	0.009	191	6	11
4	4	2	14	0.001	72.523	0.080	965	13	37
5	5	16	17	0.001	309.033	0.162	1508	18	154
6..444									
445	445	276	56	0.001	16, 690.426	0.699	3574	31	2218

Additionally, a number of categorical variables pertaining to demographic characteristics are available for the same participants. Again, please refer to the data chapter of the book (Chapter 2; [59]) for more details on these variables. With an appropriate distance measure, namely the Gower distance measure, we will incorporate some of these variables in our applications of K-medoids and agglomerative hierarchical clustering (but *not* K-means) in addition to the continuous variables contained in `df`, largely for the purpose of demonstrating clustering methods which are capable of clustering variables of mixed type. We can download and preview the auxiliary categorical data set with the commands below. Note that we select only the following variables:

- `experience` (coded as a level of experience, 1–3),
- `gender` (female or male),
- `region` (Midwest, Northeast, South, and West U.S.A., along with International),

for the sake of simplifying the demonstration of mixed-type variables clustering and reducing the computational burden. We also extract the `UID` column, a user ID which corresponds to the `name` column in `df`, which will be required for later merging these two data sets. We also ensure that all but this leading `UID` column is formatted as a `factor` (Table 2).

```
demog <- import(paste0(URL, "DLT1%20Nodes.csv"))
demog <- demog |>
  select(UID, experience, gender, region) |>
  mutate_at(vars(-UID), as.factor)
demog
```

4.1.1 Pre-processing the Data

In `df`, the first column (`name`) is the student identifier, and the remaining columns are each of the centrality measures calculated from students' interactions in the MOOC forum. We will eventually discard the `name` column from future analyses; we retain it for the time being so that `df` and `demog` can be merged appropriately. The data

Table 2 Preview of the selected additional categorical demographic variables associated with the MOOC centralities data set

	UID	Experience	Gender	Region
1	1	1	Female	South
2	2	1	Female	South
3	3	2	Female	Northeast
4	4	2	Female	South
5	5	3	Female	South
6..444				
445	445	3	Female	South

also contain a small number of observations—only three rows of df—with missing values for the variable Closeness_total, as seen by

```
df |> is.na() |> which(arr.ind=TRUE)
```

```
    row col
441 441   4
442 442   4
443 443   4
```

Given that none of the clustering methods described in this chapter are capable of accommodating missing data, we remove these three observations for future analyses too. Furthermore, one of the rows in demog has NULL recorded for the gender variable. We remove all invalid rows from both df and demog. The function complete.cases() constructs a completely observed data set by extracting the rows which contain one or more missing values and we augment the index of fully observed rows with an index of non-NULL gender values. Finally, we drop the redundant NULL level from the factor variable gender.

```
obs_ind <- complete.cases(df) & demog$gender != "NULL"
df$name[!obs_ind] # indices of observations with missing values
```

```
[1] 439 441 442 443
```

```
df <- df |> filter(obs_ind)
demog <- demog |> filter(obs_ind) |> droplevels()
```

Before proceeding any further, it would be prudent to explore the complete data visually, which we do via the matrix of pairwise scatter plots, excluding the name column, in Fig. 1:

```
pairs(df[,-1])
```

From the plots in Fig. 1, we can see that there are clearly two quite extreme outliers. Simple exploratory analyses (not shown) confirms that these are the final two rows of the complete data set. These observations are known to correspond to the two instructors who led the discussion. They have been marked using a red cross symbol in Fig. 1. Though we have argued that K-medoids is a more robust clustering method than K-means, for example, we also remove these observations in order to avoid their detrimental effects on the K-means output. The rows must be removed from both df and demog so that they can later be merged. With these observations included, K-means for instance would likely add one additional cluster containing just these two observations, about whom we already know their role differs substantially from the other observations, as they are quite far from the bulk of the data in terms of squared Euclidean distance. That is not to say, however, that there will not still be outliers in df after removing these two cases.

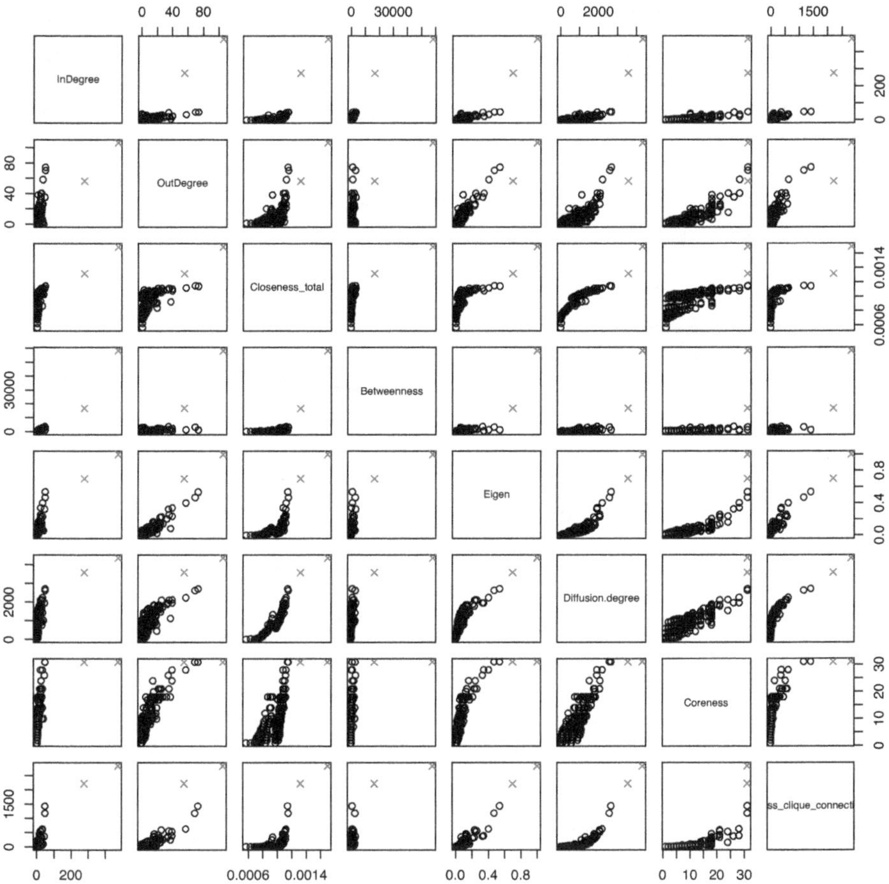

Fig. 1 Matrix of pairwise scatter plots for all variables in the MOOC centralities data set

```
keep_ind <- 1:(nrow(df) - 2)

df <- df |> slice(keep_ind)
demog <- demog |> slice(keep_ind)
```

As is good practice when using dissimilarity-based clustering algorithms, we pre-process the purely continuous data by normalising each variable to have a mean of 0 and a variance of 1, by constructing the scaled data frame `sdf` using the function `scale()`, again excluding the `name` column.

```
sdf <- scale(df[,-1], center=TRUE, scale=TRUE)
```

The object `sdf` can be used for all clustering methods described herein. To also accommodate the categorical demographic variables, we use `merge()` to combine both the scaled continuous data and the categorical data. This requires some manipulation of the name column, with which the two data sets are merged, but we ultimately discard the superfluous name column, which we do not want to contribute to any pairwise distance matrices or clustering solutions, from both merged_df and sdf.

```
merged_df <- data.frame(name=df$name, sdf) |>
  merge(demog, by=1) |>
  select(-name)
```

Finally, before proceeding to apply various clustering methods to these data, we present summaries of the counts of each level of the categorical variables experience, gender, and region. These variables imply groupings of size three, two, and five, respectively, and it will be interesting to see if this is borne out in any of the mixed-type clustering applications.

```
table(merged_df$experience)
```

```
  1   2   3
118 150 171
```

```
table(merged_df$gender)
```

```
female   male
   299    140
```

```
table(merged_df$region)
```

```
International       Midwest     Northeast         South          West
           32            77           110           168            52
```

4.2 Clustering Applications

We now show how each of the clustering methods described above can be implemented in R, using these data throughout and taking care to address all practical concerns previously raised. For each method—K-means in Sect. 4.2.1, K-medoids in Sect. 4.2.2, and agglomerative hierarchical clustering in Sect. 4.2.3—we show clustering results following the method-specific guidelines for choosing K. However, we conclude by comparing results across different methods using the average silhouette width criterion to further guide the choice of K in Sect. 4.2.4. We refrain from providing an interpretation of the uncovered clusters until Sect. 4.2.5, after the optimal clustering solution is identified.

Before we proceed, we set the seed to ensure that results relying on random number generation are reproducible.

```
set.seed(2024)
```

4.2.1 *K*-means Application

We begin by showing a naive use of the `kmeans()` function, supplying only the scaled data `sdf` and the pre-specified number of clusters K via the `centers` argument. For now, we assume for no particular reason that there are $K = 3$ clusters, just to demonstrate the use of the `kmeans()` function and its arguments. A number of aspects of the results are printed when we examine the resulting `km1` object.

```
km1 <- kmeans(sdf, centers=3)
km1
```

```
K-means clustering with 3 clusters of sizes 48, 129, 262

Cluster means:
    InDegree  OutDegree Closeness_total Betweenness       Eigen
1  2.2941694  1.9432559       1.0923551   1.9697769  2.00747791
2 -0.4088814 -0.4311073      -1.4086380  -0.3975783 -0.55451137
3 -0.2189864 -0.1437536       0.4934399  -0.1651209 -0.09475944
  Diffusion.degree   Coreness Cross_clique_connectivity
1        1.9078646  2.1923592                 2.0066265
2       -1.1038258 -0.5622732                -0.3094092
3        0.1939543 -0.1248092                -0.2152835

Clustering vector:
  [1] 1 2 2 3 1 1 1 1 3 1 1 3 1 1 1 2 1 3 1 2 2 1 2 1 3 1 1 2 1 1 2 3 3 1 1 1 2
 [38] 3 3 2 1 3 2 1 3 3 2 3 1 1 3 3 1 1 2 3 3 1 3 1 1 1 1 1 3 3 1 1 3 3 3 3 2 3
 [75] 3 3 3 3 2 2 3 3 3 2 3 2 3 1 2 3 3 1 2 3 3 3 2 1 3 1 3 3 3 3 3 2 3 2 3 3 2
[112] 3 3 3 1 1 2 2 2 3 3 2 2 3 2 3 3 3 3 2 3 2 3 3 2 3 1 3 3 2 3 3 3 3 2 3 3 3
[149] 2 3 2 3 2 3 3 2 3 3 2 3 3 3 3 2 3 3 3 2 3 3 3 3 3 2 2 3 3 3 3 2 3 2 3 3 3
[186] 3 3 3 3 3 2 3 3 3 3 3 2 1 3 3 3 3 3 3 3 3 3 2 2 2 3 3 2 3 2 2 3 3 1 2 3 2
[223] 1 2 2 3 2 2 2 2 2 3 2 1 3 3 2 2 3 3 3 3 3 2 3 3 3 3 3 3 3 3 3 2 3 3 3 3 3
[260] 2 3 2 2 2 3 2 3 2 2 2 3 2 2 2 2 3 2 3 3 3 3 2 2 3 2 2 2 2 2 2 2 3 2 2 3 2
[297] 3 3 3 3 3 3 3 3 3 2 2 3 2 1 2 2 2 3 2 2 3 3 2 3 2 3 2 3 3 3 2 2 3 3 3 2 3
[334] 2 3 3 3 3 3 3 3 3 3 3 3 3 3 2 2 3 2 2 2 3 3 3 3 3 2 3 3 3 3 3 3 3 3 3 3 2
[371] 3 3 3 3 3 3 3 3 3 3 3 3 3 3 3 3 3 3 3 3 3 3 3 3 3 3 3 3 3 3 3 3 3 3 3 3 3
[408] 3 3 3 3 3 3 3 3 2 2 2 2 2 2 2 2 3 3 3 3 3 3 3 3 1 2 2 2 2 2 2 3

Within cluster sum of squares by cluster:
[1] 858.90295  72.01655 414.30502
 (between_SS / total_SS =  61.6 %)
```

```
Available components:

[1] "cluster"      "centers"      "totss"        "withinss"
    "tot.withinss"
[6] "betweenss"    "size"         "iter"         "ifault"
```

Among other things, this output shows the estimated centroid vectors $\boldsymbol{\mu}_1, \ldots, \boldsymbol{\mu}_K$ upon convergence of the algorithm (`$centers`), an indicator vector showing the assignment of each observation to one of the $K = 3$ clusters (`$cluster`), the size of each cluster in terms of the number of allocated observations (`$size`), the within-cluster sum-of-squares *per cluster* (`$withinss`), and the ratio of the between-cluster sum-of-squares (`$betweenss`) to the total sum of squares (`$totss`). Ideally, this ratio should be large, if the total within-cluster sum-of-squares is minimised. We can access the TWCSS quantity by typing

```
km1$tot.withinss
```

```
[1] 1345.225
```

However, these results were obtained using the default values of ten maximum iterations (`iter.max=10`, by default) and only one random set of initial centroid vectors (`nstart=1`, by default). To increase the likelihood of converging to the global minimum, it is prudent to increase `iter.max` and `nstart`, to avoid having the algorithm terminate prematurely and avoid converging to a local minimum, as discussed in Sect. 3.1.2.2. We use `nstart=50`, which is reasonably high but not so high as to incur too large a computational burden.

```
km2 <- kmeans(sdf, centers=3, nstart=50, iter.max=100)
km2$tot.withinss
```

```
[1] 1339.226
```

We can see that the solution associated with the best random starting values achieves a lower TWCSS than the earlier naive attempt. Next, we see if an even lower value can be obtained using the K-means^{++} initialisation strategy, by invoking the `kmeans_pp()` function from Sect. 3.1.2.2 with `K=3` centers and supplying these centroid vectors directly to `kmeans()`, rather than indicating the number of random `centers` to use.

```
km3 <- kmeans(sdf, centers=kmeans_pp(sdf, K=3), iter.max=100)
km3$tot.withinss
```

```
[1] 1343.734
```

In this case, using the K-means^{++} initialisation strategy did not further reduce the TWCSS; in fact it is worse than the solution obtained using `nstart=50`. However, recall that K-means^{++} is itself subject to randomness and should therefore also be run several times, though the number of K-means^{++} runs need not be so high as 50. In the code below, we use `replicate()` to invoke both `kmeans_pp()` and `kmeans()` itself ten times in order to obtain a better solution.

```
KMPP <- replicate(10, list(kmeans(sdf, iter.max=100,
                               centers=kmeans_pp(sdf, K=3))))
```

Among these ten solutions, five are identical and achieve the same minimum TWCSS value.

```
TWCSS <- sapply(KMPP, "[[", "tot.withinss")
TWCSS
```

```
[1] 1345.225 1339.226 1339.226 1339.226 1340.457 1339.226 1345.225
    1345.225
[9] 1345.225 1339.226
```

Thereafter, we can extract a solution which minimises the `tot.withinss` as follows:

```
km4 <- KMPP[[which.min(TWCSS)]]
km4$tot.withinss
```

```
[1] 1339.226
```

Finally, this approach resulted in an identical solution to `km2` being obtained—with just ten runs of K-means^{++} and K-means rather than `nstart=50` runs of the K-means algorithm alone—which is indeed superior to the solution obtained with just one uninformed random start in `km1`. We will thus henceforth adopt this initialisation strategy always.

To date, the K-means algorithm has only been ran with the fixed number of clusters $K = 3$, which may be suboptimal. The following code iterates over a range of K values, storing both the `kmeans()` output itself and the TWCSS value for each K. The range $K = 1, \ldots, 10$ notably includes $K = 1$, corresponding to no group structure in the data. The reason for storing the `kmeans()` output itself is to avoid having to run `kmeans()` again after determining the optimal K value; such a subsequent run may not converge to the same minimised TWCSS and having to run the algorithm again would be computationally wasteful.

```
K <- 10 # set upper limit on range of K values

TWCSS <- numeric(K) # allocate space for TWCSS estimates

KM <- list() # allocate space for kmeans() output
```

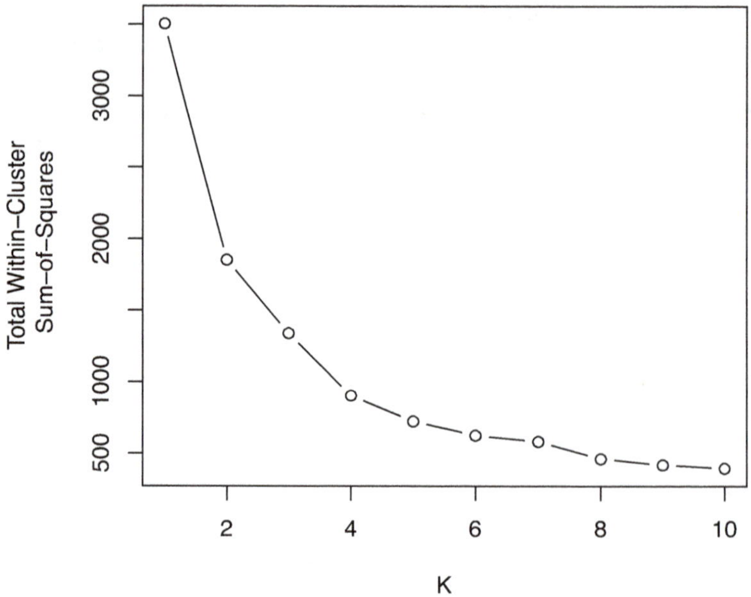

Fig. 2 Elbow plot showing a range of K values against the corresponding obtained TWCSS for K-means applied to the MOOC centralities data set using K-means with K-means^{++} initialisation

```
for(k in 1:K) { # loop over k=1,...,K

  # Run K-means using K-Means++ initialisation:
  # use the current k value and do so ten times if k > 1
  KMPP    <- replicate(ifelse(k > 1, 10, 1),
                       list(kmeans(sdf, iter.max=100,
                                   centers=kmeans_pp(sdf, K=k))))

  # Extract and store the solution which minimises the TWCSS
  KM[[k]] <- KMPP[[which.min(sapply(KMPP, "[[", "tot.withinss"))]]

  # Extract the TWCSS value for the current value of k
  TWCSS[k] <- KM[[k]]$tot.withinss
}
```

As previously stated in Sect. 3.1.2.1, the so-called "elbow method" consists of plotting the range of K values on the x-axis against the corresponding obtained TWCSS values on the y-axis and looking for an kink in the resulting curve.

```
plot(x=1:K, y=TWCSS, type="b",
     xlab="K", ylab="Total Within-Cluster\n Sum-of-Squares")
```

Figure 2 suggests that $K = 4$ would produce the best results: beyond $K = 4$, the decrease in TWCSS is minimal, which suggests that an extra cluster is not required to model the data well; between $K = 3$ and $K = 4$, there is a more substantial decrease in TWCSS, which suggests that the fourth cluster is necessary. This method is of course highly subjective, and we will further inform our choice of K for K-means using silhouette widths and the ASW criterion in Sect. 4.2.4.

For now, we can interrogate the K-means solution with $K = 4$ by examining the sizes of the clusters and their centroid vectors, using the corresponding fourth element of the list `KM` by setting `K <- 4`.

```
K <- 4

KM[[K]]$size
```

```
[1] 127  57   8 247
```

```
KM[[K]]$centers
```

```
    InDegree  OutDegree Closeness_total Betweenness      Eigen Diffusion.degree
1 -0.4086612 -0.4381057      -1.4220978  -0.3994123 -0.5596480       -1.1138889
2  1.5706416  1.1372352       0.9247804   1.3859592  1.0781087        1.4216179
3  4.1035975  5.0512503       1.5201941   3.4673658  5.4946785        3.1631065
4 -0.2852445 -0.2007813       0.4685522  -0.2267743 -0.1390054        0.1422138
    Coreness Cross_clique_connectivity
1 -0.5672245                -0.3102236
2  1.7423739                 0.9615041
3  3.4821417                 5.5033583
4 -0.2232184                -0.2406243
```

However, these centroids correspond to the *scaled* version of the data created in Sect. 4.1.1. Interpretation can be made more straightforward by undoing the scaling transformation on these centroids, so that they are on the same scale as the data `df` rather than the scaled data `sdf` which was used as input to `kmeans()`. The column-wise means and standard deviations used when `scale()` was invoked are available as the attributes `"scaled:center"` and `"scaled:scale"`, respectively and are used in the code below. We show these centroids rounded to four decimal places in Table 3.

```
rescaled_centroids <- t(apply(KM[[K]]$centers, 1, function(r) {
                           r * attr(sdf, "scaled:scale") + attr(sdf,
                           "scaled:center")
                           } ))
round(rescaled_centroids, 4)
```

Now, we can more clearly see that the first cluster, with $n_{k=1} = 127$ observations, has the lowest mean value for all $d = 8$ centrality measures, while the last cluster, the largest with $n_{k=4} = 247$ observations, has moderately larger values for all centrality measures. The two smaller clusters, cluster two with $n_{k=2} = 57$ observations and cluster three with only $n_{k=3} = 8$ observations have the second-largest and largest values for each measure, respectively. As previously stated, we

Table 3 Centroids from the $K = 4$ K-means solution on the original data scale

InDegree	OutDegree	Closeness_total	Betweenness	Eigen	Diffusion.degree	Coreness	Cross_clique_connectivity
1.1024	1.7480	0.0008	20.7010	0.0071	144.1732	2.6378	4.6850
15.3684	14.8421	0.0010	849.9328	0.0996	1370.1053	16.1053	157.5789
33.6250	47.3750	0.0011	1816.6608	0.3492	2212.1250	26.2500	703.6250
1.9919	3.7206	0.0010	100.8842	0.0308	751.5061	4.6437	13.0526

defer a more-detailed interpretation of uncovered clusters to Sect. 4.2.5, after the optimal clustering solution has been identified.

4.2.2 K-medoids Application

The function `pam()` in the `cluster` library implements the PAM algorithm for K-medoids clustering. By default, this function requires only the arguments `x` (a pre-computed pairwise dissimilarity matrix, as can be created from the functions `dist()`, `daisy()`, and more) and `k`, the number of clusters. However, there are many options for many additional speed improvements and initialisation strategies [48]. Here, we invoke a faster variant of the PAM algorithm which necessitates specification of `nstart` as a number greater than one, to ensure the algorithm is evaluated with multiple random initial medoid vectors, in a similar fashion to `kmeans()`. Thus, we call `pam()` with `variant="faster"` and `nstart=50` throughout.

Firstly though, we need to construct the pairwise dissimilarity matrices to be used as input to the `pam()` function. Unlike K-means, we are not limited to *squared* Euclidean distances. It is prudent, therefore, to explore K-medoids solutions with several different dissimilarity measures and compare the different solutions obtained for different measures. Each distance measure will yield different results at the same K value, thus the value of K and the distance measure must be considered as a pair when determining the optimal solution.

We begin by calculating the Euclidean, Manhattan, and Minkowski (with $p = 3$) distances on the scaled continuous data in `sdf`:

```
dist_euclidean <- dist(sdf, method="euclidean")

dist_manhattan <- dist(sdf, method="manhattan")

dist_minkowski3 <- dist(sdf, method="minkowski", p=3)
```

Secondly, we avail of another principal advantage of K-medoids; namely, the ability to incorporate categorical variables in mixed-type data sets, by calculating pairwise Gower distances between each row of `merged_df`, using `daisy()`.

```
dist_gower <- daisy(merged_df, metric="gower")
```

As per the K-means tutorial in Sect. 4.2.1, we can produce an elbow plot by running the algorithm over a range of K values and extracting the minimised within-cluster total distance achieved upon convergence for each value of K. We demonstrate this for the `dist_euclidean` input below.

```
K <- 10
WCTD_euclidean <- numeric(K)
pam_euclidean <- list()
for(k in 1:K) {
  pam_euclidean[[k]] <- pam(dist_euclidean, k=k,
                            variant="faster", nstart=50)
  WCTD_euclidean[k] <- pam_euclidean[[k]]$objective[2]
}
```

Equivalent code chunks for the `dist_manhattan`, `dist_minkowski3`, and `dist_gower` inputs are almost identical, so we omit them here for the sake of brevity. Suffice to say, equivalent lists `pam_manhattan`, `pam_minkowski3`, and `pam_gower`, as well as equivalent vectors `WCTD_manhattan`, `WCTD_minkowski3`, and `WCTD_gower`, can all be obtained. Using these objects, corresponding elbow plots for all four dissimilarity measures are showcased in Fig. 3.

Some of the elbow plots in Fig. 3 are more conclusive than others. As examples, there are reasonably clear elbows at $K = 3$ for the Euclidean and Minkowski distances, arguably an elbow at $K = 4$ for the Manhattan distance, and no clear, unambiguous elbow under the Gower distance. In any case, the elbow method only helps to identify the optimal K value for a given dissimilarity measure; we defer a discussion of how to choose the overall best K-medoids clustering solution to later in this tutorial.

For now, let's interrogate the $K = 3$ solution obtained using the Euclidean distance. Recall that the results are already stored in the list `pam_euclidean`, so we merely need to access the third component of that list by setting `K <- 3`. Firstly, we can examine the size of each cluster by tabulating the cluster-membership indicator vector as follows:

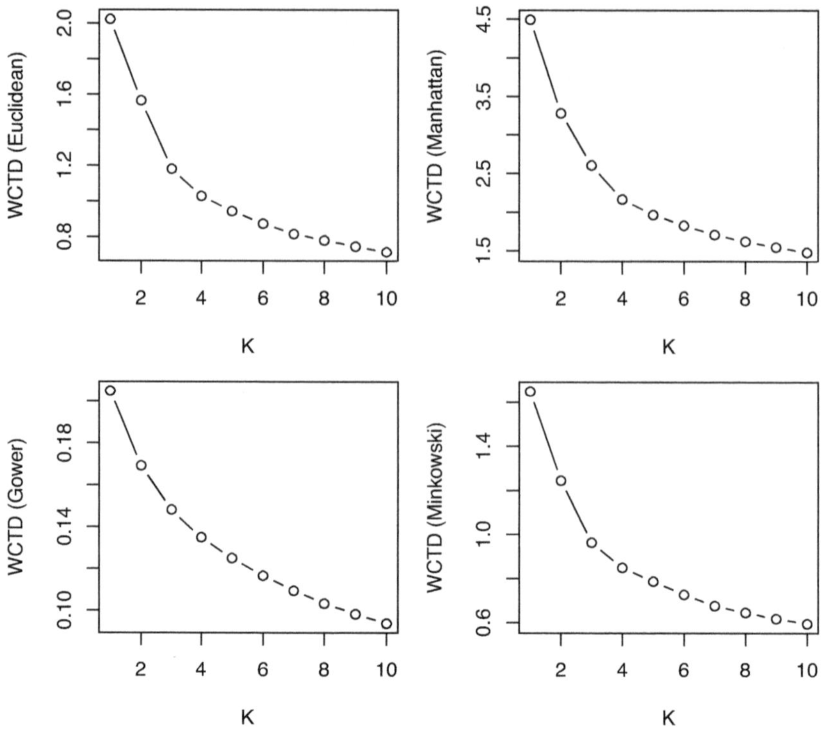

Fig. 3 Elbow plots for K-medoids clustering evaluated with different dissimilarity measures over a range of K values. In clockwise order, beginning with the top-left panel, these are the Euclidean distance, Manhattan distance, Minkowski distance and, for the merged data with additional categorical covariates, the Gower distance

```
K <- 3

table(pam_euclidean[[K]]$clustering)
```

```
  1   2   3
 67 122 250
```

Examining the medoids which serve as prototypes of each cluster is rendered difficult by virtue of the input having been a distance matrix rather than the data set itself. Though K-medoids defines the medoids to be the rows in the data set from which the distance to all other observations currently allocated to the same cluster, according to the specified distance measure, is minimised, the `medoids` component of the output instead gives the *indices* of the medoids within the data set.

```
pam_euclidean[[K]]$medoids
```

```
[1]  88 272  45
```

Table 4 Medoids for the $K = 3$ K-medoids solution obtained using the Euclidean distance on the original data scale, with the corresponding observation index in the name column

name	InDegree	OutDegree	Closeness _total	Betweenness	Eigen	Diffusion. degree	Coreness	Cross_clique _connectivity
88	16	14	0.0011	675.5726	0.1096	1415	18	157
272	1	1	0.0007	0.0000	0.0047	41	2	2
45	1	2	0.0010	29.4645	0.0194	684	3	5

Table 5 Cross-tabulation of the clusters obtained by K-means with $K = 4$ (along the columns) and K-medoids with $K = 3$ and the Euclidean distance (along the rows)

	1	2	3	4
1	0	57	8	2
2	122	0	0	0
3	5	0	0	245

Fortunately, these indices within sdf correspond to the same, unscaled observations within df. Allowing for the fact that observations with name greater than the largest index here were removed due to missingness, they are effectively the values of the name column corresponding to the medoids. Thus, we can easily examine the medoids on their original scale. In Table 4, they include the name column, which was *not* used when calculating the pairwise distance matrices, and are rounded to four decimal places.

```
df |>
  slice(as.numeric(pam_euclidean[[K]]$medoids)) |>
  round(4)
```

Thus, we can see that there is a small cluster with $n_{k=1} = 67$ observations which has the largest values for all $d = 8$ centrality measures, a slightly larger cluster with $n_{k=2} = 122$ observations and the lowest values for all variables, and the largest cluster with $n_{k=3} = 250$ and intermediate values for all variables. The cluster sizes of the K-medoids solution being more evenly balanced than the earlier K-means solution is an artefact of K-medoids being less susceptible to outlying observations by virtue of not squaring the distances. We can explore this by cross-tabulating the clusters obtained by K-means with $K = 4$ and K-medoids with $K = 3$ and the Euclidean distance in Table 5.

```
table(pam_euclidean[[3]]$clustering,
      KM[[4]]$cluster)
```

From the cross-tabulation in Table 5, we can see that the $n_k = 8$ observations in the smallest K-means cluster were absorbed into a larger cluster under the K-medoids solution, thereby demonstrating the robustness of K-medoids to outliers. Otherwise, both solutions are broadly in agreement.

4.2.3 Agglomerative Hierarchical Clustering Application

Performing agglomerative hierarchical clustering is straightforward now that the distance matrices have already been created for the purposes of running `pam()`. All that is required is to specify the distance matrix and an appropriate linkage criterion as the `method` when calling `hclust()`. We demonstrate this below for a subset of all possible distance measure and linkage criterion combinations among those described in Sect. 3.2.1. Recall that for the Ward criterion, the underlying distance measure is usually assumed to be squared Euclidean and that `"ward.D2"` is the correct `method` to use when the Euclidean distances are not already squared. For `method="centroid"`, we manually square the Euclidean distances.

```
hc_minkowski3_complete <- hclust(dist_minkowski3, method="complete")

hc_manhattan_single <- hclust(dist_manhattan, method="single")

hc_gower_average <- hclust(dist_gower, method="average")

hc_euclidean_ward <- hclust(dist_euclidean, method="ward.D2")

hc_euclidean2_centroid <- hclust(dist_euclidean^2, method="ward.D2")
```

Plotting the resulting dendrograms is also straightforward. Simply calling `plot()` on any of the items above will produce a dendrogram visualisation. We do so here for four of the hierarchical clustering solutions constructed above—the undepicted `hc_euclidean2_centroid` dendrogram is virtually indistinguishable from that of `hc_euclidean_ward`—while suppressing the observation indices along the x-axis for clarity by specifying `labels=FALSE`.

```
plot(hc_minkowski3_complete, labels=FALSE,
     main="", xlab="Minkwoski Distance (p=3) with Complete Linkage")
plot(hc_manhattan_single, labels=FALSE,
     main="", xlab="Manhattan Distance with Single Linkage")
plot(hc_gower_average, labels=FALSE,
     main="", xlab="Gower Distance with Average Linkage")
plot(hc_euclidean_ward, labels=FALSE,
     main="", xlab="Euclidean Distance with the Ward Criterion")
```

As previously alluded to in Sect. 3.2.2, some combinations of dissimilarity measure and linkage criterion are liable to produce undesirable results. The susceptibility to outliers of complete linkage clustering is visible in the top-left panel of Fig. 4, where just two observations form a cluster at a low height and are never again merged. The tendency of single linkage clustering to exhibit a "chaining" effect whereby all observations are successively merged into just one ever-larger cluster is evident in the top-right panel of Fig. 4, and similar behaviour is observed for the Gower distance with average linkage depicted in the bottom-left panel. The

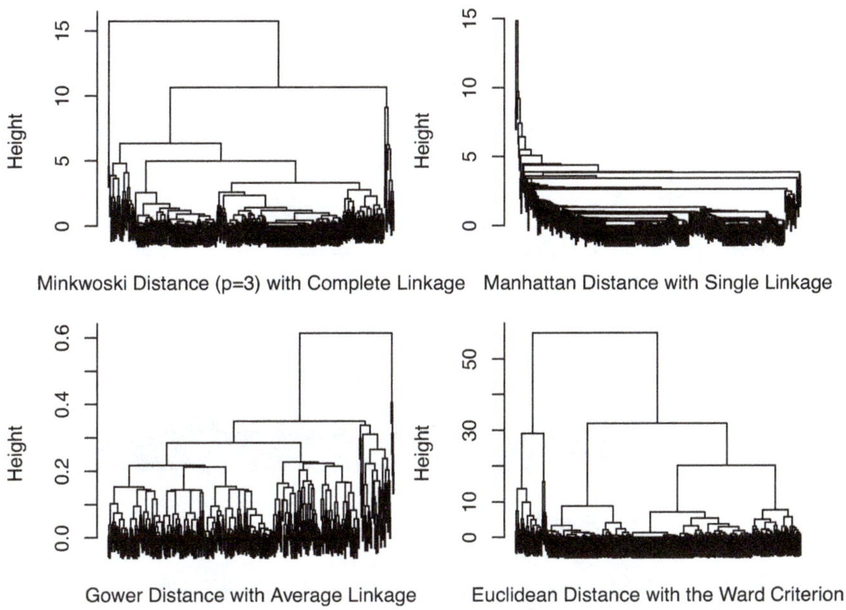

Fig. 4 Dendrograms obtained by agglomerative hierarchical clustering with a selection of dissimilarity measures and linkage criteria

most reasonable results appear to arise from using the Ward criterion in conjunction with Euclidean distances.

Taking the set of candidate partitions in `hc_euclidean_ward`, for the reasons outlined above, all that remains is to cut this dendrogram at an appropriate height. Practitioners have the freedom to explore different levels of granularity in the final partition. Figure 5 shows the dendrogram from the bottom-right panel of Fig. 4 cut horizontally at different heights, indicated by lines of different colours, as well as the corresponding implied K values.

Thereafter, one simply extracts the resulting partition by invoking `cutree()` with the appropriate height `h`. For example, to extract the clustering with $K = 3$, which we choose here because of the wide range of heights at which a $K = 3$ solution could be obtained:

```
hc_ward2 <- cutree(hc_euclidean_ward, h=45)
```

The object `hc_ward2` is now simply an vector indicating the cluster-membership of each observation in the data set. We show only the first few, for brevity, and then tabulate this vector to compute the cluster sizes. However, interpretation of these clusters is more difficult than in the case of K-means and K-medoids, as there is no centroid or medoid prototype with which to characterise each cluster.

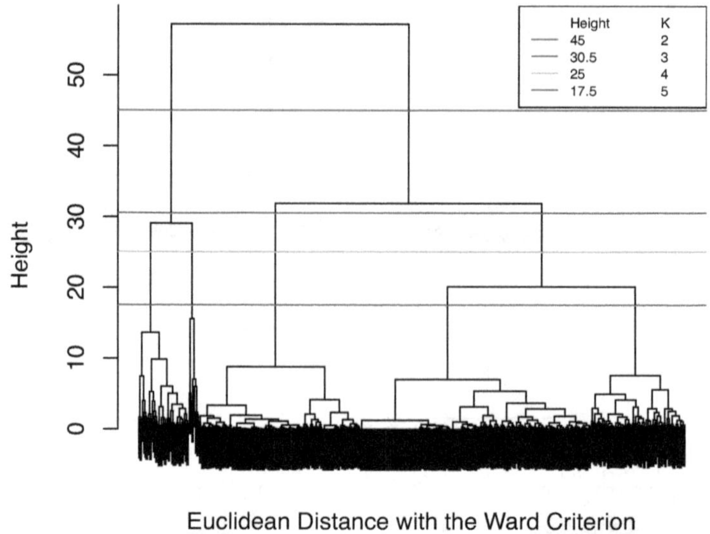

Fig. 5 Dendrogram obtained using the Euclidean distance and Ward criterion cut at different heights with the corresponding implied K indicated

```
head(hc_ward2)
```

```
[1] 1 2 2 2 1 1
```

```
table(hc_ward2)
```

```
hc_ward2
  1   2
 49 390
```

In this section, we have not presented an exhaustive evaluation of all possible pairs of dissimilarity measure and linkage criterion, but note that the code to do so is trivial. In any case, much like K-means and K-medoids, we must turn to other cluster quality indices to guide the choice of the best overall solution, be that choosing the best distance and linkage settings for agglomerative hierarchical clustering, or choosing the best clustering method in general among several competing methods. We now turn to finding the optimal clustering solution among multiple competing methods, guided by silhouette widths and plots thereof.

4.2.4 Identifying the Optimal Clustering Solution

In our application of K-means, the elbow method appeared to suggest that $K = 4$ yielded the best solution. In our applications of K-medoids, the elbow method suggested different values of K for different distance metrics. Finally, in our applications of hierarchical clustering, we noted that visualising the resulting dendrogram could be used to guide the choice of the height at which to cut to produce a single hard partition of K clusters. Now, we must determine which method yields the overall best solution. Following Sect. 3.3, we employ silhouettes for this purpose.

For K-means and agglomerative hierarchical clustering, silhouettes can be computed using the `silhouette` function in the `cluster` library, in which case the function requires two arguments; the integer vector containing the cluster membership labels and an appropriate dissimilarity matrix. Thus, for instance, silhouettes could be obtained easily for the following two examples (using `kmeans()` and `hclust()`, respectively).

```
kmeans_sil <- silhouette(kmeans(sdf, centers=kmeans_pp(sdf, K=4),
                                iter.max=100)$cluster,
                         dist_euclidean)
```

```
hclust_sil <- silhouette(cutree(hclust(dist_euclidean, method="ward.D2"), k=2),
                         dist_euclidean)
```

For K-medoids, it suffices only to supply the output of `pam()` itself, from which the cluster membership labels and appropriate dissimilarity matrix will be extracted internally, e.g.,

```
pam_sil <- silhouette(pam(dist_euclidean, k=3, variant="faster", nstart=50))
```

Thereafter, `plot()` can be called on `kmeans_sil`, `hclust_sil`, or `pam_sil` to produce a silhouette plot. For an example based on `hclust_sil`, see Fig. 6. Note that as $K = 2$ here, `col=2:3` colours the silhouettes according to their cluster.

```
plot(hclust_sil, main="", col=2:3)
```

Figure 6 shows that most silhouette widths are positive under this solution, indicating that most observations have been reasonably well-clustered. Cluster 2 appears to be the most cohesive, with a cluster-specific average silhouette width of 0.70, while cluster 1 appears to be the least cohesive, with a corresponding average of just 0.24. The overall ASW is 0.65, as indicated at the foot of the plot.

Given that we adopted a range of $K = 1, \ldots, 10$ when using K-means and K-medoids, and four different dissimilarity measures when using K-medoids, we have 50 non-hierarchical candidate solutions to evaluate, of which some seem more plausible than others according to the respective elbow plots. For agglomerative hierarchical clustering, an exhaustive comparison over a range of K values, for each dissimilarity measure and each linkage criterion, would be far too extensive for the present tutorial. Consequently, we limit our evaluation of silhouettes to the K-means and K-medoids solutions already present

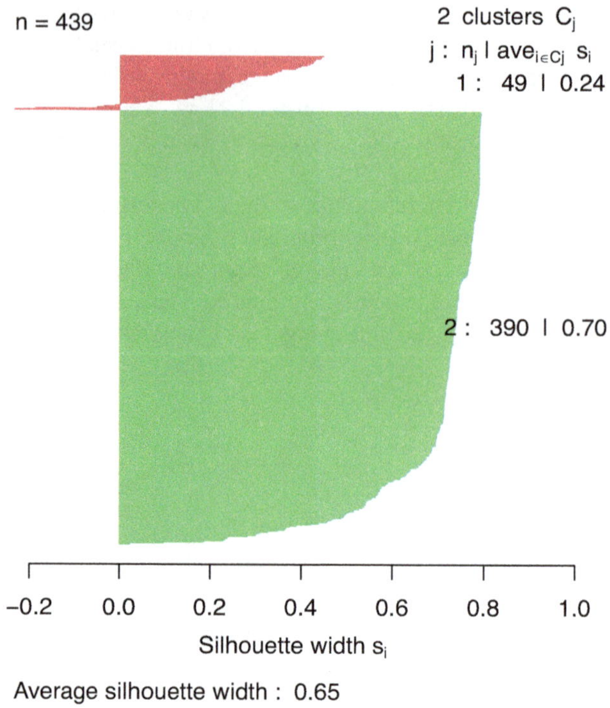

Fig. 6 Silhouette plot for the $K = 2$ hierarchical clustering solution obtained using the Ward criterion with Euclidean distances

in the objects `KM`, `pam_euclidean`, `pam_manhattan`, `pam_minkowski3`, and `pam_gower`, and the hierarchical clustering solutions which employ the Ward criterion in conjunction with Euclidean distances (of which we consider a further 10 solutions, again with $K = 1, \ldots, 10$, by considering the 10 possible associated heights at which the dendrogram can be cut). We limit the hierarchical clustering solutions to those based on the Ward criterion given that single linkage and complete linkage have been shown to be susceptible to chaining and sensitive to outliers, respectively. We use the corresponding pre-computed dissimilarity matrices `dist_euclidean`, `dist_manhattan`, `dist_minkowski3`, and `dist_gower`, where appropriate throughout.

Though the ASW associated with `hclust_sil` is given on the associated silhouette plot in Fig. 6, we can calculate ASW values for other solutions—which we must do to determine the best solution—without producing individual silhouette plots. To show how this can be done, we examine the structure of the `hclust_sil` object, showing only its first few rows.

```
head(hclust_sil)
```

```
     cluster neighbor sil_width
[1,]       1        2 0.4408007
[2,]       2        1 0.7416708
[3,]       2        1 0.7401549
[4,]       2        1 0.4696606
[5,]       1        2 0.2494557
[6,]       1        2 0.1915306
```

The columns relate to the cluster to which object i is assigned, the cluster for which the corresponding $b(i)$ was minimised, and the $s(i)$ score itself, respectively. Calculating `mean(hclust_sil[,3])` will return the ASW. Though the code is somewhat tedious, we calculate the ASW criterion values for all 60 candidate solutions—that is, six methods evaluated over $K = 1, \ldots, 10$—using a small helper function to calculate the ASW for the sake of tidying the code.

```
K <- 10
ASW <- function(x) mean(x[,3])
silhouettes <- data.frame(K=2:K,
  kmeans=sapply(2:K, function(k) ASW(silhouette(KM[[k]]$
                                  cluster, dist_euclidean))),
  kmedoids_euclidean=sapply(2:K, function(k) ASW(silhouette
                           (pam_euclidean[[k]]))),
  kmedoids_manhattan=sapply(2:K, function(k) ASW(silhouette
                           (pam_manhattan[[k]]))),
  kmedoids_minkowski3=sapply(2:K, function(k) ASW(silhouette
                            (pam_minkowski3[[k]]))),
  kmedoids_gower=sapply(2:K, function(k) ASW(silhouette
                       (pam_gower[[k]]))),
  hc_euclidean_ward=sapply(2:K, function(k) ASW(silhouette(cutree
                          (hc_euclidean_ward, k), dist_euclidean))))
```

In Fig. 7, we plot these silhouettes against K using `matplot()`, omitting the code to do so for brevity.

According to Fig. 7, there is generally little support for $K > 5$ across almost all methods considered, as most method's ASW values begin to decline after this point. The ASW values also make clear that incorporating the additional categorical demographic variables in a mixed-type clustering using the Gower distance does not lead to reasonable partitions, for any number of clusters K. Overall, the most promising solutions in terms of having the highest ASW are K-means, Ward hierarchical clustering, and K-medoids with the Manhattan distance, all with $K = 2$, and K-medoids with the Euclidean and Minkowski distances, each with $K = 3$. However, it would be wise to examine silhouette widths in more detail, rather than relying merely on the average. The silhouettes for this hierarchical clustering solution are already depicted in Fig. 6, so Fig. 8 shows individual silhouette widths for the remaining solutions.

It is notable that the silhouettes and ASW of the $K = 2$ K-means solution (top-left panel of Fig. 8) and the Ward hierarchical clustering solution (Fig. 6) appear almost identical (if one accounts for the clusters being relabelled and associated

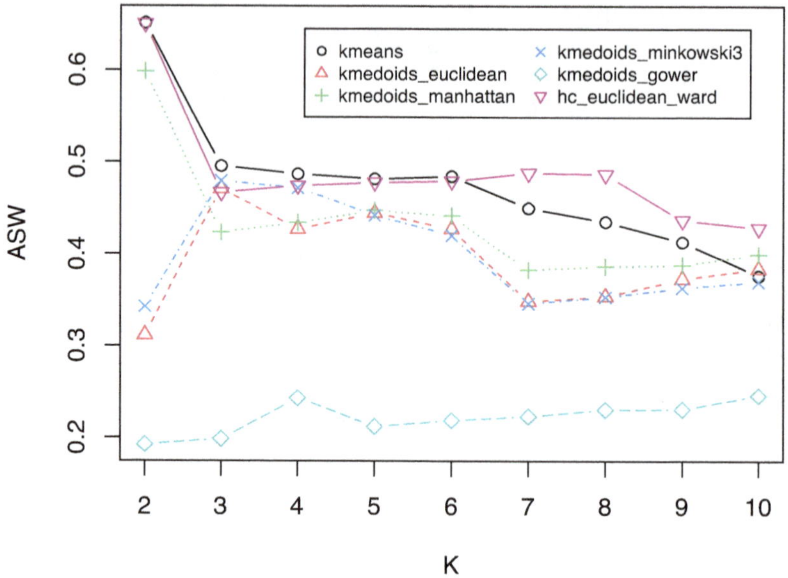

Fig. 7 ASW criterion values plotted against K for K-means, K-medoids (with the Euclidean, Manhattan, Minkowski ($p = 3$), and Gower distances), and agglomerative hierarchical clustering based on Euclidean distance and the Ward criterion

colours being swapped). Indeed, according to a cross-tabulation of their partitions (not shown), their assignments differ for just 4 out of $n = 439$ observations. Despite having the highest ASW, we can justify dismissing these solutions given that $K = 2$ was not well-supported by its corresponding elbow plot in Fig. 2. Similar logic suggests that the $K = 3$ K-means solution and the $K = 2$ K-medoids solution based on the Manhattan distance can also be disregarded. Although we stress again that an ideal analysis would more thoroughly determine an optimal solution with reference to additional cluster quality measures and note that various clustering solutions can be legitimate, for potentially different clustering aims, on the same data set [2, 3], we can judge that—among the two $K = 3$ K-medoids solutions—the one based on the Euclidean distance is arguably preferable, for two reasons. Firstly, its ASW is quite close to that of the Minkowski distance solution:

```
silhouettes |>
  filter(K == 3) |>
  select(kmedoids_euclidean, kmedoids_minkowski3)
```

```
  kmedoids_euclidean kmedoids_minkowski3
1           0.470844           0.4795972
```

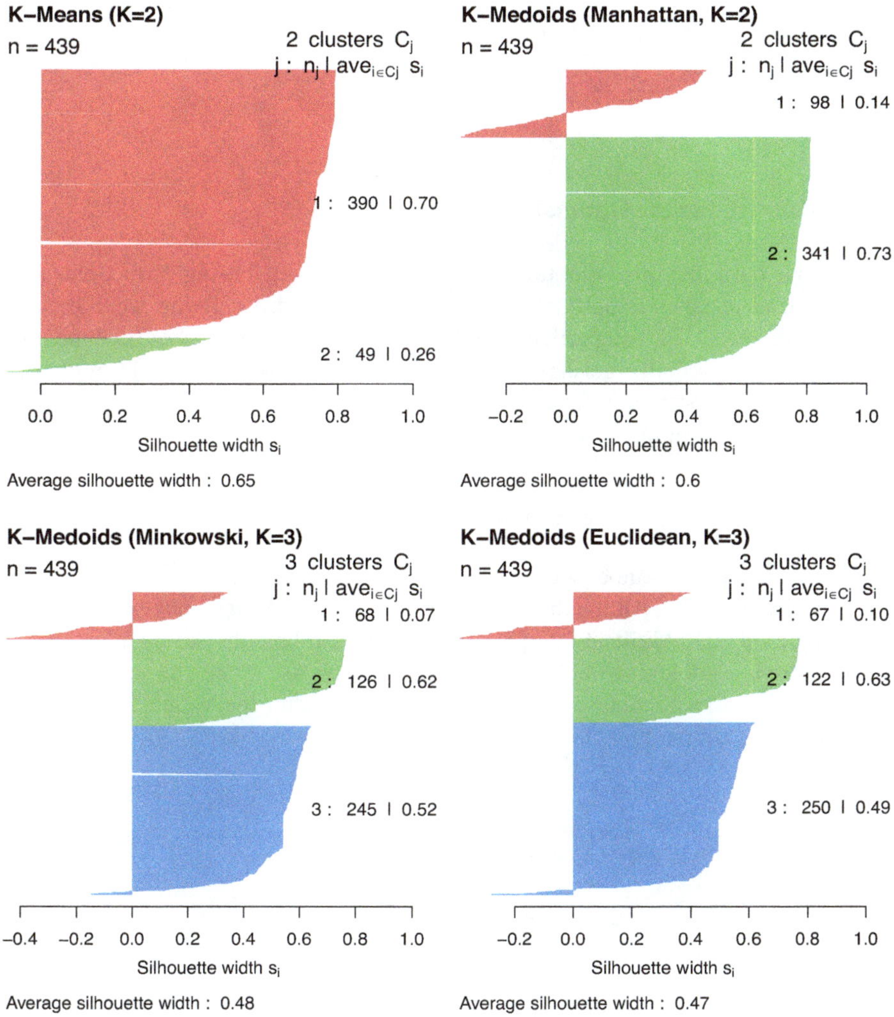

Fig. 8 Silhouette plots showing silhouette widths for a numbering of promising solutions, coloured according to cluster membership

Secondly, the first cluster has a higher cluster-specific average silhouette width under the solution based on the Euclidean distance. Indeed, this solution has fewer negative silhouette widths also:

```
sum(silhouette(pam_euclidean[[3]])[,3] < 0)
```

```
[1] 27
```

```
sum(silhouette(pam_minkowski3[[3]])[,3] < 0)
```

```
[1] 28
```

4.2.5 Interpreting the Optimal Clustering Solution

By now, we have identified that the $K = 3$ solution obtained using K-medoids and the Euclidean distance is optimal. Although aspects of this solution were already discussed in Sect. 4.2.2—in particular, Table 4 has already shown the $K = 3$ medoid vectors obtained at convergence—we now turn to examining this output in greater detail, in order to provide a fuller interpretation of each of the uncovered clusters. We extract this solution for ease of manipulation.

```
final_pam <- pam_euclidean[[3]]
```

Recall that this method yielded three clusters of sizes $n_1 = 67$, $n_2 = 122$, and $n_3 = 250$. Although the categorical variables were not used by this clustering method, additional interpretative insight can be obtained by augmenting the respective medoids in Table 4 with these cluster sizes and the `experience` values of the corresponding rows of the `merged_df` data set, which includes the additional demographic features.

```
df |>
  slice(as.numeric(final_pam$medoids)) |>
  round(4) |>
  mutate(size=table(final_pam$clustering)) |>
  left_join(slice(demog |>
                    select(UID, experience),
                  as.numeric(final_pam$medoids)),
            by=join_by(name == UID)) |>
  select(-name)
```

Interpretation and labeling of the clustering results is the step that follows, with a focus only on the medoid values of the centrality scores (had the optimal solution been obtained by `kmeans()`, we could instead examine its centroids in its `$centers` component, i.e., mean vectors). We will follow the example papers that we used as a guide for choosing the centrality measures [28] and [61]. Both papers have used traditional centrality measures (e.g., degree, closeness, and betweenness) as well as diffusion centralities (diffusion degree and coreness) to infer students' roles. Furthermore, the second paper has an extended review of the roles and how they have been inferred from centrality measures, so readers are encouraged to read this review.

Table 6 Medoids for the $K = 3$ K-medoids solution obtained using the Euclidean distance on the original data scale, augmented with the cluster sizes and the corresponding values of the `experience` variable (which was not directly used by the clustering algorithm)

InDegree	OutDegree	Closeness_total	Betweenness	Eigen	Diffusion.degree	Coreness	Cross_clique_connectivity	size	experience
16	14	0.0011	675.5726	0.1096	1415	18	157	67	1
1	1	0.0007	0.0000	0.0047	41	2	2	122	3
1	2	0.0010	29.4645	0.0194	684	3	5	250	2

As the data shows, the first cluster has the highest degree centrality measures (`InDegree` and `OutDegree`), highest betweenness centrality, as well as the highest values of the diffusion centralities (`Diffusion_degree` and `Coreness`). These values are concordant with students who were actively engaged, received multiple replies, had their contributions discussed by others, and achieved significant diffusion. All of such criteria are concordant with the role of *leaders*. It stands to reason that the *leaders* cluster would be the smallest, with $n_1 = 67$.

The third cluster has intermediate values for the degree centralities, high diffusion centrality values, as well as relatively high values of betweenness centrality. Such values are concordant with the role of an active participant who participates and coordinates the discussion. Therefore, we will use the label of *coordinators*.

Finally, the second cluster has the lowest values for all centrality measures (though its diffusion values are still fairly reasonable). Thus, this cluster could feasibly be labelled as an *isolates* cluster, gathering participants whose role in the discussions is peripheral at best. Overall, the interpretations of this $K = 3$ solution are consistent with other findings in the existing literature, e.g., [28].

We can now label the clusters accordingly to facilitate more informative cluster-specific summaries. Here, we also recall the size of each cluster with the new labels invoked, to demonstrate their usefulness.

```
final_pam$clustering <- factor(final_pam$clustering,
                               labels=c("leaders", "coordinators",
                                        "isolates"))
table(final_pam$clustering)
```

```
     leaders coordinators     isolates
          67          122          250
```

As an example, we can use these labels to guide a study of the mean vectors of each cluster (bearing in mind that these are not centroid centroid vectors obtained by K-means, but rather mean vectors obtained calculated for each group defined by the K-medoids solution), for which the interpretation of the leader, coordinator, and isolate labels are still consistent with the conclusions drawn from the medoids in Table 4. Note that, for the sake of readability, the group-wise summary below is transposed.

```
df |>
  group_by(clusters=final_pam$clustering) |>
  select(-name) |>
  summarise_all(mean) |>
  mutate_if(is.numeric, round, 2) |>
  pivot_longer(cols=-clusters,
               names_to="centrality") |>
```

```
  pivot_wider(names_from=clusters,
              values_from=value)
```

```
# A tibble: 8 x 4
  centrality                 leaders coordinators isolates
  <chr>                        <dbl>        <dbl>    <dbl>
1 InDegree                      17.1         1.1      1.98
2 OutDegree                     18.8         1.68     3.62
3 Closeness_total                0           0        0
4 Betweenness                  943.         21.2     99.1
5 Eigen                          0.13        0.01     0.03
6 Diffusion.degree            1466.        132.     742.
7 Coreness                      17.2         2.56     4.58
8 Cross_clique_connectivity  220.          4.53    12.6
```

From Table 6, we can also see that each observation which corresponds to a cluster medoid contains low, high, and medium levels of experience, respectively. However, one should be cautious not to therefore conclude that the clusters neatly map to experience levels, as the following cross-tabulation indicates little-to-no agreement between the groupings of experience levels in the data and the uncovered clusters.

```
table(final_pam$clustering,
      merged_df$experience,
      dnn=c("Clusters", "Experience"))
```

```
              Experience
Clusters        1  2  3
  leaders      17 23 27
  coordinators 40 34 48
  isolates     61 93 96
```

Finally, we can produce a visualisation of the uncovered clusters in order to better understand the solution. Visualising multivariate data with $d > 2$ is challenging and consequently such visualisations must resort to either plotting the first two principal components or mapping the pairwise dissimilarity matrix to a configuration of points in Cartesian space using multidimensional scaling. The latter is referred to as a "CLUSPLOT" [62] and is implemented in the `clusplot()` function in the same `cluster` library as `pam()` itself. This function uses classical (metric) multi-dimensional scaling [63] to create a bivariate plot displaying the partition of the data. Observations are represented by points in the scatter plot an ellipse spanning the smallest area containing all points in the given cluster is drawn around each cluster. In the code below, only `clusplot(final_pam)` is strictly necessary to produce such a plot for the optimal $K = 3$ Euclidean distance K-medoids solution; all other

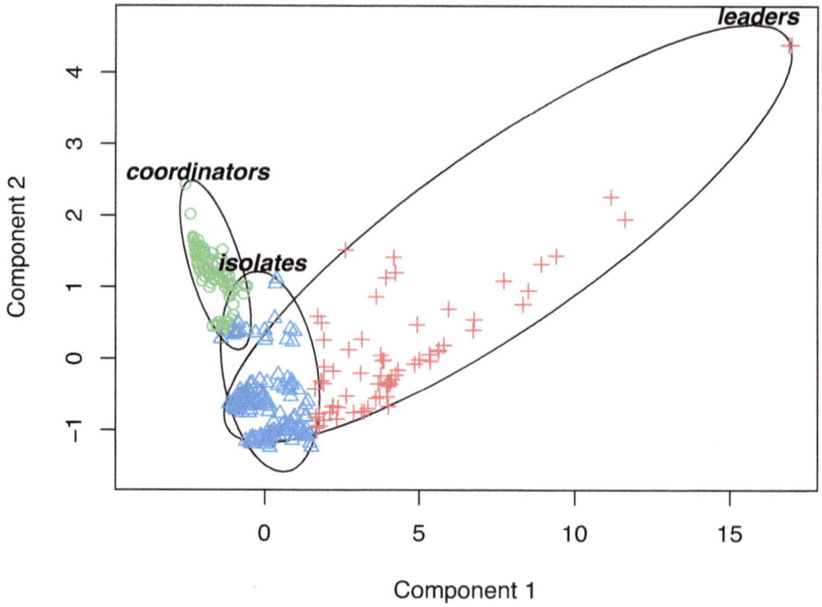

Fig. 9 Two-dimensional clustering plot for the final $K = 3$ Euclidean distance K-medoids solution obtained via classical multidimensional scaling. The ellipses around each cluster are given the associated labels of *leaders*, *coordinators*, and *isolates* and the points are coloured according to cluster membership

arguments are purely for cosmetic purposes for the sake of the resulting Fig. 9 and are described in `?clusplot`.

Figure 9 shows that the *leaders* cluster—the smallest cluster with the highest value for all centrality measures—is quite diffuse. Conversely, the larger *coordinators* cluster, with the smallest values for all centrality measures, and the *isolates* cluster, the largest of all, with intermediate values for all centrality measures, are more compact. This is consistent with the cluster-specific average silhouette widths shown in the bottom-right panel of Fig. 8. That being said, the large span of the *leaders* cluster again affirms the relative robustness of K-medoids to outliers, of which some (all of which are leaders) still remain despite the earlier pre-processing steps.

```
clusplot(final_pam,                                          # the pam() output
         main="", sub=NA,                   # remove the main title and subtitle
         lines=0,                            # do not draw any lines on the plot
         labels=4,                                      # only label the ellipses
         col.clus="black",                        # colour for ellipses and labels
         col.p=as.numeric(final_pam$clustering) + 1, # colours for points
         cex.txt=0.75,                             # control size of text labels
         xlim=c(-4, 17)             # expand x-axis to avoid trimming labels
         )
```

5 Discussion and Further Readings

The present analysis of the MOOC centralities data set incorporated three of the most commonly used dissimilarity-based clustering approaches; namely, K-means, K-medoids, and agglomerative hierarchical clustering. Throughout the applications, emphasis was placed on the sensitivity of the results to various choices regarding algorithmic inputs available to practitioners, be that the choice of how to choose initial centroids for K-means, the choice of dissimilarity measure for K-medoids, or the choice of linkage criterion for hierarchical clustering, as examples. Consequently, our analysis considered different values of K, different distances, different linkage criteria, different combinations thereof, and indeed different clustering algorithms entirely, in an attempt to uncover the "best" clustering solution for the MOOC centralities data set. We showed how elbow plots and other graphical tools can guide the choice of K for a given method but ultimately identified—via the average silhouette width criterion—an optimal solution with $K = 3$, using K-medoids in conjunction with the Euclidean distance measure. Although we note that there are a vast array of other cluster quality measures in the literature which target different definitions of what constitutes a "good" clustering [3], the solution we obtained seems reasonable, in that the uncovered clusters which we labelled as *leaders*, *coordinators*, and *isolates* are consistent, from an interpretative point of view, with existing educational research. Indeed, several published studies have uncovered similar patterns of three groups which can be labelled in the same way [28, 64, 65].

Nonetheless, there are some limitations to our applications of dissimilarity-based clustering methods in this tutorial which are worth mentioning. Firstly, we note firstly that the decision to standardise the variables—in particular, to normalise them by subtracting their means and dividing by their standard deviations—was applied across the board to each clustering method we explored. Although the standardised data were only used as inputs to each algorithm (i.e., the output was always interpreted on the original scale), we note that different groups are liable to be uncovered with different standardisation schemes. In other words, not standardising the data, or using some other standardisation method (e.g., rescaling to the [0, 1] range) may lead to different, possibly more or less meaningful clusters.

A second limitation is that all variables in Table 1 were used as inputs (either directly in the case of K-means, or indirectly as inputs to the pairwise dissimilarity matrix calculations required for K-medoids and hierarchical clustering). As well as increasing the computational burden, using all variables can be potentially problematic in cases where the clustering structure is driven by some $d^\star < d$ variables, i.e., if the data can be separated into homogeneous subgroups along fewer dimensions. In such instances where some of the variables are uninformative in terms of explaining the variability in the data, variable selection strategies tailored to the unsupervised paradigm may be of interest and again may lead to more meaningful clusters being found [66, 67].

Although dissimilarity-based clustering encompasses a broad range of flexible methodologies which can be utilised in other, diverse settings—for example, dissimilarity-based clustering is applied in the context of longitudinal categorical data in the chapter on sequence analysis methods (Chapter 10; [18])—there are other clustering paradigms which may be of interest in similar or alternative settings as the data used in the present tutorial, even if they are not yet widely adopted in educational research. We now briefly introduce some of these alternative clustering frameworks for readers interested in expanding their knowledge of the topic of clustering beyond the dissimilarity-based framework detailed herein.

A first alternative to dissimilarity-based clustering is the density-based clustering framework, most famously exemplified by the DBSCAN clustering algorithm [68, 69]. Density-based clustering broadly defines clusters as areas of higher density than the remainder of the data set, where objects are closely packed together with many nearby neighbours, while objects in the sparse areas which separate clusters are designated as outliers. This has been identified as one main advantage of DBSCAN by authors who applied it an education research context [70]—insofar as the capability to easily and effectively separate individual exceptionally poor or exceptionally outstanding students who require special attention from teachers is desirable—along with DBSCAN obviating the need to pre-specify the number of clusters K.

Another alternative is given by the model-based clustering paradigm, which is further discussed in Chapter 9 [8]. Although we direct readers to that chapter for a full discussion of model-based clustering using finite Gaussian mixture models, the relationship between this approach and latent profile analysis, and a tutorial in R using the `mclust` package [71], there are some aspects and advantages which are pertinent to discuss here. Firstly, as model-based clustering is based on an underlying generative probability model, rather than relying on dissimilarity-based heuristics, it admits the use of principled, likelihood-based model-selection criteria such as the Bayesian information criterion [72], thereby eliminating the subjectivity of elbow plots and other graphical tools for guiding the choice of K. Secondly, such models can be extended to allow covariates to guide the construction of the clusters [73], thereby enabling, for example, incorporation of the categorical demographic variables associated with the MOOC centralities data set used in the present tutorial.

A third advantage of model-based clustering is that it returns a "soft" partition, whereas dissimilarity-based approaches yield either a single "hard" partition (with each observation placed in one group only), under partitional methods like K-means or K-medoids, or a set of nested hard partitions from which a single hard partition can be extracted, under hierarchical clustering. Specifically, model-based clustering assigns each observation a probability of belonging to each cluster, such that observations can have a non-negative association with more than one cluster and the uncertainty of the cluster memberships can be quantified. This has the effect of diminishing the effect of outliers or observations which lie on the boundary of two or more natural clusters, as they are not forced to wholly belong to one cluster.

In light of these concerns, another clustering paradigm of potential interest is that of fuzzy clustering, which is notable for allowing for "soft" cluster-membership assignments while still being dissimilarity-based [74, 75]. Indeed, there are fuzzy variants of the K-means and K-medoids algorithms which relax the assumption that the latent variable $\mathbf{z}_i$ encountered in Eq. (1), for example, is binary. They are implemented in the functions `FKM()` and `FKM.med()`, respectively, in the `fclust` R package [76].

Another advantage of model-based clustering over the dissimilarity-based paradigm is the flexibility it affords in relation to the shapes, orientations, volumes, and sizes of the clusters it uncovers. At their most flexible, finite Gaussian mixture models can find clusters where all of these characteristics differ between each cluster, but intermediary configurations—whereby, as but one example, clusters can be constrained to have equal volume and orientation but varying shape and size—are permissible. This is achieved via different parsimonious parameterisations of the cluster-specific covariance matrices which control the geometric characteristics of the corresponding ellipsoids; see Chapter 9 [8] for details. By contrast, K-means is much more restrictive. The algorithm assumes that clusters are of similar size, even if the estimated clusters can vary in size upon convergence. Moreover, by relying on squared Euclidean distances to a mean vector, with no recourse to modelling covariances, K-means implicitly assumes that all clusters are spherical in shape, have equal volume, and radiate around their centroid. This can have the damaging consequence, in more challenging applications, that a larger number of spherical clusters may be required to fit the data well, rather than a more parsimonious and easily interpretable mixture model with fewer non-spherical components. Finally, in light of these concerns, we highlight the spectral clustering approach [77], implemented via the function `specc()` in the `kernlab` package in R, which is, like model-based clustering, capable of uncovering clusters with more flexible shapes, particularly when the data are not linearly separable, and shares some connections with kernel K-means [78], another flexible generalisation of the standard K-means algorithm adopted in this tutorial.

Overall, we encourage readers to further explore the potential of dissimilarity-based clustering methods, bear in mind the limitations and practical concerns of each algorithm discussed in this tutorial, and remain cognisant of the implications thereof for results obtained in educational research applications. We believe that by paying particular attention to the guidelines presented for choosing an optimal partition among multiple competing methodologies, with different numbers of clusters and/or different dissimilarity measures and/or different linkage criteria, more meaningful and interpretable patterns of student behaviour can be found. Finally, we hope that the additional references provided to other clustering frameworks will inspire a broader interest in the topic of cluster analysis among practitioners in the field of education research.

References

1. Everitt BS, Landau S, Leese M, Stahl D (2011) Cluster analysis, Fifth. John Wiley & Sons, New York, NY, U.S.A.
2. Hennig C (2015) What are the true clusters? Pattern Recognition Letters 64:53–62
3. Hennig C (2016) Clustering strategy and method selection. In: Hennig C, Meila M, Murtagh F, Rocci R (eds) Handbook of Cluster Analysis. Chapman; Hall/CRC Press, New York, N.Y., U.S.A., pp 703–730
4. MacQueen JB (1967) Some methods for classification and analysis of multivariate observations. In: Cam LML, Neyman J (eds) Proceedings of the Fifth Berkeley Symposium on Mathematical Statistics and Probability. University of California Press, June 21–July 18, 1965; December 27 1965–January 7, 1966, Statistical Laboratory of the University of California, Berkeley, CA, U.S.A., pp 281–297
5. Kaufman L, Rousseeuw PJ (1990) Partitioning around medoids (program PAM). In: Kaufman L, Rousseeuw PJ (eds) Finding Groups in Data: An Introduction to Cluster Analysis. John Wiley & Sons, New York, NY, U.S.A., pp 68–125
6. Kaufman L, Rousseeuw PJ (1990) Agglomerative nesting (program AGNES). In: Kaufman L, Rousseeuw PJ (eds) Finding Groups in Data: An Introduction to Cluster Analysis. John Wiley & Sons, New York, NY, U.S.A., pp 199–252
7. R Core Team (2023) R: A language and environment for statistical computing. R Foundation for Statistical Computing, Vienna, Austria
8. Scrucca L, Saqr M, López-Pernas S, Murphy K (2024) An introduction and R tutorial to model-based clustering in education via latent profile analysis. In: Saqr M, López-Pernas S (eds) Learning Analytics Methods and Tutorials: A Practical Guide using R. Springer
9. Bouveyron C, Celeux G, Murphy TB, Raftery AE (2019) Model-Based Clustering and Classification for Data Science: With Applications in R. Cambridge University Press, Cambridge, UK
10. Rennen-Allhoff B, Allhoff P (1983) Clusteranalysen bei psychologisch-pädagogischen Fragestellungen. Psychologie in Erziehung und Unterricht 30:253–261
11. Hickendorff M, Edelsbrunner PA, McMullen J, Schneider M, Trezise K (2018) Informative tools for characterizing individual differences in learning: Latent class, latent profile, and latent transition analysis. Learning and Individual Differences 66:4–15
12. Saqr M, López-Pernas S (2021) The longitudinal trajectories of online engagement over a full program. Computers & Education 175:104325
13. Cook CR, Kilgus SP, Burns MK (2018) Advancing the science and practice of precision education to enhance student outcomes. Journal of School Psychology 66:4–10
14. Howard MC, Hoffman ME (2018) Variable-centered, person-centered, and person-specific approaches: Where theory meets the method. Organizational Research Methods 21:846–876
15. Richters JE (2021) Incredible utility: The lost causes and causal debris of psychological science. Basic and Applied Social Psychology 43:366–405
16. Saqr M, López-Pernas S (2023) The temporal dynamics of online problem-based learning: Why and when sequence matters. International Journal of Computer-Supported Collaborative Learning 18:11–37
17. Dutt A (2015) Clustering algorithms applied in educational data mining. International Journal of Information and Electronics Engineering 5:112–116
18. Saqr M, López-Pernas S, Helske S, Durand M, Murphy K, Studer M, Ritschard G (2024) Sequence analysis: Basic principles, technique, and tutorial. In: Saqr M, López-Pernas S (eds) Learning Analytics Methods and Tutorials: A Practical Guide using R. Springer
19. Beder HW, Valentine T (1990) Motivational profiles of adult basic education students. Adult Education Quarterly 40:78–94
20. Clément R, Dörnyei Z, Noels KA (1994) Motivation, self-confidence, and group cohesion in the foreign language classroom. Language Learning 44:417–448
21. Fernandez-Rio J, Méndez-Giménez A, Cecchini Estrada JA (2014) A cluster analysis on

students' perceived motivational climate. Implications on psycho-social variables. The Spanish Journal of Psychology 17:E18
22. Cahapin EL, Malabag BA, Santiago J Cereneo Sailog, Reyes JL, Legaspi GS, Adrales KL (2023) Clustering of students admission data using K-means, hierarchical, and DBSCAN algorithms. Bulletin of Electrical Engineering and Informatics 12:3647–3656
23. Saqr M, Tuominen V, Valtonen T, Sointu E, Väisänen S, Hirsto L (2022) Teachers' learning profiles in learning programming: The big picture! Frontiers in Education 7:1–10
24. Jovanović J, Gaševič D, Dawson S, Pardo A, Mirriahi N (2017) Learning analytics to unveil learning strategies in a flipped classroom. The Internet and Higher Education 33:74–85
25. López-Pernas S, Saqr M (2021) Bringing synchrony and clarity to complex multi-channel data: A learning analytics study in programming education. IEEE Access 9:166531–166541
26. López-Pernas S, Saqr M, Viberg O (2021) Putting it all together: Combining learning analytics methods and data sources to understand students' approaches to learning programming. Sustainability 13:4825
27. Fan Y, Tan Y, Raković M, Wang Y, Cai Z, Shaffer DW, Gašević D (2022) Dissecting learning tactics in MOOC using ordered network analysis. Journal of Computer Assisted Learning 39:154–166
28. Saqr M, López-Pernas S (2021) Modelling diffusion in computer-supported collaborative learning: A large scale learning analytics study. International Journal of Computer-Supported Collaborative Learning 16:441–483
29. Perera D, Kay J, Koprinska I, Yacef K, Zaïane OR (2009) Clustering and sequential pattern mining of online collaborative learning data. IEEE Transactions on Knowledge and Data Engineering 21:759–772
30. Saqr M, López-Pernas S, Jovanović J, Gašević D (2023) Intense, turbulent, or wallowing in the mire: A longitudinal study of cross-course online tactics, strategies, and trajectories. The Internet and Higher Education 57:100902
31. Vieira Roque F, Cechinel C, Merino E, Villarroel R, Lemos R, Munoz R (2018) Using multimodal data to find patterns in student presentations. In: 2018 XIII Latin American Conference on Learning Technologies (LACLO). São Paulo, Brazil, pp 256–263
32. Lee J-E, Chan JY-C, Botelho A, Ottmar E (2022) Does slow and steady win the race?: Clustering patterns of students' behaviors in an interactive online mathematics game. Educational Technology Research and Development 70:1575–1599
33. López-Pernas S, Saqr M, Gordillo A, Barra E (2022) A learning analytics perspective on educational escape rooms. Interactive Learning Environments 1–17
34. Rosa PJ, Morais D, Gamito P, Oliveira J, Saraiva T (2016) The immersive virtual reality experience: A typology of users revealed through multiple correspondence analysis combined with cluster analysis technique. Cyberpsychology, Behavior and Social Networking 19:209–216
35. Wang X, Liu Q, Pang H, Tan SC, Lei J, Wallace MP, Li L (2023) What matters in AI-supported learning: A study of human-AI interactions in language learning using cluster analysis and epistemic network analysis. Computers & Education 194:104703
36. Maechler M, Rousseeuw P, Struyf A, Hubert M, Hornik K (2022) cluster: cluster analysis basics and extensions
37. Lloyd SP (1982) Least squares quantization in PCM. IEEE Transactions on Information Theory 28:129–137
38. Forgy EW (1965) Cluster analysis of multivariate data: Efficiency vs interpretability of classifications. Biometrics 21:768–769
39. Hartigan JA, Wong MA (1979) Algorithm AS 136: a K-means clustering algorithm. Journal of the Royal Statistical Society: Series C (Applied Statistics) 28:100–108
40. Arthur D, Vassilvitskii S (2007) K-means^{++}: The advantages of careful seeding. In: SODA '07: Proceedings of the Eighteenth Annual ACM-SIAM Symposium on Discrete Algorithms. Society for Industrial; Applied Mathematics, Philadelphia, PA, U.S.A., pp 1027–1035
41. Hamming RW (1950) Error detecting and error correcting codes. The Bell System Technical Journal 29:147–160

42. Jaccard P (1901) Distribution de la flore alpine dans le bassin des Dranses et dans quelqus régions voisines. Bulletin de la Société Vaudoise des Sciences Naturelles 37:241–272
43. Dice LR (1945) Measures of the amount of ecologic association between species. Ecology 26:397–302
44. Sørensen T (1948) A method of establishing groups of equal amplitude in plant sociology based on similarity of species and its application to analyses of the vegetation on Danish commons. Kongelige Danske Videnskabernes Selskab 5:1–34
45. Gower JC (1971) A general coefficient of similarity and some of its properties. Biometrics 27:857–871
46. Huang Z (1998) Extensions to the k-means algorithm for clustering large data sets with categorical values. Data Mining and Knowledge Discovery 2:283–304
47. Huang Z (1997) Clustering large data sets with mixed numeric and categorical values. In: Lu H, Motoda H, Luu H (eds) KDD: Techniques and Applications. World Scientific, Singapore
48. Schubert E, Rousseeuw PJ (2021) Fast and eager K-medoids clustering: $\mathcal{O}(K)$ runtime improvement of the PAM, CLARA, and CLARANS algorithms. Information Systems 101:101804
49. Kaufman L, Rousseeuw PJ (1990) Divisive analysis (program DIANA). In: Kaufman L, Rousseeuw PJ (eds) Finding Groups in Data: An Introduction to Cluster Analysis. John Wiley & Sons, New York, NY, U.S.A., pp 253–279
50. Gilpin S, Qian B, Davidson I (2013) Efficient hierarchical clustering of large high dimensional datasets. In: Proceedings of the 22nd ACM International Conference on Information & Knowledge Management. Association for Computing Machinery, New York, NY, U.S.A., pp 1371–1380
51. Bouguettaya A, Yu Q, Liu X, Zhou X, Song A (2015) Efficient agglomerative hierarchical clustering. Expert Systems with Applications 42:2785–2797
52. Ward, Jr. JH (1963) Hierarchical grouping to optimize an objective function. Journal of the American Statistical Association 58:236–244
53. Murtagh F, Legendre P (2014) Ward's hierarchical agglomerative clustering method: Which algorithms implement Ward's criterion? Journal of Classification 31:274–295
54. Rand WM (1971) Objective criteria for the evaluation of clustering methods. Journal of the American Statistical Association 66:846–850
55. Hubert L, Arabie P (1985) Comparing partitions. Journal of Classification 2:193–218
56. Rousseeuw PJ (1987) Silhouettes: A graphical aid to the interpretation and validation of cluster analysis. Computational and Applied Mathematics 20:53–65
57. Wickham H, Averick M, Bryan J, Chang W, McGowan LD, François R, Grolemund G, Hayes A, Henry L, Hester J, Kuhn M, Pedersen TL, Miller E, Bache SM, Müller K, Ooms J, Robinson D, Seidel DP, Spinu V, Takahashi K, Vaughan D, Wilke C, Woo K, Yutani H (2019) Welcome to the tidyverse. Journal of Open Source Software 4:1686
58. Chan C, Leeper TJ, Becker J, Schoch D (2023) rio: a Swiss-army knife for data file I/O
59. López-Pernas S, Saqr M, Conde J, Del-Río-Carazo L (2024) A broad collection of datasets for educational research training and application. In: Saqr M, López-Pernas S (eds) Learning Analytics Methods and Tutorials: A Practical Guide using R. Springer
60. Saqr M, López-Pernas S, Conde M Ángel, Hernández-García Ángel (2024) Social betwork analysis: A primer, a guide and a tutorial in R. In: Saqr M, López-Pernas S (eds) Learning Analytics Methods and Tutorials: A Practical Guide using R. Springer
61. Saqr M, López-Pernas S (2022) How CSCL roles emerge, persist, transition, and evolve over time: A four-year longitudinal study. Computers & Education 189:104581
62. Pison G, Struyf A, Rousseeuq PJ (1999) Displaying a clustering with CLUSPLOT. Computational Statistics and Data Analysis 30:381–392
63. Mead A (1992) Review of the development of multidimensional scaling methods. Journal of the Royal Statistical Society: Series D (The Statistician) 41:27–39
64. Kim MK, Ketenci T (2019) Learner participation profiles in an asynchronous online collaboration context. The Internet and Higher Education 41:62–76

65. Saqr M, Viberg O (2020) Using diffusion network analytics to examine and support knowledge construction in CSCL settings. In: Alario-Hoyos C, Rodríguez-Triana MJ, Scheffel M, Arnedillo-Sánchez I, Dennerlein SM (eds) Addressing Global Challenges and Quality Education: Proceedings of the 15th European Conference on Technology Enhanced Learning, EC-TEL 2020, September 14–18, 2020. Springer, Cham, Switzerland, pp 158–172
66. Witten DM, Tibshirani R (2010) A framework for feature selection in clustering. Journal of the American Statistical Association 105:713–726
67. Hancer E, Xue B, Zhang M (2020) A survey on feature selection approaches for clustering. Artificial Intelligence Review 53:4519–4545
68. Ester M, Kriegel H-P, Sander J, Xu X (1996) A density-based algorithm for discovering clusters in large spatial databases with noise. In: Simoudis E, Han Jiawei, Fayyad UM (eds) Proceedings of the Second International Conference on Knowledge Discovery and Data Mining. AAAI Press, Portland, OR, U.S.A., pp 226–231
69. Hahsler M, Piekenbrock M, Doran D (2019) dbscan: Fast density-based clustering with R. Journal of Statistical Software 91:1–30
70. Du H, Chen S, Niu H, Li Y (2021) Application of DBSCAN clustering algorithm in evaluating students' learning status. In: Proceedings of the 17th International Conference on Computational Intelligence and Security, November 19–22, 2021. Chengdu, China, pp 372–376
71. Scrucca L, Fop M, Murphy TB, Raftery AE (2016) mclust 5: Clustering, classification and density estimation using Gaussian finite mixture models. The R Journal 8:289–317
72. Schwarz GE (1978) Estimating the dimension of a model. The Annals of Statistics 6:461–464
73. Murphy K, Murphy TB (2020) Gaussian parsimonious clustering models with covariates and a noise component. Advances in Data Analysis and Classification 14:293–325
74. Kaufman L, Rousseeuw PJ (1990) Fuzzy analysis (program FANNY). In: Kaufman L, Rousseeuw PJ (eds) Finding Groups in Data: An Introduction to Cluster Analysis. John Wiley & Sons, New York, NY, U.S.A., pp 164–198
75. D'Urso P (2016) Fuzzy clustering. In: Hennig C, Meila M, Murtagh F, Rocci R (eds) Handbook of Cluster Analysis. Chapman; Hall/CRC Press, New York, NY, U.S.A., pp 245–575
76. Ferraro MB, Giordani P, Serafini A (2019) fclust: An R package for fuzzy clustering. The R Journal 11:198–210
77. Ng AY, Jordan MI, Weiss Y (2001) On spectral clustering: Analysis and an algorithm. In: Dietterich T, Becker S, Ghahramani Z (eds) Advances in Neural Information Processing Systems. MIT Press, Cambridge, MA, U.S.A., pp 849–856
78. Dhillon IS, Guan Y, Kulis B (2004) Kernel K-means: Spectral clustering and normalized cuts. In: KDD '04: Proceedings of the Tenth ACM SIGKDD International Conference of Knowledge Discovery and Data Mining, Seattle, WA, U.S.A. Association for Computing Machinery, New York, NY, U.S.A., pp 551–556

An Introduction and R Tutorial to Model-Based Clustering in Education via Latent Profile Analysis

Luca Scrucca, Mohammed Saqr, Sonsoles López-Pernas, and Keefe Murphy

1 Introduction

Education research is commonly performed with variable-centered methods (e.g., correlation, regression, comparison of means e.g., t-test) using a sample from the population to devise or compare central tendency measures or an "average" (i.e., mean or median). The average is assumed to represent the population under study and therefore could be generalised to the population at large [1, 2]. Put another way, the statistical findings of variable-centered methods are thought to apply to all learners in the same way. In doing so, variable-centered methods ignore the individual differences that are universal across all domains of human function [3]. Learners are not an exception; students vary in their behavior, attitude, and dispositions, and they rarely—if at all—conform to a common pattern or an average behavior [2, 4]. An "average" is thus a poor simplification of learners' heterogeneity; consequently, methods to capture individual differences have started to gain popularity with the increasing attention to patterns and differences among students. Our focus in this chapter is on such person-centered methods [1–3, 5].

Person-centered methods can be broadly grouped into two categories; (1) heuristic, dissimilarity-based clustering algorithms (e.g., agglomerative hierarchical clustering, and partitional clustering algorithms like k-means) on one hand and (2)

L. Scrucca (✉)
Università degli Studi di Perugia, Perugia, Italy
e-mail: luca.scrucca@unipg.it

M. Saqr · S. López-Pernas
School of Computing, University of Eastern Finland, Joensuu, Finland

K. Murphy
Department of Mathematics and Statistics, Hamilton Institute, Maynooth University, Maynooth, Ireland

M. Saqr, S. López-Pernas (eds.), *Learning Analytics Methods and Tutorials*,
https://doi.org/10.1007/978-3-031-54464-4_9

model-based clustering (MBC) approaches (e.g., finite Gaussian mixture models) on the other. Though we focus here on the MBC paradigm, we note that—contrary to variable-centered methods—person-centered methods are generally concerned with capturing heterogeneity by uncovering the latent (e.g., unobserved or hidden) patterns within data into subgroups of "clusters" or "profiles", which are assumed to be homogeneous [1, 2]. Modelling the unobserved patterns within the data could reveal the qualitative differences between learners. For instance, where students may have different patterns of approaching their learning, capturing such patterns would make sense as each different approach may benefit from a certain course design, set of instructional methods, or scaffolding approach [2]. Similarly, dispositions such as engagement, motivation, and achievement are multidimensional and vary across students; capturing such differences requires a method that could handle such nonlinear multidimensional dispositions and identify the different patterns.

This chapter deals with one of the person-centered methods; namely, latent profile analysis (from the perspective of model-based clustering via finite Gaussian mixture models). To clarify, we note that the terms finite mixture models and latent profile analysis can be used interchangeably, with the former being more common in the statistical literature and the latter being more common in the education literature and social sciences literature more broadly. The equivalence of both terminologies is further elaborated in Sect. 3.1. In any case, this framework represents a *probabilistic* approach to statistical unsupervised learning that aims at discovering clusters of observations in a data set [6]. In other words, the MBC paradigm is referred to as model-based by virtue of being based on a generative probabilistic model, unlike heuristic clustering algorithms based on dissimilarity criteria.

We also offer a walkthrough tutorial for the analysis of a data set on school engagement, academic achievement, and self-regulated learning using the popular `mclust` package [7] for R [8] which implements the approach. Whereas mixture models and `mclust` have received growing attention within social science, they have not garnered widespread utilisation in the field of educational research and their adoption in learning analytics research has been relatively slow.

2 Literature Review

Examples of mixture models being applied in educational research settings are relatively few compared to other methods of clustering. Some notable examples exist which address patterns in students' online learning [9] or patterns in students' disposition [10] or collaborative roles [11]. Most studies in education research that applied mixture models used latent profile analysis (LPA) to identify students' profiles from self-reported data. For example, [10] performed LPA on a data set of 318 students' survey responses about emotional self-efficacy, motivation, self-regulation, and academic performance, and identified four profiles: "low", "average", "above average with a low ability to handle the emotions of others", and "high emotional self-efficacy". In the work by Cheng et al. [12], the authors analyzed 615 vocational education students' achievement emotions in online learning environments, and found three profiles: "blends of negative emotions", "nonemotional", and "pure positive emotion". Hoi [13] employed LPA on self-report data on classroom engagement from 413 first-year university students in Vietnam and found three profiles: "highly engaged", "moderately engaged", and "minimally engaged". Scheidt et al. [14] collected survey responses from 2339 engineering undergraduates about 28 noncognitive and affective factors using a survey instrument and using Gaussian mixture models found four very distinct profiles of students.

The analysis of online trace log data—which is at the core of learning analytics data—with mixture models is even less common. In the study by Zhang et al. [15], the authors applied LPA to variables related to debugging derived from students' programming problems submission traces. They found a profile with higher debugging accuracy and coding speed, another profile with lower debugging performance in runtime and logic errors, and a third profile with lower performance in syntactic errors who tended to make big changes in every submission. Studies covering online collaborative learning are even more scarce. A rare example is the study by Saqr and López-Pernas [11], in which the authors used latent profile analysis to identify students' roles in collaboration based on their centrality measures. The mixture models identified three collaborative roles that represented a distinct pattern of collaboration: leaders, who contribute the most to the discussion, whose arguments are more likely to spread; mediators, who bridge others and moderate the discussions; as well as isolates, who are socially idle and do not contribute to the discussion.

A considerable number of studies that applied mixture models further investigate the association between profile membership and academic achievement. For example, in the aforementioned study by Yu et al. [10], students with high emotional self-efficacy had higher academic performance than the other profiles. In the study by Zhang et al. [15], the authors found that higher debugging accuracy was related to higher scores in all exams, whereas there were no differences between the two other identified profiles. By the same token, researchers have attempted to find reasons why a certain profile emerged, or what are the variables that are more associated with one profile more than the other. For example, [13] found that peer support, provision of choice, and task relevance are the factors more likely to predict classroom engagement profile membership. Yu et al. [10] found that self-regulation and motivation played significant roles in determining profile membership.

Clearly, there are plenty of opportunities for further exploration and investigation in this area that could augment our knowledge of learning, learners' behavior, and the variabilities of learning processes [2]. This is especially true given the numerous advantages of the MBC paradigm over more traditional, heuristic clustering algorithms, which we imminently describe. Subsequently, in the rest of this chapter we elaborate on the theoretical underpinnings of the family of Gaussian parsimonious clustering models implemented in the `mclust` R package and additionally explore some advanced features of the package, which we employ in an analysis of a real educational research application thereafter. Finally, we conclude with a brief discussion.

3 Model-Based Clustering

As stated above, clustering methods, in a general sense, are used to uncover group structure in heterogeneous populations and identify patterns in a data set which may represent distinct subpopulations. While there is no universally applicable definition of what constitutes a cluster [16], it is commonly assumed that clusters should be well separated from each other and cohesive in an ideal analysis [17]. Conversely, objects within a cluster should be more similar to each other in some sense, in such a way that an observation has a defined relationship with observations in the same cluster, but not with observations from other clusters.

Traditional clustering approaches, like the aforementioned k-means algorithm, and agglomerative hierarchical clustering, use dissimilarity-based heuristics to ultimately produce a "hard" partition of cases into groups, such that each observation is associated with exactly one cluster only. As such approaches are not underpinned by a statistical model, assessment of the optimal number of clusters is often a fraught task, lacking the guidance of principled statistical model selection criteria. However, we note that k-means can be recasted as a clustering model assuming a Gaussian mixture with equal proportions and diagonal equal covariance matrix across groups. Moreover, some (but not all) agglomerative hierarchical clustering models can be rooted in a statistical model also, as discussed in [18].

Conversely, the MBC paradigm typically assumes that data arise from a (usually finite) mixture of probability distributions, whereby each observation is assumed to be generated from a specific cluster, characterised by an associated distribution in the mixture [19]. Ideally, mixtures of distributions are supposed to provide a good model for the heterogeneity in a data set; that is, once an observation has been assigned to a cluster, it is assumed to be well-represented by the associated distribution. As such, MBC methods are based on a formal likelihood and seek to estimate parameters (e.g., means, variances, and covariances, which may or may not differ across groups) which best characterise the different distributions. Rather than yielding only a "hard" partition, each observation is assigned a probability of being associated with each mixture component—such that observations can have non-negative association with more than one cluster—from which a hard partition can be constructed. These probabilities are treated as weights when estimating the component parameters, which brings the advantages of minimising the effect of observations lying near the boundary of two natural clusters (e.g., a student with an ambiguous learning profile) and being able to quantity the uncertainty in the cluster assignment of a particular observation to provide a sense of cases for which further investigation may be warranted. Compared to other approaches, the other main advantages of this statistical modelling framework are its ability to use statistical model selection criteria and inferential procedures for evaluating and assessing the results obtained.

Inference for finite mixture models is routinely achieved by means of the expectation-maximisation (EM) algorithm [20], under which each observation's component membership is treated as a "missing" latent variable which must be estimated. This formulation assumes that the data are *conditionally* independent and identically distributed, where the conditioning is with respect to a latent variable representation of the data in which the latent variable indicates cluster membership. Given the relative familiarity of latent class and latent profile terminology in the social sciences, we now explicitly cast MBC methods in the framework of latent variable modelling.

3.1 Latent Variable Models

Latent variable models are statistical models that aim to explain the relationships between observed variables by introducing one or more unobserved or latent variables. The idea behind latent variable models is that some of the underlying constructs or concepts we are interested in cannot be measured directly, but only through their effects on observable variables. Latent variable modelling has a relatively long history, dating back from the measure of general intelligence by factor analysis [21], to the structural equation modelling approach [22], from topic modelling, such as the latent Dirichlet allocation algorithm [23], to hidden Markov models for time series [24] and longitudinal data [25]. Latent variable models are widely used in various fields, including psychology, sociology, economics, and biology, to name a few. They are particularly useful when dealing with complex phenomena that cannot be easily measured or when trying to understand the underlying mechanisms that drive the observed data.

When discussing latent variable modelling, it is useful to consider the *taxonomy* presented by Bartholomew et al. [26]. This can be particularly helpful, as the same models are sometimes referred to by different names in different scientific disciplines. Bartholomew et al. [26, Table 1.3] considered a cross-classification of latent variable methods based on the type of variable (manifest or latent) and its nature (metrical or categorical). If both the manifest and latent variables are metrical, the model is called a **factor analysis model**. If the manifest variables are categorical and the latent variables are metrical, the model is called a **latent trait model** or **item response theory model**. If the manifest variables are metrical and the latent variables are categorical, the model is called a **latent profile analysis model**. If both the manifest and latent variables are categorical, the model is called a **latent class model**.

In this scheme, finite Gaussian mixture models described in this chapter assume that the observed variables are continuous and normally distributed, while the latent variable, which represents the cluster membership of each observation, is categorical. Therefore, Gaussian mixtures belong to the family of latent profile analysis models. This connection is made apparent by the `tidyLPA` R package [27], which leverages this equivalence to provide an interface to the well-known `mclust` R package [7] used throughout this chapter, using tidyverse syntax and terminology which is more familiar in the LPA literature.

3.2 Finite Gaussian Mixture Models

As described above, finite mixture models (FMMs) provide the statistical framework for model-based clustering and allow for the modelling of complex data by combining simpler distributions. Specifically, a FMM assumes that the observed data are generated from a finite mixture of underlying distributions, each of which corresponds to a distinct subgroup or cluster within the data. Gaussian mixture models (GMMs) are a particularly widespread variant of FMMs which specifically assume that each of the underlying distributions is a (multivariate) Gaussian distribution. This means that the data within each cluster are normally distributed, but with potentially different means and covariance matrices. The relevance of GMMs stems from the well-established fact that mixtures of Gaussians can provide an accurate approximation to any continuous density.

In FMMs, the latent variable represents the cluster assignment for each observation in the data. It is a categorical variable assuming one of a finite set of possible values that correspond to different clusters. Alternatively, it can be encoded using a set of indicator variables, which take the value of 1 for the cluster to which the observation belongs, and 0 for all other clusters.

To estimate the parameters of a GMM with the associated latent variable for cluster membership, a likelihood-based approach is typically used. The likelihood function expresses the probability of observing the data, given the parameter values and the latent variable. The maximum likelihood estimation (MLE) method is commonly used to estimate the parameters and the latent variable which maximise the likelihood function. Usually, the number of clusters in a GMM is also unknown, and it is determined through a process known as model selection, which involves comparing models with different numbers of clusters and parameterisations and selecting the one which best fits the data.

In summary, model-based clustering from the perspective of latent variable modelling assumes that the data is generated from a probabilistic model with a specific number of clusters. A likelihood-based approach can be used to estimate the parameters of the model and the latent variable that represents the cluster assignment for each observation in the data, and guide the selection of the number of clusters. A GMM is a common framework for model-based clustering which assumes the data in each cluster is generated from a Gaussian distribution.

4 Gaussian Parsimonious Clustering Models

For a continuous feature vector $\boldsymbol{x} \in \mathbb{R}^d$, the general form of the density function of a Gaussian mixture model (GMM) with K components can be written as

$$f(\boldsymbol{x}) = \sum_{k=1}^{K} \pi_k \phi_d(\boldsymbol{x}; \boldsymbol{\mu}_k, \boldsymbol{\Sigma}_k), \tag{1}$$

where π_k represents the mixing probabilities, i.e., the marginal probability of belonging to the k-th cluster, such that $\pi_k > 0$ and $\sum_{k=1}^{K} \pi_k = 1$, and $\phi_d(\cdot)$ is the d-dimensional multivariate Gaussian density with parameters $(\boldsymbol{\mu}_k, \boldsymbol{\Sigma}_k)$ for $k = 1, \ldots, K$. Clusters described by such a GMM are ellipsoidal, centered at the means $\boldsymbol{\mu}_k$, and with other geometric characteristics (namely volume, shape, and orientation) determined by the covariance matrices $\boldsymbol{\Sigma}_1, \ldots, \boldsymbol{\Sigma}_K$. Parsimonious parameterisations of covariance matrices can be controlled by imposing some constraints on the covariance matrices through the following eigen-decomposition [28, 29]:

$$\boldsymbol{\Sigma}_k = \lambda_k \boldsymbol{U}_k \boldsymbol{\Delta}_k \boldsymbol{U}_k^{\top}, \tag{2}$$

where $\lambda_k = |\boldsymbol{\Sigma}_k|^{1/d}$ is a scalar which controls the *volume*, $\boldsymbol{\Delta}_k$ is a diagonal matrix whose entries are the normalised eigenvalues of $\boldsymbol{\Sigma}_k$ in decreasing order, such that $|\boldsymbol{\Delta}_k| = 1$, which controls the *shape* of the density contours, and $\boldsymbol{U}_k$ is an orthogonal matrix whose columns are the eigenvectors of $\boldsymbol{\Sigma}_k$, which controls the *orientation* of the corresponding ellipsoid. The size of a cluster is distinct from its volume and is proportional to π_k [29].

GMMs with unconstrained covariance matrices are quite flexible, but require the estimation of several parameters. To obtain a balance between model complexity and accuracy of parameter estimates, a parsimonious model parameterisation can be adopted. Constraining the geometric characteristics of cluster covariances to be equal across clusters can greatly reduce the number of estimable parameters, and is the means by which GMMs obtain intermediate covariance matrices between homoscedasticity and heteroscedasticity. A list of the 14 resulting parameterisations available in the `mclust` package [7] for R [8] is included in Table 2.1 of [30]. Of particular note is the nomenclature adopted by `mclust` whereby each model has a three-letter name with each letter pertaining to the volume, shape, and orientation, respectively, denoting whether the given component is equal (`E`) or free to vary (`V`) across clusters. Some model names also use the letter `I` in the third position to indicate that the covariance matrices are diagonal and two particularly parsimonious models have the letter `I` in the second position to indicate that the covariance matrices are isotropic. Thus, as examples, the fully unconstrained `VVV` model is one for which the volume, shape, and orientation are all free to vary across clusters, the `EVE` model constrains the clusters to have equal volume and orientation but varying shape, and the `VII` model assumes isotropic covariance matrices with cluster-specific volumes. The flexibility to model clusters with different geometric characteristics by modelling correlations according to various parameterisations represents another advantage over heuristic clustering algorithms. Taking the k-means algorithm as an example, a larger number of circular, Euclidean distance-based clusters may be required to fit the data well, rather than a more parsimonious and easily interpretable mixture model with fewer non-spherical components.

Given a random sample of observations $\{\boldsymbol{x}_1, \boldsymbol{x}_2, \dots, \boldsymbol{x}_n\}$ in d dimensions, the log-likelihood of a GMM with K components is given by

$$\ell(\boldsymbol{\theta}) = \sum_{i=1}^{n} \log \left\{ \sum_{k=1}^{K} \pi_k \phi_d(\boldsymbol{x}_i; \boldsymbol{\mu}_k, \boldsymbol{\Sigma}_k) \right\}, \tag{3}$$

where $\boldsymbol{\theta} = (\pi_1, \dots, \pi_{K-1}, \boldsymbol{\mu}_1, \dots, \boldsymbol{\mu}_K, \boldsymbol{\Sigma}_1, \dots, \boldsymbol{\Sigma}_K)$ denotes the collection of parameters to be estimated.

Maximizing the log-likelihood function in Eq. 3 directly is often complicated, so maximum likelihood estimation (MLE) of $\boldsymbol{\theta}$ is usually performed using the EM algorithm [20] by including the component membership as a latent variable. The EM algorithm consists of two steps: the E-step (Expectation step) and the M-step (Maximisation step). In the E-step, the algorithm calculates the expected membership probabilities of each data point to each of the mixture components based on the current estimates of the model parameters. Thus, though the latent variable indicating cluster membership is assumed to be categorical and represented by indicator variables taking the values 0 or 1, the model estimates assignment probabilities in the range [0, 1] at this step. In the M-step, the algorithm updates the model parameters by maximizing the likelihood of the observed data given the estimated membership probabilities. These two steps are repeated until convergence or a maximum number of iterations is reached. Details on the use of EM algorithm in finite mixture modelling is provided by McLachlan and Peel [19], while a thorough treatment and further extensions can be found in [31]. For the GMM case, see Sec. 2.2 of [30].

Following the fitting of a GMM and the determination of the MLEs of parameters, the maximum a posteriori (MAP) procedure can be used to assign the observations into the most likely cluster and recover a "hard" partition. For an observation $\boldsymbol{x}_i$ the posterior conditional probability of coming from the mixture component k is given by

$$\widehat{p}_{ik} = \frac{\widehat{\pi}_k \phi_d(\boldsymbol{x}_i; \widehat{\boldsymbol{\mu}}_k, \widehat{\boldsymbol{\Sigma}}_k)}{\displaystyle\sum_{g=1}^{K} \widehat{\pi}_g \phi_d(\boldsymbol{x}; \widehat{\boldsymbol{\mu}}_g, \widehat{\boldsymbol{\Sigma}}_g)}. \tag{4}$$

Then, an observation is assigned to the cluster with the largest posterior conditional probability, i.e., $\boldsymbol{x}_i \in \mathcal{C}_{k^\star}$ with $k^\star = \arg\max_k \widehat{p}_{ik}$.

4.1 Model Selection

Given that a wide variety of GMMs in Eq. 1 can be estimated by varying the number of mixture components and the covariance decompositions in Eq. 2, selecting the appropriate model represents a crucial decision. A popular option consists in choosing the "best" model using the Bayesian information criterion [BIC, 32], which, for a given model $\mathcal{M}$, is defined as

$$\mathrm{BIC}_{\mathcal{M}} = 2\ell_{\mathcal{M}}(\widehat{\boldsymbol{\theta}}) - \nu_{\mathcal{M}} \log(n),$$

where $\ell_{\mathcal{M}}(\widehat{\boldsymbol{\theta}})$ stands for the maximised log-likelihood of the data sample of size n under model $\mathcal{M}$, and $\nu_{\mathcal{M}}$ for the number of independent parameters to be estimated. Another option available in clustering is the Integrated Complete Likelihood [ICL, 33] criterion given by

$$\mathrm{ICL}_{\mathcal{M}} = \mathrm{BIC}_{\mathcal{M}} + 2\sum_{i=1}^{n}\sum_{k=1}^{K} c_{ik} \log(\widehat{p}_{ik}),$$

where $\widehat{p}_{ik}$ is the conditional probability that $\boldsymbol{x}_i$ arises from the k-th mixture component from Eq. 4, and $c_{ik} = 1$ if the i-th observation is assigned to cluster $\mathcal{C}_k$ and 0 otherwise.

Both criteria evaluate the fit of a GMM to a given set of data by considering both the likelihood of the data given the model and the complexity of the model itself, represented by the number of parameters to be estimated. Compared to the BIC, the ICL introduces a further entropy-based penalisation for the overlap of the clusters. For this reason, the ICL tends to select models with well-separated clusters.

Whereas there is no consensus of a standard criteria for choosing the best model, there are guidelines that the researcher could rely on. To decide on the optimal model, examining the fit indices (such as the BIC and ICL), model interpretability, and conformance to theory can be of great help. The literature recommends estimating a 1-cluster solution for each model that serves as a comparative baseline and then increasing the number of clusters one by one, evaluating if adding another cluster yields a better solution in both statistical and conceptual terms [34]. Among all fit indices, larger BIC values seems to be the preferred method for selecting the best model. However, examining other indices (e.g., AIC, ICL) is also useful. Oftentimes, fit indices do not converge to a certain model. In such cases, the interrelation between the selected models, such as whether one model is an expanded version of another, should also be taken into consideration, as well as the stability of the different models, including the relative sizes of the emergent profiles (each profile should comprise more than 5–8% of the sample) [34]. Furthermore, the elbow method could be helpful in cases where no clear number of clusters can be easily determined from the fit indices (e.g., the BIC continues to grow consistently when increasing the number of clusters). This entails plotting the BIC values and finding an elbow shape where a rise in BIC is less noticeable with increasing numbers of clusters or roughly an elbow followed by a relatively flat line. The choice of the best number of clusters can and probably should be guided by theory; that is, in cases where previous research reported a certain number of clusters or profiles, it is recommended to take this guidance into account. For instance, research on engagement has repeatedly reported three levels of engagement. Once we have chosen the most suitable model, it is suggested to compute model diagnostics (e.g., entropy and average posterior probability) to evaluate the selected model. These diagnostics are covered in Sect. 4.3.3.

4.2 mclust R Package

`mclust` is an R package [8] for model-based cluster analysis, classification, and density estimation using Gaussian finite mixture models [30, 35]. It is widely used in statistics, machine learning, data science, and pattern recognition. One of the key features of `mclust` is its flexibility in modelling quantitative data with several covariance structures and different numbers of mixture components. Additionally, the package provides extensive graphical representations, model selection criteria, initialisation strategies for the EM algorithm, bootstrap-based inference, and Bayesian regularisation, among other prominent features. `mclust` also represents a valuable tool in educational settings because it provides a powerful set of models that allows students and researchers to quickly and easily perform clustering and classification analyses on their data. We focus here on the use of `mclust` as a tool for unsupervised model-based clustering, though the package does also provide functions for supervised model-based classification.

The main function implementing model-based clustering is called `Mclust()`, which requires a user to provide at least the data set to analyze. In the one-dimensional case, the data set can be a vector, while in the multivariate case, it can be a matrix or a data frame. In the latter case, the rows correspond to observations, and the columns correspond to variables.

The `Mclust()` function allows for further arguments, including the optional argument `G` to specify the number of mixture components or clusters, and `modelNames` to specify the covariance decomposition. If both `G` and `modelNames` are not provided, `Mclust()` will fit all possible models obtained using a number of mixture components from 1 to 9 and all 14 available covariance decompositions, and it will select the model with the largest BIC. Notably, if the data set is univariate, only 2 rather than 14 models governing the scalar variance parameters are returned; that they are equal or unequal across components. Finally, computing the BIC and ICL criteria can be done by invoking the functions `mclustBIC()` and `mclustICL()`, respectively.

4.3 Other Practical Issues and Extensions

Prior to commencing the cluster analysis of a data set on school engagement, academic achievement, and self-regulated learning measures, we first provide some theoretical background on some extensions of practical interest which will be explored in the analysis.

4.3.1 Bayesian Regularisation

Including a prior distribution over the mixture parameters is an effective way to avoid singularities and degeneracies in maximum likelihood estimation. Furthermore, this can help to prevent overfitting and improve model performance. In situations where the variables of interest are discrete or take on only a few integer values, including a prior distribution can help to regularise the model.

Fraley and Raftery [36] proposed using weekly informative conjugate priors to regularise the estimation process. The EM algorithm can still be used for model fitting, but maximum likelihood estimates (MLEs) are replaced by maximum a posteriori (MAP) estimates. A slightly modified version of BIC can be used for model selection, with the maximised log-likelihood replaced by the log-likelihood evaluated at the MAP or posterior mode.

The prior distributions proposed by Fraley and Raftery [36] are:

- a uniform prior on the simplex for the mixture weights $(\pi_1, \ldots, \pi_K)$;
- a Gaussian prior on the mean vector (conditional on the covariance matrix), i.e.,

$$\boldsymbol{\mu} \mid \boldsymbol{\Sigma} \sim \mathcal{N}(\boldsymbol{\mu}_{\mathcal{P}}, \boldsymbol{\Sigma}/\kappa_{\mathcal{P}}) \tag{5}$$

$$\propto |\boldsymbol{\Sigma}|^{-1/2} \exp\left\{-\frac{\kappa_{\mathcal{P}}}{2} \operatorname{tr}\left((\boldsymbol{\mu} - \boldsymbol{\mu}_{\mathcal{P}})^{\top} \boldsymbol{\Sigma}^{-1} (\boldsymbol{\mu} - \boldsymbol{\mu}_{\mathcal{P}})\right)\right\}, \tag{6}$$

  with $\boldsymbol{\mu}_{\mathcal{P}}$ and $\kappa_{\mathcal{P}}$ being the hyperparameters controlling, respectively, the mean vector and the amount of shrinkage applied;
- an inverse Wishart prior on the covariance matrix, i.e.,

$$\boldsymbol{\Sigma} \sim \mathcal{IW}(\nu_{\mathcal{P}}, \boldsymbol{\Lambda}_{\mathcal{P}})$$

$$\propto |\boldsymbol{\Sigma}|^{-(\nu_{\mathcal{P}}+d+1)/2} \exp\left\{-\frac{1}{2} \operatorname{tr}\left(\boldsymbol{\Sigma}^{-1} \boldsymbol{\Lambda}_{\mathcal{P}}^{-1}\right)\right\}, \tag{7}$$

  with the hyperparameters $\nu_{\mathcal{P}}$ and the matrix $\boldsymbol{\Lambda}_{\mathcal{P}}$ controlling the degrees of freedom and scale of the prior distribution, respectively.

Adding a prior to GMMs estimated using the `mclust` R package is easily obtained by adding an optional `prior` argument when calling some of the fitting functions, such as `mclustBIC()` and `Mclust()`. Specifically, setting `prior = priorControl(functionName = "defaultPrior")` allows to adopt the conjugate priors described above with the following default values for the hyperparameters:

- mean vector $\boldsymbol{\mu}_{\mathcal{P}} = \bar{\boldsymbol{x}}$, the sample mean of each variable;
- shrinkage $\kappa_{\mathcal{P}} = 0.1$;
- degrees of freedom $\nu_{\mathcal{P}} = d + 2$;
- scale matrix $\boldsymbol{\Lambda}_{\mathcal{P}} = \S/(K^{2/d})$, where $\S$ is the sample covariance matrix.

Rationale for the above default values for the prior hyperparameters, together with the corresponding MAP estimates of the GMM parameters, can be found in [36, Table 2]. These values should suffice for most applications, but experienced users who want to tune the hyperparameters can refer to the documentation available in the help pages for `priorControl` and `defaultPrior()`. Further details about specifying different hyperparameter values can be found in [30].

4.3.2 Bootstrap Inference

Likelihood-based inference in mixture models is complicated because asymptotic theory applied to mixture models require a very large sample size [19], and standard errors derived from the expected or the observed information matrix tend to be unstable [37]. For these reasons, resampling approaches based on the bootstrap are often employed [38].

The *bootstrap* [39] is a general, widely applicable, powerful technique for obtaining an approximation to the sampling distribution of a statistic of interest. The bootstrap distribution is approximated by drawing a large number of samples, called *bootstrap samples*, from the empirical distribution. This can be obtained by resampling with replacement from the observed data (*nonparametric bootstrap*), or from a parametric distribution with unknown parameters substituted by the corresponding estimates (*parametric bootstrap*). A Bayesian version of the bootstrap, introduced by Rubin [40], allows posterior samples to be obtained by resampling with weights for each observation drawn from a uniform Dirichlet distribution. A strictly related technique is the *weighted likelihood bootstrap* [41], where a statistical model is repeatedly fitted using weighted maximum likelihood with weights obtained as in Bayesian bootstrap.

Let $\widehat{\boldsymbol{\theta}}$ be the estimate of a set of GMM parameters $\boldsymbol{\theta}$ for a given model $\mathcal{M}$, determined by the adopted covariance parameterisation and number of mixture components. The bootstrap distribution for the parameters of interest is obtained as follows:

- draw a bootstrap sample of size n using one of the resampling techniques described above to form the bootstrap sample $(\boldsymbol{x}_1^\star, \dots, \boldsymbol{x}_n^\star)$;
- fit a the GMM $\mathcal{M}$ to get the bootstrap estimates $\widehat{\boldsymbol{\theta}}^\star$;
- replicate the previous steps a large number of times, say B.

The bootstrap distribution for the parameters of interest, $\widehat{\boldsymbol{\theta}}_1^\star, \widehat{\boldsymbol{\theta}}_2^\star, \ldots, \widehat{\boldsymbol{\theta}}_B^\star$, can then be used for computing the bootstrap standard errors (as the square root of the diagonal elements of the bootstrap covariance matrix) or the bootstrap percentile confidence intervals. More details can be found in [30].

From a practical point of view, bootstrap resampling can be conducted in `mclust` by means of the function `MclustBootstrap()`. This function takes as arguments the fitted model object returned from e.g., `Mclust()` or `mclustBIC()`, the optional argument `type`, which allows to specify the type of bootstrap samples to draw (`"bs"` for nonparametric bootstrap, `"pb"` for parametric bootstrap, and `"wlbs"` for weighted likelihood bootstrap), and the optional argument `nboot`, which sets the number of bootstrap samples. At least 999 samples should be drawn if confidence intervals are needed.

4.3.3 Entropy and Average Posterior Probabilities

The definition of entropy in information theory [42] refers to the average amount of information provided by a random variable. Following this definition, [43] defines the entropy of a finite mixture model as follows

$$E_{\text{FMM}} = -\sum_{i=1}^{n}\sum_{k=1}^{K} \widehat{p}_{ik} \log(\widehat{p}_{ik}),$$

where $\widehat{p}_{ik}$ is the estimated posterior probability of case i to belong to cluster k (see Eq. 4). If the mixture components are well separated, $\widehat{p}_{ik} \approx 1$ for the assigned cluster $\mathcal{C}_k$ and 0 otherwise. Consequently, the entropy of the mixture model in this case is $E_{\text{FMM}} = 0$ (note that $0\log(0) = 0$ by convention). On the contrary, in the case of maximal assignment uncertainty, $\widehat{p}_{ik} = 1/K$ for all clusters $\mathcal{C}_k$ ($k = 1, \ldots, K$). As a result, the entropy of the mixture model is $E_{\text{FMM}} = n\log(K)$.

In latent class and latent profile analysis, a slightly different definition of entropy is used as a diagnostic statistic to assess how well the fitted model assigns individuals to the identified clusters based on their response patterns. Thus, a normalised version of the entropy is defined as follows

$$E = 1 - \frac{E_{\text{FMM}}}{n\log(K)} = 1 + \frac{\sum_{i=1}^{n}\sum_{k=1}^{K} \widehat{p}_{ik}\log(\widehat{p}_{ik})}{n\log(K)}.$$

Entropy takes values on the range [0, 1], with higher entropy values indicating that the model has less uncertainty in assigning cases to their respective latent classes/profiles. Thus, high entropy values typically indicate a better model which is able to distinguish between the latent components and that the components are relatively distinct. An entropy value close to 1 is ideal, while values above 0.6 are considered acceptable, although there is no agreed upon optimal cutoff for entropy.

The contribution of each observation to the overall total entropy can be defined as

$$E_i = 1 + \frac{\sum_{k=1}^{K} \widehat{p}_{ik} \log(\widehat{p}_{ik})}{\log(K)},$$

so that the overall total entropy is obtained by averaging over the individual contributions, i.e., $E = \sum_{i=1}^{n} E_i/n$. The individual contributions E_i can also be used to compute the average entropy of each latent component, which indicates how accurately the model defines components. Average posterior probabilities (AvePP) are a closely related performance assessment measure, given by the average posterior membership probabilities $\widehat{p}_{ik}$ for each component for the observations most probably assigned to that component, for which a cutoff of 0.8 has been suggested to indicate acceptably high assignment certainty and well-separated clusters [34]. The analysis below presents the necessary code to calculate entropies and average posterior probabilities thusly from a fitted `mclust` model.

5 Application: School Engagement, Academic Achievement, and Self-regulated Learning

A group of 717 primary school students from northern Spain were evaluated in terms of their school engagement, self-regulation, and academic performance through the use of various measures. The school engagement measure (SEM) was employed to assess their engagement, while their self-regulation was evaluated with the self-regulation strategy inventory—self-report. The measure for academic achievement was based on the students' self-reported grades in Spanish and mathematics, which were rated on a scale of 1–5. More information about the dataset is available in Chapter 2 [44]. This data set can be used to identify clusters of students based on their engagement and self-regulation. These clusters would represent distinct patterns or "profiles" of engagement. Finding such profiles would allow us to understand individual differences but more importantly to stratify support according to different engagement profiles.

5.1 Preparing the Data

We start by loading the packages required for the analysis. We note in particular that version 6.0.0 of `mclust` is employed here, the latest release at the time of writing.

```
library(ggplot2)
library(ggridges)
library(mclust)
```

```
library(rio)
library(tidyverse)
```

```
-- Attaching core tidyverse packages ---------------- tidyverse 2.0.0 --
v dplyr     1.1.3     v readr     2.1.4
v forcats   1.0.0     v stringr   1.5.0
v lubridate 1.9.2     v tibble    3.2.1
v purrr     1.0.2     v tidyr     1.3.0
-- Conflicts ---------------------------------- tidyverse_conflicts() --
x dplyr::filter() masks stats::filter()
x dplyr::lag()    masks stats::lag()
x purrr::map()    masks mclust::map()
i Use the conflicted package (<http://conflicted.r-lib.org/>) to force
  all conflicts to become errors
```

Then, we read the data set from an online comma-separated-value (CSV) file, followed by some data cleaning and formatting to prepare the data for subsequent analysis. Note that the CSV file to be read is not in standard format, so we have to explicitly set the separator field using the optional argument `sep = ";"`.

```
# read the data
data <- import("https://github.com/lamethods/data/raw/main/3_engSRLach/
               Manuscript_School%20Engagment.csv",
               sep = ";")

# select the variables to be analyzed
vars <- c("PRE_ENG_COND", "PRE_ENG_COGN", "PRE_ENG_EMOC")
x <- select(data, all_of(vars)) |>
  as_tibble() |>
  rename("BehvEngmnt" = "PRE_ENG_COND",  # Behavioral engagement
         "CognEngmnt" = "PRE_ENG_COGN",  # Cognitive engagement
         "EmotEngmnt" = "PRE_ENG_EMOC")  # Emotional engagement

# print the data set used in the subsequent analysis
x
```

```
# A tibble: 717 x 3
   BehvEngmnt CognEngmnt EmotEngmnt
        <dbl>      <dbl>      <dbl>
 1       3.75       3.14        4.4
 2       4          3.71        2
 3       4.25       3.86        4
 4       3.75       2.57        3
 5       4.25       3           4
 6       4          3.71        3.8
 7       3.5        2.14        3.2
 8       4.75       3.57        1.6
 9       3.25       2.71        3
10       5          4.43        4.8
# i 707 more rows
```

A table of summary statistics for the data set can be obtained as follows:

```
x |> pivot_longer(cols = colnames(x),
                  names_to = "Variable",
                  values_to = "Value") |>
  group_by(Variable) |>
  summarize(N = n(),
            Nunq = n_distinct(Value),
            Mean = mean(Value),
            SD = sd(Value),
            Min = min(Value),
            Median = median(Value),
            Max = max(Value))
```

```
# A tibble: 3 x 8
  Variable        N  Nunq  Mean    SD   Min Median   Max
  <chr>       <int> <int> <dbl> <dbl> <dbl>  <dbl> <dbl>
1 BehvEngmnt    717    17  4.17 0.627     1   4.25     5
2 CognEngmnt    717    30  2.92 0.771     1   2.92     5
3 EmotEngmnt    717    22  3.61 0.911     1   3.61     5
```

5.2 *Model Estimation and Model Selection*

To begin our latent profile analysis, we first fit a number of candidate GMMs with different numbers of latent components and covariance parameterisations, and compute the Bayesian Information Criterion (BIC) to select the "optimal" model. This model selection criterion jointly takes into account both the covariance decompositions and the number of mixture components in the model.

As mentioned earlier, given the characteristics of the data, which consists of a small number of unique values relative to the number of observations, a prior is used for regularisation. We invoke the default priors described above, summarise the BIC values of the three best models, and visualise the BIC values of all fitted models.

```
BIC <- mclustBIC(x, prior = priorControl())
summary(BIC)
```

```
Best BIC values:
             VVI,3        VVI,4        VVV,3
BIC      -4521.213 -4526.905884 -4533.57166
BIC diff     0.000    -5.693183   -12.35896
```

```
plot(BIC)
```

The selected model is a three-component GMM with diagonal covariance matrices of varying volume and shape, with axis-aligned orientation, indicated as (VVI,3). Thus, the variables are independent within each cluster (Fig. 1).

5.3 *Examining Model Output*

The fit of the optimal model is obtained using:

```
mod <- Mclust(x, modelNames = "VVI", G = 3, prior = priorControl())
summary(mod, parameters = TRUE)
```

```
----------------------------------------------------
Gaussian finite mixture model fitted by EM algorithm
----------------------------------------------------
```

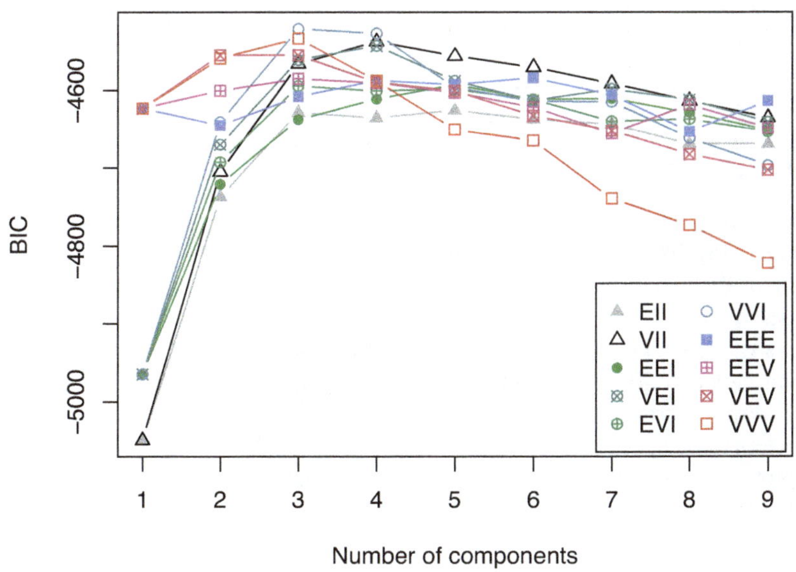

Fig. 1 BIC traces for the estimated GMMs with default priors

```
Mclust VVI (diagonal, varying volume and shape) model with 3 components:

Prior: defaultPrior()

 log-likelihood   n df       BIC       ICL
      -2194.856 717 20 -4521.213 -4769.174

Clustering table:
  1   2   3
184 119 414

Mixing probabilities:
        1         2         3
0.2895147 0.1620776 0.5484078

Means:
               [,1]     [,2]     [,3]
BehvEngmnt 3.704041 4.713234 4.257355
CognEngmnt 2.287057 3.699530 3.017293
EmotEngmnt 2.738969 4.733899 3.737286

Variances:
[,,1]
           BehvEngmnt CognEngmnt EmotEngmnt
BehvEngmnt  0.5022148  0.0000000  0.0000000
CognEngmnt  0.0000000  0.3909235  0.0000000
EmotEngmnt  0.0000000  0.0000000  0.7674268
[,,2]
           BehvEngmnt CognEngmnt EmotEngmnt
BehvEngmnt  0.0737948  0.0000000 0.00000000
CognEngmnt  0.0000000  0.4150514 0.00000000
EmotEngmnt  0.0000000  0.0000000 0.05540526
[,,3]
           BehvEngmnt CognEngmnt EmotEngmnt
BehvEngmnt  0.2048374  0.0000000  0.0000000
CognEngmnt  0.0000000  0.3327557  0.0000000
EmotEngmnt  0.0000000  0.0000000  0.2795838
```

The shown output reports some basic information about the fit, such as the maximised log-likelihood (`log-likelihood`), the number of observations (`n`), the number of estimated parameters (`df`), the BIC criterion (`BIC`), and the clustering table based on the MAP classification. The latter indicates that the clusters also vary in terms of *size*. The optional argument `parameters = TRUE` in the `summary()` function call additionally prints the estimated parameters. Observe that the `VVI` model allows variance to vary across components while fixing all covariance parameters to zero.

A plot showing the classification provided by the estimated model can be drawn as follows:

```
plot(mod, what = "classification")
```

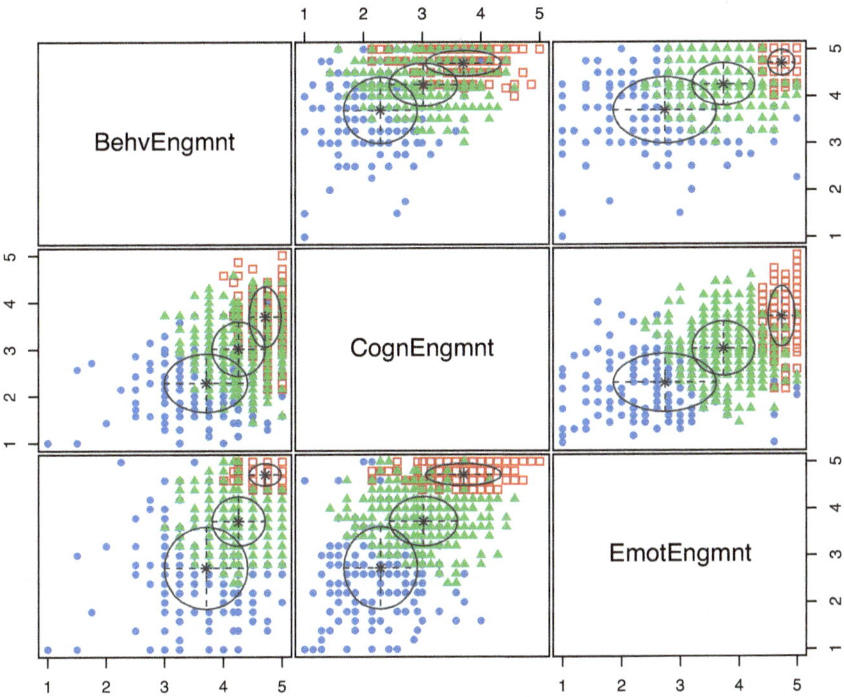

Fig. 2 Scatterplot matrix of engagement features with data points marked and coloured by the identified clusters, and ellipses corresponding to projections of the estimated cluster covariances

The estimated model identifies three clusters of varying size. The third group (shown as filled green triangles) accounts for more than 50% of the observations, while the first (shown as blue filled points) and the second (shown as red open squares) account for approximately 29% and 16%, respectively. The smallest cluster is also the group with the largest engagement scores (Fig. 2).

The different engagement behaviour of the three identified clusters can be shown using a latent profiles plot of the estimated means with point sizes proportional to the estimated mixing probabilities (see Fig. 3):

```
# collect estimated means
means <- data.frame(Profile = factor(1:mod$G),
                    t(mod$parameters$mean)) |>
  pivot_longer(cols = -1,
               names_to = "Variable",
               values_to = "Mean")

# convert variable names to factor
means$Variable <- factor(means$Variable,
                         levels = colnames(mod$data))
```

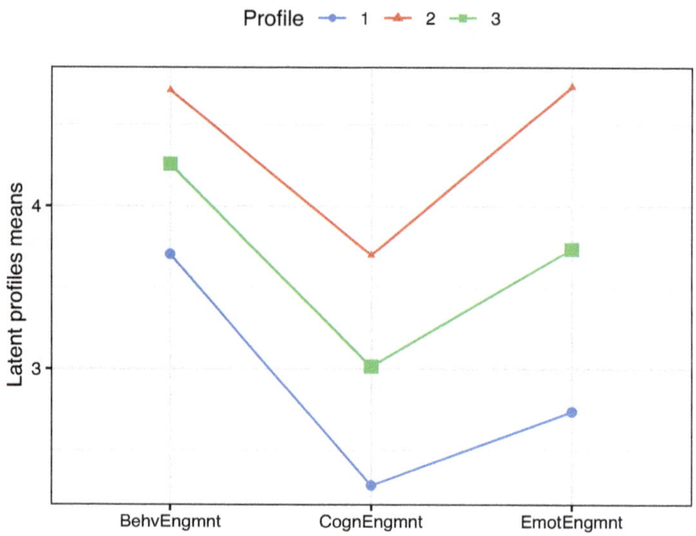

Fig. 3 Latent profiles plot showing estimated means with point sizes proportional to estimated mixing probabilities

```
# add mixing probabilities corresponding to profiles
means <- means |>
  add_column(MixPro = mod$parameters$pro[means$Profile])
means

# A tibble: 9 x 4
  Profile Variable    Mean MixPro
  <fct>   <fct>      <dbl>  <dbl>
1 1       BehvEngmnt  3.70  0.290
2 1       CognEngmnt  2.29  0.290
3 1       EmotEngmnt  2.74  0.290
4 2       BehvEngmnt  4.71  0.162
5 2       CognEngmnt  3.70  0.162
6 2       EmotEngmnt  4.73  0.162
7 3       BehvEngmnt  4.26  0.548
8 3       CognEngmnt  3.02  0.548
9 3       EmotEngmnt  3.74  0.548

# plot the means of the latent profiles
ggplot(means, aes(x = Variable, y = Mean,
                  group = Profile,
                  shape = Profile,
                  color = Profile)) +
  geom_point(aes(size = MixPro)) +
  geom_line(linewidth = 0.5) +
```

```
  labs(x = NULL, y = "Latent profiles means") +
  scale_color_manual(values = mclust.options("classPlotColors")) +
  scale_size(range = c(1, 3), guide = "none") +
  theme_bw() +
  theme(legend.position = "top")
```

The smallest cluster (Profile 2) has the highest engagement scores for all three variables. All three scores are lower for the largest cluster (Profile 3), which are all in turn lower for Profile 1. All three profiles exhibit the lowest mean scores for the cognitive engagement attribute. For Profile 2, behavioural engagement and emotional engagement scores are comparable, whereas for the other two profiles, the mean scores for this attribute are lower than those for the behaviour engagement attribute. Taken together, we could characterise profiles 1, 3, and 2 as “low”, “medium”, and “high” engagement profiles, respectively.

To provide a more comprehensive understanding of the results presented in the previous graph, it would be beneficial to incorporate a measure of uncertainty for the estimated means of the latent profiles. This can be achieved by resampling using the function `MclustBootstrap()` as described above:

```
boot <- MclustBootstrap(mod, type = "bs", nboot = 999)
```

The bootstrap distribution of the mixing weights can be visualised using histograms with the code

```
par(mfcol = c(1, 3), mar = c(2, 4, 1, 1), mgp = c(2, 0.5, 0))
plot(boot, what = "pro", xlim = c(0, 1))
```

while the bootstrap distribution of the components means can be plotted with the code

```
par(mfcol = c(3, 3), mar = c(2, 4, 1, 1), mgp = c(2, 0.5, 0))
plot(boot, what = "mean", conf.level = 0.95)
```

The resulting graphs are reported in Figs. 4 and 5, where the GMM estimates are shown as dashed vertical lines, while the horizontal segments represent the percentile confidence intervals at the 95% confidence level.

Numerical output of the resampling-based bootstrap distributions is given by:

```
sboot <- summary(boot, what = "ci")
sboot
```

```
------------------------------------------------------
Resampling confidence intervals
------------------------------------------------------
```

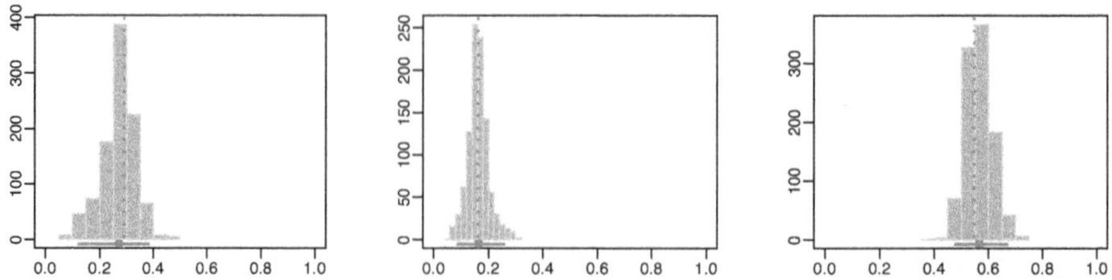

Fig. 4 Bootstrap distribution of GMM mixture weights

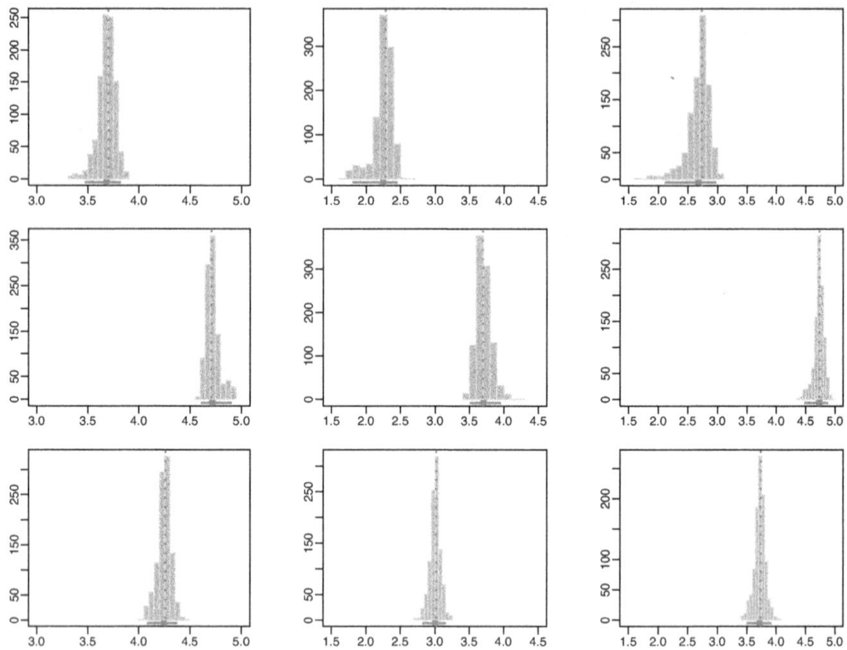

Fig. 5 Bootstrap distribution of GMM component means

```
Model                        = VVI
Num. of mixture components = 3
Replications                 = 999
Type                         = nonparametric bootstrap
Confidence level             = 0.95

Mixing probabilities:
              1          2         3
2.5%  0.1243841 0.08798243 0.4782555
97.5% 0.3835502 0.25791041 0.6687084
```

```
Means:
[,,1]
       BehvEngmnt CognEngmnt EmotEngmnt
2.5%     3.478885   1.816859   2.121577
97.5%    3.815991   2.447267   2.960526
[,,2]
       BehvEngmnt CognEngmnt EmotEngmnt
2.5%     4.620202   3.521108   4.495329
97.5%    4.900447   3.946966   4.867349
[,,3]
       BehvEngmnt CognEngmnt EmotEngmnt
2.5%     4.092707   2.835722   3.512692
97.5%    4.369398   3.149737   3.908160

Variances:
[,,1]
       BehvEngmnt CognEngmnt EmotEngmnt
2.5%   0.3925919  0.2235505  0.5453094
97.5%  0.7894949  0.4924710  0.9944741
[,,2]
       BehvEngmnt CognEngmnt EmotEngmnt
2.5%  0.01602006  0.3225734 0.01935046
97.5% 0.10643442  0.5867531 0.15708500
[,,3]
       BehvEngmnt CognEngmnt EmotEngmnt
2.5%   0.1571746  0.2842534  0.2288972
97.5%  0.2816023  0.3952229  0.4188765
```

The information above can then be used to plot the latent profile means accompanied by 95% confidence intervals represented as vertical bars, as illustrated in Fig. 6. The confidence intervals for the cognitive and emotional engagement attributes are noticeably wider for the "low" engagement profile.

```
means <- means |>
  add_column(lower = as.vector(sboot$mean[1,,]),
             upper = as.vector(sboot$mean[2,,]))
means
```

```
# A tibble: 9 x 6
  Profile Variable    Mean MixPro lower upper
  <fct>   <fct>      <dbl>  <dbl> <dbl> <dbl>
1 1       BehvEngmnt  3.70  0.290  3.48  3.82
2 1       CognEngmnt  2.29  0.290  1.82  2.45
3 1       EmotEngmnt  2.74  0.290  2.12  2.96
4 2       BehvEngmnt  4.71  0.162  4.62  4.90
```

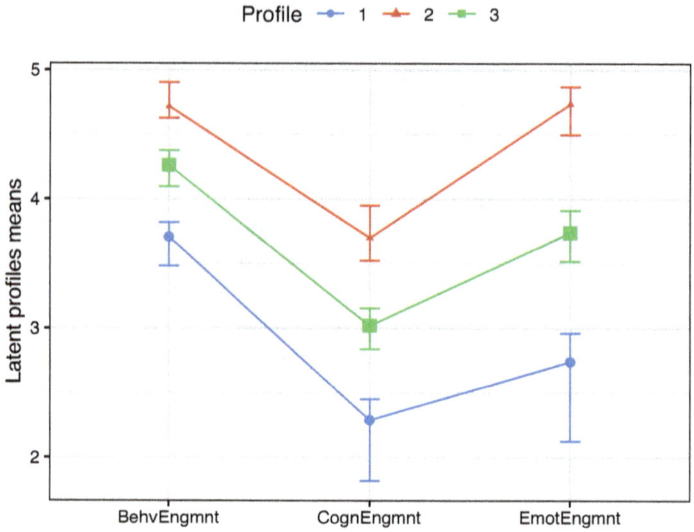

Fig. 6 Latent profiles plot showing estimated means with 95% bootstrap confidence intervals

```
5 2       CognEngmnt  3.70  0.162  3.52  3.95
6 2       EmotEngmnt  4.73  0.162  4.50  4.87
7 3       BehvEngmnt  4.26  0.548  4.09  4.37
8 3       CognEngmnt  3.02  0.548  2.84  3.15
9 3       EmotEngmnt  3.74  0.548  3.51  3.91
```

```
ggplot(means, aes(x = Variable, y = Mean, group = Profile,
                  shape = Profile, color = Profile)) +
  geom_point(aes(size = MixPro)) +
  geom_line(linewidth = 0.5) +
  geom_errorbar(aes(ymin = lower, ymax = upper),
                linewidth = 0.5, width = 0.1) +
  labs(x = NULL, y = "Latent profiles means") +
  scale_color_manual(values = mclust.options("classPlotColors")) +
  scale_size(range = c(1, 3), guide = "none") +
  theme_bw() +
  theme(legend.position = "top")
```

Finally, the entropy of the estimated partition, average entropy of each latent component, and average posterior probabilities are obtained via:

```
probs <- mod$z                      # posterior conditional probs
probs_map <- apply(probs, 1, max) # maximum a posteriori probs
clusters <- mod$classification    # cluster assignment for each obs
n <- mod$n                          # number of obs
K <- mod$G                          # number of latent profiles
```

```r
# Entropy:
E <- 1 + sum(probs * log(probs))/(n * log(K))
E
```

```
[1] 0.6890602
```

```r
# Case-specific entropy contributions:
Ei <- 1 + rowSums(probs * log(probs))/log(K)
sum(Ei)/n
```

```
[1] 0.6890602
```

```r
df_entropy  <- data.frame(clusters = as.factor(clusters), entropy = Ei)
```

```r
df_entropy |>
  group_by(clusters) |>
  summarise(count = n(),
            mean = mean(entropy),
            sd = sd(entropy),
            min = min(entropy),
            max = max(entropy))
```

```
# A tibble: 3 x 6
  clusters count  mean    sd   min   max
  <fct>    <int> <dbl> <dbl> <dbl> <dbl>
1 1          184 0.740 0.239 0.369 1.00 
2 2          119 0.690 0.225 0.187 0.993
3 3          414 0.666 0.197 0.172 0.974
```

```r
ggplot(df_entropy, aes(y = clusters, x = entropy,  fill = clusters)) +
  geom_density_ridges(stat = "binline", bins = 21,
                      scale = 0.9, alpha = 0.5) +
  scale_x_continuous(breaks = seq(0, 1 ,by=0.1),
                     limits = c(0, 1.05)) +
  scale_fill_manual(values = mclust.options("classPlotColors")) +
  geom_vline(xintercept = E, lty = 2) +
  labs(x = "Case-specific entropy contribution",
       y = "Latent profile") +
  theme_ridges(center_axis_labels = TRUE) +
  theme(legend.position = "none",
        panel.spacing = unit(1, "lines"),
        strip.text.x = element_text(size = 8))
```

```
# Average posterior probabilities by cluster:
df_AvePP <- data.frame(clusters = as.factor(clusters), pp = probs_map)

df_AvePP |>
  group_by(clusters) |>
  summarise(count = n(),
            mean = mean(pp),
            sd = sd(pp),
            min = min(pp),
            max = max(pp))
```

```
# A tibble: 3 x 6
  clusters count  mean    sd   min   max
  <fct>    <int> <dbl> <dbl> <dbl> <dbl>
1 1          184 0.864 0.160 0.513 1.00
2 2          119 0.858 0.146 0.468 0.999
3 3          414 0.850 0.135 0.502 0.996
```

```
ggplot(df_AvePP, aes(y = clusters, x = pp,  fill = clusters)) +
  geom_density_ridges(stat = "binline", bins = 21,
                      scale = 0.9, alpha = 0.5) +
  scale_x_continuous(breaks = seq(0, 1, by=0.1),
                     limits = c(0, 1.05)) +
  scale_fill_manual(values = mclust.options("classPlotColors")) +
  labs(x = "MAP probabilities", y = "Latent profile") +
  theme_ridges(center_axis_labels = TRUE) +
  theme(legend.position = "none",
        panel.spacing = unit(1, "lines"),
        strip.text.x = element_text(size = 8))
```

We note that, as shown in Figs. 7 and 8, all entropy and AvePP quantities appear satisfactory from the point of view of indicating reasonably well-separated clusters.

6 Discussion

Using a person-centered method (finite Gaussian mixture model), the present analysis uncovered the heterogeneity within the SEM engagement data by identifying three latent or unobserved clusters: low, medium, and high engagement clusters. Uncovering the latent structure could help understand individual differences among students, identify the complex multidimensional variability of a construct—engagement in our case—and possibly help personalise teaching and learning. Several studies have revealed similar patterns of engagement which—similar to the current analysis—comprise three levels that can be roughly summarised as high, moderate, and low [9, 45, 46]. The heterogeneity of engagement has been demonstrated in longitudinal studies, in both face-to-face settings as well as online engagement [9]. Furthermore, the association between engagement and

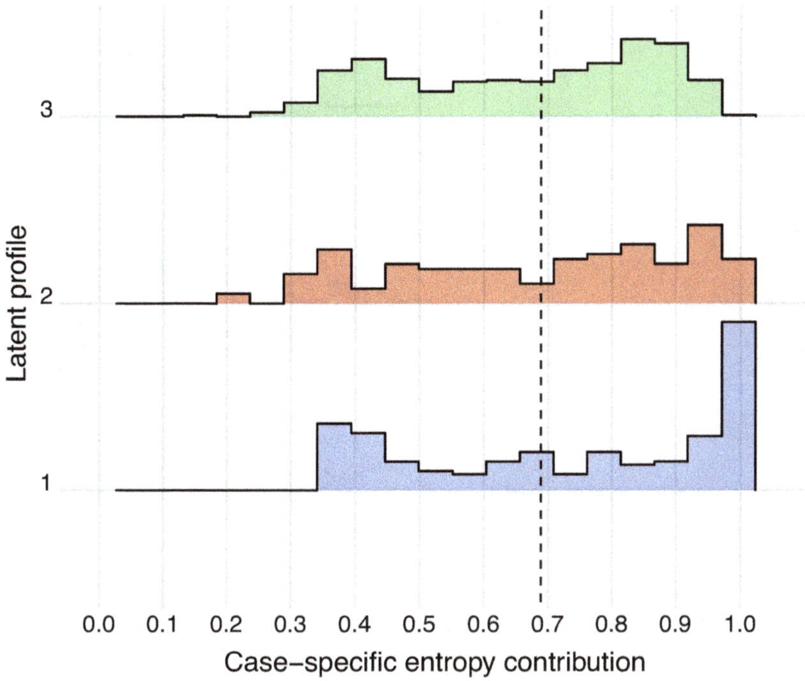

Fig. 7 Entropy contributions by cluster and total entropy (dashed line)

performance has been demonstrated to vary by achievement level, time of the year, as well as engagement state; that is, high achievers may at some point in their program descend to lower engagement states and still continue to have higher achievement [3]. Such patterns, variability, and individual differences are not limited to engagement, but has been reported for almost every major disposition in education psychology [2].

On a general level, heterogeneity has been a hot topic in recent educational literature. Several calls have been voiced to adopt methods that capture different patterns or subgroups within students' behavior or functioning. Assuming that there is "an average" pattern that represents the entirety of student populations requires the measured construct to have the same causal mechanism, same development pattern, and affect students in exactly the same way. The average assumption is of course impossible and has been proven inaccurate across a vast number of studies (e.g., [2] and [4]). Since heterogeneity is prevalent in psychological, behavioral, and physiological human data, person-centered methods will remain a very important tool for researchers [47].

Person-centered methods can be grouped into algorithmic clustering methods on one hand and the model-based clustering paradigm on the other, with the former being more traditional and the latter being more novel in the learning analytics literature. The analysis of the SEM data centered here on the model-based

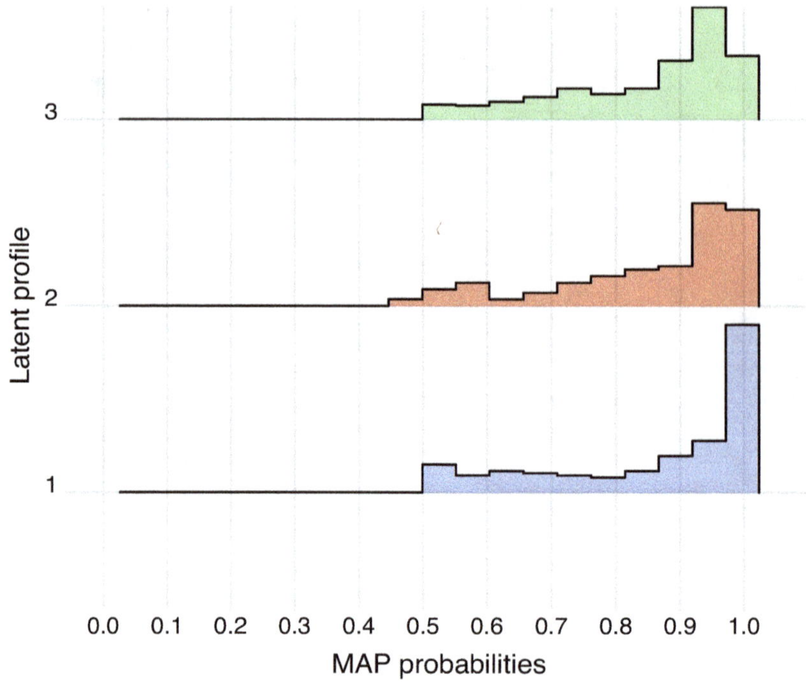

Fig. 8 Average posterior probabilities by cluster

approach, specifically the finite Gaussian mixture model framework. The `mclust` package enabled such models to be easily fitted and this framework exhibits many advantages over traditional clustering algorithms which rely on dissimilarity-based heuristics. Firstly, the likelihood-based underpinnings enable the selection of the optimal model using principled statistical model selection criteria. In particular, it is noteworthy in the present analysis that the model selection procedure was not limited to three-cluster solutions: mixtures with fewer or greater than three clusters were evaluated and the three-cluster solution—supported by previous studies in education research—was identified as optimal according to the BIC. Secondly, the parsimonious modelling of the covariance structures provides the flexibility to model clusters with different geometric characteristics. In particular, the clusters in the present analysis, whereby each group is described by a single Gaussian component with varying volume and shape, but the same orientation aligned with the coordinate axes are more flexible than the spherical, Euclidean distance-based clusters obtainable under the k-means algorithm. Thirdly, the models relax the assumption that each observation is associated with exactly one cluster and yields informative cluster-membership probabilities for each observation, which can be used to compute useful diagnostics such as entropies and average posterior probabilities which are unavailable under so-called "hard" clustering frameworks.

Finally, the `mclust` package facilitates simple summaries and visualisations of the resulting clusters and cluster-specific parameter estimates.

That being said, there are a number of methodological limitations of the GMM framework to be aware of in other settings. Firstly, and most obviously, such models are inappropriate for clustering categorical or mixed-type variables. For clustering longitudinal categorical sequences, such as those in Chapter 10 [48], model-based approaches are provided by the mixtures of exponential-distance models framework of [49] (and related `MEDseq` R package) and the mixtures of hidden Markov models framework of [50] (and related `seqHMM` package; see Chapter 12 [51]). Regarding mixed-type variables, [52] provide a model-based framework (and the related `clustMD` package).

Secondly, the one-to-one correspondence typically assumed between component distributions and clusters is not always the case [53]. This is only true if the underlying true component densities are Gaussian. When the assumption of component-wise normality is not satisfied, the performance of such models will deteriorate as more components are required to fit the data well. However, even for continuous data, GMMs tend to overestimate the number of clusters when the assumption of normality is violated. Two strategies for dealing with this are provided by the `mclust` package, one based on combining Gaussian mixture components according to an entropy criterion, and one based on a adding a so-called "noise component"—represented by a uniform distribution—to the mixture. The noise component captures outliers which do not fit the prevailing patterns of Gaussian clusters, which would otherwise be assigned to (possibly many) small clusters and minimises their deleterious effect on parameter estimation for the other, more defined clusters. Further details of combining components and adding a noise component can be found in [30]. Alternatively, mixture models which depart from normality have been an active area of research in model-based clustering in recent years. Such approaches—some of which are available in the R package `mixture` [54]—replace the underlying Gaussian component distributions with e.g., generalised hyperbolic distributions, the multivariate t distribution, and the multivariate skew-t distribution.

A third main limitation of GMMs is their ineffectiveness in high-dimensional settings, when the data dimension d is comparable to or even greater than n. Among the 14 parsimonious parameterisations available in `mclust`, only models with diagonal covariance structures are tractable when $n \leq p$. Incorporating factor-analytic covariance decompositions in so-called finite Gaussian mixtures of factor analysers have been proposed for addressing this issue [55, 56]. Imposing constraints on the parameters of such factor-analytic structures in the component covariance matrices in the spirit of `mclust` leads to another family of parsimonious Gaussian mixture models [57], which are implemented in the R package `pgmm`. Model selection becomes increasingly difficult with such models, given the need to choose both the optimal number of mixture components and the optimal number of latent factors (as well as the covariance parameterisation, in the case of `pgmm`). Infinite mixtures of infinite factor analysers—implemented in the R package

IMIFA—are a recent, Bayesian extension which enable automatic inference of the number of components and the numbers of cluster-specific latent factors [58].

Another recent extension, building directly on the 14 models from mclust, is the MoEClust model family of [59] and the associated MoEClust R package, which closely mimics its syntax. MoEClust effectively embeds Gaussian parsimonious clustering models in the mixtures of experts framework, enabling additional sources of heterogeneity in the form of covariates to be incorporated directly in the clustering model, to guide the construction of the clusters. Either, neither, or both the mixing proportions and/or component mean parameters can be modelled as functions of these covariates. The former is perhaps particularly appealing, given its analogous equivalence to latent profile *regression* [60]. Hypothetically, assuming information on the gender and age of the students in the present analysis was available, such covariates would influence the probabilities of cluster membership under such a model, while the correspondence thereafter between the parameters of the component distributions and the clusters would have the same interpretation as per standard LPA models.

References

1. Howard MC, Hoffman ME (2018) Variable-centered, person-centered, and person-specific approaches: where theory meets the method. Organ Res Methods 21:846–876. https://doi.org/10.1177/1094428117744021
2. Hickendorff M, Edelsbrunner PA, McMullen J, Schneider M, Trezise K (2018) Informative tools for characterizing individual differences in learning: Latent class, latent profile, and latent transition analysis. Learn Individ Differences 66:4–15. https://doi.org/10.1016/j.lindif.2017.11.001
3. Saqr M, López-Pernas S, Helske S, Hrastinski S (2023) The longitudinal association between engagement and achievement varies by time, students' subgroups, and achievement state: A full program study. Comput Educ 199:104787. https://doi.org/10.1016/j.compedu.2023.104787
4. Törmänen, Järvenoja, Saqr, Malmberg, others (2022) A person-centered approach to study students' socio-emotional interaction profiles and regulation of collaborative learning. Front Educ 7. https://doi.org/10.3389/feduc.2022.866612
5. Saqr M (2023) Modelling within-person idiographic variance could help explain and individualize learning. Br J Educ Technol 54:1077–1094.
6. Fraley C, Raftery AE (2002) Model-based clustering, discriminant analysis, and density estimation. J Am Stat Assoc 97:611–631. https://doi.org/10.1198/016214502760047131
7. Fraley C, Raftery AE, Scrucca L (2023) mclust: Gaussian mixture modelling for model-based clustering, classification, and density estimation
8. R Core Team (2023) R: A language and environment for statistical computing. R Foundation for Statistical Computing, Vienna, Austria
9. Saqr M, López-Pernas S (2021) The longitudinal trajectories of online engagement over a full program. Comput Educ 175:104325. https://doi.org/10.1016/j.compedu.2021.104325
10. Yu J, Huang C, He T, Wang X, Zhang L (2022) Investigating students' emotional self-efficacy profiles and their relations to self-regulation, motivation, and academic performance in online learning contexts: A person-centered approach. Educ Inf Technol 27:11715–11740. https://doi.org/10.1007/s10639-022-11099-0

11. Saqr M, López-Pernas S (2022) How CSCL roles emerge, persist, transition, and evolve over time: A four-year longitudinal study. Comput Educ 189:104581. https://doi.org/10.1016/j.compedu.2022.104581
12. Cheng S, Huang J-C, Hebert W (2023) Profiles of vocational college students' achievement emotions in online learning environments: Antecedents and outcomes. Comput Hum Behav 138:107452. https://doi.org/10.1016/j.chb.2022.107452
13. Hoi VN (2023) Transitioning from school to university: A person-oriented approach to understanding first-year students' classroom engagement in higher education. Educ Rev 1–21. https://doi.org/10.1080/00131911.2022.2159935
14. Scheidt M, Godwin A, Berger E, Chen J, Self BP, Widmann JM, Gates AQ (2021) Engineering students' noncognitive and affective factors: Group differences from cluster analysis. J Eng Educ 110:343–370. https://doi.org/10.1002/jee.20386
15. Zhang Y, Paquette L, Pinto JD, Liu Q, Fan AX (2023) Combining latent profile analysis and programming traces to understand novices' differences in debugging. Educ Inf Technol 28:4673–4701. https://doi.org/10.1007/s10639-022-11343-7
16. Hennig C (2015) What are the true clusters? Pattern Recogn Lett 64:53–62
17. Everitt BS, Landau S, Leese M, Stahl D (2011) Cluster analysis, 5th edn. John Wiley & Sons, New York
18. Fraley C (1998) Algorithms for model-based Gaussian hierarchical clustering. SIAM J Sci Comput 20:270–281. https://doi.org/10.1137/S1064827596311451
19. McLachlan GJ, Peel D (2000) Finite mixture models. John Wiley & Sons, New York
20. Dempster AP, Laird NM, Rubin DB (1977) Maximum likelihood from incomplete data via the EM algorithm (with discussion). J R Stat Soc Ser B (Stat Methodol) 39:1–38. https://doi.org/10.1111/j.2517-6161.1977.tb01600.x
21. Spearman C (1904) "General Intelligence," objectively determined and measured. Am J Psychol 15:201–292. https://doi.org/10.2307/1412107
22. Jöreskog KG (1970) A general method for analysis of covariance structures. Biometrika 57:239–251
23. Blei DM, Ng AY, Jordan MI (2003) Latent Dirichlet allocation. J Mach Learn Res 3:993–1022
24. Zucchini W, MacDonald IL, Langrock R (2016) Hidden Markov models for time series: An introduction using R. Chapman & Hall/CRC Press, London
25. Bartolucci F, Farcomeni A, Pennoni F (2012) Latent Markov models for longitudinal data. Chapman & Hall/CRC Press
26. Bartholomew DJ, Knott M, Moustaki I (2011) Latent variable models and factor analysis: A unified approach, 3rd edn. John Wiley & Sons, Chichester
27. Rosenberg JM, Beymer PN, Anderson DJ, Van Lissa CJ, Schmidt JA (2018) TidyLPA: An R package to easily carry out latent profile analysis (LPA) using open-source or commercial software. J Open Source Softw 3:978. https://doi.org/10.21105/joss.00978
28. Banfield J, Raftery AE (1993) Model-based Gaussian and non-Gaussian clustering. Biometrics 49:803–821. https://doi.org/10.2307/2532201
29. Celeux G, Govaert G (1995) Gaussian parsimonious clustering models. Pattern Recogn 28:781–793. https://doi.org/10.1016/0031-3203(94)00125-6
30. Scrucca L, Fraley C, Murphy TB, Raftery AE (2023) Model-based clustering, classification, and density estimation using mclust in R. Chapman & Hall/CRC Press, London
31. McLachlan GJ, Krishnan T (2008) The EM algorithm and extensions, 2nd edn. Wiley-Interscience, Hoboken
32. Schwarz G (1978) Estimating the dimension of a model. Ann Stat 6:461–464. https://doi.org/10.1214/aos/1176344136
33. Biernacki C, Celeux G, Govaert G (2000) Assessing a mixture model for clustering with the integrated completed likelihood. IEEE Trans Pattern Anal Mach Intell 22:719–725
34. Nylund-Gibson K, Choi AY (2018) Ten frequently asked questions about latent class analysis. Transl Issues Psychol Sci 4:440–461

35. Scrucca L, Fop M, Murphy TB, Raftery AE (2016) mclust 5: Clustering, classification and density estimation using Gaussian finite mixture models. R J 8:205–233. https://doi.org/10.32614/RJ-2016-021
36. Fraley C, Raftery AE (2007) Bayesian regularization for normal mixture estimation and model-based clustering. J Classif 24:155–181
37. Basford KE, Greenway DR, McLachlan GJ, Peel D (1997) Standard errors of fitted component means of normal mixtures. Comput Stat 12:1–18
38. O'Hagan A, Murphy TB, Scrucca L, Gormley IC (2019) Investigation of parameter uncertainty in clustering using a Gaussian mixture model via jackknife, bootstrap and weighted likelihood bootstrap. Comput Stat 34:1779–1813. https://doi.org/10.1007/s00180-019-00897-9
39. Efron B (1979) Bootstrap methods: Another look at the jackknife. Ann Stat 7:1–26
40. Rubin DB (1981) The Bayesian bootstrap. Ann Stat 9:130–134
41. Newton MA, Raftery AE (1994) Approximate bayesian inference with the weighted likelihood bootstrap (with discussion). J R Stat Soc Ser B (Stat Methodol) 56:3–48
42. Cover TM, Thomas JA (2006) Elements of information theory, 2nd edn. John Wiley & Sons, New York
43. Celeux G, Soromenho G (1996) An entropy criterion for assessing the number of clusters in a mixture model. J Classif 13:195–212
44. López-Pernas S, Saqr M, Conde J, Del-Río-Carazo L (2024) A broad collection of datasets for educational research training and application. In: Saqr M, López-Pernas S (eds) Learning analytics methods and tutorials: A practical guide using R. Springer
45. Archambault I, Dupéré V (2016) Joint trajectories of behavioral, affective, and cognitive engagement in elementary school. J Educ Res 110:188–198. https://doi.org/10.1080/00220671.2015.1060931
46. Zhen R, Liu R-D, Wang M-T, Ding Y, Jiang R, Fu X, Sun Y (2019) Trajectory patterns of academic engagement among elementary school students: The implicit theory of intelligence and academic self-efficacy matters. Br J Educ Psychol 90:618–634. https://doi.org/10.1111/bjep.12320
47. Bryan CJ, Tipton E, Yeager DS (2021) Behavioural science is unlikely to change the world without a heterogeneity revolution. Nat Hum Behav 5:980–989. https://doi.org/10.1038/s41562-021-01143-3
48. Saqr M, López-Pernas S, Helske S, Durand M, Murphy K, Studer M, Ritschard G (2024) Sequence analysis in education: Principles, technique, and tutorial with r. In: Saqr M, López-Pernas S (eds) Learning analytics methods and tutorials: A practical guide using R. Springer
49. Murphy K, Murphy TB, Piccarreta R, Gormley IC (2021) Clustering longitudinal life-course sequences using mixtures of exponential-distance models. J R Stat Soc Ser A (Stat Soc) 184:1414–1451. https://doi.org/10.1111/rssa.12712
50. Helske S, Helske J (2019) Mixture hidden Markov models for sequence data: The seqHMM package in R. J Stat Softw 88:1–32
51. Helske J, Helske S, Saqr M, López-Pernas S, Murphy K (2024) A modern approach to transition analysis and process mining with Markov models: A tutorial with R. In: Saqr M, López-Pernas S (eds) Learning analytics methods and tutorials: A practical guide using R. Springer
52. McParland D, Gormley IC (2016) Model based clustering for mixed data: clustMD. Adv Data Anal Classif 10:155–169
53. Hennig C (2010) Methods for merging Gaussian mixture components. Adv Data Anal Classif 4:3–34
54. Pocuca N, Browne RP, McNicholas PD (2022) mixture: Mixture models for clustering and classification
55. Ghahramani Z, Hinton GE (1996) The EM algorithm for mixtures of factor analyzers. Department of Computer Science, University of Toronto
56. McLachlan GJ, Peel D, Bean RW (2003) Modelling high-dimensional data by mixtures of factor analyzers. Comput Stat Data Anal 41:379–388

57. McNicholas PD, Murphy TB (2008) Parsimonious Gaussian mixture models. Stat Comput 18:285–296
58. Murphy K, Viroli C, Gormley IC (2020) Infinite mixtures of infinite factor analysers. Bayesian Anal 15:937–963
59. Murphy K, Murphy TB (2020) Gaussian parsimonious clustering models with covariates and a noise component. Adv Data Anal Classif 14:293–325. https://doi.org/10.1007/s11634-019-00373-8
60. Dayton CM, Macready GB (1988) Concomitant-variable latent-class models. Journal of the American Statistical Association 83:173–178

Part III
Temporal Methods

Sequence Analysis in Education: Principles, Technique, and Tutorial with R

Mohammed Saqr, Sonsoles López-Pernas, Satu Helske, Marion Durand, Keefe Murphy, Matthias Studer, and Gilbert Ritschard

1 Introduction

Patterns exist everywhere in our life, from the sequence of genes to the order of steps in cooking recipes. Discovering patterns, variations, regularities, or irregularities is at the heart of scientific inquiry and, therefore, several data mining methods have been developed to understand patterns. Sequence analysis—or sequence mining—was developed almost four decades ago to address the increasing needs for pattern mining [1]. Ever since, a wealth of applications, algorithms, and statistical tools have been developed, adapted, or incorporated into the array of sequence analysis. Since sequence mining has been conceptualized, it has grown in scale of adoption and range of applications across life and social sciences [2] and education research was no exception (e.g., [3]). As a data mining technique, sequence mining has been commonly implemented to identify hidden patterns that would otherwise be missed using other analytical techniques and find interesting subsequences (parts of the sequence) that have practical significance or unexpected sequences that we did not know existed [4]. For instance, by mining sequences of collaborative dialogue, we

M. Saqr (✉) · S. López-Pernas · M. Durand
School of Computing, University of Eastern Finland, Joensuu, Finland
e-mail: mohammed.saqr@uef.fi

S. Helske
INVEST Research Flagship Center & Department of Social Research, University of Turku, Turku, Finland

K. Murphy
Department of Mathematics and Statistics, Hamilton Institute, Maynooth University, Maynooth, Ireland

M. Studer · G. Ritschard
Institute of Demography and Socioeconomics & Centre LIVES, University of Geneva, Geneva, Switzerland

M. Saqr, S. López-Pernas (eds.), *Learning Analytics Methods and Tutorials*,
https://doi.org/10.1007/978-3-031-54464-4_10

could identify which sequences are followed by more argumentative interactions, and what sequences are crucial to the collaborative process. A literature review of the common applications follows in the next section.

Learning is a process that unfolds in time, a process that occurs in sequences of actions, in repeated steps, in patterns that have meanings and value for understanding learners' behavior [5]. The conceptualization of learning as a process entails two important criteria: process as a sequence of states that unfold in time and process as a transformative mechanism that drives the change from one state to another [6]. Thereupon, methods such as sequence mining have gained increasing grounds and amassed a widening repertoire of techniques in the field of education to study the learning process. In particular, sequence mining has been used to harness the temporal unfolding of learners' behavior using digital data, transcripts of conversations, or behavioral states [7]. Nevertheless, sequence mining can be used to study non-temporal sequences such as protein sequences and other types of categorical data [8].

What makes sequences in education interesting is that they have patterns of repeated or recurrent sequences. Finding such patterns has helped typify learners' behaviors and identify which patterns are associated with learning and which are associated with unfavorable outcomes [3]. Sequence mining can also describe a pathway or a trajectory of events, for example, how a student proceeds from enrolment to graduation [9], and help to identify the students who have a stable trajectory, who are turbulent, and who are likely to falter along their education [3].

2 Review of the Literature

In recent years, sequence analysis has become a central method in learning analytics research due to its potential to summarize and visually represent large amounts of student-related data. In this section we provide an overview of some of the most representative studies in the published literature. A summary of the studies reviewed in this section can be seen in Table 1. A common application of sequence analysis is the study of students' log data extracted from their interactions with online learning technologies (mostly learning management systems, LMSs) throughout a learning session [10–12]. In some studies, the session is well-delimited, such as the duration of a game [13] or moving window [14], but in most cases it is inferred from the data, considering a session as an uninterrupted sequence of events [10, 11]. Few are the studies in which longer sequences are studied, covering a whole course or even a whole study program [3, 9, 15]. In such studies, the sequences are not composed of instantaneous interactions but rather of states that aggregate students' information over a certain period, for example, we can study students' engagement [9], learning strategies [3], or collaboration roles [15] for each course in a study program.

Most of the existing research has used clustering techniques to identify distinct groups of similar sequences. Agglomerative Hierarchical Clustering (AHC) has been the most used technique, with a wealth of distance measures such as Euclidean

Table 1 Summary of the reviewed articles about sequence analysis in learning analytics

Ref.	Context	Time scheme	Actor	Alphabet	Clustering algorithm
[18]	40 students	Learning activity (5 days)	Student	LMS events	Differential sequence mining (core algorithm)
[14]	1 middle school class (40 students)	Learning activity (5 days)	Student	LMS events (e.g., Read, Linkadd)	Differential sequence mining (core algorithm, SPAMc)
[11]	1 university course (290 students)	Session	Student-session	LMS events	AHC (Optimal matching)
		Course	Student-course	Tactics obtained from previous clustering	AHC (Euclidean distance)
[10]	3 courses: one university course with 3 course offerings (1135 students), another university course with 2 course offerings (487 students), and a MOOC with a single offering (368 students)	Session	Student-session	LMS events (e.g., content_access, download)	First Order Markov Model
		Course	Student-course	Tactics obtained from previous clustering	AHC (Euclidean distance)
[9]	15 university courses (106 students)	Study program (15 courses)	Student	Engagement state (e.g., Active, Average)	Hidden Markov Models
[13]	1 educational escape room game in a university course (96 students)	Escape room game (1 h 45 min)	Team	Game activity (e.g., hint obtained, puzzle solving)	AHC (Longest Common Subsequence)

(continued)

Table 1 (continued)

Ref.	Context	Time scheme	Actor	Alphabet	Clustering algorithm
[15]	10 university courses (329 students)	Study program (10 courses)	Student	Roles in the group (e.g., Leaders, Mediators)	Mixture Hidden Markov Models
[3]	10 university courses (135 students)	Session	Student-session	LMS event (e.g., Course main view, Forum consume)	Mixture Hidden Markov Models
		Course	Student-course	Tactics obtained from previous clustering (e.g., Lecture read, Forum read)	AHC (Euclidean distance)
		Study program (10 courses)	Student	Course-level strategies from previous clustering (e.g., Light interactive, Moderate interactive)	AHC (Euclidean distance)
[16]	1 university courses, 4 Course offerings (200 students)	Week	Group of students	Interaction type on forum (e.g., Discuss, Argue)	
		Session	Student-session	Interaction type on forum (e.g., Discuss, Argue)	AHC (Longest Common Prefix)

[3, 10], Longest Common Subsequence [13], Longest Common Prefix [16], and Optimal Matching [11]. Other works have relied on Markovian Models [15, 17] or differential sequence mining [18]. Throughout the remainder of the book, we provide an introduction to sequence analysis as a method, describing in detail the most relevant concepts for its application to educational data. We provide a step-by-step tutorial of how to implement sequence analysis in a data set of student log data using the R programming language.

3 Basics of Sequences

Sequences are ordered lists of discrete elements (i.e., events, states or categories). Such elements are discrete (in contrast to numerical values such as grades) and are commonly organized chronologically. Examples include sequence of activities, sequence of learning strategies, or sequence of behavioral states [19]. A sequence of learning activities may include (*play video—solve exercise—taking quiz—access instructions*) [17], other examples include sequence of game moves e.g., (*solve puzzle—request hint—complete game)* [13], or collaborative roles, for instance, (*leader—mediator - isolate*) [15].

Before going into sequence analysis, let's discuss a basic example of a sequence inspired by Saqr and López-Pernas [9]. Let's assume we are tracking the engagement states of students from a course to the next and for a full year that has five courses. The engagement states can be either engaged (when the student is fully engaged in their learning), average (when the student is moderately engaged), and disengaged (when the student is barely engaged). Representing the sequence of engagement states of two hypothetical students may look like the example on Table 2.

The first student starts in course 1 with an *Average* engagement state, in Course 2, the student is engaged, and so in all the subsequent courses Course 3, Course 4, and Course 5. The student in row 2 has a *Disengaged* state in course 2 onwards. As we can see from the two sequences here, there is a pattern that repeats in both sequences (both students stay 4 consecutive courses in the same state). In real-life examples, sequences are typically longer and in larger numbers. For instance, the paper by Saqr and López-Pernas [9] contains 106 students for a sequence of 15 courses. Finding repeated patterns of engaged states similar to the first student or repeated patterns of disengaged states like the other student would be interesting and helpful to understand how certain subgroups of students proceed in their education and how that relates to their performance.

Table 2 An example sequence

Actor	Course 1	Course 2	Course 3	Course 4	Course 5
Student 1	Average	Engaged	Engaged	Engaged	Engaged
Student 2	Average	Disengaged	Disengaged	Disengaged	Disengaged

3.1 Steps of Sequence Analysis

Several protocols exist for sequence analysis that vary by discipline, research questions, type of data, and software used. In education, sequence analysis protocol usually follows steps that include preparing the data, finding patterns, and relating these patterns to other variables e.g., performance e.g., [11]. The protocol which will be followed in this manual includes six steps:

(1) Identifying (or coding) the elements of the sequence, commonly referred to as *alphabet*
(2) Specifying the time window or epoch (*time scheme*) or sequence alignment scheme
(3) Defining the *actor* and building the sequence object
(4) Visualization and descriptive analysis
(5) Finding similar groups or clusters of sequences,
(6) Analyzing the groups and/or using them in subsequent analyses.

3.1.1 The Alphabet

The first step of sequence analysis is defining the **alphabet** which are the elements or the possible states of the sequence [19]. This process usually entails "recoding" the states to optimize the granularity of the alphabet. In other words, to balance parsimony versus granularity and detail of the data. Some logs are overly detailed and therefore would require a careful recoding by the researcher [8]. For instance, the logs of Moodle (the LMS) include the following log entries for recoding students' access to the quiz module: *quiz_attempt*, *quiz_continue_attempt*, *quiz_close_attempt*, *quiz_view*, *quiz_view_all*, *quiz_preview.* It makes sense here to aggregate (*quiz_attempt*, *quiz_continue_attempt*, *quiz_close_attempt)* into one category with the label *attempt_quiz* and (*quiz_view*, *quiz_view all*, *quiz preview*) to a new category with the label *view_quiz*. Optimizing the alphabet into a reasonable number of states also helps reduce complexity and facilitates interpretation. Of course, caution should be exercised not to aggregate meaningfully distinct states to avoid masking important patterns within the dataset.

3.1.2 Specifying the Time Scheme

The second step is to define a **time scheme,** time epoch or window for the analysis. Sometimes the time window is fairly obvious, for instance, in case a researcher wants to study students' sequence of courses in a program, the window can be the whole program e.g., [9]. Yet, oftentimes, a decision has to be taken about the time window which might affect the interpretation of the resulting sequences. For example, when a researcher is analyzing the sequence of interactions in a collaborative task, he/she may consider the whole collaborative task as a time window or may opt to choose segments or steps within the task as time epochs.

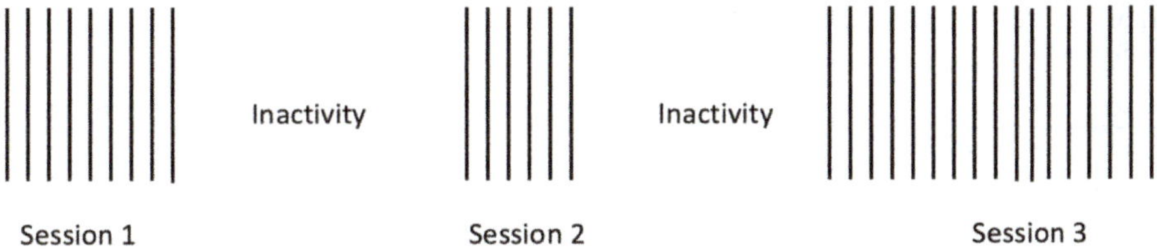

Fig. 1 Every line represents a click, a sequence of successive clicks is a session, and the session is defined by the inactivity period

Table 3 A sequence of engagement states using fictional data

Actor	Course 1	Course 2	Course 3	Course 4	Course 5
Maria	Engaged	Engaged	Engaged	Engaged	Average
Vera	Disengaged	Disengaged	Average	Engaged	Engaged
Anna	Average	Disengaged	Disengaged	Average	Average
Luis	Disengaged	Average	Average	Engaged	Engaged
Bob	Engaged	Engaged	Average	Engaged	Engaged

In the same way, analyzing the sequence of tasks in a course, one would consider the whole course to be the time window for analysis or analyze the sequence of steps in each course task e.g., [20].

In online learning, the *session* has been commonly considered the time window e.g., [11, 17]. A session is an uninterrupted sequence of online activity which can be inferred from identifying the periods of inactivity as depicted in Fig. 1. As can be seen, a user can have multiple sessions across the course. There is no standard guideline for what time window a researcher should consider, however, it is mostly defined by the research questions and the aims of analysis.

3.1.3 Defining the Actor

The third important step is to define the **actor** or the unit of analysis of the sequences (see the actor in Table 3 or User in Table 4). The actor varies according to the type of analysis. When analyzing students' sequences of actions, we may choose the student to be the actor and build a sequence of all student actions e.g., [20]. In online learning, sequence mining has always been created for "user sessions" i.e., each user session is represented as a sequence e.g., [17, 21] and therefore, a user typically has several sessions along the course. In other instances, you may be interested in the study of the sequences of the students' states, for example engagement states in [9] where the student was the actor, or a group of collaborating students' interactions as a whole such as [16] where the whole group is the actor. In the review of the literature, we have examples of such decisions.

3.1.4 Building the Sequences

This step is specific to the software used. For example, in *TraMineR* the step includes specifying the dataset on which the building of the sequences is based and telling

Table 4 The first three columns are simulated sequence data and the three gray columns are computed

User	Action	Time	Lag	Session	Order
Layla	Calendar	9.1.2023 18:44	–	Layla session 1	1
Layla	Lecture	9.1.2023 18:45	1	Layla session 1	2
Layla	Instructions	9.1.2023 18:47	2	Layla session 1	3
Layla	Assignment	9.1.2023 18:49	2	Layla session 1	4
Layla	Lecture	9.1.2023 18:50	1	Layla session 1	5
Layla	Video	9.1.2023 18:51	1	Layla session 1	6
Sophia	Lecture	9.1.2023 20:08	–	Sophia session 1	1
Sophia	Instructions	9.1.2023 20:12	4	Sophia session 1	2
Sophia	Assignment	9.1.2023 20:14	2	Sophia session 1	3
Sophia	Assignment	9.1.2023 20:18	4	Sophia session 1	4
Sophia	Assignment	9.1.2023 20:21	3	Sophia session 1	5
Carmen	Lecture	10.1.2023 10:08	–	Carmen session 1	1
Carmen	Video	10.1.2023 10:11	3	Carmen session 1	2
Layla	Instructions	10.1.2023 19:57	1506	Layla session 2	1
Layla	Video	10.1.2023 20:01	4	Layla session 2	2
Layla	Lecture	10.1.2023 20:08	7	Layla session 2	3
Layla	Assignment	10.1.2023 20:14	6	Layla session 2	4

TraMineR the alphabet, the time scheme, and the actor id variable, as well as other parameters of the sequence object. This step will be discussed in detail in the analysis section.

3.1.5 Visualizing and Exploring the Sequence Data

The fourth step is to visualize the data and perform some descriptive analysis. Visualization allows us to summarize data easily and to see the full dataset at once. *TraMineR* includes several functions to plot the common visualization techniques, each one showing a different perspective.

3.1.6 Calculating the Dissimilarities Between Sequences

The fifth step is calculating dissimilarities or distances between pairs of sequences. Dissimilarity measures are a quantitative estimation of how different—or similar—the sequences are. Since there are diverse contexts, analysis objectives and sequence types, it is natural that there are several methods to compute the dissimilarities based on different considerations.

Optimal matching (OM) may be the most commonly used dissimilarity measure used in social sciences and possibly also in education [22]. Optimal matching represents what it takes to convert or *edit* a sequence to become identical to another sequence. These edits may involve insertion, deletion (together often called *indel* operations) or substitution. For instance, in an example in Table 3, where we see a sequence of five students' engagement states, we can edit Vera's sequence and substitute the *disengaged* state with an *average* state; Vera's sequence will become identical with Luis' sequence. That is, editing Vera's sequence takes one substitution to convert her sequence to that of Luis. We can also see that it will take four substitutions to convert Anna's sequence to Maria's sequence. In other words, Anna's sequence is highly dissimilar to Maria. Different types of substitutions can be given different costs depending on how (dis)similar the two states are viewed (referred to as *substitution costs*). For example, the cost of substituting state *engaged* with state *average* might have a lower cost than substituting *engaged* with *disengaged*, since being disengaged is regarded most dissimilar to being engaged while average engagement is more similar to it. Since contexts differ, there are different ways of defining or computing the pairwise substitution costs matrix.

Optimal matching derives from bioinformatics where transformations such as indels and substitutions are based on actual biological processes such as the evolution of DNA sequences. In many other fields such a transformation process would be unrealistic. In social sciences, [22] outlined five socially meaningful aspects and compared dissimilarity measures to determine how sensitive they are to the different aspects. These similarities are particularly relevant since learning, behavior, and several related processes e.g., progress in school or transition to the labor market are essentially social processes. We explain these aspects based on an example using fictional data in Table 3 following [9].

1. **Experienced states**: how similar are the unique states forming the sequence. For instance, Maria and Bob in Table 3 have both experienced the same states *(engaged and average).*
2. **Distribution of the states**: how similar is the distribution of states. We can consider that two sequences are similar when students spend most of their time in the same states. For instance, Bob and Maria have 80% *engaged* states and 20% *average* states.
3. **Timing**: the time when each state occurs. For instance, two sequences can be similar when they have the same states occurring at the same time. For instance, Vera and Luis start similarly in a *disengaged* state, visit the *average* state in the middle, and finish in the *engaged* state.
4. **Duration**: the durations of time spent continuously in a specific state (called *spells*) e.g., the durations of *engaged* states shared by the two sequences. For instance, Vera and Anna both had spells of two successive states in the *disengaged* state while Bob had two separate spells in the *engaged* state (both of length 2).
5. **Sequencing**: The order of different states in the sequence, for instance, Vera and Luis had similar sequences starting as *disengaged*, moving to *average* and then finishing as *engaged*.

Of the aforementioned aspects, the first two can be directly determined from the last three. Different dissimilarity measures are sensitive to different aspects, and it is up to the researcher to decide which aspects are important in their specific context. Dissimilarity measures can be broadly classified in three categories [22]:

1. distance between distributions,
2. counting common attributes between sequences, and
3. edit distances.

Category 1 includes measures focusing on the distance between distributions including, e.g., *Euclidean distance* and χ^2*-distance* that compare the total time spent in each state within each sequence. The former is based on absolute differences in the proportions while the latter is based on weighted squared differences.

Category 2 includes measures based on counting common attributes. For example, *Hamming distances* are based on counting the (possibly weighted) sum of position wise mismatches between the two sequences, the *length of the longest common subsequence* (`LCS`) is the number of shared states between two sequences that occur in the same order in both, while the *subsequence vector representation-based metric* (`SVRspell`) is counted as the weighted number of matching subsequences.

Category 3 includes edit distances that measure the costs of transforming one sequence to another by using edit operations (indels and substitutions). They include (classic) OM with different cost specifications as well as variants of OM such as OM between sequences of spells (`OMspell`) and OM between sequences of transitions (`OMstran`).

Studer and Ritschard [22] give recommendations on the choice of dissimilarity measure based on simulations on data with different aspects. If the interest is on distributions of states within sequences, Euclidean and χ^2*-distance* are good choices. When timing is of importance, the Hamming distances are the most sensitive to differences in timing. With specific definitions also the Euclidean and χ^2*-distance* can be made sensitive to timing—the latter is recommended if differences in rare events are of particular importance. When durations are of importance, then `OMspell` is a good choice, and also `LCS` and classic `OM` are reasonable choices. When the main interest is in sequencing, good choices include `OMstran`, `OMspell`, and `SVRspell` with particular specifications. If the interest is in more than one aspect, the choice of the dissimilarity measure becomes more complex. By altering the specifications in measures such as `OMstran`, `OMspell`, and `SVRspell` the researcher could find a balance between the desired attributes. See [22] for more detailed information on the choice of dissimilarity measures and their specifications.

Dissimilarities are hard to interpret as such (unless the data are very small), so further analyses are needed to decrease the complexity. The most typical choice is to use cluster analysis for finding groups of individuals with similar patterns [23]. Other distance—or dissimilarity—based techniques include visualizations with multidimensional scaling [24], finding representative sequences [25], and ANOVA-type analysis of discrepancies [26].

3.1.7 Finding Similar Groups or Clusters of Sequences

The sixth step is finding similar sequences, i.e., groups or patterns within the sequences where sequences within each group or cluster are as close to each other as possible and as different from other patterns in other clusters as possible. For instance, we can detect similar groups of sequences that show access patterns to online learning which are commonly referred to as tactics e.g., [3]. Such a step is typically performed using a clustering algorithm which may—or may not—require dissimilarity measures as an input [23, 27]. Common clustering algorithms that use a dissimilarity matrix are the hierarchical clustering algorithms. Hidden Markov models are among the most non-distance based cluster algorithms. See the remaining chapters about sequence analysis for examples of these algorithms [28–30].

3.1.8 Analyzing the Groups and/or Using Them in Subsequent Analyses

Analysis of the identified patterns or subgroups of sequences is an important research question in many studies and oftentimes, it is the guiding research question. For instance, researchers may use log data to create sequences of learning actions, identify subgroups of sequences, and examine the association between the identified patterns and performance e.g., [3, 17, 18], associate the identified patterns with course and course delivery [10], examine how sequences are related to dropout using survival analysis [9], or compare sequence patterns to frequencies [16].

3.2 Introduction to the Technique

Before performing the actual analysis with R code, we need to understand how the data is processed for analysis. Four important steps that require more in-depth explanation will be clarified here, those are: defining the alphabet, the timing scheme, specifying the actor, and visualization. Oftentimes, the required information to perform the aforementioned steps are not readily obvious in the data and therefore some preparatory steps need to be taken to process the file.

The example shown in Table 4 uses fictional log trace data similar to those that come from LMSs. To build a sequence from the data in Table 4, we can use the *Action* column as an *alphabet*. If our aim here is to model the sequence of students' online actions, this is a straightforward choice that requires no preparation. Since the log trace data has no obvious timing scheme, we can use the session as a time scheme. To compute the session, we need to group the actions that occur together without a significant delay between actions (i.e., lag) that can be considered as an inactivity (see Sect. 3.1.2). For instance, Layla's actions in Table 4 started at 18:44 and ended at 18:51. As such, all Layla's actions occurred within 7 minutes. As Table 4 also shows, the first group of Layla's actions occur within 1–2 minutes of

Table 5 A sequence of engagement states using fictional data

Actor	1	2	3	4	5	6
Layla session1	Calendar	Lecture	Instructions	Assignment	Lecture	Video
Sophia session1	Lecture	Instructions	Assignment	Assignment	Assignment	
Carmen session1	Lecture	Video				
Layla session2	Instructions	Video	Lecture	Assignment		

lag. The next group of actions by Layla occur after almost one day, an hour and six minutes (1506 minutes) which constitutes a period of inactivity long enough to divide Layla's actions into two separate sessions. Layla's actions on the first day can be labeled *Layla-session1* and her actions on the second day are *Layla-session2*. The actor in this example is a composite of the student (e.g., Layla) and the session number. The same for Sophia and Carmen: their actions occurred within a few minutes and can be grouped into the sessions. Given that we have the alphabet (*Action*), the timing scheme (*session*), and the actor (*user-session*), the next step is to order the *alphabet* chronologically. In Table 4, the actions were sequentially ordered for every actor according to their chronological order. The method that we will use in our guide requires the data to be in so-called "wide format". This is performed by *pivoting* the data, or creating a wide form where the column names are the *order* and the value of the *Action* column is sequentially and horizontally listed as shown in Table 5.

The following steps are creating a sequence object using sequence mining software and using the created sequence in analysis. In our case, we use the `TraMineR` framework which has a large set of visualization and statistical functions. Sequences created with `TraMineR` also work with a large array of advanced tools, R packages, and extensions. However, it is important to understand sequence visualizations before delving into the coding part.

3.3 Sequence Visualization

Two basic plots are important here and therefore will be explained in detail. The first is the index plot (Fig. 2) which shows the sequences of stacked colored bars representing spells, with each token represented by a different color. For instance, if we take Layla's actions (in session1) and represent them as an index plot, they will appear as shown in Fig. 2 (see the arrow). Where the *Calendar* is represented as a purple bar, the *Lecture* as a yellow bar, and *instructions* as an orange bar etc. Figure 2 also shows the visualization of sequences in Table 5 and you can see each of the session sequences as stacked colored bars following their order in the table. Nevertheless, sequence plots commonly include a large number of sequences that are of the order of hundreds or thousands of sequences and may be harder to read than the one presented in the example (see examples in the next sections).

The distribution plot is another related type of sequence visualization. Distribution plots—as the name implies—represent the distribution of each alphabet at

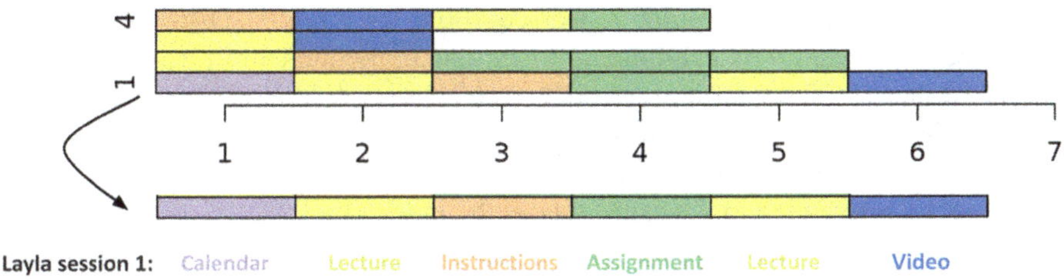

Actor	1	2	3	4	5	6
Layla session1	Calendar	Lecture	Instructions	Assignment	Lecture	Video
Sophia session1	Lecture	Instructions	Assignment	Assignment	Assignment	
Carmen session1	Lecture	Video				
Layla session2	Instructions	Video	Lecture	Assignment		

Fig. 2 A table with ordered sequences of actions and the corresponding index plot; the arrow points to Layla's actions

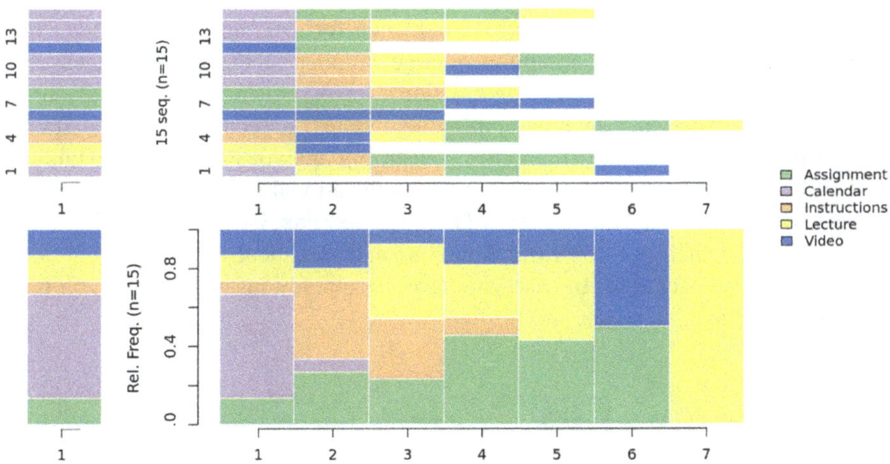

Fig. 3 Index plot (top) and distribution plot (bottom)

each time point. For example, if we look at Fig. 3 (top) we see 15 sequences in the index plot. At time point 1, we can count eight *Calendar* actions, two *Video* actions, two *Lecture* actions and one *Instruction* action. If we compute the proportions: we get 8/15 (0.53) of *Calendar* actions; for *Video*, *Assignment,* and *Lecture* we get 2/15 (0.13) in each case, and finally *Instructions* actions account for 1/15 (0.067). Figure 3 (bottom) shows these proportions. At time point 1, we see the first block *Assignment* with 0.13 of the height of the bar, followed by the *Calendar* which occupies 0.53, then a small block (0.067) for the *Instructions,* and finally two equal blocks (0.13) representing the *Video* and *Lecture* actions.

Since the distribution plot computes the proportions of activities at each time point, we see different proportions at each time point. Take for example, time point 6, we have only two actions (*Video* and *Assignment*) and therefore, the plot has 50%

for each action. At the last point 7, we see 100% for *Lecture*. Distribution plots need to be interpreted with caution and in particular, the number of actions at each time point need to be taken into account. One cannot say that at the seventh time point, 100% of actions were *Lecture*, since it was the only action at this time point. Furthermore, distribution plots do not show the transitions between sequences and should not be interpreted in the same way as the index plot.

4 Analysis of the Data with Sequence Mining in R

4.1 Important Packages

The most important package and the central framework that we will use in our analysis is the `TraMineR` package. `TraMineR` is a toolbox for creating, describing, visualizing and analyzing sequence data. `TraMineR` accepts several sequence formats, converts to a large number of sequence formats, and works with other categorical data. `TraMineR` computes a large number of dissimilarity measures and has several integrated statistical functions. `TraMineR` has been mainly used to analyze live event data such as employment states, sequence of marital states, or other life events. With the emergence of learning analytics and educational data mining, `TraMineR` has been extended into the educational field [31]. In the current analysis we will also need the packages `TraMineRextras`, `WeightedCluster`, and `seqhandbook`, which provide extra functions and statistical tools. The first code block loads these packages. In case you have not already installed them, you may need to install them.

```
library(TraMineR)
library(TraMineRextras)
library(WeightedCluster)
library(seqhandbook)
library(tidyverse)
library(rio)
library(cluster)
library(MetBrewer)
library(reshape2)
```

4.2 Reading the Data

The example that will be used here is a Moodle log dataset that includes three important fields: the User ID (`user`), the time stamp (`timecreated`), and the actions (`Event.context`). Yet, as we mentioned before, there are some steps that need to be performed to prepare the data for analysis. First, the `Event.context` is very granular (80 different categories) and needs to be re-coded as mentioned in the basics of sequence mining section. We have already prepared the file with a simpler coding scheme where, for example, all actions intended as instructions were coded as `instruction`, all group forums were coded as `group_work`, and all assignment work was coded as `Assignment`. Thus, we have a field that we can use as the alphabet titled `action`. The following code reads the original coded dataset.

```
Seqdatas <-
 import("https://github.com/lamethods/data/raw/main/1_moodleLAcourse/Events.xlsx")
```

```
# A tibble: 95,626 x 7
   Event.context      user  timecreated         Component Event.name Log   Action
   <chr>              <chr> <dttm>              <chr>     <chr>      <chr> <chr>
 1 Assignment: Fina~ 9d74~ 2019-10-26 09:37:12 Assignme~ Course mo~ Assi~ Assig~
 2 Assignment: Fina~ 9148~ 2019-10-26 09:09:34 Assignme~ The statu~ Assi~ Assig~
 3 Assignment: Fina~ 278a~ 2019-10-18 12:05:28 Assignme~ Course mo~ Assi~ Assig~
 4 Assignment: Fina~ 53d6~ 2019-10-19 13:28:37 Assignme~ The statu~ Assi~ Assig~
 5 Assignment: Fina~ aab7~ 2019-10-15 23:38:13 Assignme~ Course mo~ Assi~ Assig~
 6 Assignment: Fina~ 82ed~ 2019-10-18 17:51:43 Assignme~ Course mo~ Assi~ Assig~
 7 Assignment: Fina~ 4178~ 2019-10-18 15:22:56 Assignme~ Course mo~ Assi~ Assig~
 8 Assignment: Fina~ 82ed~ 2019-10-22 13:46:51 Assignme~ The statu~ Assi~ Assig~
 9 Assignment: Fina~ f2e9~ 2019-10-15 14:58:17 Assignme~ Submissio~ Assi~ Assig~
10 Assignment: Fina~ 53d6~ 2019-10-19 13:28:38 Assignme~ Course mo~ Assi~ Assig~
# i 95,616 more rows
```

4.3 Preparing the Data for Sequence Analysis

To create a time scheme, we will use the methods described earlier in the basis of the sequence analysis section. The timestamp field will be used to compute the `lag` (the delay) between actions, find the periods of inactivity between the actions, and mark the actions that occur without a significant lag together as a `session`. Actions that follow with a significant delay will be marked as a new session. The following code performs these steps. First, the code arranges the data according to the timestamp for each user (see the previous example in the basics section), this is why we use `arrange(timecreated, user)`. The second step (#2) is to `group_by(user)` to make sure all sessions are calculated for each user separately. The third step is to compute the lag between actions. Step #4 evaluates the lag length; if the lag exceeds 900 seconds (i.e., a period of inactivity of 900 seconds), the code marks the action as the start of a new session. Step #5 labels each session with a number corresponding to its order. The last step (#6) creates the actor variable, by concatenating the username with the string "Session_" and the session number; the resulting variable is called `session_id`.

```
sessioned_data <- Seqdatas |>
 arrange(timecreated, user) |> # Step 1
 group_by(user) |> # Step 2
```

```
mutate(Time_gap = timecreated - (lag(timecreated))) |> # Step 3
mutate(new_session = is.na(Time_gap) | Time_gap > 900) |> # Step 4
mutate(session_nr = cumsum(new_session)) |> # Step 5
mutate(session_id = paste0 (user, "_", "Session_", session_nr)) #Step 6
```

An important question here is what is the optimum lag or delay that we should use to separate the sessions. Here, we used 900 seconds (15 minutes) based on our knowledge of the course design. In the course where the data comes from, we did not have activities that require students to spend long periods of idle time online (e.g., videos). So, it is reasonable here to use a relatively short delay (i.e., 900 seconds) to mark new sessions. Another alternative is to examine the distribution of lags or compute the percentiles.

```
quantile(sessioned_data$Time_gap,
         c(0.10, 0.50, 0.75, 0.80, 0.85, 0.90, 0.95, 1.00), na.rm = TRUE)
```

```
Time differences in secs
    10%      50%      75%      80%      85%      90%      95%     100%
      1        1       61       61      181      841    12121  1105381
```

The previous code computes the proportion of lags at 10% to 100% and the results show that at 90th percentile, the length of lags is equal to 841 seconds, that is very close to the 900 seconds we set. The next step is to order the actions sequentially (i.e., create the sequence in each session) as explained in Sect. 3.1.2 and demonstrated in Table 4. We can perform such ordering using the function `seq_along(session_id)` which creates a new field called `sequence` that chronologically orders the `action` (the alphabet).

```
sessioned_data <- sessioned_data |>
  group_by(user, session_nr) |>
  mutate(sequence = seq_along(session_nr)) |>
  mutate(sequence_length = length(sequence))
```

Some sessions are outliers (e.g., extremely long or very short)—there are usually very few—and therefore, we need to trim such extremely long sessions. We do so by calculating the percentiles of session lengths.

```
quantile(sessioned_data$sequence_length,
         c(0.05, 0.1, 0.5, 0.75, 0.90, 0.95, 0.98, 0.99, 1.00),  na.rm =
         TRUE)
```

```
 5%  10%  50%  75%  90%  95%  98%  99% 100%
  3    4   16   29   42   49   59   61   62
```

We see here that 95% of sequences lie within 49 states long and therefore, we can trim these long sessions as well as sessions that are only one event long.

```
sessioned_data_trimmed <- sessioned_data |>
  filter(sequence_length > 1 & sequence <= 49)
```

The next step is to reshape or create a wide format of the data and convert each session into a sequence of horizontally ordered actions. For that purpose, we use the function `dcast` from the `reshape2` package. For this function, we need to specify the ID columns (the actor) and any other properties for the users can be specified here also. We selected the variables `user` and `session_id`. Please note that only `session_id` is necessary (`actor`) but it is always a good idea to add variables that we may use as weights, as groups, or for later comparison. We also need to specify the `sequence` column and the alphabet (`action`) column. The resulting table is similar to Table 5.

The last step is creating the sequence object using the `seqdef` function from the `TraMineR` package. To define the sequence, we need the prepared file from the previous step (similar to Table 5) and the beginning and end of the columns to consider i.e., the start of the sequence. We have started from the fourth column since the first three columns are meta-data (`user`, `session_id`, and `session_nr`). To include all columns in the data we use the `ncol` function to count the number of columns in the data. Creating a sequence object enables the full potential of sequence analysis.

```
data_reshaped <- dcast(user + session_id + session_nr ~ sequence,
                       data = sessioned_data_trimmed,
                       value.var = "Action")
Seqobject <- seqdef(data_reshaped, 4:ncol(data_reshaped))
```

```
 [>] found missing values ('NA') in sequence data
 [>] preparing 9383 sequences
 [>] coding void elements with '%' and missing values with '*'
 [>] 12 distinct states appear in the data:
     1 = Applications
     2 = Assignment
     3 = Course_view
     4 = Ethics
     5 = Feedback
     6 = General
     7 = Group_work
     8 = Instructions
     9 = La_types
     10 = Practicals
     11 = Social
```

```
    12 = Theory
[>] 9383 sequences in the data set
[>] min/max sequence length: 2/49
```

An optional—yet useful—step is to add a color palette to create a better looking plot. Choosing an appropriate palette with separable colors improves the readability of the plot by helping easily identify different alphabets.

```
Number_of_colors <- length(alphabet(Seqobject))
colors <- met.brewer(name = "VanGogh2", n = Number_of_colors)
cpal(Seqobject) <- colors
```

4.4 Statistical Properties of the Sequences

A simple way to get the properties of the sequences is through the function `summary()`. The functions show the total number of sequences in the object, the number of unique sequences, and lists the alphabet. A better way to dig deeper into the sequence properties is to use the `seqstatd()` function which returns several statistics, most notably the relative frequencies, i.e., the proportions of each state at each time point or the numbers comprising the distribution plot. The function also returns the valid states, that is, the number of valid states at each time point as well as the transversal entropy, which is a measure of diversity of states at each time point [32]. The code in the next section computes the sequence statistics and then displays the results. We show only the output of `seq_stats$Frequencies` where we see the frequency of each activity at each time point (Table 6).

```
summary(Seqobject)
seq_stats <- seqstatd(Seqobject)
seq_stats$Frequencies
seq_stats$Entropy
seq_stats$ValidStates
```

4.5 Visualizing Sequences

Visualization has a summarizing power that allows researchers to have an idea about a full dataset in one visualization. `TraMineR` allows several types of visualizations that offer different perspectives. The most common visualization type is the distribution plot (described earlier in Fig. 3). To plot a distribution plot one can use the powerful `seqplot` function with the argument `type="d"` or simply `seqdplot()`.

Table 6 Frequency of activities at each point

Activity	1	2	3	4	5	6	7	8	...	49
Applications	0.00	0.00	0.01	0.00	0.01	0.01	0.01	0.01		0.03
Assignment	0.08	0.10	0.11	0.09	0.08	0.09	0.07	0.07		0.00
Course_view	0.48	0.32	0.27	0.26	0.23	0.21	0.20	0.23		0.14
Ethics	0.01	0.01	0.01	0.01	0.01	0.01	0.01	0.01		0.07
Feedback	0.03	0.04	0.03	0.05	0.04	0.04	0.04	0.04		0.00
General	0.01	0.01	0.02	0.02	0.03	0.03	0.04	0.04		0.22
Group_work	0.21	0.28	0.31	0.33	0.36	0.37	0.38	0.37		0.31
Instructions	0.05	0.07	0.07	0.07	0.07	0.08	0.08	0.07		0.04
La_types	0.01	0.01	0.01	0.01	0.01	0.01	0.02	0.02		0.09
Practicals	0.09	0.12	0.12	0.11	0.11	0.11	0.11	0.12		0.07
Social	0.01	0.02	0.02	0.03	0.03	0.02	0.03	0.03		0.00
Theory	0.01	0.01	0.01	0.01	0.02	0.01	0.01	0.01		0.02

Fig. 4 Sequence distribution plot

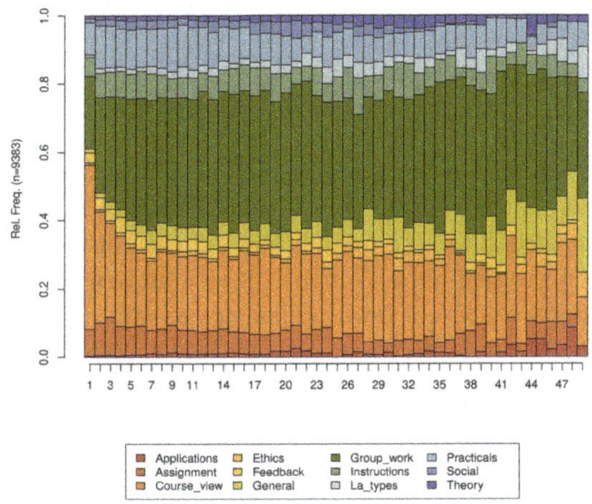


```
seqplot(Seqobject, type = "d")
```

The default distribution plot has an y-axis that ranges from 0 to 1.0 corresponding to the proportion and lists the number of sequences which in our case is 9383 However, the default output of the `seqdplot()` function is rarely satisfactory and we need to use the function arguments to optimize the resulting plot. The help file contains a rather detailed list of arguments and types of visualizations that can be consulted for more options, which can be obtained like any other R function by typing `?seqplot`. In this chapter we will discuss the most basic options. In Fig. 4, we use `cex.legend` argument to optimize the legend text size, we use the `ncol` argument to make the legend spread over six columns, the argument `legend.prop` to make the legend a bit far away from the main plot so they do not overlap and we use the argument `border=NA` to remove the borders from the plot. With such

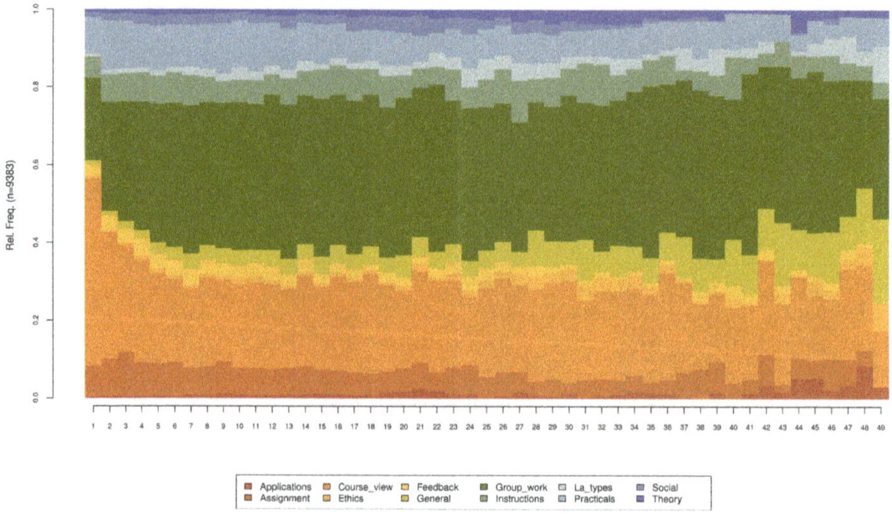

Fig. 5 Sequence distribution plot with customized arguments

small changes, we get a much cleaner and readable distribution plot. Please note, that in each case, you may need to optimize the plot according to your needs. It is important to note here that in the case of missing data or sequences with unequal lengths like ours—which is very common—the distribution plot may show results that are made of fewer sequences at later time points. As such, the interpretation of the distribution plot should take into account the number of sequences, missing data, and timing (Fig. 5). An index plot may be rather more informative in cases where missing data is prevalent.

```
seqplot(Seqobject, type = "d", cex.legend = 0.9, ncol = 6, cex.axis = 0.7,
        legend.prop = 0.1, border = NA)
```

The index plot can be plotted in the same way using `seqplot()` with the argument `type="I"` or simply using `seqIplot`. The resulting plot (Fig. 6) has each sequence of the 9383 plotted as a line of stacked colored bars. One of the advantages of index plots is that they show the transitions between states in each sequence spell. Of course, plotting more than nine thousand sequences results in very thin lines that may not be very informative. Nevertheless, index plots are very informative when the number of sequences is relatively small. Sorting the sequences could help improve the visualization. On the right side, we see the index plot using the argument "sortv ="from.start", under which sequences are sorted by the elements of the alphabet at the successive positions starting from the beginning of the time window.

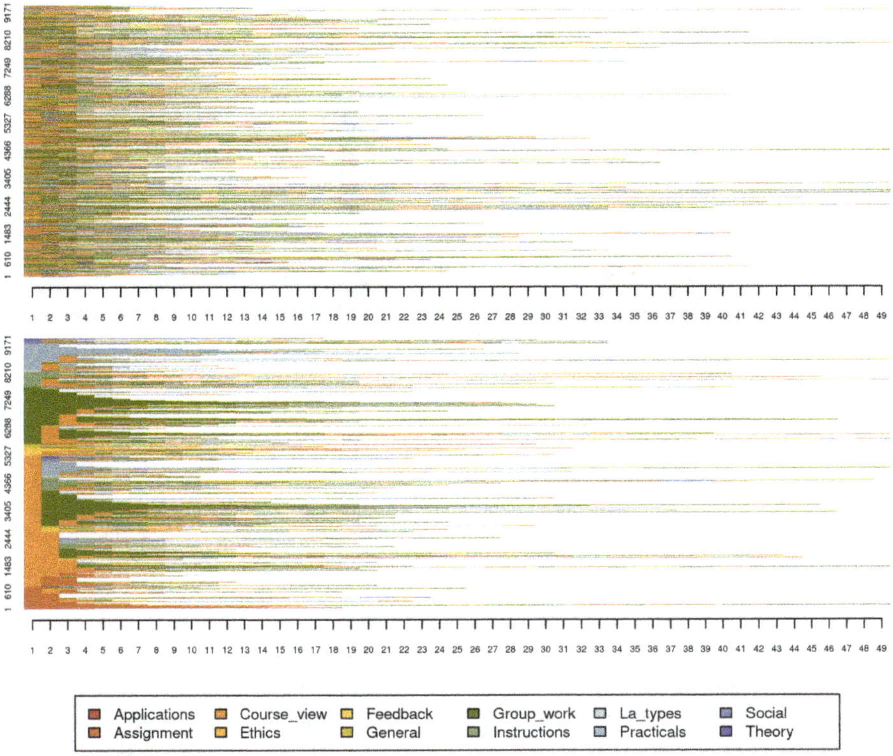

Fig. 6 Sequence index plots

```
seqplot(Seqobject, type = "I", cex.legend = 0.6, ncol = 6, cex.axis = 0.6,
        legend.prop = 0.2, border = NA)

seqplot(Seqobject, type = "I", cex.legend = 0.6, ncol = 6, cex.axis = 0.6,
        legend.prop = 0.2, border = NA, sortv = "from.start")
```

The last visualization type we discuss here is the mean time plot, which plots the total time of every element of the alphabet across all time, i.e., a frequency distribution of all states regardless of their timing. As the plot in Fig. 7 shows, group work seems to be the action that students performed the most, followed by course view.

```
seqplot(Seqobject, type = "mt", cex.legend = 0.7, ncol = 6, cex.axis = 0.6,
        legend.prop = 0.15, border = NA, ylim = c(0, 5))
```

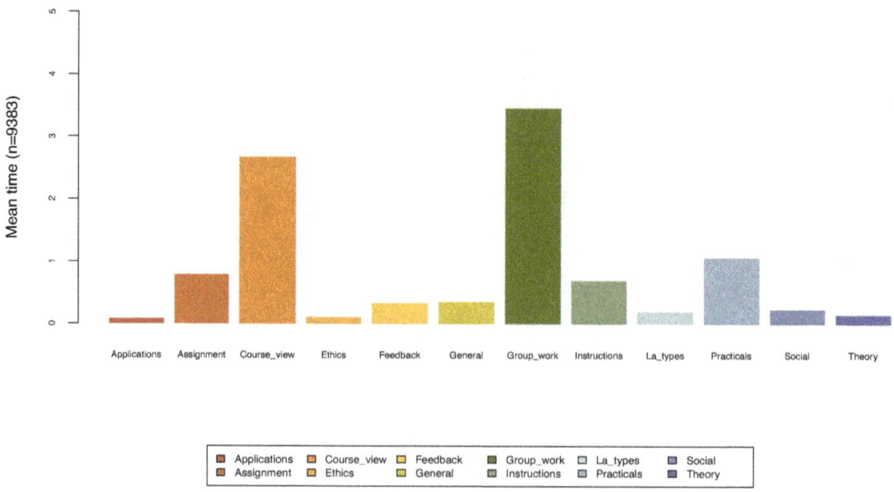

Fig. 7 Mean time plot

4.6 Dissimilarity Analysis and Clustering

Having prepared the sequences and explored their characteristics, we can now investigate if they have common patterns, recurrent sequences, or groups of similar sequences. This is a two-stage process; we will need to compute dissimilarities (along with associated substitution costs) and then perform cluster analysis on the resulting matrix. For more details on the clustering technique, please refer to Chapter 8 [33]. In the case of log trace data, clustering has always been performed to find learning tactics or sequences of students' actions that are similar to each other or, put another way, patterns of similar behavior (e.g., [3, 11]). For the present analysis, we begin with the most common method for computing the dissimilarity matrix, that is Optimal Matching (OM). OM computes the dissimilarity between two sequences as the minimal cost of converting a sequence to the other. OM requires some steps that include specifying a substitution cost matrix, indel cost. Later, we use a clustering algorithm to partition the sequences according to the values returned by the OM algorithm [34, 35].

A possible way to compute substitutions cost that has been commonly used—yet frequently criticized—in the literature is the TRATE method [36]. The TRATE method is data-driven and relies on transition rates; it assumes that pairs of states with frequent transitions between them should have "lower cost" of substitution (i.e., they are seen as being more similar). Thus, if we replace an action with another action that occurs often, it has a lower cost. This may be useful in some course designs, where some activities are very frequently visited and others are rare. The function `seqsubm()` is used to compute substitution costs with the TRATE method via:

```
substitution_cost_TRATE <- seqsubm(Seqobject, method ="TRATE")
```

If we print the substitution cost matrix (Table 7), we see that, for instance, the cost of replacing Applications with Applications is 0, whereas the cost of replacing `Applications` with `Assignment` (and vice versa) is higher (1.94). Since `Course_view` is the most common transition, replacing any action with `Course_view` tends to be the lowest in cost, which makes sense. Please note that the TRATE method is presented here for demonstration only. In fact, we do not recommend it to be used by default; readers should choose carefully what cost method best suits their data.

Nevertheless, the most straightforward way of computing the cost is to use a constant cost; that is, to assume that the states are equally distant from one another. To do so, we can use the function `seqsubm()` and supply the argument `method="CONSTANT"`. In the following example, we assign a common cost of 2 (via the argument `cval`). We also refer below to other optional arguments which are not strictly necessary for the present application but nonetheless worth highlighting as options.

```
substitution_cost_constant <- seqsubm(
  Seqobject,             # Sequence object
  method = "CONSTANT",   # Method to determine costs
  cval = 2,              # Substitution cost
  time.varying = FALSE,  # Does not allow the cost to vary over time
  with.missing = TRUE,   # Allows for missingness state
  miss.cost = 1,         # Cost for substituting a missing state
  weighted = TRUE)       # Allows weights to be used when applicable
```

To compute the OM dissimilarity matrix, the `indel` argument needs to be provided and we will use the default value 1 which is half of the highest substitution cost (2). We also need to provide the substitution cost matrix (`sm`). We opt for the matrix of constant substitution costs created above, given its straightforward interpretability.

```
dissimilarities <- seqdist(Seqobject, method = "OM",indel = 1,
                           sm = substitution_cost_constant)
```

```
[>] 9383 sequences with 12 distinct states

[>] checking 'sm' (size and triangle inequality)

[>] 4062 distinct  sequences

[>] min/max sequence lengths: 2/49

[>] computing distances using the OM metric

[>] elapsed time: 7.807 secs
```

Table 7 Substitution cost matrix for the TRATE method

From \ To	Applications	Assignment	Course_view	Ethics	Feedback	General	Group_work	Instructions	La_types	Practicals	Social	Theory
Applications	0.000	1.939	1.868	1.966	1.988	1.749	1.946	1.985	1.982	1.956	2.000	1.994
Assignment	1.939	0.000	1.750	1.996	1.970	1.936	1.961	1.960	1.922	1.962	1.993	1.982
Course_view	1.868	1.750	0.000	1.879	1.743	1.801	1.524	1.564	1.749	1.750	1.735	1.854
Ethics	1.966	1.996	1.879	0.000	1.991	1.950	1.907	1.991	1.940	1.945	1.988	1.945
Feedback	1.988	1.970	1.743	1.991	0.000	1.992	1.879	1.926	1.990	1.978	1.999	1.996
General	1.749	1.936	1.801	1.950	1.992	0.000	1.936	1.909	1.842	1.961	1.985	1.947
Group_work	1.946	1.961	1.524	1.907	1.879	1.936	0.000	1.862	1.924	1.956	1.873	1.938
Instructions	1.985	1.960	1.564	1.991	1.926	1.909	1.862	0.000	1.931	1.954	1.850	1.983
La_types	1.982	1.922	1.749	1.940	1.990	1.842	1.924	1.931	0.000	1.964	1.981	1.907
Practicals	1.956	1.962	1.750	1.945	1.978	1.961	1.956	1.954	1.964	0.000	1.978	1.948
Social	2.000	1.993	1.735	1.988	1.999	1.985	1.873	1.850	1.981	1.978	0.000	1.994
Theory	1.994	1.982	1.854	1.945	1.996	1.947	1.938	1.983	1.907	1.948	1.994	0.000

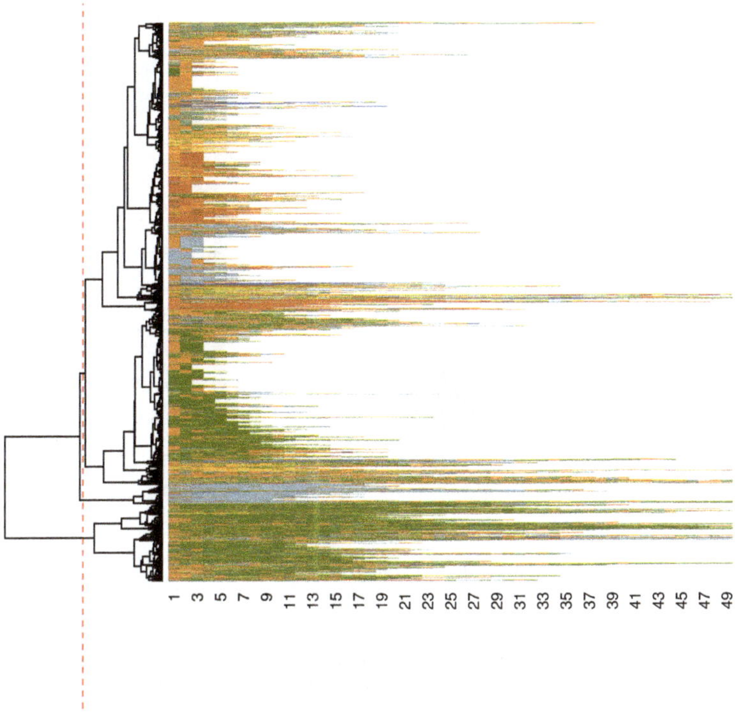

Fig. 8 Visualization of clusters using a dendrogram

In the resulting pairwise dissimilarity matrix, every sequence has a dissimilarity value with every other sequence in the dataset, and therefore, the dissimilarity matrix can be large and resource intensive in larger matrices. In our case, the dissimilarity matrix is 9383 * 9383 (i.e., 88,040,689) in size. With these dissimilarities between sequences as input, several distance-based clustering algorithms can be applied to partition the data into homogeneous groups. In our example, we use the hierarchical clustering algorithm from the package `stats` by using the function `hclust()`, but note that the choice of clustering algorithm can also affect results greatly and should be chosen carefully by the reader. For more details on the clustering technique, please refer to Chapter 8 [33]. The `seq_heatmap()` function is used to plot a dendrogram of the index plot which shows a hierarchical tree of different levels of subgrouping and helps choose the number of clusters visually.

```
clusters_sessionsh <- hclust(as.dist(dissimilarities), method = "ward.D2")
seq_heatmap(Seqobject, clusters_sessionsh)
```

To do the actual clustering, we use the function `cutree()` and with the argument `k = 3` to cluster the sequence into three clusters according to the groups highlighted in Fig. 8. The `cutree` function produces a vector of cluster numbers, we can create

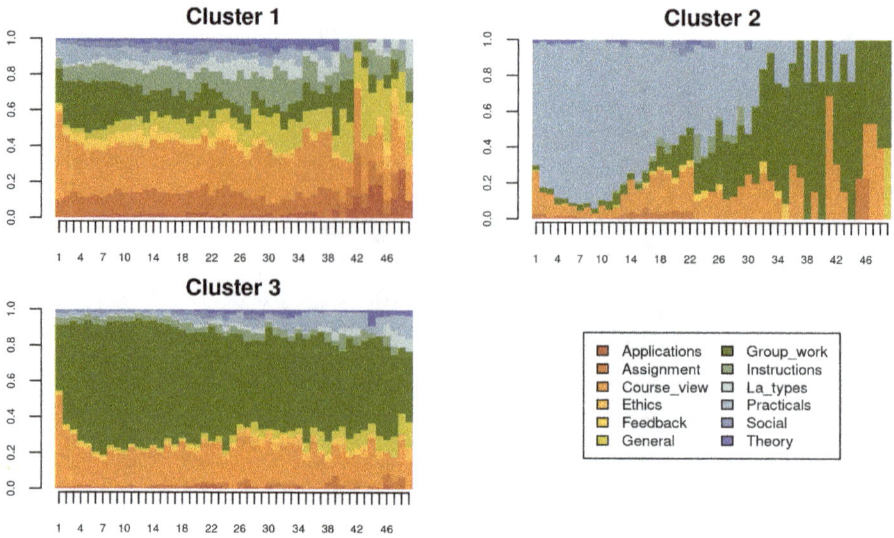

Fig. 9 Sequence distribution plot for the k=3 cluster solution

more descriptive labels as shown in the example and assign the results to an R object called Groups. Visualizations of the clustering results can be performed in a similar fashion to the earlier visualizations of the entire set of sequences: via `seqplot()`, with the desired `type` of plot, and the addition of the argument `group` (Fig. 9). Readers have to choose the arguments and parameters according to contexts, research questions, and the nature of their data.

```
Cuts <- cutree(clusters_sessionsh, k = 3)
Groups <- factor(Cuts, labels = paste("Cluster", 1:3))
seqplot(Seqobject, type = "d", group = Groups, cex.legend = 0.8, ncol = 2,
        cex.axis = 0.6, legend.prop = 0.2, border = NA)
```

However, the resulting clusters might not be the best solution and we need to try other dissimilarity measures and/or clustering algorithms, evaluate the results, and compare their fit indices. `TraMineR` provides several distance measures, the most common of which are:

- **Edit distances:** Optimal matching `"OM"` or optimal matching with sensitivity to certain factors, e.g., optimal matching with sensitivity to spell sequence (`"OMspell"`) or with sensitivity to transitions (`"OMstran"`).
- **Shared attributes:** Distance based on the longest common subsequence (`"LCS"`), longest common prefix (`"LCP"`; which prioritizes sequence common initial states), or the subsequence vectorial representation distance (`"SVRspell"`; based on counting common subsequences).

- **Distances between distributions of states:** Euclidean ("EUCLID") distance or Chi-squared ("CHI2").

Determining the distance may be done based on the research hypothesis, context, and the nature of the sequences. For instance, a researcher may decide to group sequences based on their common starting points (e.g., [16]) where the order and how a conversation starts matter. TraMineR allows the computation of several dissimilarities. The following code computes some of the most common dissimilarities and stores each in a variable that we can use later.

```
# Edit distances and sequences
dissimOMstran <- seqdist(Seqobject, method = "OMstran", otto = 0.1,
                         sm = substitution_cost_constant, indel = 1)
dissimOMspell <- seqdist(Seqobject, method = "OMspell", expcost = 0,
                         sm = substitution_cost_constant, indel = 1)
dissimSVRspell <- seqdist(Seqobject, method = "SVRspell", tpow = 0)
dissimOM <- seqdist(Seqobject, method = "OM", otto = 0.1,
                    sm = substitution_cost_constant, indel = 1)


# Distances between state distributions
dissimCHI2 <- seqdist(Seqobject, method = "CHI2", step = 1)
dissimEUCLID <- seqdist(Seqobject, method = "EUCLID", step = 49)


# Distances based on counts of common attribute e.g., duration (spell lengths)
dissimOMspell <- seqdist(Seqobject, method = "OMspell", expcost = 1,
                         sm = substitution_cost_constant, indel = 1)
dissimLCS <- seqdist(Seqobject, method = "LCS")
dissimLCP <- seqdist(Seqobject, method = "LCP")
dissimRLCP <- seqdist(Seqobject, method = "RLCP")
```

We can then try each dissimilarity with varying numbers of clusters and compute the clustering evaluation measures. The function as.clustrange from the WeightedCluster package computes several cluster quality indices including, among others, the Average Silhouette Width (ASW) which is commonly used in cluster evaluation to measure the coherence of the clusters. A value above 0.25 means that the data has some structure or patterns, whereas a value below 0.25 signifies the lack of structure in the data. The function also computes the R^2 Value which represents the ratio of the variance explained by the clustering solution. The results can be plotted and inspected. We can see that four clusters seem to be a good solution. Table 8 and Fig. 10 show that the ASW and CHsq measures are maximized for the four-cluster solution, for which other parameters such as R^2 are also relatively good. Thus, we can use the four cluster solution. We note the use of the norm="zscoremed" argument which improves the comparability of the various metrics in Fig. 10 by standardizing the values to make it easier to identify the maxima. Table 8, however, presents the values on their original scales. Finally,

Table 8 Cluster performance metrics

PBC	HG	HGSD	ASW	ASWw	CH	R2	CHsq	R2sq	HC
0.281	0.336	0.335	0.268	0.268	2230.420	0.192	3565.535	0.275	0.319
0.417	0.489	0.488	0.307	0.307	1992.266	0.298	3553.441	0.431	0.246
0.507	0.614	0.614	0.333	0.333	1952.437	0.384	3985.571	0.560	0.190
0.491	0.635	0.634	0.272	0.272	1717.644	0.423	3524.149	0.601	0.184
0.511	0.677	0.677	0.287	0.288	1580.850	0.457	3301.251	0.638	0.165
0.534	0.736	0.735	0.308	0.309	1449.482	0.481	3178.407	0.670	0.138
0.557	0.812	0.812	0.324	0.325	1362.368	0.504	3110.913	0.699	0.104
0.559	0.831	0.830	0.327	0.327	1341.529	0.534	3124.571	0.727	0.096
0.571	0.865	0.865	0.329	0.330	1274.303	0.550	3085.893	0.748	0.080

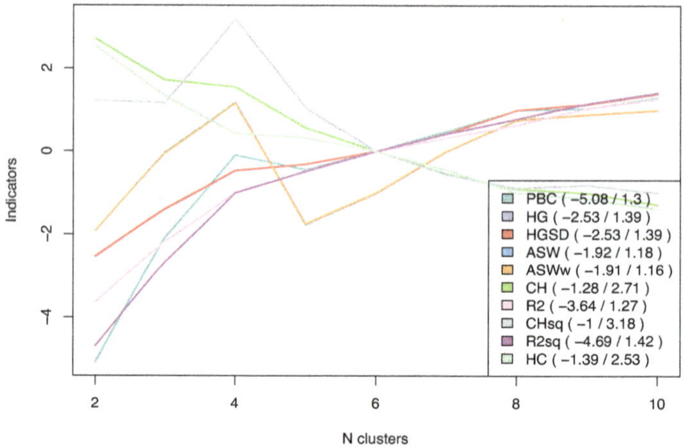

Fig. 10 Cluster performance metrics. X-axis represents the number of clusters, Y-axis represents the fit index standardized value

the ranges and other characteristics of each cluster quality metric are summarized in Table 9. For brevity, we proceed with only the Euclidean distance matrix.

```
dissimiarities_tested <- dissimEUCLID
Clustered <- hclust(as.dist(dissimiarities_tested), method = "ward.D2")
Clustered_range <- as.clustrange(Clustered, diss = dissimiarities_tested,
                                 ncluster = 10)
plot(Clustered_range, stat = "all", norm = "zscoremed", lwd = 2)
```

```
Clustered_range[["stats"]]
```

To get the cluster assignment, we can use the results from the `Clustered_range` object and plot the clusters using the previously shown distribution, index, and mean time plot types (Figs. 11, 12, and 13).

Table 9 Measures of the quality of a partition. Note: Table is based on [23] with permission from the author [23]

Name	Abrv.	Range	Min/Max	Interpretation
Point Biserial Correlation	PBC	$[-1;1]$	Max	Measure of the capacity of the clustering to reproduce the distances
Hubert's Gamma	HG	$[-1;1]$	Max	Measure of the capacity of the clustering to reproduce the distances (order of magnitude)
Hubert's Somers' D	HGSD	$[-1;1]$	Max	Measure of the capacity of the clustering to reproduce the distances (order of magnitude) taking into account ties in distances
Hubert's C	HC	[0;1]	Min	Gap between the partition obtained and the best partition theoretically possible with this number of groups and these distances
Average Silhouette Width	ASW	$[-1;1]$	Max	Coherence of assignments. High coherence indicates high between-group distances and strong within-group homogeneity
Average Silhouette Width (weighted)	ASWw	$[-1;1]$	Max	As previous, for floating point weights
Calinski-Harabasz index	CH	$[0; +\infty[$	Max	Pseudo F computed from the distances
Calinski-Harabasz index	CHsq	$[0; +\infty[$	Max	As previous, but using squared distances
Pseudo R2	R2	[0;1]	Max	Share of the discrepancy explained by the clustering solution (only to compare partitions with identical number of groups)
Pseudo R2	R2sq	[0;1]	Max	As previous, but using squared distances

```r
grouping <- Clustered_range$clustering$cluster4
seqplot(Seqobject, type = "d", group = grouping, cex.legend = 0.9, ncol = 6,
        cex.axis = 0.6, legend.prop = 0.2, border = NA)
```

```r
seqplot(Seqobject, type = "I", group = grouping, cex.legend = 0.9, ncol = 6,
        cex.axis = 0.6, legend.prop = 0.2, border = NA)
```

```r
seqplot(Seqobject, type = "mt", group = grouping, cex.legend = 1, ncol = 6,
        cex.axis = .5, legend.prop = 0.2, ylim = c(0, 10))
```

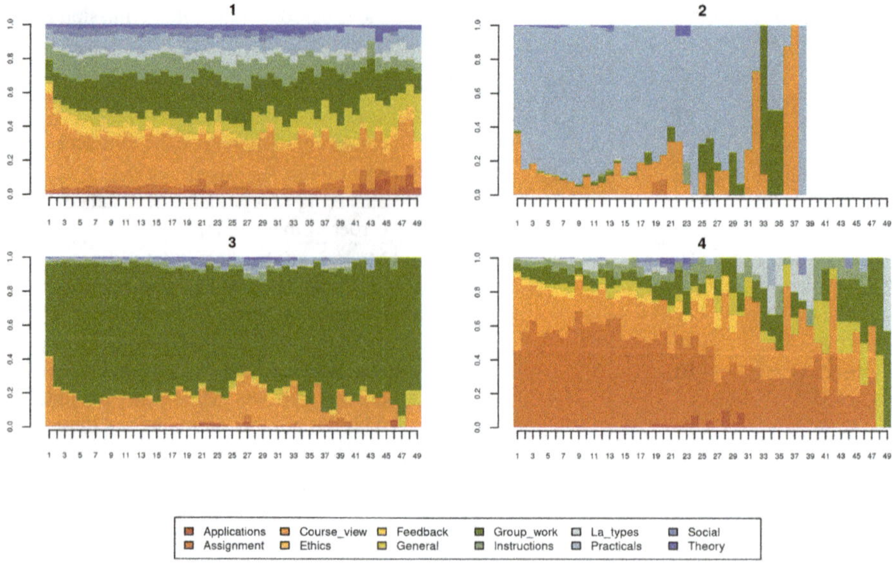

Fig. 11 Sequence distribution plot for the four clusters

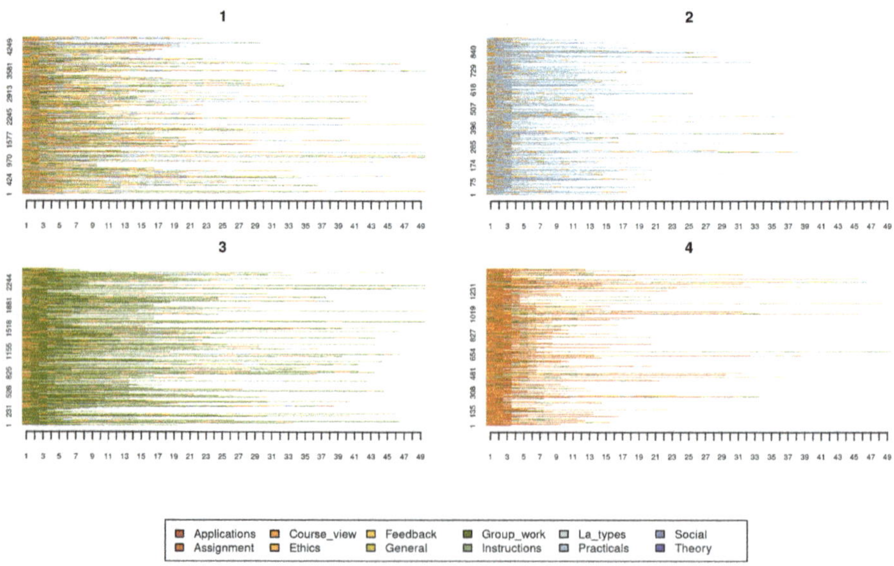

Fig. 12 Sequence index plot for the four clusters

Given the clustering structure, we also use a new plot type: the implication plot from the `TraMineRextras` package. Such a plot explicitly requires a `group` argument; in each of these plots, at each time point, "being in this group implies to be in this state at this time point". The strength of the rule is represented by a

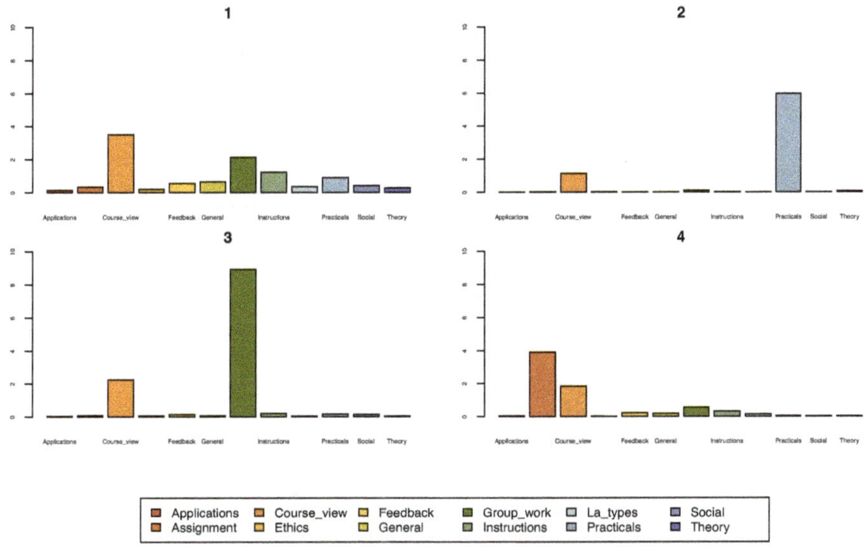

Fig. 13 Mean time plot for the four clusters

plotted line and a 95% confidence interval. Put another way, the more likely states have higher implicative values, which are more relevant when higher than the 95% confidence level (Fig. 14).

```
implicaton_plot <- seqimplic(Seqobject, group = grouping)
plot(implicaton_plot, conf.level = 0.95, cex.legend = 0.7)
```

Given the implication plot as well as the other plots, the first cluster seems to be a mixed cluster with no prominent activity. Cluster 2 is dominated by practical activities, Cluster 3 is dominated by group work activities, Cluster 4 is dominated by assignments. Researchers usually give these clusters a label e.g., for Cluster 1, one could call it a diverse cluster. See some examples here in these papers [3, 10, 11].

5 More Resources

Sequence analysis is a rather large field with a wealth of methods, procedures, and techniques. Since we have used the `TraMineR` software in this chapter, a first place to seek more information about sequence analysis would be to consult the TraMineR manuals and guides [23, 27, 37]. More tools for visualization can be found in the package `ggseqplot` [38]. The `ggseqplot` package reproduces similar plots to `TraMineR` with the `ggplot2` framework as well as other interesting visualizations [39]. This allows further personalisation using the `ggplot2` grammar, as we have

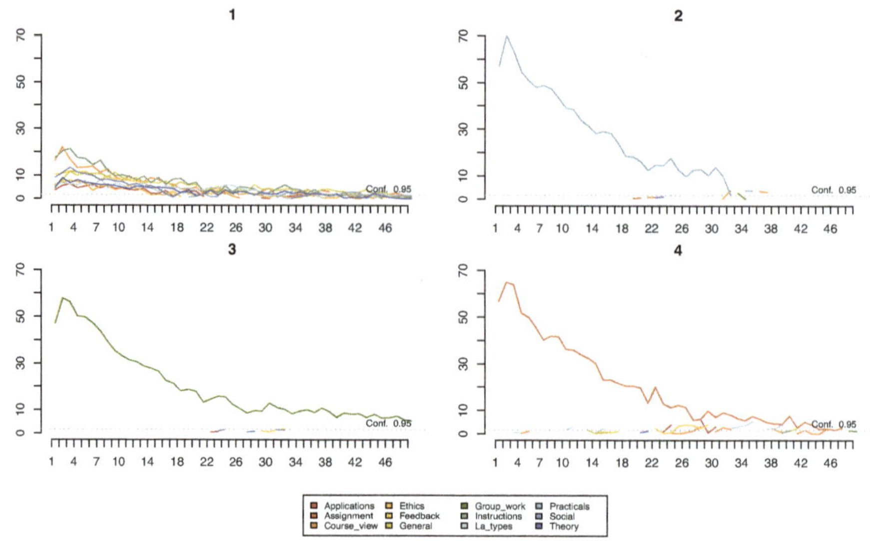

Fig. 14 Implication plot for the four clusters

learned in Chapter 6 of this book on data visualization [40]. Another important sequence analysis package is `seqHMM` [41], which contains several functions to fit hidden Markov models. In the next chapter, we see more advanced aspects of sequence analysis for learning analytics [28–30].

To learn more about sequence analysis in general, you can consult the book by Cornwell (2015), which is the first general textbook on sequence analysis in the context of social sciences. Another valuable resource is the recent textbook by Raab and Struffolino [38], which introduces the basics of sequence analysis and some recent advances as well as data and R code.

Acknowledgments This paper has been co-funded by the Academy of Finland (decision numbers 350560 and 331816), the Academy of Finland Flagship Programme (decision number 320162), the Strategic Research Council (SRC), FLUX consortium (decision numbers: 345130 and 345130), and the Swiss National Science Foundation (project "Strengthening Sequence Analysis", grant No.: 10001A_204740).

References

1. Abbott A (1983) Sequences of social events: concepts and methods for the analysis of order in social processes. Hist Methods J Quant Interdiscip Hist 16:129–147. https://doi.org/10.1080/01615440.1983.10594107
2. Piccarreta R, Studer M (2019) Holistic analysis of the life course: Methodological challenges and new perspectives. Adv Life Course Res 41:100251
3. Saqr M, López-Pernas S, Jovanović J, Gašević D (2023) Intense, turbulent, or wallowing in the mire: A longitudinal study of cross-course online tactics, strategies, and trajectories. Internet High Educ 57:100902

4. Fournier-Viger P, Lin JC-W, Kiran RU, Koh YS, Thomas R (2017) A survey of sequential pattern mining. Data Sci Pattern Recogn 1:54–77
5. Alexander PA, Schallert DL, Reynolds RE (2009) What is learning anyway? A topographical perspective considered. Educ Psychol 44:176–192
6. Schmitz B (2006) Advantages of studying processes in educational research. Learning and Instruction 16:433–449
7. Saqr M, Peeters W, Viberg O (2021) The relational, co-temporal, contemporaneous, and longitudinal dynamics of self-regulation for academic writing. Res Pract Technol Enhanced Learn 16:29
8. Liao TF, Bolano D, Brzinsky-Fay C, Cornwell B, Fasang AE, Helske S, Piccarreta R, Raab M, Ritschard G, Struffolino E, Studer M (2022) Sequence analysis: Its past, present, and future. Soc Sci Res 107:102772
9. Saqr M, López-Pernas S (2021) The longitudinal trajectories of online engagement over a full program. Comput Educ 175:104325
10. Matcha W, Gašević D, Ahmad Uzir N, Jovanović J, Pardo A, Lim L, Maldonado-Mahauad J, Gentili S, Pérez-Sanagustín M, Tsai Y-S (2020) Analytics of learning strategies: Role of course design and delivery modality. J Learn Anal 7:45–71
11. Jovanović J, Gašević D, Dawson S, Pardo A, Mirriahi N, Others (2017) Learning analytics to unveil learning strategies in a flipped classroom. Internet High Educ 33:74–85
12. Saqr M, Matcha W, Uzir NA, Jovanovic J, Gašević D, López-Pernas S (2023) Transferring effective learning strategies across learning contexts matters: A study in problem-based learning. AJET 35–57. https://doi.org/10.14742/ajet.8303
13. López-Pernas S, Saqr M, Gordillo A, Barra E (2023) A learning analytics perspective on educational escape rooms. Interact Learn Environ 31(10):6509–6525
14. Kinnebrew JS, Loretz KM, Biswas G (2013) A contextualized, differential sequence mining method to derive students' learning behavior patterns. J Educ Data Min 5:190–219
15. Saqr M, López-Pernas S (2022) How CSCL roles emerge, persist, transition, and evolve over time: A four-year longitudinal study. Comput Educ 189:104581
16. Saqr M, López-Pernas S (2023) The temporal dynamics of online problem-based learning: Why and when sequence matters. Int J Comput Support Collab Learn 18:11–37
17. Matcha W, Gašević D, Uzir NA, Jovanović J, Pardo A (2019) Analytics of learning strategies: Associations with academic performance and feedback. In: ACM international conference proceeding series, pp 461–470
18. Kinnebrew JS, Biswas G (2012) Identifying learning behaviors by contextualizing differential sequence mining with action features and performance evolution. In: Proceedings of the 5th international conference on educational data mining, EDM 2012, pp 57–64
19. Raab M, Struffolino E (2022) Sequence analysis. SAGE Publications
20. López-Pernas S, Saqr M (2021) Bringing synchrony and clarity to complex multi-channel data: A learning analytics study in programming education. IEEE Access 9: 166531–166541
21. López-Pernas S, Saqr M, Viberg O (2021) Putting it all together: Combining learning analytics methods and data sources to understand students' approaches to learning programming. Sustain Sci Pract Pol 13:4825
22. Studer M, Ritschard G (2016) What matters in differences between life trajectories: A comparative review of sequence dissimilarity measures. J R Stat Soc Ser A Stat Soc 179:481–511
23. Studer M (2013) WeightedCluster library manual. Pract Guide Creat Typol Trajectories Soc Sci 2296–1658
24. Piccarreta R, Lior O (2010) Exploring sequences: a graphical tool based on multi-dimensional scaling. J R Stat Soc Ser A (Stat Soc) 173:165–184. https://doi.org/10.1111/j.1467-985x.2009.00606.x
25. Gabadinho A, Ritschard G (2013) Searching for typical life trajectories applied to childbirth histories. Gendered life courses–Between individualization and standardization A European approach applied to Switzerland, pp 287–312
26. Studer M, Ritschard G, Gabadinho A, Müller NS (2011) Discrepancy analysis of state sequences. Sociol Methods Res 40:471–510

27. Gabadinho A, Ritschard G, Müller NS, Studer M (2011) Analyzing and visualizing state sequences in R with TraMineR. J Stat Softw 40:1–37
28. López-Pernas S, Saqr M, Helske S, Murphy K (2024, this volume) Multichannel sequence analysis in educational research using r. In: Saqr M, López-Pernas S (eds) Learning analytics methods and tutorials: A practical guide using R. Springer
29. Helske J, Helske S, Saqr M, López-Pernas S, Murphy K (2024, this volume) A modern approach to transition analysis and process mining with markov models: A tutorial with R. In: Saqr M, López-Pernas S (eds) Learning analytics methods and tutorials: A practical guide using R. Springer
30. López-Pernas S, Saqr M (2024, this volume) Modelling the dynamics of longitudinal processes in education. A tutorial with R for the VaSSTra method. In: Saqr M, López-Pernas S (eds) Learning analytics methods and tutorials: A practical guide using R. Springer
31. Bergner Y, Shu Z, Davier A von (2014) Visualization and confirmatory clustering of sequence data from a simulation-based assessment task. In: Educational data mining 2014
32. Billari FC (2001) The analysis of early life courses: Complex descriptions of the transition to adulthood. J Population Res 18:119–142. https://doi.org/10.1007/bf03031885
33. Murphy K, López-Pernas S, Saqr M (2024, this volume) Dissimilarity-based cluster analysis of educational data: A comparative tutorial using R. In: Saqr M, López-Pernas S (eds) Learning analytics methods and tutorials: A practical guide using R. Springer
34. Abbott A, Tsay A (2000) Sequence analysis and optimal matching methods in sociology: Review and prospect. Sociol Methods Res 29:3–33
35. Studer M, Ritschard G (2015) What matters in differences between life trajectories: a comparative review of sequence dissimilarity measures. J R Stat Soc Ser A Stat Soc 179:481–511. https://doi.org/10.1111/rssa.12125
36. Taub M, Banzon AM, Zhang T, Chen Z (2022) Tracking changes in students' online self-regulated learning behaviors and achievement goals using trace clustering and process mining. Front Psychol 13:813514
37. Gabadinho A, Ritschard G, Studer M, Nicolas SM (2009) Mining sequence data in R with the TraMineR package: A users guide for version 1.2. University of Geneva, Geneva, vol 1
38. Raab M (2022) ggseqplot: Render Sequence Plots using 'ggplot2'. https://maraab23.github.io/ggseqplot
39. Wickham H, Chang W, Wickham MH (2016) Package 'ggplot2'. Create elegant data visualisations using the grammar of graphics. Version 2(1):1–189
40. López-Pernas S, Misiejuk K, Kopra J, Tikka S, Heinäniemi M, Saqr M (2024, this volume) Visualizing and reporting educational data with r. In: Saqr M, López-Pernas S (eds) Learning analytics methods and tutorials: A practical guide using R. Springer
41. Helske S, Helske J (2019) Mixture hidden Markov models for sequence data: the seqHMM package in R. J Stat Softw 88(3):1–32. https://doi.org/10.18637/jss.v088.i03

Modeling the Dynamics of Longitudinal Processes in Education. A Tutorial with R for the VaSSTra Method

Sonsoles López-Pernas and Mohammed Saqr

1 Introduction

Modeling a longitudinal process brings a lot of variability over time. The modeling procedure becomes even harder when we use multivariate continuous variables to model a single construct. For example, in education research we might model students' online behavioral engagement through their number of clicks, time spent online, and frequency of interactions [1]. Most human behavioral constructs are an amalgam of interrelated features with complex fluctuations over time. Modeling such processes requires a method that takes into account the multidimensional nature of the examined construct as well as the temporal evolution. Nevertheless, despite the rich boundless information in the quantitative data, discrete patterns can be captured, modeled, and traced using appropriate methods. Such discrete patterns represent an archetype or a "state" that is typical of behavior or function [2]. For instance, a combination of frequent consumption of online course resources, long online time, interaction with colleagues and intense interactions in cognitively challenging collaborative tasks can be coined as an "engaged state" [3] The same student that shows an "engaged state" at one point, can transition to having few interactions and time spent online at the next time point, i.e., a "disengaged state".

Capturing a multidimensional construct into qualitative discrete states has several advantages. First, it avoids information overload where the information at hand is overwhelmingly hard to process accurately because of the multiplicity and lack of clarity of how to interpret small changes and variations. Second, it allows an easy way of communicating the information; it is understandable that communicating a state such as "engaged" is easier than telling the values of several activity variables.

S. López-Pernas (✉) · M. Saqr
School of Computing, University of Eastern Finland, Joensuu, Finland
e-mail: sonsoles.lopez@uef.fi

M. Saqr, S. López-Pernas (eds.), *Learning Analytics Methods and Tutorials*,
https://doi.org/10.1007/978-3-031-54464-4_11

Third, it is easier to trace or track. As we are interested in significant shifts overtime, fine-grained changes in the activities are less meaningful. We are rather interested in significant shifts between behavioral states, e.g., from engaged to disengaged. Besides, such a shift is also actionable. As [4] puts it, "reliability can sometimes be improved by tuning grain size of data so it is neither too coarse, masking variance within bins, nor too fine-grained, inviting distinctions that cannot be made reliably." More importantly, capturing the states is more tethered to reality of human nature and function. In fact, many psychological, physiological or disease constructs have been described as states with defining criteria, e.g., motivation, depression or migraine.

Existing methods for longitudinal analysis are often limited to the study of a single variable's evolution over time [5]. Some examples of such methods are longitudinal k-means [6], group-based trajectory modeling (GBTM) [7], or growth models [8]. However, when multivariate data is the target of analysis, these methods cannot be used. Multivariate methods are usually limited to one step or another of the analysis e.g., clustering of multivariate data into categorical variables (e.g., states), or chart the succession of categories into sequences. The method presented in this chapter provides an ensemble of methods and tools to effectively model, visualize and statistically analyse the longitudinal evolution of multivariate data. As such, modeling the temporal evolution of latent states, as we propose in this chapter, may not be entirely new and has been performed—at least in part—using several models, algorithms and software platforms [9–11]. For instance, the package `lmfa` can capture latent states in multivariate data, model their trajectories as well as transition probabilities [12]. Nevertheless, most of the existing methods are concerned with modeling disease states, time-to event (survival models), time-to-failure models [11] or lack a sequential analysis.

The *VaSSTra* method described in this chapter allows to summarize multiple variables into states that can be analyzed using sequence analysis across time. Then, using life-event methods, distinct trajectories of sequences that undergo a similar evolution can be analyzed in detail. *VaSSTra* consists of three steps: (1) capturing the states or patterns (from variables); (2) modeling the temporal process (from states); and (3) capturing the patterns of longitudinal development (similar sequences are grouped in trajectories). As such, the method described in this chapter is a combination of several methods. First, a person-centered method (latent class or latent profile analysis) is used to capture the unobserved "states" within the data. The states are then used to construct a "sequence of states", where a sequence represents a person's ordered states for each time point. The construction of "sequence of states" unlocks the full potential of sequence analysis visually and mathematically. Later, the longitudinal modeling of sequences is performed using a clustering method to capture the possible trajectories of progression of states. Thus, the name of the method is "from variables to states", "from states to sequences" and "from sequences to trajectories" *VaSSTra* [5].

Throughout the chapter, we discuss how to derive states from different variables related to students, how to construct sequences from students' longitudinal progression of states, and how to identify and study distinct trajectories of sequences that undergo a similar evolution. We also cover some advanced properties of sequences that can help us analyze and compare trajectories. In the next section, we explain the *VaSSTra* method in detail. Next, we review the existing literature that has used the method. After that, we present a step-by-step tutorial on how to implement the method using a dataset of students' engagement indicators across a whole program.

2 VaSSTra: From Variables to States, from States to Sequences, from Sequences to Trajectories

In Chap. 10, we went through the basics of sequence analysis in learning analytics [13]. Specifically, we learned how to construct a sequence from a series of ordered student activities in a learning session, which is a very common technique in learning analytics (e.g., [14]). In the sequences we studied, each time point represents a single instantaneous event or action by the students. In this advanced chapter, we take a different approach, where sequences are not built from a series of events but rather from states. Such states represent a certain construct (or cluster of variables) related to students (e.g., engagement, motivation) during a certain period (e.g., a week, a semester, a course). The said states are derived from a series of data variables related to the construct under study over the stipulated time period. Analyzing the sequence of such states over several sequential periods allows us to summarize large amounts of longitudinal information and to study complex phenomena across longer timespans [5]. This approach is known as the VaSSTra method. VaSSTra utilizes a combination of person-based methods (to capture the latent states) along with life events methods to model the longitudinal process. In doing so, VaSSTra effectively leverages the benefits of both families of methods in mapping the patterns of longitudinal temporal dynamics. The method has three main steps that can be summarized as (1) identifying latent **S**tates from **Va**riables, (2) modeling states as **S**equences, and (3) identifying **Tra**jectories within sequences. The three steps are depicted in Fig. 1 and described in detail below:

- **Step 1. From variables to states:** In the first step of the analysis, we identify the "states" within the data using a method that can capture latent or unobserved

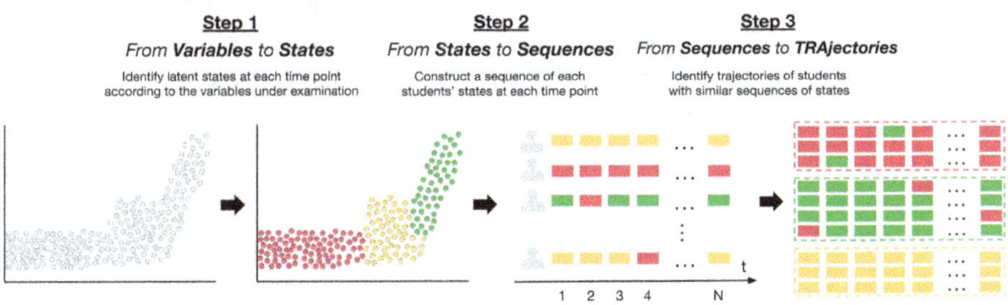

Fig. 1 Summary of the three steps of the VaSSTra method

patterns from multidimensional data (variables). The said states represent a behavioral pattern, function or a construct that can be inferred from the data. For instance, engagement is a multidimensional construct and is usually captured through several indicators. e.g., students' frequency and time spent online, course activities, cognitive activities and social interactions. Using an appropriate method, such as person-based clustering in our case, we can derive students' engagement states for a given activity or course. For instance, the method would classify students who invest significant time, effort and mental work are "engaged." Similarly, students who are investing low effort and time in studying would be classified as "disengaged." Such powerful summarization would allow us to use the discretized states in further steps. An important aspect of such states is that they are calculated for a specific timespan. Therefore, in our example we could infer students' engagement states per activity, per week, per lesson, per course, etc. Sometimes, such time divisions are by design (e.g., lessons or courses), but in other occasions researchers have to establish a time scheme according to the data and research questions (e.g., weeks or days). Computing states for multiple time periods is a necessary step to create time-ordered state sequences and prepare the data for sequence analysis.

- **Step 2. From states to sequences:** Once we have a state for each student at each time point, we can construct an ordered sequence of such states for each student. For example, if we used the scenario mentioned above about measuring engagement states, a sequence of a single student's engagement states across a six-lesson course would be like the one below. When we convert the ordered states to sequences, we unlock the potential of sequence analysis and life events methods. We are able to plot the distribution of states at each time point, study the individual pathways, the entropy, the mean time spent at each state, etc. We can also estimate how frequently students switch states, and what is the likelihood they finish their sequence in a "desirable" state (i.e., "engaged").

disengaged - average - engaged - engaged - engaged - average

- **Step 3. From sequences to trajectories:** Our last step is to identify similar trajectories—sequences of states with a similar temporal evolution—using temporal clustering methods (e.g., hidden Markov models or hierarchical clustering). Covariates (i.e., variables that could explain cluster membership) can be added at this stage to help identify why a trajectory has evolved in a certain way. Moreover, sequence analysis can be used to study the different trajectories, and not only the complete cohort. We can compare trajectories according to their sequence properties, or to other variables (e.g., performance).

3 Review of the Literature

The *VaSSTra* method has been used to study different constructs related to students' learning (Table 1), such as engagement [3, 15, 16], roles in computer-supported

Table 1 Previous literature in which *VaSSTra* has been used

Ref.	Variables	States	Alphabet	Actor	Time unit	Method Step 1	Method Step 3	Advance sequence methods
[3]	Frequency, time spent and regularity of online activities	Engagement	Active, Average, Disengaged, Withdraw	Study program	Course	LCA	HMM	Entropy, sequence implication, survival, transitions
[15]	Frequency, time spent and regularity of online activities. Grades	Engagement, achievement	Active, Average, Disengages. Achiever, Intermediate, Low	Study program	Course	LCA	MHMM	Multi-channel sequence analysis
[16]	Frequency, time spent and regularity of online activities	Engagement	Actively engaged, Averagely engaged, Disengaged, Dropout	Study program	Course	LCA	AHC	Survival analysis, discriminating subsequences, mean time
[17]	Centrality measures	CSCL roles	Influencer, Mediator, Isolate	Study program	Course	LPA	MHMM	Covariate analysis, spell duration, transitions, entropy, integrative capacity
[18]	Online activities	Learning strategies	Diverse, Forum read, Forum interact, Lecture read. Intense diverse, Light diverse, Light interactive, Moderate interactive	Course. Study program	Learning session. Course	MHMM	AHC	Entropy, implication, transitions
[19]	Online activities	Learning strategies	Video-oriented instruction, Slide-oriented instruction, Help-seeking. Assignment struggling, Assignment approaching, Assignment succeeding	Course	Learning session	AHC	HMM	Multi-channel sequence analysis

collaborative learning (CSCL) [17], and learning strategies [18, 19]. Several algorithms have been operationalized to identify latent states from students' online data. Some examples are: Agglomerative Hierarchical Clustering (AHC) [19], Latent Class Analysis (LCA) [3, 15, 16], Latent Profile Analysis (LPA) [17], and mixture hidden Markov models (MHMM) [18].

Moreover, sequences of states have mostly been used to represent each course in a program [3, 15–18], but also smaller time spans, such as each learning session in a single course [18, 19]. Different algorithms have also been used to cluster sequences of states into trajectories including HMM [3, 19], mixture hidden Markov models (MHMM) [15, 17], and AHC [16, 18]. Moreover, besides the basic aspects of sequence analysis discussed in the previous chapter, previous work have explored advanced features of sequence analysis such as survival analysis [3, 16], entropy [3, 17, 18], sequence implication [3, 18], transitions [3, 17, 18], covariates [17], discriminating subsequences [16] or integrative capacity [17]. Other studies have made used of multi-channel sequence analysis [15, 19], which is covered in Chap. 13 [20].

4 *VassTra* with R

In this section we provide a step-by-step tutorial on how to implement *VaSSTra* with R. To illustrate the method, we will conduct a case study in which we examine students' engagement throughout all the courses of the first two years of their university studies, using variables derived from their usage of the learning management system.

4.1 The Packages

In order to conduct our analysis we will need several packages besides the basic `rio` (for reading and saving data in different extensions), `tidyverse` (for data wrangling), `cluster` (for clustering features), and `ggplot2` (for plotting). Below is a brief summary of the rest of the packages needed:

- `BBmisc`: A package with miscellaneous helper functions [21]. We will use its `normalize` function to normalize our data across courses to remove the differences in students' engagement that are due to different course implementations (e.g., larger number of learning resources).
- `tidyLPA`: A package for conducting Latent Profile Analysis (LPA) with R [22]. We will use it to cluster students' variables into distinct clusters or states.
- `TraMineR`: As we have seen in Chap. 10 about sequence analysis [13], this package helps us construct, analyze and visualize sequences from time-ordered states or events [23].

- `seqhandbook`: This package complements `TraMineR` by providing extra analyses and visualizations [24].
- `Gmisc`: A package with miscellaneous functions for descriptive statistics and plots [25]. We will use it to plot transitions between states.
- `WeightedCluster`: A package for clustering sequences using hierarchical cluster analysis [26]. We will use it to cluster sequences into similar trajectories.

The code below imports all the packages that we need. You might need to install them beforehand using the `install.packages` command:

```
library(rio)
library(tidyverse)
library(ggplot2)
library(cluster)
library(BBmisc)
library(tidyLPA)
library(TraMineR)
library(seqhandbook)
library(Gmisc)
library(WeightedCluster)
```

4.2 The Dataset

For our analysis, we will use a dataset of students' engagement indicators throughout eight courses, corresponding to the first two years of a blended higher education program. The dataset is described in detail in the data chapter. The indicators or variables are calculated from students' log data in the learning management system, and include the frequency (i.e., number of times) with which certain actions have been performed (e.g., view the course lectures, read forum posts), the time spent in the learning management system, the number of sessions, the number of active days, and the regularity (i.e., consistency and investment in learning). These variables represent students' behavioral engagement indicators. The variables are described in detail in Chapter 2 about the datasets of the book [27] and Chapter 7 about predictive learning analytics [28] and in previous works [3]. Below we use `rio`'s `import` function to read the data.

```
LongitudinalData <-
  import("https://github.com/lamethods/data/raw/main/
        9_longitudinalEngagement/LongitudinalEngagement.csv")
```

```
# A tibble: 1,136 x 15
   UserID   CourseID Sequence Freq_Course_View Freq_Forum_Consume
```

```
     <chr>    <chr>        <int>            <int>             <int>
 1 D2C5F64E C6107FC4           1              150               251
 2 D2C5F64E 4C3F37F0           2               98                84
 3 D2C5F64E E54A52A3           3              254               354
 4 D2C5F64E AB7EC624           4              332               825
 5 D2C5F64E B0E95213           5              386               960
 6 D2C5F64E 3DE2A32B           6              261              1026
 7 D2C5F64E D7DF3685           7              250               652
 8 D2C5F64E ECD1AFC8           8              287               697
 9 D7E3D0DC B51E1259           1              186               597
10 D7E3D0DC C473D477           2              241               580
# i 1,126 more rows
# i 10 more variables: Freq_Forum_Contribute <int>, Freq_Lecture_View <int>,
#   Regularity_Course_View <dbl>, Regularity_Lecture_View <dbl>,
#   Regularity_Forum_Consume <dbl>, Regularity_Forum_Contribute <dbl>,
#   Session_Count <int>, Total_Duration <int>, Active_Days <int>,
#   Final_Grade <dbl>
```

4.3 From Variables to States

The first step in our analysis is to detect latent states from the multiple engagement-related variables in our dataset (e.g., frequency of course views, frequency of forum posts, etc.). For this purpose, we will use LPA, a person-based clustering method, to identify each student's engagement state at each course. We first need to standardize the variables, to account for the possible differences in course implementations (e.g., each course has a different number of learning materials and slightly different duration). This way, the mean value of each indicator will be always 0 regardless of the course. Any value above the mean will be always positive, and any value below will be negative. As such the engagement is measures on the same scale. To standardize the data we first group it by `CourseID` using `tidyverse`'s `group_by` and then we apply the `normalize` function from the `BBmisc` package to all the columns that contain the engagement indicators using `mutate_at` and specifying the range of columns. If we inspect the data now, we will see that all variables are centered around 0.

```
LongitudinalData |> group_by(CourseID) |>
  mutate_at(vars(Freq_Course_View:Active_Days),
            function(x) normalize(x, method = "standardize")) |>
  ungroup() -> df
```

```
# A tibble: 1,136 x 15
   UserID   CourseID Sequence Freq_Course_View Freq_Forum_Consume
   <chr>    <chr>       <int>            <dbl>              <dbl>
 1 D2C5F64E C6107FC4        1           -1.13              -1.87
 2 D2C5F64E 4C3F37F0        2           -2.03              -2.69
 3 D2C5F64E E54A52A3        3            0.519             -1.24
```

```
 4 D2C5F64E AB7EC624         4           1.88            1.33
 5 D2C5F64E B0E95213         5           2.78            2.11
 6 D2C5F64E 3DE2A32B         6           0.603           2.46
 7 D2C5F64E D7DF3685         7           0.434           0.357
 8 D2C5F64E ECD1AFC8         8           0.707           0.707
 9 D7E3D0DC B51E1259         1           0.939           1.22
10 D7E3D0DC C473D477         2           1.55            1.39
# i 1,126 more rows
# i 10 more variables: Freq_Forum_Contribute <dbl>, Freq_Lecture_View <dbl>,
#   Regularity_Course_View <dbl>, Regularity_Lecture_View <dbl>,
#   Regularity_Forum_Consume <dbl>, Regularity_Forum_Contribute <dbl>,
#   Session_Count <dbl>, Total_Duration <dbl>, Active_Days <dbl>,
#   Final_Grade <dbl>
```

Now, we need to subset our dataset and choose only the variables that we need for clustering. That is, we exclude the metadata about the user and course, and keep only the variables that we believe are relevant to represent the engagement construct.

```
to_cluster <- dplyr::select(df, Freq_Course_View, Freq_Forum_Consume,
                             Freq_Forum_Contribute, Freq_Lecture_View,
                             Regularity_Course_View, Session_Count,
                             Total_Duration, Active_Days)
```

Before we go on, we must choose a seed so we can obtain the same results every time we run the clustering algorithm. We can now finally use `tidyLPA` to cluster our data. We try from 1 to 10 clusters and different models that enforce different constraints on the data. For example, model 3 takes equal variances and equal covariances, whereas model 6 takes varying variances and varying covariances. You can find out more about this in the `tidyLPA` documentation [22]. Be aware that running this step may take a while. For more details about LPA, consult the model-based clustering chapter.

```
set.seed(22294)
Mclustt <- to_cluster |> ungroup() |>
  single_imputation() |>
  estimate_profiles(1:10, models = c(1, 2, 3, 6))
```

Once all the possible cluster models and numbers have been calculated, we can calculate several statistics that will help us choose which is the right model and number of clusters for our data. For this purpose, we use the `compare_solutions` function from `tidyLPA` and we use the results of calling this function to plot the BIC and the entropy of each model for the range of cluster numbers that we have tried (1–10) (Fig. 2). In the model-based clustering chapter you can find out more details about how to choose the best cluster solution.

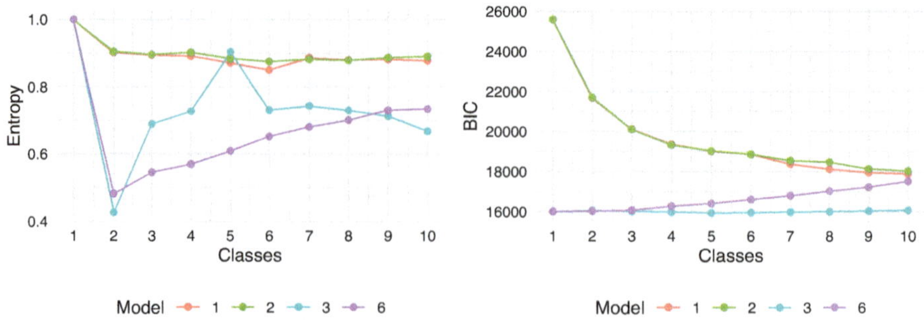

Fig. 2 Choosing the number of clusters by plotting statistics

```
cluster_statistics <- Mclustt |>
  compare_solutions(statistics = c("AIC", "BIC","Entropy"))
fits <- cluster_statistics$fits |> mutate(Model = factor(Model))

fits |>
  ggplot(aes(x = Classes,y = Entropy, group = Model, color = Model))+
  geom_line() +
  scale_x_continuous(breaks = 0:10) +
   geom_point() +
  theme_minimal() +
  theme(legend.position = "bottom")

fits |>
  ggplot(aes(x = Classes,y = BIC, group = Model, color = Model)) +
  geom_line() +
  scale_x_continuous(breaks = 0:10) +
  geom_point() +
  theme_minimal() +
  theme(legend.position = "bottom")
```

Although, based on the BIC values, Model 6 with 3 classes would be the best fit, the entropy for this model is quite low. Instead, Models 1 and 2 have a higher overall entropy and quite a large fall in BIC when increasing from 2 to 3 classes. Taken together, we choose Model 1 with 3 classes which shows better separation of clusters (high entropy) and a large drop in BIC value (elbow). We add the cluster assignment back to the data so we can compare the different variables between clusters and use the cluster assignment for the next steps. Now, for each student's course enrollment, we have assigned a state (i.e., cluster) that represents the student's engagement during that particular course.

```
df$State <- Mclustt$model_1_class_3$model$classification
```

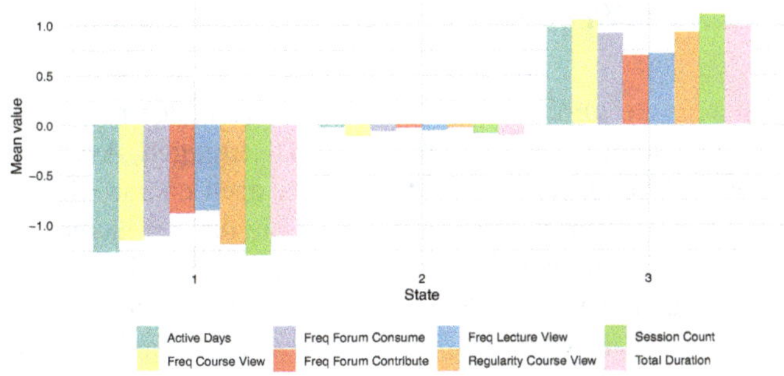

Fig. 3 Mean value of each variable for each cluster

We can plot the mean variable values for each of the three clusters (Fig. 3) to understand what each of them represents:

```
df |> pivot_longer(Freq_Course_View:Active_Days) |>
  mutate(State = factor(State)) |>
  filter(name %in% names(to_cluster)) |>
  mutate(name = gsub("_", " ", name)) |>
  group_by(name, State) |>
  summarize_if(is.double, mean) -> long_mean

long_mean |>
  ggplot(aes(fill = name, y = value, x = State)) +
  geom_col(position = "dodge") +
  scale_fill_brewer("", type = "qual", palette=8) +
  theme_minimal() +
  ylab ("Mean value") +
  theme(legend.position = "bottom")
```

We clearly see that the first cluster represents students with low mean levels of all engagement indicators; the second cluster represents students with average values, and the third cluster with high values. We can convert the `State` column of our dataset to a factor to give the clusters an appropriate descriptive label:

```
engagement_levels = c("Disengaged", "Average", "Active")
df_named <- df |>
  mutate(State = factor(State, levels = 1:3, labels = engagement_levels))
```

4.4 *From States to Sequences*

In the previous step, we turned a large amount of variables representing student engagement in a given course into a single state: **Disengaged**, **Average**, or **Active**. Each student has eight engagement states: one per each course (time point in our data) in the first two years of the program. Since the dataset includes the order of each course for each student, we can construct a sequence of the engagement states throughout all courses for each student. To do that we first need to transform our data into a wide format, in which each row represents a single student, and each column represents the student's engagement state at a given course:

```
clus_seq_df <- df_named |> arrange(UserID, Sequence) |>
  pivot_wider(id_cols = "UserID", names_from = "Sequence", values_from = "State")
```

Now we can use `TraMineR` to construct the sequence and assign colors to represent each of the engagement states:

```
colors <- c("#28a41f", "#FFDF3C", "#e01a4f")
clus_seq <- seqdef(clus_seq_df , 2:9,
                   alphabet = sort(engagement_levels), cpal = colors)
```

```
 [>] 3 distinct states appear in the data:
     1 = Active
     2 = Average
     3 = Disengaged
 [>] state coding:
       [alphabet] [label]    [long label]
     1 Active     Active     Active
     2 Average    Average    Average
     3 Disengaged Disengaged Disengaged
 [>] 142 sequences in the data set
 [>] min/max sequence length: 8/8
```

We can use the sequence distribution plot from `TraMineR` to visualize the distribution of each state at each time point (Fig. 4). We see that the distribution of states is almost constant throughout the eight courses. The 'Average' state takes the largest share, followed by the 'Engaged' state, and the 'Disengaged' state is consistently the least common. For more hints on how to interpret the sequence distribution plot, refer to Chapter 10 [13].

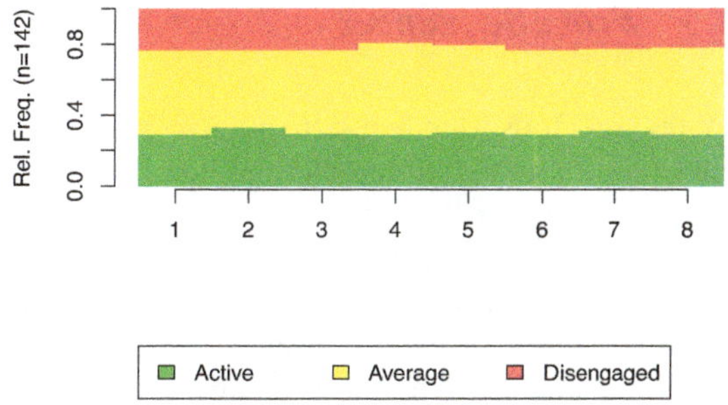

Fig. 4 Sequence distribution plot of the course states

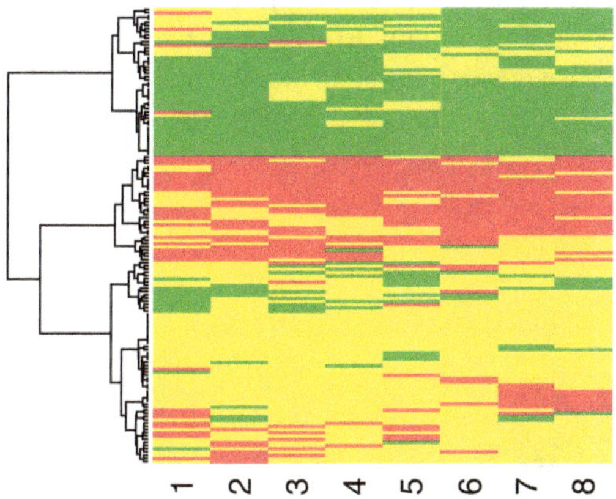

Fig. 5 Sequence index plot of the course states ordered by sequence distance

```
seqdplot(clus_seq, border = NA, use.layout = TRUE,
         with.legend = T, ncol = 3, legend.prop = 0.2)
```

We can also visualize each of the individual students' sequences of engagement states using a sequence index plot (Fig. 5). In this type of visualization, each horizontal bar represents a single student, and each of the eight colored blocks along the bar represents the students' engagement states. We can order the students' sequences according to their similarity for a better understanding. To do this, we calculate the substitution cost matrix (`seqsubm`) and the distance between the sequences according to this cost (`seqdist`). Then we use an Agglomerative Nesting Hierarchical Clustering algorithm (`agnes`) to group sequences together according to their similarity (see Chapter 10 [13]). We may now use the `seq_heatmap` function of `seqhandbook` to plot the sequences ordered by their similarity. From this plot, we already sense the existence of students that are mostly active, students that are mostly disengaged, and students that are in-between, i.e., mostly average.

```
sm <- seqsubm(clus_seq, method = "CONSTANT")
mvad.lcss <- seqdist(clus_seq, method = "LCS", sm = sm, with.missing = T)
clusterward2 <- agnes(mvad.lcss, diss = TRUE, method = "ward")
seq_heatmap(clus_seq, clusterward2)
```

4.5 From Sequences to Trajectories

In the previous step we constructed a sequence of each student's engagement states throughout eight courses. When plotting these sequences, we observed that there might be distinct trajectories of students that undergo a similar evolution of engagement. In this last step, we use hierarchical clustering to cluster the sequences of engagement states into distinct trajectories of similar engagement patterns. To perform hierarchical clustering we first need to calculate the distance between all the sequences. For more details on the clustering technique, please refer to Chapter 8 [29]. As we have seen in the Sequence Analysis chapter, there are several algorithms to calculate the distance. We choose LCS (Longest Common Subsequence), implemented in the `TraMineR` package which calculates the distance based on the longest common subsequence.

```
dissimLCS <- seqdist(clus_seq, method = "LCS")
```

```
[>] 142 sequences with 3 distinct states

[>] creating a 'sm' with a substitution cost of 2

[>] creating 3x3 substitution-cost matrix using 2 as
constant value

[>] 103 distinct  sequences

[>] min/max sequence lengths: 8/8

[>] computing distances using the LCS metric

[>] elapsed time: 0.012 secs
```

Now we can perform the hierarchical clustering. For this purpose, we use the `hclust` function of the `stats` package

```
Clustered <- hclust(as.dist(dissimLCS), method = "ward.D2")
```

We create partitions for clusters ranging from 2 to 10 and we plot the cluster statistics to be able to select the most suitable cluster number (Fig. 6).

```
Clustered_range <- as.clustrange(Clustered, diss = dissimLCS, ncluster = 10)
plot(Clustered_range, stat = "all", norm = "zscoremed", lwd = 2)
```

There seems to be a maximum for most statistics at three clusters, so we save the cluster assignment for three clusters in a variable named `grouping`.

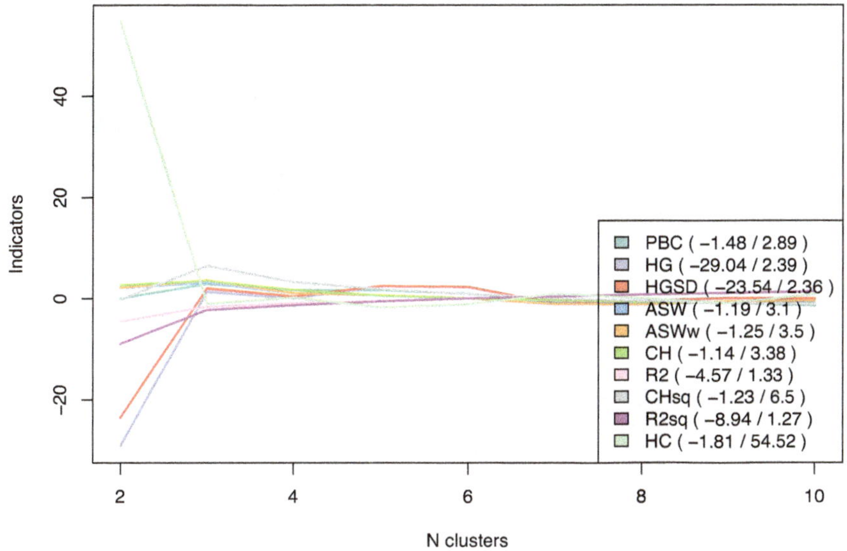

Fig. 6 Cluster statistics for the hierarchical clustering

```
grouping <- Clustered_range$clustering$cluster3
```

Now we can use the variable `grouping` to plot the sequences for each trajectory using the sequence index plot (Fig. 7):

```
seqIplot(clus_seq, group = grouping, sortv = "from.start")
```

In Fig. 7, we see that the first trajectory corresponds to mostly average students, the second one to mostly active students, and the last one to mostly disengaged students. We can rename the clusters accordingly.

```
trajectories_names = c("Mostly average", "Mostly active", "Mostly disengaged")
trajectories = trajectories_names[grouping]
```

We can plot the sequence distribution plot to see the overall distribution of the sequences for each trajectory (Fig. 8).

```
seqdplot(clus_seq, group = trajectories)
```

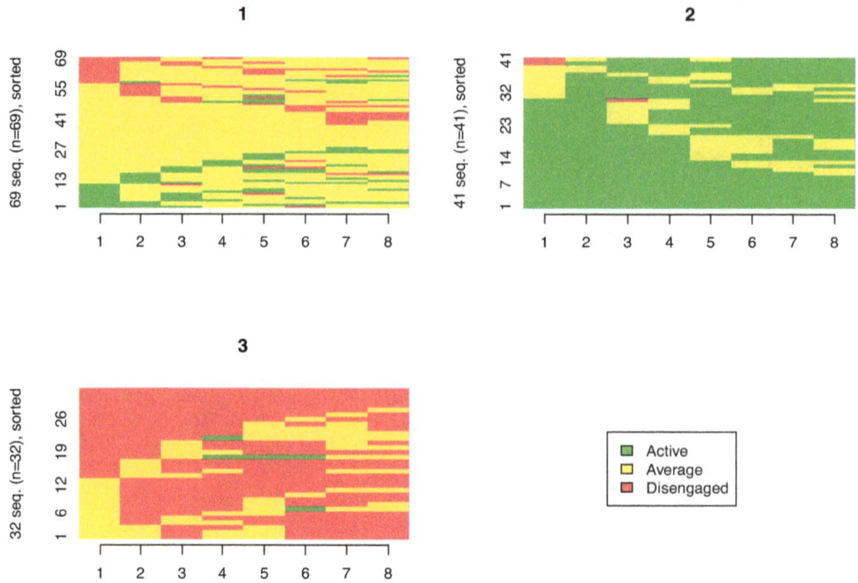

Fig. 7 Sequence index plot of the course states per trajectory

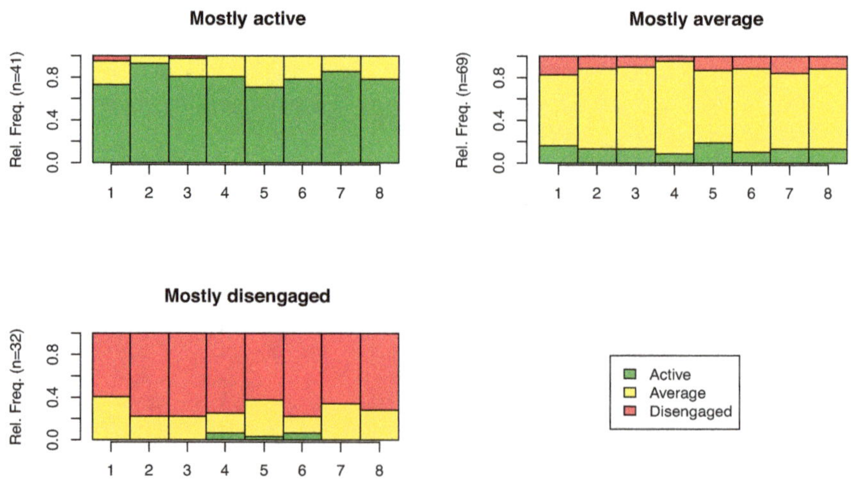

Fig. 8 Sequence distribution plot of the course states per trajectory

4.6 *Studying Trajectories*

There are many aspects that we can study about our trajectories. For example, we can use the mean time plot to compare the time spent in each engagement state for each trajectory. This plot summarizes the time distribution plot across all time points

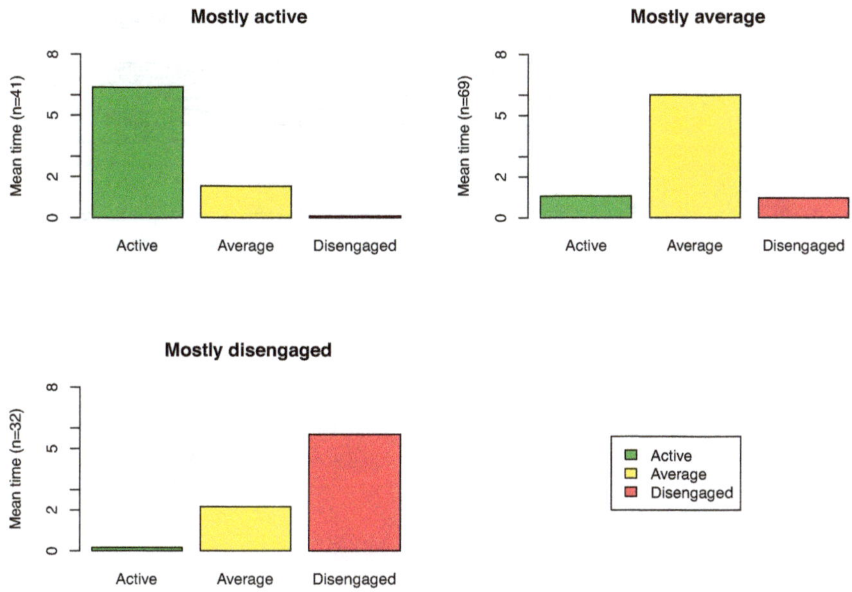

Fig. 9 Mean time plot of the course states per trajectory

(Fig. 9). As expected, we see that the mostly active students spend most of their time in an 'Active' state, the mostly average students in an Average state, and the mostly disengaged students in a 'Disengaged' state, although they spend quite some time in an 'Average' state as well.

```
seqmtplot(clus_seq, group = trajectories)
```

Another very useful plot is the sequence frequency plot (Fig. 10), that shows the most common sequences in each trajectory and the percentage of all the sequences that they represent. We see that, for each trajectory, the sequence that has all equal engagement states is the most common. The mostly active has, of course, sequences dominated by 'Active' states, with sparse average states, and one 'Disengaged' state. The mostly disengaged shows a similar pattern dominated by disengaged states with some average and one active state. The mostly average, although it is dominated by 'Average' states, shows diversity of shifts to active or disengaged.

```
seqfplot(clus_seq, group = trajectories)
```

To measure the stability of engagement states for each trajectory at each time point, we can use the between-study entropy. Entropy is lowest when all students have the same engagement state at the same time point and highest when the heterogeneity is

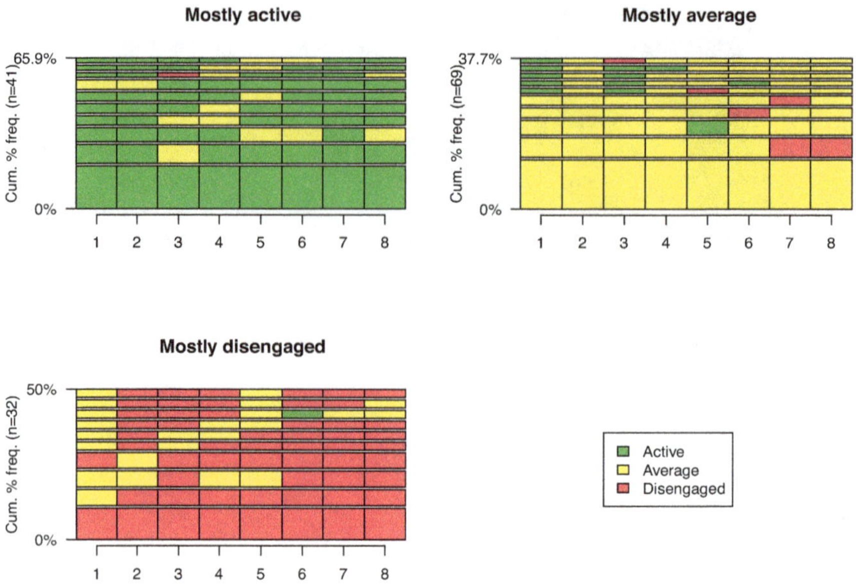

Fig. 10 The 10 most frequent sequences in each trajectory

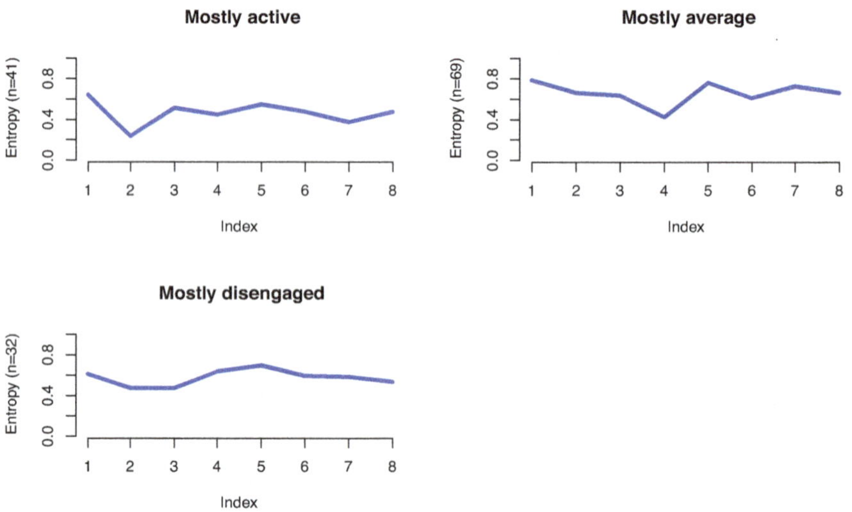

Fig. 11 Transversal entropy plot of each trajectory

maximum. We can see that the "Mostly active" and "Mostly disengaged" trajectories have a slightly lower entropy compared to the "Mostly average" one, which is a sign that the students in this trajectory are the least stable (Fig. 11).

```
seqHtplot(clus_seq, group = trajectories)
```

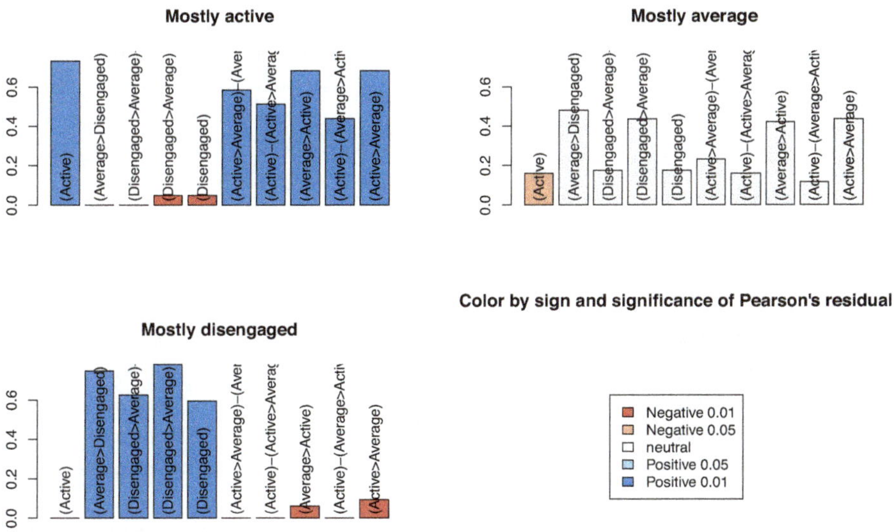

Fig. 12 Most discriminating subsequences per trajectory

Another interesting aspect to look into is the difference in the most common subsequences among trajectories (Fig. 12). We first search the most frequent subsequences overall and then compare them among the three trajectories. Interestingly enough, the most frequent subsequence is remaining 'Active', and remaining 'Disengaged' is number five. Remaining average is not among the top 10 most common subsequences, but rather the subsequences containing the 'Average' state always include transitions to other states.

```
mvad.seqe <- seqecreate(clus_seq)
fsubseq <- seqefsub(mvad.seqe, pmin.support = 0.05, max.k = 2)
discr <- seqecmpgroup(fsubseq, group = trajectories, method = "chisq")
plot(discr[1:10])
```

There are other sequence properties that we may need compare among the trajectories. The function `seqindic` calculates sequence properties for each individual sequence (Table 2). Some of these indices need additional information about the sequences, namely, which are the positive and the negative states. In our case, we might consider the *Active* state to be positive and the *Disengaged* state to be negative. Below we discuss some of the most relevant measures:

- `Trans`: Number of transitions. It represents the number of times there has been a change of state. If a sequence maintains the same state throughout its whole length, the value of `Trans` would be zero; if there were two shifts of state, the value would be 2.
- `Entr`: Longitudinal entropy or within-student entropy is a measure of the diversity of the sequence states. In contrast with the transversal or between-

Table 2 Sequence indicators

	Trans	Entr	Volat	Cplx	Integr	Prec
1	2.000	0.343	0.393	0.313	0.000	0.600
2	2.000	0.343	0.393	0.313	0.000	0.600
3	4.000	0.512	0.536	0.541	0.000	0.964
4	0.000	0.000	0.000	0.000	1.000	0.000
5	4.000	0.631	0.536	0.600	0.000	1.159
6..141						
142	2.000	0.343	0.393	0.313	0.917	0.006

students entropy that we saw earlier (Fig. 11), which is calculated per time point, longitudinal entropy is calculated per sequence (i.e., which represents a student's engagement through the courses in a program in this case). Longitudinal entropy is calculated using Shannon's entropy formula. Sequences that remain in the same state most of the time have a low entropy whereas sequences that shift states continuously with great diversity have a high entropy.

- `Cplx`: The complexity index is a composite measure of a sequence's complexity based on the number of transitions and the longitudinal entropy. It measures the variety of states within a sequence, as well as the frequency and regularity of transitions between them. In other words, a sequence with a high complexity index is characterized by many different and unpredictable states or events, and frequent transitions between them.
- `Prec`: Precarity is a measure of the (negative) stability or predictability of a sequence. It measures the proportion of time that a sequence spends in negative or precarious states, as well as the frequency and duration of transitions between positive and negative states. A sequence with a high precarity index is characterized by a high proportion of time spent in negative or precarious states, and frequent transitions between positive and negative states.
- `Volat`: Objective volatility represents the average between the proportion of states visited and the proportion of transitions (state changes). It is measure of the variability of the states and transitions in a sequence. A sequence with high volatility would be characterized by frequent and abrupt changes in the states or events, while a sequence with low volatility would have more stable and predictable patterns.
- `Integr`: Integrative capacity (potential) is the ability to reach a positive state and then stay in a positive state. Sequences with a high integrative capacity not only include positive states but also manage to stay in such positive states.

```
Indices <- seqindic(clus_seq,
            indic = c("trans","entr","cplx","prec","volat","integr"),
            ipos.args = list(pos.states = c("Active")),
            prec.args = list(c("Disengaged")))
```

We can compare the distribution of these indices between the different trajectories to study their different properties (Fig. 13). Below is an example for precarity and

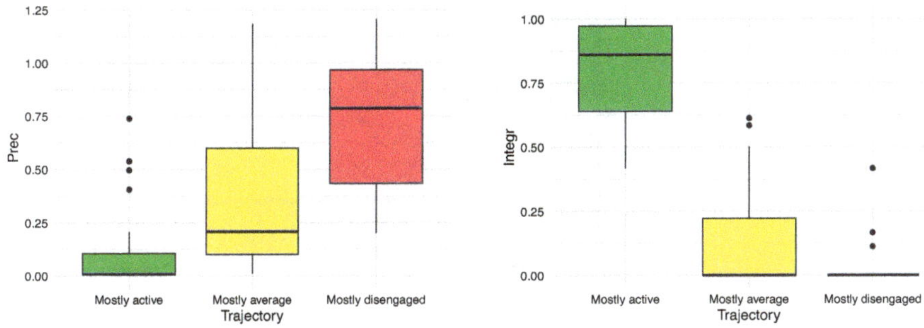

Fig. 13 Comparison of sequence indicators between trajectories

Table 3 Transition rate between states

	[-> Active]	[-> Average]	[-> Disengaged]
[Active ->]	216	82	1
[Average ->]	78	320	77
[Disengaged ->]	5	75	140

integrative capacity. We clearly see how the *Mostly disengaged* trajectory has the highest value of precarity, whereas the *Mostly active* students have the highest integrative capacity. Beyond a mere visual representation, we could also conduct statistical tests to compare whether these properties differ significantly from each other among trajectories.

```
Indices$Trajectory = trajectories

Indices |> ggplot(aes(x = Trajectory, y = Prec, fill = Trajectory)) +
  geom_boxplot() + scale_fill_manual(values = colors) +
  theme_minimal() + theme(legend.position = "none")

Indices |> ggplot(aes(x = Trajectory, y = Integr, fill = Trajectory)) +
  geom_boxplot() + scale_fill_manual(values = colors) +
  theme_minimal() + theme(legend.position = "none")
```

As we have mentioned, an important aspect of the study of students' longitudinal evolution is looking at the transitions between states. We can calculate the transitions using `seqtrate` from `TraMineR` and plot them using `transitionPlot` from `Gmisc`.

```
transition_matrix = seqtrate(clus_seq, count = T)
```

From Table 3 and Fig. 14 we can see how most transitions are between one state and the same (no change). The most unstable state is 'Average' with frequent transitions

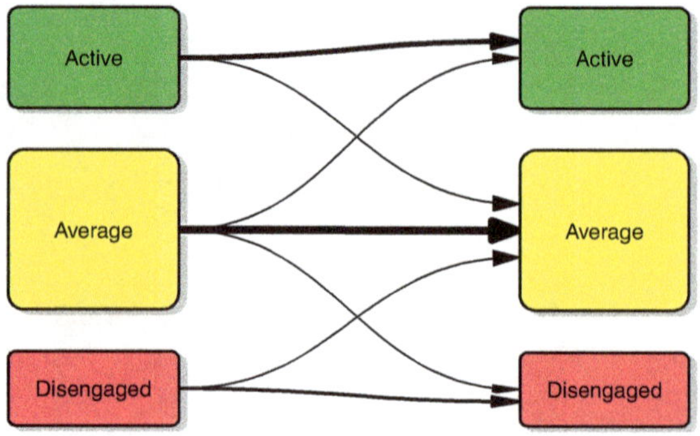

Fig. 14 Transition plot between states

both to 'Active' and 'Disengaged'. Both 'Active' and 'Disengaged' had occasional transitions to 'Average' but rarely from one another.

```
transitionPlot(transition_matrix,
                              fill_start_box = colors,
                              txt_start_clr = "black",
                              cex = 1,
                              box_txt = rev(engagement_levels))
```

5 Discussion

In this chapter, the *VaSSTra* method is presented as a person-centered approach for the longitudinal analysis of complex behavioral constructs over time. In the step-by-step tutorial, we have analyzed students' engagement states throughout all the courses in the first two years of program. First, we have clustered all the indicators of student engagement into three engagement states using model-based clustering: active, average and disengaged. This step allowed us to summarize eight continuous numerical variables representing students' online engagement indicators of each course into a single categorical variable (state). Then, we constructed a sequence of engagement states for each student, allowing us to map the temporal evolution of engagement and make use of sequence analysis methods to visualize and investigate such evolution. Lastly, we have clustered students' sequences of engagement states into three different trajectories: a mostly active trajectory which is dominated by engaged students who are stable throughout time, a mostly average trajectory with averagely engaged students who often transition to engaged or disengaged states, and a mostly disengaged trajectory with inactive students that fail to catch up and remain disengaged most of the program. As such, *VaSSTra* offers several advantages over the existing longitudinal methods for clustering (such as growth models or longitudinal k-means) which are limited to a single continuous variable

[30–32], instead of taking advantage of multiple variables in the data. Through the summarizing power of visualization, *VaSSTra* is able to represent complex behaviors captured through several variables using a limited number of states. Moreover, through sequence analysis, we can study how the sequences of such states evolve over time and differ from one another, and whether there are distinct trajectories of evolution.

Several literature reviews of longitudinal studies (e.g., [33]) have highlighted the shortcomings of existing research for using variable-centered methods or ignoring the heterogeneity of students' behavior. Ignoring the longitudinal heterogeneity means mixing trends of different types, e.g., an increasing trend in a subgroup and a decreasing trend in another subgroup exist. Another limitation of the existing longitudinal clustering methods is that cluster membership can not vary with time so one student is assigned to a single longitudinal cluster, which makes it challenging to study variability and transitions.

As we have seen in the literature review section, the *VaSSTra* method can be adapted to various scenarios beyond engagement, such as collaborative roles, attitudes, achievement, or self-efficacy, and can be used with different time points such as tasks, days, weeks, or school years. The reader should refer to Chapter 8 [29] and Chapter 9 [34] about clustering to learn other clustering techniques that may be more appropriate for transforming different types of variables into states (that is, conducting the first step of *VaSSTra*). Moreover, in Chapter 10 [13], the basics of sequence analysis are described, including how to cluster sequences into trajectories using different distance measures that might be more appropriate in different situations. The next chapter (Chapter 12) [35] presents Markovian modeling, which constitutes another way of clustering sequences into trajectories according to their state transitions. Lastly, Chapter 13 [20] presents multi-channel sequence analysis, which could be used to extend *VaSSTra* to study several parallel sequences (of several constructs) at the same time.

References

1. Henrie CR, Bodily R, Manwaring KC, Graham CR (2015) Exploring intensive longitudinal measures of student engagement in blended learning. Int Rev Res Open Distrib Learn 16. https://doi.org/10.19173/irrodl.v16i3.2015
2. Lazarus G, Song J, Jeronimus BF, Fisher AJ (2023) Delineating discrete generalizable states from intraindividual time series: towards a science of moments. https://doi.org/10.31234/osf.io/4nxqh
3. Saqr M, López-Pernas S (2021) The longitudinal trajectories of online engagement over a full program. Comput Educ 175:104325. https://doi.org/10.1016/j.compedu.2021.104325
4. Winne PH (2020) Construct and consequential validity for learning analytics based on trace data. Comput Human Behav 112:106457. https://doi.org/10.1016/j.chb.2020.106457
5. López-Pernas S, Saqr M (2023) From variables to states to trajectories (VaSSTra): a method for modelling the longitudinal dynamics of learning and behaviour. In: Proceedings of the tenth international conference on technological ecosystems for enhancing multiculturality (TEEM'22). Springer, Salamanca, pp. 1169–1178. https://doi.org/10.1007/978-981-99-0942-1_123

6. Genolini C, Falissard B (2010) KmL: K-means for longitudinal data. Comput Stat 25:317–328. https://doi.org/10.1007/s00180-009-0178-4
7. Nagin DS (2014) Group-based trajectory modeling: an overview. Ann Nutr Metab 65:205–210. https://doi.org/10.1159/000360229
8. Ram N, Grimm KJ (2009) Growth mixture modeling: a method for identifying differences in longitudinal change among unobserved groups. Int J Behav Dev 33:565–576. https://doi.org/10.1177/0165025409343765
9. Hougaard P (1999) Multi-state models: a review. Lifetime Data Anal 5:239–264. https://doi.org/10.1023/a:1009672031531
10. Jackson CH (2011) Multi-state models for panel data: the msm package for R. J Stat Softw 38. https://doi.org/10.18637/jss.v038.i08
11. McClintock BT, Langrock R, Gimenez O, Cam E, Borchers DL, Glennie R, Patterson TA (2020) Uncovering ecological state dynamics with hidden Markov models. Ecol Lett 23:1878–1903. https://doi.org/10.1111/ele.13610
12. Vogelsmeier LVDE, Vermunt JK, Roover KD (2022) How to explore within-person and between-person measurement model differences in intensive longitudinal data with the r package lmfa. Behav Res Methods. https://doi.org/10.3758/s13428-022-01898-1
13. Saqr M, López-Pernas S, Helske S, Durand M, Murphy K, Studer M, Ritschard G (2024) Sequence analysis in education: principles, technique, and tutorial with r. In: Saqr M, López-Pernas S (eds) Learning analytics methods and tutorials: a practical guide using R. Springer, Berlin
14. Jovanović J, Gašević D, Dawson S, Pardo A, Mirriahi N (2017) Learning analytics to unveil learning strategies in a flipped classroom. Internet Higher Educ 33:74–85. https://doi.org/10.1016/j.iheduc.2017.02.001
15. Saqr M, López-Pernas S, Helske S, Hrastinski S (2023) The longitudinal association between engagement and achievement varies by time, students' profiles, and achievement state: a full program study. Comput Educ 199:104787. https://doi.org/10.1016/j.compedu.2023.104787
16. Saqr M, López-Pernas S (2021) The dire cost of early disengagement: a four-year learning analytics study over a full program. In: Technology-enhanced learning for a free, safe, and sustainable world. Springer International Publishing, Cham, pp 122–136. https://doi.org/10.1007/978-3-030-86436-1_10
17. Saqr M, López-Pernas S (2022) How CSCL roles emerge, persist, transition, and evolve over time: a four-year longitudinal study. Comput Educ 104581. https://doi.org/10.1016/j.compedu.2022.104581
18. Saqr M, López-Pernas S, Jovanović J, Gašević D (2023) Intense, turbulent, or wallowing in the mire: a longitudinal study of cross-course online tactics, strategies, and trajectories. Internet High 57:100902. https://doi.org/10.1016/j.iheduc.2022.100902
19. López-Pernas S, Saqr M (2021) Bringing synchrony and clarity to complex multi-channel data: a learning analytics study in programming education. IEEE Access. https://doi.org/10.1109/ACCESS.2021.3134844
20. López-Pernas S, Saqr M, Helske S, Murphy K (2024) Multichannel sequence analysis in educational research using r. In: Saqr M, López-Pernas S (eds) Learning analytics methods and tutorials: a practical guide using R. Springer, in press
21. Bischl B, Lang M, Bossek J, Horn D, Richter J, Surmann D (2022) BBmisc: miscellaneous helper functions for b. bischl. https://CRAN.R-project.org/package=BBmisc
22. Rosenberg JM, Beymer PN, Anderson DJ, Van Lissa CJ, Schmidt JA (2018) tidyLPA: an r package to easily carry out latent profile analysis (LPA) using open-source or commercial software. J Open Source Softw 3:978. https://doi.org/10.21105/joss.00978
23. Gabadinho A, Ritschard G, Müller NS, Studer M (2011) Analyzing and visualizing state sequences in r with TraMineR. J Stat Softw 40. https://doi.org/10.18637/jss.v040.i04
24. Robette N (2023) Seqhandbook: miscellaneous tools for sequence analysis. https://CRAN.R-project.org/package=seqhandbook
25. Gordon M (2023) Gmisc: descriptive statistics, transition plots, and more. https://CRAN.R-project.org/package=Gmisc

26. Studer M (2013) WeightedCluster library manual: a practical guide to creating typologies of trajectories in the social sciences with r. LIVES. https://doi.org/10.12682/LIVES.2296-1658.2013.24
27. López-Pernas S, Saqr M, Conde J, Del-Río-Carazo L (2024) A broad collection of datasets for educational research training and application. In: Saqr M, López-Pernas S (eds) Learning analytics methods and tutorials: a practical guide using R. Springer, in press
28. Jovanovic J, López-Pernas S, Saqr M (2024) Predictive modelling in learning analytics using R. In: Saqr M, López-Pernas S (eds) Learning analytics methods and tutorials: a practical guide using R. Springer, in press
29. Murphy K, López-Pernas S, Saqr M (2024) Dissimilarity-based clustering educational data using R. In: Saqr M, López-Pernas S (eds) Learning analytics methods and tutorials: a practical guide using R. Springer, in press
30. Alhadabi A, Li J (2020) Trajectories of academic achievement in high schools: growth mixture model. J Educ Issu 6:140. https://doi.org/10.5296/jei.v6i1.16775
31. Shin S, Rachmatullah A, Ha M, Lee J-K (2018) A longitudinal trajectory of science learning motivation in Korean high school students. J Balt Sci Educ 17:674–687. https://doi.org/10.33225/jbse/18.17.674
32. Vanacore K, Dieter K, Hurwitz L, Studwell J (2021) Longitudinal clusters of online educator portal access: connecting educator behavior to student outcomes. In: LAK21: 11th international learning analytics and knowledge conference. ACM. https://doi.org/10.1145/3448139.3448195
33. Salmela-Aro K, Tang X, Symonds J, Upadyaya K (2021) Student engagement in adolescence: a scoping review of longitudinal studies 2010–2020. J Res Adolesc 31:256–272. https://doi.org/10.1111/jora.12619
34. Scrucca L, Saqr M, López-Pernas S, Murphy K (2024) An introduction and r tutorial to model-based clustering in education via latent profile analysis. In: Saqr M, López-Pernas S (eds) Learning analytics methods and tutorials: a practical guide using R. Springer, in press
35. Helske J, Helske S, Saqr M, López-Pernas S, Murphy K (2024) A modern approach to transition analysis and process mining with markov models: a tutorial with R. In: Saqr M, López-Pernas S (eds) Learning analytics methods and tutorials: a practical guide using R. Springer, in press

A Modern Approach to Transition Analysis and Process Mining with Markov Models in Education

Jouni Helske, Satu Helske, Mohammed Saqr, Sonsoles López-Pernas, and Keefe Murphy

1 Introduction

In the previous two chapters, we have learned about sequence analysis [1, 2] and its relevance to educational research. This chapter presents a closely-related method: Markovian models. Specifically, we focus on a particular type of Markovian model, where the data are assumed to be categorical and observed at discrete time intervals, as per the previous chapters about sequence analysis, although in general Markovian models are not restricted to categorical data. One of the main differences between sequence analysis and Markovian modelling is that the former relies on deterministic data mining, whereas the latter uses probabilistic models [3]. Moreover, sequence analysis takes a more holistic approach by analysing sequences as a whole, whereas Markovian modelling focuses on the transitions between states, their probability, and the reasons (covariates) which explain why these transitions happen.

J. Helske (✉)
Department of Mathematics and Statistics, University of Jyväskylä, Jyväskylän yliopisto, Finland

INVEST Research Flagship Centre, University of Turku, Turku, Finland
e-mail: jouni.helske@iki.fi

S. Helske
INVEST Research Flagship Centre, University of Turku, Turku, Finland

Department of Social Research, University of Turku, Turku, Finland

M. Saqr · S. López-Pernas
School of Computing, University of Eastern Finland, Joensuu, Finland

K. Murphy
Department of Mathematics and Statistics, Hamilton Institute, Maynooth University, Maynooth, Ireland

M. Saqr, S. López-Pernas (eds.), *Learning Analytics Methods and Tutorials*,
https://doi.org/10.1007/978-3-031-54464-4_12

We provide an introduction and practical guide to the topic of Markovian models for the analysis of sequence data. While we try to avoid advanced mathematical notations, to allow the reader to continue to other, more advanced sources when necessary, we do introduce the basic mathematical concepts of Markovian models. When doing so, we use the same notation as in the R package 'seqHMM' [4], which we also use in the examples. In particular, we illustrate first-order Markov models, hidden Markov models, mixture Markov models, and mixture hidden Markov models with applications to synthetic data on students' collaboration roles throughout a complete study program.

The chapter proceeds to describe the theoretical underpinnings on each method in turn, then showcases each method with code, before presenting some conclusions and further readings. In addition to the aforementioned applications to collaboration roles and achievement sequences, we also provide a demonstration of the utility of Markovian models in another context, namely process mining. In the process mining application, we leverage Markov models and mixture Markov models to explore learning management system logs. Finally, we conclude with a brief discussion of Markovian models in general and provide some recommendations for further reading of advanced topics in this area as a whole.

2 Methodological Background

2.1 Markov Model

The simple first-order Markov chain or Markov model (MM) can be used to model transitions between successive states. In the first-order MM, given the current observation, the next observation in the sequence is independent of the past—this is called the *Markov property* (the order of MM determines on how many previous observations the next observation depends on). For example, when predicting a student's school success in the fourth year under a first-order model, we only need to consider their success in the third year, while their success in the first and second year give no additional information for the prediction (see Fig. 1 for an illustration). As such, the model is said to be memoryless.

As an example, consider the data described in Table 1 which includes four sequences of length ten. The *alphabet*—that is, the list of all possible states appearing in the data—consists of two types of observed state; low achievement

Fig. 1 Illustration of the Markov Model. The nodes Y_1 to Y_4 refer to states at time points 1 to 4. The arrows indicate dependencies between states

Table 1 Four example sequences of school achievement with individuals A–D across the rows and years 1–10 across the columns

	1	2	3	4	5	6	7	8	9	10
A	L	L	L	H	L	H	L	H	H	H
B	L	H	H	L	H	L	H	L	L	H
C	H	H	L	H	L	L	H	L	H	H
D	H	H	L	L	L	H	L	L	L	H

Table 2 Transition matrix showing the probabilities of transitioning from one state to another (low or high achievement). The rows and columns describe the origin state and the destination state, respectively

	→ Low	→ High
Low →	8/20 = 0.4	12/20 = 0.6
High →	10/16 = 0.625	6/16 = 0.375

success (L) and high achievement success (H). Here, the individuals are assumed to be independent from one another:
Say t describes the position in the sequence, or in this example, the year (in other words, here t runs from 1 to 10). If we assume that the probability of observing L or H at any given point t depends on the previous observation only, we can estimate the *transition probabilities* a_{LL} (from state L to state L), a_{LH} (L to H), a_{HL} (H to L), and a_{HH} (H to H) by calculating the number of observed transitions from each state to all states and scaling these with the total number of transitions from that state. Mathematically, we can write the transition probability a_{rs} from state r to state s as

$$a_{rs} = P(z_t = s \mid z_{t-1} = r), \;\; s, r \in \{L, H\},$$

which simply states that the observed state z_t in year t being L or H depends on which of the two states were observed in the previous year $t-1$. For example, to compute $a_{LH} = P(z_t = H \mid z_{t-1} = L)$, the probability of transitioning from the origin state L to the destination state H, we divide the twelve observed transitions to state H from state L by 20, which is the total number of transitions from L to any state.

The basic MM assumes that the transition probabilities remain constant in time (this property is called *time-homogeneity*). This means, for example, that the probabilities of transitioning from the low-achievement state to the high-achievement state is the same in the ninth year as it was in the second year. We can collect the transition probabilities in a transition matrix (which we call A) which shows all of the possible transition probabilities between each pair of origin and destination states, as illustrated in Table 2. For example, when a student has low achievement in year t, they have a 40% probability to have low achievement in year $t+1$ and a higher 60% probability to transition to high achievement instead, regardless of the year t. Notice that the probabilities in each row must add up to 1 (or 100%).

Lastly, we need to define probabilities for the starting states of the sequences, i.e., the *initial probabilities*

$$\pi_s = P(z_1 = s), \ s \in \{L, H\}.$$

In the example, half of the students have low achievement and the other half have high achievement in the first year, so $\pi_L = \pi_H = 0.5$.

This basic MM is very simple and is often not realistic in the context of educational sciences. We can, however, extend the basic MM in several ways.

First of all, we can include covariates to explain the transition and/or initial probabilities. For example, if we think that transitioning from low to high achievement becomes more challenging as the students get older we may add time as an explanatory variable to the model, allowing the probability of transitioning from low to high achievement to decrease in time. We could also increase the order of the Markov chain, accounting for longer histories. This may be more realistic, but at the same time increasing the order makes the model considerably more complex, the more so the longer the history considered.

Secondly, one of the most useful extensions is the inclusion of hidden (or latent) states that cannot be observed directly but can be estimated from the sequence of observed states. An MM with time-constant hidden states is typically called the mixture Markov model (MMM). It can be used to find latent subpopulations, or in other words, to cluster sequence data. A model with time-varying hidden states is called the hidden Markov model (HMM), which allows the individuals to transition between the hidden states. Allowing for both time-constant and time-varying hidden states leads to a mixture hidden Markov model (MHMM). Unless otherwise specified, from now on when talking about hidden states we refer always to time-varying hidden states, while time-constant hidden states are referred to as clusters.

2.2 Mixture Markov Model

Consider a common case in sequence analysis where individual sequences are assumed to be clustered into subpopulations such as those with typically high and low achievement. In the introductory sequence analysis chapter, the clustering of sequences was performed based on a matrix of pairwise dissimilarities between sequences. Alternatively, we can use the MMM to group the sequences based on their initial and transition probabilities, for example, into those who tend to stay in and transition to high achievement states and those that tend to stay in and transition to low achievement states, as illustrated in Table 3.

In MMMs, we have a separate transition matrix A^k for each cluster k (for $k = 1, \ldots, K$ clusters/subpopulations), and the initial state distribution defines the probabilities to start (and stay) in the hidden states corresponding to a particular cluster. This probabilistic clustering provides group membership probabilities for

Table 3 Two transition matrices showing the probabilities of transitioning from one state of achievement to another in two clusters of Low achievement and High achievement. The rows and columns describe the origin state and the destination state, respectively

(a) Low achievement			(b) High achievement		
Cluster: Low achievement	→ Low	→ High	Cluster: High achievement	→ Low	→ High
Low →	0.8	0.2	Low →	0.6	0.4
High →	0.4	0.6	High →	0.1	0.9

each sequence; these define how likely it is that each individual is a member of each cluster. We can easily add (time-constant) covariates to the model to explain the probabilities of belonging to each cluster. By incorporating covariates in this way we could, for example, find that being in a high-achievement cluster is predicted by gender or family background. However, we note that this is distinct from the aforementioned potential inclusion of covariates to explain the transition and/or initial probabilities.

An advantage of this kind of probabilistic modelling approach is that we can use traditional model selection methods such as likelihood-based information criteria or cross-validation for choosing the best model. For example, if the number of subpopulations is not known in advance—as is typically the case—we can compare models with different clustering solutions (e.g., those obtained with different numbers of clusters, different subsets of covariates, or different sets of initial probabilities, for example) and choose the best-fitting model with, for example, the Bayesian information criterion (BIC) [5].

2.3 *Hidden Markov Model*

The HMM can be useful in a number of cases when the state of interest cannot be directly measured or when there is measurement error in the observations. In HMMs, the Markov chain operates at the level of hidden states, which subsequently generate or emit observed states with different probabilities. For example, think about a progression of a student's ability as a hidden state and school success as the observed state. We cannot measure true ability directly, but we can estimate the student's progress by their test scores that are emissions of their ability. There is, however, some uncertainty in how well the test scores represent students' true ability. For example, observing low test scores at some point in time does not necessarily mean the student has low ability; they might have scored lower than expected in the test due to other reasons such as being sick at that particular time. Such uncertainty can be reflected in the emission probabilities; for example, in the high-ability state students get high test scores eight times out of ten and low test scores with a 20% probability, while in the low-ability state the students get low test scores nine times out of ten and high test scores with a 10% probability. These probabilities are collected in an emission matrix as illustrated in Table 4.

Table 4 Emission matrix showing the probabilities of each hidden state (low or high ability) emitting each observed state (low or high test scores)

	Low scores	High scores
Low ability	0.9	0.1
High ability	0.2	0.8

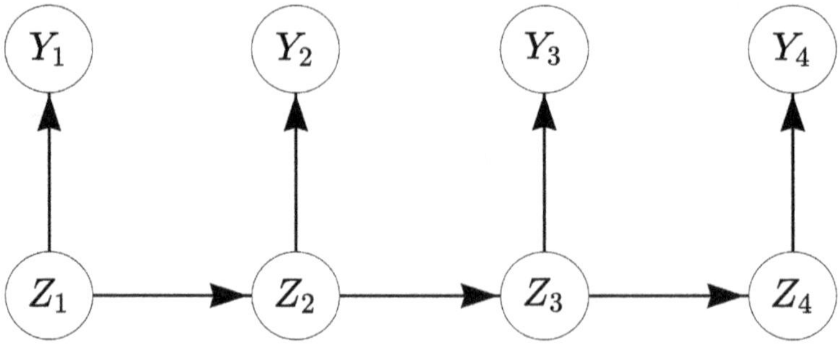

Fig. 2 Illustration of the HMM. The nodes Z_1 to Z_4 refer to hidden states at time points 1 to 4, while the nodes Y_1 to Y_4 refer to observed states. The arrows indicate dependencies between hidden and/or observed states

Again, the full HMM is defined by a set of parameters: the initial state probabilities π_s, the hidden state transition probabilities a_{rs}, and the emission probabilities of observed states $b_s(m)$. What is different to the MM is that in the HMM, the initial state probabilities π_s define the probabilities of starting from each *hidden* state. Similarly, the transition probabilities a_{rs} define the probabilities of transitioning from one *hidden* state to another hidden state. The emission probabilities $b_s(m)$ (collected in an emission matrix B) define the probability of observing a particular state m (e.g., low or high test scores) given the current hidden state s (e.g., low or high ability).

When being in a certain hidden state, observed states occur randomly, following the emission probabilities. Mathematically speaking, instead of assuming the Markov property directly on our observations, we assume that the observations are conditionally independent given the underlying hidden state. We can visualise the HMM as a directed acyclic graph (DAG) illustrated in Fig. 2. Here Z are the unobserved states (such as ability) which affect the distribution of the observed states Y (test scores). At each time point t, the state z_t can obtain one of S possible values (there are two hidden states in the example of low and high ability, so $S = 2$), which in turn defines how Y_t is distributed.

2.4 Mixture Hidden Markov Models

Combining the ideas of both time-constant clusters and time-varying hidden states leads to the concept of mixture hidden Markov model (MHMM). Here we assume

Table 5 Two transition matrices showing the probabilities of transitioning from one state of ability to another in two clusters, the Stayers and the Movers. The rows and columns describe the origin state and the destination state, respectively

(a) Stayers			(b) Movers		
Cluster: Stayers	→ Low	→ High	Cluster: Movers	→ Low	→ High
Low →	1	0	Low →	0.6	0.4
High →	0	1	High →	0.3	0.7

Table 6 Two emission matrices showing the probabilities of each hidden state (low or high ability) emitting each observed state (low or high test scores)

(a) Stayers			(b) Movers		
Cluster: Stayers	Low scores	High scores	Cluster: Movers	Low scores	High scores
Low ability	0.9	0.1	Low ability	0.7	0.3
High ability	0.1	0.9	High ability	0.2	0.8

that the population of interest consists of a finite number of subpopulations, each with their own HMM with varying transition and emission probabilities. For example, we could expect to find underlying groups which behave differently when estimating the progression of ability through the sequence of test scores, such as those that consistently stay on a low-ability or high-ability track (stayers) and those that move between low and high ability (movers). In this case, we need two transition matrices: the stayers' transition matrix allows for no transitions while the movers' transition matrix allows for transitioning between low and high ability, as illustrated in Table 5.

Similarly, we need two emission matrices that describe how the observed states are related to hidden states, as illustrated in Table 6. In this example, there is a closer match between low/high ability and low/high test scores in the Stayers cluster in comparison to the Movers cluster.

Mathematically, when estimating a MHMM we first fix the number of clusters K, and create a joint HMM consisting of K submodels (HMMs). The number of hidden states does not have to be fixed but can vary by submodel, so that the HMMs have more hidden states for some clusters and fewer for others (in our example, because the transition matrix of the Stayers cluster is diagonal, we could also split the cluster into two single state clusters, one corresponding to low and another to high ability). This can increase the burden of model selection, so often a common number of hidden states is assumed for each cluster for simplicity. In any case, the initial state probabilities of this joint model define how sequences are assigned to different clusters. We estimate this joint model using the whole data and calculate cluster membership probabilities for each individual. The idea of using mixtures of HMMs has appeared in literature under various names with slight variations, e.g., [6], [7], and [4]. Notably, MHMMs inherit from MMMs the ability to incorporate covariates to predict cluster memberships.

2.5 Multi-Channel Sequences

There are two options to analyse multi-channel (or multi-domain or multi-dimensional) sequence data with Markovian models. The first option is to combine observed states in different channels into one set of single-channel sequences with an expanded alphabet. This option is simple, and works for MMs, HMMs, MMMs, and MHMMs, but can easily lead to complex models as the number of states and channels increases considerably. The second option, which can only be used when working with HMMs and MHMMs, is to treat the observed states in each channel independently given the current hidden state. This can be easily performed by defining multiple emission probability matrices, one for each channel. The assumption of conditional independence simplifies the model, but is sometimes unrealistic, in which case it is better to resort to the first option and convert the data into single-channel sequences. Both options are discussed further in Chapter 13 [8], a dedicated chapter on multi-channel sequences, where applications of distance-based and Markovian clustering approaches are presented. In this chapter, we henceforth focus on single-channel sequences.

2.6 Estimating Model Parameters

The model parameters, i.e. the elements of the initial probability vectors π, transition probability matrices A, and emission probability matrices B, can be estimated from data using various methods. Typical choices are the Baum-Welch algorithm (an instance of the expectation-maximisation, i.e., the EM algorithm) and direct (numerical) maximum likelihood estimation. It is possible to restrict models, for example, by setting some parameters to fixed values (typically zeros), for example, to make certain starting states, transitions, or emissions impossible.

After the parameter estimation, in addition to studying the estimated model parameters upon convergence, we can, for example, compute cluster-membership probabilities for each individual and find the most probable paths of hidden state sequences using the Viterbi algorithm [9]. These can be further analysed and visualised for interpretation.

3 Review of the Literature

Markovian methods have been used across several domains in education and have gained renewed interest with the surge in learning analytics and educational data mining. Furthermore, the introduction of specialised R packages (e.g., `seqHMM` [10]) and software applications (e.g., Mplus [11, 12]) have made it easier to implement Markov models. One of the most common applications of Markovian methods is the clustering of sequence data [13–15]. Markov models offer a credible alternative to existing distance-based methods (e.g. optimal matching) and can be used with different sequence types (e.g. multi-channel sequences). Furthermore, Markovian methods offer some advantages in clustering sequential data such as the inclusion of covariates that can explain why a sequence emerged (e.g., [16]). More importantly, Markovian models are relatively scalable and can be used to cluster large sequences [17]. As Saqr et al. [17] noted, large sequences are hard to cluster using standard methods such as hierarchical clustering, which is memory inefficient, and hard to parallelise or scale [18, 19]. Furthermore, distance-based clustering methods are limited by the theoretical maximum dimension of a matrix in R which is 2,147,483,647 (i.e., a maximum of 46,430 sequences). In such a case, Markovian methods may be the solution.

Examples of Markovian methods in clustering sequences are plentiful. For example, HMMs have been used to cluster students' sequences of learning management system (LMS) trace data to detect their patterns of activities or what the authors referred to as learning tactics and strategies [15]. Another close example was that of López-Pernas and Saqr [20], who used HMMs to cluster multi-channel data of students' learning strategies of two different tools (an LMS and an automated assessment tool). Other examples include using HMM in clustering sequences of students' engagement states [21], sequences of students' collaborative roles [16], or sequences of self-regulation [13, 14].

Markovian methods are also popular in studying transitions and have therefore been used across several applications and with different types of data. One of the most common usages is what is known as stochastic processes mining which typically uses first-order Markov models to map students' transitions between learning activities. For example, Matcha et al. [22] used first-order Markov models to study students' processes of transitions between different learning tactics. Other uses include studying the transitions between tactics of academic writing [23], between self-regulated learning events [24], or within collaborative learning settings [25]. Yet, most of such work has been performed by the `pMiner` R package [26], which was recently removed from The Comprehensive R Archive Network (CRAN) due to slow updates and incompatibility with existing guidelines. This chapter offers a modern alternative that uses modern and flexible methods for fitting, plotting, and clustering stochastic process mining models as well as the possibility to add covariates to understand "why" different transitions pattern emerged.

Indeed, transition analysis in general has been a popular usage for Markovian models and has been used across several studies. For instance, for the analysis of temporal patterns of students' activities in online learning (e.g., [27]), or transitions between latent states [28], or transitions between assignment submission patterns [29].

4 Examples

As a first step, we will import all the packages required for our analyses. We have used most of them throughout the book. Below is a brief summary:

- qgraph: A package for visualising networks, which can be used to plot transition probabilities [30]. This is used only for the process mining application in Sect. 4.3.
- rio: A package for reading and saving data files with different extensions [31].
- seqHMM: A package designed for fitting hidden (latent) Markov models and mixture hidden Markov models for social sequence data and other categorical time series [4, 32].
- tidyverse: A package that encompasses several basic packages for data manipulation and wrangling [33].
- TraMineR: As seen in the introductory sequence analysis chapter, this package helps us construct, analyze, and visualise sequences from time-ordered states or events [34].

```
library(qgraph)
library(rio)
library(seqHMM)
library(tidyverse)
library(TraMineR)
```

Henceforth, we divide our examples into two parts: the first largely focuses on traditional uses of the seqHMM package to fit Markovian models of varying complexity to sequence data; the latter presents a demonstration of Markovian models from the perspective of process mining. We outline the steps involved in using seqHMM in general in Sect. 4.1, demonstrate the application of MMs, HMMs, MMMs, and MHMMs in Sect. 4.2, and explore process mining using Markovian models in Sect. 4.3, leveraging much of the steps and code from the previous two sections. We note that different datasets are used in Sects. 4.2 and 4.3; we begin by importing the data required for Sect. 4.2 and defer the importing of the data used in the process mining application to the later section.

With this in mind, we start by using the import() function from the rio package to import our sequence data. Based on the description of the MHMM in [35], we used the seqHMM package to simulate a synthetic dataset (simulated_data) consisting of students' collaboration roles (obtained from [36]) on different courses across a whole program. As the original data, the simulation was based on the two-channel model (collaboration and achievement), but we only use the collaboration sequences in the following examples, and leave the multi-channel sequence analysis to Chapter 13 [8]. While not available in the original study, we also simulated students' high school grade point average (GPA, for simplicity categorised into three levels) for each student, which will be used to predict cluster memberships. Using this data, we show how the `seqHMM` package can be used to analyse such sequences. We start with the simple MM, and then transition to HMMs and their mixtures. To be able to use the `seqHMM` functions, we need to convert the imported data to a sequence using the function `seqdef()` from the `TraMineR` package (see Chapter 10 [1] for more information about creating `stslist` objects). We subsequently assign a colour palette to each state in the alphabet for later visualisations using the function `cpal()`. Finally, we can also extract the covariate information separately (`cov_data`).

```
URL <- "https://github.com/sonsoleslp/labook-data/raw/main/"
simulated_data <- import(
   paste(URL, "12_longitudinalRoles/simulated_roles.csv"))

roles_seq <- seqdef(
  simulated_data,
  var = 3:22,
  alphabet = c("Isolate", "Mediator", "Leader"),
  cnames = 1:20
)
```

```
 [>] 3 distinct states appear in the data:

     1 = Isolate

     2 = Leader

     3 = Mediator

 [>] state coding:

       [alphabet]  [label]  [long label]

     1  Isolate     Isolate  Isolate

     2  Mediator    Mediator Mediator

     3  Leader      Leader   Leader

 [>] 200 sequences in the data set

 [>] min/max sequence length: 20/20
```

```
cpal(roles_seq) <- c("#FBCE4B", "#F67067", "#5C2262")

cov_data <- simulated_data |>
  select(ID, GPA) |>
  mutate(GPA = factor(GPA, levels = c("Low", "Middle", "High")))
```

4.1 Steps of Estimation

We will first briefly introduce the steps of the analysis with the `seqHMM` package and then show examples of estimating MMs, HMMs, MMMs, and MHMMs.

4.1.1 Defining the Model Structure

First, we need to create the model object which defines the structure of the model. This can be done by using one of the model building functions of `seqHMM`. The build functions include `build_mm()` for constructing the simple MM, `build_hmm()` for the HMM, `build_mmm()` for the MMM, and `build_mhmm()` for the MHMM. The user needs to give the build function the sequence data and the number of hidden states and/or clusters (when relevant). The user can also set restrictions on the models, for example, to forbid some transitions by setting the corresponding transition probabilities to zero. To facilitate the estimation of the parameters of more complex models, the user may also set informative starting values for model parameters.

4.1.2 Estimating the Model Parameters

After defining the model structure, model parameters need to be estimated. The `fit_model()` function estimates model parameters using maximum likelihood estimation. The function has several arguments for configuring the estimation algorithms. For simple models the default arguments tend to work well enough, but for more complex models the user should adjust the algorithms. This is because the more parameters the algorithm needs to estimate, the higher the risk of not finding the model with the optimal parameter values (the ones which maximises the likelihood).

In order to reduce the risk of being trapped in a local optimum of the likelihood surface (instead of a global optimum), we advise to estimate the model numerous times using different starting values for the parameters. The `seqHMM` package strives to automate this. One option is to run the EM algorithm multiple times with random starting values for any or all of initial, transition, and emission probabilities. These are specified in the `control_em` argument. Although not done by default, this method seems to perform very well as the EM algorithm is relatively fast. Another option is to use a global direct numerical estimation method such as the multilevel single-linkage method. See [4] for more detailed information on model estimation.

4.1.3 Examining the Results

The output of the `fit_model()` contains the estimated model (stored in `fit_hmm$model`) as well as information about the estimation of the model, such as the log-likelihood of the final model (`fit_hmm$logLik`). The `print()` method provides information about the estimated model in a written format, while the `plot()` method visualises the model parameters as a graph. For HMMs and MHMMs, we can calculate the most probable sequence of hidden states for each individual with the `hidden_paths()` function. Sequences of observed and hidden state sequences can be plotted with the `ssplot()` function for MMs and HMMs and with the `mssplot()` function for the MMMs and the MHMMs. For MMMs and MHMMs, the `summary()` method automatically computes some features of the models such as standard errors for covariates and prior and posterior cluster membership probabilities for the subjects.

4.2 Markov Models

We now follow the steps outlined above for each model in turn, starting from the most basic Markov model, proceeding through a hidden Markov model and a mixture Markov model, and finally concluding with a mixture hidden Markov model.

4.2.1 Markov Model

We focus on the sequences of collaboration roles, collected in the `roles_seq` object. The `build_mm()` function only takes one argument, `observations`, which should be an `stslist` object created with the `seqdef()` function from the `TraMineR` package as mentioned before. We can build a MM as follows:

```
markov_model <- build_mm(roles_seq)
```

For the MM, the `build_mm()` function estimates the initial probabilities and the transition matrix. Note that the `build_mm()` function is the only build function that automatically estimates the parameters of the model. This is possible because for the MM the estimation is a simple calculation while for the other types of models the estimation process is more complex. The user can access the estimated parameters by calling `markov_model$initial_probs` and `markov_model$transition_probs` or view them by using the `print` method of the model:

```
print(markov_model)
```

```
Initial probabilities :
 Isolate Mediator   Leader
   0.375    0.355    0.270

Transition probabilities :
          to
from       Isolate Mediator Leader
  Isolate   0.4231    0.478 0.0987
  Mediator  0.1900    0.563 0.2467
  Leader    0.0469    0.428 0.5252
```

We can see that the initial state probabilities are relatively uniform, with a slightly lower probability for starting in the Leader state. In terms of the transition probabilities, the most distinct feature is that that it is rare to transition directly from the Leader state to Isolate and vice versa (estimated probabilities are about 5% and 10%, respectively). It is also more common to drop from Leader to Mediator (43%) than to increase collaboration from Mediator to Leader (25%). Similarly, the probability of moving from Mediator to Isolate is only 19%, but there is a 48% chance of transitioning from Isolate to Mediator.

We can also draw a graph of the estimated model using the `plot()` method which by default shows the states as pie graphs (for the MM, the pie graphs only consist of one state), transition probabilities as directed arrows, and initial probabilities below each state (see Fig. 3).

```
plot(markov_model,
  legend.prop = 0.2, ncol.legend = 3,
  edge.label.color = "black", vertex.label.color = "black")
```

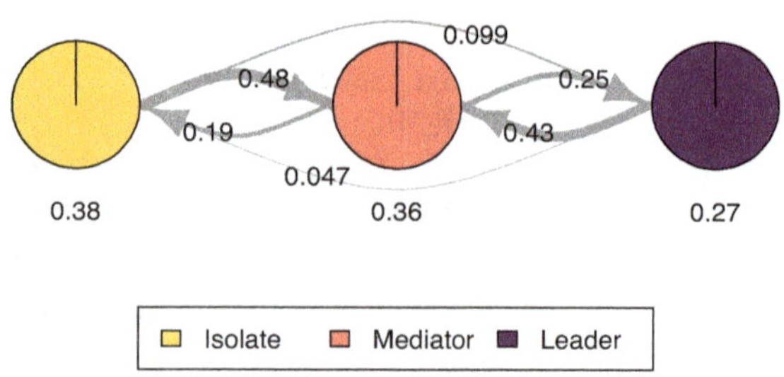

Fig. 3 Estimated Markov model as a pie charts with the transition probabilities shown as labelled edges

4.2.2 Hidden Markov Models

The structure of an HMM is set with the `build_hmm()` function. In contrast to `build_mm()`, other `build_*()` functions such as `build_hmm()` do not directly estimate the model parameters. For `build_hmm()`, in addition to observations (an `stslist`), we need to provide the `n_states` argument which tells the model how many hidden states to construct. Using again the collaboration roles sequences, if we want to estimate an HMM with two hidden states, we can write:

```
set.seed(1)
hidden_markov_model <- build_hmm(
   observations = roles_seq,
   n_states = 2
)
```

The `set.seed()` call ensures that we will always end up with the same exact initial model with hidden states in the same exact order even though we use random values for the initial parameters of the model (which is practical for reproducibility). We are now ready to estimate the model with the `fit_model()` function. The HMM we want to estimate is simple, so we rely on the default values and again use the `print()` method to provide information about the estimated model:

```
fit_hmm <- fit_model(hidden_markov_model)
fit_hmm$model
```

```
Initial probabilities :
State 1 State 2
  0.657   0.343

Transition probabilities :
         to
from      State 1 State 2
  State 1  0.9089  0.0911
  State 2  0.0391  0.9609

Emission probabilities :
           symbol_names
state_names Isolate Mediator Leader
    State 1  0.4418    0.525 0.0336
    State 2  0.0242    0.478 0.4980
```

The estimated initial state probabilities show that it is more probable to start from hidden state 1 than from hidden state 2 (66% vs. 34%). The high transition

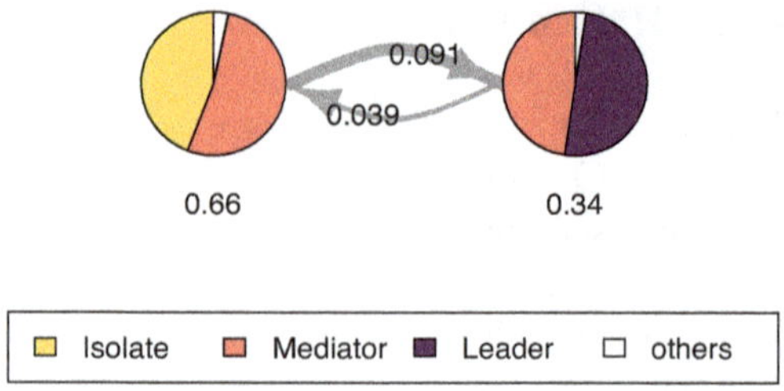

Fig. 4 HMM with two hidden states (pie charts), with transitions between hidden states shown as labelled edges

probabilities on the diagonal of the transition matrix indicate that the students typically tend to stay in the hidden state they currently are in. Transition probabilities between the hidden states are relatively low and also asymmetric: it is more likely that students move from state 1 to state 2 than from state 2 to state 1. Looking at the emission matrices, we see that the role of the students in state 2 is mostly Leader or Mediator (emission probabilities are 50% and 48%). On the other hand, state 1 captures more of those occasions where students are isolated or exhibit at most a moderate level of participation (mediators). We can also visualise this with the `plot()` method of `seqHMM` (see Fig. 4):

```
plot(fit_hmm$model,
  ncol.legend = 4, legend.prop = 0.2,
  edge.label.color = "black", vertex.label.color = "black"
)
```

The plot values mainly shows the same information. By default, to simplify the graph, the plotting method combines all states with less than 5% emission probabilities into one category. This threshold can be changed with the `combine.slices` argument (setting `combine.slices = 0` plots all states).

For simple models, using `n_states` is sufficient. It automatically draws random starting values that are then used for the estimation of model parameters. However, as parameter estimation of HMMs and mixture models can be sensitive to starting values of parameters, it may be beneficial to provide starting values manually using the `initial_probs`, `transition_probs`, and `emission_probs` arguments. This is also necessary in case we want to define structural zeros for some of these components, e.g., if we want to restrict the initial probabilities so that each sequence starts from the same hidden state, or if we want to set the lower diagonal part of the transition matrix to zero, which means that the model does not allow transitioning back to previous states (this is called a left-to-right model) [4]. It is also possible to mix random and user-defined starting values by using `simulate_*()` functions (e.g. `simulate_transition_probs()`) for some of the model components and user-defined values for others.

In the following example we demonstrate estimating a three-state HMM with user-defined starting values for the initial state probabilities and the transition matrix but simulate starting values for the emission matrices. For simulating starting values with `simulate_emission_probs()`, we need to define the number of hidden states, and the number of observed symbols, i.e., the length of the alphabet of the sequences.

```
# Set seed for randomisation
set.seed(1)

# Initial state probability vector, must sum to one
init_probs <- c(0.3, 0.4, 0.3)

# a 3x3 transition matrix, each row should sum to one
trans_probs <- rbind(
   c(0.8, 0.15, 0.05),
   c(0.2, 0.6, 0.2),
   c(0.05, 0.15, 0.8)
   )
# Simulate emission probabilities
emission_probs <- simulate_emission_probs(
  n_states = 3, n_symbols = length(alphabet(roles_seq))
)

# Build the HMM
hidden_markov_model_2 <- build_hmm(
  roles_seq, initial_probs = init_probs,
  transition_probs = trans_probs,
  emission_probs = emission_probs
)
```

Our initial probabilities suggest that it is slightly more likely to start from the second hidden state than the first and the third. Furthermore, the starting values for the transition matrices suggest that staying in hidden states 1 and 3 is more likely than staying in hidden state 2. All non-zero probabilities are, however, mere suggestions and will be estimated with the `fit_model()` function. We now estimate this model 50 times with the EM algorithm using randomised starting values:

```
set.seed(1)
fit_hmm_2 <- fit_model(hidden_markov_model_2,
  control_em = list(restart = list(times = 50))
)
```

We can get the information on the EM estimation as follows:

```
fit_hmm_2$em_results
```

```
$logLik
[1] -3546.155

$iterations
[1] 488

$change
[1] 9.947132e-11

$best_opt_restart
 [1] -3546.155 -3546.155 -3546.155 -3546.155 -3546.155 -3546.155 -3546.155
 [8] -3546.155 -3546.155 -3546.155 -3546.155 -3546.155 -3546.155 -3546.155
[15] -3546.155 -3546.155 -3546.155 -3546.155 -3546.155 -3546.155 -3546.155
[22] -3546.155 -3546.155 -3546.155 -3546.155
```

The `loglik` element gives the log-likelihood of the final model. This value has no meaning on its own, but it can be used to compare HMMs with the same data and model structure (e.g., when estimating the same model from different starting values). The `iterations` and `change` arguments give information on the last EM estimation round: how many iterations were used until the (local) optimum was found and what was the change in the log-likelihood at the final step.

The most interesting element is the last one: `best_opt_restart` shows the likelihood for 25 (by default) of the best estimation rounds. We advise to always check these to make sure that the best model was found several times from different starting values: this way we can be fairly certain that we have found the actual maximum likelihood estimates of the model parameters (global optimum). In this case all of the 25 log-likelihood values are identical, meaning that it is likely that we have found the best possible model among all HMMs with three hidden states.

```
plot(fit_hmm_2$model,
  legend.prop = 0.15, ncol.legend = 3,
  edge.label.color = "black", vertex.label.color = "black",
  combine.slices = 0, trim = 0.0001
)
```

Interpreting the results in Fig. 5 we see that the first hidden state represents about equal amounts of isolate and mediator roles, the second hidden state represents mainly Leaders and some Mediator roles, and the third hidden state represents mainly Mediator roles and partly Leader roles. Interestingly, none of the students start as Mediator/Leader, while of the other two the Isolate/Mediator state is more typical (two thirds). There are no transitions from the first to the second state nor vice versa, and transition probabilities to the third state are considerably higher than away from it. In other words, it seems that the model has two different origin states and one destination state.

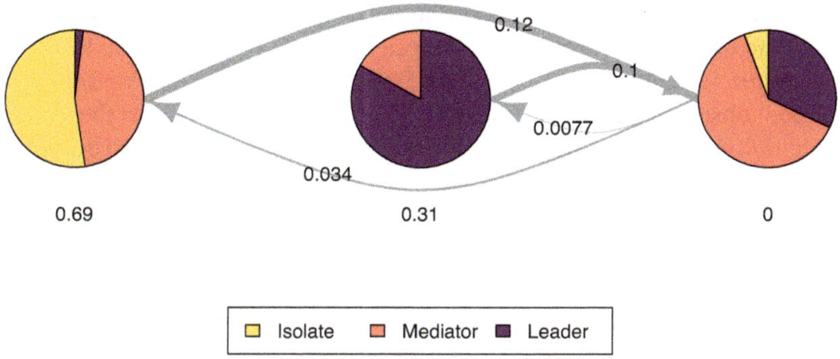

Fig. 5 HMM with three hidden states (pie charts), with transitions between hidden states shown as labelled edges

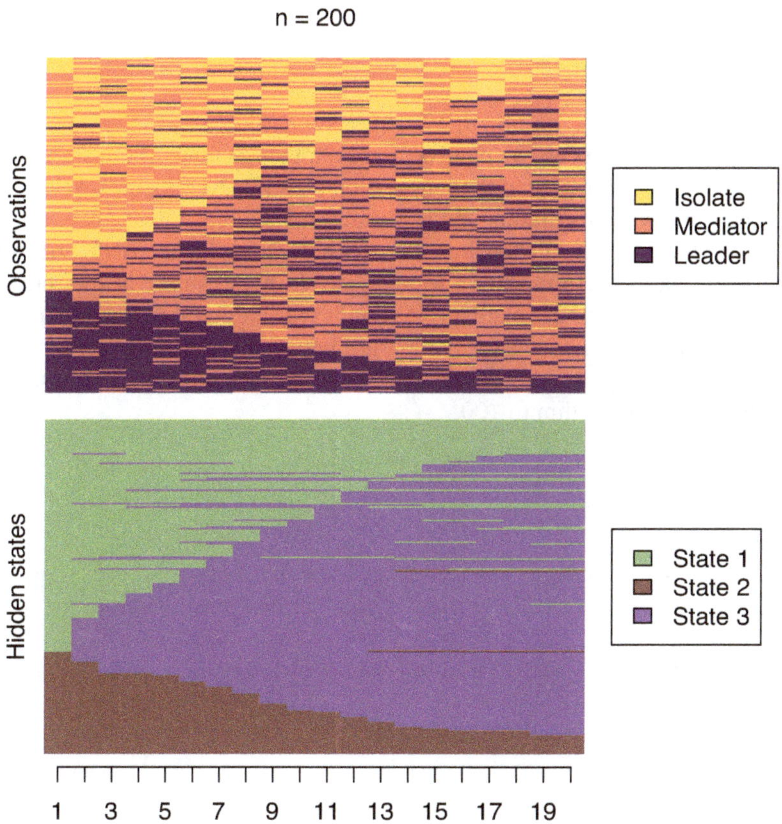

Fig. 6 Observed and hidden state sequences from the HMM with three hidden states

We can visualise the observed and/or hidden state sequences with the `ssplot()` function. The `ssplot()` function can take an `stslist` object or a model object of class `mm` or `hmm` (see Fig. 6). Here we want to plot full sequence index plots (`type = "I"`) of both observed and hidden states (`plots = "both"`) and sort the sequences using multidimensional scaling of hidden states (`sortv = "mds.hidden"`). See the `seqHMM` manual and visualisation vignette for more information on the different plotting options.

```
ssplot(fit_hmm_2$model,
  # Plot sequence index plot (full sequences)
  type = "I",
  # Plot observed and hidden state sequences
  plots = "both",
  # Sort sequences by the scores of multidimensional scaling
  sortv = "mds.hidden",
  # X axis tick labels
  xtlab = 1:20
)
```

By looking at the sequences, we can see that even though none of the students start in hidden state 3, the majority of them transition there. In the end, most students end up alternating between mediating and leadership roles.

Is the three-state model better than the two-state model? As already mentioned, we can use model selection criteria to test that. To make sure that the three-state model is the best, we also estimate a HMM with four hidden states and then use the Bayesian information criterion for comparing between the three models. Because the four-state model is more complex, we increase the number of re-estimation rounds for the EM algorithm to 100.

```
# Set seed for randomisation
set.seed(1)

# Build and estimate a HMM with four states
hidden_markov_model_3 <- build_hmm(roles_seq, n_states = 4)

fit_hmm_3 <- fit_model(hidden_markov_model_3,
  control_em = list(restart = list(times = 100))
)

fit_hmm_3$em_results$best_opt_restart
```

```
 [1] -3534.304 -3534.304 -3534.304 -3534.304 -3534.304 -3534.304 -3534.304
 [8] -3534.304 -3534.304 -3534.304 -3534.304 -3534.304 -3534.304 -3534.305
[15] -3534.305 -3534.306 -3534.308 -3534.310 -3534.332 -3534.335 -3534.335
[22] -3534.335 -3534.336 -3534.337 -3534.337
```

The best model was found only 13 times out of 101 estimation rounds from randomised starting values. A cautious researcher might be wise to opt for a higher number of estimation rounds for increased certainty, but here we will proceed to calculating the BIC values.

```
BIC(fit_hmm$model)
```

```
[1] 7430.028
```

```
BIC(fit_hmm_2$model)
```

```
[1] 7208.427
```

```
BIC(fit_hmm_3$model)
```

```
[1] 7259.37
```

Generally speaking, the lower the BIC, the better the model. We can see that the three-state model (`fit_hmm_2`) has the lowest BIC value, so three clusters is the best choice (at least among HMMs with 2–4 hidden states).

4.2.3 Mixture Markov Models

The MMM can be defined with the `build_mmm()` function. Similarly to HMMs, we need to either give the number of clusters with the `n_clusters` argument, which generates random starting values for the parameter estimates, or give starting values manually as `initial_probs` and `transition_probs`. Here we use random starting values:

```
# Set seed for randomisation
set.seed(123)
# Define model structure (3 clusters)
mmm <- build_mmm(roles_seq, n_clusters = 3)
```

Again, the model is estimated with the `fit_model()` function:

```
fit_mmm <- fit_model(mmm)
```

The results for each cluster can be plotted one at a time (interactively, the default), or in one joint figure. Here we opt for the latter (see Fig. 7). At the same time we also illustrate some other plotting options:

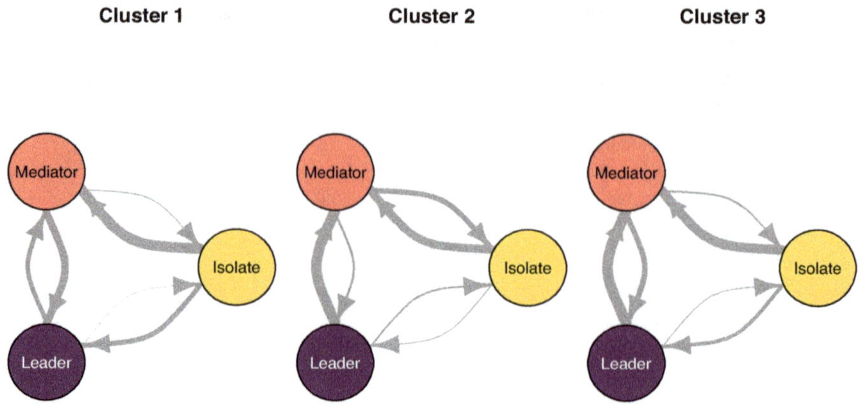

Fig. 7 MMM with three clusters

```
plot(fit_mmm$model,
  # Plot all clusters at the same time
  interactive = FALSE,
  # Set the number of rows (1) and columns (3) for cluster plots
  nrow = 1, ncol = 3,
  # Omit legends
  with.legend = FALSE,
  # Choose another layout for the vertices (see plot.igraph)
  layout = layout_in_circle,
  # Omit pie graphs from vertices
  pie = FALSE,
  # Set state colours
  vertex.label.color = c("black", "black", "white"),
  # Set state label colours
  vertex.color = cpal(roles_seq),
  # Increase the size of the circle
  vertex.size = 80,
  # Plot state labels instead of initial probabilities
  vertex.label = "names",
  # Set state label in the centre of the circle
  vertex.label.dist = 0,
  # Omit labels for transition probabilities
  edge.label = NA
)
```

The following code plots the sequence distribution plot of each cluster (Fig. 8). In Cluster 1, we see low probabilities to downward mobility and high probabilities for upward mobility, so this cluster describes leadership trajectories. In Cluster 2, we can see that the thickest arrows lead to mediator and isolates roles, so this cluster describes trajectories with less central roles in collaboration. In Cluster 3, we see the highest transition probabilities for entering the mediator role but also

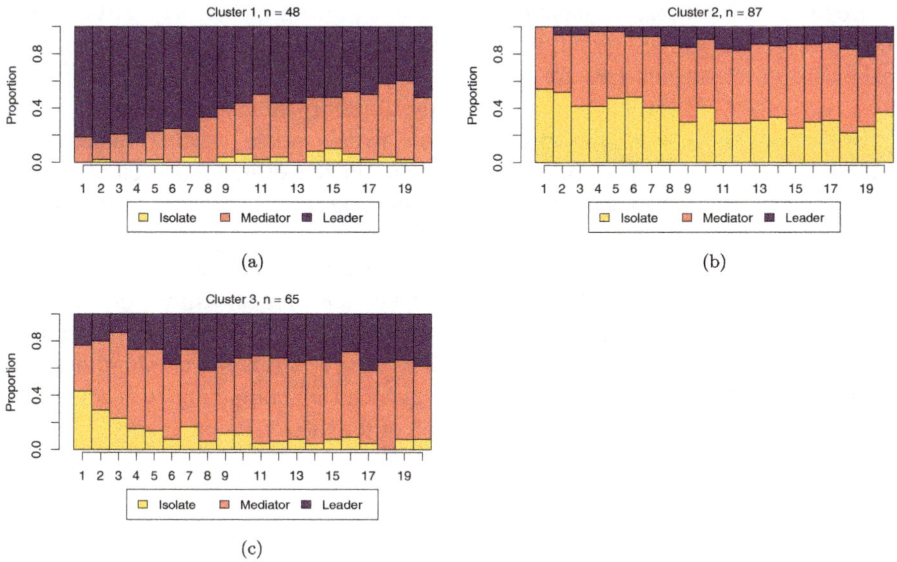

Fig. 8 State distribution plots by most probable clusters estimated with the mixture Markov model. (**a**) Cluster 1. (**b**) Cluster 2. (**c**) Cluster 3

some transitions from mediator to leader, so this cluster describes trajectories with more moderate levels of participation in comparison to cluster 1. This behavior is easier to see when visualising the sequences in their most probable clusters. The plot is interactive, so we need to hit 'Enter' on the console to generate each plot. Alternatively, we can specify which cluster we want to plot using the `which.plots` argument.

```
cl1 <- mssplot(fit_mmm$model,
  # Plot Y axis
  yaxis = TRUE,
  # Legend position
  with.legend = "bottom",
  # Legend columns
  ncol.legend = 3,
  # Label for Y axis
  ylab = "Proportion"
)
```

We can add covariates to the model to explain cluster membership probabilities. For this, we need to provide a data frame (argument `data`) and the corresponding formula (argument `formula`). In the example data we use the data frame called `cov_data` that we created at the beginning of the tutorial with columns `ID` and `GPA`, where the order of the ID variable matches to that of the sequence data roles_seq (note that the ID variable is not used in the model building, so the user needs to make sure that both matrices are sorted by ID). We can now use the information about students' GPA level as a predictor of the cluster memberships.

Numerical estimation of complex models from random starting values may lead to convergence issues and other problems in the estimation (you may, for example, get warnings about the EM algorithm failing). To avoid such issues, giving informative starting values is often helpful. This model is more complex than the model without covariates and estimation from random starting values leads to convergence issues (not shown here). To facilitate model estimation, we use the results from the previous MMM as informative starting values. Here we also remove the common intercept by adding 0 to the formula, which simplifies the interpretation of the covariate effects later (instead of comparing to a reference category, we get separate coefficients for each of the three GPA categories).

```
set.seed(98765)
mmm_2 <- build_mmm(
  roles_seq,
  # Starting values for initial probabilities
  initial_probs = fit_mmm$model$initial_probs,
  # Starting values for transition probabilities
  transition_probs = fit_mmm$model$transition_probs,
  # Data frame for covariates
  data = cov_data,
  # Formula for covariates (one-sided)
  formula = ~ 0 + GPA
)
```

Again, the model is estimated with the fit_model() function. Here we use the EM algorithm with 50 restarts from random starting values:

```
set.seed(12345)
fit_mmm_2 <- fit_model(
  mmm_2,
  # EM with randomised restarts
  control_em = list(
    restart = list(
      # 50 restarts
      times = 50,
      # Store loglik values from all 50 + 1 estimation rounds
      n_optimum = 51
    )
  )
)
```

```
Warning in fit_model(mmm_2, control_em = list(restart
= list (times = 50, : EM algorithm failed: Estimation of
gamma  coefficients  failed due to singular Hessian.
```

The model was estimated 50 + 1 times (first from the starting values we provided and then from 50 randomised values). We get one warning about the EM algorithm failing. However, 50 estimation rounds were successful. We can check that the best model was found several times from different starting values (37 times, to be precise):

```
fit_mmm_2$em_results$best_opt_restart
```

```
 [1] -3614.627 -3614.627 -3614.627 -3614.627 -3614.627 -3614.627 -3614.627
 [8] -3614.627 -3614.627 -3614.627 -3614.627 -3614.627 -3614.627 -3614.627
[15] -3614.627 -3614.627 -3614.627 -3614.627 -3614.627 -3614.627 -3614.627
[22] -3614.627 -3614.627 -3614.627 -3614.627 -3614.627 -3614.627 -3614.627
[29] -3614.627 -3614.627 -3614.627 -3614.627 -3614.627 -3614.627 -3614.627
[36] -3614.627 -3614.627 -3619.695 -3624.547 -3624.547 -3624.547 -3624.547
[43] -3624.547 -3624.547 -3624.547 -3624.547 -3624.547 -3624.547 -3631.328
[50] -3637.344      -Inf
```

We can now be fairly certain that the optimal model has been found, and can proceed to interpreting the results. The clusters are very similar to what we found before. We can give the clusters more informative labels and then show state distribution plots in each cluster in Fig. 9:

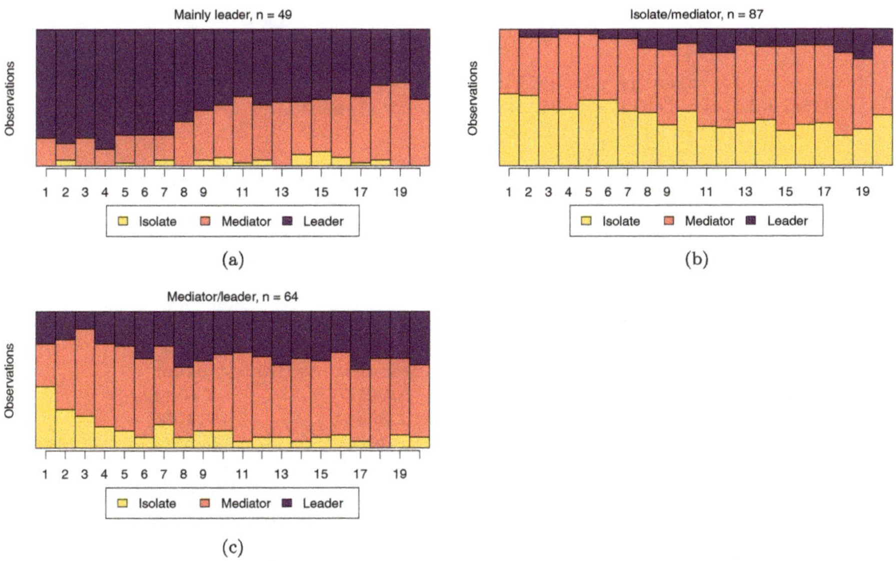

Fig. 9 State distribution plots by most probable clusters estimated with the mixture Markov model with covariates. (**a**) Mainly leader. (**b**) Isolate/mediator. (**c**) Mediator/leader

```r
cluster_names(fit_mmm_2$model) <- c(
  "Mainly leader", "Isolate/mediator", "Mediator/leader"
)
mssplot(fit_mmm_2$model, with.legend = "bottom", ncol.legend = 3)
```

The model summary shows information about parameter estimates of covariates and prior and posterior cluster membership probabilities (these refer to cluster membership probabilities before or after conditioning on the observed sequences, respectively):

```r
summary_mmm_2 <- summary(fit_mmm_2$model)
summary_mmm_2
```

```
Covariate effects :
Mainly leader is the reference.

Isolate/mediator :
           Estimate  Std. error
GPALow       1.9221       0.478
GPAMiddle    0.3901       0.314
GPAHigh     -0.0451       0.277

Mediator/leader :
           Estimate  Std. error
GPALow        1.670       0.487
GPAMiddle     0.411       0.312
GPAHigh      -0.667       0.332

Log-likelihood: -3614.627   BIC: 7461.487

Means of prior cluster probabilities :
   Mainly leader Isolate/mediator  Mediator/leader
           0.244            0.425            0.331

Most probable clusters :
             Mainly leader  Isolate/mediator  Mediator/leader
count                   49                87               64
proportion           0.245             0.435             0.32

Classification table :
Mean cluster probabilities (in columns) by the most probable cluster (rows)

                  Mainly leader Isolate/mediator Mediator/leader
Mainly leader           0.91758          0.00136          0.0811
Isolate/mediator        0.00081          0.89841          0.1008
Mediator/leader         0.05902          0.10676          0.8342
```

We will first interpret the information on prior and posterior cluster membership probabilities and then proceed to interpreting covariate effects. Firstly, the `means of prior cluster probabilities` give information on how likely each cluster is in the whole population of students (33% in Mediator, 24% in Leader, and 43% in Isolate). Secondly, `Most probable clusters` shows group sizes and proportions if each student would be classified into the cluster for which they have the highest cluster membership probability.

Thirdly, the `Classification table` shows mean cluster probabilities (in columns) by the most probable cluster (in rows). We can see that the clusters are fairly crisp (the certainty of cluster memberships are fairly high) because the membership probabilities are large in the diagonal of the table. The uncertainty of the classification is the highest for the Mediator/leader cluster (among those that had the highest membership probability in that cluster, average cluster memberships were 83% for the Mediator/leader cluster, 6% for the Mainly leader cluster, and 11% for the Isolate/mediator cluster) and the highest in the Mainly leader cluster (92% for the Mainly leader cluster, 8% for the Mediator/leader cluster, and 0.1% for the Isolate/mediator cluster).

The part titled `Covariate effects` shows the parameter estimates for the covariates. Interpretation of the values is similar to that of multinomial logistic regression, meaning that we can interpret the direction and uncertainty of the effect—relative to the reference cluster Mainly leader—but we cannot directly interpret the magnitude of the effects (the magnitudes are on log-odds scale). We can see that individuals with low GPA more often end up in the Isolate/mediator cluster and the Mediator/leader cluster in comparison to the Mainly leader cluster (i.e., the standard errors are small in comparison to the parameter estimates), while individuals with high GPA levels end up in the Mediator/leader cluster less often but are not more or less likely to end up in the Isolate/mediator cluster. For categorical covariates such as our `GPA` variable, we can also easily compute the prior cluster membership probabilities from the estimates with the following call:

```
exp(fit_mmm_2$model$coefficients)/rowSums(exp(fit_mmm_2$model$coefficients))
```

```
          Mainly leader Isolate/mediator Mediator/leader
GPALow       0.07605453        0.5198587       0.4040868
GPAMiddle    0.25090105        0.3705958       0.3785031
GPAHigh      0.40497185        0.3870997       0.2079285
```

The matrix shows the levels of the covariates in the rows and the clusters in the columns. Among the high-GPA students, 40% (0.40497) are classified as Mainly leaders, 39% as Isolate/mediators, and 21% as Mediator/leaders. Among middle-GPA students classification is relatively uniform (25% as Mainly leaders, 37% as Isolate/mediators and 38 Mediator/leaders) whereas most of the low-GPA students are classified as Isolate/mediators or Mediator/leaders (52% and 40%, respectively).

The summary object also calculates prior and posterior cluster memberships for each student. We omit them here, for brevity, but demonstrate that they can be obtained as follows:

```
prior_prob <- summary_mmm_2$prior_cluster_probabilities
posterior_prob <- summary_mmm_2$posterior_cluster_probabilities
```

4.2.4 Mixture Hidden Markov Models

Finally, we will proceed to the most complex of the models, the MHMM.

For defining a MHMM, we use the `build_mhmm()` function. Again, we can use the argument `n_states` which is now a vector showing the number of hidden states in each cluster (the length of the vector defines the number of clusters). We will begin by estimating a MHMM with three clusters, each with two hidden states:

```
set.seed(123)
mhmm <- build_mhmm(
  roles_seq,
  n_states = c(2, 2, 2),
  data = cov_data,
  formula = ~ 0 + GPA
)
fit_mhmm <- fit_model(mhmm)
```

```
Error in fit_model(mhmm): EM algorithm failed: Estimation of
gamma coefficients failed due to singular Hessian.
```

In this case, we get an error message about the EM algorithm failing. This means that the algorithm was not able to find parameter estimates from the random starting values the `build_mhmm()` function generated and we need to adjust our code.

Starting values for the parameters of the MHMM can be given with the arguments `initial_probs`, `transition_probs`, and `emission_probs`. For the MHMM, these are lists of vectors and matrices, one for each cluster. We use the same number of hidden states (two) for each cluster. We define the initial values for the transition and emission probabilities as well as regression coefficients ourselves. We also restrict the initial state probabilities so that in each cluster every student is forced to start from the same (first) hidden state.

```
set.seed(1)

# Set initial probabilities
init <- list(c(1, 0), c(1, 0), c(1, 0))

# Define own transition probabilities
trans <- matrix(c(
  0.9, 0.1,
  0.1, 0.9
), nrow = 2, byrow = TRUE)

translist <- list(trans, trans, trans)

# Simulate emission probabilities
```

```
emiss <- simulate_emission_probs(
  n_states = c(2, 2, 2),
  n_symbols = 3,
  n_clusters = 3
)

emiss <- replicate(3, matrix(1/3, 2, 3), simplify = FALSE)

# Define initial values for coefficients
# Here we start from a case where low GPA correlates with Cluster 1,
# whereas middle and high GPA has no effect
beta <- cbind(0, c(-2, 0, 0), c(-2, 0, 0))

# Define model structure
mhmm_2 <- build_mhmm(
  roles_seq,
  initial_probs = init, transition_probs = translist,
  emission_probs = emiss, data = cov_data,
  formula = ~ 0 + GPA, beta = beta
)
```

Now that we have built the MHMM, we can estimate its parameters:

```
set.seed(1)
suppressWarnings(fit_mhmm_2 <- fit_model(
  mhmm_2,
  control_em = list(restart = list(times = 100, n_optimum = 101)))
)
```

We can now check how many times the log-likelihood values occurred in the 101 estimations:

```
table(round(fit_mhmm_2$em_results$best_opt_restart, 2))
```

```
     -Inf -3672.25 -3595.82 -3588.58 -3584.14 -3526.42 -3525.06 -3519.53
       56        2        1        3        1        4        1        2
  -3519.5 -3519.24
       15       16
```

The best model was found 16 times out of 101 times, although the second best model has a log-likelihood of −3519.5 which is almost indistinguishable from the optimal model (−3519.24)

We will start to interpret the model by looking at the sequence plots in each cluster (see Fig. 10). The function call is interactive. As before, if you only want to plot one cluster you can use the `which.plots` argument:

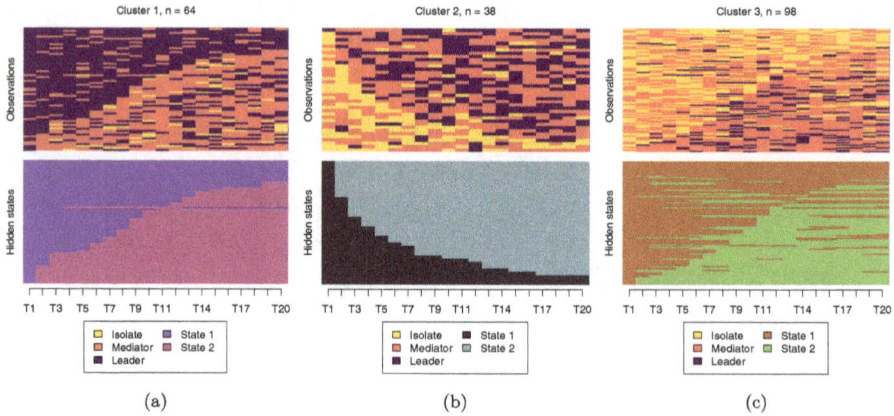

Fig. 10 MHMM estimated sequence distribution plot with hidden states. (**a**) Cluster 1. (**b**) Cluster 2. (**c**) Cluster 3

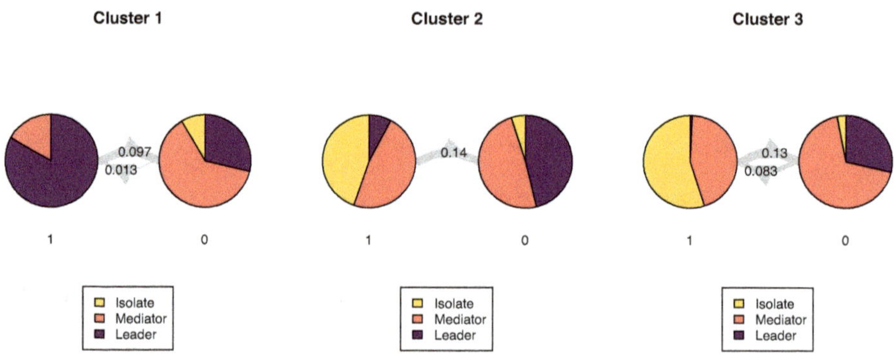

Fig. 11 Transitions between states for each trajectory

```
mssplot(fit_mhmm_2$model,
        plots = "both", type = "I", sortv = "mds.hidden",
        with.legend = "bottom.combined", legend.prop = .15)
```

We can also visualise the model parameters in each cluster (see Fig. 11):

```
plot(fit_mhmm_2$model,
  vertex.size = 60,
  label.color = "black",
  vertex.label.color = "black",
  edge.color = "lightgray",
  edge.label.color = "black",
```

```
  legend.prop = 0.4,
  ncol.legend = 1,
  ncol = 3,
  interactive = FALSE,
  combine.slices = 0
)
```

Based on the two plots, we can determine that Cluster 1 describes students who start as leaders but then transition to alternating between mediator and leader. Cluster 2 describes students who start by alternating between isolate and mediator roles and then mainly transition to alternating between mediator and leader roles. Cluster 3 describes students who start as alternating between isolate and mediator roles, after which they transition between isolate/mediator and mediator/leader.

```
cluster_names(fit_mhmm_2$model) <- c(
  "Downward transition", "Upward transition", "Alternating"
)
```

With `summary(fit_mhmm_2$model)` we get the parameter estimates and standard errors for the covariates and information about clustering:

```
summary(fit_mhmm_2$model)
```

```
Covariate effects :
Downward transition is the reference.

Upward transition :
           Estimate  Std. error
GPALow       -0.455       0.464
GPAMiddle     0.440       0.310
GPAHigh      -2.743       0.727

Alternating :
           Estimate  Std. error
GPALow       1.3560       0.324
GPAMiddle    0.3461       0.316
GPAHigh      0.0468       0.250

Log-likelihood: -3519.243   BIC: 7237.543

Means of prior cluster probabilities :
Downward transition   Upward transition        Alternating
              0.302               0.181              0.517

Most probable clusters :
            Downward transition  Upward transition  Alternating
count                        61                 30          109
proportion                0.305               0.15        0.545
```

```
Classification table :
Mean cluster probabilities (in columns) by the most probable cluster (rows)

                    Downward transition Upward transition Alternating
Downward transition             0.95727            0.0267      0.0161
Upward transition               0.03007            0.8037      0.1662
Alternating                     0.00975            0.0962      0.8940
```

We can see, that the prior probabilities of belonging to each cluster are very different: half of the students can be described as alternating, while of the rest, a downward transition is more typical (31%). Based on the classification table, the Downward transition cluster is rather crisp, while the other two are partly overlapping (see the MMM example for more information on interpreting the classification table).

The `Covariate effects` tables show that, in comparison to Alternating cluster, students with low GPA are less likely to end up in the Upward or Downward transition clusters and students with high GPA are less likely to end up in Upward transition cluster. Again, we can calculate the probabilities of belonging to each cluster by GPA levels:

```
exp(fit_mhmm_2$model$coefficients)/rowSums(exp(fit_mhmm_2$model$coefficients))
```

```
            Downward transition Upward transition Alternating
GPALow                0.1813217        0.11502283   0.7036555
GPAMiddle             0.2521406        0.39144399   0.3564154
GPAHigh               0.4734128        0.03048189   0.4961054
```

The table shows that students with low GPA typically belong to the Alternating cluster (70% probability) while students with high GPA mainly end up in the Downward transition cluster (47%) or the Alternating cluster (50%). Most students with middle GPA end up in the Upward transition cluster (39%), but the probabilities are almost as high for the Alternating cluster (36%) and also fairly high for the Downward transition cluster (25%).

In light of this, it is worth noting that the covariates do not merely explain the uncovered clusters; as part of the model, they drive the formation of the clusters. In other words, an otherwise identical model without the dependence on the `GPA` covariate may uncover different groupings with different probabilities.

If we are not sure how many clusters or hidden states we expect, or if we wish to investigate different combinations of covariates, we can estimate several models and compare the results with information criteria or cross-validation. Estimating a large number of complex models is, however, very time-consuming. Using prior information for restricting the pool of potential models is useful, and sequence analysis can also be used as a helpful first step [10, 37].

4.3 Stochastic Process Mining with Markovian Models

Process mining is a relatively recent method for the analysis of event-log data (time-stamped logs) which aims to understand the flow and dynamics of the process under study. In education, process mining has been used extensively to analyse learners' online logs collected from Learning Management Systems (LMS), to understand how they utilize learning resources and transitions between learning activities to mention a few [38, 39]. In this book, we have devoted a full chapter for process mining where we explained how process mining can be performed in R [40]. Yet, in this chapter we will present a novel method that we propose to perform stochastic process mining using MMs. While process mining can be performed using different software, techniques and algorithms, MMs offer a powerful framework for process mining with several advantages over the commonly used methods. First, it is more theoretically aligned with the idea of a transition from an action to an action and that actions are temporally dependent on each other. Second, MMs allow for data to be clustered into similar transition patterns, a possibility not offered by other process mining methods (see the process mining chapter of this book [40]). Third, contrary to other process mining methods, MMs do not require researchers to arbitrarily exclude—or trim—a large part of the data to "simplify" the model. For instance, most of the process mining analyses require an arbitrary cutoff to trim the data so that the process model is readable. This trimming significantly affects the resulting model and makes it hard to replicate. Most importantly, MMs have several fit statistics that we can use to compare and judge the model fit as we have seen before.

Several R packages can perform stochastic process mining; in this tutorial we will rely on the same package we discussed earlier and combine it with a powerful visualisation that allows us to effectively visualise complex processes. In the next example, we will analyse data extracted from the learning management system logs and offer a detailed guide to process mining. We will also use MMMs to cluster the data into latent patterns of transitions. Given that the traditional plotting function in `seqHMM` works well with a relatively short alphabet, we will use a new R package called `qgraph` for plotting. The package `qgraph` offers powerful visualisations which makes plotting easier, and more interpretable especially for larger models. Furthermore, `qgraph` allows researchers to use a fixed layout for all the plotted networks so the nodes can be compared to each other more easily.

Let us now go through the analysis. The next chunk of code imports the prepared sequence data from the sequence analysis chapter. The data belong to a learning analytics course and the events are coded trace logs of students' actions such as *Course view*, *Instructions*, *Practicals*, *Social*, etc. Then, we build a sequence object using the function `seqdef()` from `TraMineR`.

```
library(qgraph)
library(rio)
library(seqHMM)
library(tidyverse)
library(TraMineR)
seq_data <- import(paste0(URL, "1_moodleLAcourse/LMS_data_wide.xlsx"))
seq_data_all <- seqdef(seq_data, var = 7:54 )
```

Before proceeding further, it is advisable to visualise the sequences. Figure 12 shows the sequence index plot, sorted according to the first states. The data are much larger than the collaboration roles and achievement sequences analysed previously; there are 9478 observations with an alphabet of 12 states. Unlike in the previous example, the sequence lengths vary considerably. Due to this, shorter sequences contain missing values to fill the empty cells in the data frame. However, there

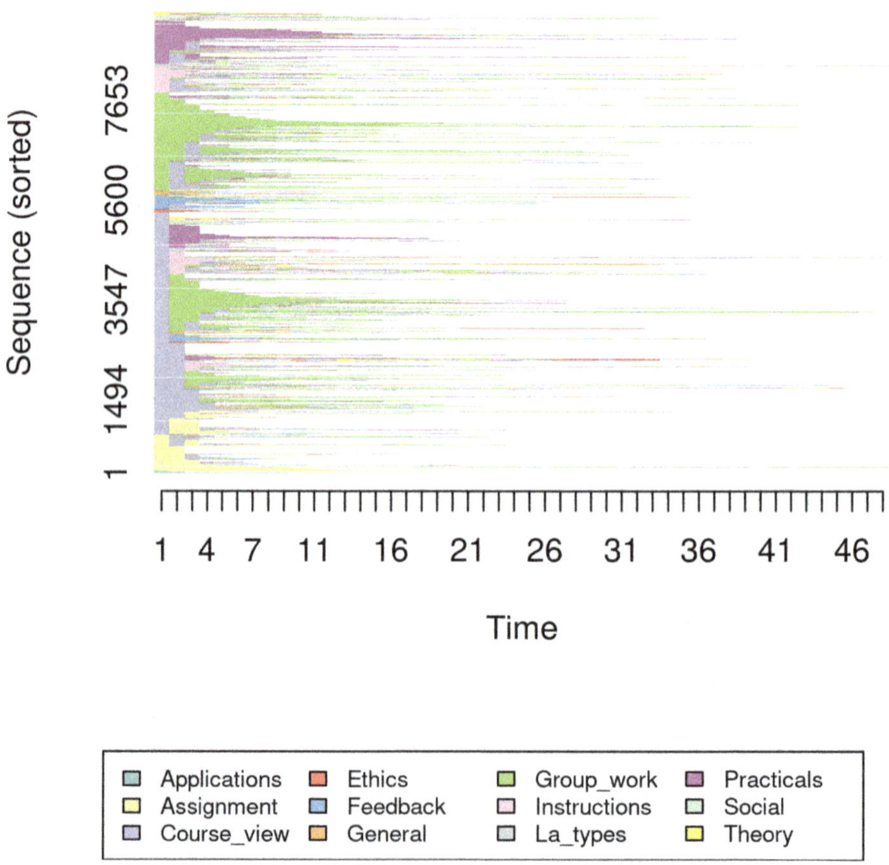

Fig. 12 Sequence index plot for the learning management system logs

are no internal gaps. When creating the sequence object with the `seqdef` function, `TraMineR` allows for distinguishing between real missing values (`NA`, where the true state is unknown) and technical missing values (void) used to pad the sequences to equal lengths. The `seqHMM` package is able to account for both types of missing values and treats them slightly differently, for example when calculating the most probable paths of hidden states.

```
seqplot(seq_data_all,
  type = "I", ncol = 4, sortv = "from.start",
  legend.prop = 0.2, cex.legend = 0.7, border = NA,
  ylab = "Sequence (sorted)", xlab = "Time"
)
```

A simple transition analysis can be performed by estimating and plotting the transition probabilities. This can be performed using the `TraMineR` package. Yet, this simple approach has drawbacks and it is advisable to estimate the MM and use their full power. The next code estimates the transition probabilities of the full dataset using the function `seqtrate()` from `TraMineR` package (shown in Table 7).

```
overalltransitions <- seqtrate(seq_data_all)
```

As we mentioned earlier, we will use a novel plotting technique that is more suitable for large process models. Below, we plot the transition probabilities with the `qgraph()` function from the `qgraph` package (Fig. 13). We use some arguments to improve the process model visualisation. First, we use the argument `cut = 0.15` to show the edges with probabilities below 0.15 in lower thickness and colour intensity. This *cut* makes the graph easier to read and less crowded, and gives emphasis to the edges which matter. The argument `minimum = 0.05` hides small edges below the probability threshold of 0.05. We use `edge.labels = TRUE` to show the transition probabilities as edge labels. The argument `color` gets the colour palette from the sequence with the function `cpal()` and the argument `curveAll = TRUE` ensures the graph shows curved edges. The `"colorblind"` theme makes sure that the colours can be seen by everyone regardless of colour vision abilities. Lastly, the `mar`

within the figure area.

```
# get the labeles to use them as node names.
Labelx <- alphabet(seq_data_all)
transitionsplot <- qgraph(
  overalltransitions, cut = 0.15, minimum = 0.05,
  labels = Labelx, edge.labels = TRUE, edge.label.cex = 0.65,
  color = cpal(seq_data_all), curveAll = TRUE,
  theme = "colorblind", mar = c(4, 3, 4, 3)
)
```

Table 7 Transition probabilities

From\To	Applications	Assignment	Course_view	Ethics	Feedback	General	Group_work	Instructions	La_types	Practicals	Social	Theory
Applications	0.46	0.07	0.13	0.01	0.01	0.19	0.05	0.01	0.01	0.05	0.00	0.00
Assignment	0.00	0.70	0.19	0.00	0.01	0.02	0.03	0.02	0.02	0.02	0.00	0.00
Course_view	0.01	0.07	0.35	0.01	0.03	0.03	0.28	0.10	0.02	0.08	0.02	0.01
Ethics	0.01	0.00	0.12	0.61	0.01	0.04	0.10	0.01	0.03	0.04	0.01	0.02
Feedback	0.00	0.02	0.23	0.00	0.56	0.00	0.11	0.04	0.01	0.02	0.00	0.00
General	0.04	0.05	0.18	0.01	0.00	0.49	0.06	0.06	0.05	0.03	0.01	0.02
Group_work	0.00	0.01	0.19	0.00	0.01	0.01	0.73	0.02	0.00	0.01	0.01	0.00
Instructions	0.00	0.02	0.33	0.00	0.03	0.04	0.12	0.37	0.02	0.03	0.04	0.00
La_types	0.01	0.06	0.24	0.01	0.00	0.10	0.07	0.05	0.38	0.03	0.01	0.03
Practicals	0.00	0.02	0.17	0.00	0.01	0.01	0.03	0.02	0.01	0.73	0.00	0.01
Social	0.00	0.01	0.25	0.00	0.00	0.01	0.12	0.11	0.01	0.02	0.48	0.00
Theory	0.00	0.02	0.15	0.03	0.00	0.02	0.06	0.01	0.05	0.05	0.00	0.60

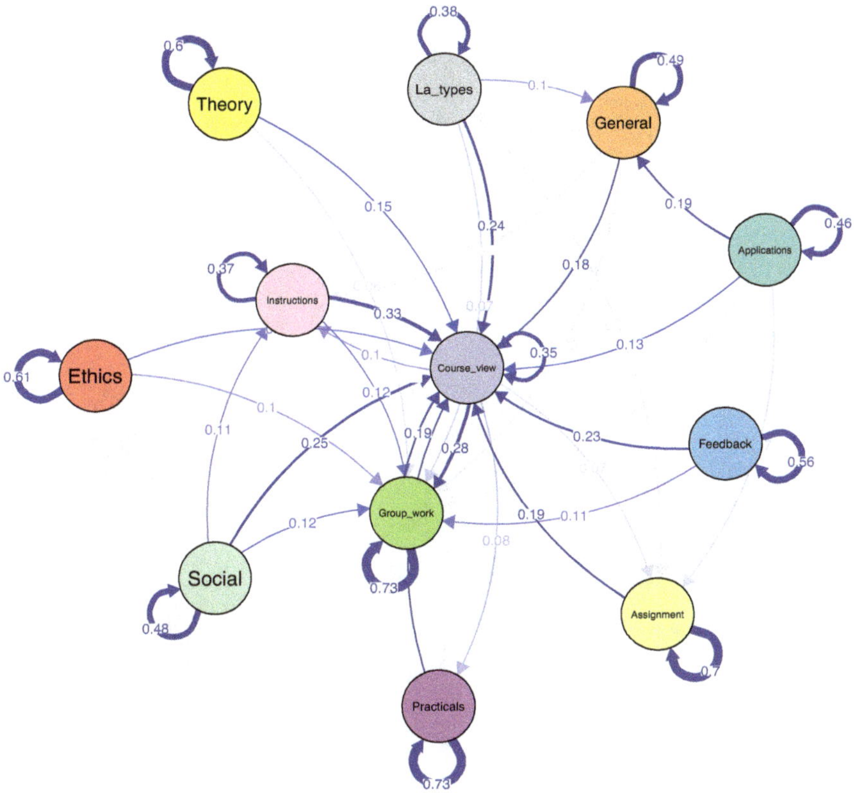

Fig. 13 Process map for the overall process

The `seqtrate()` function only computes the transition probabilities but does not compute the initial probabilities. While it is not difficult to calculate the proportions of starting in each state, we can also estimate a simple Markov model which does the same with a short command. We do so using the `build_mm()` function as per Sect. 4.2, recalling that the `build_mm()` function is distinct from `build_hmm()`, `build_mmm()`, and `build_mhmm()` in that it is the only build function that automatically estimates the parameters of the model.

The plotting now includes an extra option called `pie = overallmodel$initial_probs` which tells `qgraph` to use the initial probabilities from the fitted MM as the sizes of the pie charts in the borders of the nodes in Fig. 14. For instance, the pie around *Course view* is around half of the circle corresponding to 0.48 initial probability to start from *Course view*. Please also note that the graph is otherwise equal to the one generated via `seqtrate()` apart from these initial probabilities.

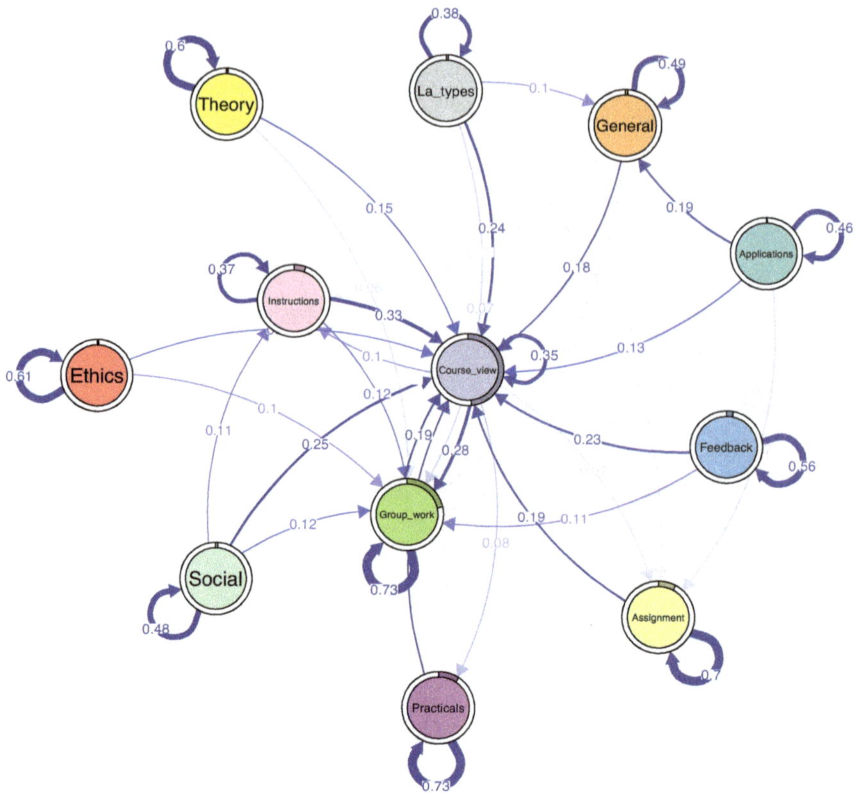

Fig. 14 Process map for the overall process with initial probabilities

```
overallmodel <- build_mm(seq_data_all)

overallplot <- qgraph(
  overallmodel$transition_probs
  cut = 0.15,
  minimum = 0.05,
  labels = Labelx,
  mar = c(4, 3, 4, 3),
  edge.labels = TRUE,
  edge.label.cex = 0.65,
  color = cpal(seq_data_all),
  curveAll = TRUE,
  theme = "colorblind",
  pie = overallmodel$initial_probs
)
```

Having plotted the transitions of the full dataset, we can now look for transition patterns, that is typical transition patterns (i.e., clusters) that are repeated within the data. The procedure is the same as before. In the next example, we use the function `build_mmm()` to build the model with four clusters as a demonstration. Ideally, researchers need to estimate several models and choose the best model based on model selection criteria (such as BIC) values as well as interpretability.

The steps involved in fitting the model are as before; we make use of the function `fit_model()` to estimate the model. The results of the running the code will be an MM for each cluster (with distinct initial and transition probabilities). Given the number of sequences in the dataset, their length, and the number of states, the computational burden is larger than for previous applications in this chapter. For illustrative purposes, instead of repeated EM runs with random starting values, we use single EM run followed by global optimisation, using the argument `global_step = TRUE`. One benefit of this global (and local) step in `fit_model` over the EM algorithm is the flexibility to define a maximum runtime (in seconds) for the optimization process (argument `maxtime` in `control_global`). This can be valuable for larger problems with predefined runtime (e.g., in a shared computer cluster). Note, however, that relying on the runtime can lead to non-reproducible results even with fixed seed if the optimisation terminates due to the time limit. Finally, we run additional local optimisation step using the results of the global optimisation, for more accurate results. The last argument `threads = 16` instructs to use parallel computing to enable faster fitting (please, customise according to the number of cores in your computer). As for the starting values, we use the transition probabilities computed from the full data for all clusters, and random values for the initial probabilities.

While in theory many of the global optimisation algorithms should eventually find the global optimum, in practice there are no guarantees that it is found in limited time. Thus, as earlier, in practice it is advisable to try different global/local optimisation algorithms and/or EM algorithm with different initial values to make it more likely that the global optimum is found (see [4] for further discussion).

```
set.seed(1)
trans_probs <- simulate_transition_probs(12, 4, diag_c = 5)
init_probs <- as.numeric(prop.table(table(seq_data_all[,1])[1:12]))
init_probs <- replicate(4, init_probs, simplify = FALSE)

builtseqLMS <- build_mmm(
  seq_data_all,
  transition_probs = trans_probs,
  initial_probs = init_probs
)

fitLMS <- fit_model(
  builtseqLMS,
  global_step = TRUE,
  control_global = list(
    maxtime = 3600,
```

```
    maxeval = 1e5,
    algorithm = "NLOPT_GD_STOGO_RAND"),
  local_step = TRUE,
  threads = 16
)

fitLMS$global_results$message
fitLMS$logLik
```

```
[1] "NLOPT_SUCCESS: Generic success return value."

[1] -114491.2
```

Before plotting the clusters, let us do some cleanups. First, we get the transition probabilities of each cluster and assign them to a variable. In that way, they are easier to manipulate and work with. In the same way, we can extract the initial probabilities for each cluster.

```
#extract transition probabilities of each cluster
Clustertp1 <- fitLMS$model$transition_probs$`Cluster 1`
Clustertp2 <- fitLMS$model$transition_probs$`Cluster 2`
Clustertp3 <- fitLMS$model$transition_probs$`Cluster 3`
Clustertp4 <- fitLMS$model$transition_probs$`Cluster 4`

#extract initial probabilities of each cluster
Clusterinitp1 <- fitLMS$model$initial_probs$`Cluster 1`
Clusterinitp2 <- fitLMS$model$initial_probs$`Cluster 2`
Clusterinitp3 <- fitLMS$model$initial_probs$`Cluster 3`
Clusterinitp4 <- fitLMS$model$initial_probs$`Cluster 4`
```

Plotting the process maps can be performed in the same way we did before. However, if we need to compare clusters, it is best if we use a unified layout. An average layout can be computed with the function `averageLayout()` which takes the transition probabilities of the four clusters as input and creates—as the name implies—an averaged layout. Another option is to use the same layout of the `overallplot` in the previous example. This can be obtained from the plot object `overallplot$layout`. This can be helpful if you would like to plot the four plots corresponding to each cluster with the same layout as the overall plot (see Fig. 15).

```
Labelx <- colnames(Clustertp1) # we need to get the labels

Averagelayout <- averageLayout(
  list(Clustertp1, Clustertp2, Clustertp3, Clustertp4)
)
```

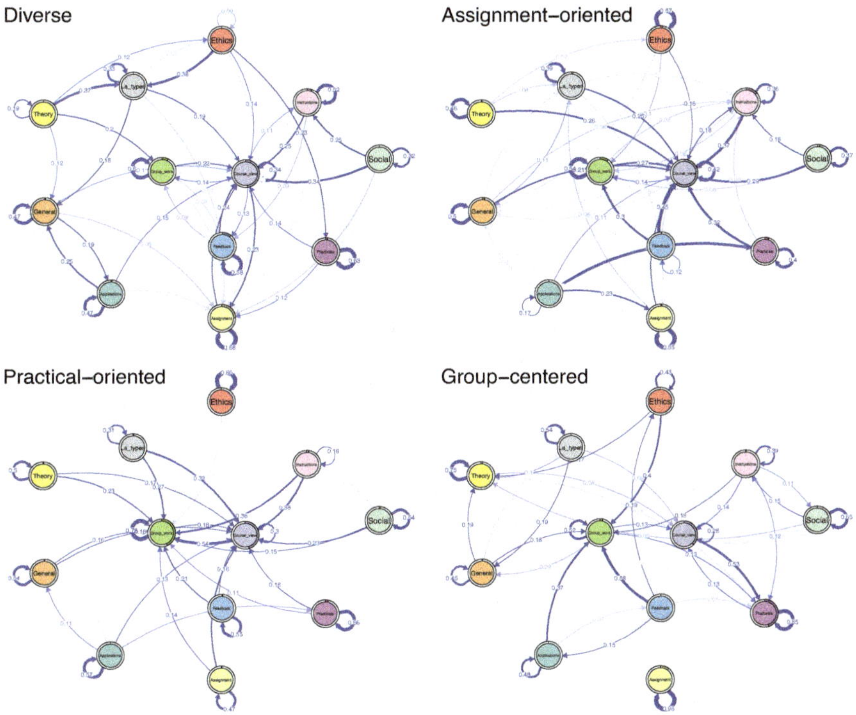

Fig. 15 Process maps for each cluster

```
# You can also try with this layout from the previous plot
Overalllayout <- overallplot$layout

qgraph(
  Clustertp1, cut = 0.15, minimum = 0.05 , labels = Labelx,
  edge.labels = TRUE, edge.label.cex = 0.65, color = cpal(seq_data_all),
  layout = Averagelayout, pie = Clusterinitp1, curveAll = TRUE,
  theme = "colorblind", title = "Diverse"
)

qgraph(
  Clustertp2, cut = 0.15, minimum = 0.05, labels = Labelx,
  edge.labels = TRUE, edge.label.cex = 0.65, color = cpal(seq_data_all),
  layout = Averagelayout, pie = Clusterinitp2, curveAll = TRUE,
  theme = "colorblind", title = "Assignment-oriented"
)

qgraph(
  Clustertp3, cut = 0.15, minimum = 0.05, labels = Labelx,
  edge.labels = TRUE, edge.label.cex = 0.65, color = cpal(seq_data_all),
  layout = Averagelayout, pie = Clusterinitp3, curveAll = TRUE,
  theme = "colorblind", title = "Practical-oriented"
)
```

```
qgraph(
  Clustertp4, cut = 0.15, minimum = 0.05 , labels = Labelx,
  edge.labels = TRUE, edge.label.cex = 0.65, color = cpal(seq_data_all),
  layout = Averagelayout, pie = Clusterinitp4, curveAll = TRUE,
  theme = "colorblind", title = "Group-centered"
)
```

Oftentimes, the researcher is interested in comparing two pre-defined fixed groups, e.g., high achievers and low achievers, rather than between the computed clusters. In the next example we will compare high to low achievers based on their achievement levels. First, we have to create a separate sequence object for each group. We do this by filtering but you can do it in other ways. For instance, you can create two sequences from scratch for each group. The next step is to build the MMs separately for each group.

```
seq_high <- seq_data_all[seq_data$Achievementlevel4 < = 2,]
seq_low <-  seq_data_all[seq_data$Achievementlevel4 > 2,]

high_mm <- build_mm(seq_high)
low_mm <- build_mm(seq_low)
```

Before plotting the groups, let us do some cleaning, like we did before. First, we get the transition and initial probabilities of each group. We also compute an average layout. Please note that you can use the layout from the previous examples if you are comparing the models against each other and you need a unified framework. The plotting is the same as before (see Fig. 16).

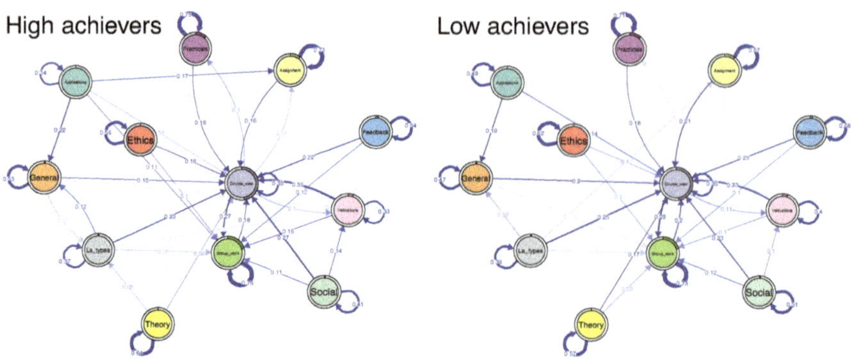

Fig. 16 Process maps for high achievers and low achievers using average layout

```
par(mfrow=c(1, 2))

#extract transition probabilities of each cluster
Highprobs <- high_mm$transition_probs
Lowprobs <- low_mm$transition_probs

#extract initial probabilities of each cluster
Highinit <- high_mm$initial_probs
Lowinit <- high_mm$initial_probs

Averagelayout <- averageLayout(list(Highprobs, Lowprobs))

Highplot <- qgraph(
  Highprobs, cut = 0.15, minimum = 0.05, labels = Labelx,
  edge.labels = TRUE, edge.label.cex = 0.65,
  color = cpal(seq_data_all), layout = Averagelayout,
  pie = Highinit, theme = "colorblind", title = "High achievers"
)

Lowplot <-  qgraph(
  Lowprobs, cut=0.15, minimum = 0.05, labels = Labelx,
  edge.labels = TRUE, edge.label.cex = 0.65,
  color = cpal(seq_data_all), layout = Averagelayout,
  pie = Lowinit, theme = "colorblind", title = "Low achievers"
)
```

We can also plot the difference plot (see Fig. 17); that is, what the low achievers do less than high achievers. In this case, red edges are negative (transitions they do less) and blue edges are positive (transitions that they do more than high achievers). As you can see, the differences are not that huge. In fact, much of the literature comparing high and low achievers uses higher thresholds e.g., top 25% to bottom 25% or even top 10% to bottom 10%.

```
diffplot <- qgraph(
  Lowprobs - Highprobs, cut = 0.15, minimum = 0.05, labels = Labelx,
  edge.labels = TRUE, edge.label.cex = 0.65, layout = Averagelayout,
  color = cpal(seq_data_all), theme = "colorblind"
)
```

5 Conclusions and Further Readings

Markovian models provide a flexible model-based approach for analysing complex sequence data. MMs and HMMs have proven useful in many application areas such as biology and speech recognition, and can be a valuable tools in analysing data in educational settings as well. Their mixture variants allow for the representation of complex systems by combining multiple MMs or HMMs, each capturing different aspects of the underlying processes, allowing probabilistic clustering, information compression (e.g. visualisation of multicategory data from multiple domains), and detection of latent features of sequence data (e.g., extraction of different learning strategies). The ability to incorporate covariates in the case of MMMs and MHMMs makes those models even more powerful, and generally MMs and MMMs represent useful tools in the field of process mining also.

The seqHMM package used in the examples supports time-constant covariates for predicting cluster memberships for each individual. In theory, covariates could be used to define transition or emission probabilities as well, leading to subject-specific and possibly time-varying transition and emission probabilities (in the case of time-varying covariates). However, at the time of writing this chapter, these are not supported in seqHMM (this may change in the future). In R, there are at least two other, potentially useful packages: for MMs, the dynamite [41] package supports covariates on the transition probabilities with potentially time-varying effects, whereas LMest [42] supports MMs and HMMs with covariates, and restricted variants of the MHMM where only the initial and transition matrices vary between clusters. Going beyond the R software, some commercial software also offers tools to analyse Markovian models, including latentGold [43] and Mplus [11].

The conditional independence assumption of observations given the hidden states in HMMs can sometimes be unrealistic. In these settings, the so-called double chain MMs can be used [44]. There the current observation is allowed to depend on both the current hidden state and the previous observation. Some restricted variants of such models are implemented in the march package in R [45]. Finally, variable-order MMs extend basic MMs by allowing the order of the MM to vary in time. A TraMineR-compatible implementation of variable-order models can be found in the PST package [46].

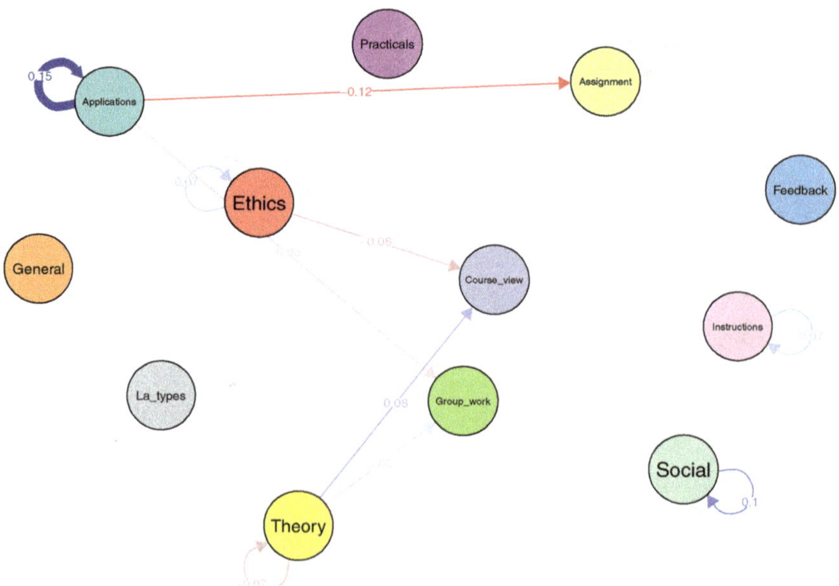

Fig. 17 Difference between process maps of high achievers and low achievers using average layout

We encourage readers to read more about how to interpret the results in the original study where the data for this chapter was drawn from [36]. We also encourage readers to learn more about Markovian models in the context of multi-channel sequence analysis in Chapter 13 [8].

Acknowledgments JH and SH were supported by Research Council of Finland (PREDLIFE: Towards well-informed decisions: Predicting long-term effects of policy reforms on life trajectories, decision numbers 331817 and 331816, and Research Flagship INVEST: Inequalities, Interventions and New Welfare State, decision number 345546). SH was also supported by the Strategic Research Council (SRC), FLUX consortium: Family Formation in Flux – Causes, Consequences, and Possible Futures (decision numbers: 345130 and 345130). MS was supported by Research Council of Finland (TOPEILA: Towards precision education: Idiographic learning analytics, decision number 350560).

References

1. Saqr M, López-Pernas-Pernas S, Helske S, Durand M, Murphy K, Studer M, Ritschard G (2024) Sequence analysis in education: principles, technique, and tutorial with r. In: Saqr M, López-Pernas-Pernas S (eds) Learning analytics methods and tutorials: a practical guide using R. Springer, Berlin
2. López-Pernas-Pernas S, Saqr M (2024) Modeling the dynamics of longitudinal processes in education. A tutorial with R for the VaSSTra method. In: Saqr M, López-Pernas-Pernas S (eds) Learning analytics methods and tutorials: a practical guide using R. Springer, in press
3. Liao TF, Bolano D, Brzinsky-Fay C, Cornwell B, Fasang AE, Helske S, Piccarreta R, Raab M, Ritschard G, Struffolino E, Studer M (2022) Sequence analysis: its past, present, and future. Soc Sci Res 107:102772. https://doi.org/10.1016/j.ssresearch.2022.102772
4. Helske S, Helske J (2019) Mixture hidden Markov models for sequence data: the seqHMM package in R. J Stat Softw 88. https://doi.org/10.18637/jss.v088.i03
5. Schwarz GE (1978) Estimating the dimension of a model. Ann Stat 6:461–464. https://doi.org/10.1214/aos/1176344136
6. Pol F van de, Langeheine R (1990) Mixed Markov latent class models. Sociol Methodol 20:213. https://doi.org/10.2307/271087
7. Vermunt JK, Tran B, Magidson J (2008) Latent class models in longitudinal research. In: Menard S (ed) Handbook of longitudinal research. Elsevier, Amsterdam, pp 373–385
8. López-Pernas-Pernas S, Murphy K, Saqr M (2024) Multichannel sequence analysis in educational research using R. In: Saqr M, López-Pernas-Pernas S (eds) Learning analytics methods and tutorials: a practical guide using R. Springer, in press
9. Rabiner L (1989) A tutorial on hidden Markov models and selected applications in speech recognition. Proc IEEE 77:257–286. https://doi.org/10.1109/5.18626
10. Helske S, Helske J, Eerola M (2018) Combining sequence analysis and hidden markov models in the analysis of complex life sequence data. In: Ritschard G, Studer M (eds) Sequence analysis and related approaches: innovative methods and applications. Springer International Publishing, Cham, pp 185–200. https://doi.org/10.1007/978-3-319-95420-2/_11
11. Muthén LK, OMB (2017) Mplus user's guide, 8th edn. Los Angeles
12. Muthén B, Muthén L Mplus: A general latent variable modeling program. https://www.statmodel.com/download/Mplus-A/%20General/%20Latent/%20Variable/%20Modeling/%20Program.pdf
13. Törmänen T, Järvenoja H, Saqr M, Malmberg J, & Järvelä S (2022) A person-centered approach to study students' socio-emotional interaction profiles and regulation of collaborative learning. Front Educ 7

14. Törmänen T, Järvenoja H, Saqr M, Malmberg J, Järvelä S (2023) Affective states and regulation of learning during socio-emotional interactions in secondary school collaborative groups. Br J Educ Psychol 93(Suppl 1):48–70
15. Fincham E, Gašević D, Jovanović J, Pardo A (2019) From study tactics to learning strategies: an analytical method for extracting interpretable representations. IEEE Trans Learn Technol 12:59–72
16. Saqr M, López-Pernas-Pernas S (2022) How CSCL roles emerge, persist, transition, and evolve over time: a four-year longitudinal study. Comput Educ 189:104581
17. Saqr M, López-Pernas-Pernas S, Jovanović J, Gašević D (2023) Intense, turbulent, or wallowing in the mire: a longitudinal study of cross-course online tactics, strategies, and trajectories. Internet Higher Educ 57:100902
18. Bouguettaya A, Yu Q, Liu X, Zhou X, Song A (2015) Efficient agglomerative hierarchical clustering. Expert Syst Appl 42:2785–2797. https://doi.org/10.1016/j.eswa.2014.09.054
19. Gilpin S, Qian B, Davidson I (2013) Efficient hierarchical clustering of large high dimensional datasets. In: Proceedings of the 22nd ACM international conference on information & knowledge management. Association for Computing Machinery, New York, pp 1371–1380. https://doi.org/10.1145/2505515.2505527
20. López-Pernas-Pernas S, Saqr M (2021) Bringing synchrony and clarity to complex multi-channel data: a learning analytics study in programming education. IEEE Access 9:166531–166541
21. Saqr M, López-Pernas-Pernas S (2021) The longitudinal trajectories of online engagement over a full program. Comput Educ 175:104325
22. Matcha W, Gasevic D, Ahmad Uzir N, Jovanovic J, Pardo A, Lim L, Maldonado-Mahauad J, Gentili S, Perez-Sanagustin M, Tsai Y-S (2020) Analytics of learning strategies: role of course design and delivery modality. J Learn Anal 7:45–71. https://doi.org/10.18608/jla.2020.72.3
23. Peeters W, Saqr M, Viberg O (2020) Applying learning analytics to map students' self-regulated learning tactics in an academic writing course. In: Proceedings of the 28th international conference on computers in education, pp 245–254
24. Lim L, Bannert M, Graaf J van der, Singh S, Fan Y, Surendrannair S, Rakovic M, Molenaar I, Moore J, Gašević D (2023) Effects of real-time analytics-based personalized scaffolds on students' self-regulated learning. Comput Human Behav 139:107547. https://doi.org/10.1016/j.chb.2022.107547
25. Saqr M, López-Pernas-Pernas S (2023) The temporal dynamics of online problem-based learning: why and when sequence matters. Int J Comput-Support Collab Learn 18:11–37. https://doi.org/10.1007/s11412-023-09385-1
26. Gatta R, Vallati M, Lenkowicz J, Rojas E, Damiani A, Sacchi L, De Bari B, Dagliati A, Fernandez-Llatas C, Montesi M, Marchetti A, Castellano M, Valentini V (2017) Generating and comparing knowledge graphs of medical processes using pMineR. In: Proceedings of the knowledge capture conference. ACM, New York. https://doi.org/10.1145/3148011.3154464
27. Boroujeni MS, Dillenbourg P (2019) Discovery and temporal analysis of MOOC study patterns. J Learn Anal 6:16–33. https://doi.org/10.18608/jla.2019.61.2
28. Andrade A, Danish JA, Maltese AV (2017) A measurement model of gestures in an embodied learning environment: accounting for temporal dependencies. J Learn Anal 4:18–46. https://doi.org/10.18608/jla.2017.43.3
29. Kokoç M, Akçapınar G, Hasnine MN (2021) Unfolding students' online assignment submission behavioral patterns using temporal learning analytics. Educ Technol Soc 24:223–235. https://www.jstor.org/stable/26977869
30. Epskamp S, Cramer AOJ, Waldorp LJ, Schmittmann VD, Borsboom D (2012) qgraph: network visualizations of relationships in psychometric data. J Stat Softw 48:1–18
31. Chan C, Chan GC, Leeper TJ, Becker J (2021) rio: a Swiss-army knife for data file I/O. https://cran.r-project.org/package=rio
32. Helske J, Helske S (2023) seqHMM: mixture hidden Markov models for social sequence data and other multivariate, multichannel categorical time series. https://cran.r-project.org/package=seqHMM

33. Wickham H, Averick M, Bryan J, Chang W, McGowan LD, François R, Grolemund G, Hayes A, Henry L, Hester J, Kuhn M, Pedersen TL, Miller E, Bache SM, Müller K, Ooms J, Robinson D, Seidel DP, Spinu V, Takahashi K, Vaughan D, Wilke C, Woo K, Yutani H (2019) Welcome to the tidyverse. J Open Source Softw 4:1686. https://doi.org/10.21105/joss.01686
34. Gabadinho A, Ritschard G, Müller NS, Studer M (2011) Analyzing and visualizing state sequences in R with TraMineR. J Stat Softw 40. https://doi.org/10.18637/jss.v040.i04
35. Saqr M, López-Pernas-Pernas S, Helske S, Hrastinski S (2023) The longitudinal association between engagement and achievement varies by time, students' profiles, and achievement state: a full program study. Comput Educ 199:104787
36. Saqr M, López-Pernas-Pernas S (2022) How CSCL roles emerge, persist, transition, and evolve over time: a four-year longitudinal study. Comput Educ 189:104581. https://doi.org/10.1016/j.compedu.2022.104581
37. Helske S, Keski-Säntti M, Kivelä J, Juutinen A, Käriälä A, Gissler M, Merikukka M, Lallukka T (2023) Predicting the stability of early employment with its timing and childhood social and health-related predictors: a mixture markov model approach. Longitud Life Course Stud 14:73–104
38. Peeters W, Saqr M, Viberg O (2020) Applying learning analytics to map students' self-regulated learning tactics in an academic writing course. In: Proceedings of the 28th international conference on computers in education. Asia-Pacific Society for Computers in Education, pp 245–254
39. Saqr M, Matcha W, Jovanovic J, Gašević D, López-Pernas-Pernas S, et al (2022) Transferring effective learning strategies across learning contexts matters: a study in problem-based learning. Australas J Educ Technol 39(3)35–57
40. López-Pernas-Pernas S, Saqr M (2024) The why, the how, and the when of educational process mining in R. In: Saqr M, López-Pernas-Pernas S (eds) Learning analytics methods and tutorials: a practical guide using R, Chap. 14. Springer, Cham
41. Tikka S, Helske J (2023) dynamite: an R package for dynamic multivariate panel models. https://doi.org/10.48550/ARXIV.2302.01607
42. Bartolucci F, Pandolfi S, Pennoni F (2017) LMest: an R package for latent Markov models for longitudinal categorical data. J Stat Softw 81:1–38. https://doi.org/10.18637/jss.v081.i04
43. Vermunt JK, Magidson J (2016) Guide for latent GOLD 5.1: basic, advanced, and syntax. Statistical Innovations Inc., Belmont
44. Berchtold A (1999) The double chain Markov model. Commun Stat Theory Methods 28:2569–2589. https://doi.org/10.1080/03610929908832439
45. Maitre O, Emery K, Oliver Buschor with contributions from, Berchtold A (2020). march: Markov chains. https://CRAN.R-project.org/package=march
46. Gabadinho A, Ritschard G (2016) Analyzing state sequences with probabilistic suffix trees: the PST R package. J Stat Softw 72:1–39. https://doi.org/10.18637/jss.v072.i03

Multi-Channel Sequence Analysis in Educational Research: An Introduction and Tutorial with R

Sonsoles López-Pernas, Mohammed Saqr, Satu Helske, and Keefe Murphy

1 Introduction

Learning is a dynamic phenomenon which follows the unidirectional forward law of time; that is, learning occurs sequentially in a time-ordered manner [1, 2]. Throughout this book, we have devoted several chapters to sequence analysis of learner-related data, due to its increasingly central role in learning analytics and education research as a whole. First, in Chapter 10 [3], we presented an introduction to sequence analysis and a discussion on the relevance of this method. Subsequently, in Chapter 11 [4], we studied how to build sequences from multiple variables and analyse more complex aspects of sequences. Both Chapter 10 and Chapter 11 also deal with clustering sequences into trajectories which undergo a similar evolution using distance-based methods. Then, in Chapter 12 [5], we learned how to cluster sequential data using Markov models. Though Chapter 12 touched briefly on simultaneous analysis of multi-channel sequences —albeit only from the Markovian point of view— we present the multi-channel perspective in greater detail here.

In this new chapter, we cover multi-channel sequence analysis, which deals with the analysis of two or more synchronised sequences, in greater detail. An example in educational research would be the analysis of the simultaneous unfolding of motivation, achievement, and engament sequences. Multi-channel sequence

S. López-Pernas-Pernas (✉) · M. Saqr
School of Computing, University of Eastern Finland, Joensuu, Finland
e-mail: sonsoles.lopez@uef.fi

S. Helske
INVEST Research Flagship Center & Department of Social Research, University of Turku, Turku, Finland

K. Murphy
Department of Mathematics and Statistics, Hamilton Institute, Maynooth University, Maynooth, Ireland

M. Saqr, S. López-Pernas (eds.), *Learning Analytics Methods and Tutorials*,
https://doi.org/10.1007/978-3-031-54464-4_13

analysis is a rather novel method in social sciences [6–9] and its applications within educational research in general remain scarce. The increasing availability of student multimodal temporal data makes multi-channel sequence analysis a relevant and timely method to tackle the challenges that these new data bring.

Throughout this chapter, we describe multi-channel sequence analysis in detail, with an emphasis on how to detect patterns within the sequences, i.e., clusters —or trajectories— of multi-channel sequences that share similar temporal evolutions (or similar trajectories). We illustrate the method with a case study in which we examine students' sequences of online engagement and academic achievement, where we analyse the longitudinal association between both constructs simultaneously. Here, we outline two perspectives on clustering multi-channel sequences and present both in the tutorial. The first approach uses distance-based clustering algorithms, very much in the spirit of the single-channel cluster analysis described in Chap. 10. We describe some limitations of this framework and then principally focus on a more efficient approach to identifying distinct trajectories using mixture hidden Markov models. We also show how covariates can be incorporated to make the Markovian framework even more powerful. This main analysis is inspired by the recently published paper by Saqr et al. [10]. In the next section, we provide a description of the multi-channel sequence analysis framework. We follow with a review of the existing literature in learning analytics that has relied on multi-channel sequence analysis. Next, we include a step-by-step tutorial on how to implement multi-channel sequence analysis with R, with particular attention paid to clustering via distance-based algorithms and mixture hidden Markov models. Finally, we conclude with a brief discussion and provide recommendations for further reading.

2 Multi-Channel Sequence Analysis

Multi-channel sequence analysis refers to the process of analysing sequential data that consists of multiple parallel channels or streams of (categorical) information. Each channel provides a different perspective or type of information. The goal of multi-channel sequence analysis is to jointly explore the dependencies and temporal interactions between the different channels and extract meaningful insights that may not be evident when considering each channel in isolation. The sources of information can be of very varied nature. For example, using video data of students working on a collaborative task, we could code students' spoken utterances in one channel, and their facial expressions in another channel throughout the whole session (see Fig. 1). In this way, we could analyse how students' expressions relate to what they are saying. Multi-channel sequence analysis often follows the following steps:

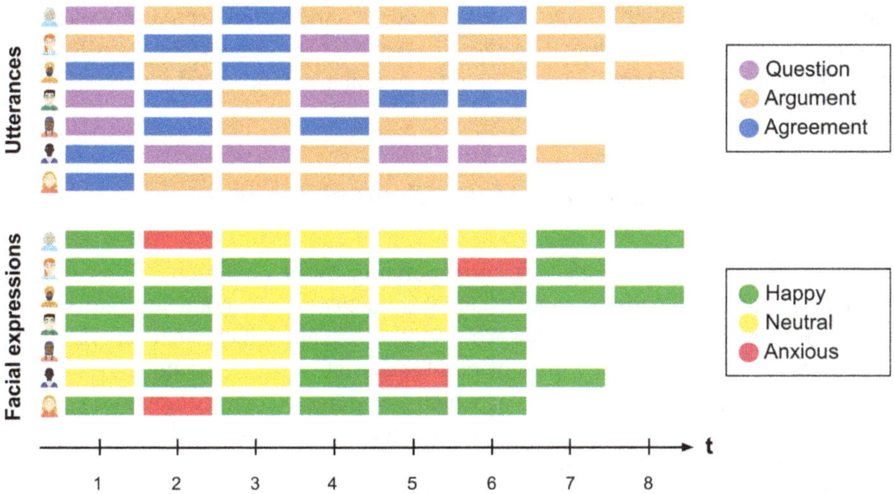

Fig. 1 An example of two channels of sequence data where the first channel (top) represents students' utterances in collaborative work and the second channel (bottom) represents their facial expressions

2.1 Step 1: Building the Channel Sequences

Just like in regular sequence analysis (see Chapter 10 [3]), the first step in multi-channel sequence analysis is to construct the sequences which constitute the different channels. A sequence is an ordered list of categorical elements (i.e., events, states or categories). These elements are discrete (as opposed to numerical values such as grades) and are ordered chronologically. In a sequence, time is not a continuum but a succession of discrete timepoints. These timepoints can be generated by sampling (e.g., each 30 seconds constitutes a new timepoint) or can emerge from an existing time scheme (e.g., each new lesson in a course is a new timepoint). In multi-channel sequence analysis, it is very important that all channels (i.e., parallel sequences) follow the same time scheme so they can be synchronised. As well as time points being aligned in this fashion, it is also typically assumed that the sequence length for each observation matches across channels (although some of the methods used can also deal with partially missing information). The elements of each sequence —typically referred to as the *alphabet*, which is unique for each channel— can be predefined events or categories from the beginning (e.g., utterances or facial expressions) or, in cases where we are dealing with numerical variables (e.g., heart rate or grades), we can convert them into a state or category by dividing the numerical variable into levels (e.g., tertiles, quartiles) or using clustering techniques. This way, we can focus on sharp changes in numerical variables and conceal small, probably insignificant changes. As Winne puts it [11], "reliability can sometimes be improved by tuning grain size of data so it is neither too coarse, masking variance within bins, nor too fine-grained, inviting distinctions that cannot be made reliably". Once we have our channels represented as sequences of ordered elements that follow the same time scheme, we can use the steps learned in the introductory sequence analysis chapter to construct the sequences.

2.2 Step 2: Visualising the Multi-Channel Sequence

When we have built the sequences corresponding to all of the channels, we can visualise the data. Visualising multi-channel sequence data is not a straightforward task, and we need to decide whether we want to present each channel separately or extend the alphabet to show a plot of combined states—or sometimes even both. The extended alphabet approach —in which the states are the combination of the states of all the channels— helps us understand the structure of the data and the typical combinations and evolution of the combined states from all of the channels. It usually works well when (1) the extended alphabet is moderate in size, (2) when there are no state combinations that are very rare, and (3) when the states are either fully observed or missing in all channels at the same time. The extended alphabet is likely to be of moderate size when there are at most 2–3 channels with a small alphabet in each, or if some combinations are not present in the data. Rare state combinations are typically impractical both when visualizing and analyzing the data—especially if there are many rare combinations or if the data are very sensitive in nature. Finally, if there are partially missing observations, i.e., observations that at some specific time point are missing from some of the channels but not all, we would need to deal with each combination of missing and observed states separately, which further extends the alphabet. If one or more of the three conditions are not met, it is often preferable to resort to presenting each channel separately or using visualization techniques that summarise information (e.g., sequence distribution plots).

Continuing with the previous example, the possible states in the extended alphabet would be all the combinations between the alphabet of the utterances channel (Question-Argument-Agreement) and the alphabet of the facial expressions channel (Happy-Neutral-Anxious). In Fig. 2, we can see that the first student starts by 'Question/Happy', then goes to 'Argument/Anxious' and so on.

2.3 Step 3: Finding Patterns (Clusters or Trajectories)

Patterns may exist in all types of sequences and, in fact, in all types of data. Discovering such patterns, variations, or groups of patterns is a central theme in analytics. In sequence analysis, similar patterns reflect common temporal pathways where the data share a similar temporal course. This could be a group of students, a typical succession of behavior, or a sequence of interactions. As such, the last typical step in the analysis is to find such patterns. Since these patterns are otherwise latent or unobservable, we need a clustering algorithm that unveils such hidden patterns. There are different possibilities to cluster multi-channel sequence data, of which two are described below: extensions to traditional distance-based clustering approaches widely used in single-channel analyses and Markovian models. The two approaches are then further distinguished in Sect. 2.4, with regard to the ability of the Markovian approach to directly accommodate covariates.

2.3.1 Traditional Sequence Analysis Extensions

Broadly speaking, the first paradigm involves extending techniques commonly adopted in single-channel analyses to the multi-channel domain, by suitably summarising the information arising from all constituent channels. Within this first framework, two different approaches have been proposed.

The first consists of flattening the multi-channel data by combining the states (such as in the example in Fig. 2) and treating the sequence as a single channel with an extended alphabet. Typically, one would then compute dissimilarities using this extended alphabet, most commonly using Optimal Matching (OM). However, this approach is not limited to distance-based clustering [12]; any of the methods, distance-based or otherwise, that we have seen in the previous chapters related to sequence analysis can be used to cluster sequences represented in this way (e.g., agglomerative hierarchical clustering using OM, or hidden Markov models). While this seems like an easier choice, the models tend to become too complex —in the sense of requiring a larger number of clusters and/or hidden states in order to adequately characterise the data— when the number of channels or numbers of per-channel states increase. Moreover, for even a moderate number of channels and/or moderate numbers of states per-channel, the size of the combined alphabet can become unwieldy. Indeed, using this approach in conjunction with OM has been criticised because of the difficulty of specifying appropriate substitution costs for such large combined alphabets [13]. As a reminder, the substitution cost refers to the penalty associated with replacing one element of a sequence with another one; these dissimilarites between all pairs of states are then used to calculate all pairwise

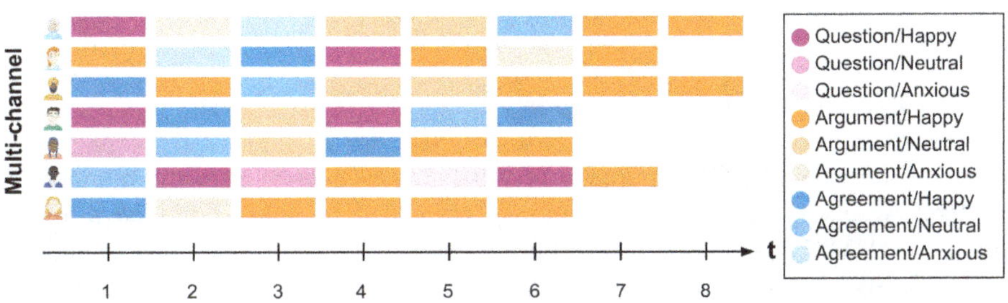

Fig. 2 An example of a multi-channel sequence object obtained by using the extended alphabet approach, i.e., combining the states of the utterances and facial expressions channels of the data shown in Fig. 1

distances between whole sequences, which are treated as the input to distance-based clustering algorithms (see Chapter 10 [3]).

The second, more common approach is explicitly distance-based and relies on an extension of the OM metric itself, by using a set of overall multi-channel costs derived from channel-specific costs, to calculate a pairwise dissimilarity matrix. The method relies on computing substitution costs for each constituent channel, combining the channels into a new multi-channel sequence object, and deriving an overall substitution cost matrix for the multi-channel object by adding (or averaging) the appropriate substitution costs across channels for a particular multi-state change, typically with equal weight given to each channel [14]. For instance, if the cost associated with substituting the states 'Question' and 'Agreement' in the utterances channel in Fig. 1 is 0.6 and the cost of substituting 'Happy' and 'Neutral' in the facial expressions channel is 1.2, the cost associated with substitutions of 'Question/Neutral' for 'Question/Happy' and 'Question/Neutral' for 'Agreement/Happy' in Fig. 2 would be 1.2 and 1.8, respectively. From such multi-channel costs, a pairwise multi-channel dissimilarity matrix can be obtained and all subsequent steps of the clustering analysis can proceed as per Chap. 10 thereafter.

That being said, adopting the extended OM approach rather than the extended alphabet approach does not resolve all limitations of the distance-based clustering paradigm. As Saqr et al. [15] noted, large sequences are hard to cluster using standard methods such as hierarchical clustering, which is memory inefficient and hard to parallelise or scale [16, 17]. Furthermore, distance-based clustering methods are limited by the theoretical maximum dimension of a matrix in R which is 2,147,483,647, corresponding to a maximum of 46,430 sequences. Using weights can fix the memory issue if the number of unique sequences remains below the threshold [18]. However, the more states (and combinations of states), the more unique the sequences tend to be, so the memory issue is even more typical with multi-channel sequence data. In such a case, Markovian methods may be the solution. Moreover, the particular choice of distance-based clustering algorithm must be chosen with care, and as ever the user must pre-specify the number of clusters, thereby often necessitating the evaluation of several computing solutions with different numbers of clusters, different hierarchical clustering linkage criteria, or different distance-based clustering algorithms entirely.

2.3.2 Mixture Hidden Markov Models

Even so, the extended OM approach is still liable to overestimate the number of clusters. A second, more sophisticated option is to use mixture hidden Markov models (MHMM) [19], which we have already encountered in some detail in Chap. 12. Such models notably support both single- and multi-channel sequences and often lead to more parsimonious representations of the latent group structure than the distance-based paradigm. MHMMs are a thus a far more efficient and powerful method for temporally aligning the multiple channels of data into homogeneous temporal patterns. As we have learned in Chap. 12, when we estimate a mixture hidden Markov model, we need to specify the number of clusters (K) and we create K submodels that are each a hidden Markov model. Each submodel can have a different number of hidden states, which allows for great flexibility in modeling the variability of the data, without unduly inflating the number of clusters as per the distance-based approaches, thereby enabling more concise and interpretable characterisation of the distinct patterns in the data. Accommodating multi-channel sequences in the MHMM framework requires treating the observed states in each channel independently given the current hidden state. This can be easily performed by defining multiple emission probability matrices (one per channel). The assumption of conditional independence simplifies the model but is sometimes unrealistic, particularly if there are strong time-dependencies between observed states even after accounting for the hidden state. In some cases, it would be better to either analyse single-channel sequences (using an extended alphabet) or resort to the distance-based paradigm described above.

We also learned in Chap. 12 that an advantage of MHMMs, as a probabilistic modelling approach, is that we can use traditional model selection methods such as likelihood-based information criteria or cross-validation for choosing the best model. For example, if the number of subpopulations is not known in advance —as is typically the case— we can compare models with different clustering solutions (be they those obtained with different numbers of clusters or different sets of initial probabilities, for example) and choose the best-fitting model with, for example, the Bayesian information criterion (BIC) [20]. Conversely, the distance-based paradigm relies on heuristic cluster quality measures for conducting model selection. In addition, an advantage of the distance-based approach is that it does not rely on making any assumptions about the data-generating mechanism. In other words, it does not assume that the data follows a generative probabilistic model, Markovian or otherwise.

The extended OM method for calculating multi-channel dissimilarities is implemented in the R package `TraMineR` and is applied in our case study, while the extended alphabet approach (i.e., combining all channels into a single one where the alphabet contains all possible state combinations) is not pursued any further here. We also cluster via the MHMM framework in our case study, using the R package `seqHMM` [21], in order to empirically compare and contrast both perspectives. It is worth noting that, of the Markovian methods we learned about in Chap. 12, only hidden Markov models and mixture hidden Markov models explicitly allow for multi-channel sequences by treating the observed states in each channel independently given the current hidden state. The more basic Markov and mixture Markov models can only accommodate multi-channel sequences by treating them as a single channel using the extended alphabet approach described above, but we do not consider this option any further here. To summarise, we do not explore the extended alphabet approach in either the distance-based or Markovian settings; we only demonstrate the multi-channel extension of the OM metric from the distance-based point of view, and only the MHMM approach from the Markovian point of view.

2.4 Step 4: Relating Clusters to Covariates

As we have seen, there are, broadly speaking, two conflicting perspectives on clustering in the sequence analysis community; algorithmic, distance-based clustering on the one hand, and model-based clustering on the other. Distance-based clustering of multi-channel sequences has already been described above, but it is useful to acknowledge that mixture Markov models and mixture hidden Markov models are precisely examples of model-based clustering methods. Moreover, `seqHMM` and the methods it implements are unique in offering a model-based approach to clustering multi-channel sequences. A key advantage of the model-based clustering framework is that it can easily be extended to directly account for information available in the form of covariates, as part of the underlying probabilistic model. Though the recently introduced mixture of exponential-distance models framework [22] attempts to reconcile the distance-based and model-based cultures in sequence analysis, while allowing for the direct incorporation of covariates, it is not based on Markovian principles, and most importantly can currently only accommodate single-channel sequences.

Otherwise, however, distance-based approaches typically cannot accommodate covariates, except as part of a post-hoc step whereby, typically, covariates are used as explanatory variables in a post-hoc multinomial regression with the cluster memberships used as the response (or as independent variables in linear or non-linear regression models). This is questionable as substituting a categorical variable indicating cluster membership disregards the heterogeneity within clusters and is clearly only sensible when the clusters are sufficiently homogeneous [23, 24]. It is often more desirable to incorporate covariates directly, i.e. to cluster sequences and relate the clusters to the covariates simultaneously. This represents an important distinction between the two approaches outlined above; this step does not apply to distance-based clustering approaches and is a crucial advantage of the Markovian approach which makes MHMMs even more powerful tools for clustering multi-channel sequences.

In theory, MHMMs can include covariates to explain the initial, transition, and/or emission probabilities. A natural use-case would be to allow subject-specific and possibly time-varying transition and emission probabilities (in the case of time-varying covariates). For example, if we think that transitioning from low to high achievement gets harder as the students get older we may add time as an explanatory variable to the model, allowing the probability of transitioning from low to high achievement to diminish in time. However, at the time of writing this chapter, these extensions are not supported in `seqHMM` (this may change in the future). Instead, the `seqHMM` package used in the examples supports time-constant covariates only. Specifically, the manner in which covariates are accommodated is similar in spirit to the aforementioned mixture of exponential-distance models framework and latent class regression [25], in that covariates are used for predicting cluster memberships for each observed sequence. Adding covariates to the model thus helps to both explain and influence the probabilities that each individual has of belonging to each cluster. By incorporating covariates in this way, we could, for example, find that being in a high-achievement cluster is predicted by gender, previous grades, or family background

In the terminology of mixtures of experts modelling, this approach corresponds to a so-called "gating network mixture of experts model", whereby the distribution of the sequences depends on the latent cluster membership variable, which in turn depends on the covariates, while the sequences are independent of the covariates, conditional on the latent cluster membership variable (see [26] and [22] for examples). This, rather than having covariates affect the component distributions, is particularly appealing, as the interpretation of state-specific parameters is the same as it would be under a model without covariates. However, it is worth noting that the covariates do not merely explain the uncovered clusters; as part of the model, they drive the formation of the clusters. In other words, an otherwise identical model without dependence on covariates may uncover different groupings with different probabilities. However, this can be easily overcome by treating the selection of the most relevant subset of covariates as an issue of model selection via the BIC or other criteria and taking the number of clusters/states, set of initial probabilities, and set of covariates which jointly optimise the chosen criterion. The incorporation of covariates in MHMMs and related practical issues are also illustrated in the case study on the longitudinal association of engagement and achievement which follows.

3 Review of the Literature

The use of multi-channel sequence analysis in learning analytics has so far been scarce. Most of the few studies that implemented this method used it to combine multiple modalities of data such as coded video data [27, 28], electrodermal activity [27, 28], clickstream/logged data [10, 29, 30], and assessment/achievement data [10, 30, 31]. The data were converted into different sequence channels using a variety of methods. For example, [27] manually coded video interactions as positive/negative/mixed to represent the emotional valence of individual interaction, and used a pre-defined threshold to discern between high and low arousal from electrodermal activity data. In the work by [30], two sequence channels were built from students' daily tactics using the learning management system and an automated assessment tool for programming assignments respectively. The learning tactics were identified through hierarchical clustering of students' trace log data in each environment. Similarly [10], created an engagement channel for each course in a study program based on students' engagement states derived from the learning management system data, and an achievement state based on their grade tertiles. Another study [29] had five separate channels: the first channel represented the interactive dimension built from the social interactions through peer communications and online behaviours; the second channel represented the cognitive dimension constructed from students' knowledge contributions at the superficial, medium, and deep levels; the third channel represented the regulative dimension which represented students' regulation of their collaborative processes, including task understanding, goal setting and planning, as well as monitoring and reflection; the fourth channel represented the behavioural dimension which analysed students' online behaviours, including resource management, concept mapping and observation, and the fifth channel represented the socio-emotional dimension, which included active listening and respect, encouraging participation and inclusion, as well as fostering cohesion.

A common step in the existing works is the use of clustering techniques to detect unobserved patterns within the multi-channel sequences. Articles have relied on hidden Markov models to identify hidden states in the data. For instance, [30] found three hidden states consisting of learning strategies which combine the use of the learning management system and an automated assessment tool: an instruction-oriented strategy where students consume learning resources to learn how to code; an independent coding strategy where students relied on their knowledge or used the learning resources available to complete the assignments, and a dependent coding strategy where students mostly relied on the help forums to complete their programming assignments. Most studies go one step further and identify distinct trajectories based on such hidden states. For example, [28] found four clusters of socio-emotional interaction episodes (positive, negative, occasional regulation, frequent regulation), which differed in terms of fluctuation of affective states and activated regulation of learning. The authors of [29] found three types of group collaborative patterns: behaviour-oriented, communication-behaviour-synergistic, and communication-oriented. To the knowledge of the authors, no article has relied on the distance-based approach (commonly used in single-channel sequence analysis) to cluster multi-channel sequences. However, this approach has been used in social sciences [9] as well as in other disciplines [e.g., 32].

4 Case Study: The Longitudinal Association of Engagement and Achievement

To learn how to implement multi-channel sequence analysis we are going to practice with a real case study: we will investigate the longitudinal association between engagement and achievement across a study program using simulated data based on the study by [10]. We begin by creating a sequence for each of the channels (engagement and achievement) and then explore visualisations thereof. We then focus on clustering these data using the methods described above specifically tailored for multi-channel sequences, in order to present an empirical view of both the distance-based and Markovian perspectives. In doing so, we will first demonstrate how to construct a multi-channel pairwise OM dissimilarity matrix in order to adapt and apply the distance-based clustering approach of Chap. 10 to these data. Secondly, we will largely focus on using a mixture hidden Markov model to detect the longitudinal clusters of students that share a similar trajectory of engagement and achievement. Finally, we will demonstrate how to incorporate covariates under the model-based Markovian framework.

4.1 The Packages

To accomplish our task, we will rely on several R packages We have use most of them throughout the book. Below is a brief summary of the packages required:

- `rio`: A package for reading and saving data files with different extensions [33].
- `seqHMM`: A package designed for fitting hidden (latent) Markov models and mixture hidden Markov models for social sequence data and other categorical time series [21].
- `tidyverse`: A package that encompasses several basic packages for data manipulation and wrangling [34].
- `TraMineR`: As seen in the introductory sequence analysis chapter, this package helps us construct, analyse, and visualise sequences from time-ordered states or events [35].
- `WeightedCluster`: A package to cluster sequences and computing quality measures [18].

If you have not done so already, install the packages using the `install.packages()` command (e.g., `install.packages("seqHMM")`). We can then import them as follows:

```
library(rio)
library(tidyverse)
library(TraMineR)
library(seqHMM)
library(WeightedCluster)
```

4.2 The Data

The data that we are going to use in this chapter is a synthetic dataset which contains the engagement states (`Active`, `Average`, or `Disengaged`) and achievement states (`Achiever`, `Intermediate`, or `Low`) for eight successive courses (each course is a timepoint). Each row contains an identifier for the student (`UserID`), an identifier for the course (`CourseId`), and the order (`Sequence`) in which the student took that course (1–8). For each course, a student has both an engagement state (`Engagement`) and an achievement state (`Achievement`). For example, a student can be `'Active'` and `'Achiever'` in the first course (`Sequence = 1`) they take, and 'Disengaged' and 'Low' achiever in the next. In addition, the dataset contains the final grade (0–100) and three time-constant covariates (they are the same for each student throughout all four courses): the previous grade (Prev_grade, 1–10), Attitude (0–20), and Gender (Male/Female). The dataset is described in more detail in the Data chapter of this book. We can import the data using the import() command from rio. A preview is shown in Table 1.

```
URL <- "https://github.com/sonsoleslp/labook-data/raw/main/"
fileName <- "9_longitudinalEngagement/SequenceEngagementAchievement.xlsx"
df <- import(paste0(URL, fileName))
df
```

4.3 Creating the Sequences

We are going to first construct and visualise each channel separately (engagement and achievement), and then study them in combination.

4.3.1 Engagement Channel

To build the engagement channel, we need to construct a sequence of students' engagement states throughout the eight courses. To do so, we need to first arrange our data by student (UserID) and course order (Sequence). Then, we pivot the data to a wide format, where the first column is the UserID and the rest of the columns are the engagement states at each of the courses (ordered from 1 to 8). For more details about this process, refer to the introductory sequence analysis chapter.

```
eng_seq_df <- df |> arrange(UserID, Sequence) |>
  pivot_wider(id_cols = "UserID", names_from = "Sequence", values_from="Engagement")
```

Now we can use TraMineR to construct the sequence object and assign colours to it to represent each of the engagement states:

```
engagement_levels <- c("Active", "Average", "Disengaged")
colours <- c("#28a41f", "#FFDF3C", "#e01a4f")
eng_seq <- seqdef(eng_seq_df , 2:9, alphabet = engagement_levels, cpal = colours)
```

We can use the sequence distribution plot from TraMineR to visualise the distributions of the states at each time point (Fig. 3 left), and the sequence index plot to visualise each student's sequence of engagement states (Fig. 3, right), here sorted

Table 1 Preview of the data

	UserID	CourseID	Sequence	Engagement	Final_Grade	Achievement	Prev_grade	Attitude	Gender
1	00050F0E	4C3F37F0	1	Average	72.27	Achiever	6.826613	12.92833	Male
2	00050F0E	E54A52A3	2	Disengaged	72.56	Achiever	6.826613	12.92833	Male
3	00050F0E	AB7EC624	3	Average	78.78	Achiever	6.826613	12.92833	Male
4	00050F0E	B0E95213	4	Average	74.19	Achiever	6.826613	12.92833	Male
5	00050F0E	0B301F55	5	Average	87.35	Achiever	6.826613	12.92833	Male
6..1135									
1136	FE2E4C85	79ED6873	8	Active	87.69	Achiever	4.189420	20.00000	Male

according to the elements of the alphabet at the successive positions starting from the beginning of the sequences:

```
seqdplot(eng_seq, border = NA,  ncol = 3, legend.prop = 0.2)
seqIplot(eng_seq, border = NA,  ncol = 3, legend.prop = 0.2, sortv = "from.start")
```

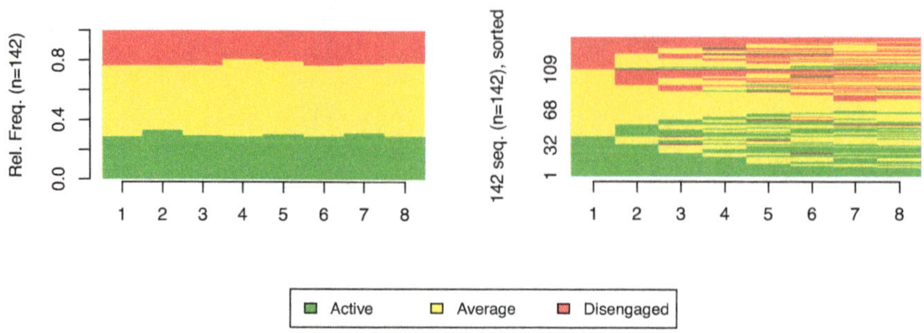

Fig. 3 Sequence distribution plot and index plot of the course engagement states

Please refer to Chapter 10 [3] for guidance on how to interpret sequence distribution and index plots.

4.3.2 Achievement Channel

We follow a similar process to construct the achievement sequences. First we arrange the data by student and course order and then we convert the data into the wider format using —this time— the achievement states as the values.

```
ach_seq_df <- df |> arrange(UserID,Sequence) |>
  pivot_wider(id_cols = "UserID", names_from = "Sequence", values_from ="Achievement")
```

We use again `TraMineR` to construct the sequence object for achievement and we again provide a suitable, distinct colour scheme:

```
achievement_levels <- c("Low", "Intermediate", "Achiever")

coloursA <- c("#a5e3ff","#309bff","#3659bf")
ach_seq <- seqdef(ach_seq_df , 2:9, cpal = coloursA, alphabet = achievement_levels)
```

We may use the same visualisations to plot our achievement sequences in Fig. 4:

```
seqdplot(ach_seq, border = NA,  ncol = 3, legend.prop = 0.2)
seqIplot(ach_seq, border = NA,  ncol = 3, legend.prop = 0.2, sortv = "from.start")
```

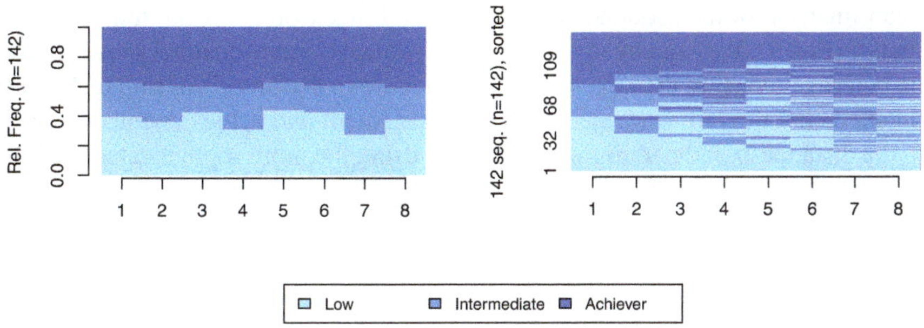

Fig. 4 Sequence distribution plot and index plot of the course achievement states

4.3.3 Visualising the Multi-Channel Sequence

Now that we have both channels, we can use the `mc_to_sc_data()` function from `seqHMM` to convert the two channels of engagement and achievement sequences to a single channel with an extended alphabet. If we had additional channels, we would just add them to the list, after `ach_seq`.

```
multi_seq <- mc_to_sc_data(list(eng_seq, ach_seq))
```

If we check the alphabet of the sequence data, we can see that it contains all combinations of engagement and achievement states.

```
alphabet(multi_seq)
```

```
[1] "Active/Achiever"          "Active/Intermediate"
[3] "Active/Low"               "Average/Achiever"
[5] "Average/Intermediate"     "Average/Low"
[7] "Disengaged/Achiever"      "Disengaged/Intermediate"
[9] "Disengaged/Low"
```

We now need to provide an appropriate colour scale to accommodate the extended alphabet. The colour scale we define below uses green colours to represent active students, blue colours represent averagely engaged students, and red colours represent disengaged students, whereas the intensity of the colour represents the achievement level: darker colours represent higher achievement, and lighter colours represent low achievement.

```
coloursM <-  c("#115a20","#24b744","#78FA94",
               "#3659bf","#309bff","#93d9ff",
               "#E01A1A","#EB8A8A","#F5BDBD")

cpal(multi_seq) <- coloursM
```

We can finally plot the sequence distribution and index plots for the multi-channel sequence (Fig. 5). We can already hint that the students' multi-channel sequences are quite heterogeneous. In the next steps, we will use two distinct clustering techniques to detect distinct combined trajectories of engagement and achievement, beginning with the distance-based paradigm and then the using the more sophisticated MHMM framework.

```
seqdplot(multi_seq, border = NA,  ncol = 3, legend.prop = 0.2)
seqIplot(multi_seq, border = NA,  ncol = 3, legend.prop = 0.2,
         sortv = "from.start")
```

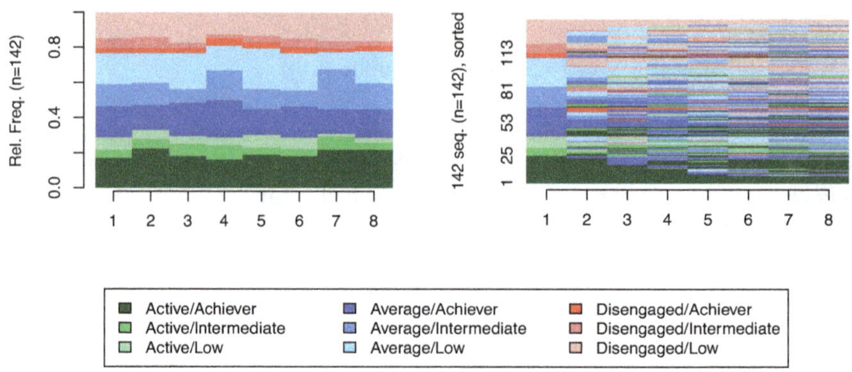

Fig. 5 Sequence distribution plot and index plot of the multi-channel sequence

4.4 *Clustering via Multi-Channel Dissimilarities*

We now describe how to construct a dissimilarity matrix for a combined multi-channel sequence object to use as input for distance-based clustering algorithms, using utilities provided in the `TraMineR` R package and relying on concepts described in Chap. 10. We begin by creating a list of substitution cost matrices for each of the engagement and achievement channels. For simplicity, we adopt the same method of calculating substitution costs for each channel, though this need not be the case. Here, we use the data-driven `"TRATE"` method, which relies on transition rates, as the `method` argument in both calls to `seqsubm()`. In other scenarios, we might want to use a constant substitution cost (same cost for all substitutions) for one or some of the channels and, for example, manually specify the substitution cost matrix for the others.

```
sub_mats <- list(engagement = seqsubm(eng_seq, method = "TRATE"),
                 achievement = seqsubm(ach_seq, method = "TRATE"))
```

Subsequently, we invoke the function `seqMD()` to compute the multi-channel dissimilarities. This requires the constituent channels to be supplied as a list of sequence objects, along with the substitution cost matrices via the argument `sm`. The argument `what = "diss"` ensures that dissimilarities are returned, according to the OM `method`, with equal weights for each channel specified via the argument `cweight`.

```
channels <- list(engagement = eng_seq, achievement = ach_seq)

mc_dist <- seqMD(channels,
                 sm = sub_mats,
                 what = "diss",
                 method = "OM",
                 cweight = c(1, 1))
```

Thereafter, `mc_dist` can be used as input to any distance-based clustering algorithm. Though we are not limited to doing so, we follow Chap. 10 in applying hierarchical clustering with Ward's criterion. However, we note that users of the distance-based paradigm are not limited to Ward's criterion or hierarchical clustering itself either.

```
Clustered <- hclust(as.dist(mc_dist), method = "ward.D2")
```

We then obtain the implied clusters by cutting the produced tree using the `cutree()` function, where the argument `k` indicates the desired number of clusters. We also create more descriptive labels for each cluster.

```
Cuts <- cutree(Clustered, k = 6)
Groups <- factor(Cuts, labels = paste("Cluster", 1:6))
```

It is worth noting that the `Groups` vector contains the clustering assignment for each student. This vector is a factor in the same order as the engagement and achievement sequences, and hence the same order as the data frames used to create the sequences. Hence, information about which student belongs to which cluster is readily accessible. Here, we show the assignments of the first 5 students only, for brevity.

```
head(Groups, 5)
```

```
         1         2         3         4         5
Cluster 1 Cluster 2 Cluster 3 Cluster 4 Cluster 3
Levels: Cluster 1 Cluster 2 Cluster 3 Cluster 4 Cluster 5 Cluster 6
```

Generally speaking, the resulting clusters for the chosen `k` value may not represent the best solution. In an ideal analysis, one would resort to one or more of the cluster quality measures given in Table 8 of Chap. 10 to inform the choice of an optimal `k` value. In doing so, we determined the chosen number of six clusters is optimal according to the average silhouette width (ASW) criterion (again, see Chap. 10). However, we omit repeating the details of these steps here for brevity and leave them as an exercise for the interested reader.

The obtained `Groups` can be used to produce visualisations of the clustering solution. We do so in two ways; first showing the combined multi-channel object with an extended alphabet (as per Fig. 2) using the `seqplot()` function from `TraMiner` in Fig. 6, and second using a stacked representation of both constituent

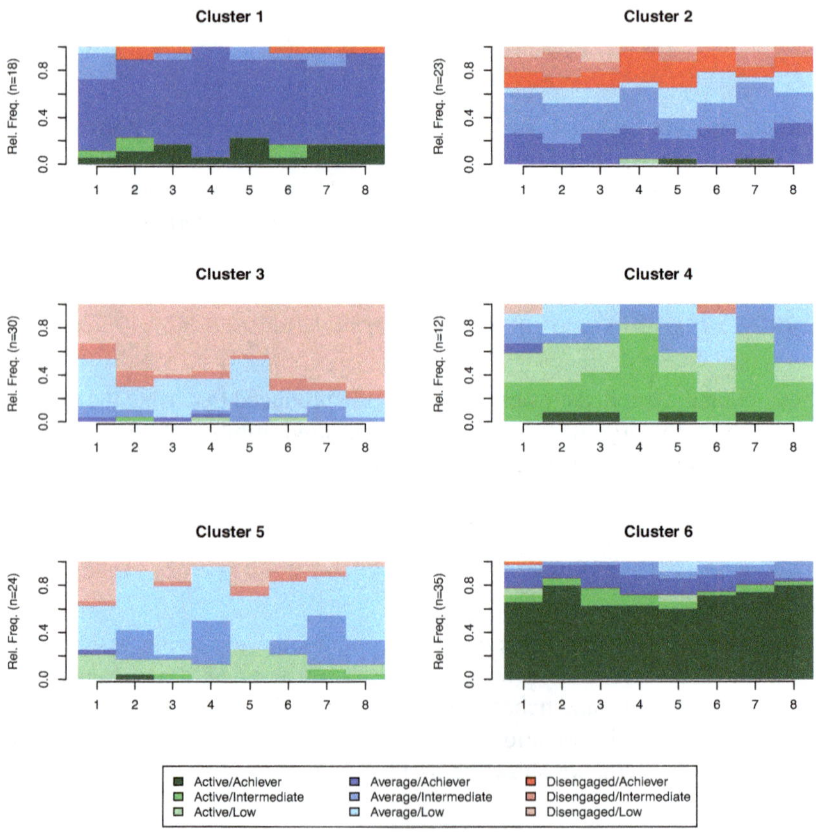

Fig. 6 State distribution plots per cluster for the combined multi-channel sequences

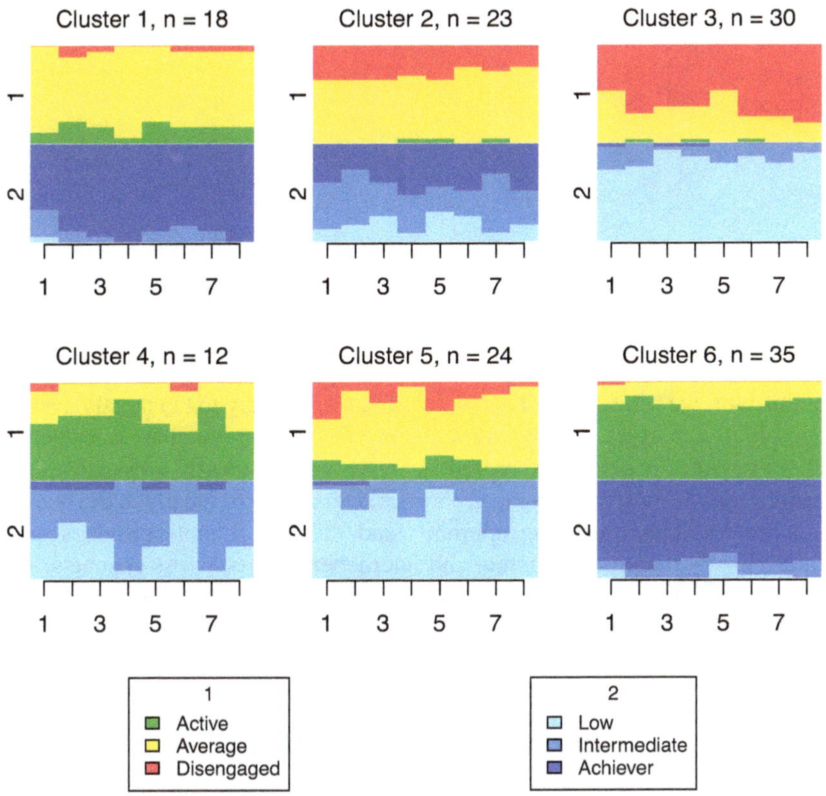

Fig. 7 Stacked state distribution plots per cluster showing both constituent channels

channels using the `ssp()` and `gridplot()` functions from `seqHMM` in Fig. 7. State distribution plots per cluster are depicted in each case.

```
seqplot(multi_seq, type = "d", group = Groups,
        ncol = 3, legend.prop = 0.1, border = NA)
```

Though there is some evidence of well-defined clusters capturing different engagement and achievement levels over time, a plot such as this can be difficult to interpret, even for a small number of channels and small numbers of per-channel states. A more visually appealing alternative is provided by the stacked state distribution plots below, which shows each individual channel with its own alphabet on the same plot, with one such plot per cluster. This requires supplying a list of the observations in each cluster in each channel to the `ssp()` function to create each plot, with each plot then arranged and displayed appropriately by the function `gridplot()`.

```
ssp_k <- list()
for(k in 1:6) {
  ssp_k[[k]] <- ssp(list(eng_seq[Cuts == k,],
                         ach_seq[Cuts == k,]),
                    type = "d", border = NA,
                    with.legend = FALSE,
                    title = levels(Groups)[k])
}
gridplot(ssp_k, ncol = 3, byrow = TRUE,
         row.prop = c(0.4, 0.4, 0.2), cex.legend = 0.8)
```

From this plot, we can see six clusters, relating respectively to (1) mostly averagely engaged high achievers, (2) mostly actively engaged intermediate achievers, (3) mostly disengaged intermediate or high achievement, (4) mostly averagely engaged low achievers, (5) mostly disengaged low achievers, and (6) mostly actively engage high achievers. This clustering approach and this set of plots provides quite a granular view of the group structure and interdependencies across channels, which differs somewhat from the clustering solution obtained by the application of MHMMs which now follows.

4.5 Building a Mixture Hidden Markov Model

The `seqHMM` package supports several types of (hidden) Markov models for single- or multi-channel sequences of data. The package provides functions for evaluating and comparing models, and methods for the visualisation of multi-channel sequence data and hidden Markov models. The models are estimated through maximum likelihood, using the Expectation Maximisation (EM) algorithm and/or direct numerical maximisation with analytical gradients. For more details on the `seqHMM` package, refer to the "Markov models" chapter. We are going to construct a mixture hidden Markov model (MHMM). For this purpose, we will rely on the `build_mhmm()` function of `seqHMM`. In its most basic form, the `build_mhmm()` function takes as arguments a list of `observations` (i.e., a list of the sequences that make up the channels), and a vector `n_states` whose length indicates the number of clusters to estimate, for which the value of each position specifies the number of hidden states in each cluster. The `build_mhmm()` function defines that structure of the model (including possible restrictions) and also assigns random or user-defined starting values to model parameters for estimation. Though we are not restricted to assuming all clusters have an equal number of hidden states, we are going to build a model with three clusters and two hidden states per cluster, inspired by the results obtained in the previous section (i.e., six clusters). Therefore, our input for `n_states` is `c(2, 2, 2)`. If we instead wanted to build a model with four clusters and three hidden states, we would need to provide `c(3, 3, 3, 3)`.

```
init_mhmm <- build_mhmm(observations = list(eng_seq,ach_seq),
                        n_states = c(2, 2, 2))
```

Now we can fit the model using the `fit_model()` function. Using the EM algorithm and the `times` argument, we can provide a number to control the number of times the model should be fit in a hope to find the globally optimal solution. We need to find a number that is not too low (or else we might be settling for a non-optimal solution too early) but also not too high (or the code will take forever to run). If we provide `100`, for example, the model will be estimated 101 times: once from the user-defined or random starting values at the build stage and then 100 re-estimations with randomised starting values. The function will give the best model as an output. Generally, the more complex the model, the more re-estimation rounds are needed.

```
set.seed(2294)
mhmm_fit <- fit_model(init_mhmm,
           control_em = list(restart = list(times = 100)))
```

Once the function has finished fitting the model, we need to check if it has likely reached the optimal solution. We can do so by checking the variable `mhmm_fit$em_results$best_opt_restart` of the fitted model. For the results to be reliable, we would expect to find the best-fitting model a number of times from different starting values. In other words, we want to see the same value repeating at the beginning of the sequence of numbers returned by `mhmm_fit$em_results$best_opt_restart` for several times if we estimate the model 100 times. In this case, we observe the same number quite many times, so the stability of our results is acceptable. If the same number does not appear several times, we can increase the number provided to the `times` argument or we can provide more informative starting values (for example, using the parameters of the newly fitted suboptimal model as starting values for the new estimation).

```
mhmm_fit$em_results$best_opt_restart
```

```
 [1] -1659.712 -1659.712 -1659.712 -1659.712 -1659.712 -1659.712 -1659.712
 [8] -1659.712 -1659.712 -1659.712 -1659.712 -1659.712 -1659.712 -1659.712
[15] -1659.712 -1659.712 -1659.712 -1659.712 -1659.712 -1659.712 -1659.712
[22] -1659.712 -1659.712 -1659.712 -1659.712
```

Now we can plot our results using the `mssplot()` function. Figure 8 shows index plots for each of the three fitted clusters. These plots show the categories at each time point within each cluster, as well as the hidden states. We can see that Cluster 1 corresponds to students who are mostly low achievers, wherein State 1 represents students who are disengaged and State 2 represents students who are averagely engaged. Cluster 2 includes the Active students who are mostly high achievers (State 2), or intermediate achievers (State 1). Cluster 3 represents mostly low achievers, where State 1 mostly represents the disengaged and State 2 the mostly averagely engaged. Therefore, the results are quite interpretable and make sense given our data.

```
HIDDEN_STATES <- c("darkred", "pink", "darkblue", "purple",
                   "lightgoldenrod1", "orange")

mssplot(mhmm_fit$model,
        plots = "both",
        type = "I",
        sortv = "mds.hidden",
        yaxis = TRUE,
        with.legend = "bottom",
        ncol.legend = 1,
        cex.title = 1.3,
        hidden.states.colors = HIDDEN_STATES)
```

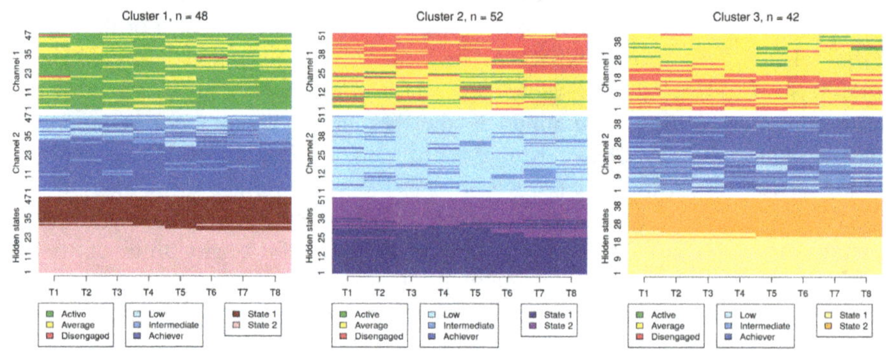

(a) Cluster 1. (b) Cluster 2. (c) Cluster 3.

Fig. 8 Multi-channel sequence index plots. (**a**) Cluster 1. (**b**) Cluster 2. (**c**) Cluster 3

We may also plot the transitions between the hidden states for each cluster using the `plot()` function (Fig. 9). Most arguments used here are purely cosmetic in nature, with the exception of `combine.slice` which corresponds to the highest probability for which emission probabilities are combined into one state. We adopt the default value of 0.05 to aid the legibility of the plots. We see that the probabilities of starting at each hidden state (initial probabilities), as shown at the bottom of each pie chart, are quite balanced: 0.61/0.39, 0.58/0.42, 0.62/0.38. The transition probabilities (i.e., the probabilities of transitioning from one hidden state to another within the same cluster) are in turn quite low (all < 0.07), meaning that in most cases students remain in the same hidden state throughout their whole academic trajectory. Lastly, the emission probabilities (i.e., the probabilities of each state within each hidden state) are quite diverse, where some hidden states are quite homogeneous (e.g, State 2 in Cluster 2 has a 0.84 probability of having Disengaged students) whereas others are more evenly distributed among all states (e.g., State 1 in Cluster 3).

```
plot(mhmm_fit$model,
     vertex.size = 60,
     cpal = coloursM,
     label.color = "black",
     vertex.label.color = "black",
     edge.color = "lightgray",
     edge.label.color = "black",
     ncol.legend = 1,
     ncol = 3,
     rescale = FALSE,
     interactive = FALSE,
     combine.slices = 0.05,
     combined.slice.label = "States with probability < 0.05")
```

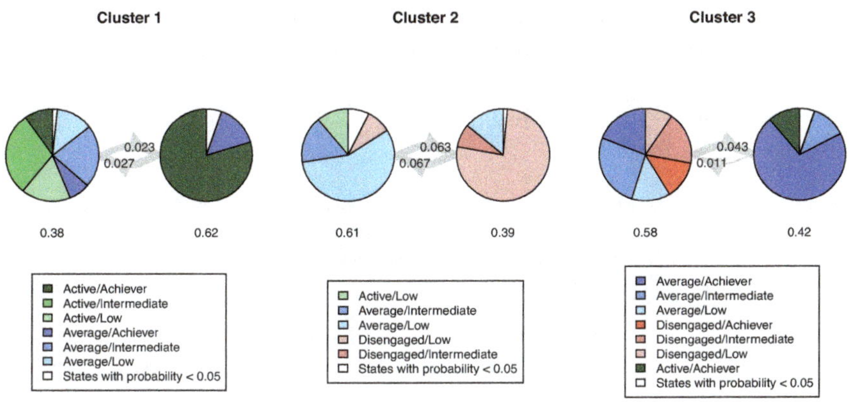

Fig. 9 Transitions between hidden states for each trajectory

In the previous example, we have used `build_mhmm()` with the default values in which we only need to provide the number of clusters and hidden states. However, the function allows much more flexibility by allowing us to constrain the different probabilities (initial, transition, and emission) that we have just talked about. An interesting case study can be to investigate how students who start in the same hidden state evolve throughout their academic trajectory; whether they remain in the same hidden state or switch to a different one. To do that, we would need to fix the initial probabilities to be 1 for the first hidden state, and 0 for the other. To do this for the three clusters, we need to create a list with the vector `c(1, 0)` repeated three times. For the initial transition and emission probabilities, we can just simulate random ones using the functions `simulate_transition_probs()` and `simulate_emission_probs()` respectively. It is worth noting that in the previous MHMM example, when we fitted the model using the default probabilities, we just needed to provide the number of clusters and numbers of hidden states per cluster, via `n_states`. However, now that we need to fix the initial probabilities, we also need to simulate the transition and emission probabilities using the corresponding functions, as `build_mhmm()` requires either `n_states` to be provided or *all three of* `initial_probs`, `transition_probs`, and `emission_probs` to be provided.

```
initial_probs <- list(c(1, 0), c(1, 0), c(1, 0))
transition_probs <- simulate_transition_probs(n_states = 2,
                                              n_clusters = 3,
                                              left_right = TRUE)
emission_probs <- simulate_emission_probs(n_states = 2,
                                          n_symbols = 3,
                                          n_clusters = 1)
```

We must now provide the probability objects to the `build_mhmm()` function along with the observations. Since the number of clusters and hidden states can be derived from the probabilities, we no longer need to provide the `n_states` argument. Specifically, this is because `emission_probs` is given as a list of length 3 with each element being a list of length 2, corresponding to three clusters each with two hidden states per cluster, as before. This new model is fitted in the same way, by using the `fit_model()` function.

```
set.seed(2294)
init_emission_probs <- list(list(emission_probs, emission_probs),
                            list(emission_probs, emission_probs),
                            list(emission_probs, emission_probs))
init_mhmm_i <- build_mhmm(observations = list(eng_seq, ach_seq),
                          initial_probs = initial_probs,
                          transition_probs = transition_probs,
                          emission_probs = init_emission_probs)

mhmm_fit_i <- fit_model(init_mhmm_i,
                        control_em = list(restart = list(times = 200)))
```

Again, we need to check whether the model has converged to a valid solution by checking if the same number is repeated at the beginning of the `best_opt_start` output:

```
mhmm_fit_i$em_results$best_opt_restart
```

```
 [1] -1726.463 -1726.463 -1726.463 -1726.463 -1726.463 -1726.463 -1726.463
 [8] -1726.463 -1726.463 -1726.463 -1726.463 -1726.463 -1726.463 -1726.463
[15] -1726.463 -1726.463 -1726.463 -1726.463 -1726.463 -1726.463 -1735.624
[22] -1735.624 -1735.624 -1735.624 -1735.624
```

Now that we have checked that our model is correct, we can give the channels and clusters representative names so that they look better when plotting:

```
mhmm_fit_i$model$cluster_names <- c("Trajectory 1",
                                    "Trajectory 2",
                                    "Trajectory 3")
mhmm_fit_i$model$channel_names <- c("Engagement",
                                    "Achievement")
```

Now we may plot our results, as before, in the form of sequence index plots (see Fig. 10a–c). We see that —due to the specified initial probabilities— all clusters start with the same hidden state and mostly remain there throughout the whole trajectory, although some students switch to the other hidden state as time advances. In Cluster 1, students are mostly low achievers and start by being mostly averagely engaged (State 1), while some transition to disengaged (State 2). In Cluster 2, students are mostly highly engaged and start being high achievers (State 1), while some students transition to being mostly intermediate achievers (State 2). Lastly, in Cluster 3, students are mostly averagely engaged or disengaged high or intermediate achievers (State 1), while a small fraction of students become more averagely engaged and more highly achieving over time (State 2).

```
HIDDEN_STATES <- c("darkred", "pink", "darkblue", "purple",
                   "lightgoldenrod1", "orange")

mssplot(mhmm_fit_i$model,
        plots = "both",
        type = "I",
        sortv = "mds.hidden",
        yaxis = TRUE,
        with.legend = "bottom",
        ncol.legend = 1,
        cex.title = 1.3,
        hidden.states.colors = HIDDEN_STATES)
```

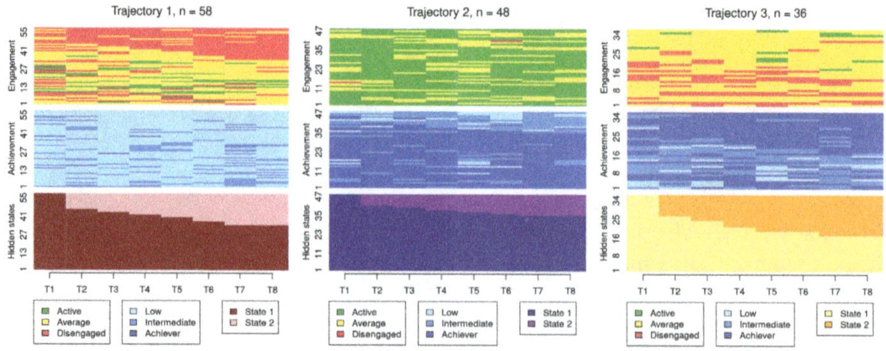

(a) Trajectory 1. (b) Trajectory 2. (c) Trajectory 3.

Fig. 10 Multi-channel sequence index plots with fixed start. (**a**) Trajectory 1. (**b**) Trajectory 2. (**c**) Trajectory 3

There are countless combinations of numbers of clusters, hidden states, and initial/transition/emission probabilities that we can use to fit our model. The choice depends on our data, our research question, the interpretability of the results, and the goodness of fit. Regarding the latter, a common way to compare the performance of several models is by using the Bayesian Information Criterion (BIC; [20]). The BIC can help us score the models which better fit our data while penalising us if we overfit it by adding too many parameters (e.g, if we specified 50 clusters instead of 3, we would better capture the variability of our data but our model would hardly generalise to other scenarios). Therefore, as a general rule, the lower the BIC the better the model, although we need to take into account the aforementioned issues (research questions, interpretability, etc.) to make the final choice. Below we show an example of how to calculate the BIC of the models we have estimated. We see that the first model with the fixed initial probabilities has a somewhat higher BIC than the second one with the restricted initial probabilities. However, we might need to try different numbers of clusters and hidden states to make our final choice.

```
BIC(mhmm_fit$model)
```

```
[1] 3565.658
```

```
BIC(mhmm_fit_i$model)
```

```
[1] 3656.949
```

4.6 Incorporating Covariates in MHMMs

The BIC can also be used to select the optimal model when different candidate models with different subsets of covariates are fitted. Recall that the clustering and the relation of clusters to covariates is performed simultaneously. Thus, otherwise identical models without dependence on covariates, or using different covariates, may uncover different groupings with different probabilities and hence a different BIC score. Here, we take the optimal model –with three clusters, each with two hidden states, using fixed initial probabilities– and demonstrate how to add and select covariates to make the model more powerful and interpretable. We begin by extracting the covariate information from the original, long format `df`. For each student, the available time-constant covariates are a numeric variable giving their previous grades from an earlier course, a numeric measure of their attitude towards learning, and their gender (male or female). The order of the rows of `cov_df` matches the order in the two `eng_seq` and `ach_seq` channels.

```
cov_df <- df |>
  arrange(UserID, Sequence) |>
  filter(!duplicated(UserID)) |>
  select(Prev_grade, Attitude, Gender)
```

The code below replicates the code used to fit the previously optimal MHMM model, and augments it by providing a data frame (via the `data` argument, using the `cov_df` just created) and the corresponding one-sided formula for specifying the covariates to include (via the `formula` argument). For simplicity, we keep the same number of clusters and hidden states and fit three models (one for each covariate in `cov_df`). To mitigate against convergence issues, we use the results of the previously optimal MHMM to provide informative starting values.

```
set.seed(2294)
init_mhmm_1 <- build_mhmm(observations = list(eng_seq, ach_seq),
                          initial_probs = mhmm_fit_i$model$initial_probs,
                          transition_probs = mhmm_fit_i$model$transition_probs,
                          emission_probs = mhmm_fit_i$model$emission_probs,
                          data = cov_df,
                          formula = ~ Prev_grade)
init_mhmm_2 <- build_mhmm(observations = list(eng_seq, ach_seq),
                          initial_probs = mhmm_fit_i$model$initial_probs,
                          transition_probs = mhmm_fit_i$model$transition_probs,
                          emission_probs = mhmm_fit_i$model$emission_probs,
                          data = cov_df,
                          formula = ~ Attitude)
```

```
init_mhmm_3 <- build_mhmm(observations = list(eng_seq, ach_seq),
                          initial_probs = mhmm_fit_i$model$initial_probs,
                          transition_probs = mhmm_fit_i$model$transition_probs,
                          emission_probs = mhmm_fit_i$model$emission_probs,
                          data = cov_df,
                          formula = ~ Gender)
```

We estimate each model 50 + 1 times (first from the starting values we provided and then from 50 randomised values).

```
mhmm_fit_1 <- fit_model(init_mhmm_1,
                        control_em = list(restart = list(times = 50)))
mhmm_fit_2 <- fit_model(init_mhmm_2,
                        control_em = list(restart = list(times = 50)))
mhmm_fit_3 <- fit_model(init_mhmm_3,
                        control_em = list(restart = list(times = 50)))
```

Note that in an ideal analysis, however, one would vary the number of clusters and hidden states along with varying sets of covariates and initial probabilities in order to find the model settings which jointly optimise the BIC. Moreover, one is not limited to including only a single covariate; one could specify for example `formula = ~ Attitude + Gender`, but we omit this consideration here for simplicity. Furthermore, one should check in an ideal analysis that the optimal model was found in a majority of the 51 runs. We did so and can be satisfied that there are no convergence issues. We select among the three models with three different, single covariates using the BIC, and include the previous model with no covariates in the comparison also.

```
BIC(mhmm_fit_i$model)
```

```
[1] 3656.949
```

```
BIC(mhmm_fit_1$model)
```

```
[1] 3614.673
```

```
BIC(mhmm_fit_2$model)
```

```
[1] 3632.004
```

```
BIC(mhmm_fit_3$model)
```

```
[1] 3649.353
```

Recalling that lower BIC values are indicative of a better model, we can see that each covariate yields a better-fitting model. Though a better model could be found by adding more covariates or modifying the numbers of clusters and hidden states, we now proceed to interrogate, interpret, and visualise the results of the best model only; namely the MHMM with `Prev_grade` as a predictor of cluster membership.

We begin by examining the clusters via sequence index plots using `mssplot()`, as before, in Fig. 11.

```
mssplot(mhmm_fit_1$model,
        plots = "both",
        type = "I",
        sortv = "mds.hidden",
        yaxis = TRUE,
        with.legend = "bottom",
        ncol.legend = 1,
        cex.title = 1.3,
        hidden.states.colors = HIDDEN_STATES)
```

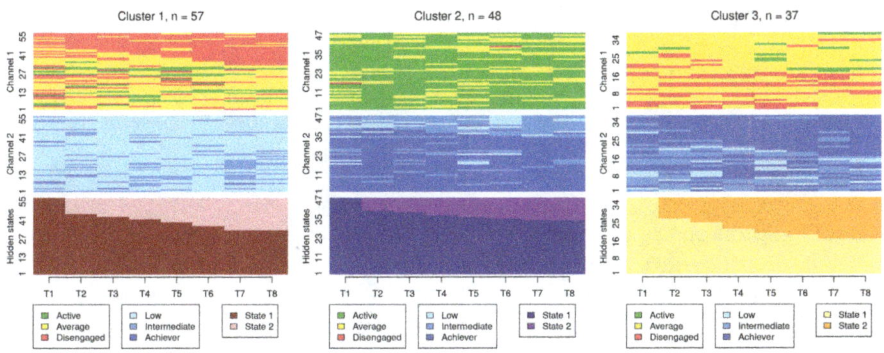

(a) Cluster 1. (b) Cluster 2. (c) Cluster 3.

Fig. 11 Multi-channel sequence index plots for the optimal covariate-dependent MHMM. (**a**) Cluster 1. (**b**) Cluster 2. (**c**) Cluster 3

Compared to Fig. 10, the results are notably similar. One observation previously assigned to the first trajectory is now assigned to the third, but the clusters retain precisely the same interpretations of Cluster 1 being mostly low achievers who move from averagely engaged to disengaged, Cluster 2 being mostly highly engaged high achievers, some of whom transition to being mostly intermediate achievers, and Cluster 3 being most averagely engaged or disengaged high or intermediate achievers, some of whom become more averagely engaged and highly achieving over time. It is worth noting that it is possible for the clusters and the interpretation thereof to change more drastically than this when covariates are added. As the clusters are very similar to what we found before, we give the clusters more informative labels as a reminder:

```
cluster_names(mhmm_fit_1$model) <- c("LowEngLowAch", "HighEngHighAch", "LowEngHighAch")
```

We now present a summary of the model which shows information about parameter estimates of covariates and prior and posterior cluster membership probabilities, which refer to cluster membership probabilities before or after conditioning on the observed sequences, respectively. In other words, prior cluster membership probabilities are calculated using each individual's observed covariate values only, while posterior cluster membership probabilities are calculated using individual's covariate values *and* their observed sequence. In each case, these represent the probabilities of belonging to a particular HMM component in the MHMM mixture.

```
summary_mhmm_1 <- summary(mhmm_fit_1$model)
summary_mhmm_1
```

```

Covariate effects :
LowEngLowAch is the reference.

HighEngHighAch :
             Estimate  Std. error
(Intercept)    -6.253       1.220
Prev_grade      0.817       0.159

LowEngHighAch :
             Estimate  Std. error
(Intercept)    -1.486       0.960
Prev_grade      0.158       0.139

Log-likelihood: -1708.843   BIC: 3614.673

Means of prior cluster probabilities :
  LowEngLowAch HighEngHighAch  LowEngHighAch
          0.40           0.34           0.26

Most probable clusters :
            LowEngLowAch  HighEngHighAch  LowEngHighAch
```

```
count                   57             48             37
proportion           0.401          0.338          0.261

Classification table :
Mean cluster probabilities (in columns) by the most probable
                                                cluster (rows)

              LowEngLowAch HighEngHighAch LowEngHighAch
LowEngLowAch       0.98577        0.00702       0.00721
HighEngHighAch     0.00316        0.98499       0.01185
LowEngHighAch      0.01335        0.01552       0.97113
```

We will first interpret the information on prior and posterior cluster membership probabilities and then proceed to interpreting covariate effects. Firstly, the section named `Means of prior cluster probabilities` gives information on how likely each cluster is in the whole population of students (40% in "LowEngLowAch", 34% in "HighEngHighAch", and 26% in "LowEngHighAch"). Secondly, the section `Most probable clusters` shows group sizes and proportions considering that each student would be classified into the cluster for which they have the highest cluster membership probability. Thirdly, the `Classification table` shows mean cluster probabilities (in columns) by the most probable cluster (in rows). We can see that the clusters are very crisp (the certainty of cluster memberships are very high) because the membership probabilities are large in the diagonal of the table.

The part titled `Covariate effects` shows the parameter estimates for the covariates. Interpretation of the values is similar to that of multinomial logistic regression, meaning that we can interpret the direction and uncertainty of the effect –relative to the reference level of "LowEngLowAch"—but we cannot directly interpret the magnitude of the effects. The magnitude of the `Prev_grade` coefficient is small with respect to its associated standard error, so we can say there is no significant relationship between previous grades and membership of the poorly engaged yet highly achieving "LowEngHighAch". However, the large positive coefficient in "HighEngHighAch" indicates that students with high previous grades are more likely to belong to the highly engaged high achieving cluster, which makes intuitive sense.

The summary object also calculates prior and posterior cluster memberships for each student. We omit them here, for brevity, but demonstrate that they can be obtained as follows:

```
prior_prob <- summary_mhmm_1$prior_cluster_probabilities
posterior_prob <- summary_mhmm_1$posterior_cluster_probabilities
```

Similarly, to obtain the most likely cluster assignment for each student according to the posterior probabilites, we simply need to access the `most_probable_cluster` element from the summary object (`summary_mhmm_1`). As per the `Groups` vector defined in the earlier distance-based clustering analysis, we show the assignments for the first 5 students only below, for brevity.

```
cluster_assignments <- summary_mhmm_1$most_probable_cluster
head(cluster_assignments, 5)
```

```
[1] LowEngHighAch  LowEngLowAch   LowEngLowAch   HighEngHighAch
LowEngLowAch Levels: LowEngLowAch HighEngHighAch LowEngHighAch
```

If we are interested in getting the most probable path of hidden states for each student, we may use the `hidden_paths()` function from `seqHMM` and pass the MHMM model as an argument (`mhmm_fit_1$model`). We then need to convert it to the long format to get a row per student and timepoint and then we extract only the `value` column to get only the hidden state. We can then assign the resulting vector to a column (e.g., `HiddenState`) in the original dataframe which follows the same data order.

```
df$HiddenState <- hidden_paths(mhmm_fit_1$model) |>
  pivot_longer(everything()) |> pull(value)
head(df$HiddenState)
```

```
[1] LowEngHighAch:State 1 LowEngHighAch:State 1 LowEngHighAch:State 2
[4] LowEngHighAch:State 2 LowEngHighAch:State 2 LowEngHighAch:State 2
8 Levels: LowEngLowAch:State 1 LowEngLowAch:State 2 ... %
```

It is good to be aware that the most probable cluster according to posterior cluster probabilities may not always match with the cluster producing the single most probable path of hidden states. This is because the posterior cluster probabilities are calculated based on all of the possible hidden state paths generated from that cluster, while the single most probable path of hidden states may sometimes come from a different cluster.

Lastly, we advise readers to consult Chap. 12 for more examples of interpreting covariate effects in the context of MHMM models.

5 Discussion

This chapter has provided an introduction to multi-channel sequence analysis, a method that deals with the analysis of two or more synchronised sequences. Throughout this chapter, we have highlighted the growing significance of multi-channel sequence analysis in educational research. While earlier chapters in this book focused on sequence analysis of learner-related data, this method presents a suitable approach to leverage the increasing availability of student multimodal temporal data. Our review of the existing literature that has utilised multi-channel sequence analysis concluded that the applications of this method within educational research are still relatively scarce. However, the potential of this framework is immense, as it opens up new possibilities to address the challenges posed by the complex and dynamic nature of educational data.

In the tutorial part of the chapter, we examined a case study on students' online engagement and academic achievement to demonstrate how multi-channel sequence analysis can reveal valuable insights into the longitudinal associations between these constructs. The identification of distinct trajectories through mixture hidden Markov models allows for a deep understanding of how both constructs evolve together over time and the distance-based clustering approach offers an alternative perspective on the group structure within these data. The step-by-step tutorial on implementing multi-channel sequence analysis with R according to both clustering frameworks provides readers with practical guidance to explore and apply these methods in their own research. Armed with this knowledge, researchers and educators can unlock valuable insights from their data and make informed decisions to support student learning and success.

Markovian models are relatively scalable and can be used to cluster large sequence data. See, e.g., [36] for scalability in terms of hidden states and time points in hidden Markov models. In terms of the number of sequences, computations can be easily parallelised; see, e.g., [19]. Then again, unlike data-mining type sequence analysis, probabilistic models require making assumptions about the data-generating mechanisms which may not always be realistic (such as time-homogeneity or the Markov property itself [19]. Furthermore, analysis of complex multi-channel sequence data can be very time consuming [6, 37]. Although we have highlighted the advantages of the model-based Markovian approach, particularly with regard to the ability to incorporate covariates as predictors of cluster membership, it is difficult to say if one approach is definitively better than the other for all potential use-cases. Although information criteria such as the BIC or cross-validation can be used to guide the choice of MHMM, particularly regarding the choice of the number of clusters (and numbers of hidden states within each cluster, and/or initial/transition/emission probabilities), we did not comprehensively do so here. Rather, we opted for three clusters, each with two hidden states, in both applications of MHMM herein, and only considered different initial probability settings and different covariates to arrive at the best model in terms of BIC. In an ideal analysis, one would compare different MHMM fits with different numbers of clusters and hidden states, different restrictions on the probabilities, and different covariates to arrive at an optimal model in terms of BIC or other model selection criteria.

Conversely, the choice of six clusters for the hierarchical clustering solution was arrived at in a different fashion, driven by the average silhouette width cluster quality measure. Indeed, the BIC cannot be used to choose the number of clusters with hierarchical clustering, nor can it be used to compare a hierarchical clustering solution with MHMM. However, it is interesting to note that the six clusters obtained by hierarchical clustering map reasonably well to the six hidden states of the MHMM solutions. As examples, considering the MHMM with fixed initial probabilities, one could say that clusters 4 and 5 of the hierarchical clustering solution, capturing averagely engaged and disengaged low achievers respectively, matches the two hidden states of the first MHMM cluster, while clusters 2 and 6, capturing actively engaged intermediate and high achievers respectively, matches the two hidden states of the second MHMM cluster.

Overall, we encourage readers to further explore the potential of multi-channel sequence analysis, the merits of both clustering approaches, and broader implications in the field of education by diving into the recommended readings. The next section provides a list of valuable resources for readers to expand their knowledge on the topic.

6 Further Readings

For more thorough presentations and comparisons of different options for the distance-based analysis of multi-channel sequences, we recommend the text book by Struffolino and Raab [38] and articles by Emery and Berchtold [8] and Ritschard et al. [13]. For assessing whether sequences in different channels are associated to the extent of needing joint analysis, Piccarreta and Elzinga [39, 40] have proposed solutions for quantifying the extent of association between two or more channels using Cronbach's α and principal component analysis.

For general presentations of the MHMM for multi-channel sequences, see the chapter by Vermunt et al. [41] and the article by Helske and Helske [19]. For further examples and tips for visualization of multi-channel sequence data and estimation of Markovian models with the `seqHMM` package, see the `seqHMM` vignettes [42, 43].

To learn how to combine distance-based sequence analysis and hidden Markov models for the analysis of complex multi-channel sequence data with partially missing observations, see the chapter by Helske et al. [6].

Acknowledgments SH was supported by the Academy of Finland (decision number 331816), the Academy of Finland Flagship Programme (decision number 320162), and the Strategic Research Council (SRC), FLUX consortium (decision numbers: 345130 and 345130). MS was supported by the Academy of Finland (TOPEILA: Towards precision education: Idiographic learning analytics, grant 350560).

References

1. Saqr M, Nouri J, Fors U (2019) Time to focus on the temporal dimension of learning: a learning analytics study of the temporal patterns of students' interactions and self-regulation. Int J Technol Enhanced Learn 11:398. https://doi.org/10.1504/ijtel.2019.102549
2. Saqr M, López-Pernas-Pernas S (2023) The temporal dynamics of online problem-based learning: why and when sequence matters. Int J Comput-Support Collaborat Learn 18:11–37. https://doi.org/10.1007/s11412-023-09385-1

3. Saqr M, López-Pernas-Pernas S, Helske S, Durand M, Murphy K, Studer M, Ritschard G (2024) Sequence analysis in education: principles, technique, and tutorial with R. In: Saqr M, López-Pernas-Pernas S (eds) Learning analytics methods and tutorials: a practical guide using R. Springer, Berlin
4. López-Pernas-Pernas S, Saqr M (2024) Modeling the dynamics of longitudinal processes in education. A tutorial with r for the VaSSTra method. In: Saqr M, López-Pernas-Pernas S (eds) Learning analytics methods and tutorials: a practical guide using R. Springer, Berlin
5. Helske J, Helske S, Saqr M, López-Pernas-Pernas S, Murphy K (2024) A modern approach to transition analysis and process mining with markov models: a tutorial with R. In: Saqr M, López-Pernas-Pernas S (eds) Learning analytics methods and tutorials: a practical guide using R. Springer, Berlin
6. Helske S, Helske J, Eerola M (2018) Combining sequence analysis and hidden Markov models in the analysis of complex life sequence data. In: Life course research and social policies. Springer, Berlin, pp 185–200
7. Eisenberg-Guyot J, Peckham T, Andrea SB, Oddo V, Seixas N, Hajat A (2020) Life-course trajectories of employment quality and health in the U.S.: a multichannel sequence analysis. Soc Sci Med 264:113327. https://doi.org/10.1016/j.socscimed.2020.113327
8. Emery K, Berchtold A (2022) Comparison of two approaches in multichannel sequence analysis using the Swiss Household Panel. Long Life Course Stud **14**, 1–32. https://doi.org/10.1332/175795921x16698302233894
9. Gauthier J-A, Widmer ED, Bucher P, Notredame C (2010) Multichannel sequence analysis applied to social science data. Sociol Methodol 40:1–38. https://doi.org/10.1111/j.1467-9531.2010.01227.x
10. Saqr M, López-Pernas S, Helske S, Hrastinski S (2023) The longitudinal association between engagement and achievement varies by time, students' profiles, and achievement state: a full program study. Comput Edu 199:104787. https://doi.org/10.1016/j.compedu.2023.104787
11. Winne PH (2020) Construct and consequential validity for learning analytics based on trace data. Comput Hum Behav 112:106457. https://doi.org/10.1016/j.chb.2020.106457
12. Murphy K, López-Pernas-Pernas S, Saqr M (2024) Dissimilarity-based cluster analysis of educational data: a comparative tutorial using R. In: Saqr M, López-Pernas-Pernas S (eds) Learning analytics methods and tutorials: a practical guide using R. Springer, Berlin
13. Ritschard G, Liao TF, Struffolino E (2023) Strategies for multidomain sequence analyis in social research. Sociol Methodol 53:288–322. https://doi.org/10.1177/00811750231163833
14. Pollock G (2007) Holistic trajectories: a study of combined employment, housing and family careers by using multiple-sequence analysis. J R Stat Soc Ser A (Stat Soc) 170:167–183
15. Saqr M, López-Pernas-Pernas S, Jovanović J, Gašević D (2023) Intense, turbulent, or wallowing in the mire: a longitudinal study of cross-course online tactics, strategies, and trajectories. Internet Higher Edu 57:100902
16. Bouguettaya A, Yu Q, Liu X, Zhou X, Song A (2015) Efficient agglomerative hierarchical clustering. Exp Syst Appl 42:2785–2797. https://doi.org/10.1016/j.eswa.2014.09.054
17. Gilpin S, Qian B, Davidson I (2013) Efficient hierarchical clustering of large high dimensional datasets. In: Proceedings of the 22nd ACM international conference on information & knowledge management. Association for Computing Machinery, New York, NY, pp 1371–1380
18. Studer M (2013) WeightedCluster library manual: a practical guide to creating typologies of trajectories in the social sciences with R. LIVES. https://doi.org/10.12682/LIVES.2296-1658.2013.24
19. Helske S, Helske J (2019) Mixture hidden Markov models for sequence data: the seqHMM package in R. J Stat Softw 88:1–32. https://doi.org/10.18637/jss.v088.i03
20. Schwarz GE (1978) Estimating the dimension of a model. Ann. Stat. 6:461–464. https://doi.org/10.1214/aos/1176344136
21. Helske J, Helske S (2023). https://cran.r-project.org/web/packages/seqHMM/index.html
22. Murphy K, Murphy TB, Piccarreta R, Gormley IC (2021) Clustering longitudinal life-course sequences using mixtures of exponential-distance models. J R Stat Soc Ser A (Stat Soc) 184:1414–1451

23. Studer M (2018) Divisive Property-Based and fuzzy clustering for sequence analysis. In: Ritschard G, Studer M (eds) Sequence analysis and related approaches: innovative methods and applications. Springer, Cham, pp 223–239
24. Helske S, Helske J, Chihaya GK (2023) From sequences to variables: rethinking the relationship between sequences and outcomes. Soc Method 54(1)27–51. https://doi.org/10.1177/00811750231177026
25. Dayton CM, Macready GB (1988) Concomitant-variable latent-class models. J Am Stat Assoc 83:173–178
26. Murphy K, Murphy TB (2020) Gaussian parsimonious clustering models with covariates and a noise component. Adv Data Anal Class 14:293–325
27. Törmänen T, Järvenoja H, Saqr M, Malmberg J, Järvelä S (2022) A person-centered approach to study students' socio-emotional interaction profiles and regulation of collaborative learning. Front Edu 7: https://doi.org/10.3389/feduc.2022.866612
28. Törmänen T, Järvenoja H, Saqr M, Malmberg J, Järvelä S (2022) Affective states and regulation of learning during socio-emotional interactions in secondary school collaborative groups. British J Edu Psychol 93:48–70. https://doi.org/10.1111/bjep.12525
29. Ouyang F, Xu W, Cukurova M (2023) An artificial intelligence-driven learning analytics method to examine the collaborative problem-solving process from the complex adaptive systems perspective. Int J Comput-Suppor Collab Learn 18:39–66. https://doi.org/10.1007/s11412-023-09387-z
30. López-Pernas S, Saqr M (2021) Bringing synchrony and clarity to complex multi-channel data: a learning analytics study in programming education. IEEE Access 9:166531–166541. https://doi.org/10.1109/access.2021.3134844
31. Bacci S, Bertaccini B (2022) A mixture hidden Markov model to mine students' university curricula. Data 7:25. https://doi.org/10.3390/data7020025
32. Liu B, Widener MJ, Smith LG, Farber S, Minaker LM, Patterson Z, Larsen K, Gilliland J (2021) Disentangling time use, food environment, and food behaviors using multi-channel sequence analysis. Geograph Anal 54:881–917. https://doi.org/10.1111/gean.12305
33. Chan C, Chan GC, Leeper TJ, Becker J (2021) Rio: a Swiss-army knife for data file I/O. https://cran.r-project.org/package=rio
34. Wickham H, Averick M, Bryan J, Chang W, McGowan LD, François R, Grolemund G, Hayes A, Henry L, Hester J, Kuhn M, Pedersen TL, Miller E, Bache SM, Müller K, Ooms J, Robinson D, Seidel DP, Spinu V, Takahashi K, Vaughan D, Wilke C, Woo K, Yutani H (2019) Welcome to the tidyverse. J Open Source Softw 4:1686. https://doi.org/10.21105/joss.01686
35. Gabadinho A, Ritschard G, Müller NS, Studer M (2011) Analyzing and visualizing state sequences in R with TraMineR. J Stat Softw 40:1–37. https://doi.org/10.18637/jss.v040.i04
36. Rabiner LR (1989) A tutorial on hidden Markov models and selected applications in speech recognition. Proc IEEE 77:257–286. https://doi.org/10.1109/5.18626
37. Berchtold A (2004) Optimization of mixture models: comparison of different strategies. Comput Stat 19:385–406. https://doi.org/10.1007/bf03372103
38. Raab M, Struffolino E (2022) Sequence analysis. SAGE, London
39. Piccarreta R, Elzinga CH (2013) Mining for association between life course domains. In: Contemporary issues in exploratory data mining in the behavioral sciences. Routledge, New York, pp 212–242
40. Piccarreta R (2017) Joint sequence analysis: association and clustering. Sociol Methods Res 46:252–287
41. Vermunt JK, Tran B, Magidson J (2008) Latent class models in longitudinal research. In: Handbook of longitudinal research: design, measurement, and analysis. Elsevier, Amsterdam, pp 373–385
42. Helske S (2017). https://cran.r-project.org/web/packages/seqHMM/vignettes/seqHMM_visualization.pdf
43. Helske S (2017). https://cran.r-project.org/web/packages/seqHMM/vignettes/seqHMM_estimation.pdf

The Why, the How and the When of Educational Process Mining in R

Sonsoles López-Pernas and Mohammed Saqr

1 Introduction

Nowadays, almost all learning platforms generate vast amounts of data that include every interaction a student has with the learning environment. Such large amounts of data offer a unique opportunity to analyze and understand the dynamics of the learning process. In the previous chapters of the book, we covered several methods for analyzing the temporal aspects of learning, such as sequence analysis [1, 2], Markovian modeling [3] or temporal network analysis [4]. In this chapter, we present an analytical method that is specifically oriented towards analyzing time-stamped event log data: process mining. Process mining is a technique that allows us to discover, visualize, and analyze the underlying process from time-stamped event logs. Through process mining, we may uncover hidden patterns, bottlenecks, and inefficiencies in students' learning journeys. Tracking students' actions step-by-step, allows us to identify which resources are most effective, which topics are more challenging, and even predict possible problems before they may occur.

Process mining emerged as a business tool that allows organizations to analyze and improve their operational processes. The field has rapidly expanded with several modelling methods, algorithms, tools and visualization techniques. Further, the method has been met with enthusiasm from several researchers leading to a rapid uptake by other disciplines such as health care management and education. As the field currently stands, it is a blend of process management and data science with less emphasis on statistical methods. The field has found its place in educational research with the recent surge of trace log data generated by students' activities and the interest that learning analytics and educational data mining have kindled.

S. López-Pernas (✉) · M. Saqr
School of Computing, University of Eastern Finland, Joensuu, Finland
e-mail: sonsoles.lopez@uef.fi

M. Saqr, S. López-Pernas (eds.), *Learning Analytics Methods and Tutorials*,
https://doi.org/10.1007/978-3-031-54464-4_14

This tutorial chapter will introduce the reader to the fundamental concepts of process mining technique and its applications in learning analytics and education research at large. We will first describe the method, the main terminology, and the common steps of analysis. Then, we will provide a review of related literature to gain an understanding of how this method has been applied in learning analytics research. We then provide a step-by-step tutorial on process mining using the R programming language and the `bupaverse` framework. In this tutorial, we analyze a case study of students' online activities in an online learning management system using the main features of process mining.

2 Basic Steps in Process Mining

The goal of process mining is to extract process models from event data. The resulting models can then be used to portray students' pathways in learning, identify common transitions, and find issues in their approach. In doing so, process mining promises to find deviations from the norm, suggest corrective actions, and optimize processes as an ultimate goal [5]. Process mining starts by the extraction of event data. In the case of educational process mining, event data often reflects students' activities in learning management systems (LMSs), or in other types of digital tools that record time-stamped events of students' interactions with the digital tools, such as automated assessment tools or online learning games. The said data is used to construct what is known as an event log. An event log has three necessary parts:

- **Case identifier**: A case represents the subject of the process. For example, if we are analyzing students' enrollment process, each student would represent a different case. If we are analyzing students' online activities in the LMS, we can also consider each student as a separate case; alternatively, if we want a greater level of granularity, each student's learning session can be then considered a separate case. All event logs need to have a case identifier that unequivocally identifies each case and that allows to group together all the events that belong to the same case.
- **Activity identifier**: Activities represent each action or event in the event data. Continuing with the previous examples, an activity would represent each step in the enrollment process (e.g, application, revision, acceptance, payment etc.), or each action in the LMS (e.g., watch video, read instructions, or check calendar).
- **Timestamp**: The timestamp is a record of the moment each event has taken place. It allows to establish the order of the events. In the case of online activity data, for instance, the timestamp would be the instant in which a student clicks on a learning resource. In some occasions, activities are not instantaneous, but rather have a beginning and an end. For example, if we are analyzing student's video watching, we might have an event record when they start watching and when they finish. If we want to treat these events as parts of the same activity, we need to provide additional information when constructing an event log. As

such, we need to specify an *activity instance identifier*, which would allow us to unequivocally identify and group together all instances of the same overarching activity (watching a video, in our example). Moreover, we would need to provide a *lifecycle* identifier (e.g., start and end), to differentiate between all stages of the same activity. A common limitation of LMS data is that only one click is recorded per activity so this type of analysis is often not possible.

Once we have our event log defined, we can calculate multiple metrics that allow us to understand the data. For example, we can see the most frequent activities and the most frequent transitions. We can also see the case coverage for each activity, e.g., how many cases contain each activity, and the distribution of the length of each case (how many activities they have). We can also calculate performance metrics, such as idle time (i.e., time spent without doing any activities) or throughput time (i.e., overall time taken).

From the event log, we often construct what is known as the Directly-Follows Graph (DFG), in which we graphically represent all the activities in the event log as nodes and all the observed transitions between them as edges [5]. Figure 1 shows an example with event log data from students, where each case identifier represents a student's session. First, we build the sequence of activities for each case. As such, Layla's path for her first learning session would be: Calendar → Lecture → Video → Assignment. The path for Sophia's first session would be: Calendar → Instructions → Video → Assignment. We put both paths together and construct a partial DFG that starts from Calendar, then it transitions either to Lecture or Instructions and then it converges back into Video and ends in Assignment. We create the paths for the remaining student sessions. Then, we combine them together through an iterative process until we have the complete graph with all the possible transitions between activities. Our final graph with the four sessions shown in Fig. 1 would start by Calendar, then transition either to Lecture or Instructions. Then the

User	Activity	Time	Case (session)
Layla	Calendar	8.1.2023 18:44	Layla_Session_1
Sophia	Calendar	8.1.2023 18:45	Sophia_Session_1
Layla	Lecture	8.1.2023 18:48	Layla_Session_1
Carmen	Calendar	8.1.2023 18:50	Carmen_Session_1
Sophia	Instructions	9.1.2023 18:51	Sophia_Session_1
Layla	Video	8.1.2023 18:52	Layla_Session_1
Sophia	Video	8.1.2023 18:53	Sophia_Session_1
Layla	Assignment	8.1.2023 18:54	Layla_Session_1
Carmen	Lecture	8.1.2023 19:00	Carmen_Session_1
Sophia	Assignment	8.1.2023 19:01	Sophia_Session_1
Layla	Calendar	8.1.2023 19:11	Layla_Session_2
Carmen	Instructions	8.1.2023 19:12	Carmen_Session_1
Layla	Instructions	8.1.2023 19:14	Layla_Session_2
Carmen	Calendar	8.1.2023 19:16	Carmen_Session_1
Layla	Lecture	8.1.2023 19:19	Layla_Session_2
Layla	Lecture	8.1.2023 19:20	Layla_Session_2

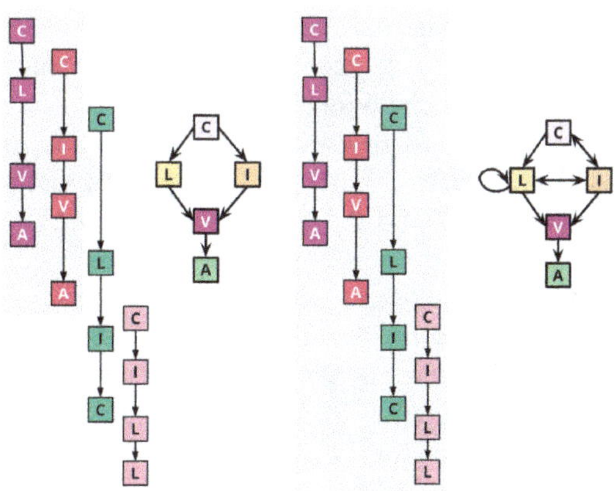

Fig. 1 Step-by-step construction of a DFG

process could transition from Lecture to Instructions or viceversa, or to Video. In addition, Lecture has a self-loop because it can trigger another lecture (see Layla's session 2). From Video, the only possible transition is to Assignment.

In real-life scenarios, building the DFG for a complete —or large— event log may turn to be overcrowded and hard to visualize or interpret [5]. Therefore, it is "common to trim activities or transitions" that do not occur often. Other options include splitting the event logs by group (e.g., class sections) to reduce the granularity of the event log to be able to compare processes between groups. We can also zoom into specific parts of the course (e.g, a specific assignment or lecture) to better understand students' activities at that time. Moreover, we can filter the event log to see cases that contain specific activities or that match certain conditions.

The DFGs are often enhanced with labels that allow us to understand the graph better. These labels are often based on the frequency of activities and transitions. For example, the nodes (representing each activity) can be labeled with the number of times (or proportion) they appear in the event data, or with the case coverage, i.e., how many cases (or what percentage thereof) they appear in. The edges (representing transitions between activities) can be labeled with the frequency of the transitions or the case coverage as well, among others. Another common way of labeling the graph is using performance measures that indicate the mean time (or median, maximum, etc.) taken by each activity and/or transition.

The DFG gives us an overarching view of the whole event log. However, in some occasions, we would like to understand the underlying process that is common to most cases of our event log. This step is called **process discovery** [5] and there are several algorithms to perform it such as the alpha algorithm [6], inductive techniques [7] or region-based approaches [8]. The discovered processes are then represented using specialized notation, such as Business Process Model and Notation (BPMN) [9] or Petri nets [10].

In some occasions, there are certain expectations regarding how the process should go. For instance, if we are analyzing students' activities during a lesson, the teacher might have given a list of instructions that students are expected to follow in order. To make sure that the discovered process (i.e., what students' data reveal) aligns with the expected process (i.e., what students were told to do in our example), **conformance checking** is usually performed. Conformance checking deals with comparing the discovered process with an optimal model or theorized process (i.e., an ideal process) [5]. The idea is to find similarities and differences between the model process and the real observed process, identify unwanted behavior, detect outliers, etc. As seen from our example, in educational settings, this can be used, for instance, to detect whether students are following the learning materials in the intended order or whether they implement the different phases of self-regulated learning. However, given that students are rarely asked to access learning materials in a strict sequential way, this feature has been rarely used. In the next section, we present a review of the literature on educational process mining where we discuss more examples.

3 Review of the Literature

A review of the literature by Bogarín et al. [11] mapped the landscape of educational process mining and found a multitude of applications of this technique in a diversity of contexts. The most common application was to investigate the sequence of students' activities in online environments such as MOOCs and other online or blended courses, as well as in computer-supported collaborative learning contexts [11]. One of the main aims was to detect learning difficulties to be able to provide better support for students. For example, López-Pernas et al. [12] used process mining to explore how students' transition between a learning management system and an automated assessment tool, and identified how struggling students make use of the resources to solve their problems. Arpasat et al. [13] used process mining to study students' activities in a MOOC, and compared the behavior of high- and low-achieving students in terms of students' activities, bottlenecks and time performance. A considerable volume of research has studied processes where the events are instantaneous, such as clicks in online learning management systems (e.g., [14–16]). Fewer are the studies that have used activities with a start time and an end time due to the limitations in data collection in online platforms. However, this limitation has been often overcome by grouping clicks into learning sessions, as is often done in the literature on students' learning tactics and strategies, or self-regulated learning (e.g., [12, 17–19]).

Regarding the methods used, much of the existing research is limited to calculating performance metrics and visualizing DFGs, whereby researchers attempt to understand the most commonly performed activities and the common transitions between them. For example, Vartiainen et al. [20] used video coded data of students' participation in an educational escape room to visualize the transitions between in-game activities using DFGs. Oftentimes, researchers use DFGs to compare (visually) across groups, for example high vs. low achievers, or between clusters obtained through different methods. For instance, Saqr et al. [21] implemented k-means clustering to group students according to their online activity frequency, and used DFGs to understand the strategies adopted by the different types of learners and how they navigate their learning process. Using a different approach, Saqr and López-Pernas [22] clustered students groups according to their sequence of interactions using distance-based clustering, and then compared the transitions between different interactions among the clusters using DFGs.

Going one step further, several studies have used process discovery to detect the underlying overall process behind the observed data [11]. A variety of algorithms have been used in the literature for this purpose, such as the alpha algorithm [23], the heuristic algorithm [24], or the fuzzy miner [18]. Less often, research on educational proecss mining has performed conformance checks [11], comparing the observed process with an "ideal" or "designed" one. An example is the work by Pechenizkiy et al. [25], who used conformance checking to verify whether students answered an exam's questions in the order specified by the teacher.

When it comes to the tools used for process mining, researchers have relied on various point-and-click software tools [11]. For example, Disco [26] has been used for DFG visualization by several articles (e.g., [27]). ProM [28] is the dominant technology when it comes to process discovery (e.g., [19, 29]) and also conformance checking (e.g., [25]). Many articles have used the R programming language to conduct process mining, relying on the `bupaverse` [30] framework for basic metrics and visualization (the one covered in the present chapter), although not for process discovery (e.g., [12]) since the algorithm support is scarce.

4 Process Mining with R

In this section, we present a step-by-step tutorial with R on how to conduct process mining of learners' data. First, we will install and load the necessary libraries. Then, we will present the data that we will use to illustrate the process mining method.

4.1 The Libraries

As a first step, we need two basic libraries that we have used multiple times throughout the book: `rio` for importing the data [31], and `tidyverse` for data manipulation [32]. As for the libraries used for process mining, we will first rely on `bupaverse`, a meta-package that contains many relevant libraries for this purpose (e.g., `bupaR`), which will help us with the frequentist approach [30]. We will use `processanimateR` to see a play-by-play animated representation of our event data. You can install the packages with the following commands:

```
install.packages("rio")
install.packages("tidyverse")
install.packages("bupaverse")
install.packages("processanimateR")
```

You can then load the packages using the `library()` function.

```
library(bupaverse)
library(tidyverse)
library(rio)
library(processanimateR)
```

4.2 Importing the Data

The dataset that we are going to analyze using process mining contains logs of students' online activities in an LMS during their participation on a course about learning analytics. We will also make use of students' grades data to compare activities between high and low achievers. More information about the dataset can be found in the data chapter of this book [33]. In the following code chunk, we download students' event and demographic data and we merge them together into the same dataframe (`df`).

```
df <- import("https://github.com/lamethods/data/raw/main/1_moodleLAcourse/
       Events.xlsx")
all <- import("https://github.com/lamethods/data/raw/main/1_moodleLAcourse/
       AllCombined.xlsx") |>
  select(User, AchievingGroup)
df <- df |> merge(all, by.x = "user", by.y = "User")
```

When analyzing students' learning event data, we are often interested in analyzing each learning session separately, rather than considering a longer time span (e.g., a whole course). A learning session is a sequence (or episode) of un-interrupted learning events. To do such grouping, we define a threshold of *inactivity*, after which, new activities are considered to belong to a new episode of learning or *session*. In the following code, we group students' logs into learning sessions considering a threshold of 15 minutes (15 min. × 60 sec./min. = 900 seconds), in a way that each session will have its own session identifier (`session_id`). For a step-by-step explanation of the sessions, code and rationale, please refer to the sequence analysis chapter [1]. A preview of the resulting dataframe can be seen in Table 1. We see that each group of logs that are less than 900 seconds (15 minutes) apart (`Time_gap` column) are within the same session (`new_session = FALSE`) and thus have the same `session_id`. Logs that are more than 900 seconds apart are considered a new session (`new_session = TRUE`) and get a new `session_id`.

```
sessioned_data <- df |>
  group_by(user) |>
  arrange(user, timecreated) |>
  mutate(Time_gap = timecreated - (lag(timecreated)))
                                                |>
  mutate(new_session = is.na(Time_gap) | Time_gap >
                                          900) |>
  mutate(session_nr = cumsum(new_session)) |>
  mutate(session_id = paste0 (user, "_", "Session_",
                    session_nr)) |> ungroup()
```

Table 1 Preview of the sessioned data

	User	Timecreated	Component	Event.name	Action	AchievingGroup	Time_ gap	New_ session	Session_ nr	Session_id
1	00a05cc62	1568053081	System	Course viewed	Course_view	Low achiever	NA	TRUE	1	00a05cc62_Session_1
2	00a05cc62	1568053142	System	Course viewed	Course_view	Low achiever	61	FALSE	1	00a05cc62_Session_1
3	00a05cc62	1568053383	Forum	Course module viewed	Instructions	Low achiever	241	FALSE	1	00a05cc62_Session_1
4	00a05cc62	1568053384	System	Course viewed	Course_view	Low achiever	1	FALSE	1	00a05cc62_Session_1
5	00a05cc62	1568061010	System	Course viewed	Course_view	Low achiever	7626	TRUE	2	00a05cc62_Session_2
6	00a05cc62	1568061071	System	Course viewed	Course_view	Low achiever	61	FALSE	2	00a05cc62_Session_2
7	00a05cc62	1568062780	Forum	Course module viewed	Group_work	Low achiever	1709	TRUE	3	00a05cc62_Session_3
8	00a05cc62	1568062781	System	Course viewed	Course_view	Low achiever	1	FALSE	3	00a05cc62_Session_3
9	00a05cc62	1568062842	System	Course viewed	Course_view	Low achiever	61	FALSE	3	00a05cc62_Session_3
10	00a05cc62	1568062843	Page	Course module viewed	Instructions	Low achiever	1	FALSE	3	00a05cc62_Session_3
11	00a05cc62	1568161913	System	Course viewed	Course_view	Low achiever	99070	TRUE	4	00a05cc62_Session_4
12	00a05cc62	1568161914	Page	Course module viewed	Instructions	Low achiever	1	FALSE	4	00a05cc62_Session_4
13	00a05cc62	1568161975	File	Course module viewed	General	Low achiever	61	FALSE	4	00a05cc62_Session_4
14	00a05cc62	1568161976	System	Course viewed	Course_view	Low achiever	1	FALSE	4	00a05cc62_Session_4
15	00a05cc62	1568191205	File	Course module viewed	General	Low achiever	29229	TRUE	5	00a05cc62_Session_5
16..95625										
95626	fc95def14	1571909890	System	Course viewed	Course_view	High achiever	1	FALSE	163	fc95def14_Session_163

4.2.1 Creating an Event Log

Now, we need to prepare the data for process mining. This is performed by converting the data into an event log. At a minimum, to construct our event log we need to specify the following:

- `case_id`: As we have explained before, the case identifier allows to differentiate between all cases (i.e, subjects) of the process. In our dataset, since we are analyzing activities in each learning session, the `case_id` would represent the identifier of each session (the column `session_id` of our dataframe that we have just created in the previous step).
- `activity_id`: It indicates the type of activity of each event in the event log. In our dataset, we will use the column `Action` that can take the values: "Course_view", "Assignment", "Instructions", "Group_work", etc. LMS logs often contain too much granularity so it is useful to recode the events to give them more meaningful names [33].
- `timestamp`: It indicates the moment in which the event took place. In our dataset, this is represented by the `timecreated` column which stores the time where each 'click' on the LMS took place.

Now that we have defined all the required parameters, we can create our event log. We will use the `simple_eventlog()` function from `bupaR` and supply the arguments to the function (corresponding to the aforementioned process elements) to the respective columns in our dataset.

```
event_log <- simple_eventlog(sessioned_data,
                case_id = "session_id",
                activity_id = "Action",
                timestamp = "timecreated")
```

In our example, each row in our data represents a single activity. This is often the case in online trace log data, as each activity is a mere instantaneous click with no duration. As explained earlier, in other occasions, activities have a beginning and an end, and we have several rows to represent all instances of an activity. For example, we might have a row to represent that a student has begun to solve a quiz, and another one to represent the quiz's submission. If the data looks like that, we might need to use the `activitylog()` function from `bupaR` to create our activity log, and indicate, using the `lifecycle_id` argument, which column indicates whether it is the start or end of the activity, or even intermediate states.

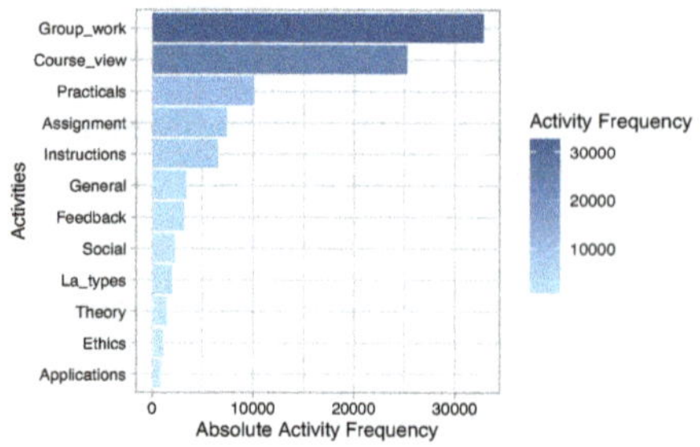

(a) Overall frequency

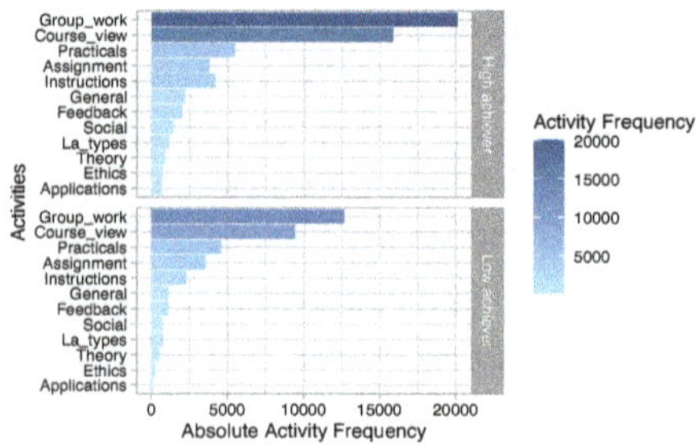

(b) Frequency by group

Fig. 2 Activity frequency. (**a**) Overall frequency. (**b**) Frequency by group

4.2.2 Inspecting the Logs

Now that we have created our event log, we can use the `summary()` function to get the descriptive statistics. The results show that we have a total of 95,626 events for 9560 cases (in our example, student sessions) and 12 distinct activities (the actions in the learning management system). We have 4076 distinct traces (i.e., unique cases), which in our example means that there are sessions that have the exact same sequence of events. These traces are, on average, 10 events long, and span from September 9th 2019 to October 27th 2019.

```
event_summary <- summary(event_log)
```

```
Number of events:  95626
Number of cases:  9560
Number of traces:  4076
Number of distinct activities:  12
Average trace length:  10.00272

Start eventlog:  2019-09-09 14:08:01
End eventlog:  2019-10-27 19:27:41
```

We may now inspect what the most common activities are in our event log (Table 2). The `absolute_frequency` column represents the total count of activities of each type, whereas the `relative_frequency` represents the percentage (as a fraction of 1) that each activity represents relative to the total count of activities. We see that the working on the group project (`Group_work`) is the most frequent activity (with 32,748 instances), followed by viewing the course main page (`Course_view`) with 25,293 instances.

```
activities(event_log)
```

Table 2 Frequency of each activity

Action	Absolute_frequency	Relative_frequency
Group_work	32748	0.34245916
Course_view	25293	0.26449919
Practicals	10020	0.10478322
Assignment	7369	0.07706063
Instructions	6474	0.06770125
General	3346	0.03499048
Feedback	3114	0.03256437
Social	2191	0.02291218
La_types	1891	0.01977496
Theory	1377	0.01439985
Ethics	1028	0.01075021
Applications	775	0.00810449

We can also view the most frequent activities in graphical form. We can even visually compare the frequency of activities between different groups (e.g., high achievers vs. low achievers) (Fig. 2).

```
event_log |> activity_frequency("activity") |> plot()
event_log |> group_by(AchievingGroup) |>
        activity_frequency("activity") |> plot()
```

We may be also interested in looking at *activity presence*, also known as case coverage. This refers to the percentage of cases that contain each of the activities at least once. Most sessions (85.8%) include visiting the main page of the course (`Course_view`). Slightly over half (54.6%) involve working on the group project (`Group_work`). Around one quarter include working on the `Practicals` (25.7%), `Instructions` (25.2%), and `Assignment` (23.2%). Other activities are less frequent (Fig. 3).

```
event_log |> activity_presence() |> plot()
event_log |> group_by(AchievingGroup) |>
        activity_presence() |> plot()
```

If we are interested in the transition between activities, we can plot the antecedent-consequent matrix. This visualization tells us how many transitions there has been from the activities on the left side (antecedent) to the activities on the bottom side (consequent).

```
event_log |> process_matrix() |> plot()
```

Figure 4 shows that there are 6416 transitions from `Course_view` to `Group_work` within the same session. The most common transition is from

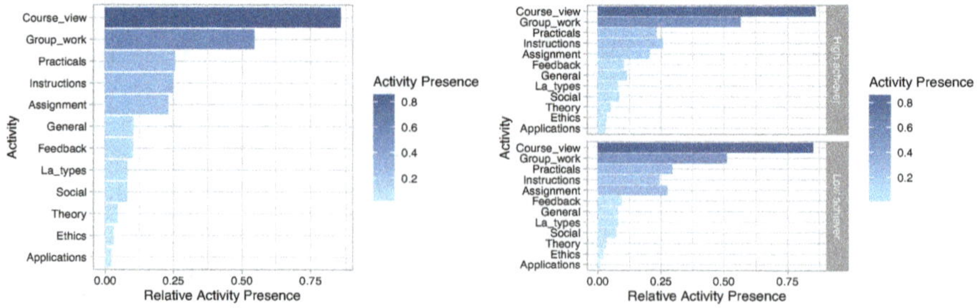

(a) Overall presence (b) Presence by group

Fig. 3 Activity presence. (**a**) Overall presence. (**b**) Presence by group

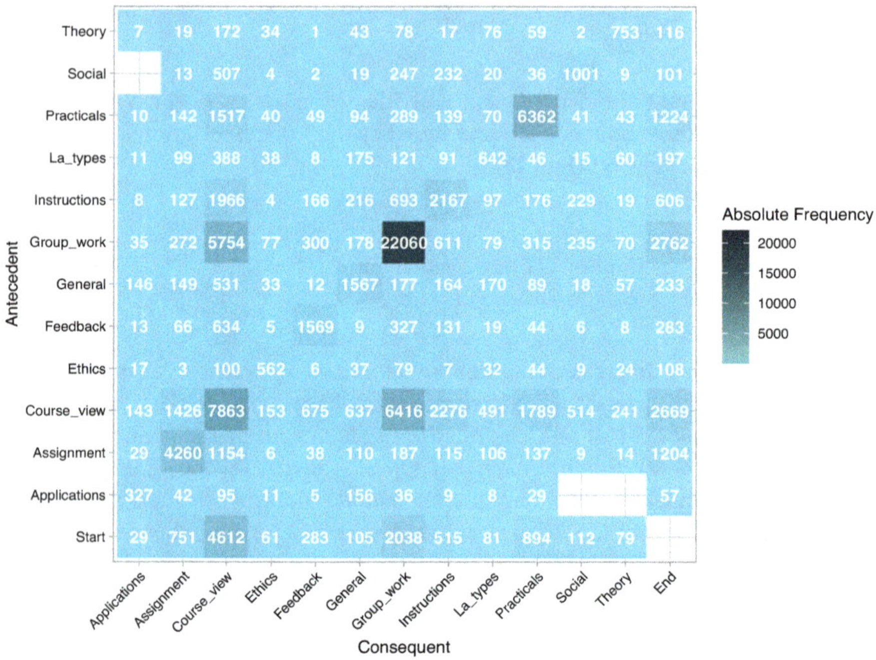

Fig. 4 Antedecent-consequent matrix

Group_work to more Group_work, with 22060 instances. Note that, on the antecedent side (left), there is an additional row for Start. This allows to represent the activities that occur right at the beginning as a transition from Start, since they do not have any other activity as antecedent. In this case, we see that most sessions start with Course_view (4612). On the consequent side (bottom), we see that there is an additional column on the right side for End, representing the end of a session. This allows to represent the activities that occur at the very end of a session, since they do not have any other activities following. We see that sessions are most likely

Fig. 5 Default process map

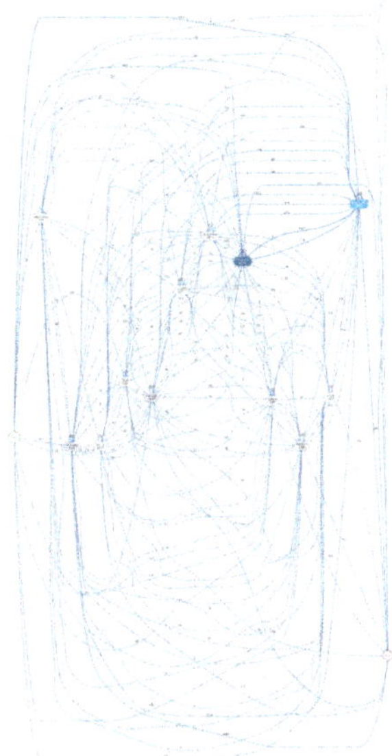

to end by Group_work (2762), closely followed by Course_view (2669). Another aspect worth noting is that some cells in our matrix are empty, which represents the fact that a transition between the activity on the left, and the activity on the bottom is not present in any of the cases in our dataset. For example, we see that there are no transitions between Applications and Theory. Another —obvious— example is the cell from Start to End, as sessions should have at least one event and therefore they can never go from Start to End without any other event in the middle.

4.2.3 Visualizing the Process

Now, it is time to visualize the event log using the DFG. Within the bupaverse framework this graph is referred to as process map. We can use the process_map() function for this purpose (Fig. 5).

```
event_log |> process_map()
```

By default, the process map contains every single activity and transition between activities. As the number of activities and cases increase, this can become unreadable, as is the case with our example. Therefore, we might need to trim some less frequent activities to be able to better visualize the process. We will choose to cover 80% of the activities. This means that we will select the most frequent activities in order, up to when we have accounted for 80% of all the activities in the event log. If we go back to Table 2, we can see that all activities ranging from the most frequent activity (Group_work) to Instructions account roughly to 80%, so we are left with the top five activities and exclude the rest. There is no standard for choosing a threshold for trimming and it is up to the researcher to define the threshold. As discussed in the introduction, such approach may yield different process maps according to the trimming threshold. A good rule of thumb for choosing this threshold is that the process map should be readable but at the same time no essential information should be removed. In our case, the activities filtered out are very infrequent and specific to certain lessons in the course, so they are not representative to the course as a whole. Another strategy would be to combine these activities into a single one such as `Reading`, which would allow to simplify our analysis while keeping all the information. For demonstration processes, we proceed by filtering the log and to operate with it in subsequent steps.

```
event_log |> filter_activity_frequency(percentage =
                          0.8) -> event_log_trimmed
```

If we now plot the process map of the trimmed event log, we will be able to see the process much more clearly (Fig. 6). When interpreting a process map it is common to start by commenting on the nodes (the activities) and their frequency. For example, we would say that `Group_work` is the most frequent activity with 32,748 instances, followed by `Group_work`, etc., as we have already done in Table 2. Next, we would describe the transitions between activities, with emphasis on how sessions start. Most sessions (5,059) —in the trimmed dataset— start by viewing the course main page `Course view`. We can see this information on the edge (arrow) that goes from `Start` to `Course_view`. The second most common first step is `Group_work` (2,129). Most instances of `Group_work` are followed by more `Group_work` (22,429, as can be seen from the self-loop). The most common transitions between activities are between `Course_view` and `Group_work`, with 6,000+ instances between one another both ways, as we can infer from the labels of the edges between one another. Viewing the course main page (`Course view`) is the most central activity in the process map as it has multiple transitions with all the other activities, since students often use the main page to go from one activity to the next.

```
event_log_trimmed |> process_map()
```

As we have seen, each node of the process contains the activity name along with its overall frequency, that is, the number of instances of that activity in our event log. Similarly, the edges (the arrows that go from activity to activity) are labeled with the number of times each transition has taken place. These are the default process map settings. Another way of labeling our process map is using the `frequency("absolute-case")` option, which counts the number of cases (i.e., sessions in our event log) that contain each given activity and transition (Fig. 7). Instead of seeing the count of activities and transitions, we are seeing the count of sessions that contain each activity and transition. For example, we could say that 8,203 sessions contain the activity `Course_view`.

Fig. 6 Process map trimmed

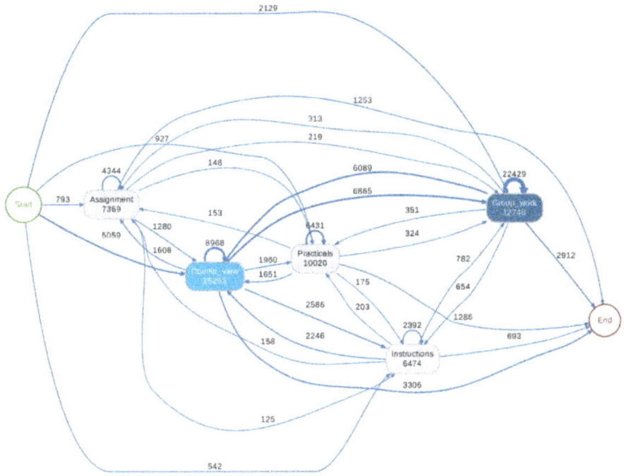

Fig. 7 Absolute case process map

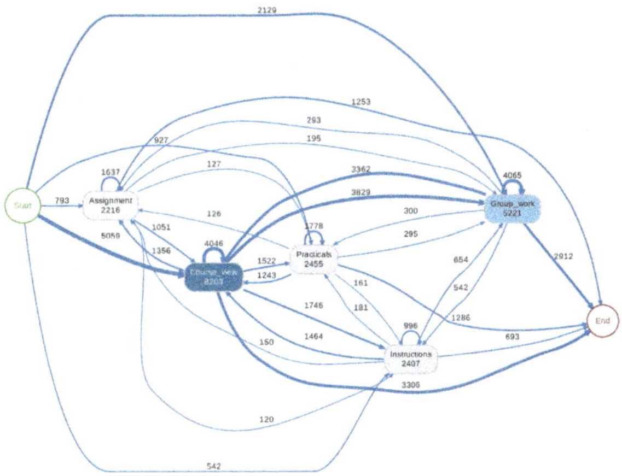


```
event_log_trimmed |> process_map(frequency("absolute-case"))
```

Sometimes we may be interested in the percentage (relative proportion) rather than the overall counts because it may be easier to compare and interpret. In such a case, we can use the `frequency("relative")` and `frequency("relative-case")` options. In the former, the labels will be relative to the total number of activities and transitions (Fig. 8a), whereas in the latter, the labels will be relative to the number of cases (or proportion thereof) (Fig. 8b). For example, in Fig. 8a, we see that `Group_work` accounts for 39.98% of the activities, whereas in Fig. 8b we see that it is present in 55.25% of the cases (i.e., sessions).

```
event_log_trimmed |> process_map(frequency("relative"))
event_log_trimmed |> process_map(frequency("relative-case"))
```

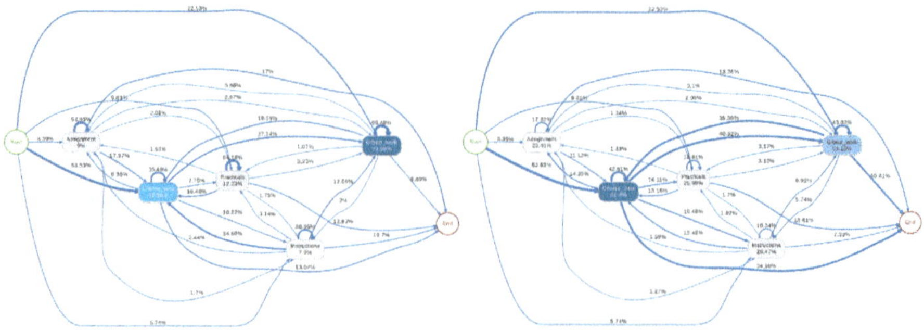

(a) Relative (b) Relative case

Fig. 8 Relative process maps. (**a**) Relative. (**b**) Relative case

We can combine several types of labels in the same process map, and even add labels related to aspects other than frequency. In the next example, we label the nodes based on their absolute and relative frequency and we label the edges based on their performance, that is, the time taken in the transition (Fig. 9). We have chosen the mean time to depict the performance of the edges, but we could also choose median, minimum, maximum, etc. We see, for example, that the longest transition seems to be between `Assingment` and `Group_work`, in which students take on average 1.59 minutes (as seen on the edge label). The shortest transition is from `Practicals` to `Practicals` (self-loop) with 0.22 minutes on average.

```
event_log_trimmed |> process_map(
                type_nodes = frequency("absolute"),
                sec_nodes = frequency("relative"),
                type_edges = performance(mean))
```

Fig. 9 Relative nodes and performance edges process map

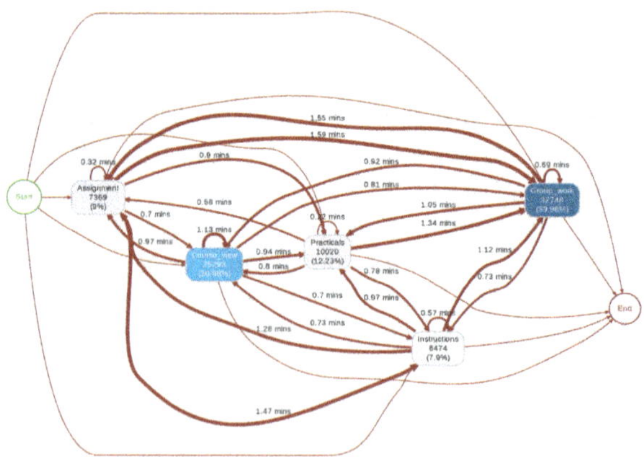

Another very useful feature is to be able to compare processes side by side. We can do so by using the `group_by()` function before calling `process_map()`. We

see that the process map does not capture any meaningful difference between the two groups (Fig. 10).

```
event_log_trimmed |> group_by(AchievingGroup) |>
                     process_map(frequency("relative"))
```

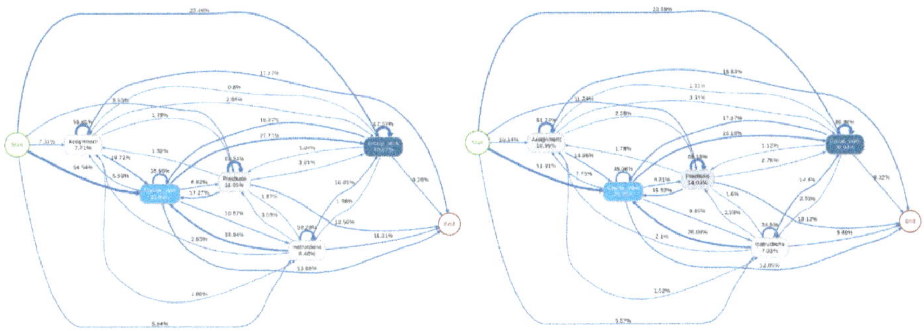

(a) High achievers (b) Low achievers

Fig. 10 Comparison between two process maps. (**a**) High achievers. (**b**) Low achievers

We can even reproduce the whole event log as a video using the function `animate_process` from `processanimateR` (Fig. 11).

```
animate_process(event_log_trimmed)
```

5 Discussion

We have demonstrated how to analyze event log data with process mining and how the visualization of process models can summarize a large amount of activities and produce relevant insights. Process models are powerful, summarizing, visually appealing and easy to interpret and understand. Therefore, process mining as a method has gained large scale adoption and has been used in a large number of studies to analyze all sorts of data. This chapter has offered an easy to understand overview of the basic concepts including the composition of the event log and the modelling approach that process mining undertakes. We also offered a brief overview of some examples of literature that have used the method in the analysis of students' data. Then, a step-by-step tutorial has shown different types of visualization in the form of process maps that are based on raw frequency, relative frequency and performance (time). Later, we have shown how process maps can be used to compare between groups and show their differences. The last part of the chapter offered a guide of how to animate the process maps and see how events

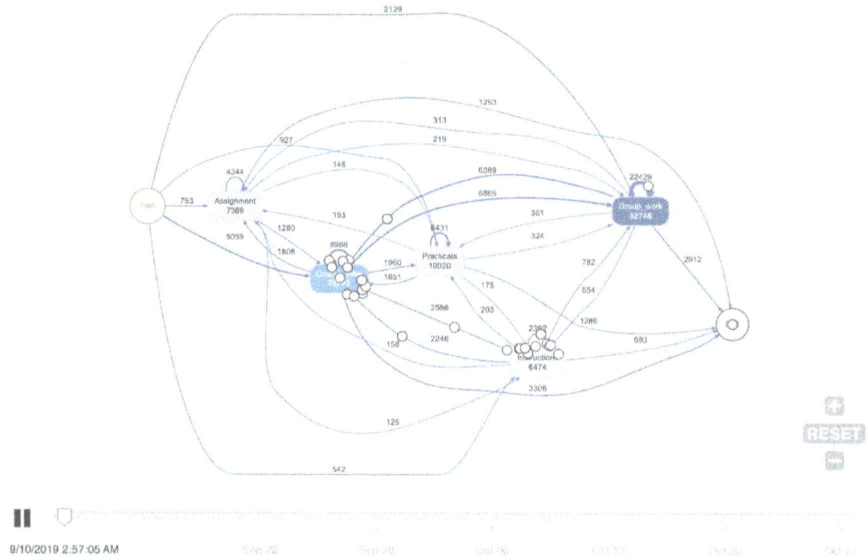

Fig. 11 Process animation

occur in real-time. Please note that our approach is restricted to `bupaverse`, which is the most popular and well-maintained framework for process mining within the R environment.

Whereas we share the enthusiasm for process mining as a powerful method, we also need to discuss the method from the rigor point of view. In many ways —as we will discuss later— process mining is largely descriptive and often times exploratory. While this is acceptable in the business domain, it needs to be viewed here from the viewpoint of scientific research and what claims we can make or base on a process map. As the tutorial has already shown, several parameters and decisions have to be made by the researcher and such decisions affect the output visualizations in considerable ways (e.g., filtering the event log). The said decisions are often arbitrary with no guidance or standards to base such decisions on. So, caution has to be exercised from over interpreting the process maps that have undergone extensive filtering. In fact, most researchers have used process mining as an auxiliary or exploratory method hand-in-hand with other tools such as sequence analysis or traditional statistics. In most cases, the conclusions and inferences were based on the other rigorous methods that are used, for instance, to statistically compare two groups. One can think of the process mining described in this chapter more of a visualization tool for the most common transitions or a method for visualizing the big picture of an event log in an intuitive way.

Another known problem of process mining is that most algorithms are black-boxed (i.e., the data processing is unclear to the researcher), with no fit statistics or randomization techniques to tell if the estimated process model is different from random. The absence of confirmatory tests makes the answer to the question of whether process mining actually recover the underlying process largely speculative. Moreover, all of the tools, algorithms, and visualization techniques have been imported verbatim from the business world and therefore, their validity for educational purposes remains to be verified. In the same vein, comparing across groups is rather done descriptively, without a statistic to tell whether the observed differences are actually significant or just happened by chance. In other methods such as psychological network analysis [34], several methods are available to create randomized models that assess the credibility of each edge, the stability of centrality measures, the differences between edges as well as to compare across networks regarding each of these parameters. Given that the algorithms, and the modelling technique necessitate trimming (discarding a considerable amount of data), it is bold to assume that the remaining process map (after trimming) is a true representation of the underlying process.

It is also important to emphasize here that, given that the field name is "process mining", the method does not translate to efficient analysis of the learning process. In fact, it is unrealistic to assume that process mining as a method —or any other method at large— can capture the full gamut of the complexity of the learning process as it unfolds across various temporal scales in different ways (e.g., transitions, variations, co-occurrence). A possible remedy for the problem lies in the triangulation of process mining models with other methods, e.g., statistics or better use a Markovian process modeling approach [3]. Stochastic process mining models (covered in detail in Chapter 12 [3]) are more advantageous theoretically and methodologically. Stochastic process mining models are theoretically robust, account for time and variable dependencies, offer fit statistics, clustering methods, several other statistics to assess the models and a wealth of possibilities and a growing repertoire of inter-operable methods.

6 Further Readings

There are multiple resources for the reader to advance their knowledge about process mining To learn more about the method itself, the reader should refer to the book *Process Mining: Data Science in Action* [35] and the *Process Mining Handbook* [36]. For more practical resources, the reader can refer to the `bupaverse` [30] documentation as well as find out more about the existing `bupaverse` extensions. The tutorial presented in this chapter has dealt with process analysis and visualization but not process discovery or conformance checking as the tools available in R are limited for these purposes. For process discovery in R, the reader can use the `heuristicsMineR` package [37], and for conformance checking, the `pm4py` package [38], a wrapper for the Python library of the same name. A practical guide with Python is also available in *A Primer on Process Mining: Practical Skills with Python and Graphviz* [39]. For specific learning resources on educational process mining, the reader can refer to the chapter "Process Mining from Educational Data" [40] in the *Handbook of Educational Data Mining*, and

"Educational process mining: A tutorial and case study using Moodle data sets" [41] in the book *Data Mining and Learning Analytics: Applications in Educational Research*. Moreover, the reader is encouraged to read the existing literature reviews on educational process mining (e.g., [11, 42]).

References

1. Saqr M, López-Pernas S, Helske S, Durand M, Murphy K, Studer M, Ritschard G (2024) Sequence analysis in education: principles, technique, and tutorial with R. In: Saqr M, López-Pernas S (eds) Learning analytics methods and tutorials: a practical guide using R. Springer, Berlin
2. López-Pernas S, Saqr M (2024) Modelling the dynamics of longitudinal processes in education. A tutorial with R for the VaSSTra method. In: Saqr M, López-Pernas S (eds) Learning analytics methods and tutorials: a practical guide using R. Springer, Berlin
3. Helske J, Helske S, Saqr M, López-Pernas S, Murphy K (2024) A modern approach to transition analysis and process mining with markov models: a tutorial with R. In: Saqr M, López-Pernas S (eds) Learning analytics methods and tutorials: a practical guide using R. Springer, Berlin
4. Saqr M (2024) Temporal network analysis: introduction and methods and analysis with R. In: Saqr M, López-Pernas S (eds) Learning analytics methods and tutorials: a practical guide using R. Springer, Berlin
5. van der Aalst WMP (2022) Process mining: a 360 degree overview. In: Lecture notes in business information processing. Springer, Berlin, pp 3–34
6. van der Aalst WMP, Weijters T, Maruster L (2004) Workflow mining: discovering process models from event logs. IEEE Trans Knowl Data Eng 16:1128–1142. https://doi.org/10.1109/tkde.2004.47
7. Leemans SJJ, Fahland D, van der Aalst WMP (2014) Discovering block-structured process models from event logs containing infrequent behaviour. In: Business process management workshops. Springer, Berlin, pp 66–78
8. Bergenthum R, Desel J, Lorenz R, Mauser S Process mining based on regions of languages. Lecture notes in computer science. Springer, Berlin, pp 375–383
9. Dumas M, Rosa ML, Mendling J, Reijers HA (2018) Fundamentals of business process management. Springer, Berlin
10. Peterson JL (1977) Petri nets. ACM Comput Surv 9:223–252. https://doi.org/10.1145/356698.356702
11. Bogarín A, Cerezo R, Romero C (2017) A survey on educational process mining. Data Min Knowl Discov 8:e1230. https://doi.org/10.1002/widm.1230
12. López-Pernas S, Saqr M, Viberg O (2021) Putting it all together: combining learning analytics methods and data sources to understand students' approaches to learning programming. Sustainability 13:4825. https://doi.org/10.3390/su13094825
13. Arpasat P, Premchaiswadi N, Porouhan P, Premchaiswadi W (2021) Applying process mining to analyze the behavior of learners in online courses. Int J Inf Edu Technol 11:436–443. https://doi.org/10.18178/ijiet.2021.11.10.1547
14. Hachicha W, Ghorbel L, Champagnat R, Zayani CA, Amous I (2021) Using process mining for learning resource recommendation: a moodle case study. Proc Comput Sci 192:853–862. https://doi.org/10.1016/j.procs.2021.08.088
15. Romero C, Cerezo R, Bogarín A, Sánchez-Santillán M (2016) Educational process mining: a tutorial and case study using moodle data sets. In: Data mining and learning analytics. Wiley, Hoboken, pp 1–28
16. Dolak R (2019) Using process mining techniques to discover student's activities, navigation paths, and behavior in LMS moodle. Lecture notes in computer science. Springer, Berlin, pp 129–138

17. Matcha W, Gašević D, Uzir NA, Jovanović J, Pardo A, Lim L, Maldonado-Mahauad J, Gentili S, Pérez-Sanagustín M, Tsai Y-S (2020) Analytics of learning strategies: role of course design and delivery modality. J Learn Anal 7:45–71. https://doi.org/10.18608/jla.2020.72.3
18. Beheshitha SS, Gašević D, Hatala M (2015) A process mining approach to linking the study of aptitude and event facets of self-regulated learning. In: Proceedings of the fifth international conference on learning analytics and knowledge. ACM, New York
19. Cerezo R, Bogarín A, Esteban M, Romero C (2020) Process mining for self-regulated learning assessment in e-learning. J Comput Higher Edu 32:74–88. https://doi.org/10.1007/s12528-019-09225-y
20. Vartiainen H, López-Pernas S, Saqr M, Kahila J, Parkki T, Tedre M, Valtonen T (2023) Mapping students' temporal pathways in a computational thinking escape room. In: Hirsto L, López-Pernas S, Saqr M, Sointu E, Valtonen T, Väisänen S (eds) Proceedings of the finnish learning analytics and artificial intelligence in education conference (FLAIEC22). CEUR, Joensuu, pp 77–88
21. Saqr M, Tuominen V, Valtonen T, Sointu E, Väisänen S, Hirsto L (2022) Teachers' learning profiles in learning programming: the big picture! Front Edu 7. https://doi.org/10.3389/feduc.2022.840178
22. Saqr M, López-Pernas S (2023) The temporal dynamics of online problem-based learning: why and when sequence matters. Int J Comput-Support Collab Learn 18:11–37. https://doi.org/10.1007/s11412-023-09385-1
23. Nafasa P, Waspada I, Bahtiar N, Wibowo A (2019) Implementation of alpha miner algorithm in process mining application development for online learning activities based on MOODLE event log data. In: 2019 3rd international conference on informatics and computational sciences (ICICoS). IEEE, Piscataway
24. Bogarín A, Romero C, Cerezo R, Sánchez-Santillán M (2014) Clustering for improving educational process mining. In: Proceedings of the fourth international conference on learning analytics and knowledge. ACM, New York
25. Pechenizkiy M, Trcka N, Vasilyeva E, van der Aalst WM, De Bra P (2009) Process mining online assessment data. In: Educational data mining 2009: 2nd international conference on educational data mining: proceedings [EDM'09], Cordoba, Spain. July 1–3, 2009. International Working Group on Educational Data Mining, pp 279–288
26. Fluxicon Disco
27. Juhaňák L, Zounek J, Rohlíková L (2019) Using process mining to analyze students' quiz-taking behavior patterns in a learning management system. Comput Human Behav 92:496–506. https://doi.org/10.1016/j.chb.2017.12.015
28. Van der Aalst WM, Dongen BF van, Günther CW, Rozinat A, Verbeek H, Weijters A (2009) ProM: the process mining toolkit. In: Proceedings of the BPM 2009 demonstration track (BPMDemos 2009). CEUR, Ulm, pp 1–4
29. Ariouat H, Cairns AH, Barkaoui K, Akoka J, Khelifa N (2016) A two-step clustering approach for improving educational process model discovery. In: 2016 IEEE 25th international conference on enabling technologies: infrastructure for collaborative enterprises (WETICE), pp 38–43
30. Janssenswillen G, Depaire B, Swennen M, Jans MJ, Vanhoof K (2019) bupaR: enabling reproducible business process analysis. Knowl Based Syst 163:1857. https://doi.org/10.1016/j.knosys.2018.10.018
31. Chan C, Chan GC, Leeper TJ, Becker J (2021) Rio: a swiss-army knife for data file i/o. https://cran.r-project.org/web/packages/rio/index.html
32. Wickham H, Averick M, Bryan J, Chang W, McGowan LD, François R, Grolemund G, Hayes A, Henry L, Hester J, Kuhn M, Pedersen TL, Miller E, Bache SM, Müller K, Ooms J, Robinson D, Seidel DP, Spinu V, Takahashi K, Vaughan D, Wilke C, Woo K, Yutani H (2019) Welcome to the tidyverse. J Open Source Softw 4:1686. https://doi.org/10.21105/joss.01686
33. López-Pernas S, Saqr M, Conde J, Del-Río-Carazo L (2024) A broad collection of datasets for educational research training and application. In: Saqr M, López-Pernas S (eds) Learning analytics methods and tutorials: a practical guide using R. Springer, Berlin

34. Saqr M, Beck E, López-Pernas S (2024) Psychological networks. In: Saqr M, López-Pernas S (eds) Learning analytics methods and tutorials: a practical guide using R. Springer, Berlin
35. Van der Aalst WMP (2016) Process mining: data science in action. Springer, Berlin
36. Van der Aalst WMP, Carmona J (2022) Process mining handbook. Springer, Berlin
37. Mannhardt F (2023) heuristicsmineR: discovery of process models with the heuristics miner
38. Mannhardt F (2023) pm4py: interface to the 'PM4py' process mining library
39. Ferreira DR (2020) A primer on process mining: practical skills with python and graphviz. Springer, Berlin
40. Trcka N, Pechenizkiy M, Van der Aalst WMP (2010) Process mining from educational data. In: Handbook of educational data mining. CRC Press, Boca Raton, pp 123–142
41. Romero C, Cerezo R, Bogarín A, Sánchez-Santillán M (2016) Educational process mining. In: Data mining and learning analytics. Wiley, Hoboken, pp 1–28
42. Ghazal MA, Ibrahim O, Salama MA (2017) Educational process mining: a systematic literature review. In: 2017 European conference on electrical engineering and computer science (EECS), pp 198–203

Part IV
Network Analysis

Social Network Analysis: A Primer, a Guide and a Tutorial in R

Mohammed Saqr, Sonsoles López-Pernas, Miguel Ángel Conde-González, and Ángel Hernández-García

1 Introduction

Social Network Analysis (SNA) has emerged as a computational method for studying interactions, relationships, and connections among a vast array of different entities that include humans, animals, and cities, just to name a few. Two related and largely overlapping fields are also concerned with the network as a concept: network science and network analysis. Network science is concerned with the study of the structure of networks, finding patterns and universal laws that may explain or underpin such structure in a large variety of phenomena. Network analysis is a very closely related field that is concerned with the analysis of networks that are not necessarily "social". In this chapter, we will simply use the terms social network analysis and network analysis interchangeably.

1.1 What Are Networks?

A quintessential concept in most analytical methods is that observations are —or should be—independent from each other whereas, in network analysis, observations can be related and interdependent and may interact with or influence each other [6]. As such, network analysis offers a more realistic view of the interconnected

M. Saqr (✉) · S. López-Pernas
School of Computing, University of Eastern Finland, Joensuu, Finland
e-mail: mohammed.saqr@uef.fi

M. Á. Conde-González
Robotics Research Group, Engineering School, University of León, León, Spain

Á. Hernández-García
Universidad Politécnica de Madrid, Madrid, Spain

M. Saqr, S. López-Pernas (eds.), *Learning Analytics Methods and Tutorials*,
https://doi.org/10.1007/978-3-031-54464-4_15

world around us and allows us to paint an accurate picture of the relationships and interactions that underpin our world [1, 2]. For instance, when we study students' engagement at a school, we may use a survey to measure each individual student's engagement and compute statistics such as the correlation between engagement and grades. In doing so, we ignore that students interact with peers, teachers, and the environment around them [3, 4]. We also ignore that students get influenced by an engaged student, get supported by their friends, or face a problematic social environment that may hinder their engagement [5, 6]. Network analysis offers a rich set of methods for modeling and addressing such issues [7].

A network is simply a group of entities (often called *vertices*, nodes, or *actors*) connected through a relationship (often called *edges*, *links, or arcs*) [2]. In this chapter, we will use the terms vertices and edges for simplicity. Vertices can be humans (students, teachers, or families), ideas, keywords, behaviors, emotions, feelings, concepts, schools, countries, or any entity that can be hypothesized to have a relationship with other entities. Vertices may be connected through a vast array of relations. For instance, students may be connected to each other by being friends, teammates, classmates, group members, neighbors, sporting club fans, competitors, sharing a desk, or working together on an assignment. There is virtually no limit to how a researcher can hypothesize a network. Nevertheless, the interpretation of network analysis relies heavily on how the network was constructed [2, 8].

In learning contexts, the most common type of networks comes from digital data, and in particular online communications, the most common of which are discussion forums, where the conversation occurs between forum participants (vertices). Each reply from one participant to another forms a relationship between the two vertices, creating an edge in the network. In such a situation, we are talking about a *directed* network, where the interaction has a source (the person who replies) and a target (the replied-to) [9]. Other examples of directed networks are citation networks (where documents cite other documents), or users that follow other users in social media [10]. In turn, an *undirected* network contains non-directional relationships such as a network of siblings, friends, teammates, husband and wife, or co-workers [2]. Representing interactions as a network equips the researchers with a rich toolset to harness the power of social network analysis methods. *Please note that the researcher can choose to model a directed network as undirected in case direction is deemed inappropriate according to theory, connect or research question.*

2 Analysis of Social Networks

Two types of analysis methods are commonly used: network visualization and mathematical network analysis [2]. Visualization summarizes all the interactions in the network, giving the researcher a bird's eye view of the structure of relationships, who is interacting with whom, who is leading the conversation, and who is isolated. Visualization is so powerful that it can summarize a whole semester of interactions in one visualization and still give meaningful information that can be used for intervention e.g., [11]. Mathematical network analysis offers quantification of all the network properties and the individual actors. The network properties are commonly known as graph-level measures. The mathematical measures of individual nodes are known node-level measures (often referred to as centrality measures) [2, 8, 12] Visualization is not discussed in details in this chapter, but interested readers are encouraged to see the excellent visualization tutorial maintained by Kateto Ognyanova [13].

2.1 Mathematical Analysis

Three types of mathematical analysis can be obtained with SNA: graph-level measures, node-level measures (centrality measures) and edge-level measures. Graph level measures describe the network as a whole in terms of size, interactivity and components. *Centrality measures* quantify the connectivity and importance of individual actors such as the degree of involvement in interactions, the distance to others, or the importance of position among other interacting actors [12, 14, 15]. However, each of these "importance" examples can be measured in different ways. For example, students can be considered central if they frequently contribute to discussions, reply to multiple unique threads of discussions, their contributions receive multiple replies, or their contributions trigger several threads of discussions. As such, there are several centralities with diverse methods that allow us to quantify the degree of connectedness. The website centiserver.org counts more than 400 centrality measures to this date and counting. In this book chapter, we will review and learn the most commonly used centralities in education. We rely on the recent meta-analysis 2022 for listing the used centrality measures, their classification, and operationalization in the literature. Edge measures describe the edge strength, edge weights or edges or edge centralities.

2.1.1 Graph-Level Measures

Graph-level measures are a type of analysis used in social network analysis that describes the overall structure and characteristics of a network. They can be used to compare different networks or track changes in a network over time.

Size or **vertex count** is the number of vertices (individuals or groups) in the network.

Edge count is the number of edges (interactions) between vertices in the network.

Density represents the number of edges that are present in the network divided by the number of all possible connections. High density indicates that many vertices in the network are connected to one another.

Reciprocity is another important graph-level measure that also reflects group cohesion. A reciprocated edge is an edge where two vertices have a reciprocal relationship (e.g., they are simultaneously source and target) [16]. The higher the ratio of the reciprocated edges, the more collaborative the network is, less centralized (dominated by few) and more participatory.

Transitivity (or global clustering coefficient) measures the tendency of the vertices to cluster together or form tightly knit cliques. There is a large volume of research that associates the ratio of cliques with cohesion in collaborative groups, strong ties and productive knowledge construction [16]

Other graph-level measures are aggregations of vertex-level or edge-level measures. For example, the **mean degree** is the mean number of edges of all the vertices in the network. In a collaborative group, the higher the mean degree, the more interactive the group is.

2.1.2 Local Centrality Measures

Local centrality measures are centralities that map the direct or immediate connections of an actor. Put another way, the local centralities quantify the neighbors of a certain vertex where each of these neighbors is directly connected to the target vertex without any intermediates (Figure 1).

In-degree centrality represents the total number of incoming interactions or relationships of a vertex [12, 14]. Several examples exist in the literature with different operationalization and interpretations. In the case of collaborative learning, in-degree centrality has been interpreted as the worthiness of a participant's contributions to receive replies, popularity, or influence [17–19].

Out-degree centrality represents the total number of outgoing interactions or links from an actor to other actors in the network [12, 14, 20]. Out-degree has been commonly interpreted as an indicator of participation, effort, and activity [21, 22].

Degree centrality refers to the total number of connections or interactions (incoming or outgoing) a vertex has. In undirected networks, it is simply the number of all connections of the vertex. Degree centrality can be interpreted in a similar way to the previous similar centralities as an indication of interactivity, communication, and social role in the collaborative process [23–25].

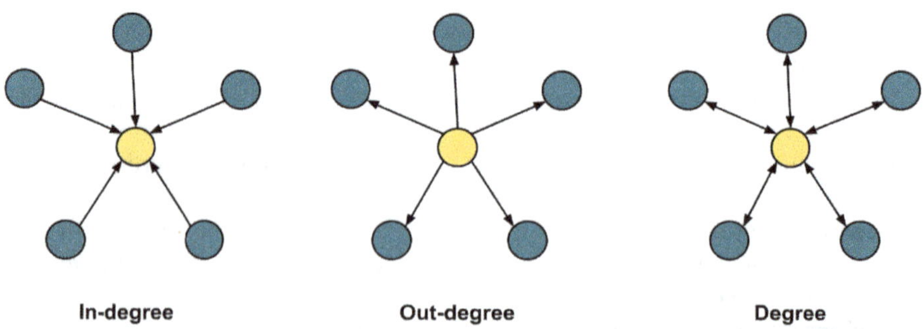

Fig. 1 Representation of a vertex in-degree (left), out-degree (middle), and degree (right)

2.1.3 Measures Based on Shortest Paths

Other relevant measures in assessing social network graphs are the shortest paths. These are based on the shortest distance between a pair of points (vertex) in the graph and represent how easy it is for a vertex to access others' resources [12, 14, 26]. Shortest paths can be used to better measure and understand centrality, as it is based on the distribution and distance of different points in a network which can help understand other educational insights [8, 27] The most relevant measures in this sense are represented in Fig. 2 and described below.

Closeness centrality is a measure based on the farness between the vertices of a network; more specifically between a specific vertex and the rest of the network [12, 14]. Small closeness values indicate greater proximity to other vertices, whereas larger values indicate greater distances from other vertices. In the field of education, it can be seen as a way to measure: (1) closure in interaction [11, 28–32], (2) ease of interaction [11, 24, 29, 33, 34], (3) time for accessing information [23, 31, 35], (4) dependencies [30], (5) control over resources [25, 28, 33], and (6) awareness of opportunities [28].

Betweenness centrality is a measure to show the frequency of a vertex lying on the shortest path between two other vertices. In the field of education, it indicates how actors mediate communication among themselves [8, 27]. It can be a way to understand who are the leaders of the interactions, that is, those who can moderate interactions, reach unconnected groups (inter-group connection) , influence the information flow (information brokering) and manage that information to solve problems effectively [28, 29, 36].

Eccentricity can help to see the vertices not involved in the interactions. It is calculated as the distance to the farthest other vertices in the network. It can be used in the educational field to understand which are the students at risk of dropout or those that are not participating in the activities [11, 34, 36].

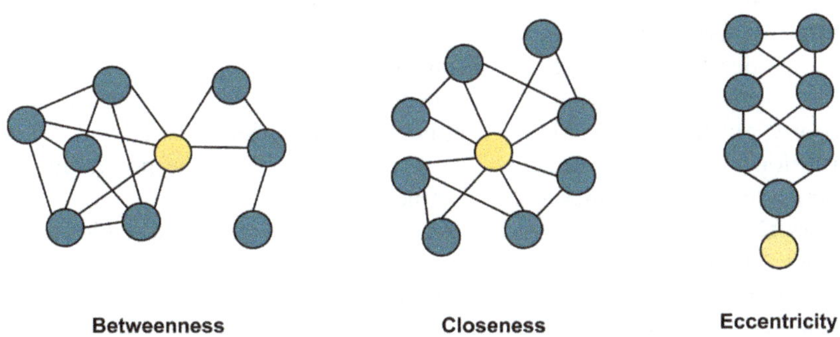

Fig. 2 Examples of networks where the highlighted vertex has a high value of betweenness (left), closeness (middle), and eccentricity (right)

2.1.4 Eigenvector-Based Centralities

Other useful metrics in social network graphs are those based on eigenvector centralities, i.e., those related to the value of a connection between vertices. It is based on the idea that it is preferable to have fewer connections to strong well-connected vertices than to have many connections to weak isolated vertices. There are several measures based on this principle:

Eigenvector centrality assesses the importance of a vertex based on the centralities of the vertex to which it is connected, that is, how many connections it has to influential vertices. It has been used to understand social capital, ego, and connection strength [21, 23, 34, 37].

Pagerank is based on link analysis and allows for defining the popularity of an individual in a network. It can be used in education to understand the reputation or influence upon a group of stakeholders based on his/her contacts [36, 38].

Hub centrality is based on Kleinberg's HITS algorithm [39] that measures who interacts more with the most influential vertices in a network. In the educational field, it could help to understand the type of students' interaction [23, 38, 40].

Authority is also based on Kleinberg's HITS algorithm but it is used to identify which vertices might be considered an authority on a specific topic score. It is calculated taking into account if its incoming links connect the vertex to others that have an important number of outgoing links [23, 38].

2.1.5 Other Measures

There are other possible metrics to be applied in social network graphs:

Clustering coefficient is a measure of how individuals, represented as vertices, are embedded in their neighborhood [36], with interactions with other individuals forming triangles. In the educational field, it shows how an individual works with peers in groups or clusters, which shows network cohesion [11, 25, 34, 38].

Diffusion Centrality belongs to the diffusion measures that explore the structural properties that facilitate the diffusion and uptake process; that is, the diffusion resulting from the interaction. It tries to measure how well a vertex can diffuse a property given the semantics and structure of a social network and a model of diffusion [41, 42]. In education, it can be used to show the possibility of an interaction to generate replies and these other replies [19].

Cross Clique Centrality assesses the number of cliques or triangles a vertex belongs to. It is related to the degree of embeddedness, connectivity with other vertices, and strength of ties among vertices. In education, it allows understanding, for example, whether a post is going to be replied to and spread throughout the network [19].

Coreness (k-core or linkage) is similar to h-index metrics for publications, and is based on networks where all vertices have at least a k degree [43]. It is interesting because if an individual may produce promising contributions, he or she is going to attract others with similar contributions and establish strong connections. In the field

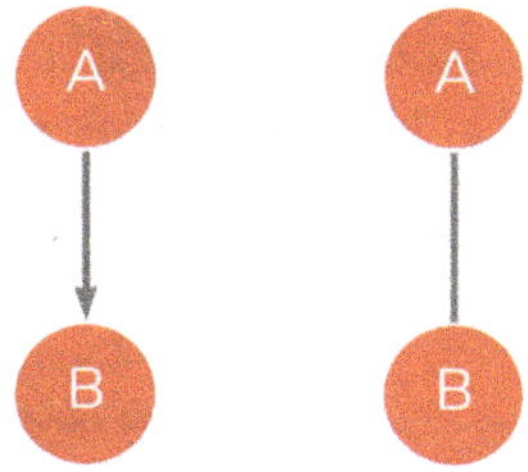

Fig. 3 Left: a directed edge where an interaction happens from A->B. Right: an undirected edge where A-B are connected

of education, those students involved in active interactions/discussions will attract other active users, which will enrich the discussion and therefore the interaction, collaboration, the quality of content, etc. [19].

2.2 Network Visualization

As we discussed earlier, networks represent the relationships (edges) between actors (vertices). A vertex is commonly represented as a circle —although other shapes are also used— and the edge as an arrow (in the case of a directed network) from the source of interaction to the target of the interaction (Fig. 3). For example, a phone call from the caller to the call receiver is represented by an arrow from the caller to the receiver. In the case of an undirected network, the edge is represented by a line connecting both vertices, for instance, two siblings, where the relationship is mutual.

2.3 Network Analysis

Let us start with a simple example that uses a common analysis scenario. Our example is a fictional discussion among students where we model interactions between students as a network. Building a network from interactions in discussion forums has been commonly performed by building a post-reply network where an edge is constructed from the author (source of the reply) to the replied-to (target of the reply). For instance, in Fig. 4, **B** replies to **A** by saying *"I agree, this could slow the progression of the disease but does it prevent the spread completely or terminate the pandemic?"*. This can be represented as an edge from B to A. In the same token, **C** replies to **B** which represents another edge C to B. We can compile all of these interactions in a list of edges as shown in Table 1 which is often referred to as an *edge list*. An edge list is a simple way —among many others— to represent the network and therefore we will use it in this chapter. Constructing the network can be performed by aggregating all edges as a network as Fig. 4.

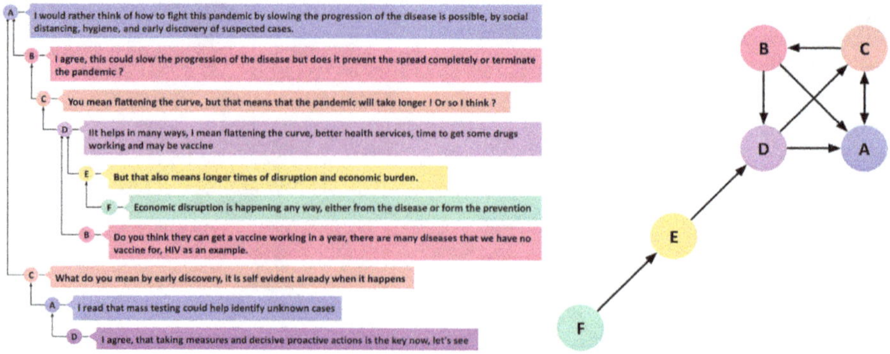

Fig. 4 A conversation between students in a discussion forum

Table 1 Edge list of students' interactions in the discussion forum

Source	Target
B	A
C	B
D	C
E	D
F	E
B	D
C	A
A	C
D	A

There are other ways to construct the network that depend on the researcher's conceptualization, context, and research question. For instance, researchers may consider all students present in a discussion (co-present) and therefore linked together, which means that all vertices will have connections to each other [9, 22, 44]. Similarly, several ways exist to aggregate the network. Figure 5 shows three identical networks where we have duplicate ties (i.e., repeated interactions between the same pair of vertices). In Fig. 5a, all interactions are represented. We can see, for instance, that there are six edges between A and F vertices. We can also see that there is a "loop" or "self-loop" around C and E. A loop exists when an interaction occurs between the vertex and itself, as is the case when one replies to their own post. This type of network configuration where multiple edges and self-loops are allowed is often referred to as *multigraph*. Another possible way to aggregate the network is to create a weight for each edge which represents the frequency of interactions between each pair of vertices (Fig. 5b). For instance, the edge between A and F will have a weight of 6 which is the number of edges between A and F, whereas the weight of the edge between H and B is 1 since this interaction only happened once. The network in Fig. 5b has an edge with a thickness that corresponds to the weight of 6, i.e., six times as thick as the edge between H and B. In other instances, we may disregard these repeated edges when they do not make a difference to our

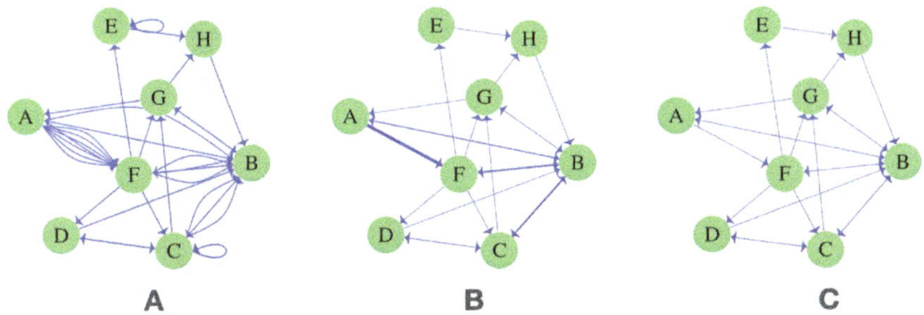

Fig. 5 Several ways of constructing a network: (**a**) Multigraph, (**b**) Weighted, (**c**). Simplified

conceptualization of the network (Fig. 5c). Such a type of network is often referred to as a simplified network. For a discussion about network configurations and how they influence research results see [22].

3 Network Analysis in R

The R language has a large number of libraries for the analysis of networks. The `igraph` package —the one used in our chapter— seems to be the preferred package by the R community given the number of dependencies, i.e., the number of other packages that rely on `igraph` or work with the `igraph` format [45]. The `igraph` package is fast, efficient, and well-respected within the academic community. The `igraph` package is also well maintained, continuously updated, has a large community, and has been released for other platforms besides R, e.g., Python. Other packages, such as sna and network, have a large user base, especially among those who are interested in statistical network modeling. Any of these packages —`sna`, `network` or `igraph`— can effectively perform the analysis described in this chapter. However, we will use `igraph` based on its relative ease of use and convenience for the chapter objectives.

Example 1 Let us start with a simple example where we analyze the network created for Fig. 6. Before doing anything else, we need to import the necessary packages. We will use `igraph` to construct and represent networks, and we will use rio to download and import the data files that we need to use as an input for `igraph`.

```
library(igraph)
library(rio)
```

We can now use the import function from rio to download the data for the example (the data shown in Table 1), and assign it to a variable named SNA_example1.

```
URL<-
"https://github.com/lamethods/data/raw/main/8_examples/"
SNA_example1 <- import(paste0(URL, "SNA_example1.xlsx"))
```

The function `graph_from_data_frame` from `igraph` converts the edge list in Table 1 into a network. R expects a dataframe where the first two columns are used as edges (the first column is used as source column, and the second is used as source column). Please also note that the two columns can have any name. Also, all extra columns —if they are there— will be used as edge attributes. We can print it to see if it has been created correctly. The print function is commonly used to test if the graph creation has been successful. In other words, does the created network have the expected number of vertices, edges, and attributes?

```
Net <- graph_from_data_frame(SNA_example1)
print(Net)
```

```
IGRAPH f3fa5a7 DN-- 6 9 --
+ attr: name (v/c)
+ edges from f3fa5a7 (vertex names):
[1] B->A C->B D->C E->D F->E B->D C->A A->C D->A
```

The output of the print function gives a glimpse of the network properties. First, `igraph` states that the object is an `igraph` object (a network). Then, `igraph` gives a unique seven-letter identifier for the network (not usually needed for analysis). Next, `igraph` tells us that the network was directed (D) and named (N), i.e., vertices have a name attribute. Then, `igraph` lists the attributes and the edges of the network. We can also visualize the created network (Fig. 6) by using the function `plot`.

```
plot(Net)
```

We have seen here the most basic functions we can use in a graph with no arguments. As shown, networks work with little effort with R. In the next section, we will take a deeper look into these functions and others using another network from a published paper.

Example 2 The next example is a larger network that comes from the interactions of a group of teachers in a Massive Open Online Course (MOOC). The MOOC included 445 participants from different places in the United States. The dataset has an edges file where the first two columns are the sender (`source`) and receiver (`target`). There is also a file for the vertices that contains demographic and other relevant information about each vertex: gender, location, and their role, etc. For more information about this dataset, please refer to the data chapter [47]. To get the data into R, we first need to read the data, store it in a dataframe and then build a network with the appropriate arguments.

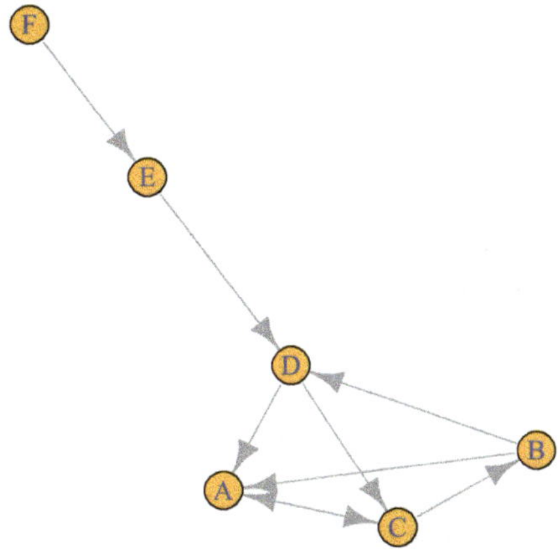

Fig. 6 Example of a simple network plotted with `igraph`

The first line of the code reads the edges list data with their attributes into a dataframe with the name `net_edges`. The second line imports the vertex data with their attributes into a dataframe with the name net_nodes.

```
URL <-
"https://github.com/lamethods/data/raw/main/6_snaMOOC/"
net_edges <- import(paste0(URL, "DLT1%20Edgelist.csv"))
net_nodes <- (import(paste0(URL, "DLT1%20Nodes.csv"))
```

To create the network, we again use the function `graph_from_data_frame`. This time we have to specify the edges dataframe using the argument d=net_edges. The second argument (optional) tells `igraph` that the network should be directed, if not provided, the network is created directed by default. The third argument which is also optional vertices = net_nodes tells `igraph` to use the dataframe net_nodes for vertex attributes. If the vertices argument is not provided, `igraph` will extract vertex names from the edges data. In case there are important vertex attributes for the analysis, providing the vertices data can be useful. Building the network and explicitly setting all arguments —as we did— helps avoid the problems that could happen from the default settings of the function. For instance, the network could be created as directed where we aim at creating an undirected network. Note that `igraph` generates a multigraph network by default (see Fig. 7).

```
DLT1 <- graph_from_data_frame(d=net_edges, directed =
        TRUE, vertices = net_nodes)
```

Let us now explore the network and see if it was built correctly using the function print. The print function output shows that the network is an `igraph` object, directed and named (*DN*) has 445 vertices, 2529 edges and then `igraph` lists the attributes of the vertices and the edges. Vertex attributes are listed along with their type.

For instance, name (v/c) means the name attribute is a (v)ertex attribute and a (c_)haracter. Edge attributes are listed in the same way. For instance, timestamp (e/c) means that it is an (e)_dge and a (c)haracter.

```
print(DLT1)
```

```
IGRAPH 97c4a14 DN-- 445 2529 --
+ attr: name (v/c), Facilitator (v/n), role1 (v/c), experience (v/n),
| experience2 (v/c), grades (v/c), location (v/c), region (v/c),
| country (v/c), group (v/c), gender (v/c), expert (v/c), connect
| (v/c), Timestamp (e/c), Discussion Title (e/c), Discussion Category
| (e/c), Parent Category (e/c), Category Text (e/c), Discussion
| Identifier (e/c), Comment ID (e/n), Discussion ID (e/n)
+ edges from 97c4a14 (vertex names):
 [1] 360->444 356->444 356->444 344->444 392->444 219->444 318->444 4   ->444
 [9] 355->356 355->444 4   ->444 310->444 248->444 150->444 19 ->310 216->19
[17] 19 ->444 19 ->4   217->310 385->444 217->444 393->444 217->19  256->219
+ ... omitted several edges
```

A network can also be plotted with the function `plot()` (Fig. 7). However, plotting with R is a vast field and will not be discussed in detail here.

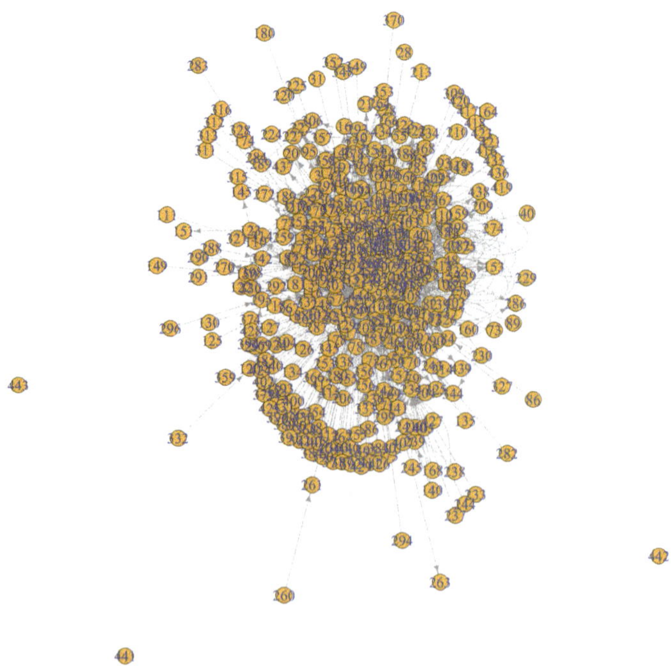

Fig. 7 Example of a complex network plotted with `igraph`

```
plot(DLT1,  layout = layout.fruchterman.reingold,
     vertex.size = 5, vertex.label.cex = 2)
```

3.1 Graph Level Analysis

Now that we have seen how to build a network from edge and vertex data, we are ready to understand some of the most commonly performed analyses in learning settings. The first type of analysis will look at the network level, or the whole group of collaborators. Analyzing the network level can tell us how interactive the group is, how cohesive, and how distributed the interactions are. We will go through each of these graph-level measures with a brief explanation of what they actually mean. We will use the data from example 1 (`DLT1`).

Let us first start by calculating the basic measures of the network. The number of vertices can be queried using the function `vcount`, which adds up to 445, and the number of edges can be queried using the function `ecount`, which is 2,529. We can get the average number of interactions by a participant by dividing the number edges by the number of vertices which is 5.68.

```
vcount(DLT1) # 445
ecount(DLT1) # 2529
ecount(DLT1) / vcount(DLT1) # 5.683146
```

The density of a graph is an important parameter of a collaborative network that refers to the ratio of existing edges to the maximum possible among all participants. Density is maximum (1) when every vertex has interacted with every other vertex in the network. Graph density can be measured using the function `graph.density`.

```
graph.density(DLT1) # 0.01279988
```

However, the `graph.density` function may result in erratic results if the network is multigraph; this is because the `igraph` algorithm will count the repeated edges and loops. Thus, we need to *simplify* the network (delete all repeated edges and loops) before computing the density and use the simplified network to compute the graph density. The results of the density of the graph of 0.0097 which is rather a low value.

```
graph.density(simplify(DLT1)) # 0.009798563
```

Reciprocity is another important graph-level measure that also reflects group cohesion. A reciprocated edge is an edge where two vertices have a reciprocal relationship (e.g., they are simultaneously source and target) [16]. The higher the ratio of the reciprocated edges, the more collaborative the network is, less centralized (dominated by few) and more participatory. Reciprocity can be computed using the function `reciprocity`, which automatically removes the loops (i.e., does not consider when a person replies to oneself). The reciprocity by `igraph` definition is the fraction of reciprocated edges in a directed graph. The value here is 0.1997544 which means that only 20% of all edges were reciprocated.

```
reciprocity(DLT1) # 0.1997544
```

We can also compute the dyad.census which returns the number of mutual interactions (reciprocated between a pair of vertices), the number of asymmetric interactions (interactions that are not reciprocated), and the number of non-connected pairs. The number of mutual interactions in our network is 212, which is relatively small given the asymmetric (1512) and non-connected pairs (97066).

```
dyad.census(DLT1) # $mut [1] 212      $asym [1]
                    1512     $null [1] 97066
```

Transitivity (or global clustering coefficient) measures the tendency of the vertices to cluster together or form tightly knit cliques. In `igraph`, transitivity is measured as the probability that the neighboring vertices of a vertex are also connected to each other or, more accurately, the ratio of triangles in the network to the total count of triplets (all occurrences of three vertices connected by two edges). There is a large volume of research that associates the ratio of cliques with cohesion in collaborative groups, strong ties and productive knowledge construction Block_2015. There are several methods for the estimation of transitivity. Here, we are going to focus on global transitivity (i.e., at the network-level) using the `igraph` method. The transitivity can be calculated by the function `transitivity`; the default function returns the global transitivity measure by default. The transitivity of our network here is 0.08880774.

```
transitivity(DLT1) # 0.08880774
```

Another possible way is to use the related function `triad_census` which reports the numbers of triangles and their different types. The reader may need to refer to the package manual to dig deeper in the results.

```
triad_census(DLT1)
```

```
[1] 13901588   486626   124805     4227    35745    11186    15929     3668
[9]      932       81     1857      376      223      334      345       68
```

Group productivity or intensity of interactivity can be explored using the *degree* measures and its variants. The average *degree* of the network measures how much on average each group member has contributed and received interactions. To compute the *average degree*, we first have to compute the *degree* for each member and then compute the *mean*.

In directed networks (like the one in this example), we can also compute the average in-degree and out-degree. For the same set of vertices, the network average in-degree should be equal to out-degree and both combined should be equal to the total degree. However, if we, for instance, compute a subset of vertices (only students excluding the teachers), in-degree and out-degree may be different. The code below computes the mean and median of the three measures, using the function *degree* with the argument `mode="total"` for the total degree, `mode="in"` for the *in-degree*, and mode="out" for *out-degree*.

```
Mean_degree <-  mean(degree(DLT1, mode = "total"))
                        # 11.36629
Mean_in_degree <-  mean(degree(DLT1, mode = "in"))
                        # 5.683146
Mean_out_degree <-  mean(degree(DLT1, mode = "out"))
                        # 5.683146
Median_degree <-  median(degree(DLT1, mode = "total"))
                        # 4
Median_in_degree <-  median(degree(DLT1, mode = "in"))
                        # 1
Median_out_degree <-  median(degree(DLT1, mode = "out"))
                        # 2
```

The mean degree is 11.36629 and the mean in-degree and out-degree are 5.683146. The median degree is 4, the median in-degree is 1, and the median out-degree is 2. The median differs significantly from the mean and may be more relevant here in this large network, where participation may not be well-distributed (see next section).

Collaboration is participatory by design but, oftentimes, some students may dominate and contribute disproportionately more than others. In the same vein, some may prefer to be isolated and thus rarely participate. Several measures allow us to measure the distribution of interactions across the network and how skewed the network contribution patterns are. An obvious method that comes straight from statistics is the standard deviation (SD) of the degree centrality. We can compute the SD like we calculated the mean and median in the previous step. The SD of *degree* centrality in our case is 34.2, SD for in-degree centrality is 26.7, and SD for *out-degree* centrality is 9.8. The SD is higher than the mean which suggests that calculation and inspection of the median was justified. We can also see that the SD of the *in-degree* centrality is much higher than the SD of the *out-degree,* which means that the variability in receiving replies is higher than that of contributions. This variability is rather common since students are selective about whom they respond to and choose the reply-worthy contributions.

```
SD_degree <-  sd(degree(DLT1, mode = "total"))
                                        # 34.20511
SD_in_degree <-  sd(degree(DLT1, mode = "in"))
                                        # 26.73596
SD_out_degree <-  sd(degree(DLT1, mode = "out"))
                                        # 9.84249
```

SNA has dedicated indices for measuring dominance in networks, known as centralization indices. Centralization indices are 0 when every vertex contributes equally and reaches the maximum of 1 when a single vertex dominates. A centralization index exists for many centralities —e.g., degree, closeness, and betweenness. Nevertheless, most of the literature has reported degree centralization, which we will demonstrate here.

The code below computes the degree, in-degree and out-degree centralization. Please note that we use a simplified network to avoid the loops and repeated edges. The results of degree centralization confirm the previous results. The degree centralization is 0.38, out-degree centralization is 0.16, and in-degree centralization is 0.60. In our network, we see that the in-degree centralization is the highest index (0.60), which means that only a few students received replies.

```
Centralization_degree <- centralization.degree(simplify(DLT1),
mode =  "all", loops = FALSE)$centralization  # 0.3826871
Centralization_in_degree <-  centralization.degree(simplify(DLT1),
mode = "in", loops = FALSE)$centralization   # 0.6064291
Centralization_out_degree <- centralization.degree(simplify(DLT1),
mode =  "out", loops = FALSE)$centralization  # 0.1572214
```

Another way to see how the interactions are distributed is to plot the degree distribution using the hist function, as demonstrated in Fig. 8.

```
par(mfrow=c(1,2))
hist(degree(DLT1, mode = "in"), breaks = 100)
hist(degree(DLT1, mode = "out"), breaks = 100)
```

3.2 *Network Connectivity*

We can also examine how connected the whole group is; this can be performed using the function is.connected which returns `FALSE` in our case, meaning that the graph has some disconnected components or subgroups of vertices that are isolated. We

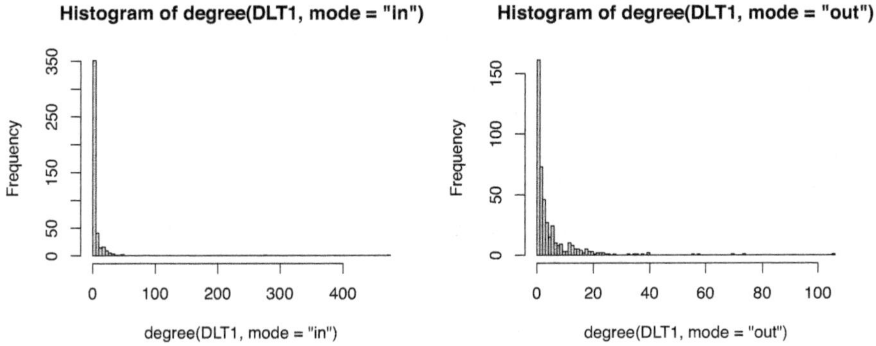

Fig. 8 Distribution of degree. Left: in-degree. Right: out-degree

can check these subgroups by the function components which tells us that there are four components:

```
is.connected(DLT1)
[1] FALSE
```

```
Components <- components(DLT1)
print(Components)
```

```
$membership
  1   2   3   4   5   6   7   8   9  10  11  12  13  14  15  16  17  18  19  20
  1   1   1   1   1   1   1   1   1   1   1   1   1   1   1   1   1   1   1   1
 21  22  23  24  25  26  27  28  29  30  31  32  33  34  35  36  37  38  39  40
  1   1   1   1   1   1   1   1   1   1   1   1   1   1   1   1   1   1   1   1
 41  42  43  44  45  46  47  48  49  50  51  52  53  54  55  56  57  58  59  60
  1   1   1   1   1   1   1   1   1   1   1   1   1   1   1   1   1   1   1   1
 61  62  63  64  65  66  67  68  69  70  71  72  73  74  75  76  77  78  79  80
  1   1   1   1   1   1   1   1   1   1   1   1   1   1   1   1   1   1   1   1
 81  82  83  84  85  86  87  88  89  90  91  92  93  94  95  96  97  98  99 100
  1   1   1   1   1   1   1   1   1   1   1   1   1   1   1   1   1   1   1   1
101 102 103 104 105 106 107 108 109 110 111 112 113 114 115 116 117 118 119 120
  1   1   1   1   1   1   1   1   1   1   1   1   1   1   1   1   1   1   1   1
121 122 123 124 125 126 127 128 129 130 131 132 133 134 135 136 137 138 139 140
  1   1   1   1   1   1   1   1   1   1   1   1   1   1   1   1   1   1   1   1
141 142 143 144 145 146 147 148 149 150 151 152 153 154 155 156 157 158 159 160
  1   1   1   1   1   1   1   1   1   1   1   1   1   1   1   1   1   1   1   1
161 162 163 164 165 166 167 168 169 170 171 172 173 174 175 176 177 178 179 180
  1   1   1   1   1   1   1   1   1   1   1   1   1   1   1   1   1   1   1   1
181 182 183 184 185 186 187 188 189 190 191 192 193 194 195 196 197 198 199 200
  1   1   1   1   1   1   1   1   1   1   1   1   1   1   1   1   1   1   1   1
201 202 203 204 205 206 207 208 209 210 211 212 213 214 215 216 217 218 219 220
  1   1   1   1   1   1   1   1   1   1   1   1   1   1   1   1   1   1   1   1
221 222 223 224 225 226 227 228 229 230 231 232 233 234 235 236 237 238 239 240
  1   1   1   1   1   1   1   1   1   1   1   1   1   1   1   1   1   1   1   1
241 242 243 244 245 246 247 248 249 250 251 252 253 254 255 256 257 258 259 260
```

```
  1   1   1   1   1   1   1   1   1   1   1   1   1   1   1   1   1   1   1   1
261 262 263 264 265 266 267 268 269 270 271 272 273 274 275 276 277 278 279 280
  1   1   1   1   1   1   1   1   1   1   1   1   1   1   1   1   1   1   1   1
281 282 283 284 285 286 287 288 289 290 291 292 293 294 295 296 297 298 299 300
  1   1   1   1   1   1   1   1   1   1   1   1   1   1   1   1   1   1   1   1
301 302 303 304 305 306 307 308 309 310 311 312 313 314 315 316 317 318 319 320
  1   1   1   1   1   1   1   1   1   1   1   1   1   1   1   1   1   1   1   1
321 322 323 324 325 326 327 328 329 330 331 332 333 334 335 336 337 338 339 340
  1   1   1   1   1   1   1   1   1   1   1   1   1   1   1   1   1   1   1   1
341 342 343 344 345 346 347 348 349 350 351 352 353 354 355 356 357 358 359 360
  1   1   1   1   1   1   1   1   1   1   1   1   1   1   1   1   1   1   1   1
361 362 363 364 365 366 367 368 369 370 371 372 373 374 375 376 377 378 379 380
  1   1   1   1   1   1   1   1   1   1   1   1   1   1   1   1   1   1   1   1
381 382 383 384 385 386 387 388 389 390 391 392 393 394 395 396 397 398 399 400
  1   1   1   1   1   1   1   1   1   1   1   1   1   1   1   1   1   1   1   1
401 402 403 404 405 406 407 408 409 410 411 412 413 414 415 416 417 418 419 420
  1   1   1   1   1   1   1   1   1   1   1   1   1   1   1   1   1   1   1   1
421 422 423 424 425 426 427 428 429 430 431 432 433 434 435 436 437 438 439 440
  1   1   1   1   1   1   1   1   1   1   1   1   1   1   1   1   1   1   1   1
441 442 443 444 445
  2   3   4   1   1

$csize
[1] 442   1   1   1

$no
[1] 4
```

Using the function decompose, we can look at each of the components. The largest component has 442 vertices, and three others have one vertex. These isolated vertices are simply students who did not contribute at all and do not represent a real subgroup.

```
Decomposed <- decompose(DLT1)
Decomposed[[1]]
```

```
IGRAPH 57821fd DN-- 442 2529 --
+ attr: name (v/c), Facilitator (v/n), role1 (v/c), experience (v/n),
| experience2 (v/c), grades (v/c), location (v/c), region (v/c),
| country (v/c), group (v/c), gender (v/c), expert (v/c), connect
| (v/c), Timestamp (e/c), Discussion Title (e/c), Discussion Category
| (e/c), Parent Category (e/c), Category Text (e/c), Discussion
| Identifier (e/c), Comment ID (e/n), Discussion ID (e/n)
+ edges from 57821fd (vertex names):
 [1] 360->444 356->444 356->444 344->444 392->444 219->444 318->444 4   ->444
 [9] 355->356 355->444 4   ->444 310->444 248->444 150->444 19 ->310 216->19
[17] 19 ->444 19 ->4   217->310 385->444 217->444 393->444 217->19  256->219
+ ... omitted several edges
```

We can also look at the network diameter or largest number of steps between vertices to see how far distant vertices are (using the distance function). A more representative variable would be the average distance between vertices, which we can obtain from the mean_distance function. The network diameter is 8 and the mean distance is 3. Both numbers are relatively high. Global efficiency is another network-level measure that examines how effective is the network structure as a conduit for information exchange using the distances between vertices. When all vertices are *close* to each other, reachable with a few number of steps, the network is said to be efficient. The value of efficiency is high in well connected groups, and low otherwise. We can examine the efficiency using the function `global_efficiency`.

```
diameter(DLT1) # 8
```

```
[1] 8
```

```
mean_distance(DLT1) # 3.030694
```

```
[1] 3.030694
```

```
global_efficiency(DLT1) # 0.1961034
```

```
[1] 0.1961034
```

3.3 Network Operations

There are many functions and tools to manipulate networks in `igraph`, which makes a comprehensive discussion of all of them beyond the scope of this introductory chapter. Nonetheless, we will discuss the most important functions. Oftentimes, we need to set an attribute to the vertices —e.g., setting the gender attribute for vertices— to be used in the analysis. Setting an attribute can be performed by using the V function followed by the $ character and the attribute that we want to set. Similarly, setting edge attributes can be done using the function E. In the next example, we define an attribute called `weight` for the vertices and we do the same for the edges. Using the skills we learnt, we can use them to create a simplified weighted network. We do so by concatenating all repeated edges into a single edge with the weight as the frequency. For that, we start by first assigning weights of 1 to each node and edge. Lastly, we use the function `simplify` to remove the duplicated edges and aggregate the weights (`edge.attr.comb = list(weight = "sum", "ignore")`) while all other edge attributes will be *ignored*.

```
V(DLT1)$weight <- 1
E(DLT1)$weight <- 1
simple.DLT1 <- simplify(DLT1,
```

```
    remove.multiple = TRUE, remove.loops = TRUE,
     edge.attr.comb = list(weight ="sum", "ignore"))
```

There are two important functions that we may need if we want to divide or create a subnet of the network. The function `subgraph.edges` allows us to create a subset of a network based on edge characteristics. In the following example, we create a subgraph with the discussions involving *Curriculum & Instruction* by using the argument `E(DLT1)$'Discussion Category'== 'Curriculum & Instruction'`. In the same way, the induced_subgraph function allows us to specify a subgraph based on the vertex characteristics. In the next example, we create a network for North Carolina teachers using `V(DLT1)$location== "NC"`.

```
k <- subgraph.edges(DLT1,
    eids = which(E(DLT1)$'Discussion Category' ==
                    'Curriculum & Instruction'))
NC_network <- induced_subgraph(DLT1, vids =
                    which(V(DLT1)$location == "NC"))
```

3.4 Individual Vertex Measures (Centrality Measures)

We have discussed the centrality measures in the introductory section and how they can be used in an educational context. Centrality measures may serve many functions, —e.g., indicators of performance in collaborative environments [48] or indicators for roles in collaboration [21]. The `igraph` package allows the calculation of several centrality measures, many of which have been used in educational research and some of which may not be relevant. Other packages, such as `centiserve`, allow an even larger number of centrality measures [49]. In this section, we will focus on the common centrality measures according to the recent meta-analysis by Saqr et al. [48] and other recently used measures, such as diffusion centrality measures.

Degree centrality measures can be computed using the function *degree* and the argument *mode* specifies the type of *degree* where `mode="in"` returns *in-degree*, `mode="out"` returns *out-degree*, `mode="total"` returns total *degree* centrality. In case of undirected networks, the *mode* argument is ignored and the function returns only the *degree* centrality, since there is no direction. We can combine all the centralities that we calculate together in a dataframe using the tibble function from the tibble package.

```
InDegree  <- degree(DLT1, mode = "in")
OutDegree <- degree(DLT1, mode = "out")
Degree    <- degree(DLT1, mode = "total")
```

```
Degree_df <- tibble::tibble(
     name = V(DLT1)$name, InDegree, OutDegree, Degree)
     print(Degree_df)
```

```
# A tibble: 445 x 4
   name  InDegree OutDegree Degree
   <chr>    <dbl>     <dbl>  <dbl>
 1 1           20        33     53
 2 2            2         5      7
 3 3            2         4      6
 4 4            2        14     16
 5 5           16        17     33
 6 6            9        24     33
 7 7           32        26     58
 8 8           13        18     31
 9 9            2        12     14
10 10           8        12     20
# i 435 more rows
```

Note that `igraph` has another function called graph.strength that computes the *degree* centrality and takes the edge weight attribute into account. In multigraph networks —like ours— *degree* centrality and `graph.strength` should return the same result. However, in networks where the edges have a `weight` attribute both functions (`degree` and `graph.strength`) return different results. In such weighted networks —like the simplified network we created in the previous example— the `degree` function will return the unique connections of every vertex or the size of the vertex direct collaborators —known as the size of the ego network. The `graph.strength` function will return the number of interactions a vertex has made. See the next example and compare the results. For more information about the different calculation methods of degree centrality of weighted networks, readers are advised to refer to the seminal article by Opsahl et al. [50].

```
InStrength  <- graph.strength(DLT1, mode = "in")
OutStrength <- graph.strength(DLT1,mode = "out")
Strength    <- graph.strength(DLT1, mode = "total")
Strength_df <- tibble::tibble(name=V(DLT1)$
                 name,InStrength,OutStrength,Strength)
print(Strength_df)
```

```
# A tibble: 445 x 4
   name  InStrength OutStrength Strength
   <chr>      <dbl>       <dbl>    <dbl>
 1 1             20          33       53
```

```
2 2                 2         5       7
3 3                 2         4       6
4 4                 2        14      16
5 5                16        17      33
6 6                 9        24      33
7 7                32        26      58
8 8                13        18      31
9 9                 2        12      14
10 10               8        12      20
# i 435 more rows
```

Closeness and *betweenness* centralities are the most commonly used centrality measures according to [48] and both rely on the position of the vertex on the shortest paths between others. *Closeness* centrality can be calculated using the function closeness, which is directional; this means that we can compute *in-closeness*, *out-closeness* and total *closeness* centrality. *Betweenness* centrality can be computed using the function *betweenness* and the function is directional: the argument `directed=TRUE` computes the directional version, and vice versa. Another commonly used centrality measure is eigenvector centrality, which can be computed using the function `eigen_centrality`. The `eigen_centrality` function default is `directed=FALSE`, since it is less suited for directed networks [14]. *Pagerank* is another closely related centrality measure that uses a similar algorithm and is more suitable for directed networks. The *Pagerank* centrality can be calculated using the function `pagerank`. Please note that to obtain the value of the centrality, you need to use `$vector` at the end as demonstrated in the code.

An important question here is whether to compute these centralities with a simplified network, weighted network or a multigraph network. The answer depends on the context, the network structure and the research question. However, evidence suggests that multigraph configuration may render the most accurate results when centralities are used as indicators for performance [22]. The code below computes the aforementioned centralities, you may need to read the help of each centrality function for more options and arguments for customization:

```
Closeness_In <- closeness(DLT1, mode = c("in"))
Closeness_Out <- closeness(DLT1, mode = c("out"))
Closeness_total <- closeness(DLT1, mode = c("total"))

Betweenness <- betweenness(simple.DLT1, directed = FALSE)
Eigen <- eigen_centrality(simple.DLT1, directed = FALSE)$
                        vector
Pagerank <- page_rank(DLT1, directed = FALSE)$vector
```

Diffusion centralities have been introduced recently in several studies and seem to offer a more robust estimation of a vertex role in spreading information [19, 51]. *Diffusion* centrality can be computed in the same way as degree centrality. However, there is no function in the `igraph` package to calculate this centrality, so we rely on the `diffusion.degree` function from the `centiserve` package. The function `diffusion.degree` accepts the mode argument to compute different variants, i.e., `"in"`, `"out"` and `"total"` diffusion degrees.

```
library(centiserve)
Diffusion.degree_in  <- diffusion.degree(DLT1, mode =c("in"))
Diffusion.degree_out  <- diffusion.degree(DLT1, mode =c("out"))
Diffusion.degree  <- diffusion.degree(DLT1, mode =c("all"))
```

Coreness and *cross-clique connectivity* are related centralities that estimate the embeddedness of the vertex in the network can be calculated using the functions `coreness` and `crossclique`. Both `coreness` and `crossclique` centralities have been shown to better correlate with performance as well as with productive and reply-worthy content [19].

```
Coreness <- coreness(DLT1)
Cross_clique_connectivity <- crossclique(DLT1)
```

```
Warning in cliques(graph): At core/cliques/cliquer_wrapper.c:57 : Edge
directions are ignored for clique calculations.
```

We can also combine the rest of the centralities together in a single dataframe:

```
Centdf  <-tibble::tibble(
name=V(DLT1)$name,Closeness_total,Betweenness,
Eigen,Pagerank,Diffusion.degree,Coreness,Cross_clique_
connectivity)print(Centdf)
```

```
# A tibble: 445 x 8
   name  Closeness_total Betweenness   Eigen Pagerank Diffusion.degree Coreness
   <chr>           <dbl>       <dbl>   <dbl>    <dbl>            <dbl>    <dbl>
 1 1            0.00110       1258.  0.206    0.00840             1865       18
 2 2            0.000808        26.5 0.0107   0.00140              218        6
 3 3            0.000799        30.6 0.00862  0.00130              191        6
 4 4            0.00102         72.5 0.0803   0.00273              965       13
 5 5            0.00106        309.  0.162    0.00525             1508       18
 6 6            0.00108        250.  0.155    0.00539             1607       18
 7 7            0.00111       1935.  0.230    0.00931             2088       21
 8 8            0.00106        164.  0.136    0.00503             1483       18
 9 9            0.00104         69.5 0.119    0.00251             1216       13
10 10           0.00106        716.  0.0875   0.00343             1432       17
# i 435 more rows
# i 1 more variable: Cross_clique_connectivity <int>
```

The calculation of graph level measures and centrality measures are usually a step in the analysis to answer a research question. For instance, density can tell how distributed the interactions between students are and therefore, how it is collaborative [5]. Centrality may be calculated to identify who are the most important students in the discussions or used to infer the roles e.g., who are the leaders who drive the discussion [21, 23]. Several studies have calculated centrality measures to investigate their relationship with performance [18]. All of such types of analysis can be performed using the analysis we have demonstrated. Of course, there are no limits to the potentials of SNA and researchers have a wide range of possibilities and potentials that they can achieve by building on the aforementioned tutorials.

4 Discussion

The present chapter offered a primer on social network analysis as well as a tutorial on the most common types of SNA analysis. SNA is a vast field with diverse applications that are far beyond a chapter or even a whole book. Readers who are interested in expanding their knowledge about SNA are advised to read the literature cited in this chapter. Furthermore, several systematic reviews have tried to offer a synthesis of the extant literature and can help the readers get an idea about the status of SNA research in education. Two systematic reviews, [27, 52]. –despite being relatively old– they give a useful review on the uses and applications of SNA in learning settings. For instance, the methods used by SNA researchers have been addressed in a dedicated systematic review by Dado and Bodemer [8], where the authors offered a detailed review of methodological approaches used in SNA research. Centrality measures were the topic of a recent systematic review and meta-analysis that synthesized the literature and offered evidence of the association of centrality measures with academic achievement [48].

A more recent scientometric study by Saqr et al. [6] offers a comprehensive review of all research on network analysis and network science across the past five decades. The study also offers a review of authors, countries, research themes and research foundations. Whereas not a traditional systematic review, the recent paper by Poquet et al. [9] offers a review of the seminal papers of SNA with a methodological approach. The paper also offers recommendations for a reporting scheme for research using SNA. It is also important to mention that our chapter covered only static networks. Readers who are interested in the more advanced time varying networks, the temporal network chapter offers a great starting point [53]. Also several guides and empirical papers demonstrate examples of temporal network analysis [20, 54, 55]. Readers who want to go deeper in analysis of learning communities, the community detection chapter can be a good place [56]. Also, for readers interested in the novel methods of psychological networks, they are encouraged to read the psychological network chapter [57].

5 More Reading Resources

Books related to SNA that the readers can consult are:

- Kolaczyk, E. D., & Csárdi, G. (2014). *Statistical analysis of network data with R* (Vol. 65). New York: Springer.
- Luke, D. A. (2015). *A user's guide to network analysis in R* (Vol. 72, No. 10.1007, pp. 978-3). New York: Springer.
- Newman, Mark. *Networks*. Oxford university press, 2018.
- Hanneman, R. A., & Riddle, M. (2005). Introduction to social network methods. (Link)
- Carolan, B. V. (2013). *Social network analysis and education: Theory, methods & applications*. Sage Publications.
- Network Science by Albert-László Barabási (Link)

Acknowledgments This chapter is co-funded by the Academy of Finland the project Towards precision education: Idiographic learning analytics (TOPEILA), Decision Number 350560

References

1. Barabási A-L (2013) Network science. Philos Trans A Math Phys Eng Sci 371:20120375. https://doi.org/10.1098/rsta.2012.0375
2. Borgatti SP, Mahra A, Brass DJ, Labianca G (2009) Network analysis in the social sciences. Science 323:892–895. https://doi.org/10.1126/science.1165821
3. Saqr M, López-Pernas S (2022) How CSCL roles emerge, persist, transition, and evolve over time: a four-year longitudinal study. Comput Edu 189:104581. https://doi.org/10.1016/j.compedu.2022.104581
4. Saqr M, Nouri J, Fors U, Viberg O, Alsuhaibani M, Alharbi A, Alharbi M, Alamer A (2021) How networking and social capital influence performance: the role of long-term ties. In: Antonyuk A, Basov N (eds) Lecture notes in networks and systems. Springer, Cham, pp 335–346
5. Saqr M, Nouri J, Vartiainen H, Tedre M (2020) Robustness and rich clubs in collaborative learning groups: a learning analytics study using network science. Sci Rep 10:14445–14461. https://doi.org/10.1038/s41598-020-71483-z
6. Saqr M, Poquet O, López-Pernas S (2022) Networks in education: a travelogue through five decades. IEEE Access Practical Innov Open Solutions 10:32361–32380. https://doi.org/10.1109/access.2022.3159674
7. Saqr M, Alamro A (2019) The role of social network analysis as a learning analytics tool in online problem based learning. BMC Med Edu 19:1–11. https://doi.org/10.1186/s12909-019-1599-6
8. Dado M, Bodemer D (2017) A review of methodological applications of social network analysis in computer-supported collaborative learning. Edu Res Rev 22:159–180. https://doi.org/10.1016/j.edurev.2017.08.005
9. Poquet O, Saqr M, Chen B (2021) Recommendations for network research in learning analytics: to open a conversation. In: Proceedings of the NetSciLA21 workshop
10. López-Pernas S, Saqr M, Apiola M (2023) Scientometrics: a concise introduction and a detailed methodology for mapping the scientific field of computing education research. In: Apiola M, López-Pernas S, Saqr M (eds) Past, present and future of computing education research: a global perspective. Springer, Cham, pp 79–99

11. Saqr M, Fors U, Tedre M (2018) How the study of online collaborative learning can guide teachers and predict students' performance in a medical course. BMC Med Edu 18:24. https://doi.org/10.1186/s12909-018-1126-1
12. Borgatti SP, Brass DJ (2019) Centrality: concepts and measures. In: Social networks at work. Routledge, New York, pp 9–22
13. Ognyanova K (2023) Static and dynamic network visualization with R. https://kateto.net/network-visualization
14. Liao H, Mariani MS, Medo M, Zhang Y-C, Zhou M-Y (2017) Ranking in evolving complex networks. Phys Rep 689:1–54. https://doi.org/10.1016/j.physrep.2017.05.001
15. Saqr M, López-Pernas S (2022) The curious case of centrality measures: a large-scale empirical investigation. J Learn Anal 9:13–31. https://doi.org/10.18608/jla.2022.7415
16. Block P (2015) Reciprocity, transitivity, and the mysterious three-cycle. Soc Netw 40:163–173. https://doi.org/10.1016/j.socnet.2014.10.005
17. Jo I, Park Y, Lee H (2017) Three interaction patterns on asynchronous online discussion behaviours: a methodological comparison. J Comput Assist Learn 33:106–122. https://doi.org/10.1111/jcal.12168
18. Romero C, López M-I, Luna J-M, Ventura S (2013) Predicting students' final performance from participation in on-line discussion forums. Comput Edu 68:458–472. https://doi.org/10.1016/j.compedu.2013.06.009
19. Saqr M, López-Pernas S (2021) Modelling diffusion in computer-supported collaborative learning: a large scale learning analytics study. Int J Comput Support Collab Learn 16:441–483. https://doi.org/10.1007/s11412-021-09356-4
20. Saqr M, López-Pernas S (2022) Instant or distant: a temporal network tale of two interaction platforms and their influence on collaboration. In: Educating for a new future: making sense of technology-enhanced learning adoption. Springer, Berlin, pp 594–600
21. Marcos-García J-A, Martínez-Monés A, Dimitriadis Y (2015) DESPRO: a method based on roles to provide collaboration analysis support adapted to the participants in CSCL situations. Comput Edu 82:335–353. https://doi.org/10.1016/j.compedu.2014.10.027
22. Saqr M, Viberg O, Vartiainen H (2020) Capturing the participation and social dimensions of computer-supported collaborative learning through social network analysis: which method and measures matter? Int J Comput-Support Collab Learn 15:227–248. https://doi.org/10.1007/s11412-020-09322-6
23. Hernández-García Á, González-González I, Jiménez-Zarco AI, Chaparro-Peláez J (2015) Applying social learning analytics to message boards in online distance learning: a case study. Comput Hum Behav 47:68–80. https://doi.org/10.1016/j.chb.2014.10.038
24. Joksimovic S, Manataki A, Gaševic D, Dawson S, Kovanovic V, De Kereki IF (2016) Translating network position into performance: importance of centrality in different network configurations. ACM international conference proceeding series 25–29 Apri, pp 314–323. https://doi.org/10.1145/2883851.2883928
25. Reychav I, Raban DR, McHaney R (2018) Centrality measures and academic achievement in computerized classroom social networks: an empirical investigation. ACM J Edu Resour Comput 56:589–618. https://doi.org/10.1177/0735633117715749
26. Freeman LC (1978) Centrality in social networks conceptual clarification. Soc Netw 1:215–239. https://doi.org/10.1016/0378-8733(78)90021-7
27. Cela KL, Sicilia MÁ, Sánchez S (2015) Social network analysis in e-learning environments: a preliminary systematic review. Edu Psychol Rev 27:219–246. https://doi.org/10.1007/s10648-014-9276-0
28. Cadima R, Ojeda J, Monguet JM (2012) Social networks and performance in distributed learning communities. Edu Technol Soc 15:296–304
29. Cho H, Gay G, Davidson B, Ingraffea A (2007) Social networks, communication styles, and learning performance in a CSCL community. Comput Edu 49:309–329. https://doi.org/10.1016/j.compedu.2005.07.003
30. Liu Z, Kang L, Domanska M, Liu S, Sun J, Fang C (2018) Social network characteristics of learners in a course forum and their relationship to learning outcomes. In: CSEDU

2018 - proceedings of the 10th international conference on computer supported education. SciTePress, Setúbal[National Engineering Research Center for E-Learning, Central China Normal University, Luoyu Road 152, Wuhan, 430079, China, Department of Computer Science, Humboldt University of Berlin, Rudower Chaussee 25, Berlin, 12489, Germany], pp 15–21
31. Osatuyi BJ, Passerini K (2016) Twittermania: Understanding how social media technologies impact engagement and academic performance of a new generation of learners. Commun Assoc Inf Syst 39:509–528
32. Putnik G, Costa E, Alves C, Castro H, Varela L, Shah V (2016) Analysing the correlation between social network analysis measures and performance of students in social network-based engineering education. Int J Technol Des Edu 26:413–437. https://doi.org/10.1007/s10798-015-9318-z
33. Gašević D, Joksimović S, Eagan BR, Shaffer DW (2019) SENS: network analytics to combine social and cognitive perspectives of collaborative learning. Comput Hum Behav 92:562–577. https://doi.org/10.1016/j.chb.2018.07.003
34. Saqr M, Fors U, Nouri J (2018) Using social network analysis to understand online problem-based learning and predict performance. PloS One 13:e0203590. https://doi.org/10.1371/journal.pone.0203590
35. Liu S, Chai H, Liu Z, Pinkwart N, Han X, Hu T (2019) Effects of proactive personality and social centrality on learning performance in SPOCs. In: CSEDU 2019 - Proceedings of the 11th International Conference on Computer Supported Education vol 2, pp 481–487. https://doi.org/10.5220/0007756604810487
36. De-Marcos L, Garciá-López E, Garciá-Cabot A, Medina-Merodio J-AJA, Domínguez A, Martínez-Herraíz JJJ-J, Diez-Folledo T (2016) Social network analysis of a gamified e-learning course: small-world phenomenon and network metrics as predictors of academic performance. Comput Hum Behav 60:312–321. https://doi.org/10.1016/j.chb.2016.02.052
37. Wise AF, Cui Y (2018) Unpacking the relationship between discussion forum participation and learning in MOOCs: content is key. In: ACM international conference proceeding series. Association for computing machinery, learning analytics research Network, New York University, New York, pp 330–339
38. Liu Z, Kang L, Su Z, Liu S, Sun J (2018) Investigate the relationship between learners' social characteristics and academic achievements. In: Journal of physics: conference series. Institute of Physics Publishing, ["National Engineering Research Center for E-Learning, Central China Normal University, Wuhan, China", "National Engineering Laboratory for Technology of Big Data Applications in Education, Central China Normal University, Wuhan, China"]
39. Kleinberg JM (1999) Authoritative sources in a hyperlinked environment. J ACM 46:604–632. https://doi.org/10.1145/324133.324140
40. García-Saiz D, Palazuelos C, Zorrilla M (2014) Data mining and social network analysis in the educational field: an application for non-expert users. Stud Comput Intell 524:411–439
41. Banerjee A, Chandrasekhar AG, Duflo E, Jackson MO (2013) The diffusion of microfinance. Science 341:1236498. https://doi.org/10.1126/science.1236498
42. Kang C, Molinaro C, Kraus S, Shavitt Y, Subrahmanian VS (2012) Diffusion centrality in social networks. In: 2012 IEEE/ACM international conference on advances in social networks analysis and mining, pp 558–564
43. Kitsak M, Gallos LK, Havlin S, Liljeros F, Muchnik L, Stanley HE, Makse HA (2010) Identification of influential spreaders in complex networks. Nat Phys 6:888–893. https://doi.org/10.1038/nphys1746
44. Fincham E, Gašević D, Pardo A (2018) From social ties to network processes: do tie definitions matter? J Learn Anal 5:9–28. https://doi.org/10.18608/jla.2018.52.2
45. Csardi G, Nepusz T, et al (2006) The igraph software package for complex network research. Int J Complex Syst 1695:1–9
46. Kolaczyk ED, Csárdi G (2020) Statistical analysis of network data with R. Springer, Berlin

47. López-Pernas S, Saqr M, Del Rio L (2024) A broad collection of datasets for educational research training and application. In: Saqr M, López-Pernas S (eds) Learning analytics methods and tutorials: a practical guide using R. Springer, Berlin
48. Saqr M, Elmoazen R, Tedre M, López-Pernas S, Hirsto L (2022) How well centrality measures capture student achievement in computer-supported collaborative learning? – a systematic review and meta-analysis. Edu Res Rev 35:100437. https://doi.org/10.1016/j.edurev.2022.100437
49. Jalili M, Salehzadeh-Yazdi A, Asgari Y, Arab SS, Yaghmaie M, Ghavamzadeh A, Alimoghaddam K (2015) CentiServer: a comprehensive resource, web-based application and R package for centrality analysis. PloS One 10:e0143111–e0143111. https://doi.org/10.1371/journal.pone.0143111
50. Opsahl T, Agneessens F, Skvoretz J (2010) Node centrality in weighted networks: generalizing degree and shortest paths. Soc Netw 32:245–251. https://doi.org/10.1016/j.socnet.2010.03.006
51. Saqr M, Viberg O (2020) Using diffusion network analytics to examine and support knowledge construction in CSCL settings. In: Alario-Hoyos C, Rodríguez-Triana MJ, Scheffel M, Arnedillo-Sánchez I, Dennerlein SM (eds) Addressing global challenges and quality education. EC-TEL 2020. Lecture notes in computer science, vol 12315. Springer, Cham. https://doi.org/https://doi.org/10.1007/978-3-030-57717-9_12
52. Sie RLL, Ullmann DT, Rajagopal K, Cela K, B. MB-R, Sloep PB (2012) Social network analysis for technology-enhanced learning: review and future directions. Int J Technol Enhanced Learn 4:172–190. https://doi.org/10.1504/IJTEL.2012.051582
53. Saqr M (2023) Temporal network analysis: Introduction, methods and detailed tutorial with R. arXiv [cs.SI]
54. Saqr M, López-Pernas S (2022) The why, the what and the how to model a dynamic relational learning process with temporal networks. In: Proceedings of the NetSciLA22 workshop
55. Saqr M, Nouri J (2020) High resolution temporal network analysis to understand and improve collaborative learning. In: Proceedings of the tenth international conference on learning analytics & knowledge. ACM, New York, pp 314–319
56. Hernández-García Á, Cuenca-Enrique C, Traxler A, López-Pernas S, Conde MÁ, Saqr M (2024) Community detection in learning networks using R. In: Saqr M, López-Pernas S (eds) Learning analytics methods and tutorials: a practical guide using R. Springer, Berlin
57. Saqr M, Beck E, López-Pernas S (2024) Psychological networks. In: Saqr M, López-Pernas S (eds) Learning analytics methods and tutorials: a practical guide using R. Springer, Berlin

Community Detection in Learning Networks Using R

Ángel Hernández-García, Carlos Cuenca-Enrique, Adrienne Traxler, Sonsoles López-Pernas, Miguel Ángel Conde-González, and Mohammed Saqr

1 Introduction to Community Detection in Social Networks

The world is essentially social. Interactions, relationships and connections form the social fabric of the structure that makes it "social" [1]. Nevertheless, the world is far from uniform or random. The social actors (e.g., humans) tend to aggregate, coalesce, or work together in teams, groups, and communities [2, 3]. In humans, communities form neighborhoods, villages, cities, and even counties. On a smaller scale, communities can be a group of friends, teams, or work colleagues. These are —more or less— organized communities. However, communities can also emerge spontaneously among, for instance, people who happen to work together more than they work with others. Other examples include groups of universities that collaborate with each other more than they collaborate with other universities. In learning situations, communities can be groups of students within a whole cohort who collaborate with each other to a larger extent than with other students in a discussion forum. In the broader context of network analysis, community structure, or simply 'community', denotes those nodes within the network that can be categorized into distinct sets, ensuring that nodes within each set exhibit a high degree of internal connectivity.

Á. Hernández-García (✉) · C. Cuenca-Enrique
Universidad Politécnica de Madrid, Madrid, Spain
e-mail: angel.hernandez@upm.es

A. Traxler
University of Copenhagen, Copenhagen, Denmark

S. López-Pernas · M. Saqr
School of Computing, University of Eastern Finland, Joensuu, Finland

M. Á. Conde-González
Robotics Research Group, Engineering School, University of León, León, Spain

M. Saqr, S. López-Pernas (eds.), *Learning Analytics Methods and Tutorials*,
https://doi.org/10.1007/978-3-031-54464-4_16

Finding these communities is integral to understanding the interaction process, the structure of the formed groups and how they contribute to the overall discussion. Community detection is not limited to discussion boards but extends to all types of networks where we would like to understand their structure. For instance, communities of key terms in a semantic network, where each concept is represented as a node, would allow us to understand the main themes of the conversation [4]. Communities of behaviors or strategies would allow us to understand the structure of human behavior and discern a better classification. The power of community detection has made it one of the most commonly used network approaches in education, social sciences and, in fact, in scientific method at large [3].

As such, the main aim of community detection is to identify different groups or clusters of nodes within the network that show strong interconnectivity. Generally, most community detection algorithms are based on the idea that some nodes are more densely connected to each other than to nodes outside of the group. As a result, the application of community detection algorithms partitions the network into different groups, clusters or communities, based on the strength of their connections. An alternate branch of community detection, blockmodeling, adopts the perspective that people with similar patterns of ties are likely to act in similar ways within the network. This approach seeks to identify those similarly connected actors to abstract the network into different positions or roles [5, 6]. A special type of communities are the rich club communities [7, 8]. A rich club is a subset of nodes in the network who are working together —and who are therefore strongly connected— far more than with others in the network. The rich club entails a few doing most of the interaction and a majority who are far less involved and less favorable in collaborative networks. This chapter explores the first and more common approach, that of identifying densely connected subgraphs.

As mentioned above, a basic approach understands communities in social networks as subgraphs where the number of internal edges is larger than the number of external edges, and therefore they usually group nodes that have a higher probability of being connected to each other than to members of other groups [9]. The goal of identifying communities in social networks depends on the research question and context and, same as in the case of network analysis in general [10], the communities found in the analysis depend on how the network was constructed, and therefore their meaning must be explained by the researcher in each different context of application. For instance, a researcher might want to find communities to understand a social shift in some larger group [11, 12], to map political polarization [13], to interrupt disease transmission or to spread health-related behaviors [14], to study how students in a course self-segregate and stabilize into groups [15], or for many other reasons.

Many partitioning algorithms use a quantitative measure called modularity index, which is, up to a multiplicative constant, the number of edges falling within groups minus the expected number in an equivalent network with edges placed at random

[16]. Modularity, which ranges between −0.5 and 1[1] [18] and of which positive values indicate the possible presence of community structure [16] is not itself a clustering method, but a measure for comparing different partitionings to judge the best. There are no established values/thresholds for optimum modularity, but in practice it is found that a value above about 0.3 is a good indicator of significant community structure in a network [19].

In the previous chapter on Social Network Analysis [10], we learned how to build a network from a set of vertices (or nodes) and edges, how to visualize it, and how to calculate centrality measures that help us understand and describe the structure of the network and the position of the nodes with respect to others. In this new chapter, we focus on detecting communities (groups of highly connected nodes) within a wider network, and how to visualize them using R.

2 Community Detection in Social Networks Based on Educational Data

The application of community detection techniques in social network analysis of educational data (for further information about educational data and sources of educational data, we refer the reader to Chapter 2 in this book [20]) dates back to the emergence of learning analytics as a discipline. For example, in a 2012 study, Rabbany et al. [21] included community mining in their implementation of Meerkat-ED to monitor how student interactions and collaborations occurred and changed over time.

Community detection can serve multiple purposes in learning analytics. For example, Pham et al. [22] explored the community structure of the eTwinning project collaboration network and how the quality of the contributors' work relates to their level of participation. Similarly, Suthers and Chu [23] unveiled communities emerging within the Tapped In network of educational professionals from the associations between members of this network distributed across media (chat rooms, discussion forums and file sharing). Orduña et al. [24] applied modularity analysis to identify collaborative behaviors in a mobile remote laboratory. Skrypnyk et al. [25] and Gruzd et al. [26] identified emerging communities from Twitter-based interactions in a cMOOC. Joksimovic et al. [27] extracted clusters of concepts into topics (concept clusters) exchanged through social media also in the scope of a cMOOC. Hernández-García et al. [28] used connected components to analyze group cohesion. Adraoui et al. [29] identified student groups through their social interactions in Facebook groups. Nistor et al. [30] used cohesion network analysis to predict newcomer integration in online learning communities. López Flores et al. [31] performed community detection analysis to identify the learning behavior profiles of undergraduate computer science students. Abal Abas et al. [32]

[1] Depending on the formula used to calculate modularity, it can also range from −1 to 1 [17].

determined study groups during peer learning interaction. Li et al. [33] examined whether students formed communities dominated by a specific gender or race, and Nguyen [34] explored youth's perspectives to frame climate change education by using community detection to find hashtag groups and identify different discourse themes in TikTok. The reader may find more examples of community detection in online learning environments in the recent review by Yassine et al. [4]. Another important aspect of community detection relates to the algorithms used to perform the detection, which will be discussed in the next section.

3 Algorithms for Community Detection

Community detection in social network analysis involves the application of specific algorithms (the number of community detection algorithms is rather high and will not be addressed in this chapter; for further information on the nature and performance of community detection algorithms, we suggest reading Lancichinetti and Fortunato [35], Fortunato [36], Fortunato and Hric [9] and Chunaev [37]), the most popular being Louvain [38], the one implemented in the software applications *Gephi*, *Pajek* and *visone* by default, and Girvan-Newman's edge betweenness [39] and clique percolation [40], used in *CFinder*.

In this chapter, we will focus on how to perform community detection with the R programming language. In R, community detection algorithms are implemented as functions of the `igraph` library [41], which was already introduced in the Social Network Analysis chapter [10]. The library includes functions to apply the following methods:

- **Louvain** (`cluster_louvain`): A multi-level modularity optimization algorithm that aims to discover community structure [42].
- **Girvan-Newman** (`cluster_edge_betweenness`): The concept behind this method is that edges connecting distinct communities tend to have a high edge betweenness and all the shortest paths from one community to another must pass through these edges [43].
- **Fast greedy optimization** (`cluster_fast_greedy`): A fast greedy modularity optimization method to identify communities within a network [19].
- **Fluid communities** (`cluster_fluid_communities`): This method is based on the concept of multiple fluids interacting in a non-uniform environment (represented by the graph topology). The method identifies communities by observing their expansion and contraction patterns, driven by their interactions and density [44].
- **Random walk-based**:
  - **Infomap** (`cluster_infomap`): This method discovers the community structure that minimizes the expected description length of a random walker's trajectory [45].

  - **Walktrap** (`cluster_walktrap`): This method is based on the underlying concept that short random walks have a tendency to remain within the same community [46].

- **Label propagation** (`cluster_label_propagation`): This method for detecting community structure in networks is efficient, with a nearly linear time complexity [47]. It operates by assigning unique labels to the vertices and subsequently updating these labels through majority voting within the vertex's neighborhood.
- **Leading eigenvector** (`cluster_leading_eigen`): This method identifies densely connected subgraphs within a graph by computing the principal non-negative eigenvector of the modularity matrix associated with the graph [48].
- **Leiden** (`cluster_leiden`): The Leiden algorithm is similar to the Louvain algorithm, but offers superior speed and delivers higher-quality solutions. It has the capability to optimize both modularity and the Constant Potts Model, which overcomes the resolution limit challenge [49].
- **Optimal modularity clustering** (`cluster_optimal`): This method computes the optimal community structure of a graph by maximizing the modularity measure across all potential partitions [18].
- **Simulated annealing** (`cluster_spinglass`): This method leverages the spin-glass model and simulated annealing techniques to explore the graph's structure and identify cohesive communities [50].

Other available libraries to perform community detection include `nett`, which allows for spectral clustering (function `spec_clust()`) and implements methods for hypothesis testing.

When to choose one algorithm over another usually depends on the characteristics of the network data, whether or not directed edges are allowed and the goals of the analysis. For example, if the network is very large, Louvain or Label Propagation algorithms may offer satisfactory results with a small computational effort, whereas Girvan-Newman might be more appropriate if the network has a hierarchical structure or when networks are large. If the network has a flow of information, Infomap may be better suited, and spectral clustering might be preferrable when community structures are complex. Guidance about which algorithms to use is not robust, and issues such as computational efficiency are more often discussed than the match with research questions [9, 51]. Ideally, the choice should be made to align with the processes believed to drive community formation and the research purpose in seeking this structure [51]. In any case, we would recommend trying different algorithms and comparing their results before choosing one.

It is important to note the limitations of some community detection algorithms in `igraph`. For example, fast greedy optimization, leading eigenvector, Leiden, Louvain and simulated annealing only work with undirected graphs —even though some of these algorithms, such as Leiden, work in directed networks, their implementation in `igraph` does not allow for community detection in directed networks—, whereas fluid communities or simulated annealing only work with simple and connected graphs.

Finally, it is worth mentioning that a limitation of some methods is that they are non-overlapping community detection algorithms; that is, they consider that a node belongs only to one group, partition or community. However, that is not often the case, which is why some overlapping community detection algorithms have been proposed (for example, random walk-based algorithms may handle overlapping), and the reader might want to check whether the chosen algorithm finds overlapping or non-overlapping communities when choosing what algorithm to use. Fortunato and Hric [9] offer additional guidance if overlapping communities are required. Now that we have covered the `igraph` library along with its features and limitations, in the next section we will present an example of how to use this library.

4 Community Detection in R: An Annotated Example Using `igraph`

To illustrate the use of `igraph` to perform community detection algorithms in R, we will use data from one of the courses reported in an article by Hernández-García and Suárez-Navas [52]. More specifically, this example focuses on the data from a "Programming Basics" undergraduate course. The data set includes forum activity of 110 students from a BSc Degree on Biotechnology, which follows the Comprehensive Training Model of the Teamwork Competence (CTMTC) methodology [53]. In this course, students work in groups of between five and seven members each. The activity in the forum includes Q&As, technical and academic support in general forums, and message exchanges between group members in group-exclusive forums.

The original data set including all forum activity was divided into three different subsets, with the help of GraphFES [54]: views (how many times user *a* read a message posted by user *b*), replies (which user replies to a different user, and how many times) and messages (which message is a reply to another message).

In this example, we will focus on the *Replies* data set, a directed graph. The node list of the *Replies* data set includes a total of 124 nodes with three attributes: `initPosts` (number of first posts of a discussion sent by a user), `replyPosts` (number of posts replying to a previous post in the discussion) and `totalPosts` (the sum of `initPosts` and `replyPosts`). Weights in the edge list (attribute *w*) represent the number of times that user `Target` replied to user `Source`. The data set includes a total of 662 weighed edges. First off, we load the required libraries for data loading (`rio`) and analysis (`igraph`). Install them first if you have not done so before.

```
library(igraph)
library(rio)
```

Then, we import the node and edges data, and we build the graph object, indicating that the network is directed (with the argument `directed = TRUE` in the call to `graph_from_data_frame()`):

```
repo <- "https://github.com/lamethods/data/raw/main/10_snaProgramming/"
ds.nodes <- import(paste0(repo,"hg_data_nodes.xlsx"))
ds.edges <- import(paste0(repo,"hg_data_edges.xlsx"))
ds <- graph_from_data_frame(d = ds.edges, directed = TRUE,
                            vertices = ds.nodes)
```

We can now observe the structure of the graph:

```
str(ds)
```

```
Class 'igraph'  hidden list of 10
 $ : num 124
 $ : logi TRUE
 $ : num [1:662] 101 73 48 51 84 58 48 101 62 27 ...
 $ : num [1:662] 73 1 51 1 1 1 101 62 27 51 ...
 $ : NULL
 $ : NULL
 $ : NULL
 $ : NULL
 $ :List of 4
  ..$ : num [1:3] 1 0 1
  ..$ : Named list()
  ..$ :List of 5
  .. ..$ name       : chr [1:124] "54" "55" "56" "57" ...
  .. ..$ User       : chr [1:124] "user_54" "user_55" "user_56" "user_57" ...
  .. ..$ initPosts  : num [1:124] 0 23 0 0 0 0 1 0 3 3 ...
  .. ..$ replyPosts: num [1:124] 0 2 0 0 0 0 3 0 19 53 ...
  .. ..$ totalPosts: num [1:124] 0 25 0 0 0 0 4 0 22 56 ...
  ..$ :List of 1
  .. ..$ weight: num [1:662] 1 2 1 2 5 4 1 1 1 1 ...
 $ :<environment: 0x5565f3418c28>
```

We can also inspect the main attributes of the graph, which is shown as a directed, named and weighted network with 124 nodes and 662 edges:

```
print(ds)
```

```
IGRAPH 29b3383 DNW- 124 662 --
+ attr: name (v/c), User (v/c), initPosts (v/n), replyPosts (v/n),
| totalPosts (v/n), weight (e/n)
+ edges from 29b3383 (vertex names):
 [1] 192->164 164->55  139->142 142->55  175->55  149->55  139->192 192->153
 [9] 153->118 118->142 160->160 158->55  163->175 152->55  182->161 161->55
[17] 210->55  149->138 138->55  117->178 178->55  127->55  160->55  197->55
[25] 155->55  122->55  189->55  145->55  135->55  207->55  203->55  140->55
[33] 159->55  126->55  123->55  139->55  133->55  153->55  201->55  139->139
```

```
[41] 139->180 192->134 134->206 206->192 192->205 205->205 205->192 192->113
[49] 113->192 134->192 192->206 113->205 205->206 192->192 113->134 206->205
+ ... omitted several edges
```

In addition, we may use the plot function to visualize the network, A first glance at the graph does not offer much information about the underlying community structure of the graph:

```
plot(ds)
```

At any moment, we can apply a community detection algorithm to the graph. In Fig. 1, we observe that the graph is not connected and is directed. Therefore, we can only apply a subset of community finding algorithms, such as Girvan-Newman, Infomap, Label propagation, or Walktrap (note that R will trigger a warning for Girvan-Newman, because edge weights have different meaning in modularity calculation and edge betweenness community detection). For the purpose of this example, we will apply the Infomap algorithm —because it is a random walk-based algorithm, we provide a random seed for reproducibility.

```
set.seed(1234)
comm.ds <- cluster_infomap(ds)
```

The basic call to the `cluster_infomap()` function takes the graph as an argument (other arguments include edge weights, node weights, number of attempts

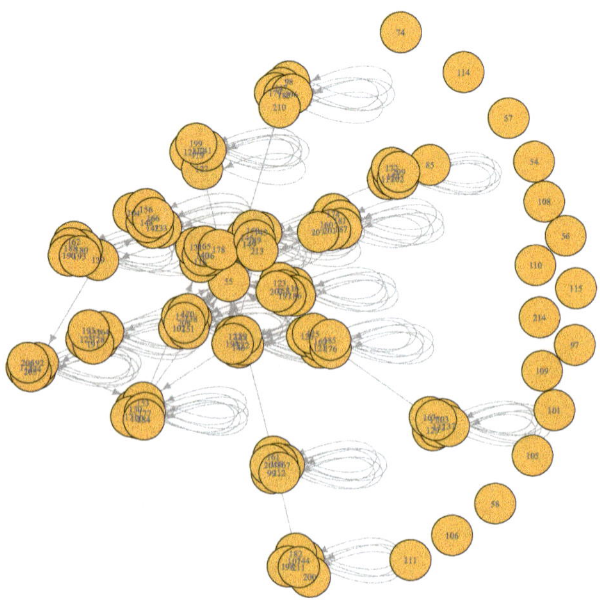

Fig. 1 Graph of the Replies from the data set of [52] using the plot function

to partition the network and whether modularity needs to be calculated; in this case, the function takes the 'weight' edge attribute as default edge weight) and returns a *community* object. We can observe its structure:

```
str(comm.ds)
```

```
Class 'communities'  hidden list of 6
 $ membership: num [1:124] 1 2 3 4 5 6 7 8 9 10 ...
 $ codelength: num 3.95
 $ names     : chr [1:124] "54" "55" "56" "57" ...
 $ vcount    : int 124
 $ algorithm : chr "infomap"
 $ modularity: num 0.927
```

The attributes of the community object include the group each node belongs to, the code length or average length of the code describing a step of the random walker (this parameter only adopts a value in random walk algorithms), node names, number of nodes and algorithm used to partition the network in communities. It is also possible to access values of the community object using different functions such as *length()*, *sizes()* or *membership()*, which return the number of communities, sizes of each community and membership of each node, respectively. At this point, it is possible to plot the graph and the communities (Fig. 2):

```
plot(comm.ds, ds)
```

See `help(plot.communities)` for more details on this method, which takes a *communities* object as its first argument and an `igraph` object as its second

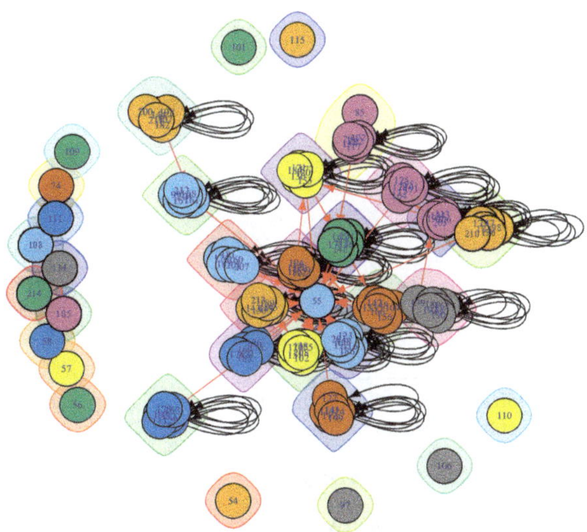

Fig. 2 Communities emerging from the Replies data set using the Infomap community finding algorithm

argument. By default, it colors the nodes and their surrounding "bubbles" with a different color for each community, and marks community-bridging edges in red. In many networks this produces a very overlapping and indistinct picture.[2]

Because Fig. 2 does not provide useful information (yet), we have to proceed to clean the graph. First, we simplify the graph by removing multiple edges between nodes and turning them into a single weight (this is not strictly necessary in this case because the original edge data already included calculated weights) and self-edges. Additionally, and because isolated nodes do not belong to any community (they are their own community), we can remove them from the graph. We then re-calculate the Infomap clustering for the simplified graph and plot it again (Fig. 3). We see that the self-loops and repeated edges have disappeared.

```
simple.ds <- simplify(ds, remove.multiple = TRUE, remove.loops = TRUE,
                      edge.attr.comb = list(weight = "sum", "ignore"))
simple.ds <- delete.vertices(simple.ds, which(degree(simple.ds) == 0))
comm.simple.ds <- cluster_infomap(simple.ds)
plot(comm.simple.ds, simple.ds)
```

We can now further refine the graph visualization to better highlight the communities:

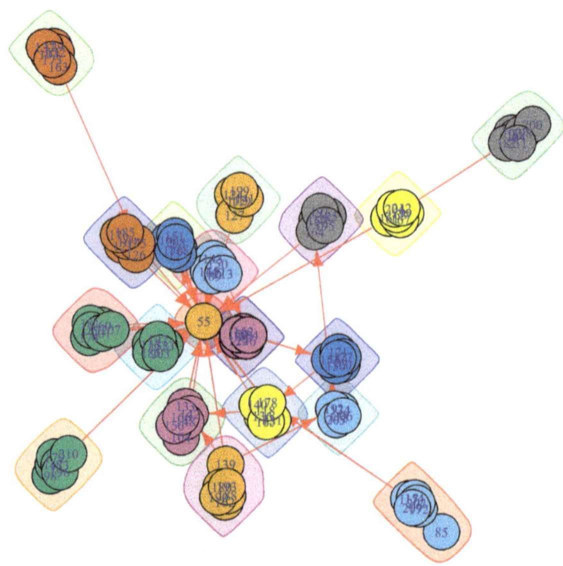

Fig. 3 Visualization of the communities emerging from the simplified Replies data set using the Infomap community finding algorithm

[2] Customization is available through other `igraph` plotting parameters, such as filtering out the colored bubbles around isolates:

```
plot(comm.ds, ds, mark.groups = communities(comm.ds)[sizes(comm.ds) >= 2])
```

However, we will shortly introduce more advanced visualization tools.

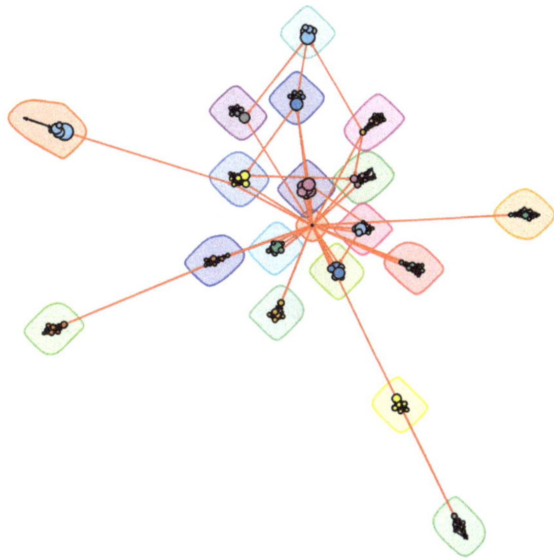

Fig. 4 Fine-tuned visualization of the simplified graph

```
lo <- layout_with_fr(simple.ds, niter = 50000,
                     weights = E(simple.ds)$weight * 0.05)
plot(comm.simple.ds,
     simple.ds,
     layout = lo,
     vertex.size = V(simple.ds)$totalPosts * 0.025,
     vertex.label = NA,
     edge.arrow.size = 0.1
)
```

From Fig. 4, we can clearly observe 19 communities, of which 18 correspond to the student groups and the remaining one (in the center of the graph) corresponds to the course instructor. Even in a network such as this with a relatively clear modular structure, different community detection algorithms can return different results. For example, `cluster_spinglass()` (which allows for adjustment of the importance of present vs. absent edges through a gamma parameter) returns a different partitioning:

```
set.seed(4321)
comm.simple2.ds <- cluster_spinglass(simple.ds, gamma = 1.0)
plot(comm.simple2.ds,
     simple.ds,
     layout = lo,
     vertex.size = V(simple.ds)$totalPosts * 0.025,
     vertex.color = membership(comm.simple2.ds),
     vertex.label = NA,
```

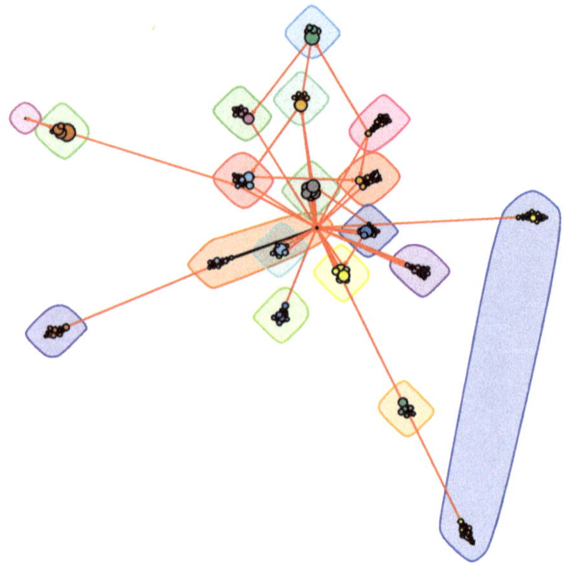

Fig. 5 Visualization of simplified graph with spinglass communities

```
        edge.arrow.size = 0.1
)
```

Figure 5 shows that the instructor is now included in one of the student communities, and a weakly connected member of another student group has split into their own community. Two groups with no direct bridging edges are also clustered together. This behavior is more common in blockmodeling, which looks for similarity of ties rather than direct links to identify groups; however, it can occur in standard community detection methods as well.

Additionally, it is possible to further simplify the network graph, by plotting only the communities and their inter-relationships. To do so, we build a condensed graph using the Infomap clustering where each node summarizes the information from all the members of the community (Fig. 6).

```
comms <- simplify(contract(simple.ds, membership(comm.simple.ds)))
plot(comms,
     vertex.size = 2.5 * sizes(comm.simple.ds),
     vertex.label = 1:length(comm.simple.ds),
     vertex.cex = 0.8,
     edge.arrow.size = 0.1
)
```

It is also worth noting that igraph incorporates a function, compare(), that takes different community objects from different partitioning methods, and allows for their comparison, based on different methods, such as variation of information, normalized mutual information, split-join distance or Rand and

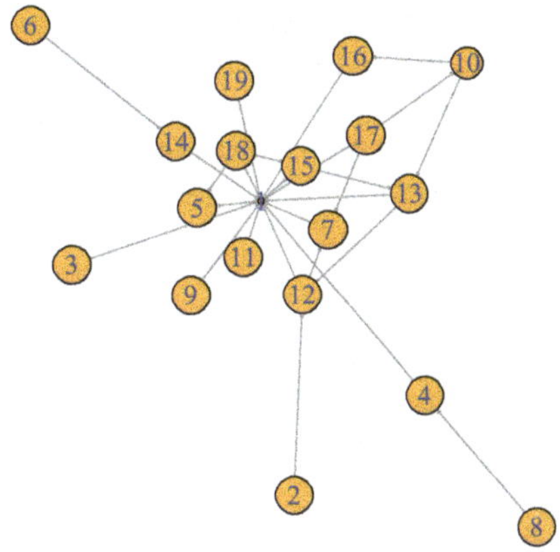

Fig. 6 Network graph of the links between Infomap communities using the simplified version of the Replies data set

adjusted Rand indices. We have not included an example here because additional knowledge of the comparison metrics is needed to interpret the results, but see `help(igraph::compare)` for references.

4.1 *Interactive Visualization of Communities in R*

In the first sections of this chapter, we have highlighted the uses and applications of community finding using educational data, as well as the main principles and methods, complemented with an example in `igraph`. However, the last section also highlights the limitations of the `igraph` library to provide advanced graphic features, such as interactive plotting. To overcome these limitations, we will further explore interactive visualization of communities using two different libraries: `visNetwork` and `networkD3`.

4.1.1 `visNetwork`

`visNetwork` is an R package for network visualization that uses the *vis.js* javascript library. It is based on `htmlwidgets`, and therefore it is compatible with Shiny, R Markdown documents, and RStudio viewer. To access its functions, we must first load the `visNetwork` package:

```
library(visNetwork)
```

Then, we need to build a data set that `visNetwork` can read. To do so, we need to create a data frame with all the original data (in this example, the data set corresponding to the simplified graph), to which we add the group that each node belongs to.

The first step to create this data frame is to build a data frame with the community of the node (in the column *group*) and a column *id* with a list of all nodes. The last line resets the row columns to a sequence starting in 1. In this step, it is critical to rename the column that represents the group assignment to "group", because this field is internally interpreted by `visNetwork` as the different communities of the network.

```
memberships <- as.data.frame(as.matrix(membership(comm.simple.ds)))
colnames(memberships)[1]<- "group"
memberships$id <- rownames(memberships)
rownames(memberships) <- 1:nrow(memberships)
```

The memberships data frame now has a column of group (community) numbers and a column of the original node id numbers:

```
head(memberships)
```

```
  group  id
1     1  55
2     2  85
3     3  98
4     4  99
5     5 102
6     6 103
```

Next, we retrieve the original node and edge list as data sets, using the `as_data_frame` function. While we could extract both data sets in a single step using the argument `what = "both"`, in this example we extract them separately for clarity of the manipulations required for each data set.

```
simple.ds.nodes <- as_data_frame(simple.ds, what = "vertices")
simple.ds.edges <- as_data_frame(simple.ds, what = "edges")
```

In the node list, and while it is not absolutely necessary, we reset the row columns to a sequence starting in 1. After that, we need to rename the original 'name' and 'User' columns to 'id' and 'title'. The former manipulation will allow us to add the group number to the dataset with the information included in the memberships object, while the latter is used by `visNetwork` to identify the different nodes.

```
rownames(simple.ds.nodes) <- 1:nrow(simple.ds.nodes)
colnames(simple.ds.nodes)[1] <- "id"
colnames(simple.ds.nodes)[2] <- "title"
```

Finally, we combine the node data set with the membership data set.

```
vis.comm <- merge(simple.ds.nodes, y = memberships, by = "id",
                  all.x = TRUE)
```

For visualization purposes, we add a column with the size of the nodes in the visualization.

```
vis.comm$size <- vis.comm$totalPosts * 0.2
```

Finally, we proceed to visualize the graph. To do so, we call the `visNetwork` function, and we pipe different visualization options (|>,[3]) which we apply to the visualization object: (1) the manual random seed ensures reproducibility; (2) the legend displays the different communities in the right part of the viewer; (3) we highlight connected nodes on selection, and allow for selection in a dropdown menu by 'id' and 'group'; and (4) we allow drag/zoom on the network, with navigation buttons that are displayed on the lower part of the viewer (see Fig. 7).

```
visNetwork(vis.comm, simple.ds.edges, width = "100%", height = "800px",
           main = "Interactive Communities") |>
 visLayout(randomSeed = 1234) |>
 visLegend(position = "right", main = "group") |>
 visOptions(highlightNearest = TRUE, nodesIdSelection = TRUE,
            selectedBy = "group") |>
 visInteraction(hideEdgesOnDrag = TRUE, dragNodes = TRUE, dragView = TRUE,
                zoomView = TRUE, navigationButtons = TRUE)
```

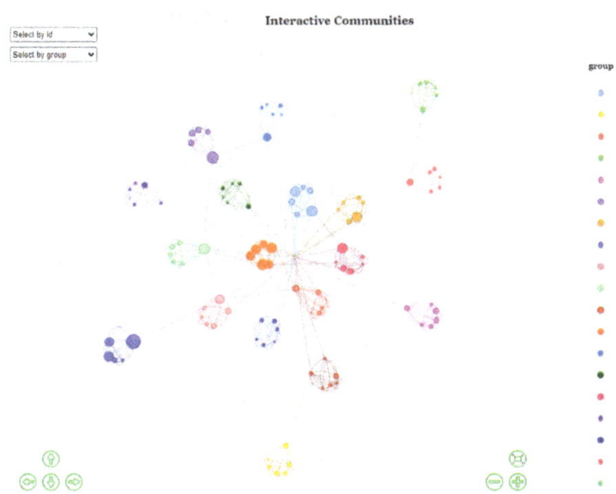

Fig. 7 Interactive visualization of the simplified version of the Replies data set using `visNetwork`

[3] The native R pipe (|>) was introduced in R version 4.1.0. If you are using an older version of R, you can use the `magrittr` pipe (%>%).

4.1.2 networkD3

The networkD3 is an advanced library for the interactive visualization of networks. It creates D3 network graphs and it is also based on the htmlwidgets framework, therefore simplifying the package's syntax for exporting the graphs and allowing integration with RStudio, RMarkdown and Shiny web apps. To access its functions, we first load the library.

```
library(networkD3)
```

An advantage of the networkD3 library is that it provides a function, igraph_to_networkD3(), that allows direct loading of an igraph network as a networkD3 compatible graph. In this case, analogously to the previous example in visNetwork, we will visualize the simplified network. We provide two arguments to the igraph_to_networkD3 function: the original igraph object and the node membership list, also obtained before with the help of igraph.

```
graph.d3 <- igraph_to_networkD3(simple.ds,
                                group = membership(comm.simple.ds))
```

Analogously to the example in visNetwork, we add node sizes for improved visualization.

```
graph.d3$nodes$size <- simple.ds.nodes$totalPosts * 0.2
```

And finally, we use the forceNetwork function to display the graph. In this case, we will store the interactive visualization in an object to enhance the behavior of the legend later by using the htmlwidgets framework. From the example below, no additional manipulation of the original data sets was required.[4]

The forceNetwork() function requires the edge (Links) and node (Nodes) data frames, as well as the name of the source and target columns in the edge data frame, the node id and group columns in the node data frame. In the following code, it is also worth noting that the argument provided to display the size of the

[4] However, it is important to note that in networkD3, the source and target vectors in the edge (links) data frame must be numeric and, most importantly, that their values are relative to the index of the node in the nodes data frame they represent. This may have implications, given that the nodes data set is based on JavaScript and it is therefore zero-indexed, unlike in R. The user may input the following code to reset the rows to a sequence starting in 0 before executing the call to forceNetwork:

```
row.names(graph.d3$nodes) <- 0:(nrow(graph.d3$nodes)-1)
row.names(graph.d3$links) <- 0:(nrow(graph.d3$links)-1).
```

nodes refers to the column number —and not the column name—, and that we can provide different repulsion values using the charge parameter —the strength of node repulsion (negative values) or attraction (positive values).

```
d3.comm <- forceNetwork(Links = graph.d3$links, Nodes = graph.d3$nodes,
                        Source = 'source', Target = 'target',
                        NodeID = 'name', Group = 'group',
                        linkColour = "#afafaf", fontSize = 12, zoom = T,
                        legend = T, Nodesize = 3, opacity = 0.8,
                        charge = -25,  width = 800, height = 800)
d3.comm
```

In the previous visualization, the legend moves when we zoom in or out, or drag the graph. A possible workaround to fix the legend is to use the `htmlwidgets` library [55]. The final result of the interactive visualization is shown in Fig. 8.

```
library(htmlwidgets)
htmlwidgets::onRender(d3.comm, jsCode = '
  function (el, x) {
    d3.select("svg").append("g").attr("id", "legend-layer");
    var legend_layer = d3.select("#legend-layer");
    d3.selectAll(".legend")
      .each(function() { legend_layer.append(() => this); });
  }
')
```

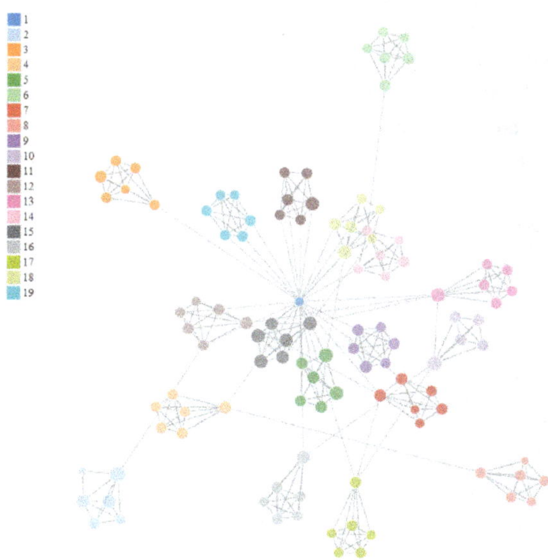

Fig. 8 Interactive visualization of the simplified version of the Replies data set using `networkD3`

5 Concluding Notes

In this chapter, we have introduced the literature on community detection in social network analysis, highlighted its uses in learning analytics, and worked through an example of finding and visualizing communities in R. The process begins outside of R, by identifying possible mechanisms of community formation in this network and why they are of interest. In educational settings, researchers may be concerned with information flow, dissemination of norms or attitudes, or other social forces. It is also important to consider whether the directionality and frequency (or other strength measure) of the interactions is important and if these considerations align with the theory and contextual peculiarity [56, 57]. Once these factors are thought out, there will still probably be a few options for community detection algorithms. It is worth trying more than one algorithm and comparing their groupings, as well as reading up on the method to see if it has tunable parameters, such as the tightness of a random walk or the relative importance of missing links. Moreover, once the communities have been detected they can be explored in several ways. For example, one can investigate the demographic differences between the communities and determine whether they are formed based on shared characteristics between the nodes (e.g., gender, race, and nationality). Other aspects to look into are the content of the interactions, the difference in performance (e.g., final grade) between communities, and their temporal evolution to understand how they formed. Furthermore, each community can be visualized and analyzed as a network of its own using the methods explained in the Social Network Analysis chapter of this book.

Though we began with the visualization tools available in the `igraph` package, in many cases researchers will want to go further. In these cases, the results of clustering algorithms can be used with libraries like `visNetwork` or `networkD3`. In the end, the goal of the visualization is to explore or present insights about network subgroups that speak to the original research questions, and it is helpful to be familiar with a range of tools for this purpose.

This chapter can be considered an introduction to the topic of community detection. However, interested users can resort to our cited papers and, for further readings, the selected papers and books provided in the following section can be a good start.

6 Further Readings

Interested readers can refer to the following resources about community detection in general:

- Fortunato, S., & Hric, D. (2016). Community detection in networks: A user guide. *Physics Reports*, 659, 1-44.

- Traag, V. A., Waltman, L., & Van Eck, N. J. (2019). From Louvain to Leiden: Guaranteeing well-connected communities. *Scientific Reports*, 9(1), 5233.
- Xie, J., Kelley, S., & Szymanski, B. K. (2013). Overlapping community detection in networks: The state-of-the-art and comparative study. *ACM Computing Surveys*, 45(4), 1-35.

For specific resources using R, the reader can consult the following:

- Borgatti, S. P., Everett, M. G., Johnson, J. C., & Agneessens, F. (2022). *Analyzing social networks using R*. SAGE.
- Kolaczyk, E. D., & Csárdi, G. (2014). *Statistical analysis of network data with R* (Vol. 65). New York: Springer.
- Luke, D. A. (2015). *A user's guide to network analysis in R* (Vol. 72, No. 10.1007, pp. 978-3). Cham: Springer.

References

1. Saqr M, Poquet O, López-Pernas S (2022) Networks in education: a travelogue through five decades. IEEE Access 10:32361–32380. https://doi.org/10.1109/access.2022.3159674
2. Dey AK, Tian Y, Gel YR (2022) Community detection in complex networks: from statistical foundations to data science applications. Comput Stat 14:e1566. https://doi.org/10.1002/wics.1566
3. Newman MEJ (2004) Detecting community structure in networks. Eur Phys J B 38:321–330. https://doi.org/10.1140/epjb/e2004-00124-y
4. Yassine S, Kadry S, Sicilia M-A (2022) Detecting communities using social network analysis in online learning environments: systematic literature review. Wiley interdisciplinary reviews Data mining and knowledge discovery 12:1–37. https://doi.org/10.1002/widm.1431
5. Prell C (2011) Social network analysis. SAGE, Thousand Oaks
6. Wasserman S, Faust K (1994) Social network analysis: methods and applications. Cambridge University Press, New York
7. Saqr M, Nouri J, Vartiainen H, Tedre M (2020) Robustness and rich clubs in collaborative learning groups: a learning analytics study using network science. Sci Rep 10:14445. https://doi.org/10.1038/s41598-020-71483-z
8. Vaquero LM, Cebrian M (2013) The rich club phenomenon in the classroom. Sci Rep 3:1174. https://doi.org/10.1038/srep01174
9. Fortunato S, Hric D (2016) Community detection in networks: a user guide. Phys Rep 659:1–44. https://doi.org/10.1016/j.physrep.2016.09.002
10. Saqr M, López-Pernas S, Conde MÁ, Hernández-García Á (2024) Social network analysis: a primer, a guide and a tutorial in R. In: Saqr M, López-Pernas S (eds) Learning analytics methods and tutorials: a practical guide using R. Springer, Berlin
11. White HC, Boorman SA, Breiger RL (1976) Social structure from multiple networks. I. Blockmodels of roles and positions. Am J Sociol 81:730–780. https://doi.org/10.1086/226141
12. Zachary WW (1977) An information flow model for conflict and fission in small groups. J Anthropol Res 33:452–473. https://doi.org/10.1086/jar.33.4.3629752
13. Adamic LA, Glance N (2005) The political blogosphere and the 2004 U.S. election. In: Proceedings of the 3rd international workshop on link discovery. ACM, New York
14. Shelton RC, Lee M, Brotzman LE, Crookes DM, Jandorf L, Erwin D, Gage-Bouchard EA (2019) Use of social network analysis in the development, dissemination, implementation, and

sustainability of health behavior interventions for adults: a systematic review. Soc Sci Med 220:81–101. https://doi.org/10.1016/j.socscimed.2018.10.013
15. Bruun J, Bearden IG (2014) Time development in the early history of social networks: link stabilization, group dynamics, and segregation. PloS One 9:e112775. https://doi.org/10.1371/journal.pone.0112775
16. Newman MEJ (2006) Modularity and community structure in networks. Proc Natl Acad Sci USA 103:8577–8582. https://doi.org/10.1073/pnas.0601602103
17. Fornito A, Zalesky A, Bullmore ET (2016) Fundamentals of Brain Network Analysis. Chapter 9 – Modularity (pp. 303–354). https://doi.org/10.1016/C2012-0-06036-X
18. Brandes U, Delling D, Gaertler M, Gorke R, Hoefer M, Nikoloski Z, Wagner D (2008) On modularity clustering. IEEE Trans Knowl Data Eng 20:172–188. https://doi.org/10.1109/tkde.2007.190689
19. Clauset A, Newman MEJ, Moore C (2004) Finding community structure in very large networks. Phys Rev E Stat Nonlinear Soft Matter Phys 70:066111. https://doi.org/10.1103/PhysRevE.70.066111
20. López-Pernas S, Saqr M, Conde J, Del-Río-Carazo L (2024) A broad collection of datasets for educational research training and application. In: Saqr M, López-Pernas S (eds) Learning analytics methods and tutorials: a practical guide using R. Springer, Berlin
21. Rabbany R, Takaffoli M, Zaïane OR (2012) Social network analysis and mining to support the assessment of on-line student participation. ACM SIGKDD Explor Newsl 13:20–29. https://doi.org/10.1145/2207243.2207247
22. Pham MC, Cao Y, Petrushyna Z, Klamma R (2012) Learning analytics in a teachers' social network. In: Hodgson V, Jones C, de Laat M, McConnell D, Ryberg T, Sloep P (eds) Proceedings of the 8th International Conference on Networked Learning 2012
23. Suthers D, Chu K-H (2012) Multi-mediated community structure in a socio-technical network. In: Proceedings of the 2nd international conference on learning analytics and knowledge. ACM, New York
24. Orduña P, Almeida A, Ros S, López-De-Ipiña D, Garcia-Zubia J (2014) Leveraging non-explicit social communities for learning analytics in mobile remote laboratories. J Univ Comput Sci 20:2043–2053. https://doi.org/10.3217/JUCS-020-15-2043
25. Skrypnyk O, Joksimović S, Kovanović V, Gasević D, Dawson S (2015) Roles of course facilitators, learners, and technology in the flow of information of a cMOOC. Int Rev Res Open Distrib Learn 16(3): 188–217
26. Gruzd A, Paulin D, Haythornthwaite C (2016) Analyzing social media and learning through content and social network analysis: a faceted methodological approach. J Learn Anal 3:46–71. https://doi.org/10.18608/jla.2016.33.4
27. Joksimović S, Kovanović V, Jovanović J, Zouaq A, Gašević D, Hatala M (2015) What do cMOOC participants talk about in social media? In: Proceedings of the fifth international conference on learning analytics and knowledge. ACM, New York
28. Hernández-García Á, González-González I, Jiménez-Zarco AI, Chaparro-Peláez J (2016) Visualizations of online course interactions for social network learning analytics. Int J Emerg Technol Learn 11(7):6–15. https://doi.org/10.3991/ijet.v11i07.5889
29. Adraoui M, Retbi A, Idrissi MK, Bennani S (2018) Evaluate learning communities in the online social media. In: Proceedings of the 12th international conference on intelligent systems: theories and applications. ACM, New York
30. Nistor N, Dascalu M, Tarnai C, Trausan-Matu S (2020) Predicting newcomer integration in online learning communities: automated dialog assessment in blogger communities. Comput Hum Behav 105:106202. https://doi.org/10.1016/j.chb.2019.106202
31. López Flores N, Islind AS, Oskarsdottir M (2022) Exploring study profiles of computer science students with social network analysis. In: Proceedings of the annual hawaii international conference on system sciences. Hawaii International Conference on System Sciences
32. Abal Abas Z, Norizan MN, Zainal Abidin Z, Abdul Rahman AFN, Rahmalan H, Ahmed Tharbe IH, Wan Fakhruddin WFW, Mohd Zaki NH, Ahmad Sobri S (2022) Modeling physical interaction and understanding peer group learning dynamics: graph analytics approach perspective. Mathematics 10:1430. https://doi.org/10.3390/math10091430

33. Li C, Xing W, Leite WL (2022) Do gender and race matter? Supporting help-seeking with fair peer recommenders in an online algebra learning platform. In: LAK22: 12th international learning analytics and knowledge conference. ACM, New York
34. Nguyen H (2023) TikTok as learning analytics data: framing climate change and data practices. In: LAK23: 13th international learning analytics and knowledge conference. ACM, New York
35. Lancichinetti A, Fortunato S (2009) Community detection algorithms: a comparative analysis. Phys Rev E 80:056117. https://doi.org/10.1103/physreve.80.056117
36. Fortunato S (2010) Community detection in graphs. Phys Rep 486:75–174. https://doi.org/10.1016/j.physrep.2009.11.002
37. Chunaev P (2020) Community detection in node-attributed social networks: a survey. Comput Sci Rev 37:100286. https://doi.org/10.1016/j.cosrev.2020.100286
38. Blondel VD, Guillaume J-L, Lambiotte R, Lefebvre E (2008) Fast unfolding of communities in large networks. J Stat Mech Theory Exp 2008:P10008. https://doi.org/10.1088/1742-5468/2008/10/p10008
39. Girvan M, Newman MEJ (2002) Community structure in social and biological networks. Proc Natl Acad Sci 99:7821–7826. https://doi.org/10.1073/pnas.122653799
40. Derényi I, Palla G, Vicsek T (2005) Clique percolation in random networks. Phys Rev Lett 94:160202. https://doi.org/10.1103/PhysRevLett.94.160202
41. Csardi G, Nepusz T (2006). The igraph software package for complex network research. Int J Complex Syst 1695, 1–9.
42. Blondel VD, Guillaume J-L, Lambiotte R, Lefebvre E (2008) Fast unfolding of communities in large networks. J Stat Mech 2008:P10008. https://doi.org/10.1088/1742-5468/2008/10/p10008
43. Newman MEJ, Girvan M (2004) Finding and evaluating community structure in networks. Phys Rev E Stat Nonlinear Soft Matter Phys 69:026113. https://doi.org/10.1103/PhysRevE.69.026113
44. Parés F, Gasulla DG, Vilalta A, Moreno J, Ayguadé E, Labarta J, Cortés U, Suzumura T (2018) Fluid communities: a competitive, scalable and diverse community detection algorithm. In: Complex networks & their applications VI. Springer, Cham, pp 229–240
45. Rosvall M, Bergstrom CT (2008) Maps of random walks on complex networks reveal community structure. Proc Natl Acad Sci USA 105:1118–1123. https://doi.org/10.1073/pnas.0706851105
46. Pons P, Latapy M (2005) Computing communities in large networks using random walks. In: Computer and information sciences - ISCIS 2005. Springer, Berlin, pp 284–293
47. Raghavan UN, Albert R, Kumara S (2007) Near linear time algorithm to detect community structures in large-scale networks. Phys Rev E Stat Nonlinear Soft Matter Phys 76:036106. https://doi.org/10.1103/physreve.76.036106
48. Newman MEJ (2006) Finding community structure in networks using the eigenvectors of matrices. Phys Rev E Stat Nonlinear Soft Matter Phys 74:036104. https://doi.org/10.1103/PhysRevE.74.036104
49. Traag VA, Van Dooren P, Nesterov Y (2011) Narrow scope for resolution-limit-free community detection. Phys Rev E Stat Nonlinear Soft Matter Phys 84:016114. https://doi.org/10.1103/PhysRevE.84.016114
50. Reichardt J, Bornholdt S (2006) Statistical mechanics of community detection. Phys. Rev. E, 74:1, 016110, 1–14. https://doi.org/10.1103/PhysRevE.74.016110
51. Smith NR, Zivich PN, Frerichs LM, Moody J, Aiello AE (2020) A guide for choosing community detection algorithms in social network studies: the question alignment approach. Am J Prevent Med 59:597–605. https://doi.org/10.1016/j.amepre.2020.04.015
52. Hernández-García Á, Suárez-Navas I (2017) GraphFES: a web service and application for moodle message board social graph extraction. In: Big data and learning analytics in higher education. Springer, Cham, pp 167–194
53. Lerís D, Fidalgo Á, Echaluce MLS (2014) A comprehensive training model of the teamwork competence. Int J Learn Intell Capital 11:1. https://doi.org/10.1504/ijlic.2014.059216

54. Chaparro-Peláez J, Acquila-Natale E, Iglesias-Pradas S, Suárez-Navas I (2015) A web services-based application for LMS data extraction and processing for social network analysis. In: New information and communication technologies for knowledge management in organizations. Springer, Cham, pp 110–121
55. Yetman C (2022) Answer to: R forceNetwork - how do I keep the legend in the top left corner when zooming is enabled? Stack Overflow. https://stackoverflow.com/a/71748184
56. Poquet O, Saqr M, Chen B (2021) Recommendations for network research in learning analytics: to open a conversation. In: Proceedings of the NetSciLA21 workshop
57. Saqr M, Viberg O, Vartiainen H (2020) Capturing the participation and social dimensions of computer-supported collaborative learning through social network analysis: which method and measures matter? Int J Comput-Support Collab Learn 15:227–248. https://doi.org/10.1007/s11412-020-09322-6

Temporal Network Analysis: Introduction, Methods and Analysis with R

Mohammed Saqr

1 Introduction

Learning is social and therefore, involves relations, interactions and connections between learners, teachers and the world at large. Such interactions are essentially temporal and unfold in time [1]; that is, facilitated, curtailed or influenced at different temporal scales [2, 3]. Therefore, time has become a quintessential aspect in several learning theories, frameworks and methodological approaches to learning [3–5]. Modeling learning as a temporal and relational process is, nevertheless, both natural, timely and more tethered to reality [4, 6]. Traditionally, relations have been modeled with Social Network Analysis (SNA) and temporal events have been modeled with sequence analysis or process mining [3, 7]. Yet, researchers have rarely combined the two aspects (the temporal and relational aspects) in an analytics framework [1]. Considering how important the timing and order of the learning process are, it is all-important that our analysis lens is not time-blind [8, 9]. Using time-blind methods flattens an essentially temporal process where the important details of progression are lost or distorted [10, 11]. In doing so, we miss the rhythm, the evolution and devolution of the process, we overlook the regularity and we may fail to capture the events that matter [9–11].

Temporal Networks

Recent advances in network analysis have resulted in the emergence of the new field of temporal network analysis which combines both the relational and temporal dimensions into a single analytical framework: temporal networks, also referred to as time-varying networks, dynamic networks or evolving networks [10]. Today, temporal networks are increasingly adopted in several fields to model dynamic

M. Saqr (✉)
School of Computing, University of Eastern Finland, Joensuu, Finland
e-mail: mohammed.saqr@uef.fi

M. Saqr, S. López-Pernas (eds.), *Learning Analytics Methods and Tutorials*,
https://doi.org/10.1007/978-3-031-54464-4_17

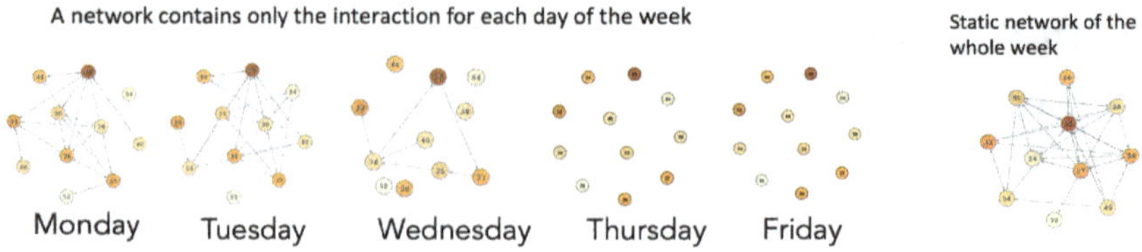

Fig. 1 Interactions are aggregated per day showing that some days show an active group e.g., Monday, and other days have an inactive group e.g., Friday. Meanwhile, the static network on the right appears dense [15]

phenomena, e.g., information exchange, the spread of infections, or the reach of viral videos on social media [12]. Whereas temporal networks are concerned with the modeling of relationships similar to traditional social networks (i.e., static or aggregate networks), they are conceptually fundamentally different [10, 11, 13]. Additionally, temporal networks are not a simple extension of social networks, nor are they time-augmented social networks or time-weighted networks. In that, temporal networks are based on different representations of data, have a different mathematical underpinning, and use distinct visualization methods. In temporal networks, edges emerge (get activated or born) and dissolve (get deactivated or die) compared to always present edges in static social networks. Also, in temporal networks, an edge represents temporary interaction, contact, co-presence, or concurrency between two nodes interacting at a specific time. The fact that static networks represent nodes as being connected together all the time exaggerates connectivity [14, 15]. For instance, in Fig. 1, we have five network visualizations, each network belonging to a weekday. We see that Monday, Tuesday, and Wednesday networks are relatively connected, whereas Thursday and Friday networks are disconnected. The corresponding aggregated or static network on the right is densely connected. The example in Fig. 1 shows how a static network both conflates connectivity and obfuscates dynamics, you can read more about this example in [15]. Similarly, network measures calculated in static networks are inflated and biased -skewed towards higher values - because they ignore the temporal direction of edges allowing the edges to run back in time. Another characteristic of temporal networks is that edges have a starting time point and ending time point, the end of each edge is understandably later than the start, i.e., follows the forward-moving direction of time. Therefore, the paths in the temporal network are unidirectional or time-restricted [10, 11]. The next section discusses the temporal networks in detail.

2 The Building Blocks of a Temporal Network

2.1 Edges

In temporal networks, edges are commonly referred to as events, links, or dynamic edges. Two types of temporal networks are commonly described based on their edge type [12].

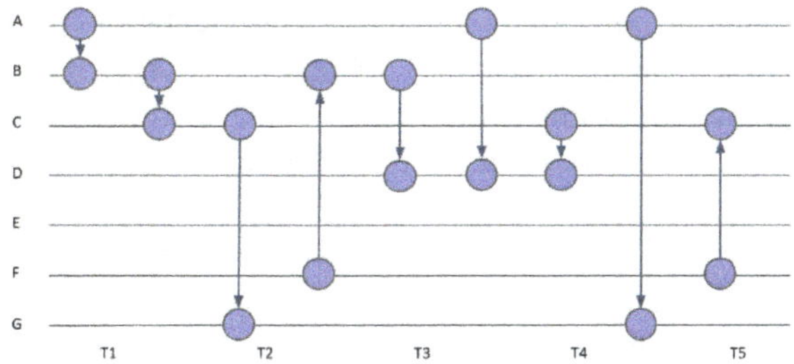

Fig. 2 Example of a contact temporal network. Edges form momentarily and have no obvious duration

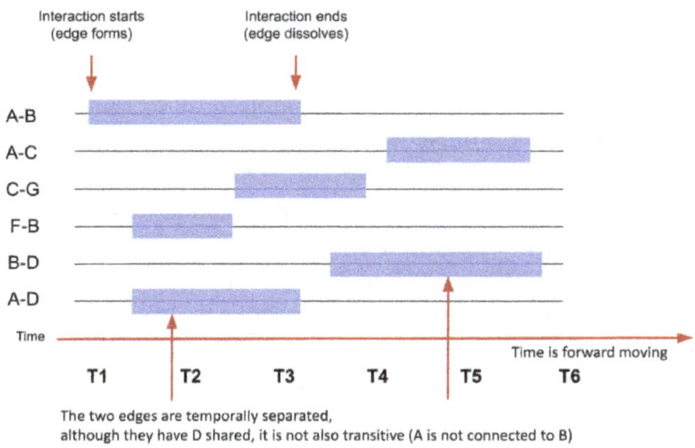

Fig. 3 An example of an interval network. On the top, we see an example of an edge. In the bottom arrows pointing to two edges B–D and A–D, we can see that the two edges do not overlap and therefore, although they share a connection (A), we can't assume that B is connected to A through D

- **Contact temporal networks:** In contact temporal networks, edge duration is very brief, undefined, or negligible. For example, instant messages have no obvious duration but have a clear source (sender), target (receiver), and timestamp. Figure 2 shows a contact temporal network where the edges are represented as sequences of contacts between nodes with no duration.
- **Interval temporal networks:** In interval temporal networks, each interaction has a duration. An example of such a network would be a conversation where each of the conversants talks for a certain length of time. In the interval temporal network, the duration of interactions matters and the modeling thereof helps understand the process. In Fig. 3, we see an interval temporal network where each edge has a clear start and clear end. For example, an edge forms between node A and node B at time point 1 and dissolves at time point 3, i.e., lasts for two time points.

2.2 Paths, Concurrency, and Reachability

Paths represent the pathways that connect edges, the identification of which can help solve essential problems like the shortest path between two places in a route planning application, e.g., Google maps. In a dynamic process, the paths represent a time-respecting sequence of edges i.e., where the timing of each edge follows one another according to time passage, that is, the timestamps are incrementally increasing [10, 11]. For instance, let's assume we have a group of students interacting about a problem, starting by defining the problem, argumenting, debating, and finding a solution. The temporal path that would represent the sequence of interactions among students in this process will be a *defining->argumenting->debating->solving*. We expect that the timestamp of *defining* precedes *argumenting* and *argumenting* precedes *debating* and so on. In that way, the path is unidirectional, follows a time-ordered sequence, and requires that each node is temporally connected, i.e., the two nodes coexist or interact with each other at the same time [11]. Such temporal co-presence is known as concurrent. Concurrency defines the duration of the nodes where they were co-present together and therefore can be a measure of the magnitude of contact between the two nodes. This is particularly important when we are modeling processes where the path length matters e.g., social influence. A student is more likely to be influenced by an idea when the student discusses the idea with another for a longer period of time. Similarly, self-regulation could be more meaningful when phases are more concurrent rather than disconnected [3]. Reachability is the proportion of nodes that can be reached from a node using time-respecting paths. A node is more influential or central, if it can reach a larger number of nodes [12].

2.3 Nodes

Nodes in temporal networks are similar to static networks at large. Such nodes can be humans, objects, semantics, historical events or chemical reactions to mention a few. Perhaps, the possible difference—if it at all exists—is that temporal network tend to be studied in fields where temporal order is consequential e.g., epidemics, linguistics and spread of ideas.

3 Previous Work and Examples of Temporal Network Analysis

Few studies have addressed temporal network analysis. Yet, some examples exist that may shed light on the novel framework and how it can be harnessed in education. In a study by Saqr and Nouri [15], the authors investigated how students interact in a problem-based learning environment using temporal networks. The study estimated temporal centrality measures, used temporal network visualization, and examined the predictive power of temporal centrality measures. The study reported rhythmic changes in centrality measures, network properties as well as the way students mix online. The study also found that temporal centrality measures were predictive of students" performance from as early as the second day of the course. Models that included temporal centrality measures have performed consistently better and from as early as the first week of the course. Another study by Saqr and Peeters [9] analyzed students' interactions in an online collaborative environment where students interacted in Facebook groups. The authors compared centrality measures from traditional social networks to temporal centrality measures and found that temporal centralities are more predictive of performance. Another study from the same group has used chat messages to study how students interact online and how temporal networks can shed light on different dynamics of students interacting using Discord instant messaging platform compared to students interacting using the forums in Moodle. Temporal networks were more informative in capturing the differences in dynamics and how such dynamics affected students" way of communicating [16].

4 Tutorial: Building a Temporal Network

Temporal network is a relatively new field with an emerging repertoire of methods that are continuously expanding. As we currently stand, a coherent tutorial that combines all possible steps of the analysis does not exist, and that is what this chapter aims to fill. The tutorial will introduce the R packages, visualization and mathematical analysis e.g., graph and node level centrality measures.

The first step is to load the needed packages. Unlike the SNA chapter [17] where we relied on the `igraph` framework, we will rely on the `statnet` framework that has a rich repertoire of temporal network packages. We will use three main packages, namely `tsna` (Temporal Social Network Analysis) which provides most functions for dealing with temporal networks as an extension for the popular `sna` package. The package `networkDynamic` offers several complementary functions for the network manipulation, whereas the package `ndtv` (Network Dynamic Temporal Visualization) offers several functions for visualizing temporal networks. To learn more about these packages, please visit their help pages. The next code chunk loads these packages as well as `tidyverse` packages to process the network dataframe [18]. We also need `tidyverse` for manipulating the file and preparing the data.

```
library(tsna)
library(ndtv)
library(networkDynamic)
library(tidyverse)
library(rio)
library(scatterplot3d)
library(ergm)
```

To create a temporal network, we need a timestamped file with interactions. The essential fields are the `source`, `target` and `time`, and perhaps also some information about the interactions or the nodes (but these constitute extra information that is good to have). A temporal network is created by combining a base static network (that has the network base information) and a dynamic network with time information. As such, we need to prepare the Massive Open Online Course (MOOC) dataset described in detail here [19] and prepare it for creating a static network that will serve as a base network.

The next code chunk loads the dataset files (edges and nodes data) from the MOOCs dataset. Some cleaning of the data is necessary.

```
URL<- "https://github.com/lamethods/data/raw/main/6_snaMOOC/"
net_edges <- import(paste0(URL, "DLT1%20Edgelist.csv"))
net_nodes <- import(paste0(URL, "DLT1%20Nodes.csv"))
```

First, we have to clean the column names from extra spaces using the function `clean_names` from the `janitor` package. Next, we have to remove loops, or instances where the source and target of the interaction are the same since it makes little sense that a person responds to oneself in a temporal network (this is not essential). Third, we need to create a dataframe where we replace duplicate edges with a weight equal to the frequency of repeated interactions, we will need this file for the creation of the base network (see later). Fourth, we recode the expertise level in the nodes file to meaningful codes (from its original numerical coding as 1,2,3) so that we can use them later in the analysis. The fifth step is to convert the timestamp to sequential days starting from the first day of the course; this makes sense for easy interpretation. Also, time works better in networkDynamic when it is numeric. The final step is to remove discussions where there are no replies. This cleaning is necessary since we have a dataset that was not essentially prepared for temporal networks.

```
net_edges <- net_edges |> janitor::clean_names() #1
                              cleaning column names
net_edges_NL <- net_edges |> filter(sender != receiver)
                                  #2 removing loops

# Removing duplicates and replacing them with weight
net_edges_NLW <- net_edges_NL |>
                 group_by(sender, receiver) |>
                 tally(name = "weight") #3

# Recoding expertise
net_nodes <- net_nodes |>
  mutate(expert_level = case_match(experience, #4
        1 ~"Expert",
        2 ~ "Student",
```

```
          3 ~ "Teacher"))

# A function to create serial days
dayizer = function(my_date) {
  numeric_date = lubridate::parse_date_time(my_date,
                 "mdy HM")
  Min_time = min(numeric_date)
  my_date = (numeric_date - Min_time) / (24*60*60)
  my_date = round(my_date,2)
  return(as.numeric(my_date))
}

net_edges_NL$new_date = dayizer(net_edges_NL$timestamp)
                                                    #5

# Remove discussions with no interactions
net_edges_NL <- net_edges_NL |>
 group_by (discussion_title) |>
 filter(n() > 1)
```

As mentioned before, the first step in creating a temporal network is creating a static base network (base network) which carries all the information about the network, e.g., the nodes, edges as well as their attributes. The base network is typically a static weighted network. Here we define the base network file (the weighted edge file we created before), we use `directed = TRUE` to create our network as directed and we tell the `network` function that the vertices attributes are in the net_nodes file.

```
NetworkD <- network(net_edges_NLW, directed = TRUE,
                    matrix.type = "edgelist",
                    loops = FALSE, multiple = FALSE,
                    vertices = net_nodes)
```

For creating a temporal network, we need more than the source and the target commonly needed for the static network. In particular, the following variables are required to be defined.

- `tail`: the source of the interaction
- `head`: the target of the interaction
- `onset`: The starting time of the interaction
- `terminus`: the end time of the interaction
- `duration`: the duration of the interaction

Our dataset—which comes from forum MOOC interactions, see Fig. 4—has an obvious starting time (which is the timestamp of each interaction) but has no

Description: df [2,529 × 10]

	sender	receiver	timestamp	discussion_title	discussion_category
	<int>	<int>	<chr>	<chr>	<chr>
1	360	444	4/4/13 16:32	Most important change for your school or district?	Group N
2	356	444	4/4/13 18:45	Most important change for your school or district?	Group D-L
3	356	444	4/4/13 18:47	DLT Resources—Comments and Suggestions	Group D-L
4	344	444	4/4/13 18:55	Most important change for your school or district?	Group O-T
5	392	444	4/4/13 19:13	Most important change for your school or district?	Group U-Z
6	219	444	4/4/13 19:16	Most important change for your school or district?	Group M
7	318	444	4/4/13 19:26	Most important change for your school or district?	Group M
8	4	444	4/4/13 19:44	Most important change for your school or district?	Group N
9	355	356	4/4/13 20:12	DLT Resources—Comments and Suggestions	Group D-L
10	355	444	4/4/13 20:13	Most important change for your school or district?	Group D-L

Fig. 4 Screenshot of the edges file. The edge file has source (sender), target (receiver) and a timestamp

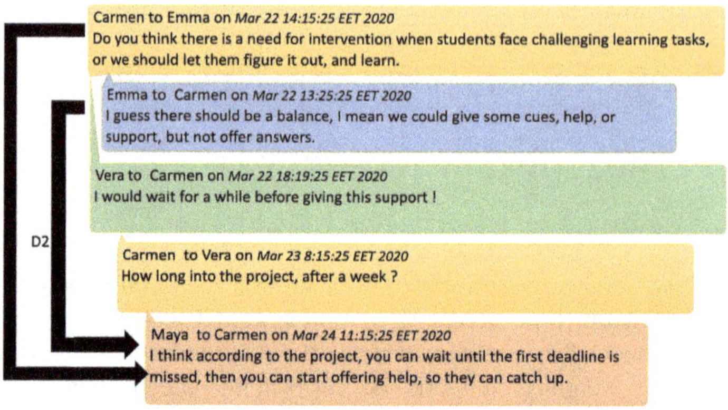

Fig. 5 A sample discussion demonstrating a way to compute the duration of the post. The first post lasts active in the discussion from starting time to the last reply it stimulated in the same thread (D1)

clear end time. There is no straightforward answer to this question. Nonetheless, a possible way to consider the duration of every post is the duration the post was active in the discussion or continued to be discussed. That is the time from the post in a discussion thread to the last post in the same threads of replies. Such a method—while far from perfect—offers a rough method for estimating the time during which this interaction has been "active" in the discussion [15, 20]. For an illustration, see Fig. 5 which shows the duration for the first and second posts.

The next code chunk creates a variable for the starting time of each interaction, computes the ending time where this post was part of an active discussion, and then computes the duration.

```
# Create the required variables (start, end, and
    duration) defined by

net_edges_NL <- net_edges_NL |>
```

```
                        group_by (discussion_title) |>
mutate(start = min(new_date), end = max(new_date),
                        duration = end - start)
```

In the same way, the second duration is computed in the same way (D2). The next step of the analysis creates a dataframe where all the needed information for the network is specified and in the next step we simply use the `networkDynamic` with two arguments, the base network and the dataframe with all the temporal network information created in the previous step. Nonetheless, dealing with around 450 nodes in a network is hard, it becomes impossible to visualize or get insights from a large number of crowded nodes. So, for the sake of simplicity of demonstration in this tutorial, we will create a smaller subset of the network of people who had a reasonable number of interactions (degree more than 20) using `get.inducedSubgraph` argument. The resulting network is then called `Active_Network` which we will analyze.

```
# Creating a dataframe with needed variables

edge_spells <- data.frame("onset" = net_edges_NL$start ,
                         "terminus" = net_edges_NL$end,
                         "tail" = net_edges_NL$sender,
                         "head" = net_edges_NL$receiver,
                         "onset.censored" = FALSE,
                         "terminus.censored" = FALSE,
                         "duration" = net_edges_NL$
                         duration)

# Creating the dynamic network network
Dynamic_network <- networkDynamic(NetworkD, edge.spells =
         edge_spells)
```

```
Edge activity in base.net was ignored
Created net.obs.period to describe network
 Network observation period info:
  Number of observation spells: 1
  Maximal time range observed: 0 until 72.01
  Temporal mode: continuous
  Time unit: unknown
  Suggested time increment: NA
```

```
Active_Network <- get.inducedSubgraph(Dynamic_network,
                               v = which(degree
                               (Dynamic_network) >
                                      20))
```

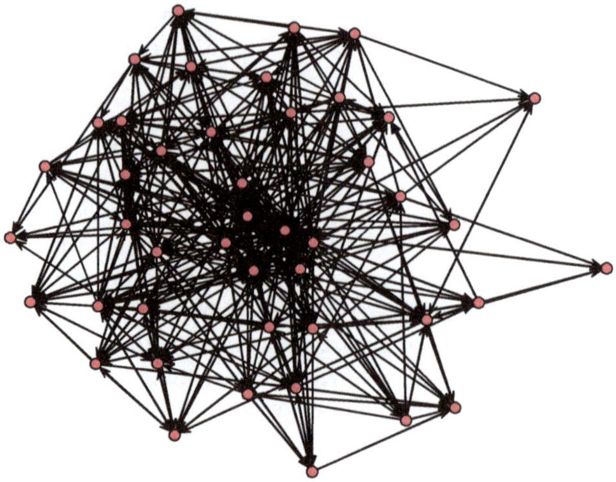

Fig. 6 A plot of the full network as plotted by the plot function

We can then confirm that the network has been created correctly using the `print` function. As the output shows, we have 521 distinct time changes, 72 days, 445 vertices and 1936 edges. We can also use the function `plot` to see how the network looks. The argument `pad` helps us remove the additional whitespace around the network. Plotting a temporal network helps summarize all the interactions in the network. As we can see in Fig. 6, the network is dense with several edges between interacting students.

```
print(Dynamic_network)
```

```
NetworkDynamic properties:
  distinct change times: 495
  maximal time range: 0 until  72.01

Includes optional net.obs.period attribute:
 Network observation period info:
  Number of observation spells: 1
  Maximal time range observed: 0 until 72.01
  Temporal mode: continuous
  Time unit: unknown
  Suggested time increment: NA

 Network attributes:
  vertices = 445
  directed = TRUE
  hyper = FALSE
  loops = FALSE
  multiple = FALSE
```

```
  bipartite = FALSE
  net.obs.period: (not shown)
  total edges= 1936
    missing edges= 0
    non-missing edges= 1936

 Vertex attribute names:
    connect country experience experience2 expert
    expert_level Facilitator gender grades group location region role1
    vertex.names

 Edge attribute names not shown
```

```
plot.network(Active_Network, pad = -0.5)
```

4.1 Visualization of Temporal Networks

To take advantage of the temporal network, we can use a function to extract the network at certain times to explore the activity. In the next example in Fig. 7, we chose the first four weeks one by one and plotted them alongside each other. The function `filmstrip` can create a similar output with a snapshot of the network at several time intervals. A similar result, yet with three-dimensional placing, can also be obtained with the `timePrism` function, as shown in Fig. 8. You may need to consult the package manual to get more information about the arguments and options for the plots.

```
plot.network(network.extract(Active_Network, onset = 1
        , terminus = 7))
plot.network(network.extract(Active_Network, onset = 8
        , terminus = 14))
plot.network(network.extract(Active_Network, onset = 15
        , terminus = 21))
plot.network(network.extract(Active_Network, onset = 22
        , terminus = 28))
```

```
compute.animation(Active_Network)
```

```
slice parameters:
  start:0
  end:72.01
  interval:1
  aggregate.dur:1
```

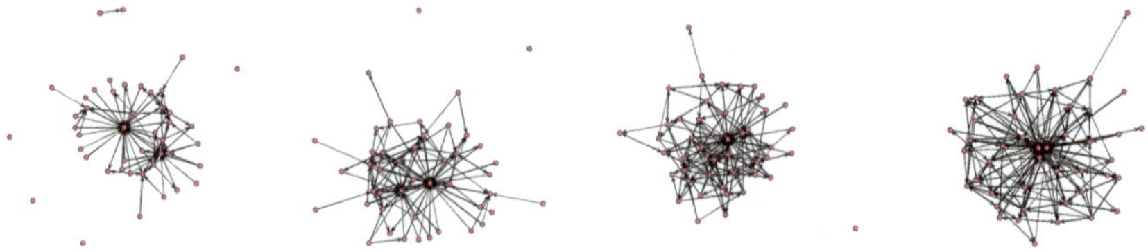

Fig. 7 A plot of the network at weeks 1, 2, 3, and 4

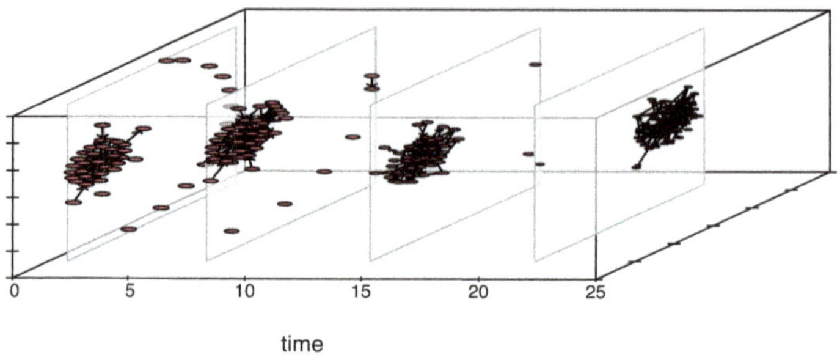

Fig. 8 A three-dimensional visualization of the network as a sequence of snapshots in space and time prism

```
rule:latest
```

```
timePrism(Active_Network, at = c(1, 7, 14, 21),
  spline.lwd = 1,
  box = TRUE,
  angle = 60,
  axis = TRUE,
  planes = TRUE,
  plane.col = "#FFFFFF99",
  scale.y = 1,
  orientation = c("z", "x", "y"))
```

However, a better way is to take advantage of the capabilities of the package `ndtv` and the network temporal information by rendering a full animated movie of the network as in Fig. 9 and explore each and every event as it happens.

```
render.d3movie(Active_Network)
```

As we mentioned in the introduction section, in temporal networks, edges or relationships form "get activated" and dissolve "get deactivated". We can plot such the dynamic edge formation and dissolution process using the functions `tEdgeFormation` which as the name implies plots the edges forming at the given

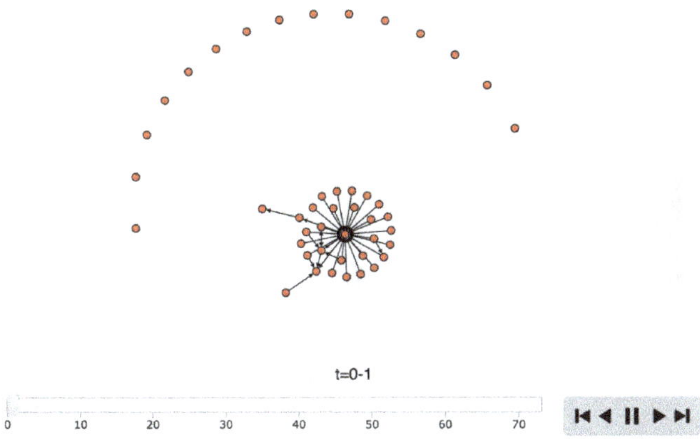

Fig. 9 A screenshot of the animated movie of the temporal network

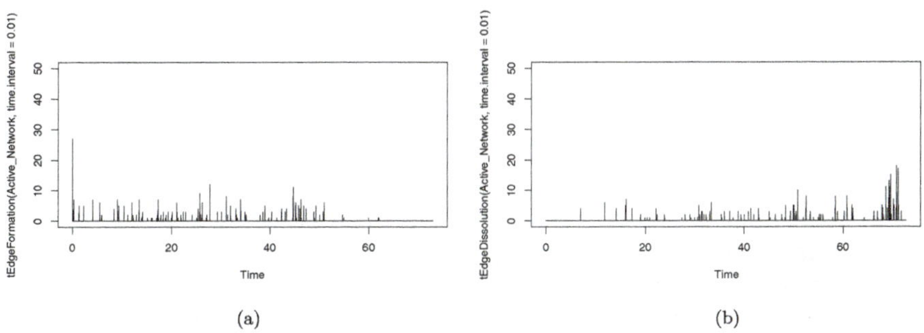

Fig. 10 Edge formation and dissolution. We see that edge formation occurs towards the start and increases till almost t=50 and edge dissolution starts scanty at the beginning and increases with time, peaking at t = 70. (**a**) Edge formation. (**b**) Edge dissolution

time point. The function `tEdgeDissolution` returns the edges terminating and can be plotted in the same way as seen in Fig. 10. Obviously, at the beginning of the MOOC, we see new relationships form and, at the end, most relationships dissolve.

```
plot(tEdgeFormation(Active_Network, time.interval =
        0.01), ylim = c(0,50))
plot(tEdgeDissolution(Active_Network, time.interval =
        0.01), ylim = c(0, 50))
```

Another way to visualize a temporal network is to use the proximity timeline, the `proximity.timeline` function tries to draw the temporal network in two dimensions, that is, it draws nodes at each time point taking into account how closely connected they are and renders them accordingly, nodes that are interacting are rendered close to each other, and nodes that not interacting are rendered apart. Technically as described in the function manual: "The passed network dynamic

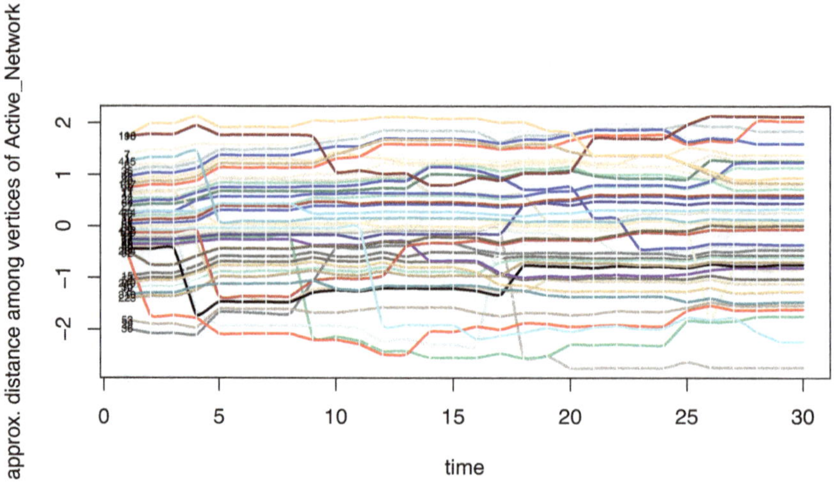

Fig. 11 A proximity timeline shows connected interacting nodes closer to each other

object is sliced up into a series of networks. It loops over the networks, converting each to a distance matrix based on geodesic path distance with `layout.distance`. The distances are fed into an MDS algorithm (specified by mode) that lays them out in one dimension: essentially trying to position them along a vertical line. The sequence of 1D layouts are arranged along a timeline, and a spline is drawn for each vertex connecting its positions at each time point. The idea is that closely-linked clusters form bands of lines that move together through the plot" [18]. The result is a timeline of temporal proximity. The next code draws the proximity timeline but also adds some colors, and a start and end for the plot, see the result in Fig. 11.

```
proximity.timeline(Active_Network, default.dist = 1,
            mode = "sammon",
                   labels.at = 1, vertex.col =
                         grDevices::colors(),
                   start = 1, end = 30, label.cex = 0.5)
```

4.2 *Statistical Analysis of Temporal Networks*

4.2.1 Graph Level Measures

Graph properties in temporal networks are dynamic and vary by time. When graph measures are computed we get a time series of the computed measures. Such fine-grained measures allow us to understand how the networks and their structure evolve or unfold in time and thus, such information can help us understand collaboration or interaction dynamics as they occur [9, 15].

The function `tSnaStats` from the package `tsna` has a large number of measures that can be computed by specifying the argument `snafun`. In the next example, we compute the graph level density with the argument `snafun=gden`. The function allows the choice of a range of time, for instance, from the end of the second week to the end of second month by supplying the arguments `start = 14` to `end = 60`. We can also use the `time.interval` to specify the granularity of the calculation. The argument `aggregate.dur` specifies the period of the aggregation, for instance, `aggregate.dur = 7` will compute the density for every seven days. We can also plot the density time series by simply using the function `plot`.

```
Density <- tSnaStats(
  nd = Active_Network,
  snafun = "gden",
  start = 14,
  end = 60,
  time.interval = 1,
  aggregate.dur = 7)
plot(Density)
```

You can see, in the resulting graph in Fig. 12, that the density increases until day 50 and then starts to drop. Of note, another type of density can be computed, known as temporal density, which computes the observed total duration of all edges and divides it by the maximum duration possible. Temporal density can be computed using the command `tEdgeDensity`.

```
tEdgeDensity(Active_Network)
```

```
[1] 0.3901841
```

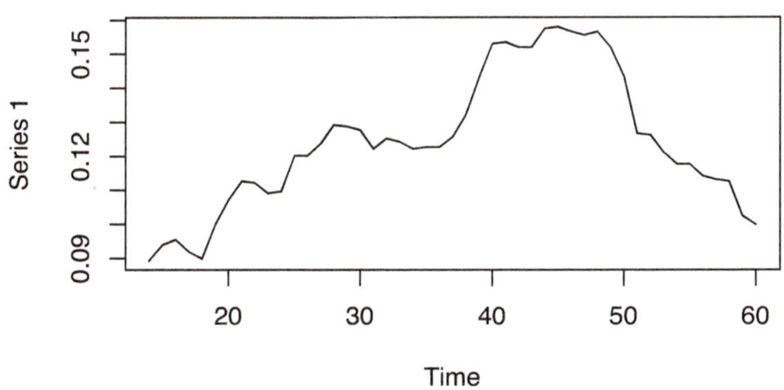

Fig. 12 A plot of temporal density from t = 14 to t = 60

```
gden(Active_Network)
```

```
[1] 0.2186869
```

Similar to density, we can compute reciprocity, i.e., the ratio of reciprocated edges to asymmetric edges. Note, that we calculate reciprocity here from day 1 to day 73 on a daily basis (this is just for demonstration of different periods). Since the function default is to calculate the reciprocated dyads, we specify the argument *measure* = *"edgewise"* to calculate the proportion of reciprocated edges. As the graph in Fig. 13a shows, reciprocity increases steadily for the first 50 days pointing to a build up of trust between collaborators. The `dyad.census` function offers a more granular view of the dyads and their reciprocity as shown in Fig. 13b. Similarly, `mutuality` is a very similar function and returns the number of complete dyads (reciprocated dyads), plotted in Fig. 13c. All of the aforementioned functions deal with reciprocity, for differences and usages, readers are encouraged to read the functions' help to explore the differences, arguments as well as the equation for each function.

```
Reciprocity <- tSnaStats(
    nd=Dynamic_network,
    snafun = "grecip" ,
    start = 1,
    end = 73,
    measure = "edgewise",
    time.interval = 1,
    aggregate.dur = 1)
plot(Reciprocity)

Dyad.census <- tSnaStats(Active_Network,
                         snafun = "dyad.census")
plot(Dyad.census)
```

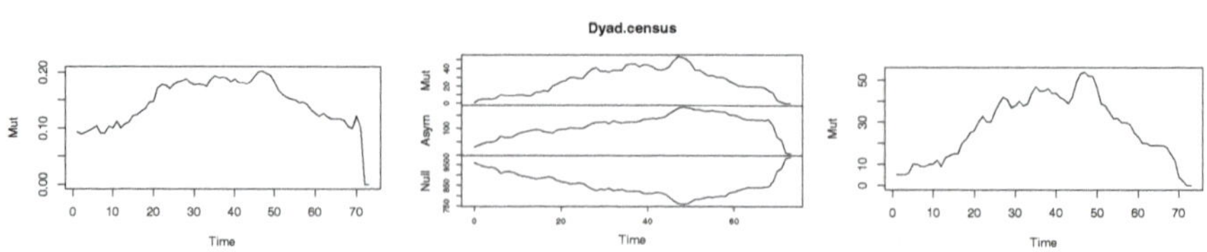

(a) (b) (c)

Fig. 13 Network descriptive statistics over time. (**a**) Reciprocity over time. (**b**) Dyad census over time. (**c**) Mutuality over time

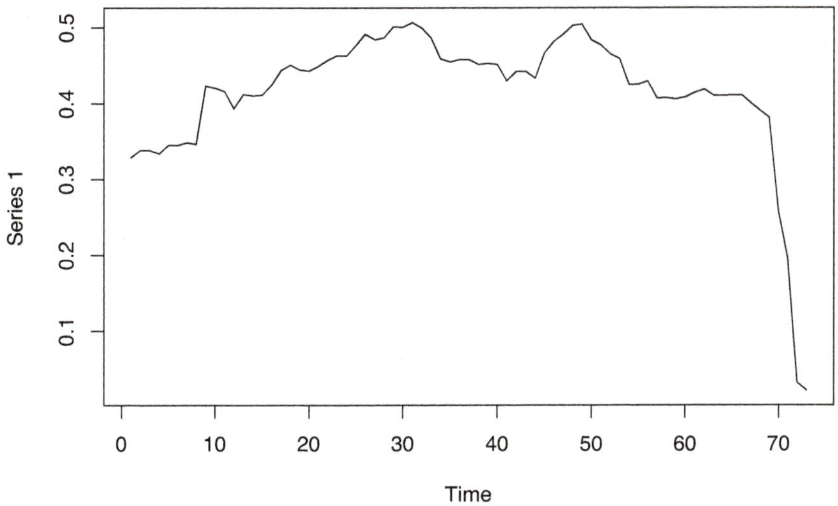

Fig. 14 A plot of degree centralization over time

```
    dynamicmutuality <- tSnaStats(
  Active_Network,
  snafun = "mutuality",
  start = 1,
  end = 73,
  time.interval = 1,
  aggregate.dur = 1
  )
plot(dynamicmutuality)
```

Centralization measures the dominance of members within the network and can be traced temporally using the function snafun = "centralization". Please note that we can choose the period, the interval and the aggregation periods using function arguments as mentioned before. The next example computes the degree centralization as demonstrated in Fig. 14. Also note that we can also compute other centralization measures such as centralization indegree, centralization outdegree, centralization betweenness, centralization closeness and eigenvector.

```
Degree_centralization <- tSnaStats(
  Active_Network,
  snafun = "centralization",
  start = 1,
  end = 73,
  time.interval = 1,
  aggregate.dur = 1,
  FUN = "degree")
```

```
plot(Degree_centralization)
```

Several other graph-level measure can be computed in the same way using the following arguments passed to the `snafun`: - `components`: count of Components within the graph over time - `triad.census`: the triad census and types of triads over time - `connectedness`: the connectedness score of the network efficiency network efficiency over time - `gtrans`: network transitivity over time - `hierarchy`: network Hierarchy over time - `lubness`: network LUBness over time - `efficiency`: network efficiency over time - `hierarchy`: network hierarchy over time

4.2.2 Node-Level Measures (Temopral Centrality Measures)

Centrality measures has been used to identify important actors, as a proxy indicator for academic achievement or to identify students' collaborative roles [21–24]. In temporal networks, centrality measures are fine-grained estimates of students" real-time centralities or importance, i.e., shows who were central and when considering the temporarily of their interaction. In doing so, we can see exactly when and for how long, at what pace, and with which rhythm a behavior happens (compare this to traditional social network analysis where centralities are computed as a single number). There are several possible uses of temporal centrality measures. For instance [15] have used them to create predictive models of students' performance. In another study by the same authors, they demonstrated that temporal centrality measures were more predictive of performance compared to traditional static centrality measures [9]. Since the temporal centralities are computed as time series, their temporal characteristics can also be used to compute other time series properties e.g., stability, variabilities, and pace, you can see for example [15]. There is a growing list of temporal centrality measures e.g., [13]. We will study here the most commonly used ones according to the latest review [23], but readers are encouraged to explore the `tsna` manual for more centrality measures.

Temporal degree centrality measures can be computed in the same way as we computed the graph level properties shown before. The next code defines the function `snafun = "degree"`, the start, the end date, and aggregation (which you can modify). An important argument here is the `cmode` argument which defines the type of centrality: "freeman", "indegree" or "outdegree" for the calculation of total in or out degree centralities. The result is a time series with the 73 values for each day from start to end, each day having a unique value for the degree centrality for each node. The rest of the code is intended to organize the results. We convert the time series to a dataframe, create a variable for the day number to make it easier to identify the day and create a variable to define the type of centrality, and then combine all centrality measures into a single data frame as below.

```
Degree_Centrality <- tSnaStats(
  Active_Network,
```

```
    snafun = "degree",
    start = 1,
    end = 73,
    time.interval = 1,
    aggregate.dur = 1,
    cmode = "freeman")

  inDegree_Centrality <- tSnaStats(
    Active_Network,
    snafun = "degree",
    start = 1,
    end = 73,
    time.interval = 1,
    aggregate.dur = 1,
    cmode = "indegree")

  OutDegree_Centrality <- tSnaStats(
    Active_Network,
    snafun = "degree",
    start = 1,
    end = 73,
    time.interval = 1,
    aggregate.dur = 1,
    cmode = "outdegree")

  Degree_Centrality_DF <- Degree_Centrality |>
          as.data.frame() |>
    mutate(Day = 1:73, centrality = "Degree_Centrality",
                         .before = 1L)


  inDegree_Centrality_DF <- inDegree_Centrality |>
              as.data.frame() |>
    mutate(Day = 1:73,centrality = "inDegree_Centrality",
                         .before = 1L)


  OutDegree_Centrality_DF <- OutDegree_Centrality |>
                  as.data.frame() |>
    mutate(Day = 1:73,centrality = "OutDegree_Centrality",
                     .before = 1L)

  rbind(Degree_Centrality_DF, inDegree_Centrality_DF,
                  OutDegree_Centrality_DF)
```

```
# A tibble: 219 x 47
     Day centrality   `1`   `5`   `6`   `7`  `11`  `13`  `15`  `17`  `19`  `24`
   <int> <chr>      <dbl> <dbl> <dbl> <dbl> <dbl> <dbl> <dbl> <dbl> <dbl> <dbl>
 1     1 Degree_Cen~     4     2     4     4     1     0     2     1     9     1
 2     2 Degree_Cen~     4     2     5     4     2     0     2     1     9     2
 3     3 Degree_Cen~     4     2     5     4     2     0     2     1     9     2
 4     4 Degree_Cen~     5     2     5     4     2     0     2     1     9     2
 5     5 Degree_Cen~     7     2     6     4     2     0     3     2    10     2
 6     6 Degree_Cen~     7     2     6     4     2     0     3     2    10     2
 7     7 Degree_Cen~     7     2     6     4     2     0     3     2     7     2
 8     8 Degree_Cen~     7     2     6     4     3     0     3     2     8     2
 9     9 Degree_Cen~     8     2     6     4     4     0     3     2     8     2
10    10 Degree_Cen~     9     2     6     4     4     0     3     2     9     2
# i 209 more rows
# i 35 more variables: `26` <dbl>, `27` <dbl>, `29` <dbl>, `30` <dbl>,
#   `34` <dbl>, `35` <dbl>, `36` <dbl>, `41` <dbl>, `44` <dbl>, `49` <dbl>,
#   `50` <dbl>, `53` <dbl>, `54` <dbl>, `58` <dbl>, `60` <dbl>, `61` <dbl>,
#   `62` <dbl>, `63` <dbl>, `64` <dbl>, `67` <dbl>, `68` <dbl>, `88` <dbl>,
#   `92` <dbl>, `98` <dbl>, `100` <dbl>, `116` <dbl>, `137` <dbl>, `198` <dbl>,
#   `219` <dbl>, `223` <dbl>, `234` <dbl>, `310` <dbl>, `432` <dbl>, ...
```

In the same way, closeness, betweenness and eigenvector centralities can be computed. Please note that we have a new argument here `gmode="graph"` which tells `snafun` that we would like to compute these centralities considering the network as undirected. You can also use the `gmode="digraph"` to compute the measures on a directed basis. You may need to use the previous code to convert each of the resulting time series into a data frame and add the day of the centralities. The `snafun` can compute other centrality measures in the same way e.g., information centrality, Bonacich Power Centrality, Harary Graph Centrality, Bonacich Power Centrality among others.

```
Closeness_Centrality <- tSnaStats(
  Active_Network,
  snafun = "closeness",
  start = 1,
  end = 73,
  time.interval = 1,
  aggregate.dur = 1,
  gmode = "graph")

Betweenness_Centrality <- tSnaStats(
  Active_Network,
  snafun = "betweenness",
  start = 1,
  end = 73,
  time.interval = 1,
  aggregate.dur = 1,
  gmode = "graph")
```

```
Eigen_Centrality <- tSnaStats(
  Active_Network,
  snafun = "evcent",
  start = 1,
  end = 73,
  time.interval = 1,
  aggregate.dur = 1,
  gmode = "graph")
```

Reachability is another important measure that temporal networks offer. You can calculate who, when, and after how many steps an interaction reaches from a certain vertex to another or to every other node in the network (if it at all does). In the next code, we compute the forward pathway (earliest reachable node) from node 44 (one of the course facilitators) using the function `tPath`. The output is a time series with all the time distance values and the number of steps.

```
FwdPathway <- tPath(
  Active_Network,
  v = 44,
  start = 0,
  graph.step.time = 7,
  end = 30,
  direction = "fwd")

FwdPathway |> as_tibble()
```

```
# A tibble: 45 x 7
   tdist previous gsteps start   end direction type
   <dbl>    <dbl>  <dbl> <dbl> <dbl> <chr>     <chr>
 1  7.09       44      1     0    30 fwd       earliest.arrive
 2 32.9        45      4     0    30 fwd       earliest.arrive
 3 34.8        31      3     0    30 fwd       earliest.arrive
 4 14.1         1      2     0    30 fwd       earliest.arrive
 5 16.1        44      1     0    30 fwd       earliest.arrive
 6 30.9        13      2     0    30 fwd       earliest.arrive
 7 28.1        23      4     0    30 fwd       earliest.arrive
 8 Inf          0    Inf     0    30 fwd       earliest.arrive
 9  7.28       44      1     0    30 fwd       earliest.arrive
10 24.2         9      2     0    30 fwd       earliest.arrive
# i 35 more rows
```

More importantly, we can plot the hierarchical path from the given node, by simply using `plot` function, see Fig. 15.

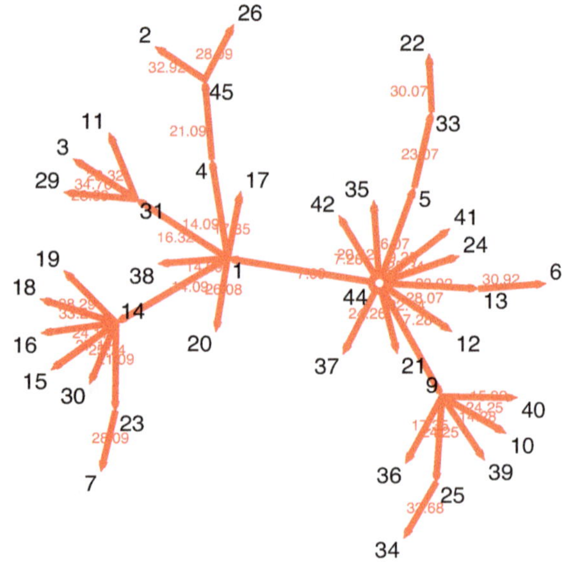

Fig. 15 A forward pathway (earliest reachable node) pathway of interactions with node 44

```
plot(FwdPathway)
```

We may also use `Graphviz` to make the plot look better and hierarchical (Fig. 16). Please follow the instructions on http://graphviz.org/download/ to download and install Graphviz.

```
plot(FwdPathway, edge.lwd = 0.1, vertex.col= "blue",
           pad = -4,
     coord=network.layout.animate.Graphviz(as.network
              (FwdPathway),
                 layout.par = list(gv.engine='dot',
                           gv.args = '-Granksep=2')))
```

Another option is to plot the network diffusion or transmission hierarchical tree with generation time vs. clock/model time using `transmissionTimeline` function as in Fig. 17. For more on these functions, readers are invited to read the function manuals and for usage, these papers offer a starting point [3, 9].

```
transmissionTimeline(FwdPathway, jitter = F,
            displaylabels = TRUE,
                 main = "Earliest forward path" )
```

Another useful package that offers a wealth of graph level measures is `tErgmStats`, you may need to consult the help files which are available by using the command `?tErgmStats`. One of the important functions that we can try here

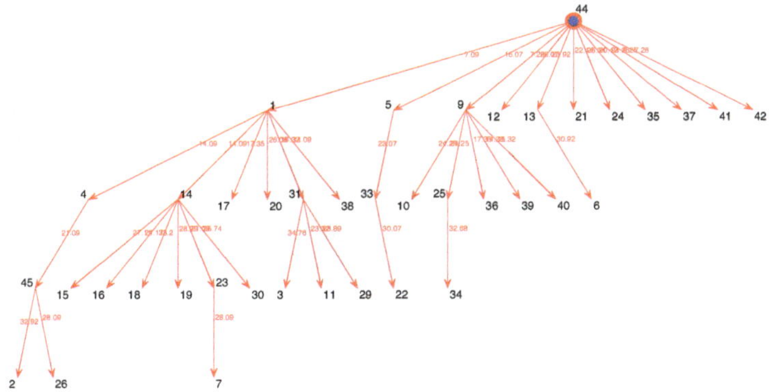

Fig. 16 An improved forward pathway with Graphviz

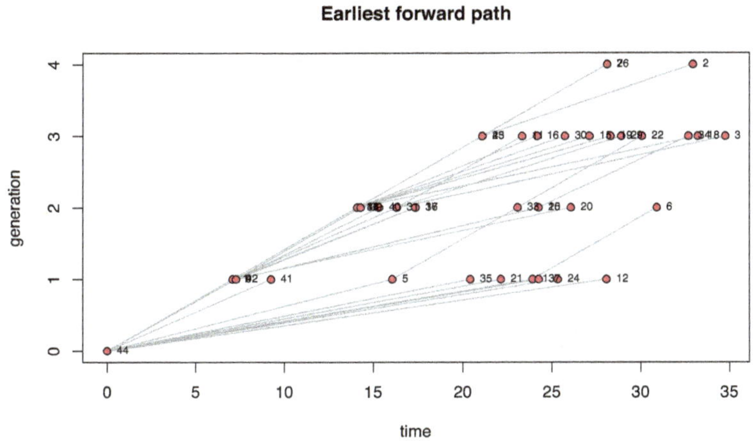

Fig. 17 Transmission hierarchical tree showing the earliest forward interaction with node 44

is the `nodemix`. We can use nodemix to examine who and when different actors interact with each other, see here for an example where the authors examined how low and high achievers mix with each other and with the teachers [15]. The next code demonstrates how to compute the mixing patterns between expertise levels. We then convert the time series as a dataframe, clean the names and then plot the results as demonstrated in Fig. 18.

```
Mix_experience <- tErgmStats(Active_Network,
  "nodemix('expert_level')",
  start = 1,
  end = 73,
  time.interval = 1,
  aggregate.dur = 1)
```

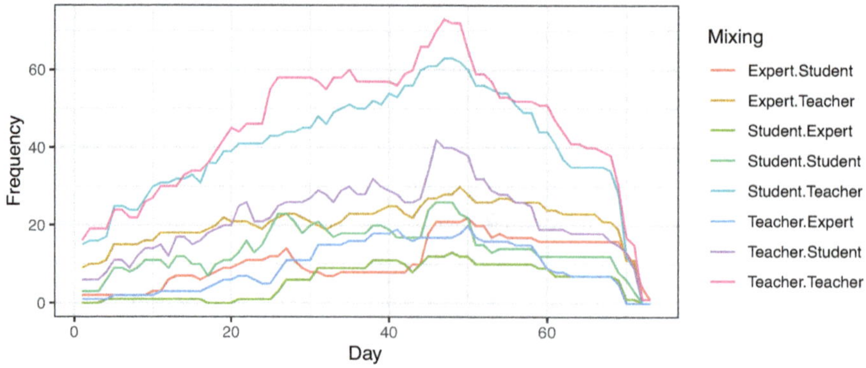

Fig. 18 Mixing patterns between expertise levels and the teachers

```
Mixing <- as.data.frame(Mix_experience)

colnames(Mixing) <- gsub("mix.expert_level.", "",
                         colnames(Mixing))

Mixing$Day <- 1:73

Mixing_long= pivot_longer(Mixing, contains("."))

names(Mixing_long)= c("Day", "Mixing", "Frequency")

ggplot(Mixing_long, aes(Day, Frequency, group =
                          Mixing, color = Mixing)) +
  geom_line(alpha = .95) + theme_bw()
```

5 Discussion

Learning can be viewed as relational, interdependent, and temporal and therefore, methods that account for such multifaceted dynamic processes are required [1, 3]. We have shown the main advantages of temporal networks and the potentials it offers for modeling dynamic learning processes. These potentials or features can facilitate the modeling of the complex natural processes–including the emergence, evolution, diffusion or disappearance of learners' activities, communities or social processes that unfold over time. Such features can augment the existing analytics method and help shed light on many learning phenomena [16]. Taking advantage of time dynamics allows us temporal evolution of co-construction of knowledge, the flow of information and the building of social relationships, to name a few examples.

What is more, temporal networks allow the longitudinal modeling and analysis of interactions across longer periods of time e.g., full duration of a course, project or meeting using time-respecting paths [9, 15].

There are several methods that can dissect the temporal dimensions of a learning process, e.g., process and sequence mining, time series methods and epistemic network analysis [25, 26]. While such methods have given a wealth of information and insights about learning processes, they fall short when it comes to the relational aspects [9]. We review here and in short the main differences between such methods of temporal networks. Process mining is a method for the discovery and modeling of a temporal process [1, 27].Yet, the relational aspect is completely ignored. The case is similar for sequence mining where the time-ordered sequences are modeled regardless of the their interactions [25, 28, 29]. Epistemic network analysis is another method that allows the study of co-temporal interactions. However, the "temporal aspect" is limited to combining data within a temporal window and later modeling the interactions as a static network. Put another way, Epistemic network analysis is a special type of static networks where edges are defined based on co-occurrence [30]. For a comparison between the various methods see [3].

6 Learning Resources

A good place to start is to get acquainted with the cited research in the literature review section. Other good references could be the methodological paper that gives a detailed overview of temporal networks which some parts of this chapter has been built around it [31]. There are few, yet very informative tutorials that we can suggest, most notable are the tutorials by Brey [32] and Bender-deMoll [33]. The packages used in this chapter have very informative manuals: TSNA[34], NDTV[18] and networkDynamic[35]. Some seminal papers can be recommended here, especially the following papers and books.

- Holme, P. (2015). Modern temporal network theory: a colloquium. European Physical Journal B, 88(9).
- Holme, P., & Saramäki, J. (2019). A Map of Approaches to Temporal Networks (pp. 1–24).
- Holme, P., & Saramäki, J. (Eds.). (2019). Temporal network theory (Vol. 2). New York: Springer.

Acknowledgments This chapter is co-funded by the Academy of Finland the project "Towards precision education: Idiographic learning analytics (TOPEILA)", Decision Number 350560

References

1. Saqr M, López-Pernas S (2023) The temporal dynamics of online problem-based learning: why and when sequence matters. Int J Comput-Support Collab Learn 18:11–37

2. Johnson AM, Azevedo R, D'Mello SK (2011) The temporal and dynamic nature of self-regulatory processes during independent and externally assisted hypermedia learning. Cogn Instr 29:471–504
3. Saqr M, Peeters W, Viberg O (2021) The relational, co-temporal, contemporaneous, and longitudinal dynamics of self-regulation for academic writing. Res Pract Technol Enhanc Learn 16:29
4. Chen B, Wise AF, Knight S, Cheng BH (2016) Putting temporal analytics into practice: the 5th international workshop on temporality in learning data. In: Proceedings of the sixth international conference on learning analytics & knowledge. ACM, New York, pp 488–489
5. Reimann P (2009) Time is precious: variable- and event-centred approaches to process analysis in CSCL research. Int J Comput-Support Collab Learn 4:239–257
6. Chen B, Poquet O (2020) Socio-temporal dynamics in peer interaction events. ACM international conference proceeding series, pp 203–208
7. López-Pernas S, Saqr M, Gordillo A, Barra E (2022) A learning analytics perspective on educational escape rooms. Interact Learn Environ 1–17
8. Saqr M, López-Pernas S (2022) How CSCL roles emerge, persist, transition, and evolve over time: A four-year longitudinal study. Comput Educ 189:104581
9. Saqr M, Peeters W (2022) Temporal networks in collaborative learning: a case study. Br J Educ Technol. https://doi.org/10.1111/bjet.13187
10. Holme P (2015) Modern temporal network theory: a colloquium. Eur Phys J B Condensed Matter Complex Syst 88. https://doi.org/10.1140/epjb/e2015-60657-4
11. Holme P, Saramäki J (2012) Temporal networks. Phys Rep 519:97–125
12. Holme P, Saramäki J (2019) A map of approaches to temporal networks. Temporal network theory. Springer, Cham, pp 1–24
13. Nicosia V, Tang J, Mascolo C, Musolesi M, Russo G, Latora V (2013) Graph metrics for temporal networks. In: Temporal networks. Springer, Berlin, pp 15–40
14. Li A, Cornelius SP, Liu Y-Y, Wang L, Barabási A-L (2017) The fundamental advantages of temporal networks. Science 358:1042–1046
15. Saqr M, Nouri J (2020) High resolution temporal network analysis to understand and improve collaborative learning. In: Proceedings of the tenth international conference on learning analytics & knowledge. ACM, New York, pp 314–319
16. Saqr M, López-Pernas S (2022) Instant or distant: a temporal network tale of two interaction platforms and their influence on collaboration. In: Educating for a new future: making sense of technology-enhanced learning adoption. Springer International Publishing, Berlin, pp 594–600
17. Saqr M, López-Pernas S, Conde MÁ, Hernández-García Á (2024) Social network analysis: a primer, a guide and a tutorial in r. In: Saqr M, López-Pernas S (eds) Learning analytics methods and tutorials: a practical guide using R. Springer, Berlin. https://doi.org/10.1007/978-3-031-54464-4
18. Bender-deMoll S (2018) Ndtv: network dynamic temporal visualizations. https://cran.r-project.org/package=ndtv
19. López-Pernas S, Saqr M, Del Rio L (2024) A broad collection of datasets for educational research training and application. In: Saqr M, López-Pernas S (eds) Learning analytics methods and tutorials: a practical guide using R. Springer, Berlin. https://doi.org/10.1007/978-3-031-54464-4
20. Vu D, Pattison P, Robins G (2015) Relational event models for social learning in MOOCs. Soc Netw 43:121–135
21. Hernández-García Á, González-González I, Jiménez-Zarco AI, Chaparro-Peláez J (2015) Applying social learning analytics to message boards in online distance learning: a case study. Comput Human Behav 47:68–80
22. Poquet O, Saqr M, Chen B (2021) Recommendations for network research in learning analytics: To open a conversation. In: Proceedings of the NetSciLA21 workshop
23. Saqr M, Elmoazen R, Tedre M, López-Pernas S, Hirsto L (2022) How well centrality measures capture student achievement in computer-supported collaborative learning? – a systematic review and meta-analysis. Educ Res Rev 35:100437

24. Saqr M, López-Pernas S (2022) The curious case of centrality measures: a large-scale empirical investigation. J. Learn Anal 9:13–31
25. López-Pernas S, Saqr M (2021) Bringing synchrony and clarity to complex multi-channel data: a learning analytics study in programming education. IEEE Access 1–1
26. Peeters W, Saqr M, Viberg O (2020) Applying learning analytics to map students' self-regulated learning tactics in an academic writing course. In: Proceedings of the 28th international conference on computers in education, vol 1, No. September. Asia-Pacific Society for Computers in Education, pp 245–254
27. Vartiainen H, López-Pernas S, Saqr M, Kahila J, Parkki T, Tedre M, Valtonen T (2022) Mapping students' temporal pathways in a computational thinking escape room. Proceedings. ISSN 1613:0073. http://ceur-wsorg
28. Heikkinen S, López-Pernas S, Malmberg J, Tedre M, Saqr M. How do business students self-regulate their project management learning? A sequence mining study. In: Saqr M, López-Pernas S (eds) Learning analytics methods and tutorials: a practical guide using R. Springer, Berlin. https://doi.org/10.1007/978-3-031-54464-4
29. Saqr M, López-Pernas S, Jovanović J, Gašević D (2023) Intense, turbulent, or wallowing in the mire: a longitudinal study of cross-course online tactics, strategies, and trajectories. Internet High Educ 57:100902
30. Elmoazen R, Saqr M, Tedre M, Hirsto L (2022) A systematic literature review of empirical research on epistemic network analysis in education. IEEE Access: Practical Innovations, Open Solutions 10:17330–17348
31. Saqr M, López-Pernas S (2022) The why, the what and the how to model a dynamic relational learning process with temporal networks. In: Proceedings of the NetSciLA22 workshop. https://www.researchgate.net/profile/Mohammed-Saqr/publication/364997941_The_Why_the_What_and_the_How_to_Model_a_Dynamic_Relational_Learning_Process_with_Temporal_Networks/links/636271222f4bca7fd0270b74/The-Why-the-What-and-the-How-to-Model-a-Dynamic-Relational-Learning-Process-with-Temporal-Networks.pdf
32. Brey A (2018) Temporal network analysis with r. The programming historian. https://doi.org/10.46430/phen0080
33. Bender-deMoll S (2016) Temporal network tools in statnet: networkDynamic, ndtv and tsna. statnet. https://web.archive.org/web/20180423112846/http://statnet.csde.washington.edu/workshops/SUNBELT/current/ndtv/ndtv_workshop.html
34. Bender-deMoll S, Morris M (2016) Tsna: Tools for temporal social network analysis. R package version 02 0. https://CRANR-projectorg/package=tsna
35. Butts C, Leslie-Cook A, Krivitsky P, Bender-deMoll S (2023) networkDynamic: dynamic extensions for network objects. R package version 0.11.4. https://statnet.org/

Epistemic Network Analysis and Ordered Network Analysis in Learning Analytics

Yuanru Tan, Zachari Swiecki, A. R. Ruis, and David Shaffer

1 Introduction

This chapter provides a tutorial on conducting *epistemic network analysis* (ENA) and *ordered network analysis* (ONA) using R. We introduce these two techniques together because they share similar theoretical foundations, but each addresses a different challenge for analyzing large-scale qualitative data on learning processes.

ENA and ONA are methods for quantifying, visualizing, and interpreting network data. Taking coded data as input, ENA and ONA represent associations between codes in undirected or directed weighted network models, respectively. Both techniques measure the strength of association among codes and illustrate the structure of connections in network graphs, and they quantify changes in the composition and strength of those connections over time. Importantly, ENA and ONA enable comparison of networks both visually and via summary statistics, so they can be used to explore a wide range of research questions in contexts where patterns of association in coded data are hypothesized to be meaningful and where comparing those patterns across individuals or groups is important.

In the following sections, we will (1) briefly review literature relevant to the application of ENA and ONA, (2) provide a step-by-step guide to implementing ENA and ONA in R, and (3) suggest additional resources and examples for further exploration. By the end of this chapter, readers will be able to apply these techniques in their own research.

Y. Tan (✉) · A. R. Ruis · D. Shaffer
University of Wisconsin, Madison, WI, USA
e-mail: yuanru.tan@wisc.edu

Z. Swiecki
Monash University, Clayton, Australia

M. Saqr, S. López-Pernas (eds.), *Learning Analytics Methods and Tutorials*,
https://doi.org/10.1007/978-3-031-54464-4_18

2 Literature Review

2.1 Epistemic Network Analysis (ENA)

ENA is a method for identifying and quantifying connections in coded data and representing them in undirected weighted network models [1]. There are two key features that differentiate ENA from other networks analysis tools or multivariate analyses: (1) ENA produces summary statistics that can be used to compare the differences in the content of networks rather than just their structure; and (2) ENA network visualizations provide information that is mathematically consistent with those summary statistics, which facilitates meaningful interpretation of statistical differences [2]. These features enable researchers to analyze a wide range of phenomena in learning analytics, including complex thinking and knowledge construction [3, 4], collaborative problem solving [5, 6], socio-emotional aspects of learning [7], mentoring [8], and teacher professional development [9–11].

One key feature that makes ENA an effective method in modeling collaborative interaction is that ENA can model individuals' unique contributions to collaborative discourse *while accounting for* group context, and thus both individuals *and* groups can be analyzed in the same model. This feature is particularly valuable in collaborative learning environments, where the interactions and contributions of each individual are related and should not be treated as a series of isolated events. For example, Swiecki et al. [6] analyzed the communications of air defense warfare teams in training exercises and found that ENA was not only able to reveal differences in individual performance identified in a qualitative analysis of the collaborative discourse, but also to test those differences statistically.

2.2 Ordered Network Analysis (ONA)

Ordered Network Analysis (ONA) extends the theoretical and analytical advantages of ENA to account for the order of events by producing *directed* weighted networks rather than undirected models [12]. Like ENA, ONA takes coded data as input, identifies and measures connections among coded items, and visualizes the structure of connections in a metric space that enables both statistical and visual comparison of networks. However, ONA models the order in which codes appear in the data, enabling analysis of phenomena in which the order of events is hypothesized to be important.

For example, Tan et al. [12] used ONA to model the performance of military teams learning to identify, assess, and respond to potential threats detected by radar. The findings demonstrate that ONA could detect qualitative differences between teams in different training conditions that were not detected with unordered models and show that they are statistically significant. In their work, Tan et al. [12] argued that ONA possesses an advantage over methods such as Sequential Pattern Mining

(SPM), which is widely used to identify frequent sequential patterns. In contrast to SPM, which prioritizes the specific micro-sequential order of events, ONA models processes by accounting for the co-temporal order of interactions between the units of analysis in response and what they are responding to. Consequently, ONA is a more appropriate methodological choice when modeling processes in ill-formed problem-solving scenarios, where collaborative interactions do not follow a prescribed sequence of steps but where the order of activities is still important.

ONA has also been used to analyze log data from online courses. For example, Fan et al. [13] analyzed self-regulated learning tactics employed by learners in Massive Open Online Courses (MOOC) using ONA and process mining. The authors found that ONA provided more nuanced interpretations of learning tactics compared to process mining because ONA models learning tactics across four dimensions: frequency, continuity, order, and the role of specific learning actions within broader tactics.

Like ENA, ONA produces summary statistics for network comparison and mathematically consistent network visualizations that enable interpretation of statistical measures. Unlike ENA, ONA models the order in which codes appear in data, enabling researchers to investigate whether and to what extent the order of events is meaningful in a given context.

In the following sections, we provide a step-by-step guide to conducting ENA and ONA analyses in R.

3 Epistemic Network Analysis in R

In this section, we demonstrate how to conduct an ENA analysis using the `rENA` package. If you are not familiar with ENA as an analytic technique, we recommend that you first read Shaffer [1], Shaffer and Ruis [14], and Bowman et al. [2] to familiarize yourself with the theoretical and methodological foundations of ENA.

3.1 Install the rENA Package and Load the Library

Before installing the `rENA` package, be sure that you are using R version 4.1 or newer. To check your R version, type `R.version` in your console. To update your R version (if needed), download and install R from the official R website: https://cran.r-project.org/

First, install the `rENA` package and then load the `rENA` library after installation is complete.

```
install.packages("rENA", repos = c("https://cran.qe-libs.org", "https://cran.
rstudio.org"))

library(rENA)
```

We also install the other package that is required for accessing the view() function Sect. 3.7.3 in rENA.

```
install.packages("tma", repos = c("https://cran.qe-libs.org", "https://cran.r
studio.org"))

library(tma )
```

3.2 Dataset

The dataset we will use as an example, `RS.data`, is included in the `rENA` package. Note that the `RS.data` file in the package is only a subset of the full dataset, and is thus intended for demonstration purposes only.

To start, pass `RS.data` from the rENA package to a data frame named **data**.

```
data = rENA::RS.data
```

Use the head() function in R to subset and preview the first three rows present in the input data frame to familiarize yourself with the data structure.

```
head(data,3)

##        UserName Condition CONFIDENCE.Pre CONFIDENCE.Post CONFIDENCE.Change
## 1      steven z FirstGame              7               8                 1
## 2       akash v FirstGame              6               8                 2
## 3 alexander b FirstGame              5               7                 1
##   C.Level.Pre NewC.Change    C.Change      Timestamp ActivityNumber GroupNa
me
## 1    High.Pre  Pos.Change Pos.Change 9/17/2013 9:43              1  Electr
ic
## 2    High.Pre  Pos.Change Pos.Change 9/17/2013 9:44              1  Electr
ic
## 3     Low.Pre  Pos.Change Pos.Change 9/17/2013 9:46              1  Electr
ic
##   GameHalf GameDay             text Data Technical.Constraints
## 1    First       1           Steven    0                     0
## 2    First       1 Hey, I am Akash    0                     0
## 3    First       1         I'm Alex    0                     0
##   Performance.Parameters Client.and.Consultant.Requests Design.Reasoning
## 1                      0                              0                0
## 2                      0                              0                0
## 3                      0                              0                0
##   Collaboration
## 1             0
## 2             0
## 3             0
```

`RS.data` consists of discourse from *RescuShell*, an online learning simulation where students work as interns at a fictitious company to solve a realistic engineering design problem in a simulated work environment. Throughout the internship, students communicate with their project teams and mentors via online chat, and these chats are recorded in the "text" column. A set of qualitative codes were applied to the data in the "text" column, where a value of 0 indicates the absence of the code and a value of 1 indicates the presence of the code in a given line.

Further details about the `RS.data` dataset can be found in Shaffer and Arastoopour [15]. Analyses of data from *RescuShell* and other engineering virtual internships can be found in Arastoopour et al. [16] and Chesler et al. [17].

3.3 Construct an ENA Model

To construct an ENA model, there is a function called `ena` which enables researchers to set the parameters for their model. This function wraps two other functions—`ena.accumulate.data` and `ena.make.set`—which can be used together to achieve the same result.

In the following sections, we will demonstrate how to set each parameter and explain how different choices affect the resulting ENA model.

3.3.1 Specify Units

In ENA, *units* can be individuals, ideas, organizations, or any other entity whose structure of connections you want to model. To set the units parameter, specify which column(s) in the data contain the variables that identify unique units.

For this example, choose the "Condition" column and the "UserName" column to define the units. The "Condition" column has two unique values: FirstGame, and SecondGame, representing novice users and relative expert users, respectively, as some students participated in *RescuShell* after having already completed a different engineering virtual internship. The "UserName" column includes unique user names for all students ($n = 48$). This way of defining the units means that ENA will construct a network for each student in each condition.

```
unitCols = c("Condition", "UserName")
```

To verify that the units are correctly specified, subset and preview the unique values in the units columns. There are 48 units from two conditions, which means that the ENA model will produce 48 individual-level networks for each of the units, and each unit is uniquely associated with either the novice group (FirstGame) or the relative expert group (SecondGame).

```
head(unique(data[, unitCols]),3)
```

```
##   Condition    UserName
## 1 FirstGame    steven z
## 2 FirstGame     akash v
## 3 FirstGame alexander b
```

3.3.2 Specify Codes

Next, specify the columns that contain the *codes*. Codes are concepts whose pattern of association you want to model for each unit. ENA represent codes as nodes in the networks and co-occurrences of codes as edges. Most researchers use binary coding in ENA analyses, where the values in the code columns are either 0 (indicating that the code is not present in that line) or 1 (indicating that the code is present in that line). `RS.data` contains six code columns, all of which will be used here.

To specify the code columns, enter the code column names in a vector.

```
codeCols = c('Data', 'Technical.Constraints', 'Performance.Parameters', 'Clie
nt.and.Consultant.Requests', 'Design.Reasoning', 'Collaboration')
```

To verify that the codes are correctly specified, preview the code columns selected.

```
head(data[,codeCols],3)
```

```
##   Data Technical.Constraints Performance.Parameters
## 1    0                     0                      0
## 2    0                     0                      0
## 3    0                     0                      0
##   Client.and.Consultant.Requests Design.Reasoning Collaboration
## 1                              0                0             0
## 2                              0                0             0
## 3                              0                0             0
```

3.3.3 Specify Conversations

The conversation parameter determines which lines in the data *can* be connected. Codes in lines that are not in the same conversation cannot be connected. For example, you may want to model connections within different time segments, such as days, or different steps in a process, such as activities.

In our example, choose the “Condition”, “GroupName”, and “ActivityNumber” columns to define the conversations. These choices indicate that connections can only happen between students who were in the same condition (FirstGame or SecondGame) and on the same project team (group), and within the same activity. This definition of conversation reflects what actually happened in the simulation: in a given condition, students only interacted with those who were in the same group, and each activity occurred on a different day.

To specify the conversation parameter, enter the column names in a vector.

```
conversationCols = c("Condition", "GroupName", "ActivityNumber")
```

To verify that the conversations are correctly specified, subset and preview the unique values in the conversation columns.

```
head(unique(data[, conversationCols]),3)

##    Condition GroupName ActivityNumber
## 1  FirstGame  Electric              1
## 12 FirstGame  Electric              3
## 15 FirstGame  Electric              4
```

3.3.4 Specify the Window

Once the conversation parameter is specified, a window method needs to be specified. Whereas the conversation parameter specifies which lines *can be* related, the window parameter determines which lines within the same conversation *are* related. The most common window method used in ENA is called a moving stanza window, which is what will be used here.

Briefly, a moving stanza window is a sliding window of fixed length that moves through a conversation to detect and accumulate code co-occurrences in recent temporal context. The lines within a designated stanza window are considered related to each other. For instance, if the moving stanza window is 7, then each line in the conversation is linked to the six preceding lines. See Siebert-Evenstone et al. [18] and Ruis et al. [19] for more detailed explanations of windows in ENA models.

Here, set the window.size.back[1]parameter equal to 7. User can specify a different moving stanza window size by passing a different numerical value to the 'window.size.back' parameter.

```
window.size.back = 7
```

3.3.5 Specify Groups and Rotation Method

When specifying the units, we chose a column that indicates two conditions: FirstGame (novice group) and SecondGame (relative expert group). To enable comparison of students in these two conditions, three additional parameters need to be specified: `groupVar`, `groups`, and `mean`.

[1] The ENA package also enables use of an infinite stanza window. The infinite stanza window works the same way as a moving stanza window, but there is no limit on the number of previous lines that are included in the window besides the conversation itself. The infinite stanza window is less commonly used in ENA, but is specified as follows: window.size.back = "INF".

```
groupVar = "Condition" # "Condition" is the column used as our grouping varia
ble
groups = c("FirstGame", "SecondGame") # "FirstGame" and "SecondGame" are the
two unique values of the "Condition" column
mean = TRUE
```

These three parameters indicate that when building the ENA model, the first dimension will maximize the difference between the two conditions: FirstGame and SecondGame. This difference maximization is achieved through `mean = TRUE`, which specifies that a *means rotation* will be performed at the dimensional reduction stage. If the means rotation is set to FALSE or there aren't two distinct groups in your data and you still set mean as TRUE, ENA will by default use singular value decomposition (SVD) to perform the dimensional reduction. Bowman et al. [2] provide a mathematical explanation of the methods used in ENA to perform dimensional reductions.

3.3.6 Specify Metadata

The last parameter to be specified is metadata. Metadata columns are not required to construct an ENA model, but they provide information that can be used to subset units in the resulting model.

Specify the metadata columns shown below to include data on student outcomes related to reported self-confidence before and after participating in engineering virtual internships. We will use this data to demonstrate a simple linear regression analysis that can be done using ENA outputs as predictors.

```
metaCols = c("CONFIDENCE.Change","CONFIDENCE.Pre","CONFIDENCE.Post","C.Change
") # optional
```

3.3.7 Construct an Model

Now that all the essential parameters have been specified, the ENA model can be constructed.

The `ena` function constructs the ENA model, and we recommend that you store the output in an object (in this case, **set.ena**).

```
set.ena =
  ena(
    data = data,
    units = unitCols,
    codes = codeCols,
    conversation = conversationCols,
    window.size.back = 7,
    metadata = metaCols, # optional
    groupVar = groupVar,
    groups = groups,
    mean = TRUE
    )
```

As noted above, the ena helper function combines the functions `ena.accumulate.data` and `ena.make.set`. The following code will construct the same ENA model specified above using these two functions.

```
accum.ena =
  ena.accumulate.data(
    text_data = RS.data[, 'text'],
    units = data[,unitCols],
    conversation = data[,conversationCols],
    metadata = data[,metaCols], # optional
    codes = data[,codeCols],
```

```
    window.size.back = 7
)

set.ena =
  ena.make.set(
    enadata = accum.ena, # the accumulation ran above
    rotation.by = ena.rotate.by.mean, # equivalent of mean=TRUE in the ena fu
nction
    rotation.params = list(
      accum.ena$meta.data$Condition=="FirstGame", # equivalent of groups in t
he ena function
      accum.ena$meta.data$Condition=="SecondGame" # equivalent of groups in t
he ena function
    )
)
```

3.4 *Summary of Key Model Outputs*

Users can explore what is stored in the object **set** by typing `set$` and select items from the drop down list. Here, we briefly describe the top-level items in `set` that are often of interest.

3.4.1 Connection Counts

Connection counts are the frequencies of unique connections a unit made. For each unit, ENA creates a cumulative adjacency vector that contains the sums of all unique code co-occurrences for that unit across all stanza windows. Here, there are 48 units in the ENA model, so there are 48 adjacency vectors. Each term in an ENA adjacency vector represents a unique co-occurrence of codes. Thus with six codes, each vector has 15 terms (*n* choose two). This is because ENA models are undirected and do not model co-occurrences of the same code.

To access ENA adjacency vectors, use `set.ena$connection.counts`.

```
head(set.ena$connection.counts,3)
```

```
##                   ENA_UNIT Condition    UserName CONFIDENCE.Change CONFIDENC
E.Pre
## 1:     FirstGame.steven z FirstGame    steven z                 1
7
## 2:      FirstGame.akash v FirstGame     akash v                 2
6
## 3: FirstGame.alexander b FirstGame alexander b                 1
5
##    CONFIDENCE.Post   C.Change Data & Technical.Constraints
## 1:               8 Pos.Change                           22
## 2:               8 Pos.Change                           47
## 3:               7 Pos.Change                            9
##    Data & Performance.Parameters Technical.Constraints & Performance.Param
eters
## 1:                            18
20
## 2:                            34
42
## 3:                             5
8
##    Data & Client.and.Consultant.Requests
## 1:                                     5
## 2:                                    10
## 3:                                     5
##    Technical.Constraints & Client.and.Consultant.Requests
## 1:                                                      6
## 2:                                                     14
## 3:                                                      3
##    Performance.Parameters & Client.and.Consultant.Requests
## 1:                                                       5
## 2:                                                      13
## 3:                                                       3
##    Data & Design.Reasoning Technical.Constraints & Design.Reasoning
## 1:                      21                                       26
## 2:                      45                                       59
## 3:                       5                                        8
##    Performance.Parameters & Design.Reasoning
## 1:                                         19
## 2:                                         38
## 3:                                          5
##    Client.and.Consultant.Requests & Design.Reasoning Data & Collaboration
## 1:                                                 6                    7
## 2:                                                 9                   12
## 3:                                                 2                    4
##    Technical.Constraints & Collaboration Performance.Parameters & Collabor
ation
## 1:                                     9
7
## 2:                                    21
11
## 3:                                     6
4
##    Client.and.Consultant.Requests & Collaboration
## 1:                                              1
## 2:                                              2
## 3:                                              0
##    Design.Reasoning & Collaboration
## 1:                                6
## 2:                               19
## 3:                                5
```

3.4.2 Line Weights

To compare networks in terms of their *relative* patterns of association, researchers can spherically normalize the cumulative adjacency vectors by diving each one by its length. The resulting normalized vectors represent each unit's relative frequencies of code co-occurrence. In other words, the sphere normalization controls for the fact that different units might have different amounts of interaction or different numbers of activities than others.

Notice that in `set.ena$connection.counts`, the value for each unique code co-occurrence is an integer equal or greater than 0, because they represent the raw connection counts between each unique pair of codes. In `set."ena"$line.weights`, those raw counts are normalized, and therefore the values are rational numbers between 0 and 1.

To access the normalized adjacency vectors, use `set.ena$line.weights`.

```
head(set.ena$line.weights,3)
```

```
##                 ENA_UNIT Condition     UserName CONFIDENCE.Change CONFIDENC
E.Pre
## 1:     FirstGame.steven z FirstGame     steven z                 1
7
## 2:      FirstGame.akash v FirstGame      akash v                 2
6
## 3: FirstGame.alexander b FirstGame alexander b                 1
5
##    CONFIDENCE.Post   C.Change Data & Technical.Constraints
## 1:               8 Pos.Change                    0.4000661
## 2:               8 Pos.Change                    0.4016067
## 3:               7 Pos.Change                    0.4370786
##    Data & Performance.Parameters Technical.Constraints & Performance.Param
eters
## 1:                     0.3273268                                      0.36
36965
## 2:                     0.2905240                                      0.35
88826
## 3:                     0.2428215                                      0.38
85143
##    Data & Client.and.Consultant.Requests
## 1:                            0.09092412
## 2:                            0.08544824
## 3:                            0.24282147
##    Technical.Constraints & Client.and.Consultant.Requests
## 1:                                              0.1091089
## 2:                                              0.1196275
## 3:                                              0.1456929
##    Performance.Parameters & Client.and.Consultant.Requests
## 1:                                               0.09092412
## 2:                                               0.11108271
## 3:                                               0.14569288
```

```
##     Data & Design.Reasoning Technical.Constraints & Design.Reasoning
## 1:               0.3818813                                0.4728054
## 2:               0.3845171                                0.5041446
## 3:               0.2428215                                0.3885143
##     Performance.Parameters & Design.Reasoning
## 1:                                  0.3455117
## 2:                                  0.3247033
## 3:                                  0.2428215
##     Client.and.Consultant.Requests & Design.Reasoning Data & Collaboration
## 1:                                         0.10910895            0.1272938
## 2:                                         0.07690342            0.1025379
## 3:                                         0.09712859            0.1942572
##     Technical.Constraints & Collaboration Performance.Parameters & Collabor
ation
## 1:                              0.1636634                             0.127
29377
## 2:                              0.1794413                             0.093
99306
## 3:                              0.2913858                             0.194
25717
##     Client.and.Consultant.Requests & Collaboration
## 1:                                      0.01818482
## 2:                                      0.01708965
## 3:                                      0.00000000
##     Design.Reasoning & Collaboration
## 1:                         0.1091089
## 2:                         0.1623517
## 3:                         0.2428215
```

3.4.3 ENA Points

As the product of a dimensional reduction, for each unit, ENA produces an ENA point in a two-dimensional space. Since there are 48 units, ENA produces 48 ENA points.

By default, rENA visualizes ENA points on an x-y coordinate plane defined by the first two dimensions of the dimensional reduction: for a means rotation, MR1 and SVD2, and for an SVD, SVD1 and SVD2.

To access these points, use `set.ena$points`.

```
head(set.ena$points,3)
```

```
##                  ENA_UNIT Condition     UserName CONFIDENCE.Change CONFIDENC
E.Pre
## 1:     FirstGame.steven z FirstGame     steven z                 1
7
## 2:      FirstGame.akash v FirstGame      akash v                 2
6
## 3: FirstGame.alexander b FirstGame alexander b                 1
5
##     CONFIDENCE.Post    C.Change            MR1           SVD2           SVD3
```

```
## 1:               8 Pos.Change -0.05423338 -0.008491458  0.06551249
## 2:               8 Pos.Change -0.07742095  0.031134440  0.03362490
## 3:               7 Pos.Change -0.30594927 -0.098348499 -0.01105519
##            SVD4         SVD5        SVD6        SVD7        SVD8
SVD9
## 1:  2.034477e-02  0.011885463 -0.02427483 -0.023161244 -0.01643227 -0.0128
85771
## 2: -2.531589e-05  0.006465571  0.01336324  0.001215593 -0.01390223  0.0040
71313
## 3:  9.816549e-02 -0.003662261  0.07609149  0.077059745 -0.09198643 -0.0624
49678
##           SVD10       SVD11        SVD12        SVD13       SVD14       SV
D15
## 1: -0.01169155  0.005448827 -0.027880312 -0.006140204  0.01230336  0.01989
532
## 2: -0.04017379  0.035710843 -0.011559722  0.002192745  0.02967617 -0.01100
021
## 3: -0.01146149 -0.031675014  0.003382794  0.026665665 -0.01779259 -0.01083
445
```

ENA points are thus summary statistics that researchers can use to conduct statistical tests, and they can also be used in subsequent analyses. For example, statistical differences between groups in the data can be tested using ENA dimension scores, and those scores can also be used in regression analyses to predict outcome variables, which we will demonstrate later.

3.4.4 Rotation Matrix

The rotation matrix used during the dimensional reduction can be accessed through `set.ena$rotation`. This is mostly useful when you want to construct an ENA metric space using one dataset and then project ENA points from different data into that space, as in Sect. 5.1.

```
head(set.ena$rotation.matrix,3)
```

```
##                                              codes          MR1          SVD
2
## 1:                    Data & Technical.Constraints -0.297140113  0.25542803
7
## 2:                  Data & Performance.Parameters  0.146745148 -0.40786334
0
## 3: Technical.Constraints & Performance.Parameters  0.006251295 -0.00672433
8
##           SVD3      SVD4        SVD5        SVD6       SVD7        SVD8
## 1:  0.4027380 0.1829314 -0.18841712 -0.22238612  0.3539426 -0.04618008
## 2:  0.4998255 0.1655853 -0.01271575  0.02831357 -0.5461929 -0.19336181
## 3: -0.0357799 0.4901334  0.29999562  0.27301947  0.1197999  0.45958816
##           SVD9      SVD10       SVD11         SVD12      SVD13      SVD14
## 1: 0.39593034  0.3556577 -0.03738872 -0.0005144122 0.30025177 0.24188117
## 2: 0.27237864 -0.1917366  0.18450254 -0.1247520324 0.17795795 0.06161493
```

```
## 3: 0.01614337 -0.2375881 -0.25630008 -0.3668342035 0.05404011 0.25435787
##          SVD15
## 1: 0.07161604
## 2: 0.04655934
## 3: 0.20907031
```

3.4.5 Metadata

`set$meta.data` returns a data frame that includes all the columns of the ENA set except for the columns representing code co-occurrences.

```
head(set.ena$meta.data,3)
```

```
##                  ENA_UNIT Condition    UserName CONFIDENCE.Change CONFIDENC
E.Pre
## 1:     FirstGame.steven z FirstGame    steven z                 1
7
## 2:      FirstGame.akash v FirstGame     akash v                 2
6
## 3: FirstGame.alexander b FirstGame alexander b                 1
5
##    CONFIDENCE.Post   C.Change
## 1:               8 Pos.Change
## 2:               8 Pos.Change
## 3:               7 Pos.Change
```

3.5 *ENA Visualization*

Once an ENA set is constructed, it can be visualized, which facilitates interpretation of the model. Here, we will look at the two conditions, "FirstGame" (novices) and "SecondGame" (relative experts), by plotting their mean networks.

3.5.1 Plot a Mean Network

To plot a network, use the `ena.plot.network` function. This function requires the `network` parameter (a character vector of line weights), and the line weights come from `set$line.weights`.

First, subset line weights for each of the two groups.

```
# Subset lineweights for FirstGame
first.game.lineweights = as.matrix(set.ena$line.weights$Condition$FirstGame)

# Subset lineweights for SecondGame
second.game.lineweights = as.matrix(set.ena$line.weights$Condition$SecondGame)
```

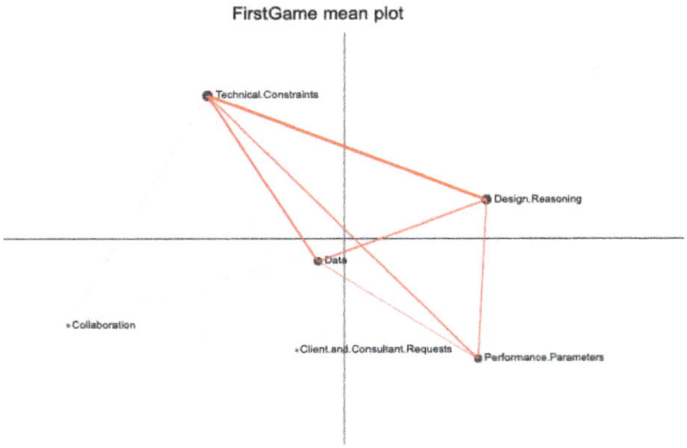

Fig. 1 ENA mean network for FirstGame group

Next, calculate the mean networks for the two groups, and store the line weights as vectors.

```
first.game.mean = as.vector(colMeans(first.game.lineweights))
second.game.mean = as.vector(colMeans(second.game.lineweights))
```

During plotting, use a pipe | > to send the output of one function into the first parameter of the subsequent function. To distinguish the two mean networks, set the color of the FirstGame mean network to red (Fig. 1) .

```
ena.plot(set.ena, title = "FirstGame mean plot")  |>
  ena.plot.network(network = first.game.mean, colors = c("red"))
```

and the color of the SecondGame mean network to blue (Fig. 2).

```
ena.plot(set.ena, title = "SecondGame mean plot")  |>
  ena.plot.network(network = second.game.mean, colors = c("blue"))
```

As you can see from the two network visualizations above, their node positions are exactly same. All ENA networks from the same model have the same node positions, which are determined by an optimization routine that attempts to place the nodes such that the centroid of each unit's network and the location of the ENA point in the reduced space are co-located.

Because of the fixed node positions, ENA can construct a subtracted network, which enables the identification of the most salient differences between two networks. To do this, ENA subtracts the weight of each connection in one network from the corresponding weighted connection in another network, then visualizes the differences in connection strengths. Each edge is color-coded to indicate which of

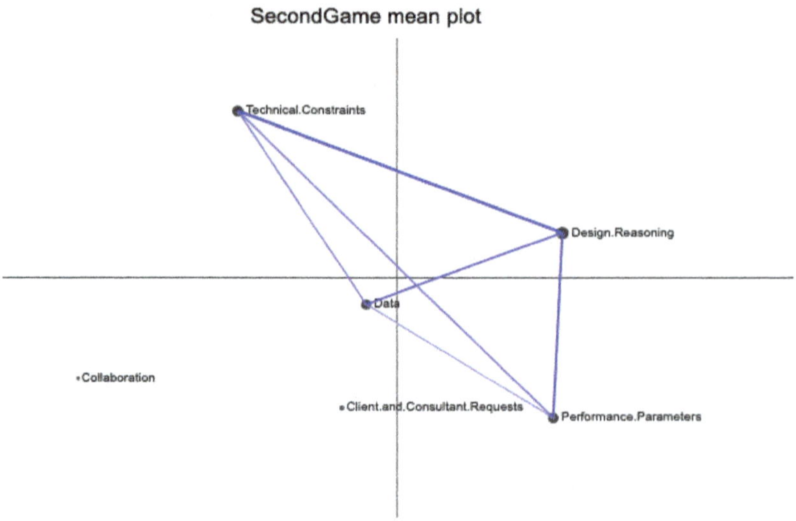

Fig. 2 ENA mean network for SecondGame group

the two networks contains the stronger connection, and the thickness and saturation of the edges corresponds to the magnitude of the difference.

To plot a subtracted network, first calculate the subtracted network line weights by subtracting one group's line weights from the other. (Because ENA computes the absolute values of the differences in edge weights, the order of the two networks in the subtraction doesn't matter.)

```
subtracted.mean = first.game.mean - second.game.mean
```

Then, use the ena.plot function to plot the subtracted network. If the differences are relatively small, a multiplier can be applied to rescale the line weights, improving legibility (Fig. 3).

```
ena.plot(set.ena, title = "Subtracted: FirstGame (red) - SecondGame (blue)")
|>
  ena.plot.network(network = subtracted.mean * 5, # Optional rescaling of the
line weights
                   colors = c("red", "blue"))
```

Here, the subtracted network shows that on average, students in the FirstGame condition (red) made more connections with Technical.Constraints and Collaboration than students in the SecondGame condition (blue), while students in the SecondGame condition made more connections with Design.Reasoning and Performance.Parameters than students in the FirstGame condition. This is because

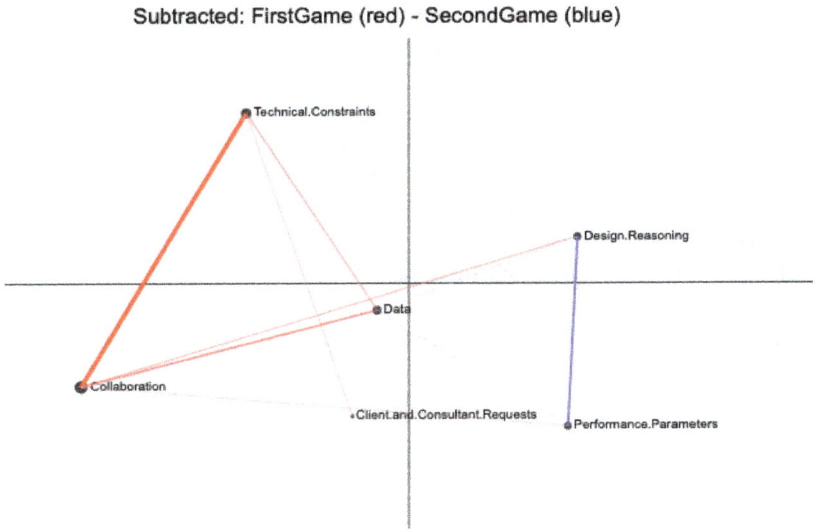

Fig. 3 ENA subtracted mean network for FirstGame (red) and SecondGame (blue)

students with more experience of engineering design practices did not need to spend as much time and effort managing the collaborative process and learning about the basic technical elements of the problem space, and instead spent relatively more time focusing on more complex analysis and design reasoning tasks.

Note that this subtracted network shows no connection between Technical.Constraints and Design.Reasoning, simply because the strength of this connection was similar in both conditions. Thus, subtraction networks should always be visualized along with with the two networks being subtracted.

3.5.2 Plot a Mean Network and its Points

The ENA point or points associated with a network or mean network can also be visualized.

To visualize the points associated with each of the mean networks plotted above, use `set$points` to subset the rows that are in each condition and plot each condition as a different color.

```
# Subset rotated points for the first condition
first.game.points = as.matrix(set.ena$points$Condition$FirstGame)

# Subset rotated points for the second condition
second.game.points = as.matrix(set.ena$points$Condition$SecondGame)
```

Then, plot the FirstGame mean network the same as above using `ena.plot.network`, use | > to pipe in the FirstGame points that we want to include, and plot them using `ena.plot.points`.

Each point in the space is the ENA point for a given unit. The red and blue squares on the x-axis are the means of the ENA points for each condition, along with the 95% confidence interval on each dimension. (might need to zoom in for better readability).

Since we used a means rotation to construct the ENA model, the resulting space highlights the differences between FirstGame and SecondGame by constructing a rotation that places the means of each condition as close as possible to the x-axis of the space and maximizes the distance between them (Figs. 4, 5, 6, 7, and 8).

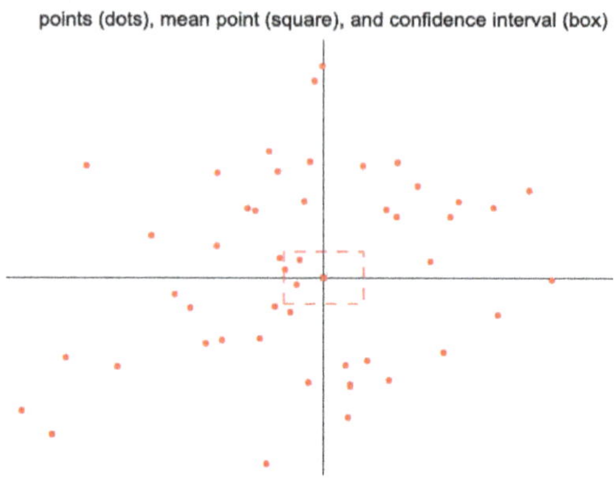

Fig. 4 FirstGame ENA points (dots), mean point (square), and confidence interval (box)

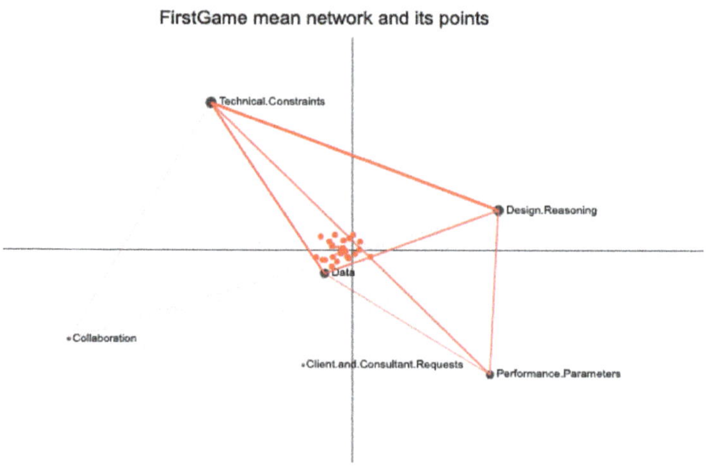

Fig. 5 FirstGame ENA mean network and points

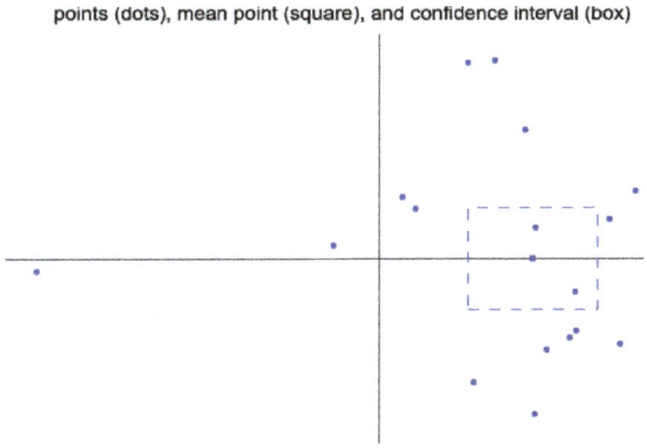

Fig. 6 SecondGame ENA points (dots), mean point (square), and confidence interval (box)

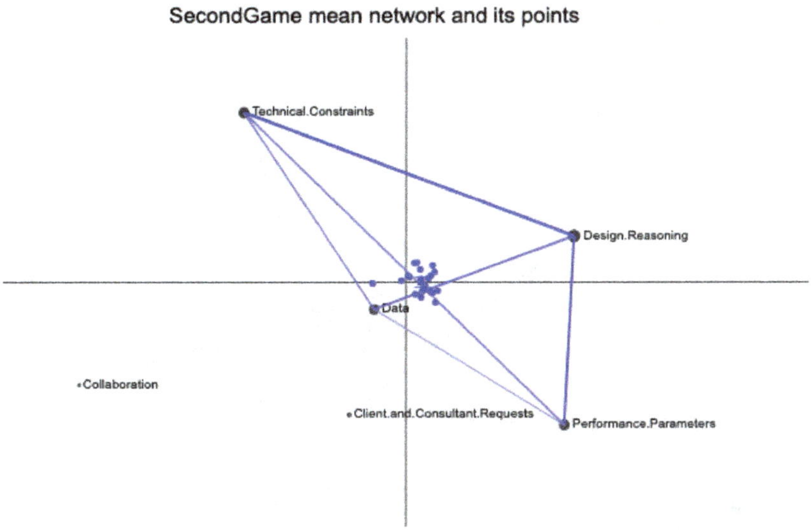

Fig. 7 SecondGame ENA mean network and points

```
ena.plot(set.ena, title = "points (dots), mean point (square), and confidenc
e interval (box)") |>
        ena.plot.points(points = first.game.points, colors = c("red")) |>
        ena.plot.group(point = first.game.points, colors =c("red"),
                    confidence.interval = "box")
```

```
ena.plot(set.ena, title = "FirstGame mean network and its points") |>
        ena.plot.network(network = first.game.mean, colors = c("red")) |>
        ena.plot.points(points = first.game.points, colors = c("red")) |>
        ena.plot.group(point = first.game.points, colors =c("red"),
                    confidence.interval = "box")
```

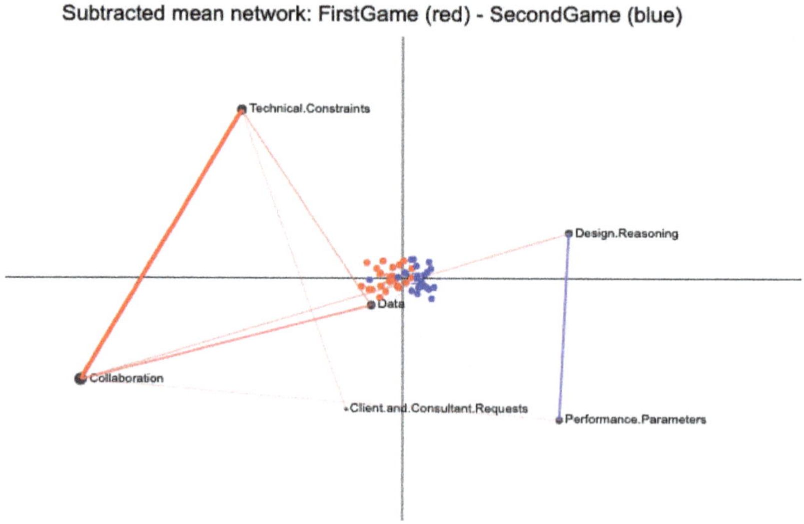

Fig. 8 ENA subtracted mean network for FirstGame (blue) and SecondGame (red)

Then, do the same for the SecondGame condition.

```
ena.plot(set.ena, title = " points (dots), mean point (square), and confidenc
e interval (box)") |>
          ena.plot.points(points = second.game.points, colors = c("blue")) |>
          ena.plot.group(point = second.game.points, colors =c("blue"),
                         confidence.interval = "box")
```

```
ena.plot(set.ena, title = "SecondGame mean network and its points") |>
          ena.plot.network(network = second.game.mean, colors = c("blue")) |>
          ena.plot.points(points = second.game.points, colors = c("blue")) |>
          ena.plot.group(point = second.game.points, colors =c("blue"),
                         confidence.interval = "box")
```

Lastly, do the same for subtraction as well.

```
ena.plot(set.ena, title = "Subtracted mean network: FirstGame (red) - SecondG
ame (blue)")  |>
          ena.plot.network(network = subtracted.mean * 5,
          colors = c("red", "blue")) |>
          ena.plot.points(points = first.game.points, colors = c("red")) |>
          ena.plot.group(point = first.game.points, colors =c("red"),
                         confidence.interval = "box") |>
          ena.plot.points(points = second.game.points, colors = c("blue")) |>
          ena.plot.group(point = second.game.points, colors =c("blue"),
                         confidence.interval = "box")
```

Note that the majority of the red points (FirstGame) are located on the left side of the space, and the blue points (SecondGame) are mostly located on the right side of the space. This is consistent with the line weights distribution in the mean network: the FirstGame units make relatively more connections with nodes on the left side of the space, while the SecondGame units make relatively more connections with nodes on the right side of the space. The positions of the nodes enable interpretation of the dimensions, and thus interpretation of the locations of the ENA points.

3.5.3 Plot an Individual Unit Network and its Point

Plotting the network and ENA point for a single unit uses the same approach. First, subset the line weights and point for a given unit.

```
unit.A.line.weights = as.matrix(set.ena$line.weights$ENA_UNIT$`FirstGame.stev
en z`) # subset line weights
unit.A.point = as.matrix(set.ena$points$ENA_UNIT$`FirstGame.steven z`) # subs
et ENA point
```

Then, plot the network and point for that unit (Fig. 9).

```
ena.plot(set.ena, title = "Individual network: FirstGame.steven z") |>
          ena.plot.network(network = unit.A.line.weights, colors = c("red"))
|>
          ena.plot.points(points = unit.A.point, colors = c("red"))
```

Following the exact same procedure, we can, for example, choose a unit from the other condition to plot and also construct a subtracted plot for those two units (Fig. 10).

```
unit.B.line.weights = as.matrix(set.ena$line.weights$ENA_UNIT$`SecondGame.sam
uel o`) # subset line weights
unit.B.point = as.matrix(set.ena$points$ENA_UNIT$`SecondGame.samuel o`) # sub
set ENA point

ena.plot(set.ena, title = "Individual network: SecondGame.samuel o") |>
          ena.plot.network(network = unit.B.line.weights, colors = c("blue"))
|>
          ena.plot.points(points = unit.B.point, colors = c("blue"))
```

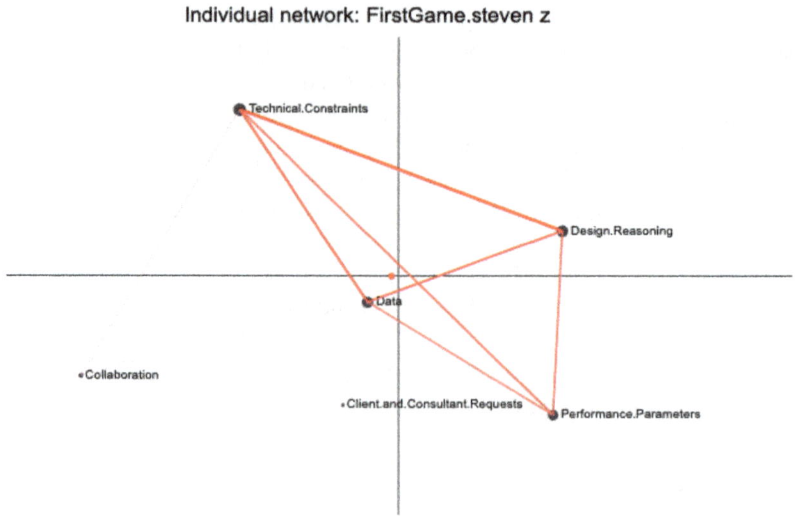

Fig. 9 EAN network for a student from FirstGame and its corresponding ENA point

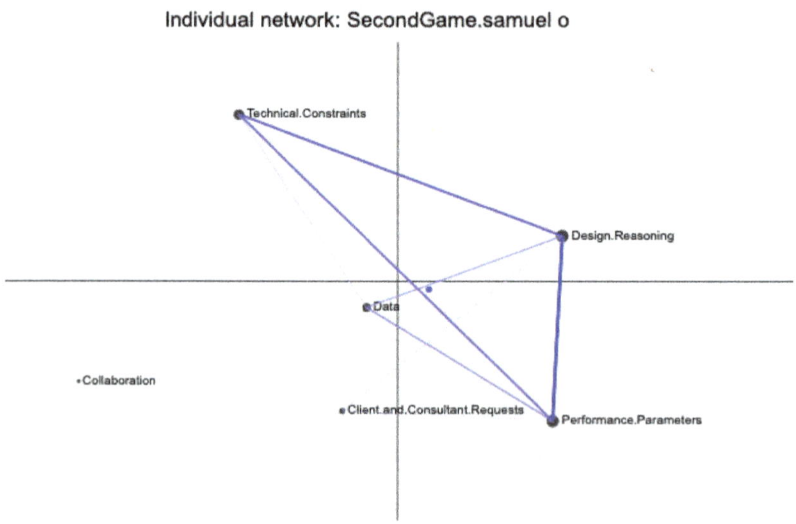

Fig. 10 ENA network for a student from SecondGame and its corresponding ENA point

To visually analyze the differences between the two individual networks, plot their subtracted network (Fig. 11).

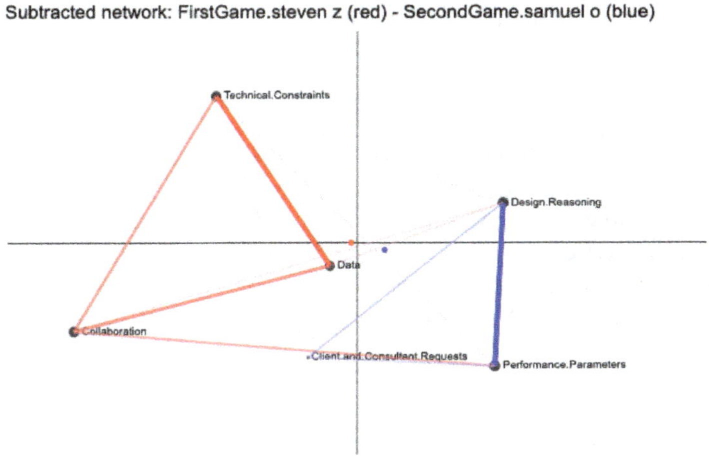

Fig. 11 ENA subtracted network showing the differences between one student from FirstGame (red) and another student from SecondGame (blue)

```
ena.plot(set.ena, title = "Subtracted network: FirstGame.steven z (red) - Sec
ondGame.samuel o (blue)") |>
          ena.plot.network(network = (unit.A.line.weights - unit.B.line.weigh
ts) * 5,
          colors = c("red", "blue")) |>
          ena.plot.points(points = unit.A.point, colors = c("red")) |>
          ena.plot.points(points = unit.B.point, colors = c("blue"))
```

In this unit-level subtracted network, Unit A (red) made relatively more connections with codes such as Technical.Constraints, Data, and Collaboration, while Unit B (blue) made relatively more connections with Design.Reasoning and Performance.Parameters.

3.5.4 Plot Everything, Everywhere, All at Once

The helper function `ena.plotter` enables users to plot points, means, and networks for each condition at the same time. This gives the same results as above more parsimoniously. However, this approach does not enable customization of edge and point colors.

```
#with helper function
plot = ena.plotter(set.ena,
                      points = T,
                      mean = T,
                      network = T,
                      print.plots = T,
                      groupVar = "Condition",
                      groups = c("SecondGame","FirstGame"),
                      subtractionMultiplier = 5)
```

3.6 Compare Groups Statistically

In addition to visual comparison of networks, ENA points can be analyzed statistically. For example, here we might test whether the patterns of association in one condition are significantly different from those in the other condition.

To demonstrate both parametric and non-parametric approaches to this question, the examples below use a Student's *t* test and a Mann-Whitney *U* test to test for differences between the FirstGame and SecondGame condition. For more on differences between parametric and non-parametric tests, see Kaur and Kumar [20].

First, install the `lsr` package to enable calculation of effect size (Cohen's *d*) for the *t* test.

```
install.packages('lsr')
library(lsr)
```

Then, subset the points to test for differences between the points of the two conditions.

```
ena_first_points_d1 = as.matrix(set.ena$points$Condition$FirstGame)[,1]
ena_second_points_d1 = as.matrix(set.ena$points$Condition$SecondGame)[,1]

ena_first_points_d2 = as.matrix(set.ena$points$Condition$FirstGame)[,2]
ena_second_points_d2 = as.matrix(set.ena$points$Condition$SecondGame)[,2]
```

Conduct the *t* test on the first and second dimensions.

```
# parametric tests
t_test_d1 = t.test(ena_first_points_d1, ena_second_points_d1)
t_test_d1

##
##  Welch Two Sample t-test
##
## data:  ena_first_points_d1 and ena_second_points_d1
## t = -6.5183, df = 45.309, p-value = 5.144e-08
## alternative hypothesis: true difference in means is not equal to 0
## 95 percent confidence interval:
##  -0.2687818 -0.1419056
## sample estimates:
##   mean of x   mean of y
## -0.09411588  0.11122786

t_test_d2 = t.test(ena_first_points_d2, ena_second_points_d2)
t_test_d2

##
##  Welch Two Sample t-test
##
## data:  ena_first_points_d2 and ena_second_points_d2
## t = 1.9334e-16, df = 43.175, p-value = 1
## alternative hypothesis: true difference in means is not equal to 0
```

```
## 95 percent confidence interval:
##  -0.07768526  0.07768526
## sample estimates:
##     mean of x     mean of y
##  1.935145e-19 -7.254914e-18
```

Compute any other statistics that may be of interest. A few examples are given below.

```
mean(ena_first_points_d1)
## [1] -0.09411588
mean(ena_second_points_d1)
## [1] 0.1112279
mean(ena_first_points_d2)
## [1] 1.935145e-19
mean(ena_second_points_d2)
## [1] -7.254914e-18
sd(ena_first_points_d1)
## [1] 0.1115173
sd(ena_second_points_d1)
## [1] 0.1063515
sd(ena_first_points_d2)
## [1] 0.1267104
sd(ena_second_points_d2)
## [1] 0.1380851
length(ena_first_points_d1)
## [1] 26
length(ena_second_points_d1)
## [1] 22
length(ena_first_points_d2)
## [1] 26
length(ena_second_points_d2)
## [1] 22
```

```
cohensD(ena_first_points_d1, ena_second_points_d1)
```

```
## [1] 1.880622
```

```
cohensD(ena_first_points_d2, ena_second_points_d2)
```

```
## [1] 5.641688e-17
```

Here, along the x axis (MR1), a two-sample t test assuming unequal variance shows that the FirstGame (mean = −0.09, SD = 0.11, N = 26) condition is statistically significantly different for alpha = 0.05 from the SecondGame condition (mean = 0.11, SD = 0.10, N = 22; t(45.31) = −6.52, p = 0.00, Cohen's d = 1.88). Along the y axis (SVD2), a two-sample t test assuming unequal variance shows that the FirstGame condition (mean = 0.11, SD = 0.13, N = 26) is not statistically significantly different for alpha = 0.05 from the SecondGame condition (mean = 0.00, SD = 1.3, N = 22; t(43.17) = 0, p = 1.00).

The Mann-Whitney U test is a non-parametric alternative to the independent two-sample t test.

First, install the `rcompanion` package to calculate the effect size (r) for a Mann-Whitney U test.

```
install.packages('rcompanion')
library(rcompanion)
```

Then, conduct a Mann-Whitney U test on the first and second dimensions.

```
# non parametric tests
w_test_d1 = wilcox.test(ena_first_points_d1, ena_second_points_d1)
w_test_d2 = wilcox.test(ena_first_points_d2, ena_second_points_d2)

w_test_d1
```

```
##
##  Wilcoxon rank sum exact test
##
## data:  ena_first_points_d1 and ena_second_points_d1
## W = 50, p-value = 8.788e-08
## alternative hypothesis: true location shift is not equal to 0
```

```
w_test_d2
```

```
##
##  Wilcoxon rank sum exact test
##
## data:  ena_first_points_d2 and ena_second_points_d2
## W = 287, p-value = 0.9918
## alternative hypothesis: true location shift is not equal to 0
```

Compute any other statistics that may be of interest. A few examples are given below.

```
median(ena_first_points_d1)
```

```
## [1] -0.08464154
```

```
median(ena_second_points_d1)
```

```
## [1] 0.1300029
```

```
median(ena_first_points_d2)
```

```
## [1] -0.007252397
```

```
median(ena_second_points_d2)
```

```
## [1] 0.0003031848
```

```
length(ena_first_points_d1)
```

```
## [1] 26
```

```
length(ena_second_points_d1)
```

```
## [1] 22
```

```
length(ena_first_points_d2)
```

```
## [1] 26
```

```
length(ena_second_points_d2)
```

```
## [1] 22
```

```
abs(wilcoxonR(ena_first_points_d1, ena_second_points_d1))
```

```
##     r
## 0.863
```

```
abs(wilcoxonR(ena_first_points_d2, ena_second_points_d2))
```

```
##     r
## 0.863
```

Here, along the x axis (MR1), a Mann-Whitney U test shows that the FirstGame condition (Mdn $= -0.08$, N $= 26$) was statistically significantly different for alpha $= 0.05$ from the SecondGame condition (Mdn $= -0.007$, N $= 22$; $U = 50$, $p = 0.00$, $r = 0.86$). Along the y axis (SVD2), a Mann-Whitney U test shows that the FirstGame condition (Mdn $= 0.13$, N $= 26$) is not statistically significantly different for alpha $= 0.05$ from the SecondGame condition (Mdn $= 0.00$, N $= 22$; U $= 287$, $p = 0.99$). The absolute value of `r` value in Mann-Whitney U test varies from 0 to close to 1. The interpretation values for r commonly in published literature is: `0.10 - < 0.3` (small effect), `0.30 - < 0.5` (moderate effect) and > = `0.5` (large effect).

3.7 *Model Evaluation*

In this section, we introduce three ways users can evaluate the quality of their ENA models.

3.7.1 Variance Explained

Briefly, variance explained (also called explained variation) refers to the proportion of the total variance in a dataset that is accounted for by a statistical model or set of predictors.

In ENA, to represent high-dimensional vectors in a two-dimensional space, ENA uses either singular value decomposition or means rotation combined with SVD. For each of the reduced dimensions, the variance in patterns of association among units explained by that dimension can be computed.

```
head(set.ena$model$variance,2)
```

```
##       MR1      SVD2
## 0.3204602 0.2445006
```

Here, the first dimension is MR1 and the second dimension is SVD2. The MR1 dimension has the highest variance explained at 32%.

As with any statstical model, greater explained variance does not necessarily indicate a better model, as it may be due to overfitting, but it provides one indicator of model quality.

3.7.2 Goodness of Fit

Briefly, a model's goodness of fit refers to how well a model fits or represents the data. A model with a high goodness of fit indicates that it accurately represents the data and can make reliable predictions.

In ENA, a good fit means that the positions of the nodes in the space—and thus the network visualizations—are consistent with the mathematical properties of the model. In other words, we can confidently rely on the network visualizations to interpret the ENA model. The process that ENA uses to achieve high goodness of fit is called co-registration. The mathematical details of co-registration are beyond the scope of this chapter and can be found in Bowman et al. [2].

To test a model's goodness of fit, use `ena.correlations`. The closer the value is to 1, the higher the model's goodness of fit is. Most ENA models have a goodness of fit that is well above 0.90.

```
ena.correlations(set.ena)
```

```
##      pearson  spearman
## 1  0.993766  0.994119
## 2 0.9850392 0.9850519
```

3.7.3 Close the Interpretative Loop

Another approach to evaluate an ENA model is to confirm the alignment between quantitative model (in our case, our ENA model) and the original qualitative data. In other words, we can return to the original data to confirm that quantitative findings give a fair representation of the data. This approach is an example of what's called as closing the interpretative loop in Quantitative Ethnography field [1].

For example, based on our visual analysis of the network of "SecondGame.samuel o" in previous section, we are interested in what the lines are in the original data that contributed to the connection between Design.Reasoning and Performance.Parameters.

Let's first review what "SecondGame.samuel o" ENA network looks like (Fig. 12).

```
ena.plot(set.ena, title = "Individual network: SecondGame.samuel o") |>
          ena.plot.network(network = as.matrix(set.ena$line.weights$ENA_UNIT$
`SecondGame.samuel o`), colors = c("blue")) |>
          ena.plot.points(points = as.matrix(set.ena$points$ENA_UNIT$`SecondG
ame.samuel o`), colors = c("blue"))
```

To do so, we use `view()` function and specify required parameters as below.

This is going to activate a window shows up in your `Viewer` panel. If it is too small to read, you can click on the "Show in new window" button to view it in your browser for better readability.

In the `Viewer` panel, hover over your cursor on any of the lines that are in bold, a size of 7 lines rectangle shows up, representing that in a moving stanza window of size 7, the referent line (the line in bold) and its preceding 6 lines. The 1 and 0 in Technical.Constraints column and Design.Reasoning column shows where the connections happened (Fig. 13).

For example, line 2477 Samuel shared his [Design.Reasoning] about "mindful of (the) how one device scores relative to other ones", to reference back to what Casey said in line 2476 about [Performance.Parameters] "not one source/censor can be the best in every area so we had to sacrifice certain attributes", as well as what Jackson said in line 2475 about safety as one of the [Performance.Parameters] "when it came to the different attributes, i think that all were important in their own way but i think safety is one of the most important".

This is a qualitative example of a connection made between Performance.Parameters and Design.Reasoning.

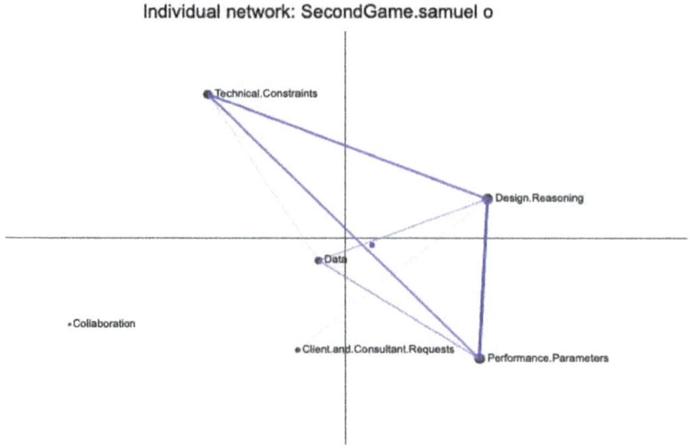

Fig. 12 ENA network for Samuel O from SecondGame

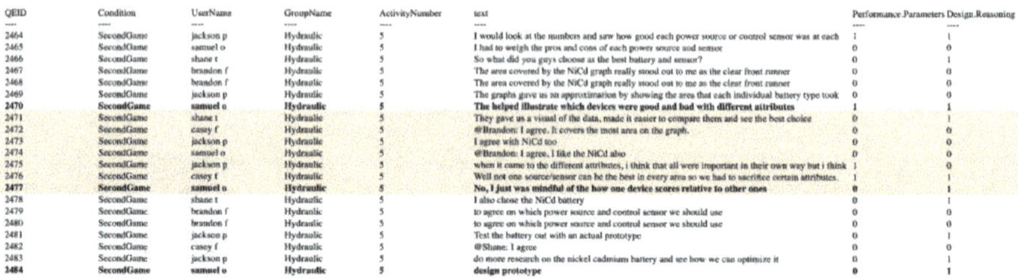

QEID	Condition	UserName	GroupName	ActivityNumber	text	Performance.Parameters	Design.Reasoning
----	----	----	----	----	----	----	----
2464	SecondGame	jackson p	Hydraulic	5	I would look at the numbers and saw how good each power source or control sensor was at each	1	1
2465	SecondGame	samuel o	Hydraulic	5	I had to weigh the pros and cons of each power source and sensor	0	0
2466	SecondGame	shane t	Hydraulic	5	So what did you guys choose as the best battery and sensor?	0	1
2467	SecondGame	brandon f	Hydraulic	5	The area covered by the NiCd graph really stood out to me as the clear front runner	0	0
2468	SecondGame	brandon f	Hydraulic	5	The area covered by the NiCd graph really stood out to me as the clear front runner	0	0
2469	SecondGame	jackson p	Hydraulic	5	The graphs gave us an approximation by showing the area that each individual battery type took	0	0
2470	**SecondGame**	**samuel o**	**Hydraulic**	**5**	**The helped illustrate which devices were good and bad with different attributes**	**1**	**1**
2471	SecondGame	shane t	Hydraulic	5	They gave us a visual of the data, made it easier to compare them and see the best choice	0	1
2472	SecondGame	casey f	Hydraulic	5	@Brandon: I agree. It covers the most area on the graph.	0	0
2473	SecondGame	jackson p	Hydraulic	5	I agree with NiCd too	0	0
2474	SecondGame	samuel o	Hydraulic	5	@Brandon: I agree, I like the NiCd also	0	0
2475	SecondGame	jackson p	Hydraulic	5	when it came to the different attributes, i think that all were important in their own way but i think	1	0
2476	SecondGame	casey f	Hydraulic	5	Well not one source/sensor can be the best in every area so we had to sacrifice certain attributes.	1	1
2477	**SecondGame**	**samuel o**	**Hydraulic**	**5**	**No, I just was mindful of the how one device scores relative to other ones**	**0**	**1**
2478	SecondGame	shane t	Hydraulic	5	I also chose the NiCd battery	0	1
2479	SecondGame	brandon f	Hydraulic	5	to agree on which power source and control sensor we should use	0	0
2480	SecondGame	brandon f	Hydraulic	5	to agree on which power source and control sensor we should use	0	0
2481	SecondGame	jackson p	Hydraulic	5	Test the battery out with an actual prototype	0	1
2482	SecondGame	casey f	Hydraulic	5	@Shane: I agree	0	0
2483	SecondGame	jackson p	Hydraulic	5	do more research on the nickel cadmium battery and see how we can optimize it	0	1
2484	**SecondGame**	**samuel o**	**Hydraulic**	**5**	**design prototype**	**0**	**1**

Fig. 13 A screenshot of the view() function result. The highlighted lines represent lines within the same stanza window

3.8 *Using ENA Model Outputs in Other Analyses*

It is often useful to use the outputs of ENA models in subsequent analyses. The most commonly used outputs are the ENA points, i.e., `set$points`. For example, we can use a linear regression analysis to test whether ENA points on the first two dimensions are predictive of an outcome variable, in this case, change in confidence in engineering skills.

```
regression_data = set.ena$points
regression_data$CONFIDENCE.Change = as.numeric(regression_data$CONFIDENCE.Cha
nge)

## Warning: NAs introduced by coercion

condition_regression = lm(CONFIDENCE.Change ~ MR1 + SVD2 + Condition,
                          data = regression_data,
                          na.action = na.omit)
summary(condition_regression)

##
## Call:
## lm(formula = CONFIDENCE.Change ~ MR1 + SVD2 + Condition, data = regression
_data,
##     na.action = na.omit)
##
## Residuals:
##      Min       1Q   Median       3Q      Max

## -1.18092 -0.24324 -0.08171  0.30716  1.88404
##
## Coefficients:
##                     Estimate Std. Error t value Pr(>|t|)
## (Intercept)           1.1111     0.1490   7.457 2.82e-09 ***
## MR1                  -0.4540     0.8616  -0.527    0.601
## SVD2                  0.3268     0.7154   0.457    0.650
## ConditionSecondGame  -0.3484     0.2566  -1.358    0.182
## ---
## Signif. codes:  0 '***' 0.001 '**' 0.01 '*' 0.05 '.' 0.1 ' ' 1
##
## Residual standard error: 0.6374 on 43 degrees of freedom
##   (1 observation deleted due to missingness)
## Multiple R-squared:  0.1228, Adjusted R-squared:  0.0616
## F-statistic: 2.007 on 3 and 43 DF,  p-value: 0.1273
```

The results of this regression analysis show that ENA points are not a significant predictor of the students' pre-post change in confidence (MR1: $t = -0.53, p = 0.60$; SVD2: $t = 0.46, p = 0.65$; Condition: $t = -1.36, p = 0.18$). The overall model was also not significant (F(3, 43) = 2.01, $p = 0.13$) with an adjusted r-squared value of 0.06.

Recall that the dataset we are using is a small subset of the full `RS.data`, and thus results that are significant for the whole dataset may not be for this sample.

4 Ordered Network Analysis with R

This section demonstrates how to conduct an ONA analysis using the `ona` R package. If you are new to ONA as an analytic technique, Tan et al. [12] provides a more detailed explication of its theoretical and methodological foundations.

Because ONA shares some conceptual and procedural similarities with ENA, you may also want to read the recommended papers from the ENA section [1, 2, 14].

4.1 Install the ONA Package and Load the Library

Install the ona package and load the ona library after installing.

```
install.packages("ona", repos = c("https://cran.qe-libs.org",
"https://cran.rstudio.org"))

library(ona)
```

Then, install the other package that is required for ONA analysis.

```
install.packages("tma", repos = c("https://cran.qe-libs.org",
"https://cran.rstudio.org"))

library(tma)
```

4.2 Dataset

(Refer to Sect. 3.2 for a detailed description of the dataset used here.)

Load the `RS.data` dataset.

```
data = ona::RS.data
```

4.3 Construct an ONA Model

To construct an ONA model, identify which columns in the data to use for the parameters required by the ONA modeling function. The parameters are defined identically in both ENA and ONA; see Sect. 3.3 for detailed explanations.

4.3.1 Specify Units

Select the *units* as in Sect. 3.3.1.

```
my_units <- c("Condition", "UserName")
```

4.3.2 Specify Codes

Select the *codes* as in Sect. 3.3.2.

```
my_codes = c(
          'Data',
          'Technical.Constraints',
          'Performance.Parameters',
          'Client.and.Consultant.Requests',
          'Design.Reasoning',
          'Collaboration')
```

4.3.3 Specify Conversations

The parameter to specify *conversations* in rENA is called "conversation"; in ONA, the equivalent is called "my_hoo_rules", where "hoo" is an abbreviation of "horizon of observation."

Choose the combination of "Condition" column, "GroupName" column, and "ActivityNumber" column to define the conversation parameter.

The syntax to specify conversations using `my_hoo_rules` in ONA is slightly different from the syntax to specify `conversation` in ENA, but the conceptual definition is the same.

```
my_hoo_rules <- conversation_rules(
                  (Condition %in% UNIT$Condition &
                   GroupName %in% UNIT$GroupName &
                   ActivityNumber %in% UNIT$ActivityNumber))
```

4.3.4 Specify the Window

Specify a moving stanza window size by passing a numerical value to the `window_size` parameter.

```
window_size = 7
```

4.3.5 Specify Metadata

As in ENA, metadata columns can be included if desired. Metadata columns are not required to construct an ONA model, but they provide information that can be used to subset units in the resulting model.

```
metaCols = c("CONFIDENCE.Change","CONFIDENCE.Pre","CONFIDENCE.Post","C.Change
")
```

4.3.6 Accumulate Connections

Now that all the parameters are specified, connections can be accumulated. For each unit, the ONA algorithm uses a moving stanza window to identify connections formed from a current line of data (e.g., a turn of talk), or *response*, to the preceding lines within the window (the *ground*).

Unlike in ENA, where connections among codes are recorded in a symmetric adjacency matrix, ONA accounts for the order in which the connections occur by constructing an *asymmetric adjacency matrix* for each unit; that is, the number of connections from code A to code B may be different than the number of connections from B to A.

To accumulate connections, pass the parameters specified to the `contexts` and `accumulate_contexts` functions, and store the output in an object (in this case, **accum.ona**).

```
accum.ona <-
  contexts(data,
           units_by = my_units,
           hoo_rules = my_hoo_rules) |>
  accumulate_contexts(codes = my_codes,
                      decay.function = decay(simple_window, window_size = 7),
                      meta.data = metaCols,
                      return.ena.set = FALSE) # keep this as FALSE to get an
ONA model, otherwise it will return an undirected model)
```

4.3.7 Construct an ONA Model

After accumulation, call the `model` function to construct an ONA model. ONA currently implements *singular value decomposition* (SVD) and *means rotation* (MR) to perform dimensional reduction.

To create an ONA model using SVD, pass the `accum.ona` object to the `model` function.

```
set.ona <-
  model(accum.ona)
```

When there are two discrete groups to compare, a means rotation can be used, as described in Sect. 3.3.5.

A means rotation is specified using `rotate.using ="mean"` in the `model` function. Additionally, the `model` function expects `rotation.params` to be a `list` with two named elements, each containing a logical vector representing the rows of units to be included in each group.

```
set.ona <-
  model(accum.ona,                        # the previously run accumulation above
        rotate.using ="mean",             # means rotation method
        rotation.params =                 # two groups for means rotation in a list
              list(FirstGame=accum.ona$meta.data$Condition=="FirstGame",
                   SecondGame=accum.ona$meta.data$Condition=="SecondGame")
        )
```

Here, construct the ONA model as shown below.

4.4 Summary of Key Model Outputs

Information about an ONA model is now stored in the R object `set.ona`.

As in rENA, users can explore the data stored in the object by typing `set.ona$` and select items from the drop down list. Here, we briefly explain the top-level items in `set.ona$`.

4.4.1 Connection Counts

Because ONA accounts for the order in which the connections occur by constructing an *asymmetric adjacency matrix* for each unit, connection counts *from* code A *to* code B and *from* B *to* A, as well as self-connections for each code (from A to A) are recorded. Thus, because six codes were included in the model, the cumulative adjacency vector for each unit contains 36 terms (*n*^2).

```
head(set.ona$connection.counts,3)
```

```
##    Condition     UserName                  ENA_UNIT Data to Data
## 1: FirstGame     steven z     FirstGame::steven z           26
## 2: FirstGame      akash v      FirstGame::akash v           72
## 3: FirstGame alexander b FirstGame::alexander b           11
##    Technical.Constraints to Data Performance.Parameters to Data
## 1:                          44.0                             32
## 2:                         102.5                             66
## 3:                          21.5                             15
##    Client.and.Consultant.Requests to Data Design.Reasoning to Data
## 1:                                     12                       52
## 2:                                     10                       88
```

```
## 3:                                    4                       14
##    Collaboration to Data Data to Technical.Constraints
## 1:                  8                        27.0
## 2:                 12                       106.5
## 3:                  1                        11.5
##    Technical.Constraints to Technical.Constraints
## 1:                                          63
## 2:                                         193
## 3:                                          40
##    Performance.Parameters to Technical.Constraints
## 1:                                         29.5
## 2:                                         90.5
## 3:                                         15.5
##    Client.and.Consultant.Requests to Technical.Constraints
## 1:                                                 8.5
## 2:                                                23.5
## 3:                                                 1.0
##    Design.Reasoning to Technical.Constraints
## 1:                                    62.0
## 2:                                   160.5
## 3:                                    23.0
##    Collaboration to Technical.Constraints Data to Performance.Parameters
## 1:                                 11.0                             24
## 2:                                 29.0                             77
## 3:                                 15.5                             10
##    Technical.Constraints to Performance.Parameters
## 1:                                         35.5
## 2:                                         91.5
## 3:                                         17.5
##    Performance.Parameters to Performance.Parameters
## 1:                                            34
## 2:                                            72
## 3:                                            10
##    Client.and.Consultant.Requests to Performance.Parameters
## 1:                                                  5.5
## 2:                                                 20.5
## 3:                                                  3.0
##    Design.Reasoning to Performance.Parameters
## 1:                                       42
## 2:                                       77
## 3:                                       14
##    Collaboration to Performance.Parameters Data to Client.and.Consultant.R
equests
## 1:                                   7.5
6
## 2:                                  14.5
18
## 3:                                   1.0
5
##    Technical.Constraints to Client.and.Consultant.Requests
```

```
## 1:                                                     7.5
## 2:                                                    19.5
## 3:                                                     7.0
##     Performance.Parameters to Client.and.Consultant.Requests
## 1:                                                      11.5
## 2:                                                      18.5
## 3:                                                       3.0
##     Client.and.Consultant.Requests to Client.and.Consultant.Requests
## 1:                                                               5
## 2:                                                              12
## 3:                                                               1
##     Design.Reasoning to Client.and.Consultant.Requests
## 1:                                                15.5
## 2:                                                 9.0
## 3:                                                 3.0
##     Collaboration to Client.and.Consultant.Requests Data to Design.Reasonin
g
## 1:                                               2                     19
## 2:                                               3                     86
## 3:                                               0                     10
##     Technical.Constraints to Design.Reasoning
## 1:                                       50.0
## 2:                                      152.5
## 3:                                       17.0
##     Performance.Parameters to Design.Reasoning
## 1:                                          20
## 2:                                          69
## 3:                                           9
##     Client.and.Consultant.Requests to Design.Reasoning
## 1:                                                 5.5
## 2:                                                16.0
## 3:                                                 2.0
##     Design.Reasoning to Design.Reasoning Collaboration to Design.Reasoning
## 1:                                    59                              6.0
## 2:                                   136                             33.5
## 3:                                    19                              4.0
##     Data to Collaboration Technical.Constraints to Collaboration
## 1:                      0                                    7.0
## 2:                     15                                   27.0
## 3:                      6                                   18.5
##     Performance.Parameters to Collaboration
## 1:                                      0.5
## 2:                                      8.5
## 3:                                      6.0
##     Client.and.Consultant.Requests to Collaboration
## 1:                                               0
## 2:                                               2
## 3:                                               0
##     Design.Reasoning to Collaboration Collaboration to Collaboration
## 1:                                 5.0                            0
## 2:                                42.5                           14
## 3:                                 7.0                            9
```

4.4.2 Line Weights

To compare networks in terms of their *relative* patterns of association, researchers can spherically normalize the cumulative adjacency vectors by diving each one by its length. The resulting normalized vectors represent each unit's relative frequencies of code co-occurrence. In other words, the sphere normalization controls for the fact that different units might have different amounts of interaction or different numbers of activities than others.

In `set.ona$connection.counts`, the value for each unique co-occurrence of codes is an integer equal or greater than 0, because they represent the directional connection counts between each pair of codes. In `set.ona$line.weights`, the connection counts are sphere normalized, and so the values are between 0 and 1.

```
head(set.ona$line.weights,3)

##     Condition    UserName               ENA_UNIT ENA_DIRECTION Data to Data
## 1: FirstGame    steven z    FirstGame::steven z      response    0.1543564
## 2: FirstGame     akash v     FirstGame::akash v      response    0.1619657
## 3: FirstGame alexander b FirstGame::alexander b      response    0.1424314
##     Technical.Constraints to Data Performance.Parameters to Data
## 1:                      0.2612185                      0.1899771
## 2:                      0.2305762                      0.1484686
## 3:                      0.2783886                      0.1942246
##     Client.and.Consultant.Requests to Data Design.Reasoning to Data
## 1:                              0.07124140                0.3087127
## 2:                              0.02249524                0.1979581
## 3:                              0.05179323                0.1812763
##     Collaboration to Data Data to Technical.Constraints
## 1:             0.04749427                     0.1602931
## 2:             0.02699429                     0.2395743
## 3:             0.01294831                     0.1489055
##     Technical.Constraints to Technical.Constraints
## 1:                                      0.3740173
## 2:                                      0.4341581
## 3:                                      0.5179323
##     Performance.Parameters to Technical.Constraints
## 1:                                       0.1751351
## 2:                                       0.2035819
## 3:                                       0.2006988
##     Client.and.Consultant.Requests to Technical.Constraints
## 1:                                               0.05046266
## 2:                                               0.05286381
## 3:                                               0.01294831
##     Design.Reasoning to Technical.Constraints
## 1:                                 0.3680806
## 2:                                 0.3610486
```

```
## 3:                              0.2978111
##    Collaboration to Technical.Constraints Data to Performance.Parameters
## 1:                           0.06530462                        0.1424828
## 2:                           0.06523619                        0.1732133
## 3:                           0.20069875                        0.1294831
##    Technical.Constraints to Performance.Parameters
## 1:                                       0.2107558
## 2:                                       0.2058314
## 3:                                       0.2265954
##    Performance.Parameters to Performance.Parameters
## 1:                                        0.2018506
## 2:                                        0.1619657
## 3:                                        0.1294831
##    Client.and.Consultant.Requests to Performance.Parameters
## 1:                                               0.03265231
## 2:                                               0.04611524
## 3:                                               0.03884492
##    Design.Reasoning to Performance.Parameters
## 1:                                 0.2493449
## 2:                                 0.1732133
## 3:                                 0.1812763
##    Collaboration to Performance.Parameters Data to Client.and.Consultant.R
equests
## 1:                              0.04452587                            0.035
62070
## 2:                              0.03261810                            0.040
49143
## 3:                              0.01294831                            0.064
74153
##    Technical.Constraints to Client.and.Consultant.Requests
## 1:                                              0.04452587
## 2:                                              0.04386571
## 3:                                              0.09063815
##    Performance.Parameters to Client.and.Consultant.Requests
## 1:                                               0.06827301
## 2:                                               0.04161619
## 3:                                               0.03884492
##    Client.and.Consultant.Requests to Client.and.Consultant.Requests
## 1:                                                       0.02968392
## 2:                                                       0.02699429
## 3:                                                       0.01294831
##    Design.Reasoning to Client.and.Consultant.Requests
## 1:                                        0.09202014
## 2:                                        0.02024571
## 3:                                        0.03884492
##    Collaboration to Client.and.Consultant.Requests Data to Design.Reasonin
g
## 1:                                     0.011873567              0.1127989
## 2:                                     0.006748571              0.1934590
## 3:                                     0.000000000              0.1294831
```

```
##    Technical.Constraints to Design.Reasoning
## 1:                                 0.2968392
## 2:                                 0.3430524
## 3:                                 0.2201212
##    Performance.Parameters to Design.Reasoning
## 1:                                  0.1187357
## 2:                                  0.1552171
## 3:                                  0.1165348
##    Client.and.Consultant.Requests to Design.Reasoning
## 1:                                         0.03265231
## 2:                                         0.03599238
## 3:                                         0.02589661
##    Design.Reasoning to Design.Reasoning Collaboration to Design.Reasoning
## 1:                            0.3502702                        0.03562070
## 2:                            0.3059352                        0.07535905
## 3:                            0.2460178                        0.05179323
##    Data to Collaboration Technical.Constraints to Collaboration
## 1:            0.00000000                             0.04155748
## 2:            0.03374286                             0.06073714
## 3:            0.07768984                             0.23954367
##    Performance.Parameters to Collaboration
## 1:                             0.002968392
## 2:                             0.019120952
## 3:                             0.077689840
##    Client.and.Consultant.Requests to Collaboration
## 1:                                     0.000000000
## 2:                                     0.004499048
## 3:                                     0.000000000
##    Design.Reasoning to Collaboration Collaboration to Collaboration
## 1:                        0.02968392                     0.00000000
## 2:                        0.09560476                     0.03149333
## 3:                        0.09063815                     0.11653476
```

4.4.3 ONA Points

For each unit, ONA produces an ONA point in a two-dimensional space formed by the first two dimensions of the dimensional reduction.

Here, the MR1 column represents the *x*-axis coordinate for each unit, and the SVD2 column represents the *y*-axis coordinate for each unit.

```
head(set.ona$points,3)
```

```
##     Condition     UserName                  ENA_UNIT ENA_DIRECTION         MR1
## 1: FirstGame    steven z     FirstGame::steven z      response  0.00753635
## 2: FirstGame     akash v      FirstGame::akash v      response -0.07719283
## 3: FirstGame alexander b FirstGame::alexander b      response -0.20600855
##            SVD2        SVD3        SVD4       SVD5         SVD6          SVD
7
## 1: -0.05350532  0.02308722  0.03365899 0.18576251 -0.064647347 -0.01638733
9
```

```
## 2:  0.01840812 -0.01485049 -0.03634380 0.02065882 -0.003406081 -0.00802670
5
## 3: -0.05806135 -0.08409658  0.13340227 0.03017168 -0.046102014 -0.09567464
9
##           SVD8         SVD9       SVD10           SVD11          SVD12        SVD
13
## 1:  0.02504138 -0.04662639 0.01654204 -0.0050339193  0.001089911  0.020150
19
## 2: -0.01584017 -0.01879391 0.03338419 -0.0082095228 -0.004596587  0.050346
82
## 3: -0.10337234  0.14247946 0.01749875  0.0005982879 -0.073284812 -0.019597
35
##          SVD14          SVD15        SVD16        SVD17          SVD18         S
VD19
## 1: -0.03170014  0.014028551 -0.03292583 -0.01286482  0.0002353795 -0.01426
1325
## 2: -0.02280942  0.042727200  0.02373320 -0.02829064 -0.0208970211  0.01012
0401
## 3: -0.01633710 -0.002697103 -0.03717024  0.02019079 -0.0119060565 -0.00651
8936
##          SVD20        SVD21        SVD22        SVD23        SVD24
## 1:  0.01159296 0.0009366777 -0.033851942 -0.003230414 -0.016553686
## 2: -0.01752303 0.0045814224 -0.006533689 -0.007668149  0.017903544
## 3: -0.03746771 0.0075254949  0.038768078  0.008012232  0.009589488
##             SVD25         SVD26        SVD27          SVD28        SVD29
## 1:  2.124235e-03  0.008351416  0.024600338 -0.0003563738  0.006848068
## 2: -4.643724e-05 -0.013146155 -0.003761067 -0.0172067645 -0.009072690
## 3:  1.347210e-03  0.014229380  0.006423058  0.0020076670  0.003970270
##           SVD30        SVD31        SVD32         SVD33         SVD34
SVD35
## 1:  0.009214306 0.000689622 0.001926662 -0.0001196152 0.0057286766  0.0031
16433
## 2: -0.006574454 0.001518995 0.006561024  0.0037200464 0.0019185441  0.0064
47044
## 3: -0.003266165 0.004738366 0.003057179  0.0018318401 0.0008428842 -0.0048
84616
##           SVD36
## 1:  0.003361801
## 2: -0.001230173
## 3:  0.001080194
```

4.4.4 Rotation Matrix

The rotation matrix used during the dimensional reduction can be accessed through `set.ona$rotation`. This is mostly useful when you want to construct an ONA metric space using one dataset and then project ONA points from different data into that space, as in Sect. 5.2.

```
head(set.ona$rotation.matrix,3)
##                                  codes         MR1       SVD2         SVD3
## 1:                        Data to Data -0.06930733 -0.4261566  0.041261221
## 2:   Technical.Constraints to Data -0.18483941 -0.3305414 -0.001705309
## 3: Performance.Parameters to Data  0.16511484 -0.4098705 -0.057056137
##           SVD4        SVD5       SVD6       SVD7        SVD8       SVD9
## 1: -0.24813254 -0.12386178 -0.1212851 -0.1813800 -0.02634818 0.02463562
## 2: -0.09652822  0.29776415 -0.2487902  0.4530700 -0.08131861 0.19027807
## 3: -0.06943836  0.02894105 -0.1666548 -0.1512543  0.26221091 0.15918103
##         SVD10      SVD11      SVD12       SVD13      SVD14       SVD15
## 1: -0.1685089 0.02779782  0.1634913  0.03515259 -0.1307011  0.11108864
## 2: -0.2985731 0.03356852 -0.1788196  0.24614774 -0.1162268 -0.06526278
## 3:  0.2634069 0.30265567  0.0364710 -0.16880948  0.2246206  0.04608908
##         SVD16       SVD17      SVD18        SVD19       SVD20       SVD21
## 1:  0.1678676 -0.21955326  0.4390146  0.263975884 -0.03791243 -0.26637411
## 2:  0.2427948  0.31464439 -0.1658976 -0.023955700 -0.04756904  0.05450570
## 3: -0.1521164 -0.06523085 -0.3436932  0.005934216 -0.21114098  0.07531336
##          SVD22       SVD23       SVD24       SVD25       SVD26       SVD27
## 1: -0.16378180 -0.04863544 -0.03046897  0.01129295  0.07990007 -0.10906748
## 2:  0.09542229  0.04913771  0.09391774 -0.07213795 -0.01391693 -0.03060512
## 3:  0.18550364 -0.20153993  0.11835360 -0.08759355 -0.06613629  0.01739627
##         SVD28       SVD29       SVD30       SVD31        SVD32        SVD3
3
## 1:  0.1218933  0.05135484 -0.08227500  0.09675121 -0.205485302 -0.15944684
3
## 2: -0.0431860 -0.04667436 -0.03855961 -0.06592640 -0.006227139 -0.00619161
0
## 3:  0.1294605  0.01342884 -0.17369350  0.07209455 -0.153340746 -0.00982495
5
##          SVD34       SVD35      SVD36
## 1: -0.12907089  0.14685408 0.01703190
## 2:  0.02971639  0.11327562 0.07985307
## 3:  0.17078477 -0.02710075 0.06842069
```

4.4.5 Metadata

`set.ona$meta.data` gives a data frame that includes all the columns except for the code connection columns.

```
head(set.ona$meta.data,3)

##    Condition     UserName                ENA_UNIT
## 1: FirstGame     steven z     FirstGame::steven z
## 2: FirstGame      akash v      FirstGame::akash v
## 3: FirstGame alexander b FirstGame::alexander b
```

4.5 ONA Visualization

Once an ONA model is constructed, ONA networks can be visualize. The plotting function in ONA is called `plot`, and it works similarly to the same function in ENA.

Before plotting, you can set up several global parameters to ensure consistency across plots. These parameters will be clearer in subsequent sections.

```
node_size_multiplier = 0.4 # scale up or down node sizes
node_position_multiplier = 1 # zoom in or out node positions
point_position_multiplier =1.5 # zoom in or out the point positions
edge_arrow_saturation_multiplier = 1.5 # adjust the chevron color lighter or
darker
edge_size_multiplier = 1 # scale up or down edge sizes
```

4.5.1 Plot a Mean Network

Mean ONA networks can be plotted for each of the conditions along with their subtracted network.

First, plot the mean network for the FirstGame condition. Use a pipe | > to connect the `edges` function and the `nodes` function. Users are only required to specify the `weights` parameter, as the remaining parameters have default values unless specified otherwise (Fig. 14).

```
ona::plot.ena.ordered.set(set.ona, title = "FirstGame (red) mean network") |
>
  edges(
    weights =set.ona$line.weights$Condition$FirstGame,
    edge_size_multiplier = edge_size_multiplier,
    edge_arrow_saturation_multiplier = edge_arrow_saturation_multiplier,
    node_position_multiplier = node_position_multiplier,
    edge_color = c("red")) |>
  nodes(
    node_size_multiplier = node_size_multiplier,
    node_position_multiplier = node_position_multiplier,
    self_connection_color = c("red"))
```

Since this is the first ONA network visualization in this chapter, we briefly explain how to read an ONA network.

Node size: In ONA, the node size is proportional to the number of occurrences of that code as a *response* to other codes in the data, with larger nodes indicating more responses. For example, in this plot, students in the FirstGame condition responded most frequently with discourse about Technical.Constraints.

Self-connections: The color and saturation of the circle within each node is proportional to the number of *self-connections* for that code: that is, when a code is both what students *responded to* and what they *responded with*. Colored circles that are larger and more saturated reflect codes with more frequent self-connections.

Edges: Note that unlike most directed network visualizations, which use arrows or spearheads to indicate direction, ONA uses a "broadcast" model, where the source of a connection (what students responded to) is placed at the apex side of the triangle and the destination of a connection (what students responded with) is placed at its base.

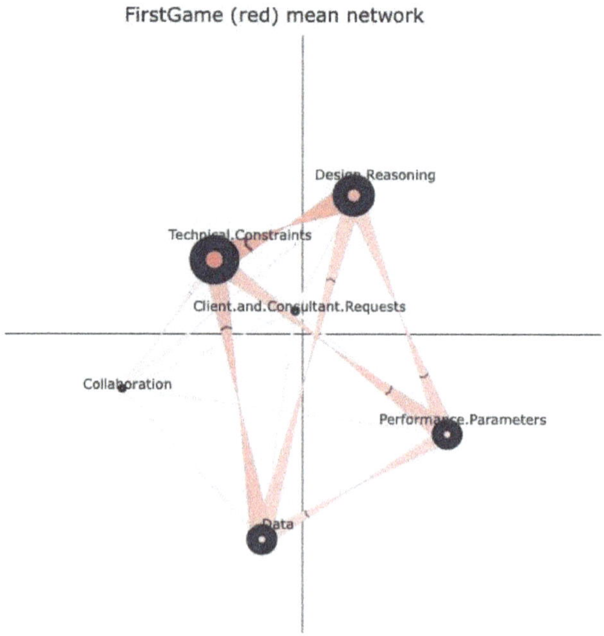

Fig. 14 ONA mean network for FirstGame group

Chevrons on edges: The chevrons point in the direction of the connection. Between any pair of nodes, if there is a bidirectional connection, the chevron only appears on the side with the stronger connection. This helps viewers differentiate heavier edges in cases such as between Technical.Constraints and Data, where the connection strengths from both directions are similar. When the connection strengths are identical between two codes, the chevron will appear on both edges.

Now, plot the mean network for SecondGame (Fig. 15).

```
ona::plot.ena.ordered.set(set.ona, title = "SecondGame (blue) mean network")
|>
  edges(
    weights = set.ona$line.weights$Condition$SecondGame,
    edge_size_multiplier = edge_size_multiplier,
    edge_arrow_saturation_multiplier = edge_arrow_saturation_multiplier,
    node_position_multiplier = node_position_multiplier,
    edge_color = c("blue")) |>
  nodes(
    node_size_multiplier = node_size_multiplier,
    node_position_multiplier = node_position_multiplier,
    self_connection_color = c("blue"))
```

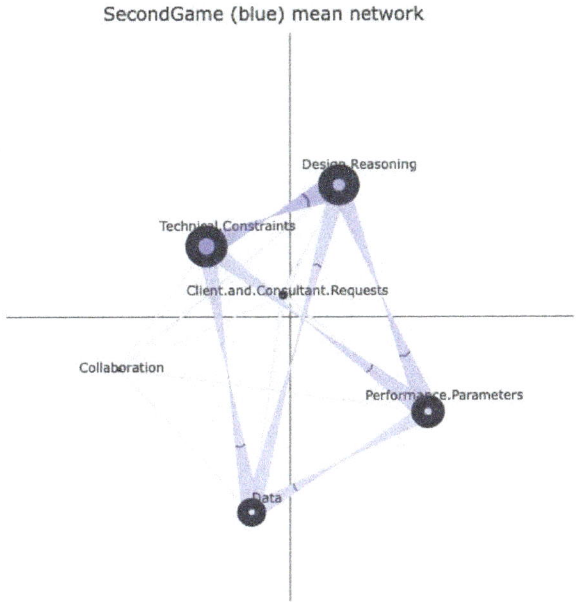

Fig. 15 ONA mean network for SecondGame group

Then, plot the subtracted network to show the differences between the mean networks of the FirstGame and SecondGame conditions (Fig. 16).

```
ona::plot.ena.ordered.set(set.ona, title = "Subtracted mean network: FirstGa
me (red) vs SecondGame (blue)") |>
  edges(
    weights = (colMeans(set.ona$line.weights$Condition$FirstGame) - colMeans(
set.ona$line.weights$Condition$SecondGame))*4, # optional weights multiplier
to adjust readability
    edge_size_multiplier = edge_size_multiplier,
    edge_arrow_saturation_multiplier = edge_arrow_saturation_multiplier,
    node_position_multiplier = node_position_multiplier,
    edge_color = c("red","blue")) |>
  nodes(
    node_size_multiplier = node_size_multiplier,
    node_position_multiplier = node_position_multiplier,
    self_connection_color = c("red","blue"))
```

4.5.2 Plot a Mean Network and its Points

Besides plotting the mean network for each condition and the subtracted network, we can also plot the individual units within each condition.

Use `set.ona$points` to subset the rows that are in each condition and plot the units in each condition as a different color.

The points are specified in the `units` function. The `edges` and `nodes` functions remain the same as above (Figs. 17, 18, 19, and 20).

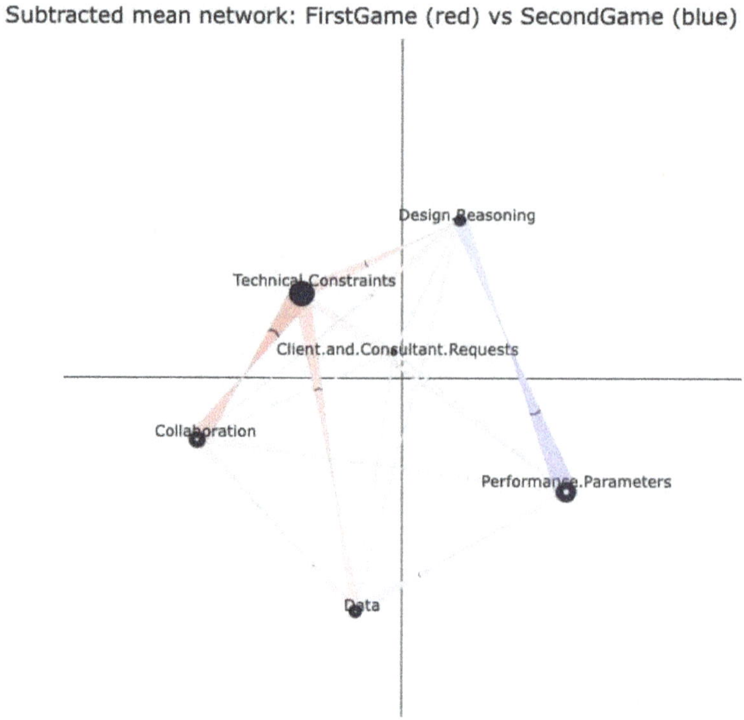

Fig. 16 ONA subtracted network showing the differences between FirstGame (red) and SecondGame (blue)

```
ona:::plot.ena.ordered.set(set.ona, title = "points (dots), mean point (squar
e), and confidence interval") |>
  units(
    points=set.ona$points$Condition$FirstGame,
    points_color = c("red"),
    show_mean = TRUE, show_points = TRUE, with_ci = TRUE)
```

```
ona:::plot.ena.ordered.set(set.ona, title = "FirstGame (red) mean network") |
>
  units(
    points=set.ona$points$Condition$FirstGame,
    points_color = c("red"),
    show_mean = TRUE, show_points = TRUE, with_ci = TRUE) |>
  edges(
    weights =set.ona$line.weights$Condition$FirstGame,
    edge_size_multiplier = edge_size_multiplier,
    edge_arrow_saturation_multiplier = edge_arrow_saturation_multiplier,
    node_position_multiplier = node_position_multiplier,
    edge_color = c("red")) |>
  nodes(
    node_size_multiplier = node_size_multiplier,
    node_position_multiplier = node_position_multiplier,
    self_connection_color = c("red"))
```

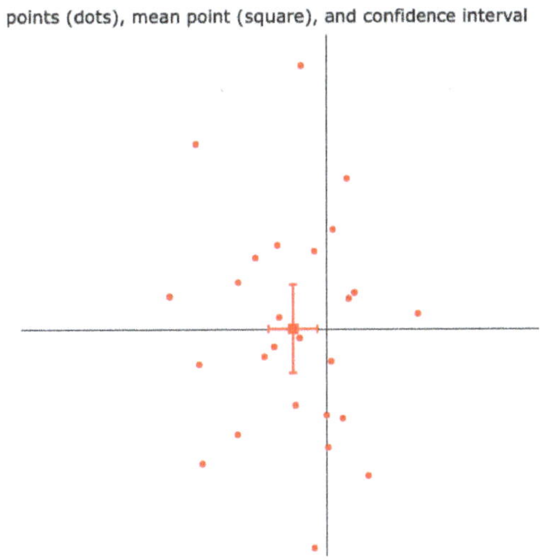

Fig. 17 ONA points (dots) and their mean point (square) for FirstGame group

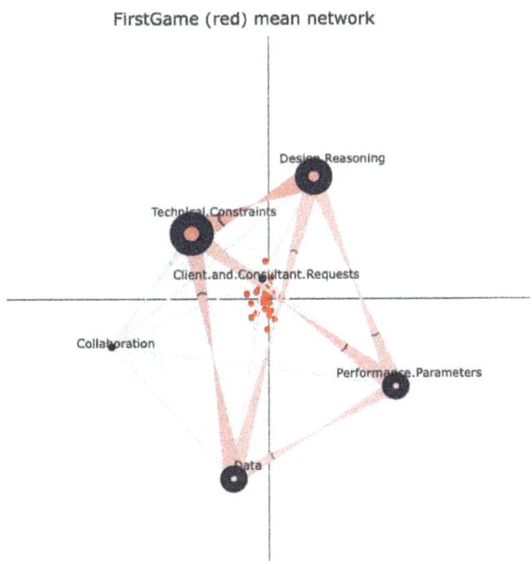

Fig. 18 ONA mean network for FirstGame group

```
ona::plot.ena.ordered.set(set.ona, title = "points (dots), mean point (squar
e), and confidence interval") |>
  units(
    points=set.ona$points$Condition$SecondGame,
    points_color = c("blue"),
    show_mean = TRUE, show_points = TRUE, with_ci = TRUE)
```

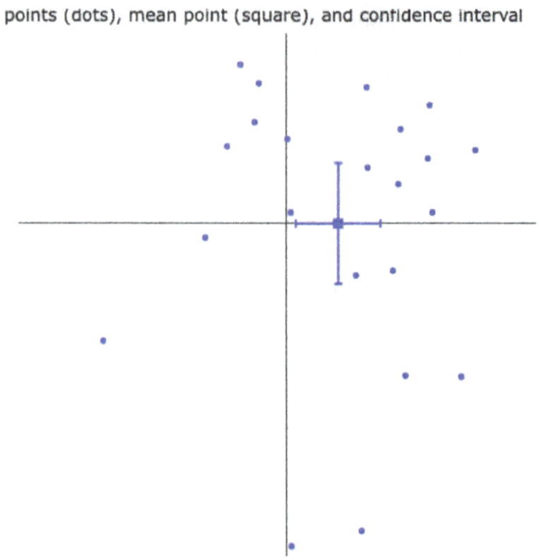

Fig. 19 ONA points (dots) and their mean point (square) for SecondGame group

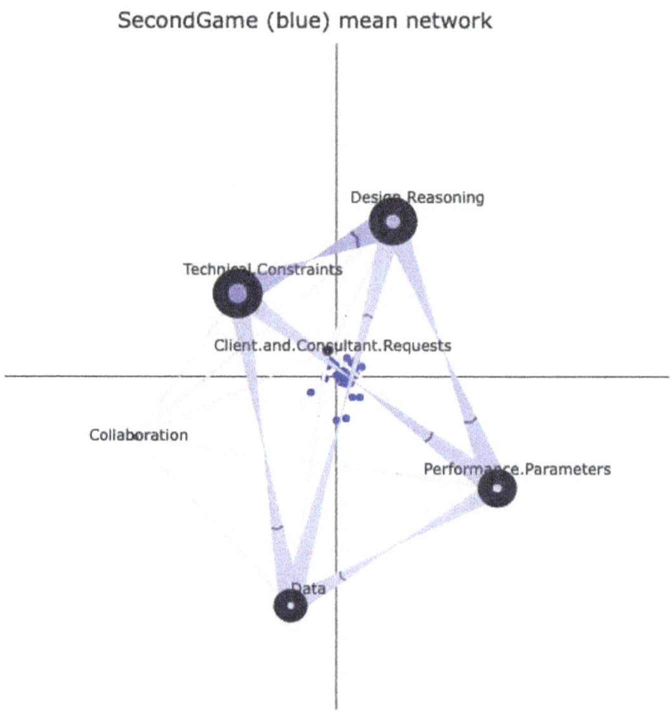

Fig. 20 ONA mean network and points for SecondGame

```
ona::plot.ena.ordered.set(set.ona, title = "SecondGame (blue) mean network")
|>
  units(
    points=set.ona$points$Condition$SecondGame,
    points_color = "blue",
    show_mean = TRUE, show_points = TRUE, with_ci = TRUE) |>
  edges(
    weights = set.ona$line.weights$Condition$SecondGame,
    edge_size_multiplier = edge_size_multiplier,
    edge_arrow_saturation_multiplier = edge_arrow_saturation_multiplier,
    node_position_multiplier = node_position_multiplier,
    edge_color = c("blue")) |>
  nodes(
    node_size_multiplier = node_size_multiplier,
    node_position_multiplier = node_position_multiplier,
    self_connection_color = c("blue"))
```

Plot the subtracted network as follows (Fig. 21).

```
# FirstGame and SecondGame subtracted plot
ona::plot.ena.ordered.set(set.ona, title = "Difference: FirstGame (red) vs S
econdGame (blue)") |>
  units(
    points = set.ona$points$Condition$FirstGame,
    points_color = "red",
    show_mean = TRUE, show_points = TRUE, with_ci = TRUE) |>
  units(
    points = set.ona$points$Condition$SecondGame,
    points_color = "blue",
    show_mean = TRUE, show_points = TRUE, with_ci = TRUE) |>
  edges(
    weights = (colMeans(set.ona$line.weights$Condition$FirstGame) - colMeans(
```

```
set.ona$line.weights$Condition$SecondGame))*4, # optional multiplier to adjus
t for readability
    edge_size_multiplier = edge_size_multiplier,
    edge_arrow_saturation_multiplier = edge_arrow_saturation_multiplier,
    node_position_multiplier = node_position_multiplier,
    edge_color = c("red","blue")) |>
  nodes(
    node_size_multiplier = node_size_multiplier,
    node_position_multiplier = node_position_multiplier,
    self_connection_color = c("red","blue"))
```

4.5.3 Plot an Individual Network and its Points

To plot an individual student's network and ONA point, use `set.ona$points`.

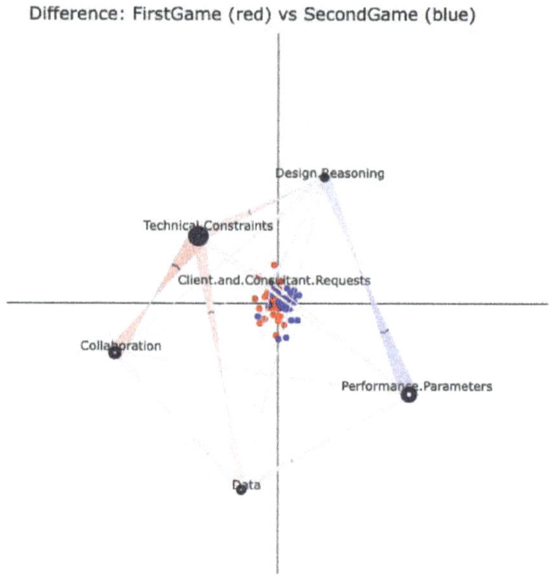

Fig. 21 ONA subtracted network showing the differences between FirstGame (red) and SecondGame (blue)

Here, we choose the same two units we compared in the ENA analysis (Sect. 3.5.3) (Figs. 22 and 23).

```
# first game
ona:::plot.ena.ordered.set(set.ona, title = "FirstGame::steven z") |>
  units(
    points=set.ona$points$ENA_UNIT$`FirstGame::steven z`,
    points_color = "red",
    show_mean = FALSE, show_points = TRUE, with_ci = FALSE) |>
  edges(
    weights = set.ona$line.weights$ENA_UNIT$`FirstGame::steven z`,
    edge_size_multiplier = edge_size_multiplier,
    edge_arrow_saturation_multiplier = edge_arrow_saturation_multiplier,
    node_position_multiplier = node_position_multiplier,
    edge_color = c("red")) |>
  nodes(
    node_size_multiplier = node_size_multiplier,
    node_position_multiplier = node_position_multiplier,
    self_connection_color = c("red"))
```

```
# second game
ona:::plot.ena.ordered.set(set.ona, title = "SecondGame::samuel o") |>
  units(
    points=set.ona$points$ENA_UNIT$`SecondGame::samuel o`,
    points_color = "blue",
    show_mean = FALSE, show_points = TRUE, with_ci = FALSE) |>
  edges(
```

```
    weights = set.ona$line.weights$ENA_UNIT$`SecondGame::samuel o`,
    edge_size_multiplier = edge_size_multiplier,
    edge_arrow_saturation_multiplier = edge_arrow_saturation_multiplier,
    node_position_multiplier = node_position_multiplier,
    edge_color = c("blue")) |>
  nodes(
    node_size_multiplier = node_size_multiplier,
    node_position_multiplier = node_position_multiplier,
    self_connection_color = c("blue"))
```

In this case, both units make relatively strong connections between Design.Reasoning and Data. However, for Unit A (red), the connection is relatively more *from* Design.Reasoning *to* Data than the other way around. This indicates that more often this unit responded with Data. In contrast, Unit B (blue) responded more frequently to Data with Design.Reasoning.

A subtracted network can make such differences more salient (Fig. 24).

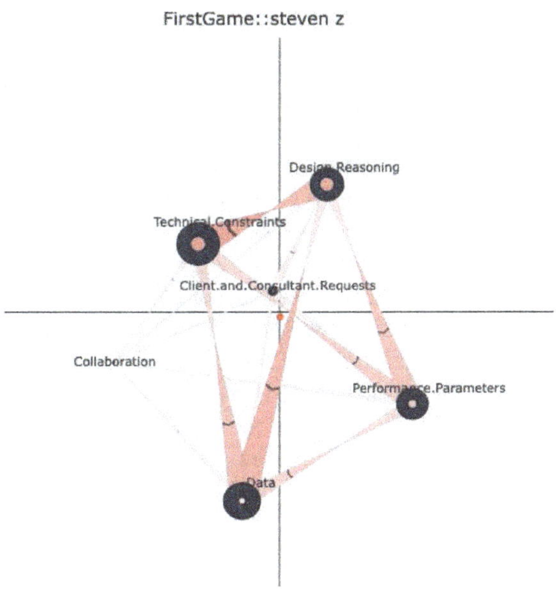

Fig. 22 ONA network for a student from FirstGame

```
# units difference
mean1 = as.vector(as.matrix(set.ona$line.weights$ENA_UNIT$`FirstGame::steven
z`))
mean2 = as.vector(as.matrix(set.ona$line.weights$ENA_UNIT$`SecondGame::samuel
o`))

subtracted.mean = mean1 - mean2

ona:::plot.ena.ordered.set(set.ona, title = "subtracted network of steven z.F
irstGame.Electric and SecondGame.luke u") |>
  units(
    points = set.ona$points$ENA_UNIT$`FirstGame::steven z`, points_color = "r
ed",
    point_position_multiplier = point_position_multiplier,
    show_mean = FALSE, show_points = TRUE, with_ci = FALSE) |>
  units(
    points = set.ona$points$ENA_UNIT$`SecondGame::samuel o`, points_color = "
blue",
    point_position_multiplier = point_position_multiplier,
    show_mean = FALSE, show_points = TRUE, with_ci = FALSE) |>
  edges(
    weights = subtracted.mean*2,
    edge_size_multiplier = edge_size_multiplier,
    edge_arrow_saturation_multiplier = edge_arrow_saturation_multiplier,
    node_position_multiplier = node_position_multiplier,
    edge_color = c("red", "blue")) |>
  nodes(
    node_size_multiplier = node_size_multiplier,
    node_position_multiplier = node_position_multiplier,
    self_connection_color = c("red", "blue"))
```

The connection between Design.Reasoning and Data consists of two triangles, one in blue pointing from Data to Design.Reasoning, the other in red pointing from Design.Reasoning to Data. This indicates that although both units made strong connections between these two codes, the relative directed frequencies are different. Recall that in the ENA subtracted network for the same two units, the connections between Data and Design.Reasoning were basically the same. ONA, by accounting for the order of events, shows that while the undirected relative frequencies were similar, there was a difference in the order in which the two students made the connection.

4.6 Compare Groups Statistically

In addition to visual comparison of networks, ENA points can be analyzed statistically. For example, here we might test whether the patterns of association in one condition are significantly different from those in the other condition.

To demonstrate both parametric and non-parametric approaches to this question, the examples below use a Student's t test and a Mann-Whitney U test to test for differences between the FirstGame and SecondGame condition.

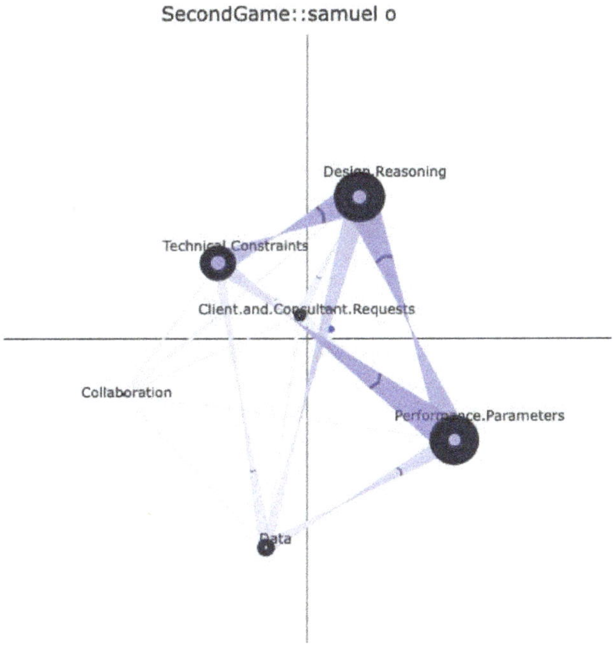

Fig. 23 ONA network for a student from SecondGame

First, install the `lsr` package to enable calculation of effect size (Cohen's *d*) for the *t* test.

```
install.packages('lsr')
library(lsr)
```

Then, subset the points to test for differences between the points of the two conditions.

```
ona_first_points_d1 = as.matrix(set.ona$points$Condition$FirstGame)[,1]
ona_second_points_d1 = as.matrix(set.ona$points$Condition$SecondGame)[,1]

ona_first_points_d2 = as.matrix(set.ona$points$Condition$FirstGame)[,2]
ona_second_points_d2 = as.matrix(set.ona$points$Condition$SecondGame)[,2]
```

Conduct the *t* test on the first and second dimensions.

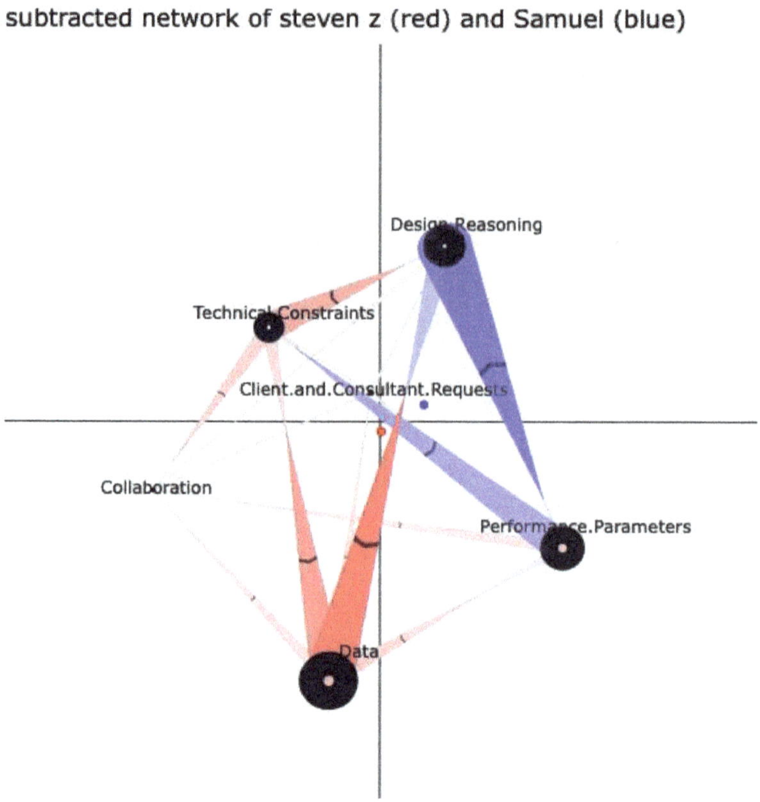

Fig. 24 ONA subtracted network showing the differences between one student from FirstGame (red) and another student from SecondGame (blue)

```
parametric tests
t_test_d1 = t.test(ona_first_points_d1, ona_second_points_d1)
t_test_d1

##
##  Welch Two Sample t-test
##
## data:  ona_first_points_d1 and ona_second_points_d1
## t = -3.7729, df = 41.001, p-value = 0.0005111
## alternative hypothesis: true difference in means is not equal to 0
## 95 percent confidence interval:
##  -0.18227713 -0.05517572
## sample estimates:
##   mean of x   mean of y
## -0.05441628  0.06431015

t_test_d2 = t.test(ona_first_points_d2, ona_second_points_d2)
t_test_d2

##
##  Welch Two Sample t-test
##
## data:  ona_first_points_d2 and ona_second_points_d2
## t = -6.9301e-16, df = 45.45, p-value = 1
## alternative hypothesis: true difference in means is not equal to 0
## 95 percent confidence interval:
```

```
##  -0.1008208  0.1008208
## sample estimates:
##     mean of x     mean of y
## -1.727628e-17  1.742362e-17
```

Compute any other statistics that may be of interest. A few examples are given below.

```
mean(ona_first_points_d1)

## [1] -0.05441628

mean(ona_second_points_d1)

## [1] 0.06431015

mean(ona_first_points_d2)

## [1] -1.727628e-17

mean(ona_second_points_d2)

## [1] 1.742362e-17

sd(ona_first_points_d1)

## [1] 0.09754142

sd(ona_second_points_d1)

## [1] 0.1171941

sd(ona_first_points_d2)

## [1] 0.1784777

sd(ona_second_points_d2)

## [1] 0.1679372

length(ona_first_points_d1)

## [1] 26

length(ona_second_points_d1)

## [1] 22

length(ona_first_points_d2)

## [1] 26

length(ona_second_points_d2)

## [1] 22

cohensD(ona_first_points_d1, ona_second_points_d1)
```

```
## [1] 1.109985
```

```
cohensD(ona_first_points_d2, ona_second_points_d2)
```

```
## [1] 1.997173e-16
```

Here, along the *x* axis (MR1), a two-sample *t* test assuming unequal variance shows that the FirstGame (mean = −0.05, SD = 0.09, N = 26) condition is statistically significantly different for alpha = 0.05 from the SecondGame condition (mean = 0.06, SD = 0.12, N = 22; t(41.001) = −3.77, p = 0.00, Cohen's d = 1.1). Along the *y* axis (SVD2), a two-sample *t* test assuming unequal variance shows that the FirstGame condition (mean = −1.73, SD = 0.17, N = 26) is not statistically significantly different for alpha = 0.05 from the SecondGame condition (mean = 1,74, SD = 0.17, N = 22; t(45.45) = 0, p = 1.00, Cohen's d = 0.00).

The Mann-Whitney *U* test is a non-parametric alternative to the independent two-sample *t* test.

First, install the `rcompanion` package to calculate the effect size (*r*) for a Mann-Whitney *U* test.

```
# install.packages('rcompanion')
library(rcompanion)
```

Then, conduct a Mann-Whitney *U* test on the first and second dimensions.

```
# non parametric tests
w_test_d1 = wilcox.test(ona_first_points_d1, ona_second_points_d1)
w_test_d2 = wilcox.test(ona_first_points_d2, ona_second_points_d2)
```

```
w_test_d1
```

```
##
##  Wilcoxon rank sum exact test
##
## data:  ona_first_points_d1 and ona_second_points_d1
## W = 130, p-value = 0.0009533
## alternative hypothesis: true location shift is not equal to 0
```

```
w_test_d2
```

```
##
##  Wilcoxon rank sum exact test
##
## data:  ona_first_points_d2 and ona_second_points_d2
## W = 264, p-value = 0.6593
## alternative hypothesis: true location shift is not equal to 0
```

Compute any other statistics that may be of interest. A few examples are given below.

```
median(ona_first_points_d1)

## [1] -0.04307778

median(ona_second_points_d1)

## [1] 0.09596238

median(ona_first_points_d2)

## [1] 0.001753116

median(ona_second_points_d2)

## [1] 0.05862436

length(ona_first_points_d1)

## [1] 26

length(ona_second_points_d1)

## [1] 22

length(ona_first_points_d2)

## [1] 26

length(ona_second_points_d2)

## [1] 22

abs(wilcoxonR(ona_first_points_d1, ona_second_points_d1))

## r
## 0

abs(wilcoxonR(ona_first_points_d2, ona_second_points_d2))

##     r
## 0.707
```

Here, along the *x* axis (MR1), a Mann-Whitney *U* test shows that the FirstGame condition (Mdn = −0.04, N = 26) was statistically significantly different for alpha = 0.05 from the SecondGame condition (Mdn = 0.10, N = 22 U = 130, p = 0.001, r = 0.00). Along the *y* axis (SVD2), a Mann-Whitney *U* test shows that the FirstGame condition (Mdn = 0.001, N = 26) is not statistically significantly different for alpha = 0.05 from the SecondGame condition (Mdn = 0.00, N = 22, U = 264, p = 0.66, r = 0.71). The absolute value of `r` value in Mann-Whitney U test varies from 0 to close to 1. The interpretation values for r commonly in published literature is: `0.10 - < 0.3` (small effect), `0.30 - < 0.5` (moderate effect) and > = `0.5` (large effect).

4.7 *Model Evaluation*

4.7.1 Variance Explained

Briefly, variance explained (also called explained variation) refers to the proportion of the total variance in a dataset that is accounted for by a statistical model or set of predictors.

In ONA, to represent high-dimensional vectors in a two-dimensional space, ONA uses either singular value decomposition or means rotation combined with SVD. For each of the reduced dimensions, the variance in patterns of association among units explained by that dimension can be computed.

```
head(set.ona$model$variance,2)
```

```
##       MR1      SVD2
## 0.1367940 0.2736079
```

In our example above, since we used means rotation method, the first dimension is labeled as MR1 and the second dimension is labeled as SVD2.The two dimensions in combination explained more than 40% of the variance.

Here, the first dimension is MR1 and the second dimension is SVD2. The two dimensions in combination explained more than 40% of the variance.

As with any statistical model, greater explained variance does not necessarily indicate a better model, as it may be due to overfitting, but it provides one indicator of model quality.

4.7.2 Goodness of Fit

Briefly, a model's goodness of fit refers to how well a model fits or represents the data. A model with a high goodness of fit indicates that it accurately represents the data and can make reliable predictions.

In ONA, a good fit means that the positions of the nodes in the space—and thus the network visualizations—are consistent with the mathematical properties of the model. In other words, we can confidently rely on the network visualizations to interpret the ONA model. The process that ONA uses to achieve high goodness of fit is called co-registration, the same as the one used in ENA. The mathematical details of co-registration are beyond the scope of this chapter and can be found in Bowman et al. [2].

To test a model's goodness of fit, use `ona::correlations`. The closer the value is to 1, the higher the model's goodness of fit is. Most ENA models have a goodness of fit that is well above 0.90.

```
ona::correlations(set.ona)

##     pearson  spearman
## 1 0.9801173 0.9801799
## 2 0.9801431 0.9759160
```

4.7.3 Close the Interpretative Loop

Another approach to evaluate an ONA model is to confirm the alignment between quantitative model (in our case, our ONA model) and the original qualitative data. In other words, we can return to the original data to confirm that quantitative findings give a fair representation of the data. This approach is an example of what's called as closing the interpretative loop in Quantitative Ethnography field [1].

For example, based on our visual analysis of the network of "SecondGame::samuel o" in previous section, we are interested in what the lines are in the original data that contributed to the connection *from* Performance.Parameters *to* Design.Reasoning.

Let's first review what "SecondGame::samuel o" ONA network looks like. Based on the connection direction and strength from Technical.Constraints to Performance.Parameters, we would expect to see more examples of Samuel responded with "Design.Reasoning" to "Performance.Parameters", than the other way around (Fig. 25).

```
ona:::plot.ena.ordered.set(set.ona, title = "SecondGame::samuel o") |>
  units(
    points=set.ona$points$ENA_UNIT$`SecondGame::samuel o`,
    points_color = "blue",
    show_mean = FALSE, show_points = TRUE, with_ci = FALSE) |>
  edges(
    weights = set.ona$line.weights$ENA_UNIT$`SecondGame::samuel o`,
    edge_size_multiplier = edge_size_multiplier,
    edge_arrow_saturation_multiplier = edge_arrow_saturation_multiplier,
    node_position_multiplier = node_position_multiplier,
    edge_color = c("blue")) |>
  nodes(
    node_size_multiplier = node_size_multiplier,
    node_position_multiplier = node_position_multiplier,
    self_connection_color = c("blue"))
```

To do so, we use `view()` function and specify required parameters as below.

This is going to activate a window shows up in your `Viewer` panel. If it is too small to read, you can click on the "Show in new window" button to view it in your browser for better readability.

In the `Viewer` panel, hover over your cursor on any of the lines that are in bold, a size of 7 lines rectangle shows up, representing that in a moving stanza window of size 7, the referent line (the line in bold) and its preceding 6 lines. The 1 and 0 in Technical.Constraints column and Design.Reasoning column shows where the connections happened (Fig. 26).

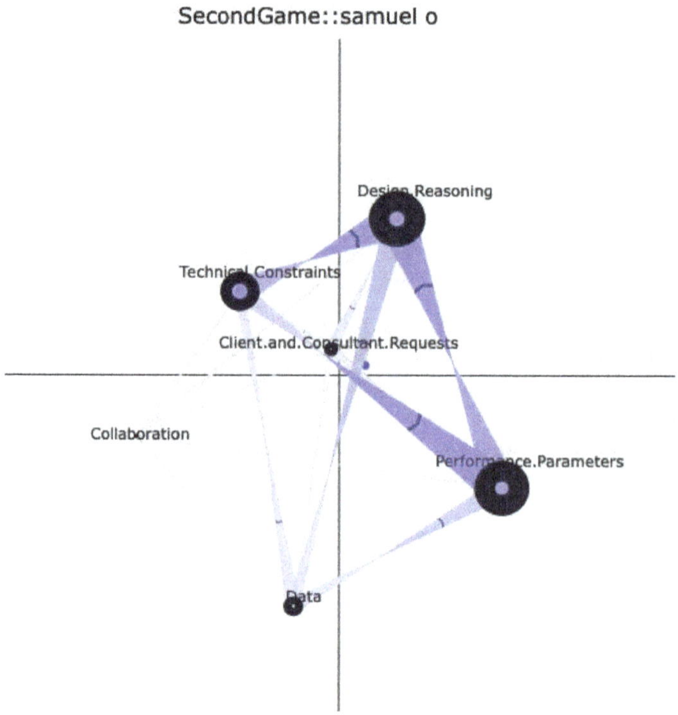

Fig. 25 ONA network for a student from SecondGame

QEID	Condition	UserName	GroupName	ActivityNumber	text	Performance.Parameters	Design.Reasoning
2464	SecondGame	jackson p	Hydraulic	5	I would look at the numbers and saw how good each power source or control sensor was at each	1	1
2465	SecondGame	samuel o	Hydraulic	5	I had to weigh the pros and cons of each power source and sensor	0	0
2466	SecondGame	shane t	Hydraulic	5	So what did you guys choose as the best battery and sensor?	0	1
2467	SecondGame	brandon f	Hydraulic	5	The area covered by the NiCd graph really stood out to me as the clear front runner	0	0
2468	SecondGame	brandon f	Hydraulic	5	The area covered by the NiCd graph really stood out to me as the clear front runner	0	0
2469	SecondGame	jackson p	Hydraulic	5	The graphs gave us an approximation by showing the area that each individual battery type took	0	0
2470	**SecondGame**	**samuel o**	**Hydraulic**	**5**	**The helped illustrate which devices were good and bad with different attributes**	**1**	**1**
2471	SecondGame	shane t	Hydraulic	5	They gave us a visual of the data, made it easier to compare them and see the best choice	0	1
2472	SecondGame	casey f	Hydraulic	5	@Brandon: I agree. It covers the most area on the graph.	0	0
2473	SecondGame	jackson p	Hydraulic	5	I agree with NiCd too	0	0
2474	SecondGame	samuel o	Hydraulic	5	@Brandon: I agree, I like the NiCd also	0	0
2475	SecondGame	jackson p	Hydraulic	5	when it came to the different attributes, i think that all were important in their own way but i think	1	0
2476	SecondGame	casey f	Hydraulic	5	Well not one source/sensor can be the best in every area so we had to sacrifice certain attributes.	1	1
2477	**SecondGame**	**samuel o**	**Hydraulic**	**5**	**No, I just was mindful of the how one device scores relative to other ones**	**0**	**1**
2478	SecondGame	shane t	Hydraulic	5	I also chose the NiCd battery	0	1
2479	SecondGame	brandon f	Hydraulic	5	to agree on which power source and control sensor we should use	0	0
2480	SecondGame	brandon f	Hydraulic	5	to agree on which power source and control sensor we should use	0	0
2481	SecondGame	jackson p	Hydraulic	5	Test the battery out with an actual prototype	0	1
2482	SecondGame	casey f	Hydraulic	5	@Shane: I agree	0	0
2483	SecondGame	jackson p	Hydraulic	5	do more research on the nickel cadmium battery and see how we can optimize it	0	1
2484	**SecondGame**	**samuel o**	**Hydraulic**	**5**	**design prototype**	**0**	**1**

Fig. 26 A screenshot of the view() function result. The highlighted lines represent lines within the same stanza window

Notice that here we are viewing the same qualitative example as in Sect. 3.7.3 in ENA. In line 2477 Samuel shared his [Design.Reasoning] about "mindful of (the) how one device scores relative to other ones", ***as a response to*** what Casey said in line 2476 about [Performance.Parameters] "not one source/censor can be the best in every area so we had to sacrifice certain attributes", as well as what Jackson said in line 2475 about safety as one of the [Performance.Parameters] "when it came to the different attributes, i think that all were important in their own way but i think safety is one of the most important".

Here, ONA was able to not only capture the occurrence between code Design.Reasoning and Performance.Parameters as ENA did, but also represent the connection direction *from* Design.Reasoning *to* Performance.Parameters.

4.8 Using ONA Model Outputs in Other Analyses

As with ENA, the outputs of ONA models can be used as inputs in other statistical models. See Sect. 3.8 for an example using ENA points.

5 Additional Features

In the sections above, we demonstrated how to do an ENA analysis and an ONA analysis. In this section, we show how to project new data into a space constructed with different data. This can be done as long as the same codes are used in both sets.

5.1 Projections in ENA

To project the ENA points from one model into a space constructed with different data, replace the `rotation.set` parameter of `ena.make.set`. In the example below, an "expert" model is developed using the SecondGame units and the FirstGame (novice) units are projected into that space. By projecting novice model's units into expert model's space, users can interpret the projected novice units' networks based on the two dimensions defined by expert model's node positions. In other words, interpreting novice's networks in the context of experts' space (Figs. 27 and 28).

```
data = rENA::RS.data

#expert data
exp.data = subset(data, Condition == "SecondGame")

#novice data
nov.data = subset(data, Condition == "FirstGame")

#expert model
units_exp = exp.data[,c("Condition","UserName")]
conversation_exp = exp.data[,c("Condition","GroupName","ActivityNumber")]
codes_exp = exp.data[,codeCols]
meta_exp = exp.data[,c("CONFIDENCE.Change",
                    "CONFIDENCE.Pre","CONFIDENCE.Post","C.Change")]

set_exp =
  ena.accumulate.data(
  text_data = exp.data[, 'text'],
  units = units_exp,
  conversation = conversation_exp,
  codes = codes_exp,
  metadata = meta_exp,
  window.size.back = 7,
) |>
  ena.make.set()

#novice model
units_nov = nov.data[,c("Condition","UserName")]
conversation_nov = nov.data[,c("Condition","GroupName","ActivityNumber")]
codes_nov = nov.data[,codeCols]
meta_nov = nov.data[,c("CONFIDENCE.Change",
                    "CONFIDENCE.Pre","CONFIDENCE.Post","C.Change")]

set_nov =
  ena.accumulate.data(
  text_data = nov.data[, 'text'],
  units = units_nov,
  conversation = conversation_nov,
  codes = codes_nov,
  metadata = meta_nov,
  window.size.back = 7,
) |>
  ena.make.set(rotation.set = set_exp$rotation)


# plot expert model (what we projected into) Using plotting wrapper to save t
ime
plot_exp = ena.plotter(set_exp,
                        points = T,
```

```
                                   mean = T,
                                   network = T,
                                   print.plots = F
                                  )

# plot test model (points from test model in training model space)
plot_nov = ena.plotter(set_nov,
                                   points = T,
                                   mean = T,
                                   network = T,
                                   print.plots = F)

#compare plots
plot_exp$plot
```

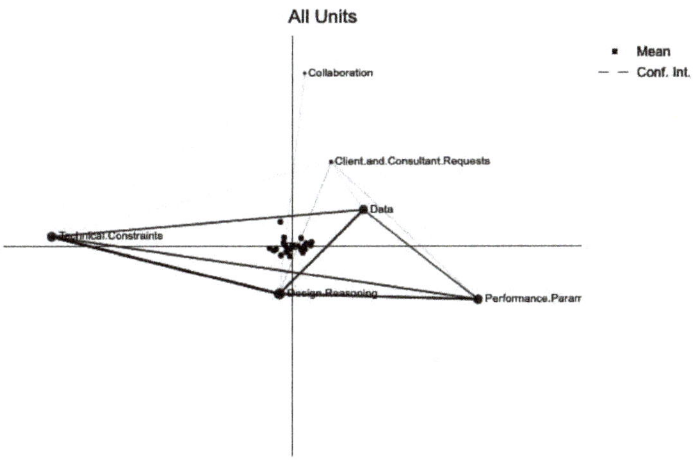

Fig. 27 Mean network for all units in the expert model

```
plot_nov$plot
```

5.2 *Projections in ONA*

Projection works similarly in ONA (Fig. 29).

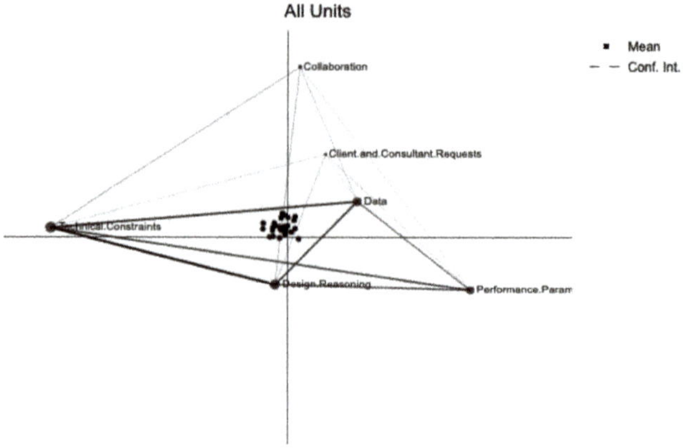

Fig. 28 Mean network for novice students projected into expert space

```
data = ona::RS.data

#expert data
exp.data = subset(data, Condition == "SecondGame")

#novice data
nov.data = subset(data, Condition == "FirstGame")

#shared unit cols
units = c("UserName","Condition","GroupName")

#shared code cols
codes = c(
          'Data',
          'Technical.Constraints',
          'Performance.Parameters',
          'Client.and.Consultant.Requests',
          'Design.Reasoning',
          'Collaboration')
```

```
#shared hoo
hoo = conversation_rules(
  (Condition %in% UNIT$Condition & GroupName %in% UNIT$GroupName))


#expert accum
accum.exp = contexts(exp.data, units_by = units, hoo_rules = hoo) |>
  accumulate_contexts(codes = codes,
                      decay.function = decay(simple_window, window_size = 7),
                      return.ena.set = FALSE, norm.by = NULL)
#expert model
set.exp = model(accum.exp)

#novice accum
accum.nov = contexts(nov.data, units_by = units, hoo_rules = hoo) |>
  accumulate_contexts(codes = codes,
                      decay.function = decay(simple_window, window_size = 7),
                      return.ena.set = FALSE, norm.by = NULL)
#novice model
set.nov = model(accum.nov)

# projecting novice data into expert space
set = model(accum.nov, rotation.set = set.exp$rotation)

ona:::plot.ena.ordered.set(set, title = "novice data into expert space") |>
  units(
    points = set$points,
    show_mean = TRUE, show_points = TRUE, with_ci = TRUE) |>
  edges(
    weights = set$line.weights) |>
  nodes(
    self_connection_color = "red",
    node_size_multiplier = 0.6)
```

6 Discussion

In this chapter, we introduced two techniques, ENA and ONA, for quantifying, visualizing, and interpreting networks using coded data. Through the use of a demonstration dataset that documents collaborative discourse among students collaborating to solve an engineering design problem, we provided step-by-step instructions on how to model complex, collaborative thinking using ENA and ONA in R. The chapter combines theoretical explanations with tutorials, intended to be of aid to researchers with varying degrees of familiarity with network analysis techniques and R. This chapter mainly showcased the standard and most common use of these two tools. The ENA and ONA R packages, akin to other R packages, offer flexibility to researchers to tailor their analyses to their specific needs. For example, users with advanced R knowledge can supply their own adjacency matrices and use ENA or ONA solely as a visualization tool rather than an integrated modeling and visualization tool.

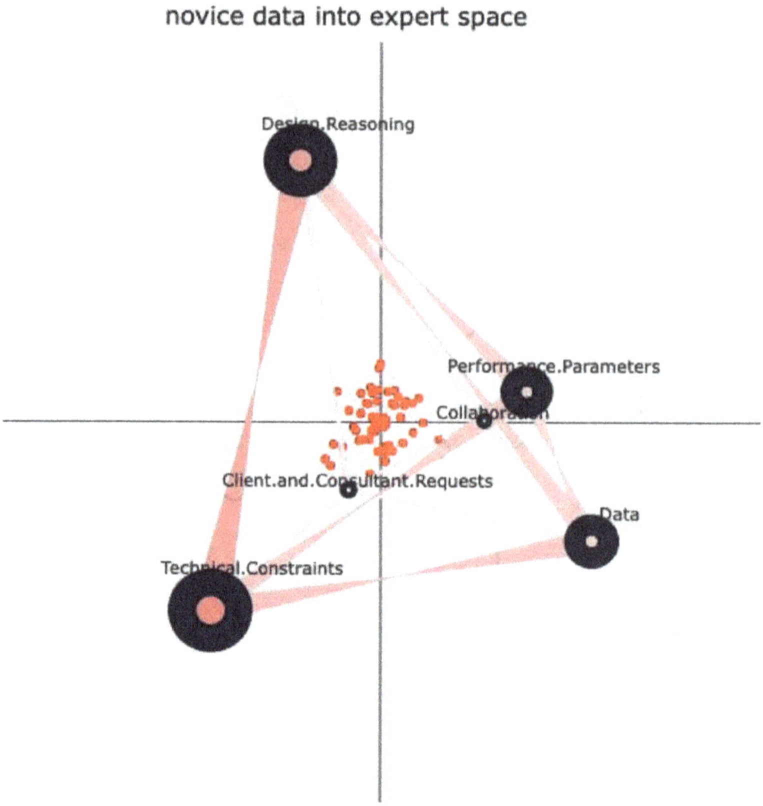

Fig. 29 Projecting novice data into expert space

Due to the technical and practical focus of this chapter, we omitted detailed explanations of the theoretical, methodological, and mathematical foundations of ENA and ONA that are crucial for informed, theory-based learning analytics research using these techniques. Consult the Further Reading section for papers that explain these aspects of ENA and ONA in greater detail.

References

1. Shaffer DW (2017) Quantitative ethnography. Cathcart Press
2. Bowman D, Swiecki Z, Cai Z, Wang Y, Eagan B, Linderoth J, Shaffer DW (2021) The mathematical foundations of epistemic network analysis. In: Advances in Quantitative Ethnography: Second International Conference, ICQE 2020, Malibu, CA, USA, February 1–3, 2021, Proceedings 2, pp 91–105
3. Csanadi A, Eagan B, Shaffer DW, Kollar I, Fischer F (2018) When coding-and-counting is not enough: using epistemic network analysis (ENA) to analyze verbal data in CSCL research. Int J Comput-Support Collab Learn 13(4):419–438

4. Oshima J, Oshima R, Fujita W (2018) A mixed-methods approach to analyze shared epistemic agency in Jigsaw instruction at multiple scales of temporality. J Learn Anal *5*(1):10–24. https://doi.org/10.18608/jla.2018.51.2
5. Bressler DM, Bodzin AM, Eagan B, Tabatabai S (2019) Using epistemic network analysis to examine discourse and scientific practice during a collaborative game. J Sci Educ Technol *28*:553–566
6. Swiecki Z, Ruis AR, Farrell C, Shaffer DW (2020) Assessing individual contributions to collaborative problem solving: a network analysis approach. Comput Hum Behav 104:105876
7. Prieto LP, Rodríguez-Triana MJ, Ley T, Eagan B (2021) The value of epistemic network analysis in single-case learning analytics: a case study in lifelong learning. In: Advances in quantitative ethnography: second international conference, ICQE 2020, Malibu, CA, USA, February 1–3, 2021, Proceedings 2, Springer International Publishing, pp. 202–217
8. Zhang S, Gao Q, Sun M, Cai Z, Li H, Tang Y, Liu Q (2022) Understanding student teachers' collaborative problem solving: Insights from an epistemic network analysis (ENA). Comput Educ 183:104485
9. Bauer E, Sailer M, Kiesewetter J, Shaffer DW, Schulz C, Pfeiffer J, Gurevych I, Fischer MR, Fischer F (2020) Pre-Service teachers' diagnostic argumentation: what is the role of conceptual knowledge and epistemic activities?. In: Proceedings of the fifteenth international conference of the learning sciences, pp 2399–2400.
10. Fernandez-Nieto GM, Martinez-Maldonado R, Kitto K, Buckingham Shum S (2021) Modelling spatial behaviours in clinical team simulations using epistemic network analysis: methodology and teacher evaluation. In: LAK21: 11th international learning analytics and knowledge conference, April, pp 386–396
11. Phillips M, Trevisan O, Carli M, Mannix T, Gargiso R, Gabelli L, Lippiello S (2023, March) Uncovering patterns and (dis) similarities of pre-service teachers through Epistemic Network Analysis. In: Society for information technology & teacher education international conference. Association for the Advancement of Computing in Education (AACE), pp 1021–1030
12. Tan Y, Ruis AR, Marquart C, Cai Z, Knowles MA, Shaffer DW (2022, October) Ordered network analysis. In: International conference on quantitative ethnography. Springer Nature Switzerland, Cham, pp 101–116
13. Fan Y, Tan Y, Raković M, Wang Y, Cai Z, Shaffer DW, Gašević D (2022) Dissecting learning tactics in MOOC using ordered network analysis. J Comput Assist Learn:1–13
14. Shaffer DW, Collier W, Ruis AR (2016) A tutorial on epistemic network analysis: analyzing the structure of connections in cognitive, social, and interaction data. J Learn Anal 3(3):9–45
15. Shaffer DW, Arastoopour G (2014) Guide to RSdata.csv sample ENA data set. Madison, WI: Games and Professional Simulations Technical Report 2014-3
16. Arastoopour G, Shaffer DW, Swiecki Z, Ruis AR, Chesler NC (2016) Teaching and assessing engineering design thinking with virtual internships and epistemic network analysis. Int J Eng Educ 32(3B):1492–1501
17. Chesler NC, Ruis AR, Collier W, Swiecki Z, Arastoopour G, Shaffer DW (2015) A novel paradigm for engineering education: Virtual internships with individualized mentoring and assessment of engineering thinking. J Biomech Eng *137*(2)
18. Siebert-Evenstone AL, Arastoopour Irgens G, Collier W, Swiecki Z, Ruis AR, Shaffer DW (2017) In search of conversational grain size: Modelling semantic structure using moving stanza windows. J Learn Anal 4(3):123–139
19. Ruis AR, Siebert-Evenstone AL, Pozen R, Eagan B, Shaffer DW (2019) Finding common ground: a method for measuring recent temporal context in analyses of complex, collaborative thinking. In: Lund K, Niccolai G, Lavoué E, Hmelo-Silver C, Gwon G, Baker M (eds) A wide lens: combining embodied, enactive, extended, and embedded learning in collaborative settings: 13th international conference on Computer Supported Collaborative Learning (CSCL), I, pp 136–143
20. Kaur A, Kumar R (2015) Comparative analysis of parametric and non-parametric tests. J Comput Math Sci *6*(6):336–342

Further Reading

21. Brohinsky J, Marquart C, Wang J, Ruis AR, Shaffer DW (2021) Trajectories in epistemic network analysis. In Ruis AR, Lee SB (eds) Advances in quantitative ethnography: second international conference, ICQE 2020, Malibu, CA, USA, February 1–3, 2021, Proceedings, Springer, pp 106–121
22. Gasevic D, Greiff S, Shaffer DW (2022) Towards strengthening links between learning analytics and assessment: challenges and potentials of a promising new bond. Comput Hum Behav:1–7
23. Shaffer DW (2017) Quantitative ethnography. Cathcart Press
24. Shaffer DW (2018) Big data for thick description of deep learning. In: Millis K, Long D, Magliano J, Weimer K (eds) Deep learning: multi-disciplinary approaches. Routledge, New York, pp 262–275
25. Shaffer DW (2018) Epistemic network analysis: understanding learning by using big data for thick description. In: Fischer F, Hmelo-Silver CE, Goldman SR, Reimann P (eds) International handbook of the learning sciences. Routledge, New York, pp 520–531
26. Shaffer DW, Eagan B, Knowles M, Porter C, Cai Z (2022) Zero re-centered projection: an alternative proposal for modeling empty networks in ENA. In: Wasson B, Zörgő S (eds) Advances in quantitative ethnography: third international conference, ICQE 2021, Virtual Event, November 6–11, 2021, Proceedings, Springer, pp 66–79
27. Swiecki Z, Lian Z, Ruis AR, Shaffer DW (2019) Does order matter? Investigating sequential and cotemporal models of collaboration. In: Lund K, Niccolai G, Lavoué E, Hmelo-Silver C, Gwon G, Baker M (eds) A wide lens: combining embodied, enactive, extended, and embedded learning in collaborative settings: 13th international conference on Computer Supported Collaborative Learning (CSCL), I, pp 112–119
28. Wang Y, Swiecki Z, Ruis AR, Shaffer DW (2021) Simplification of epistemic networks using parsimonious removal with interpretive alignment. In Ruis AR, Lee SB (eds) Advances in quantitative ethnography: second international conference, ICQE 2020, Malibu, CA, USA, February 1–3, 2021, Proceedings, Springer, pp 137–151

Part V
Psychometrics

Psychological Networks: A Modern Approach to Analysis of Learning and Complex Learning Processes

Mohammed Saqr, Emorie Beck, and Sonsoles López-Pernas

1 Introduction

Learning has long been described as a complex system that involves the interactions of several elements across time, persons and contexts [1–3]. Such descriptions include learning theories, constructs, classroom, learners, education as a whole and educational policies. For instance, Zimmerman describes self-regulation as a "complex system of interdependent processes" [4]. Winne describes the SRL process as "complex, dynamically changeable and contextually sensitive" [5]. Similar descriptions and operationalizations exist for engagement [6], motivation [7, 8], metacognition [9], agency [10] and achievement goals [11], to mention a few. Similarly, the students [12], the classroom [13], collaborative groups [14] have all been conceptualized from the perspective of complex systems. Nevertheless, despite the long history and the solid theoretical grounds, methodological approaches that capture the complexities of learning and learners have been lagging behind both in adoption or applications [1, 15]. This chapter introduces the complex systems and offers tutorial on one of the most important and promising methods of analyzing complex systems with a real-life dataset.

1.1 Complex Systems

A complex system is an ensemble of interdependent elements that interact with one another, evolve, adapt or self-organize in a nonlinear way leading to the emergence

M. Saqr (✉) · S. López-Pernas
School of Computing, University of Eastern Finland, Joensuu, Finland
e-mail: mohammed.saqr@uef.fi

E. Beck
UC Davis, Davis, CA, USA

M. Saqr, S. López-Pernas (eds.), *Learning Analytics Methods and Tutorials*,
https://doi.org/10.1007/978-3-031-54464-4_19

of new phenomenon or behavior [16, 17]. Let's for example take engagement as an example. Engagement is understood as a multidimensional construct with three dimensions: behavioral, cognitive and emotional dimensions [18, 19]. Research has shown that each of the engagement dimensions influence (or interact with) each other; for example, enjoying school (emotions) stimulates school attendance (behavior) and investment in learning (cognitive) [18, 20]. Such interactions may lead to the emergence of resilience (new behavior) [21]. We also know that such interactions are nonlinear—that is, we can not combine engagement dimensions together in a single score (e.g., behavioral + cognitive + emotional $\neq$ engagement). In other words, the sum of the parts does not equal the whole. The same can be said about self-regulated learning, the Zimmerman cyclical phases model describes three phases: forethought (task analysis and goal-setting), performance phase (task enactment and strategies) and self-reflection (evaluation and adaptation). Each of these phases are explicitly modeled as influencing each other in a cyclical way; they interact in a non-linear way and such interactions lead to the emergence of learning strategies [22].

The study of these complexities of learning requires methods that account for such interactions, interdependencies and heterogeneity [1, 15]. Linear methods such as correlations, regression or comparison of means are essentially reductionist, that is, disregard the inter-relationships between the variables. For instance, in regression analysis, all variables are included as predictors of one predicted variable (dependable variable). By design, such variables are assumed to be independent and not correlated. While this view can offer understanding through simplification, it does so by compromising the understanding of the big picture. In contrast to the reductionist view of human behavior, embracing the view that learning is complex is more tethered to reality and promises for renewal of our knowledge [1, 15, 23].

1.2 Network Analysis

Network analysis as a framework affords researchers a powerful tool set to chart relations, map connections, and discover clusters or communities between interacting components and therefore, has become one of the most important methods for understanding complex systems [24, 25]. In education, social network analysis has been used for decades [25], and we covered it in detail in previous chapters of the book [26]; [27]; [28]}. Yet, to go beyond social interactions, one needs a different type of networks: probabilistic networks [29, 30]. In probabilistic networks, the variables (often indicators of constructs or scale scores) are the nodes or the vertices of the networks, the relationships or magnitude of interactions (i.e. probabilistic associations) between such variables form the edges. In particular, we will focus in this chapter on psychological networks: Gaussian graphical models (GGM) [30]. In psychological networks, the variables may be constructs, behaviors, attitudes, etc., and the interactions are partial correlations among the nodes [3]. A partial correlation measures the relationship (or the conditions dependence) between two variables after removing (controlling or adjusting) for all other variables in the network or what is known as ceteris paribus [30, 31]. For instance, if we are modeling a network where motivation, achievement, engagement, self-regulation and well-being are nodes, and we observed a relationship between well-being and achievement, such relationships means that well-being is associated with achievement beyond what can be explained by their associations with motivation, engagement, self-regulation (all other variables in the network). The absence of a relationship between two variables signifies a conditional independence between the two variables after controlling for all other variables in the network [31]. As such, the presence or lack of interactions are both interpretable and carry a meaning. Psychological networks offer rigorous methods for the assessment of the accuracy of estimated networks, edge weights and centrality measures through bootstrapping, assessment of sampling variability through simulation of "expected replicability" of the studied networks) [29].

2 Related Work

As the term *psychological networks* suggests, most studies operationalizing this method are rooted in psychological research. Consequently, its connection to educational research is primarily established through educational psychology. For example, a study by Liu et al. [32] sought to identify a link between students' personality and their experience of psychological distress during and after the COVID-19 lockdown. Their results suggest that neuroticism was linked to heightened psychological distress both during and after the lockdown, whereas extraversion and conscientiousness were associated with the alleviation of psychological distress. Another study [33] investigated the relationship among various aspects influencing nursing students' psychological well-being, and found that perceived social support and one's professional self-concept were the most central predictors.

The cited studies have relied on self-reported data. However, a recent trend is to take advantage of the intensive trace log data collected from digital educational tools. An example is the work by Saqr et al. [34], who used psychological networks to discover the interplay between self-regulated learning tactics among foreign language students. Their findings reveal a strong correlation between writing text and social bonding, as well as between acknowledging others and social bonding. López-Pernas et al. [35] clustered students into player profiles according to their performance in an educational game and used psychological networks to map the relationships between game performance indicators for each profile. They found that certain indicators (e.g., the use of hints for help) are decisive for the success of certain player profiles and superfluous for others.

A recent trend in the use of psychological networks in education is to understand and visualize within-person (i.e., idiographic) phenomena, in which networks are constructed for data from a single individual at a time [36, 37]. For instance, Malmberg et al. [38] examined the association between monitoring events and phases of regulation in collaborative learning. Their results showed that cyclical phases of regulation do not occur simultaneously but rather monitoring motivation predicts the monitoring of task definition, ultimately leading to task enactment. A recent study by Saqr [39] constructed a network of engagement indicators using between-person data and another using within-person data. The results revealed that group-level inferences hardly generalize to individuals. For example, regularity in study and frequency of accessing resources were positively correlated at the group level but negatively so at the individual level. Such findings highlighted the need for person-specific insights to leverage personalized interventions to increase learner engagement.

3 Tutorial with R

In this section, we present a step-by-step tutorial on how to use psychological networks in cross-sectional survey data. The dataset that we are using contains the results of 6071 students' responses to a survey investigating students' psychological characteristics related to their well-being during the COVID-19 pandemic in Finland and Austria. The survey questions cover students' basic psychological needs (relatedness, autonomy, and experienced competence), self-regulated learning, positive emotion and intrinsic learning motivation. Moreover, the dataset contains demographic variables, such as country, gender, and age and is well described in the dataset chapter [40]. In the tutorial, we will construct and visualize a network that represents the relationship between the different psychological characteristics, we will interpret and evaluate these relationships, and we will compare how the networks differ among demographic groups.

3.1 The Libraries

We start by importing the necessary packages We know `rio` [41] and `tidyverse` [42] from previous chapters for importing and manipulating data respectively. The `bootnet` [43] package provides methods to estimate and assess accuracy and stability of estimated network structures and centrality indices. The package `networktools` [44] includes a selection of tools for network analysis and plotting. `NetworkToolbox` [45] implements network analysis and graph theory measures used in neuroscience, cognitive science, and psychology. `NetworkComparisonTest` [46] allows to assess the difference between two networks based on several invariance measures. The package `qgraph` [47] offers network visualization and analysis, as well as Gaussian graphical model computation. The package `mgm` [48] provides estimation of k-Order time-varying Mixed Graphical Models and mixed VAR(p) models via elastic-net regularized neighborhood regression. Lastly, `matrixcalc` [49] offers a collection of functions to support matrix calculations for probability, econometric and numerical analysis.

```
library(rio)
library(tidyverse)
library(bootnet)
library(networktools)
library(NetworkToolbox)
library(NetworkComparisonTest)
library(qgraph)
library(mgm)
library(matrixcalc)
```

3.2 Importing and Preparing the Data

The first step is importing the data and doing the necessary preparation, that is, removing the missing and incomplete responses.

```
URL<-("https://raw.githubusercontent.com/lamethods/data/main/
                              11_universityCovid/data.sav")  df  <- import(URL)  |>
  drop_na()
```

To represent each of the constructs that the survey aimed at capturing, we combine all the columns representing items from the same construct into one by averaging the responses. The next code calculates the mean for each construct by averaging all the related items:

```
aggregated <- df |> rowwise() |> mutate(
       Competence = rowMeans(cbind
                (comp1.rec , comp2.rec, comp3.rec),
                 na.rm = T),
       Autonomy = rowMeans(cbind
                 (auto1.rec , auto2.rec, auto3.rec),
                 na.rm = T),
       Motivation = rowMeans(cbind
                    (lm1.rec , lm2.rec, lm3.rec),
                    na.rm = T),
       Emotion = rowMeans(cbind
                    (pa1.rec , pa2.rec, pa3.rec),
                    na.rm = T),
       Relatedness = rowMeans(cbind
                        (sr1.rec , sr2.rec, sr3.rec),
                        na.rm = T),
```

Table 1 Preview of the data

	Relatedness	Competence	Autonomy	Emotion	Motivation	SRL
1	3.000	2.333	2.333	2.000	1.333	2.667
2	5.000	3.000	1.667	2.333	2.000	4.333
3	4.667	4.000	2.667	4.000	3.000	4.333
4	4.333	3.333	3.000	3.333	2.667	2.667
5	5.000	4.000	3.000	4.000	3.000	4.333
6..7159						
7160	4.333	4.000	4.000	3.000	5.000	4.000

```
    SRL = rowMeans(cbind
                  (gp1.rec , gp2.rec, gp3.rec),
                  na.rm = T))
```

We can now keep only the newly created columns. We also create subsets of the data based on gender (a dataset for males and another for females) and country (a data set for Austria and another for Finland). We will use these datasets later for comparison across genders and countries.

```
cols <- c("Relatedness", "Competence", "Autonomy",
    "Emotion", "Motivation", "SRL")

filter(aggregated, country == 1) |>
select(all_of(cols))  -> finlandData
filter(aggregated, country == 0) |>
select(all_of(cols))  -> austriaData
filter(aggregated, gender  == 1) |>
select(all_of(cols)) -> femaleData
filter(aggregated, gender  == 2) |>
select(all_of(cols)) -> maleData

select(aggregated, all_of(cols)) -> allData
```

The `allData` dataframe should look as follows (Table 1) :

3.3 *Assumption Checks*

As a first step of the analysis, we need to check some assumptions to make sure that the dataset and the estimated network are appropriate. First, we need to ensure that the correlation matrix is **positive definite** i.e., the included variables are not a linear combination of each other and therefore, so similar they do not offer new information. This is performed by using the function `is.positive.definite()` from the package `matrixcalc`. Please note that in cases where the correlation matrix is non-positive definite, we can use option `cor_auto` to search for possible positive definite matrices (see later). In our case, the matrix is already positive definite. Note that we also set the `use` argument to `"pairwise.complete.obs"`, or pairwise complete observations to include maximal observations.

```
correlationMatrix <- cor(x = allData, use = c(
            "pairwise.complete.obs"))
is.positive.definite(correlationMatrix)
```

```
[1] TRUE
```

The second assumption that we need to check is whether some variables are highly correlated and therefore **redundant**. In doing so we make sure that each variable is sufficiently distinct from all other variables and captures a unique construct. The `goldbricker` algorithm compares the variables' correlation patterns with all other variables in the dataset. Below, we search for items which are highly inter-correlated using the default values: ($r > 0.50$) with 0.25 as the significant fraction of variables and p-value of 0.05. The results show that the variables are significantly distinct from each other.

```
goldbricker(allData, p = 0.05, method = "hittner2003",
            threshold = 0.25, corMin = 0.5, progressbar =
                     FALSE)
```

```
Suggested reductions: Less than 25 % of correlations are significantly
different for the following pairs:
[1] "No suggested reductions"
```

3.4 Network Estimation

Given that we made sure that the data satisfy the necessary assumptions, we can now build or estimate the network. The estimation means that we quantify the associations between the different variables. Several types of associations can be estimated. The most common in psychological networks are *dependency measures*, such as correlation and regression. In these networks, we are interested in how the levels or values of the variables in the network vary together in a similar way (e.g., if and to what extent higher levels of motivation are associated with higher levels of engagement). Such patterns can be estimated using covariance, simple correlation, partial correlation or relative importance (regression). In this tutorial, we focus on the the most commonly used estimation method, which is regularized partial correlation.

Regularized partial correlations have been shown to (1) retrieve the true structure of the network in most situations, and (2) offer an interpretable sparse network that shows conditional association between variables. A partial correlation means that the association between each two variables is above and beyond all other variables in the network, or conditioning on all other variables in the network or holding all other variables constant often referred to as ***ceteris paribus***. This allows us to estimate, for example, the association between motivation and engagement beyond other variables that are included in the network, e.g., achievement, anxiety or enjoyment. Regularization is the process of applying an extra penalty for the complexity of network model. A growing body of research recommends the procedure for several reasons. Regularization helps eliminate spurious edges that results from influence of other nodes, and shrinks trivial edges to zero and thus help eliminates Type 1 error or "false positive" edges. In doing so, the resulting network model is less complex, sparser, simpler to interpret with only the strong meaningful correlations. In doing so, regularization helps retrieve a true—and conservative—structure of the network. The penalty is commonly applied using the *least absolute shrinkage and selection operator* "LASSO".

Network estimation can be performed using several packages. We will use the package `bootnet` and the function `estimateNetwork()`. To estimate the network we need to pass the data as an argument, and select the option `default = "EBICglasso"` to estimate a regularized network. By default, the correlation type is set to `cor_auto` which can detect the distribution of the variables and set the appropriate correlation coefficient: polychoric, polyserial or Pearson correlation. Other options can be `"cor"` which will compute a correlation network and `"npn"` which will apply non-paranormal transformation—to normalize the data—and then compute correlations. By default, `estimateNetwork()` computes 100 models with various degrees of sparsity. The best model is selected based on the lowest Extended Bayesian Information Criterion value (EBIC) given a hyperparameter gamma (γ) which balances a trade off between false positive edges and suppressing of the true edges [30]. Gamma ranges from 0 (favors models with more edges [i.e. uses no regularization]) to 0.5 (favors model with fewer edges). The default for the hyper-parameter gamma (γ) is set to 0.5 to ensure that edges included in the model are true and part of the network. The next code estimates the network and assigns the estimated network to an R object `allNetwork`. The estimated network can be accessed from `allNetwork$graph`. We can also see the details about the network using the function `summary`.

```
allNetwork <- estimateNetwork(allData, default =
                "EBICglasso",
                              corMethod = "cor_auto",
                                    tuning = 0.5)
summary(allNetwork)
```

```
=== Estimated network ===
Number of nodes: 6
Number of non-zero edges: 15 / 15
Mean weight: 0.1397203
Network stored in object$graph

Default set used: EBICglasso
```

3.5 Plotting the Network

The network can be plotted by using the function `plot()`. The resulting plot uses a color blind theme by default where blue edges are positive correlations and red edges are negative correlations. The thickness of the edges is proportional to the magnitude of the regularized partial correlation. As the network shows, we see a very strong correlation between motivation, autonomy and competence. We also observe that emotion is strongly correlated with competence. As we mentioned before, each of the correlations in the network are conditioned—or above and beyond—all other nodes in the network same as regression.

Please also note that we assign the plot to an R object under the name `allDataPlot`. This facilitates further plotting as well as retrieving other data, e.g., the `allDataPlot` object contains the correlation matrix, the plotting parameters, the call arguments etc. For instance, `allDataPlot$layout` returns the network layout which we can get and use in further plotting so that we have a fixed layout for all network to enable us to compare them easily against each other. Plotting in `bootnet` is mostly handled by the `qgraph` package and most options work with few differences. As such, the plot object works with all functions of the `qgraph` package as we will see later. The next code plots the network using the default settings (Fig. 1). The second line retrieves the layout and stores it in an object so we can later reuse it to plot our networks.

```
allDataPlot <- plot(allNetwork)
LX <- allDataPlot$layout
```

However, in most situations, we will need to customize the plot. First, we may need to add a title using the argument `title = "Both countries combined"`. Second, we define the node size using the argument `vsize=9` (you may need to adjust according to your preference). Third, we choose to show the edge weight (i.e., the magnitude of the partial correlation) by setting `edge.labels = TRUE`; this is very important as it shows exactly how strong the correlation is. Fourth, we can choose `cut = 0.10`, which emphasizes edges higher than a threshold of 0.10. Edges below that threshold are shown with less color intensity. Whereas the `cut` argument is optional and relies on the network, it can be used to emphasize the

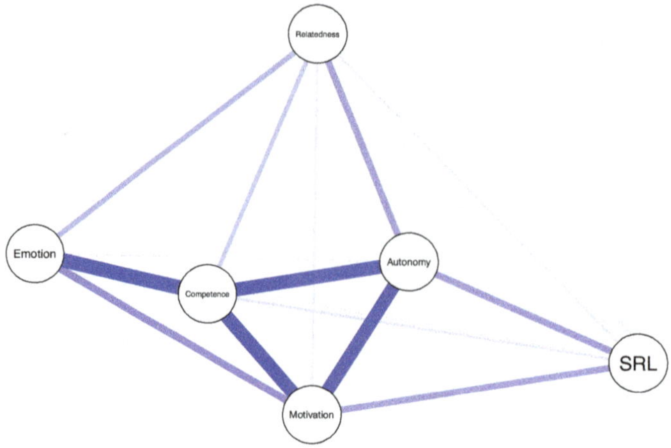

Fig. 1 Network with default settings

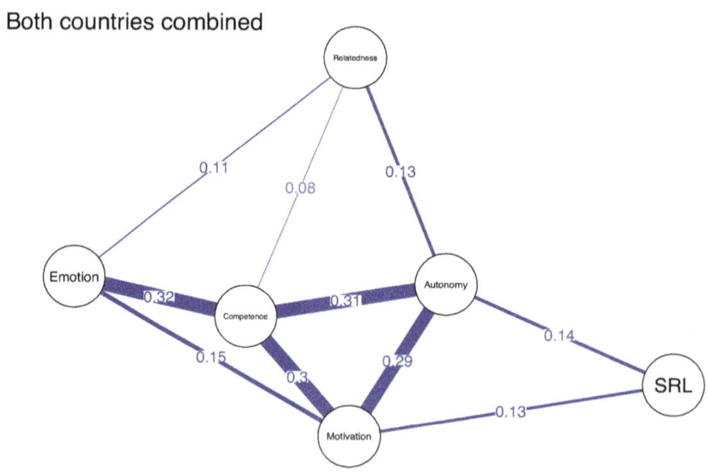

Fig. 2 Network with customized settings

important strong correlations and downsize the "clutter". Another cosmetic option is to hide (but not remove) edges below 0.05 with the argument `minimum = 0.05`. Both `cut` and `minimum` make the network easy to read, interpret and less crowded. For more options, you need to consult the plot function manual `bootnet.plot()` or `qgraph()`. The final result can be seen in Fig. 2.

```
allDataPlot <- plot(
  allNetwork,
  title = "Both countries combined",
  vsize = 9,
```

```
    edge.labels = TRUE,
    cut = 0.10,
    minimum = 0.05,
    layout = "spring")
```

3.6 Explaining Network Relationships

It may be useful—and all the more advisable—to compute the predictability of all nodes which simply means to calculate to which extent each node is explained by its relationships or the proportion of explained variance when regressing all nodes over the said node. In other words, the function computes regression for each node (where all other nodes are the predictors) and returns the R^2. This process is repeated for each node. When the predictability (R^2) is zero, the node is not explained at all by it connections and one should think whether the node belongs to this network or whether the measurement was accurate. When predictability is 1, then 100% of the variance is explained by the relationships of the node. This of course is unlikely and warrants serious checks. Predictability has also been linked to controlability, or the extent to which one can control all other nodes if acted on this node e.g. intervention targeting this node.

For the calculation of predictability, we will need the function `mgm()` from the package `mgm` to estimate the model. The only required parameter for the function is the type of each variable which we specify as `type = rep ('g', 6)` or `c("g" ,"g", "g", "g" ,"g" ,"g")`, which means all variables are Gaussian.

```
fitAllData <- mgm(as.matrix(allData), type = c("g", "g"
            , "g", "g", "g", "g"))
```

The predictability of each variable can be obtained by querying the resulting object as the code shows below. We can also obtain the mean predictability to see how far our network nodes are explained by their relationships on average. Using predictability in plots can enhance the readability of the network and give more information about the extent the node is explained by the current network. We do so by assigning the pie the value of R^2 `pie= predictAll$errors$R2`. We can also see the the RMSE (Root Mean Square Error) which measures the difference between predicted values and the actual values using `predictAll$errors$RMSE`.

```
# Compute the predictability
predictAll <- predict(fitAllData, na.omit(allData))
predictAll$errors$R2
                    # Check  predictability for all variables
```

```
[1] 0.139 0.518 0.442 0.315 0.458 0.126
```

Table 2 Predictability dataframe

Var	R2	RMSE
Relatedness	0.139	0.928
Competence	0.518	0.694
Autonomy	0.442	0.747
Emotion	0.315	0.828
Motivation	0.458	0.736
SRL	0.126	0.935

```
mean(predictAll$errors$R2)    # Mean predictability
```

```
[1] 0.333
```

```
mean(predictAll$errors$RMSE) # Mean RMSE
```

```
[1] 0.8113333
```

We can also create a data frame of the predictability (Table 2).

```
data.frame(
  var = predictAll$errors$Variable,
  R2 = predictAll$errors$R2,
  RMSE = predictAll$errors$RMSE
  )
```

As the network plot (Fig. 3) and the data frame show, the highest predictability belongs to the *competence*, *motivation* and *autonomy*. We also see that SRL and relatedness have low predictability and therefore, are less explained by their connections.

```
allDataPlot <- plot(
  allNetwork,
  title = "Both countries combined",
  vsize = 9,
  edge.labels = TRUE,
  cut = 0.10,
  minimum = 0.05,
  pie = predictAll$errors$R2
  )
```

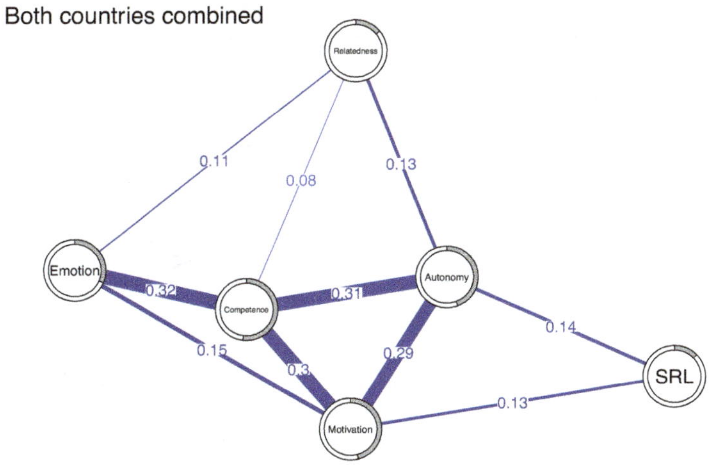

Fig. 3 Network with predicatability as pie charts in the nodes

3.7 Network Inference

Similar to traditional networks, centrality measures can be computed for psychological networks. Here—as in any network—the interpretation depends on the network variables, the estimation method, the weights of edges and of course, the theoretical underpinning that underpins the network structure. In the general sense, centrality measures estimate important, influential or central nodes. Early research has shown that centrality measures can be potential targets for intervention among others. Whereas many centrality measures exist, two centralities have gained traction: degree (or strength) centrality and expected influence. Others, e.g., betweenness, closeness and eigenvector centralities can be calculated but have not so far being "well understood" to be routinely recommended for analysis.

Degree centrality is the tally of connections a node has regardless of weight. Strength centrality is the sum of absolute weights of all connections (the sum of correlations weights positive or negative). Expected influence is similar, however, expected influence sums the raw values. For instance, if a node has connections of 0.3, -0.1 and 0.5, the degree centrality is 3, the strength centrality is $0.3 + 0.1 + 0.5 = 0.9$ and the expected influence centrality is $0.3 - 0.1 + 0.5 = 0.7$. Given that the network we have does not have any negative edges, strength and expected influence centralities are equivalent. To compute the centrality measures, we use the function `centralityPlot()` and provide the network as the main argument. We also need to provide the centralities that we wish to compute. The function plots the centralities by default (Fig. 4).

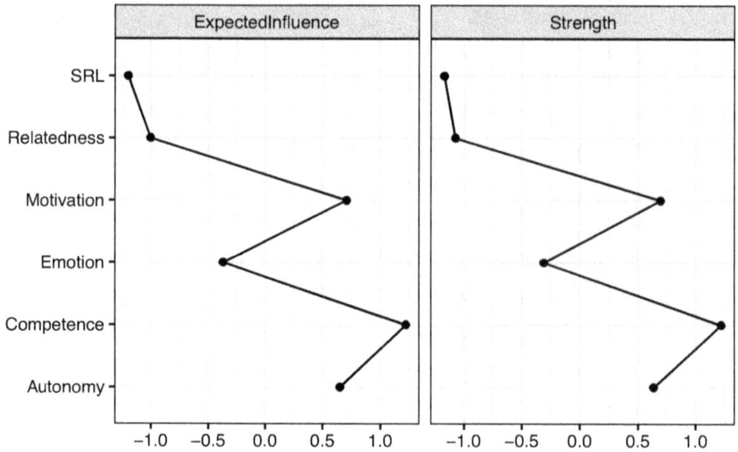

Fig. 4 Centrality plot

Table 3 Centrality table

	Graph	Type	Node	Measure	Value
1	Graph 1	NA	Relatedness	Betweenness	−0.6172134
2	Graph 1	NA	Competence	Betweenness	0.7715167
3	Graph 1	NA	Autonomy	Betweenness	1.6973368
4..21					
22	Graph 1	NA	Emotion	ExpectedInfluence	−0.3710702
23	Graph 1	NA	Motivation	ExpectedInfluence	0.7079431
24	Graph 1	NA	SRL	ExpectedInfluence	−1.2029339

```
centralityPlot(allNetwork, include = c("ExpectedInfluence",
                    "Strength"),
                   scale = "z-scores" )
```

```
Note: z-scores are shown on x-axis rather than raw centrality indices.
```

If we wanted only the values, we could obtain the centralities by using the function `centralityTable()` (Table 3).

```
centralityTable(allNetwork)
```

Another possible way to compute several types of centralities is to use the `NetworkToolbox` package (Table 4) which offers a large collection of centralities, e.g., degree, strength, closeness, eigenvector or leverage centralities. `NetworkToolbox` has even more centrality measures that the reader can try. Please refer to the manual of this package for more information about the functions and

Table 4 Centrality measures calculated with NetworkToolbox

Var	Degree	Strength	Betweenness	Closeness	Eigenvector	Leverage
Relatedness	5.00	0.40	0.00	1.89	0.22	−3.43
Competence	5.00	1.07	6.00	3.27	0.55	1.35
Autonomy	5.00	0.90	10.00	3.63	0.47	1.07
Emotion	5.00	0.62	0.00	2.58	0.36	−0.42
Motivation	5.00	0.91	0.00	3.16	0.49	1.11
SRL	5.00	0.37	0.00	1.88	0.21	−5.38

usage. Nevertheless, as mentioned above, the interpretations of these centralities are not clearly understood or straightforward as strength or expected influence.

```
Degree <- degree(allNetwork$graph)
Strength <- strength(allNetwork$graph)
Betweenness <- betweenness(allNetwork$graph)
Closeness <- closeness(allNetwork$graph)
Eigenvector <- eigenvector(allNetwork$graph)
Leverage <- leverage(allNetwork$graph)
data.frame(Var = names
      (Degree), Degree, Strength, Betweenness, Closeness,
      Eigenvector, Leverage)
```

As we mentioned above, besides the regularized partial correlation networks, several other estimation options are possible. We will demonstrate some here (Fig. 5). However, interested readers can refer to the manual pages of `estimateNetwork()` function. First, we can fit a model that is simply an association (i.e. correlation) network to explore the correlations between different variables. The association network is not recommended except for data exploration and is shown here for comparison. Another very interesting estimation method is the `ggmModSelect()`. This is recommended in large datasets with a low number of nodes. The `ggmModSelect` algorithm starts by estimating a regularized network as a starting baseline network, then estimates all possible un-regularized networks and selects the best model based on the lowest EBIC criteria. The last model we show here is the relative importance model `relimp` which estimates a directed network where edges are magnitude of the relative importance of the predictors in a linear regression model. You may notice that the `ggmModSelect` network is rather similar to the regularized network we estimated above whereas the correlation network is very dense. The relative importance network is directed and shows how each variable is expected to be influencing the other as per the regression results.

```
allNetwork_cor <- estimateNetwork(allData, default =
        "cor", verbose = FALSE)
allNetwork_mgm <- estimateNetwork(allData, default =
```

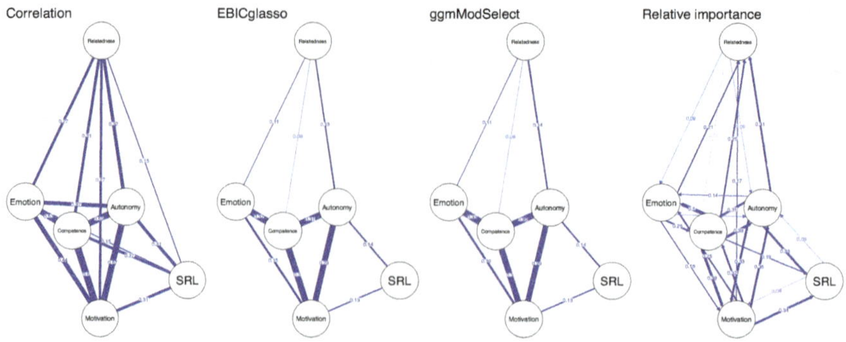

Fig. 5 Correlation networks with several estimation options

```
                "ggmModSelect", verbose = FALSE)
allNetwork_relimp <- estimateNetwork(allData, default =
                "relimp", verbose = FALSE)


plot(allNetwork_cor, title = "Correlation", vsize = 18,
                edge.labels = TRUE,
     cut = 0.10, minimum = 0.05, layout = LX)
plot(allNetwork, title = "EBICglasso", vsize = 18,
                edge.labels = TRUE,
     cut = 0.10, minimum = 0.05, layout = LX)
plot(allNetwork_mgm, title = "ggmModSelect",
                vsize = 18, edge.labels = TRUE,
     cut = 0.10, minimum = 0.05, layout = LX)
plot(allNetwork_relimp, title = "Relative importance",
vsize = 18, edge.labels = TRUE,
     cut = 0.10, minimum = 0.05, layout = LX)
```

3.8 *Comparing Networks*

Having shown the basic steps of estimation of a single network, we now proceed with comparing across different networks. Psychological networks offer rigorous methods to compare networks as a whole as well as edge weights and centralities using robust methods. Given that our data has two countries, we can estimate two networks for each country and test how they differ.

First, to do the comparison we need to estimate the networks as we did before. The next section repeats the estimation steps for each country. We start with Finland

and then Austria. First, the code performs the basic steps of assumption checking for each network. As the results show that the matrix of each of the networks is not positive definite and the `goldbricker` algorithm does not suggest that there are highly similar nodes that need to be reduced. Then, the next code chunk estimates two regularized partial correlation networks. As we have done before, we estimate the predictability of both networks. The results show that in general the mean predictability is similar in the two networks.

```
### Check the assumptions
## Finland
# check for positive definitiveness
correlationMatrix <- cor(x = finlandData, use = c(
                "pairwise.complete.obs"))
is.positive.definite(correlationMatrix)

# check for redundancy
goldbricker(finlandData, p = 0.05, method = "hittner2003",
            threshold = 0.25, corMin = 0.5, progressbar =
                        FALSE)
```

```
Suggested reductions: Less than 25 % of correlations are significantly
different for the following pairs:
[1] "No suggested reductions"
```

```
## Austria
# check for positive definitiveness
correlationMatrix <- cor(x = austriaData, use = c(
                    "pairwise.complete.obs"))
is.positive.definite(correlationMatrix)

# check for redundancy
goldbricker(austriaData, p = 0.05, method = "hittner2003",
            threshold = 0.25, corMin = 0.5, progressbar =
                            FALSE)
```

```
Suggested reductions: Less than 25 % of correlations are significantly
different for the following pairs:
[1] "No suggested reductions"
```

```
#Estimate the networks
finlandNetwork <- estimateNetwork(finlandData,
    default = "EBICglasso",
                        corMethod = "cor_auto",  tuning = 0.5)
austriaNetwork <- estimateNetwork(austriaData,
```

```
    default = "EBICglasso",
                              corMethod = "cor_auto",   tuning = 0.5)


# Compute the predictability
fitFinland <- mgm(
  as.matrix(finlandData), # data
  c("g" ,"g", "g", "g" ,"g" ,"g"),
                            # distribution for each var
  verbatim = TRUE, # hide warnings and progress bar
  signInfo = FALSE # hide message about signs
  )
predictFinland <- predict(fitFinland, na.omit(finlandData))

mean(predictFinland$errors$R2)
                    # Mean predictability of Finland: 0.3085
mean(predictFinland$errors$RMSE)
                    # Mean RMSE of Finland: 0.8283333

fitAustria <- mgm(
  as.matrix(austriaData), # data
  c("g" ,"g", "g", "g" ,"g" ,"g"),
                                    # distribution for each var
  verbatim = TRUE, # hide warnings and progress bar
  signInfo = FALSE # hide message about signs
  )
predictAustria <- predict(fitAustria, na.omit(austriaData))

mean(predictAustria$errors$R2)
                # Mean predictability of Austria: 0.3436667
mean(predictAustria$errors$RMSE)
                # Mean RMSE of Austria: 0.8036667
```

```
[1] 0.3085
[1] 0.8283333
[1] 0.3436667
[1] 0.8036667
```

Now as the networks were estimated, we plot the networks side by side in the same way we used before (Fig. 6). It is always a good idea to have a common layout to facilitate interpretation. We can use the function `averageLayout()` to generate an average layout from the two networks as the first line of the code below shows. However, we will use the layout of the first network (`LX`) to maintain comparability. Please also note that we use the function `qgraph()`

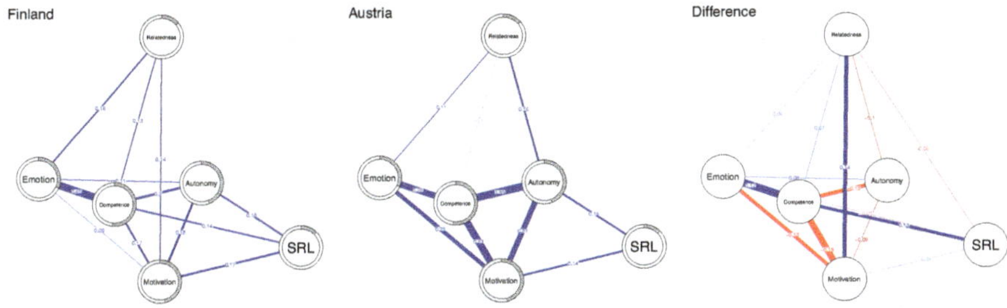

Fig. 6 Country networks and comparison. The weight of the edges represents the magnitude of the correlation while the color represents the sign: blue for positive and red for negative

which is very similar to `plot()` with more options and arguments. Yet, two small differences: in `qgraph()` you need to supply the labels as an argument, otherwise `qgraph()` will use shortened labels. Second, `qgraph()` requires either an estimated network or matrix, to plot the difference network we subtract the two matrices `finlandNetwork$graph-austriaNetwork$graph` (not the networks) to get the matrix. The following code plots the three networks using the same layout side by side as well as computes a simple difference network. It is obvious that the two network differ substantially regarding several interactions. Finland has stronger connection between competence and emotion and between motivation and relatedness. Whereas in Austria, there is a stronger connection between motivation and competence, motivation and emotion as well as competence and autonomy and autonomy and relatedness.

```
AverageLayout <-  averageLayout
                      (finlandNetwork, austriaNetwork)

plot(finlandNetwork,                # input network
     title = "Finland",             # plot title
     vsize = 19,                    # size of the nodes
     edge.labels = TRUE,            # label the edge weights
     cut = 0.10,                    # saturate edges > .10
     minimum = 0.05,                # remove edges < .05
     pie = predictFinland$errors$R2, # put R2 as pie
     layout = LX)                   # set the layout

plot(austriaNetwork,                # input network
     title = "Austria",             # plot title
     vsize = 19,                    # size of the nodes
     edge.labels = TRUE,            # label the edge weights
     cut = 0.10,                    # saturate edges > .10
     minimum = 0.05,                # remove edges < .05
     pie = predictAustria$errors$R2, # put R2 as pie
```

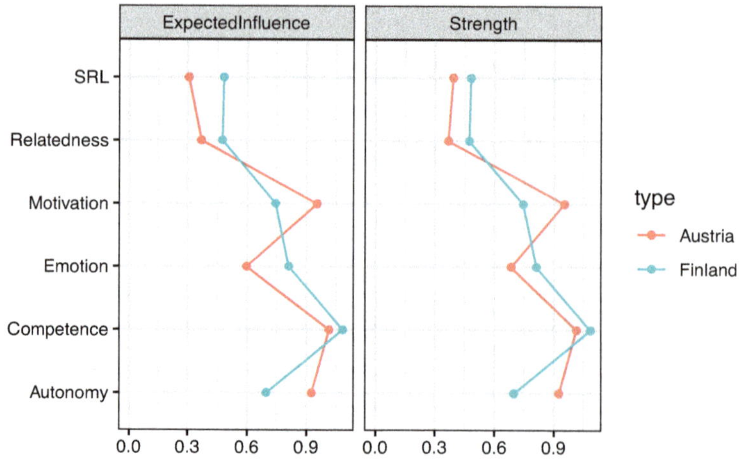

Fig. 7 Comparison of centralities plot between countries

```
    layout = LX)                        # set the layout

qgraph(finlandNetwork$graph - abs(austriaNetwork$graph),
       title = "Difference",                   # plot title
       theme = allDataPlot$Arguments$theme,
       vsize = 19,                      # size of the  nodes
       edge.labels = TRUE,           # label the edge  weights
       labels = allDataPlot$Arguments$labels, # node labels
       cut = 0.10,                     # saturate edges  > .10
       layout = LX)                            # set the layout
```

A visual comparison of centralities can be performed in the same way we did before (Fig. 7). Here, we supply the networks that we want to compare as a list, and we specify the centralities. As the results show, Motivation has the highest centrality value in the Austria network, meaning that motivation is the factor that is expected to drive the connectivity. In the Finland network, competence is the most central variable that drives the network connectivity.

```
centralityPlot(
  list(Finland = finlandNetwork,
       Austria = austriaNetwork),
  include = c("ExpectedInfluence", "Strength"))
```

Yet, to compare the networks in a rigorous way, we need a statistical test that tells which edges or centrality measures are actually different and not due to chance.

For such a comparison, NCT (short for Network Comparison Test) is an effective method that allows for a detailed comparison regarding the network structure, edges, and centralities. To do so, NCT uses permutation to generate a large number of networks (based on the original network) as a references distribution and later compares the original networks against the permuted. To perform the test, we supply the networks and the number of iterations (a large number, at least 1000 iteration is recommended). In addition, it is good practice to specify `test.edges = TRUE, edges = 'all'` to test all edges and `test.centrality = TRUE` to test the centralities since they are not tested by default. The code and the output are in the next chunk. To check the results, we can obtain the global strength (sum of all edge weights) by the using `Compared$glstrinv.sep` which is 2.148451 for Finland and 2.1725 for Austria. The difference between global strength `Compared$glstrinv.real` is 0.02404855 and is statistically insignificant (`Compared$glstrinv.pval` = 0.735). The maximum difference in any of the edges between the networks (`Compared$nwinv.real`) is 0.1713064. `Compared$einv.real` returns the difference matrix (as the difference network we computed above). The Holm-Bonferroni adjusted p-value (which adjusts for multiple comparisons) for each edge can be obtained using `Compared$einv.pvals`. The results show that most of the edges differed significantly except e.g., Relatedness-Emotion, Relatedness-SRL. Similarly, the difference in centralities can be obtained using `Compared$diffcen.real` and the p-values using `Compared$diffcen.pval` which shows for example, that the difference in expected influence centrality of competence was not statistically significant.

```
set.seed(1337)
Compared <- NCT(
  finlandNetwork,          # network 1
  austriaNetwork,          # network 2
  verbose = FALSE,         # hide warnings  and progress bar
  it = 1000,               # number of iterations
  abs = T,
  binary.data = FALSE,     # set data distribution
  test.edges = TRUE,       # test edge differences
  edges = 'all',           # which edges to test
  test.centrality   = TRUE, # test centrality
  progressbar = FALSE      # progress bar
  )

Compared$glstrinv.sep  # Separate global strength values
                       of the individual networks
```

```
[1] 2.148451 2.172500

Compared$glstrinv.real # Difference in global strength
                                    between the networks

[1] 0.02404855

Compared$glstrinv.pval # p-value of strength difference

[1] 0.7352647

Compared$nwinv.real # Maximum difference in any of the edges
                      between networks

[1] 0.1713064

Compared$einv.real # Difference in edge weight of the
                     observed networks

            Relatedness Competence   Autonomy    Emotion Motivation        SRL
Relatedness  0.00000000 0.06812689 0.09520436 0.04142711 0.13714460 0.04672616
Competence   0.06812689 0.00000000 0.13478602 0.17130636 0.16045438 0.12581888
Autonomy     0.09520436 0.13478602 0.00000000 0.07667907 0.08758119 0.01228288
Emotion      0.04142711 0.17130636 0.07667907 0.00000000 0.12873747 0.05299023
Motivation   0.13714460 0.16045438 0.08758119 0.12873747 0.00000000 0.03062823
SRL          0.04672616 0.12581888 0.01228288 0.05299023 0.03062823 0.00000000

Compared$einv.pvals # Holm-Bonferroni adjusted p-values
                                    for each edge

          Var1       Var2     p-value Test statistic E
7  Relatedness Competence 0.017982018       0.06812689
13 Relatedness   Autonomy 0.001998002       0.09520436
14  Competence   Autonomy 0.000999001       0.13478602
19 Relatedness    Emotion 0.165834166       0.04142711
20  Competence    Emotion 0.000999001       0.17130636
21    Autonomy    Emotion 0.009990010       0.07667907
25 Relatedness Motivation 0.000999001       0.13714460
26  Competence Motivation 0.000999001       0.16045438
27    Autonomy Motivation 0.003996004       0.08758119
28     Emotion Motivation 0.001998002       0.12873747
```

```
31 Relatedness         SRL 0.077922078         0.04672616
32  Competence         SRL 0.000999001         0.12581888
33    Autonomy         SRL 0.688311688         0.01228288
34     Emotion         SRL 0.073926074         0.05299023
35  Motivation         SRL 0.301698302         0.03062823
```

```
Compared$diffcen.real # Difference in centralities
```

```
               strength expectedInfluence
Relatedness  0.10476808        0.10476808
Competence   0.07001173        0.07001173
Autonomy    -0.22860961       -0.22860961
Emotion      0.12670208        0.21366531
Motivation  -0.20900022       -0.20900022
SRL          0.08803084        0.17499406
```

```
Compared$diffcen.pval # Holm-Bonferroni adjusted
                      p-values for each centrality
```

```
               strength expectedInfluence
Relatedness 0.006993007       0.006993007
Competence  0.127872128       0.127872128
Autonomy    0.000999001       0.000999001
Emotion     0.009990010       0.000999001
Motivation  0.000999001       0.000999001
SRL         0.132867133       0.000999001
```

3.9 The Variability Network

The variability network offers a good indication of how the edge weights vary across the networks (Fig. 8). In other words, the range of variability (i.e. the degree of individual differences) across the included population. Edges with low variability are expected to be similar across the networks and vice versa. The following code creates two matrices and then loops across the two networks to compute the standard deviation.

```
# Construct a network where edges are standard deviations
                          across edge weights of networks
edgeMeanJoint <- matrix(0, 6, 6)
edgeSDJoint <- matrix(0, 6, 6)
```

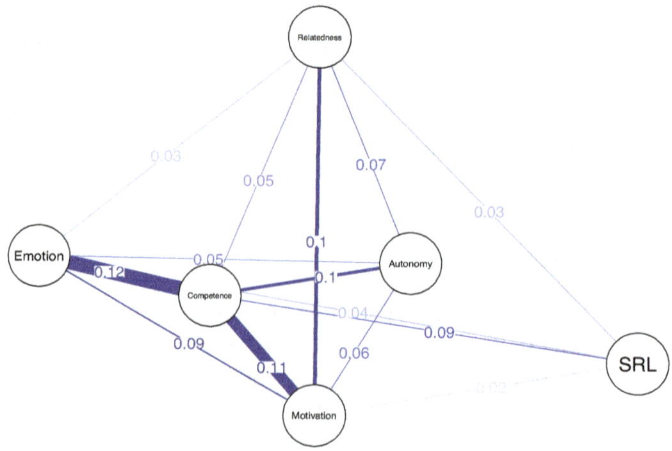

Fig. 8 Variability plot

```
for(i in 1:6){
  for(j in 1:6) {
    vector <- c(getWmat(finlandNetwork)[i, j], getWmat
                                (austriaNetwork)[i, j])
    edgeMeanJoint[i, j] <- mean(vector)
    edgeSDJoint[i, j] <- sd(vector)
  }
}
```

We then plot the networks where the edge weights are the standard deviations of all edges (Fig. 8).

```
qgraph(edgeSDJoint, layout = LX, edge.labels = TRUE,
       labels = allDataPlot$Arguments$labels, vsize = 9,
       cut = 0.09, minimum = 0.01, theme = "colorblind")
```

In the same way we compared countries, we can compare across genders, and as we see in the next code chunk, we estimated the male network, the female network and the difference network (Fig. 9). Nonetheless, the differences are really small or even trivial.

```
maleNetwork <- estimateNetwork(maleData, default =
                          "EBICglasso")
femaleNetwork <- estimateNetwork(femaleData, default =
                            "EBICglasso")

plot(maleNetwork, title = "Male", vsize = 9, edge.labels =
```

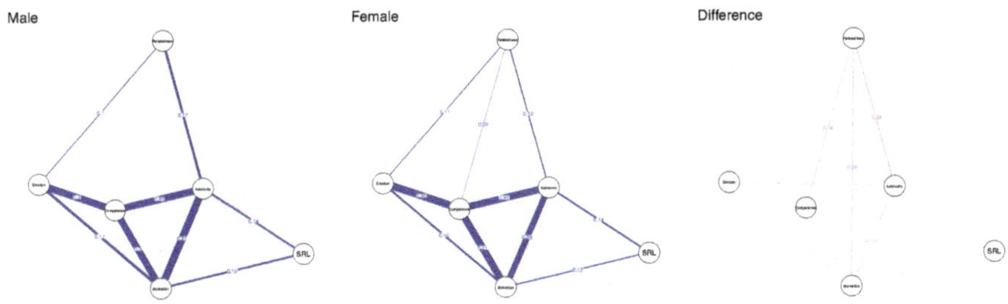

Fig. 9 Gender networks and comparison

```
                                            TRUE,
      cut = 0.10, minimum = 0.05, layout = LX)

plot(femaleNetwork, title = "Female", vsize = 9,
                            edge.labels = TRUE,
      cut = 0.10, minimum = 0.05, layout = LX)

qgraph(femaleNetwork$graph - maleNetwork$graph, title =
                            "Difference", cut = 0.1,
        labels = allDataPlot$Arguments$labels, vsize = 9
                                    ,minimum = 0.01,
        edge.labels = TRUE, layout = LX, theme =
                                    "colorblind")
```

Below we perform the network comparison test and we see that the p-values of differences between all edges is statistically insignificant.

```
ComparedGender <- NCT(
  maleNetwork, # network 1
  femaleNetwork, # network 2
  verbose = FALSE,  # hide warnings and progress bar
  it = 1000, # number of iterations
  abs = T, # test strength or expected influence?
  binary.data = FALSE, # set data distribution
  test.edges = TRUE, # test edge differences
  edges = 'all', # which edges to test
  progressbar = FALSE) # progress bar

ComparedGender$einv.pvals # Holm-Bonferroni adjusted
                          p-values for each edge
```

```
          Var1       Var2    p-value Test statistic E
```

```
7  Relatedness Competence 0.10889111      0.04185543
13 Relatedness   Autonomy 0.05394605      0.05347524
14  Competence   Autonomy 0.99400599      0.00010546
19 Relatedness    Emotion 0.72127872      0.01087833
20  Competence    Emotion 0.42757243      0.01989769
21    Autonomy    Emotion 0.18981019      0.03229342
25 Relatedness Motivation 0.08791209      0.04844325
26  Competence Motivation 0.63736264      0.01271230
27    Autonomy Motivation 0.20779221      0.03390618
28     Emotion Motivation 0.47552448      0.01965116
31 Relatedness        SRL 0.73426573      0.00892731
32  Competence        SRL 0.56543457      0.01648147
33    Autonomy        SRL 0.92907093      0.00250644
34     Emotion        SRL 1.00000000      0.00000000
35  Motivation        SRL 0.34965035      0.02428596
```

3.10 Evaluation of Robustness and Accuracy

The most common procedure to evaluate the stability and accuracy of the of the estimated networks is bootstrapping, in which a large number (1000 or more) of bootstrapped networks are created based on the original data. The resulting edge weights of the bootstrapped networks are then used to create confidence intervals to assess the accuracy of the edges. Each edge weight in the estimated networks is contrasted to the confidence intervals of the bootstrapped edges. Edges where the upper and lower bounds do not cross zero are considered statistically significant and edges that either the upper or lower bound of the confidence interval crosses the 0 line are considered not significant.

```
nCores <- parallel::detectCores() - 1
# Non-parametric bootstrap for stability of edges
                             and of edge differences

allBoot <-  bootnet(
  allNetwork,                  # network input
  default = "EBICglasso",      # method
  nCores = nCores,             # number of cores for
                                         parallelization
  computeCentrality = FALSE, # estimate centrality?
  statistics = "edge"          # what statistics do we
                             want?
  )
```

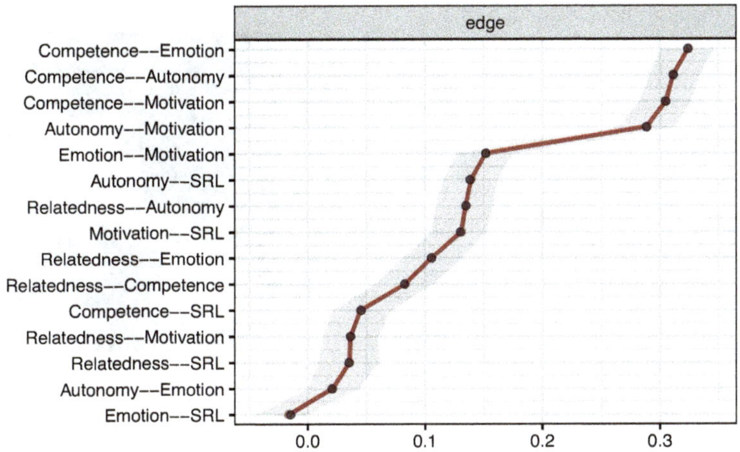

Fig. 10 Edge weights

```
plot(allBoot, plot = "area", order = "sample", legend =
                             FALSE)
```

As Fig. 10 shows, only the edges of autonomy-emotion and emotion-SRL are crossing the 0 line and therefore, are insignificant. We can also plot the edge difference plot which tests if the edge weights are different from each other (Fig. 11). As the figure shows, edges where the 95% bootstrapped confidence interval of the difference between any pair of edges crosses the zero line, the square is grey. The square is black if the edge difference does not cross the 0 (significant). For instance, autonomy-emotion and emotion-SRL have a grey square indicating non-significant difference whereas emotion-SRL and relatedness-SRL has a black square indicating that the two nodes are statistically significantly different.

```
plot(allBoot, plot = "difference", order = "sample",
                    onlyNonZero = FALSE, labels = TRUE)
```

The accuracy of centrality is assessed by the case dropping test. In the case dropping test, several proportions of cases are dropped from the data and the correlation between the observed centrality measure and those obtained from the subsetted data is calculated. If the correlation dropped markedly after dropping a small subset of the cases, then the centrality measure is unreliable.

```
set.seed(1)
centBoot <- bootnet(
  allNetwork,                  # network input
  default = "EBICglasso",      # method
```

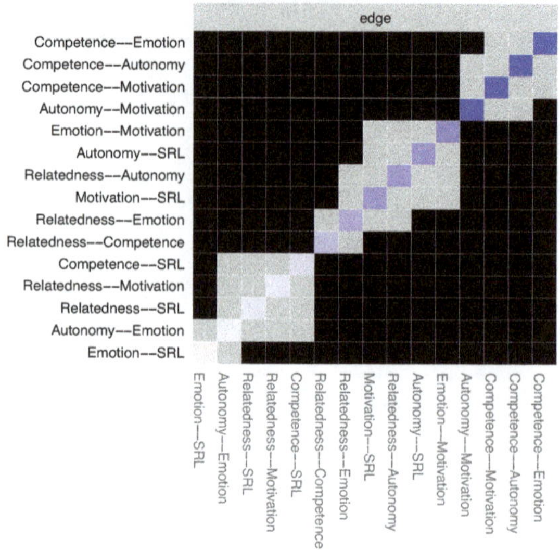

Fig. 11 Differences between edges

```
    type = "case",                # method for testing
                                          centrality stability
    nCores = nCores,              # number of cores
    computeCentrality = TRUE,     # compute centrality
    statistics = c("strength", "expectedInfluence"),
    nBoots = 19000,               # number of bootstraps
    caseMin = .05,                # min cases to drop
    caseMax = .95                 # max cases to drop
    )
```

The correlation stability coefficient is a metric that is used to judge the stability of the centrality measure using the case dropping test and is estimated as the maximum drop to retain retain 0.7 of the sample.

```
corStability(centBoot)
```

If we plot the results (Fig. 12), we can see that the correlation stability coefficient is 0.95 which is very high indicating the stability of the edges.

```
plot(centBoot)
```

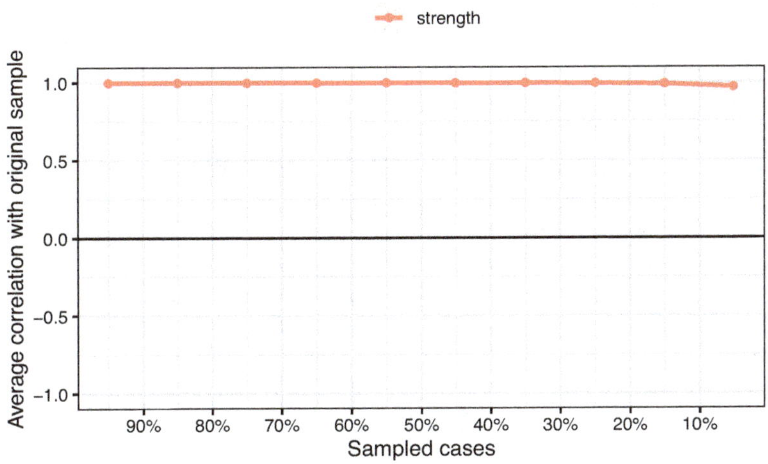

Fig. 12 Average correlation stability coefficient

3.11 Discussion

The field of psychological networks is growing fast and methods are refined at a fast speed. In the current chapter, we have tried to show the basic steps of analyzing a psychological network, visualizing the results and comparing different networks. We have also shown how networks can be compared using robust statistical methods. Furthermore, we also showed how to test the accuracy of the estimated networks using the bootstrapping methods.

An important question here is how psychological networks compare to other methods that are prevalent in the educational field e.g., Epistemic Network Analysis (ENA). In ENA, there is no way to test if the network edges are different from random, there is no rigorous method for comparison of networks or verifying the edge weights. Also, there are no centrality measures or network measures. In fact, ENA loses all connection to network methods and therefore, the usual methods for verifying, randomization or computing of network measures do not apply in ENA [50]. The same can be said about process mining which produces transition networks. In process mining, there are few confirmatory tests that verify the resulting model or rigorously compare across process models. Perhaps social networks analysis (SNA) is the closest to psychological networks. However, SNA is limited to—or has been commonly used with—-limited types of edges (e.g., co-occurrence, reply or interactions); these edges are almost always unsigned (i.e., are always positive) and have been limited to either social interactions or semantic interactions [25, 51, 52].

In sum, psychological networks offer far more wider perspectives into interactions—or interdependencies—among variables with a vast number of possible estimation methods and optimization techniques. Furthermore, psychological networks have a vibrant community who refine and push the boundary of the existing methods [53]. All of such advantages make psychological networks a promising method for modeling complex systems, understanding interactions and structure of psychological constructs as we demonstrated here [3, 54]. Furthermore, psychological networks require no prior theory or strong assumptions about the modeled variables and can serve as powerful analytical methods. As Borsboom et al. states psychological networks "form a natural bridge from data analysis to theory formation based on network science principles" and therefore can be used "to generate causal hypotheses" [29].

The book is a comprehensive reference for psychological networks from theory, to methods and estimation techniques. The following papers offer excellent guides and tutorials:

- Epskamp, S., Borsboom, D., & Fried, E. I. (2018). Estimating psychological networks and their accuracy: A tutorial paper. *Behavior research methods*, 50, 195-212.
- Epskamp, S., & Fried, E. I. (2018). A tutorial on regularized partial correlation networks. *Psychological methods*, 23(4), 617.
- Van Borkulo, C. D., van Bork, R., Boschloo, L., Kossakowski, J. J., Tio, P., Schoevers, R. A., … & Waldorp, L. J. (2022). Comparing network structures on three aspects: A permutation test. *Psychological methods*.
- Borsboom, D., Deserno, M. K., Rhemtulla, M., Epskamp, S., Fried, E. I., McNally, R. J., … & Waldorp, L. J. (2021). Network analysis of multivariate data in psychological science. *Nature Reviews Methods Primers*, 1(1), 58.
- Bringmann, L. F., Elmer, T., Epskamp, S., Krause, R. W., Schoch, D., Wichers, M., … & Snippe, E. (2019). What do centrality measures measure in psychological networks?. *Journal of abnormal psychology*, 128(8), 892.

References

1. Koopmans M (2020) Education is a complex dynamical system: challenges for research. J Exp Educ 88:358–374
2. Koopmans M, Stamovlasis D (2016) Complex dynamical systems in education: concepts, methods and applications. Springer International Publishing, Cham
3. Malmberg J, Saqr M, Järvenoja H, Haataja E, Pijeira-Díaz HJ, Järvelä S (2022) Modeling the complex interplay between monitoring events for regulated learning with psychological networks. In: The multimodal learning analytics handbook. Springer International Publishing, Cham, pp 79–104
4. Zimmerman BJ, Risemberg R (1997) Becoming a self-regulated writer: a social cognitive perspective. Contemp Educ Psychol 22:73–101
5. Winne PH, Zhou M, Egan R (2011) Designing assessments of self-regulated learning. Assessment Higher Order Thinking Skills 418:89–118
6. Wang M-T, Fredricks JA (2014) The reciprocal links between school engagement, youth problem behaviors, and school dropout during adolescence. Child Dev 85:722–737
7. Papi M, Hiver P (2020) Language learning motivation as a complex dynamic system: a global perspective of truth, control, and value. Mod Lang J 104:209–232

8. Yuan Y, Zhen H (2021) Teaching and researching motivation. Front Psychol 12:804304
9. Vollmeyer R, Rheinberg F (1999) Motivation and metacognition when learning a complex system. Eur J Psychol Educ 14:541–554
10. Deakin Crick R, Huang S, Ahmed Shafi A, Goldspink C (2015) Developing resilient agency in learning: The internal structure of learning power. Br J Educ Stud 63:121–160
11. Urdan T, Kaplan A (2020) The origins, evolution, and future directions of achievement goal theory. Contemp Educ Psychol 61:101862
12. Brown JS (1997) On becoming a learning organization. About Campus 1:5–10
13. Smit N, Dijk M van, Bot K de, Lowie W (2022) The complex dynamics of adaptive teaching: observing teacher-student interaction in the language classroom. IRAL, International review of applied linguistics in language teaching: Revue internationale de linguistique appliquee enseignement des langues Internationale Zeitschrift fur angewandte Linguistik in der Spracherziehung 60:23–40
14. Mennin S (2007) Small-group problem-based learning as a complex adaptive system. Teach Teach Educ 23:303–313
15. Hilpert JC, Marchand GC (2018) Complex systems research in educational psychology: aligning theory and method. Educ Psychol 53:185–202
16. Ladyman J, Lambert J, Wiesner K (2013) What is a complex system? Eur J Philos Sci 3:33–67
17. Simon HA (1962) The architecture of complexity. Proc Am Philos Soc 106:467–482
18. Fredricks JA, Blumenfeld PC, Paris AH (2004) School engagement: potential of the concept, state of the evidence. Rev Educ Res 74:59–109
19. Reschly AL, Christenson SL (2022) Jingle-jangle revisited: History and further evolution of the student engagement construct. In: Reschly AL, Christenson SL (eds) Handbook of research on student engagement. Springer International Publishing, Cham, pp 3–24
20. Tinto V (2022) Exploring the character of student persistence in higher education: the impact of perception, motivation, and engagement. In: Reschly AL, Christenson SL (eds) Handbook of research on student engagement. Springer International Publishing, Cham, pp 357–379
21. Skinner EA (2016) Engagement and disaffection as central to processes of motivational resilience and development. Handbook of motivation at school. Routledge, London, pp 145–168
22. Zimmerman BJ, Moylan AR (2009) Self-regulation: where metacognition and motivation intersect. Handbook of metacognition in education. Routledge/Taylor & Francis Group, London. https://books.google.com/books?hl=en&lr=&id=JpWOAgAAQBAJ&oi=fnd&pg=PA299&dq=Zimmerman+2009&ots=eyy5XecnF5&sig=935OMu05qYyTSfogzo4QStnI1G0
23. Yoon SA (2008) An evolutionary approach to harnessing complex systems thinking in the science and technology classroom. Int J Sci Educ 30:1–32
24. Barabási A-L (2013) Network science. Philos Trans Series A Math Phys Eng Sci 371:20120375
25. Saqr M, Poquet O, López-Pernas S (2022) Networks in education: a travelogue through five decades. IEEE Access Pract Innovations, Open Solutions 1–1
26. Saqr M, López-Pernas S, Conde MÁ, Hernández-García Á (2024) Social network analysis: a primer, a guide and a tutorial in R. In: Saqr M, López-Pernas S (eds) Learning analytics methods and tutorials: a practical guide using R. Springer, Berlin, in press
27. Hernández-García Á, Cuenca-Enrique C, Traxler A, López-Pernas S, Conde MÁ, Saqr M (2024) Community detection in learning networks using R. In: Saqr M, López-Pernas S (eds) Learning analytics methods and tutorials: a practical guide using R. Springer, Berlin, in press
28. Saqr M (2024) Temporal network analysis: introduction, methods, and analysis with R. In: Saqr M, López-Pernas S (eds) Learning analytics methods and tutorials: a practical guide using R. Springer, Berlin, in press
29. Borsboom D, Deserno MK, Rhemtulla M, Epskamp S, Fried EI, McNally RJ, Robinaugh DJ, Perugini M, Dalege J, Costantini G, Isvoranu A-M, Wysocki AC, Borkulo CD van, Bork R van, Waldorp LJ (2021) Network analysis of multivariate data in psychological science. Nat Rev Methods Primers 1:1–18
30. Epskamp S, Waldorp LJ, Mõttus R, Borsboom D (2018) The gaussian graphical model in cross-sectional and time-series data. Multivariate Behav Res 53:453–480

31. Epskamp S, Fried EI (2018) A tutorial on regularized partial correlation networks. Psychol Methods 23:617–634
32. Liu T-H, Xia Y, Ma Z (2022) Multifarious linkages between personality traits and psychological distress during and after COVID-19 campus lockdown: a psychological network analysis. Frontiers in Psychiatry 13. https://doi.org/10.3389/fpsyt.2022.816298
33. Zhou L, Sukpasjaroen K, Wu Y, Wang L, Chankoson T, Cai E (2022) Predicting nursing students' psychological well-being: network analysis based on a model of thriving through relationships. BMC Med Educ 22. https://doi.org/10.1186/s12909-022-03517-1
34. Saqr M, Viberg O, Peteers W (2021) Using psychological networks to reveal the interplay between foreign language students' self-regulated learning tactics. In: STELLA2020 proceedings, pp 12–23
35. López-Pernas S., Gordillo A., Barra E., Saqr M. (2023) The dynamics of students' playing profiles in a programming educational escape room. In: Proceedings TEEM 2023: eleventh international conference on technological ecosystems for enhancing multiculturality. TEEM 2023. Lecture notes in computer science. Brangança, Portugal, in press
36. Saqr M, López-Pernas S (2021) Idiographic learning analytics: a single student (n= 1) approach using psychological networks. http://ceur-ws.org/Vol-2868/article_4.pdf
37. Saqr M, López-Pernas S (2021) Idiographic learning analytics: A definition and a case study. In: Proceedings of the 2021 international conference on advanced learning technologies (ICALT). IEEE, Piscataway, pp 163–165. https://doi.org/10.1109/icalt52272.2021.00056
38. Malmberg J, Saqr M, Järvenoja H, Järvelä S (2022) How the monitoring events of individual students are associated with phases of regulation. J. Learn. Anal. 9:77–92. https://doi.org/10.18608/jla.2022.7429
39. Saqr M (2024) Group-level analysis of engagement poorly represents individual students: why we need idiographic precision learning analytics. Comput Human Behav
40. López-Pernas S, Saqr M, Conde J, Del-Río-Carazo L (2024) A broad collection of datasets for educational research training and application. In: Saqr M, López-Pernas S (eds) Learning analytics methods and tutorials: a practical guide using R. Springer, Berlin, in press
41. Chan C, Leeper TJ, Becker J, Schoch D (2021) Rio: A swiss-army knife for data file i/o https://cran.r-project.org/package=rio
42. Wickham H, Averick M, Bryan J, Chang W, McGowan LD, François R, Grolemund G, Hayes A, Henry L, Hester J, Kuhn M, Pedersen TL, Miller E, Bache SM, Müller K, Ooms J, Robinson D, Seidel DP, Spinu V, Takahashi K, Vaughan D, Wilke C, Woo K, Yutani H (2019) Welcome to the tidyverse. J. Open Source Softw 4:1686. https://doi.org/10.21105/joss.01686
43. Epskamp S, Borsboom D, Fried EI (2018) Estimating psychological networks and their accuracy: a tutorial paper. Behav Res Methods 50:195–212
44. Jones P (2022). Networktools: Tools for identifying important nodes in networks. https://CRAN.R-project.org/package=networktools
45. Christensen AP (2018) NetworkToolbox: Methods and measures for brain, cognitive, and psychometric network analysis in R. R J 422–439. https://doi.org/10.32614/RJ-2018-065
46. van Borkulo CD, Boschloo L, Kossakowski JJ, Tio P, Schoevers RA, Borsboom D, Waldorp LJ (2017) Comparing network structures on three aspects: a permutation test. J Stat Softw. https://doi.org/10.13140/RG.2.2.29455.38569
47. Epskamp S, Cramer AOJ, Waldorp LJ, Schmittmann VD, Borsboom D (2012) qgraph: network visualizations of relationships in psychometric data. J Stat Softw 48:1–18
48. Haslbeck JMB, Waldorp LJ (2020) mgm: estimating time-varying mixed graphical models in high-dimensional data. J Stat Softw 93:1–46. https://doi.org/10.18637/jss.v093.i08
49. Novomestky F (2022) Matrixcalc: collection of functions for matrix calculations. https://CRAN.R-project.org/package=matrixcalc
50. Elmoazen R, Saqr M, Tedre M, Hirsto L (2022) A systematic literature review of empirical research on epistemic network analysis in education. IEEE Access: Practical Innovations, Open Solutions 10:17330–17348. https://doi.org/10.1109/access.2022.3149812
51. López-Pernas S, Saqr M (2024) The why, how and when of process mining in learning analytics: a guided tutorial in r. In: Saqr M, López-Pernas S (eds) Learning analytics methods and tutorials: a practical guide using R. Springer, Berlin, in press

52. Saqr M, López-Pernas S, Conde MÁ, Hernández-García Á (2024) Social network analysis: a primer, a guide and a tutorial in r. In: Saqr M, López-Pernas S (eds) Learning analytics methods and tutorials: a practical guide using R. Springer, Berlin, in press
53. Isvoranu A-M, Epskamp S, Waldorp L, Borsboom D (2022) Network psychometrics with r: A guide for behavioral and social scientists. Routledge, London. https://play.google.com/store/books/details?id=-3ddEAAAQBAJ
54. Epskamp S, Borsboom D, Fried EI (2018) Estimating psychological networks and their accuracy: a tutorial paper. Behav Res Methods 50:195–212. https://doi.org/10.3758/s13428-017-0862-1

Factor Analysis in Education Research Using R

Leonie V. D. E. Vogelsmeier, Mohammed Saqr, Sonsoles López-Pernas, and Joran Jongerling

1 Introduction

Educational scientists are usually after the invisible: sense of belonging, intelligence, math ability, general aptitude, student engagement... the list of crucial, multifaceted, but not directly observable student characteristics goes on. But how do researchers measure and study what they cannot see? One solution is not looking for the invisible cause itself but for its consequences. Just like one cannot see the wind but can tell it is there from the moving leaves, researchers can indirectly infer student engagement by looking at observable aspects of students' online behavior. For example, the more engaged students are, the more effort they invest in learning (i.e., longer online time), and the more regular and frequent they post on fora. These observable variables are the gateway to the underlying engagement construct that is theorized to drive students to be engaged, and factor analysis is the key to opening that gate. If scores on a set of observed variables are all caused by the same underlying construct of interest, the relations between these variables (i.e., the covariances that indicate how increases/decreases in one behavior are related to increases/decreases in another) are an expression of this underlying construct. This is exactly the core tenet of factor analysis. The method examines the covariances between the observed variables and, from this information, extracts the underlying unobserved or *latent* construct (as well as students' relative scores on it).

The inventor of factor analysis is Spearman, who was trying to make sense of the fact that tests of widely varying cognitive abilities all positively correlated

L. V. D. E. Vogelsmeier (✉) · J. Jongerling
Tilburg University, Tilburg, Netherlands
e-mail: l.v.d.e.vogelsmeier@tilburguniversity.edu

M. Saqr · S. López-Pernas
School of Computing, University of Eastern Finland, Joensuu, Finland

M. Saqr, S. López-Pernas (eds.), *Learning Analytics Methods and Tutorials*,
https://doi.org/10.1007/978-3-031-54464-4_20

with each other. He reasoned that the cause of this was an underlying construct (or factor) called "general intelligence" that was causing people's performance on all of those tests [1]. Spearman's work on factor analysis was later extended by Thurstone, who believed that people's performance was influenced by more than just one latent dimension. Thurstone [2], therefore, expanded factor analysis to enable the extraction of multiple underlying constructs based on the covariances between variables. The extension allowed performance on a math test to, for example, be influenced both by math ability (the key ability that should be measured) and reading ability (the ability to accurately understand questions on the math test in order to answer them correctly).

Jöreskog introduced the final major addition to factor analysis. Although Spearman's and Thurstone's versions of factor analysis already allowed for *exploring* the factor structure for a given dataset, it was not yet possible to *confirm* if the factor structure fit well to the data and, thus, if the covariances between variables implied by the factor structure match the observed covariances in the dataset. Jöreskog [3] figured out how to estimate factor models in a way that made this possible. An added benefit of this estimation method was that it allowed for factor models in which observed variables (e.g., behaviors, tasks, or questionnaire items) were not influenced by all the assumed underlying dimensions but, for example, only by one, which was not possible in the methods of Spearman and Thurstone. This extension allowed adding theory and/or practical experience to the factor analysis. For instance, one could test the hypothesis that students' math ability consists of the separate sub-dimensions *addition*, *subtraction*, *division*, and *multiplication* by letting all tasks of a math test that involve addition belong to just one underlying *addition* factor, all the tasks involving multiplication to just one underlying *multiplication* factor, et cetera. If this multidimensional model of math ability fits the data well, students can subsequently be evaluated on each of these dimensions separately (instead of on an overall math ability factor). The approach in which researchers confirm if a specific factor model (i.e., for which both the number of underlying dimensions and the pattern of relations between these dimensions and the observed variables) fits to the data is nowadays called confirmatory factor analysis (CFA). In contrast, the more data-driven approach in which the number of underlying constructs is inferred from the data and all underlying constructs are assumed to influence all observed variables is called exploratory factor analysis (EFA).

Nowadays, both EFA and CFA are readily available in modern statistical (open-source) software and applied regularly across the social sciences in general and educational research in particular. For example, Krijnen et al. [4] used EFA to refine a typology of home literacy activities. They cautiously anticipated four hypothetical categories of home literacy activities (*oral language exposure*, *oral language teaching*, *code skills exposure*, and *code skills teaching activities*). However, since the authors did not have a strong theory or prior factor-analytical results to support this anticipated factor structure, they refrained from testing if that specific factor structure fit to the data with CFA and instead used the more data-driven EFA approach. Their results suggested that there were actually three factors underlying the observed variables in their dataset (*oral language teaching*, *oral language exposure*, and *general code activities*).

In contrast, Hofhuis et al. [5] used factor analysis to validate a shortened version of the Multicultural Personality Questionnaire (MPQ; [6]), a measure of intercultural competence, among a sample of students in an international university program. The authors wanted to determine if this short form of the MPQ (MPQ-SF) could be used instead of the longer original and if item scores of both versions of the questionnaire were influenced by the same five theorized aspects of inter-cultural competence (*cultural empathy*, *emotional stability*, *flexibility*, *openmindedness*, and *social initiative*). Since previous research on the original MPQ provided insight into both the number of underlying factors and the specific relations between these factors and the set of items retained on the MPQ-SF, the authors used CFA in this study. They found that the factor structure of the MPQ-SF fit well to the data and that the structure was thus comparable to that of the original MPQ.[1]

In the above, it was stressed that CFA is used when researchers have an idea about the factor structure underlying the data that they want to confirm, while EFA is used more exploratively when researchers have less concrete guiding insights for specifying the model. In practice, CFA and EFA are both used in confirmatory as well as exploratory settings and often even in the same study. Even if researchers have a well-substantiated idea about the number of constructs underlying their observations, they can use EFA to see if the number of factors found by this analysis matches their hypothesis. Similarly, researchers with competing theories about the factors underlying their observed behaviors can still use CFA to explicitly compare these competing models in terms of fit. Flora and Flake [8] discuss how neither EFA nor CFA is purely confirmatory or exploratory in more detail, arguing that, in essence, it comes down to one's specific research context. This will not further be discussed in this chapter. Instead, an integrated use of EFA and CFA often encountered in Educational Sciences is presented. The presentation is kept applied and focuses on conducting factor analysis in R. For more (technical) details, see the readings listed at the end of the chapter.

2 Literature Review

Several examples of factor analysis exist in learning analytics, which can be grouped broadly under two categories: factor analysis of self-reported data instruments (e.g., surveys) and factor analysis to explore students' online data. Analysis of self-reported instruments seems to be most widely used within the emerging research

[1] Moreover, they determined that the factor structure of the MPQ-SF was identical for Western and non-Western students and that the MPQ-SF could therefore be used to study and compare both these groups. This comparing factor structures for different groups in terms of similarity is called measurement invariance testing and will be discussed more extensively in the next chapter on SEM [7].

field of learning analytics. For instance, Whitelock-Wainwright et al. [9] used EFA and CFA to validate an instrument that measures students' expectations of ethics, privacy, and usage of their learning analytics data. The analysis suggested a two-factor model that represented two categories of variables of ideal and predicted expectations of learning analytics. Similarly, Oster et al. [10] used EFA to validate an instrument that measures learning analytics readiness among institutions. The authors found that a five-factor model best represents the data (with the dimensions *data management expertise*, *data analysis expertise*, *communication and policy application*, and *training*). Similarly, factor analysis has been used to create an instrument for measuring variables influencing learning analytics dashboard success. For instance, Park and Jo [11] measured learning analytics dashboard success through an instrument based on Kirkpatrick's four evaluation levels (i.e., Reaction, Learning, Behavior, and Results). Through EFA, they found that five dimensions were more appropriate than four, as suggested by the original instrument, which CFA later confirmed. Similarly, Kokoç and Kara [12] used the Evaluation Framework for Learning Analytics to evaluate the impact of a learning analytics dashboard on learner performance. After conducting CFA, they found that the three-factor model of the Evaluation Framework for Learning Analytics for learners provided the best model fit for the collected data, confirming the structure of the original instrument.

Besides the aforementioned traditional examples, factor analysis has also been used in learning analytics with students' online data logs. For instance, Hernández-García et al. [13] used factor analysis to identify predictors derived from data about students' interactions. The authors found that a three-factor model best represented students' interaction variables (with dimensions *groups' team message exchanges*, *distribution of postings*, and *reciprocity of interactions*). In that case, factor analysis helps find groups of predictors, understand their underlying structure, and reduce dimensionality. Similarly, Fincham et al. [14] used EFA and CFA to build a theorized model of engagement based on three groups of variables derived from online log data, analysis of sentiments, and metrics derived from the discourse of online posts. Others have applied similar approaches to study the structure of log data. For instance, Baikadi et al. [15] applied EFA to learner activities within four online courses to identify emergent behavior factors. The findings suggested a four-factor model including *activity*, *quiz activity*, *forum participation*, and *participation in activities*. Another example is the work by Wang [16], who used factor analysis for dimensionality reduction of log data, deriving predictors of learning performance from students' fine-grained actions on the learning management systems.

The remainder of this chapter first provides a brief description of the factor analysis model, the model that is at the basis of both EFA and CFA. Second, the steps needed to perform factor analysis in R are presented by applying EFA and CFA to education data that is openly available. The application begins with data preparation, requirements, and initial checks. Additionally, it is shown how to set aside a holdout sample from the original dataset (which is necessary for establishing the generalizability of results from an EFA and/or CFA to future samples, as explained below). After these preliminary steps, it is shown how to run an EFA and interpret the outcomes. The part ends with a thorough description of how to do a CFA and assess generalizability. The chapter ends with a discussion and recommendation for further reading.

3 Recap of the Factor Analysis Model

Factor analysis can be seen as a set of (multiple) linear regression models where each observed variable is regressed on one or more latent factors (see Figure 1 in [17]). Like in regression, researchers get regression weights, intercepts, and error terms. They just get a set of these parameters for each observed variable in the analysis. The regression weights are referred to as loadings in factor analysis (the straight arrows in Fig. 1) and indicate how strongly the observed variables are related to the underlying factors. The intercepts (not explicitly indicated in Fig. 1) are the expected scores on the observed variables when the factor means are equal to zero. Finally, the error terms (the ε's in Fig. 1) capture the variance unexplained by the factors and, thus, the unique variance of the individual observed variables. The difference between the factor analysis model and a regular regression model is that the values of the factors are unobserved. Therefore, estimating a factor analysis model is more complicated than estimating a regular regression analysis model with observed predictors and requires specialized software.

As mentioned in the introduction, there are two types of factor analysis: Exploratory factor analysis (EFA; Fig. 2) and confirmatory factor analysis (CFA; Fig. 3). The key difference is that all loadings (i.e., all variable-factor relations) are estimated in EFA. As a result, variables may load on more than one factor (referred to as cross-loadings). In contrast, in CFA, several loadings are set to zero (and thus not estimated) based on researchers' a priori hypotheses about which variables are unrelated to which underlying factors. Therefore, the CFA model is considered a restricted version of the EFA model.

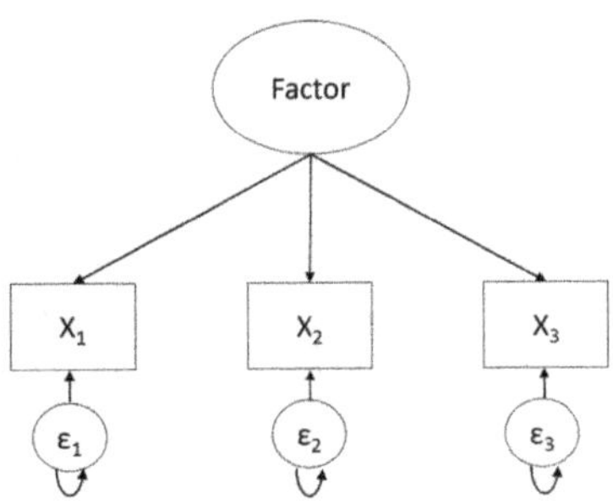

Fig. 1 A basic 1-factor model. Each observed variable (X_1–X_3) is regressed on the factor. Straight arrows indicate factor loadings (the regression weights obtained by regressing the observed variable on the factor). The ε's represent errors. Curved single-headed arrows indicated variances. Each observed variable also has an intercept, but these are not explicitly indicated in the figure

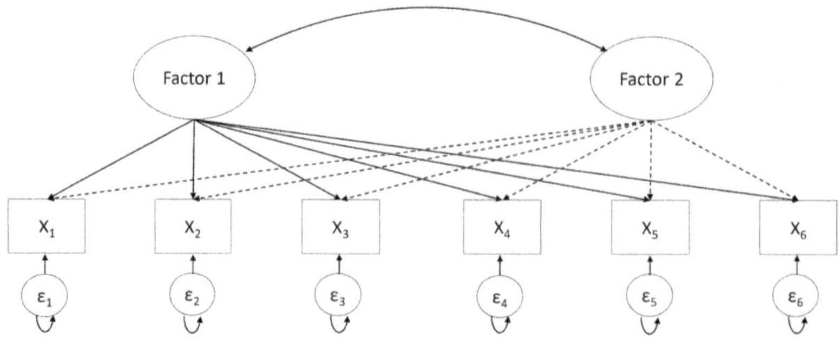

Fig. 2 An EFA Model with two latent factors. Each observed variable (X_1–X_6) is influenced by both factors. Straight arrows indicate factor loadings. Loadings for Factor 2 are depicted with dashed lines for visual clarity. The ε's represent errors. Curved single-headed arrows indicate variances and curved double-headed arrows indicate covariances

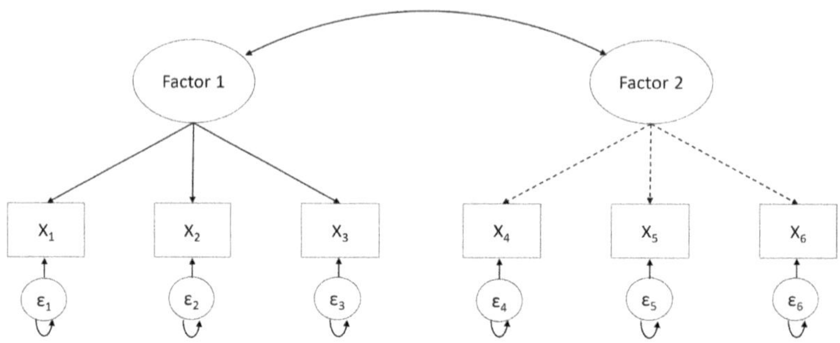

Fig. 3 A CFA Model with two latent factors. Each observed variable (X_1–X_6) is influenced by only one of the underlying factors. Straight arrows indicate factor loadings. Loadings for Factor 2 are depicted with dashed lines for visual clarity. The ε's represent errors. Curved single-headed arrows indicate variances and curved double-headed arrows indicate covariances

Another difference is the determination of the number of factors. In EFA, the number of underlying factors is determined using a data-driven approach; that is, the likely number of underlying factors for the sample data is first estimated, for example, using parallel analysis [18]. Researchers then fit a set of factor models, with the number of underlying factors in these models based on the parallel analysis result. For example, if the parallel analysis suggests 5 factors, researchers can fit models with 4, 5, and 6 underlying factors. These three models are then compared in terms of both model fit—using the Bayesian information criterion [19]—and in terms of the interpretability of the models (i.e., "*Do the relations between the factors and the observed variables they load on make sense?*"). All this will be discussed in the Sect. 5 of this chapter.

In contrast, the number of factors in CFA is determined based on strong a priori hypotheses. When researchers have multiple competing hypotheses, each can be translated into a separate CFA model. These models can then again be compared in terms of how well they fit the data. The hypothesis underlying the best-fitting model can be considered the most likely explanation of the data according to the sample at hand.

4 Integrated Strategy for a Factor Analysis

As described in the introduction, there is no clear delineation between when to use either EFA or CFA, and both methods often co-occur within the same study. Therefore, this section provides a detailed description of a principled modeling strategy that integrates both EFA and CFA. More specifically, the three steps that researchers should go through whenever latent constructs are part of their research (either because the instrument is the main focus of the study or because the latent construct is a predictor or outcome in an analysis) are discussed: (1) exploring the factor structure, (2) building the factor model and assessing fit, and (3) assessing generalizability. As a starting point, it is assumed that the researcher has already completed the initial instrument development phase for a construct of interest such as inter-cultural competence (i.e., that the researcher is using an instrument with variables/tasks/behaviors from previous research, has adjusted an instrument from previous research (e.g., by shortening, extending, translating, or revising content), or has newly developed an instrument (e.g., based on theory)). Furthermore, it is assumed that the researcher has gathered data from their population of interest (e.g., students).

4.1 Step 1: Exploring the Factor Structure

Once the variables/tasks/behaviors have been selected and data on them have been obtained using a sample from the population of interest, you, the researcher, should start with an EFA. If a *previously validated instrument* is used, or if strong prior hypotheses about the underlying factor structure of the instruments are available, you should investigate whether the number of factors and the way the variables load on the factors are in line with anticipated results. Thus, the key questions are "*Do variable scores believed to be caused by the same underlying factor indeed all load on the same factor?*" and, if only a single underlying factor is assumed to cause the scores on a set of variables, "*Do these variables indeed have strong factor loadings with only a single factor?*"

If a *new instrument* is the starting point, you should determine if the factors and the pattern of loadings can be interpreted. The key questions are "*Do all variables (primarily) loading on the same factor indeed have something in common substantively?*" and "*Are variables that load on different factors indeed qualitatively different somehow?*"

Referring back to the example about a math test, you could see that tasks involving addition, subtraction, division, and multiplication loaded on 4 distinct factors (which could then preliminarily be identified as *addition-*, *subtraction-*, *division-*, and *multiplication* ability), with each set of tasks being primarily influenced by a single factor. At this stage, you may have to make some adjustments, like removing variables without substantial loadings (e.g., loadings smaller than an absolute value of .3) on any dimensions and reestimating the EFA. Note that you should always think about the reasons for low loadings (e.g., because of an ambiguously formulated item) and not just remove variables without good reason.

4.2 Step 2: Building the Factor Model and Assessing Fit

After choosing a model with EFA, you need to refine this model and use CFA to assess the model fit (and thus how well the covariances between variables implied by the factor structure match the observed covariances in the dataset). In the EFA in the previous step, all variables were allowed to load on all factors, but often, you have theoretical or (prior) empirical information that you want to include in your analyses, and that restricts the number of these cross-loadings. In this model-building step, you could remove all factor-variable relations (i.e., factor loadings) that do not fit your theory or make substantive sense, but attention must be given to the size of the factor loadings. Close-to-zero factor loadings can be removed relatively risk-free, but larger factor loadings require more careful consideration. Even if these factor-variable relations do not make substantive sense (straight away), the data tell you they should be there. So consider them carefully; if, after closer inspection, they can be incorporated into your prior theory or assumptions, you might want to keep them; if not, you can always remove them and see if the model still fits your data.

After selecting which variable-factor relationships should be removed, you can build the corresponding CFA model and fit it to the data to determine if it fits sufficiently well. If the model does not fit well, you can return to your EFA results and see if you might have to allow additional non-zero loadings (or apply other modifications discussed further below). Note that you should only add relationships to the model that make substantive sense. This process may require several rounds of adjustments until your model fits well and is substantively sound.

4.3 Step 3: Assessing Generalizability

After the previous step, you have a preliminary model that fits well with both your new or existing theory and the current data. However, since the ultimate goal should be to present instruments that can be used to measure constructs in future studies, it is essential that you also establish that your preliminary model fit new, as-yet-unknown, data and, thus, that your model does not describe only your current sample properly but also other samples from your population. Therefore, in the final step, the preliminary model must be fitted to a new dataset from the same population as the data used to build the preliminary model. This final validation step can be referred to as cross-validation.

To assess the generalizability, one could collect a second dataset. However, in practice, gathering data more than once for instrument development purposes is often unfeasible or undesirable. A suitable alternative is to randomly split one dataset into two parts: one sample on which you perform Steps 1 (exploring the factor structure using EFA) and 2 (building the factor model and assessing fit with CFA) and one so-called holdout sample, which you set aside for Step 3 (assessing generalizability). If the CFA model fits the holdout sample, you can be more confident that your instrument can be used in future studies and settle on it as your final model. On the other hand, if the preliminary model does not fit the holdout sample well, you have to conclude that you have not found an adequate instrument yet. In that case, the sources of misfit between the CFA and the holdout sample need to be investigated (more on this below when local fit and modification indices are discussed), and findings from this inspection need to be used to update your theory/model. This updated theory/model then needs to be investigated again by going through all the steps discussed above on a new (split) dataset.

The above steps present a suitable factor modeling strategy for any study using instruments to measure latent dimensions. The only situation in which you could theoretically fit a CFA model immediately and assess its fit is when using an existing instrument on a sample from a population for which the instrument had already been validated in previous research. However, even in that situation, going through all the above steps is advisable because your sample might differ in important ways from the ones on which the instrument was validated, which could bias your results.

5 Factor Analysis in R

In the following, you are taken through the essential steps of investigating the factor structure using both EFA and CFA in the open-source software R. In describing the steps, the following is addressed: checking data characteristics for suitability for EFA/CFA, deciding on the number of factors, assessing global and local model fit, and evaluating the generalizability of the final factor model. To this end, a dataset is used to which factor analysis has been applied before [20]. The dataset contains survey data about teacher burnout in Indonesia. In total, 876 respondents have answered questions on five domains: *Teacher Self-Concept* (TSC, 5 questions), *Teacher Efficacy* (TE, 5 questions), *Emotional Exhaustion* (EE, 5 questions), *Depersonalization* (DP, 3 questions), and *Reduced Personal Accomplishment* (RPA, 7 questions). Thus, the total number of variables equals 25. The questions were assessed on a 5-point Likert scale (ranging from 1 = "never" to 5 = "always"). For more information on the dataset, see the data chapter of this book [21].

The first section shows the necessary preparation, which involves reading in the data, evaluating whether the data are suited for factor analysis, and setting apart the holdout sample needed for assessing the generalizability. The next two sections show how to conduct an EFA to arrive at a preliminary factor model/factor structure (Step 1), and how to refine this model using CFA (Step 2). The final section shows how to test the generalizability of the refined factor model using cross-validation (Step 3).

5.1 Preparation

In order to follow all the steps, you have to install the following packages with `install.packages()`(you only have to install packages the first time you want to use them; therefore, the commands are hashtagged below) and load them with `library()` whenever you open the R script.

```
library(lavaan) # install.packages("lavaan")
library(psych) # install.packages("psych")
library(semTools) # install.packages("semTools")
library(effectsize) # install.packages("effectsize")
library(rio) # install.packages("rio")
library(tidyr) # install.packages("tidyr")
library(ggplot2) # install.packages("ggplot2")
library(devtools) # install.packages("devtools")
library(sleasy) # install_github("JoranTiU/sleasy") is
      sufficient install_github("JoranTiU/sleasy")
```

5.1.1 Reading in the Data

The data can be read in, and the variable names can be extracted with the following commands:

```
dataset <- import("https://github.com/lamethods/data/raw/main/4_
                    teachersBurnout/2.%20Response.xlsx")
var_names <- colnames(dataset)
```

If not all variables in the dataset should be used for the factor analysis, you should only add the relevant variables to the `var_name` object. The commands below will then take this into account automatically.

5.1.2 Are the Data Suited for Factor Analysis?

Several data characteristics are necessary for both EFA and CFA. First, the variables must be continuous. Variables are seldom truly continuous, but they can be treated as such if they were assessed on a scale with at least five response categories and the responses are reasonably symetrically distributed [22]. If the variables are not continuous, factor analysis can still be conducted, but a specific type of estimation for categorical data is required. Note that this is beyond the scope of this chapter (interested readers are referred to [23]). Moreover, the scale on which the observed variables are assessed should be the same, which may not hold in certain educational data. If the variables have been measured on different scales, or if the variables are measured on the same scale, but the range of observed scores on the variables differs substantially between variables (e.g., some variables have scores ranging from 1 to 5, while others have scores ranging only from 2 to 4), the variables should be transformed before the factor analysis to make their scales more comparable. The following command can be used to inspect each variables' range (Table 1):

Table 1 Descriptive statistics of the dataset variables

Vars	n	Mean	sd	Median	Trimmed	Mad	Min	Max	Range	Skew	Kurtosis	se
1	876	3.65	0.68	4	3.62	0.00	1	5	4	−0.09	0.06	0.02
2	876	3.81	0.64	4	3.78	0.00	2	5	3	−0.07	−0.14	0.02
3	876	3.73	0.64	4	3.71	0.00	2	5	3	−0.17	−0.02	0.02
4	876	3.71	0.67	4	3.67	0.00	2	5	3	−0.03	−0.25	0.02
5	876	3.82	0.65	4	3.79	0.00	2	5	3	−0.10	−0.13	0.02
6	876	4.06	0.71	4	4.10	0.00	1	5	4	−0.47	0.38	0.02
7	876	4.04	0.70	4	4.07	0.00	2	5	3	−0.22	−0.45	0.02
8	876	4.12	0.71	4	4.17	0.00	1	5	4	−0.72	1.60	0.02
9	876	4.11	0.69	4	4.15	0.00	1	5	4	−0.47	0.51	0.02
10	876	3.90	0.75	4	3.92	0.00	1	5	4	−0.41	0.16	0.03
11	876	3.81	0.76	4	3.81	0.00	1	5	4	−0.35	0.23	0.03
12	876	3.73	0.85	4	3.75	1.48	1	5	4	−0.37	0.12	0.03
13	876	3.88	0.83	4	3.91	1.48	1	5	4	−0.31	−0.40	0.03
14	876	3.69	0.80	4	3.67	1.48	1	5	4	−0.03	−0.41	0.03
15	876	3.99	0.81	4	4.03	1.48	1	5	4	−0.43	−0.27	0.03
16	876	3.92	0.68	4	3.93	0.00	1	5	4	−0.53	1.25	0.02
17	876	3.60	0.68	4	3.58	1.48	1	5	4	−0.22	0.64	0.02
18	876	3.82	0.70	4	3.79	0.00	1	5	4	−0.14	0.01	0.02
19	876	3.93	0.83	4	3.97	1.48	1	5	4	−0.59	0.50	0.03
20	876	3.94	0.80	4	3.99	0.00	1	5	4	−0.79	1.22	0.03
21	876	3.88	0.79	4	3.91	0.00	1	5	4	−0.59	0.75	0.03
22	876	3.87	0.76	4	3.89	0.00	1	5	4	−0.48	0.33	0.03
23	876	3.84	0.79	4	3.86	0.00	1	5	4	−0.53	0.67	0.03

```
describe(dataset)
```

All variables were assessed on 5-point Likert scales, and from the output, you can see that all variables have very similar observed score ranges. Therefore, you can treat them as continuous, and transformation is not necessary (for information on how to transform data in R, see [24]).

Second, the sample size needs to be sufficiently large. There are several rules of thumb in the literature. Some simply state that a sample size of about 200 should be targeted, although smaller samples may be sufficient for simpler models (e.g., models with fewer factors and/or stronger relations between the factors and observed variables), while more complicated models (e.g., models with more factors and/or weaker relations between the factors and observed variables) will require larger samples. Other rules are based on the ratio between sample size and the number of estimated parameters (i.e., factor loadings, intercepts, and error variances). Bentler and Chou [25] recommend having 5 observations per estimated parameter, while Jackson [26] recommends having 10, and preferably 20 observations, for each parameter you want to estimate (e.g., for a one-factor model with 10 variables, one should aim for 30 (10 factor-loadings + 10 intercepts + 10 error-variances) $\times 10 = 300$ cases). Remember that these recommendations are for the data the model is fitted to. Since you also need a holdout sample to assess the generalizability of your model, you need to have about twice the number of observations! The sample size can be assessed by asking for the number of rows in your dataset with the following command:

```
nrow(dataset)
```

```
[1] 876
```

For the example data with 25 variables that are assumed to measure 5 latent constructs, you have to estimate 25 intercepts, 25 residual variances, and $5 \times 25 = 125$ factor loadings. This results in a total of 175 parameters. Looking at the output, you can conclude that the sample size is sufficiently large for both EFA and CFA according to the guidelines by Bentler and Chou [25] ($5 \times 175 = 875$) but not those of Johnson [27]. Since this dataset does not have twice the recommended sample size, you should not set aside a holdout sample and save the validation of the model for a future study. However, for illustration purposes, it will nevertheless be shown how to create a holdout subset for evaluating the generalizability of the final factor model.

Next, there need to be sufficiently large correlations between the variables. Otherwise, there is no point in looking at the factor structure. You can rule out that the variables in the dataset are uncorrelated with Bartlett's test [28], which tests whether the correlation matrix is an identity matrix (a matrix with off-diagonal elements equal to zero) and, thus, whether the variables are uncorrelated. The null hypothesis of this test is that the correlation matrix is an identity matrix. If the null hypothesis is rejected, it can be concluded that the variables are correlated and, thus, that you can continue with the factor analysis. With the following command, you test whether the p-value for the Bartlett's test is smaller than an alpha level of 0.05 and, thus, if the null hypothesis of "no correlation between variables" can be rejected:

```
(cortest.bartlett(R = cor(dataset[, var_names]), n =
                    nrow(dataset))$p.value) < 0.05
```

```
[1] TRUE
```

With the argument "R", you provide the correlation matrix for the data (specifically for the variables that shall be part of the factor analysis), and with the argument "n" you determine the sample size, which is equal to the number of rows. The p-value is indeed smaller than 0.05. Thus, the variables correlate.

In addition to checking for correlations between variables, it is also relevant to determine if there is enough common variance among the variables. You can assess this with the Kaiser-Meyer-Olkin (KMO) test [29]. The KMO statistic measures what proportion of the total variance among variables might be common variance. The higher this proportion, the higher the KMO value, and the more suited the data are for factor analysis. Kaiser [29] indicated that the value should be at least .8 to have good data for factor analysis (and at least .9 to have excellent data). With the following command, you obtain the results:

```
KMO(dataset)
```

```
Kaiser-Meyer-Olkin factor adequacy
Call: KMO(r = dataset)
Overall MSA =  0.94
MSA for each item =
TSC1 TSC2 TSC3 TSC4 TSC5  TE1  TE2  TE3  TE4  TE5  EE1  EE2  EE3  EE4  EE5  DE1
0.96 0.96 0.95 0.94 0.96 0.93 0.96 0.94 0.94 0.96 0.95 0.94 0.95 0.94 0.97 0.87
 DE2  DE3 RPA1 RPA2 RPA3 RPA4 RPA5
0.86 0.92 0.91 0.91 0.95 0.94 0.96
```

The overall KMO value equals 0.94. Thus, the data are excellent for factor analysis.

Next to the necessary data characteristics, you need to be aware of the non-normality of the variables and missing data. A robust estimation method is required if the variables are not normally distributed. If the data contain missing values for one or more variables, this must also be accounted for in the estimation. How to do this will be described below. Normality can be assessed by inspecting the variables' histograms:[2]

[2] There are many different ways to obtain histograms in R and this code is just one possible example. To learn more about how to create a histogram step by step, see the chapter on Data Visualization in this book [30].

```
dataset |> pivot_longer(2:ncol(dataset),
 names_to = "Variable", values_to="Score") |>
   ggplot(aes(x=Score)) + geom_histogram(bins=6) +
     scale_x_continuous(limits=c(0,6), breaks =
   c(1,2,3,4,5)) + facet_wrap("Variable", ncol = 4, scales =
   "free") + theme_minimal()
```

Looking at Fig. 4, you can see that the distribution of the variables is somewhat left-skewed. Therefore, an estimation method that is robust against non-normality should be used.

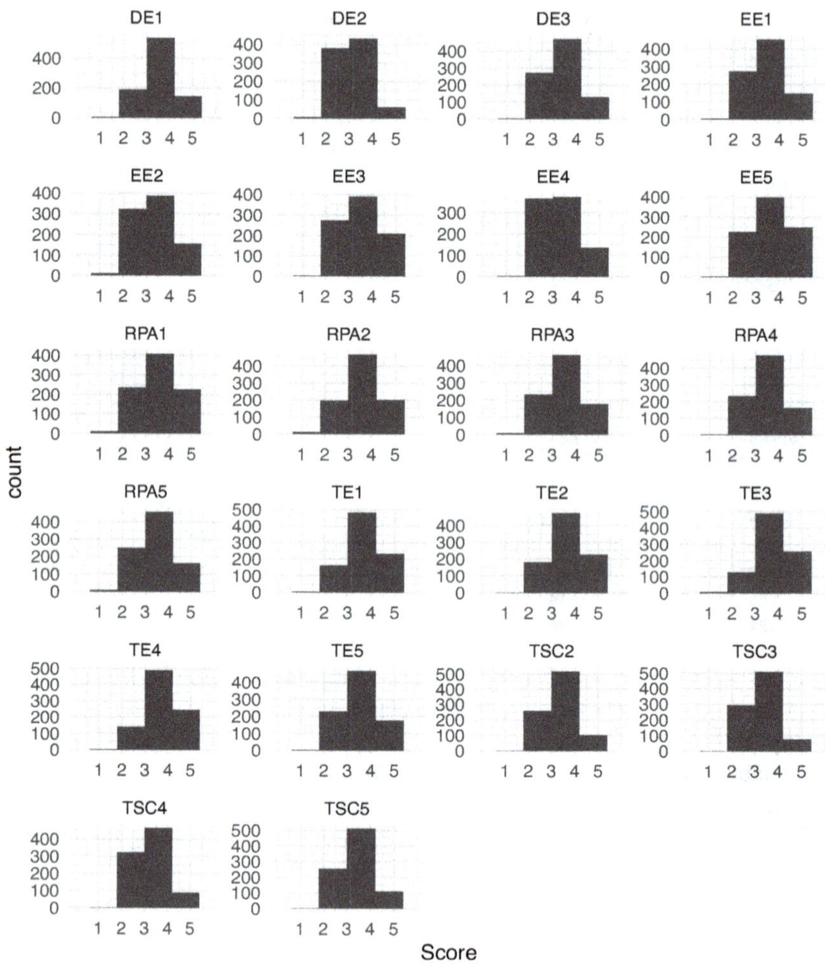

Fig. 4 Histogram of the responses to each of the items

Next, you can check for missing data with the following command:

```
colSums(is.na(dataset))
```

```
TSC1 TSC2 TSC3 TSC4 TSC5  TE1  TE2  TE3  TE4  TE5  EE1  EE2  EE3  EE4  EE5  DE1
   0    0    0    0    0    0    0    0    0    0    0    0    0    0    0    0
 DE2  DE3 RPA1 RPA2 RPA3 RPA4 RPA5
   0    0    0    0    0    0    0
```

There are no missing observations for any of the variables in this dataset.

5.1.3 Setting a Holdout Sample Apart

Once you know that the data are suited for factor analysis, you can consider setting a holdout sample apart to assess the generalizability of your findings, as explained before. However, it is important to consider the sample size in your decision. As indicated above, the minimum required sample size should at least be 5 (and preferably 10 or 20) times the number of parameters you will estimate. Do not set a holdout sample apart unless your sample size is approximately twice as large as the minimum required sample size (or larger). Otherwise, you will not have enough data to build an appropriate model in the first place. The validation of the final model then needs to be done in future research. Note, however, that the number of parameters for a CFA model is generally smaller than that for an EFA model. Therefore, it is okay if the holdout sample is somewhat smaller than the model building sample.

As was already determined above, the sample size was not twice the minimum required sample size for a model with 25 variables and 5 latent factors, but for illustrative purposes, a holdout sample is set apart nevertheless. To this end, you can randomly assign 438 rows to a model-building and holdout dataset. For this, you can use the following commands:

```
set.seed(19)
ind <- sample (c (rep("model.building", 438), rep (
                    "holdout", 438)))
tmp <- split (dataset, ind)
model.building <- tmp$model.building
holdout <- tmp$holdout
```

With the first line of code, you set a seed. Setting a seed ensures that you get the same random sequence of elements every time you rerun the code, which is crucial for replicating your results. Then, you create a vector "ind" that contains 438 times the terms "model.building" and "holdout", respectively, in random order. This gives you a total of 876 classifications, one for each participant (i.e., row) in your data. Subsequently, you create a temporal list termed "tmp", which contains two datasets: for each row number, the function `split()` checks whether it is assigned the label "model.building" or "holdout" and assigns this row to the respective dataset accordingly. For example, suppose the vector "ind" has as the first three elements "model.building", "model.building", and "holdout". In that case, the first two rows of the dataset are assigned to the model-building dataset, and the third observation is assigned to the holdout dataset. In the last step, the two new datasets are extracted from the list and stored in objects named "model.building" and "holdout". You will use the model-building data for all the following analyses until Sect. 4.3.

5.2 Step 1: Exploring the Factor Structure

The first step in exploring the factor structure is to determine how many dimensions are likely underlying the construct of interest. In this tutorial, you will see how to determine this using a combination of two commonly used data-driven approaches: parallel analysis [18] and the Bayesian information criterion (BIC; [19]). These two methods complement each other greatly: Parallel analysis indicates a range for the number of underlying dimensions, and the BIC tells us which specific number of dimensions from this range fits the data best. Parallel analysis is a simulation-based method that chooses the number of factors by comparing the amount of variance explained by a certain number of factors in *your data* (with this amount of variance explained by each factor being called the factor's eigenvalue) to the amount of information that the same number of factors would explain on average across many parallel simulated datasets (i.e., with the same number of variables and observations), but with *zero correlations between all variables*. This comparison allows for testing if the amount of explained variance in your data is larger than expected based on purely random sampling error alone. The number of factors chosen by parallel analysis is the number for which the explained variance in your data is larger than the explained variance for the simulated data. Details about how this method works are beyond the scope of this chapter but can be found in the help file[3] of the `fa.parallel()` function that performs the analysis.

With the argument `x`, you specify that you want to use your model-building data and, more specifically, all the columns corresponding to your variables. With the second argument `fa = "fa"`, you specify that you assess the best number of factors for factor analysis and not the best number of components for principal components analysis, which is a related yet different method for finding dimensions in data (e.g., [31]). The output consists of two parts: a message and a figure:

```
fa.parallel(x = model.building[, var_names], fa = "fa")
```

```
Parallel analysis suggests that the number of factors =  5  and the
                                       number of components =  NA
```

[3] Note that you can open the helpfile for any function of interest by typing `?[functionname]` (e.g., `?fa.parallel`).

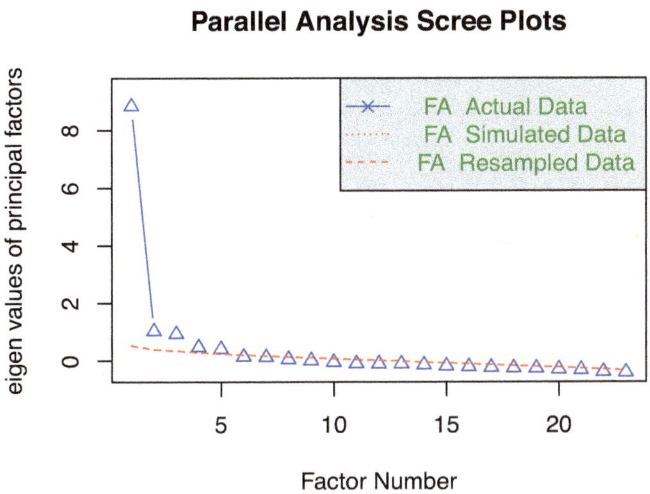

Fig. 5 Parallel analysis

The message indicates that five factors are likely underlying the data. Fig. 5 shows that with more than five factors (x-axis), the amount of information explained by the factors (on the y-axis) in your data is lower than for the simulated data.

It is important to note that parallel analysis is a purely data-driven method, and the solution depends on the specific sample. Therefore, you should see the 5 factors only as a starting point (such that the number of underlying factors is likely around five) and treat parallel analysis as giving you a plausible range for the number of underlying factors, equal to the solution plus and minus one factor (or more, if you want to). The final decision for the number of underlying factors is made by running several factor models—one for each plausible number of underlying factors as determined using the parallel analysis—and comparing these models in terms of interpretability (i.e., does the pattern of variable-factor relations make substantive sense and/or does it fit prior knowledge/assumptions) and fit (using the BIC). The BIC can be used for model selection for many analyses, including factor analysis, and the criterion balances the fit of the factor model to the data on the one hand and model parsimony (i.e., simplicity) on the other by penalizing models for each parameter they contain. The lower the BIC value, the better, so if the smallest BIC value corresponds to the model with five factors, you have stronger support for the five-factor solution. Still, the final decision should take interpretability into account. If the parallel analysis and BIC disagree, your final decision should consider interpretability even more.

The exploratory factor analysis can be performed with the following command:

```
EFA <- efa(data = model.building[, var_names],
                    nfactors = 4:6,
                    rotation = "geomin", estimator = "MLR",
                        meanstructure = TRUE)
```

The efa() function is part of lavaan [32], which is a very popular, open-source R package that can be used for estimating various latent variable models, including factor analysis and structural equation modeling (the Structural Equation Modeling Chapter [39],[21]). With the first argument data, you again specify the dataset on which you want to perform factor analysis, nfactors indicates the range of factors for which you want to retrieve the results. Next, the argument rotation pertains to identifying the model, which is only necessary for EFA and not CFA because, in the latter, you specify restrictions (i.e., restricting loadings to be equal to zero) that automatically lead to the identification of the model. In EFA, where you do not have restrictions, there are an infinite number of solutions identical in fit; that is, your factor matrix with loadings can be transformed or "rotated" in infinite ways. To clarify, consider the following: with (continuous) *observed* variables, you can visualize your data using scatterplots in which each point represents an individual's combination of scores on two variables that are represented by the x-axis and the y-axis. With observed variables, you have individuals' actual scores on the variables in your dataset. Therefore, the locations of the points relative to the x-axis and y-axis (and, therefore, the axes' position) are fixed. With factors, the situation is different. The latent variables are *unobserved*, which means that the position of the axes (which now represent latent variables) is not "pinned down" by the data. Considering Fig. 6, for example, both axis orientations are equally likely with latent variables. Note that in both plots, the data points are in the exact same location; only the orientations of the axes representing the (in this case, two) latent factors are different. Rotation is about choosing one out of all the many possible orientations of the axes.

The choice of the orientation of the axes is generally made based on how easy it makes the interpretation of the factors. Typically, the aim is to find a so-called

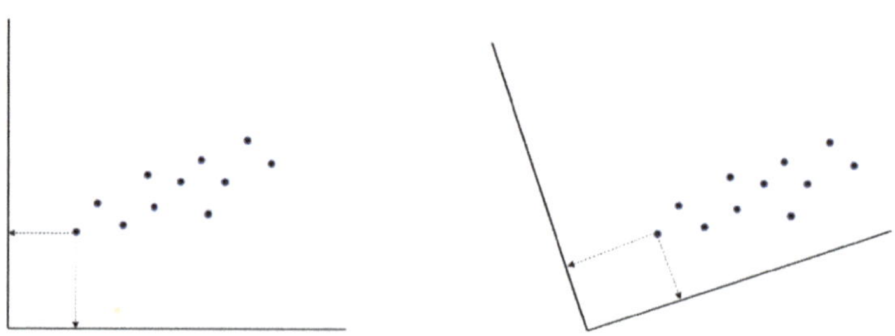

Fig. 6 The two possible axis orientations (out of an infinite number of plausible orientations) show that with rotation, the data points stay in exactly the same place, but the axes representing the latent factors (here, two factors) change

simple structure [33] in which the orientation of the axes ensures that each observed variable is strongly related to one factor (i.e., has a large loading on one factor) and is as unrelated as possible to all others (i.e., has small cross-loadings, for example, smaller in absolute value than 0.3). This simple structure makes interpreting the factors (which is done by looking for a common theme in the variables that load on them) easier.

In short, rotation is like taking a good selfie of yourself in front of the Eiffel Tower. You want both yourself and the tower to be clearly visible in the picture, so you move your camera until you get a clear shot of both. You and the Eiffel Tower both stay in exactly the same place! What changes is the vantage point, or angle, from which you look at both. Similarly, rotation moves your vantage point (the axes) to make the factors stand out clearly in your results, while the position of your observed data does not change.

Going back to the `rotation` argument in the code above, you can use `geomin`, a method of rotation that allows your factors to be correlated with each other, which is a reasonable assumption in educational settings. For other rotation options, see [23]. The argument `estimator` allows you to choose different estimation procedures. The default is standard maximum likelihood ("ML") estimation. However, for the current data, a robust maximum likelihood ("MLR") estimation was chosen to account for small violations of the normality assumption. If the data would contain missing values, you could add the argument "missing" and specify it to be equal to "fiml", which corresponds to a full information maximum likelihood approach and is a sensible approach if you have at least missing at random (MAR) data (details about missing data mechanisms are beyond the scope of this tutorial but can be found in the lavaan tutorial [23]). The final argument `meanstructure = TRUE` is necessary if you want to estimate intercepts for the observed variables as well, as opposed to just estimating variances and covariances. Note that if you add the "missing" argument to your model, `meanstructure` will be set to TRUE automatically.

With the following command, you can extract and sort the BIC values in ascending order.

```
sort(fitMeasures(EFA)["bic",])
```

```
nfactors = 5 nfactors = 4 nfactors = 6
    18142.38     18167.49     18189.29
```

The output indicates that the model with five factors is the best according to the BIC. Thus, the two techniques to determine the number of factors agree. From the original article from which the data for this tutorial was obtained, it is known that the expected number of factors was also five. Therefore, continuing the model building with the five-factor solution for this tutorial makes the most sense.

With the following command, you obtain the factor loadings for the five factors. Note that, by default, lavaan gives you standardized loadings, which means that the loadings can be interpreted as correlations between variables and factors.

```
EFA$nf5
```

```
         f1      f2      f3      f4      f5
TSC1  0.584*                               .
TSC2  0.487*                               .
TSC3  0.637*                       .
TSC4  0.578*      .                .
TSC5  0.547*                               .
TE1           0.728*               .
TE2        .  0.672*
TE3           0.708*       .
TE4           0.651*               .
TE5           0.337*       .       .
EE1               .   0.469*       .
EE2        .          0.689*
EE3                   0.768*
EE4        .          0.732*               .
EE5               .   0.479*       .
DE1                  -0.353*  0.744*       .
DE2               .           0.821*
DE3                        .  0.755*
RPA1                                  0.851*
RPA2                                  0.906*
RPA3                                  0.624*
RPA4                       .       .  0.350*
RPA5                       .       .  0.338*
```

In the output, all loadings between the variables (rows) and factors (columns) larger than an absolute value of .3 are shown by default. Inspecting the results, you can clearly see a simple structure such that every variable loads on only one factor. An exception is variable `DE1`, which positively loads on factor 4 with the other DE variables and negatively on factor 3 with the EE variables. Besides this cross-loading, the results align with the theoretical model as all `TSC` variables, all TE variables, all `EE` variables, all `DE` variables, and all RPA variables load on the same factor, respectively. In the next step, the model can be further refined based on fit. Since the model without the cross-loading is entirely in line with theory, the loading of DE1 on factor 3 will be set equal to zero in the CFA in the next section. However, if the CFA model does not fit well, putting back this cross-loading would be the first logical step.

5.3 Step 2: Building the Factor Model and Assessing Fit

The first step in building the model is to describe the model you want to estimate using special lavaan syntax. The arguments relevant for this tutorial are "=~", which can be read as "is measured by", and "~~", which translates into "is correlated with". In the following, a model is specified in which the 5 factors TSC, TE, EE, DE, and RPA are measured by different sets of variables (in line with theory and the EFA results from the previous step), separated by "+". Moreover, it is explicitly stated that correlations between factors should be estimated. Intercepts are not explicitly included in the model, but these are included again by adding the argument `meanstructure = TRUE` to the command when estimating the CFA model.

```
CFA_model <-'
#  Regressing items on factors
TSC =~ TSC1 + TSC2 + TSC3 + TSC5
TE  =~ TE1  + TE2  + TE3  + TE5
EE  =~ EE1  + EE2  + EE3  + EE4
DE  =~ DE1  + DE2  + DE3
RPA =~ RPA1 + RPA2 + RPA3 + RPA4

# Correlations between factors
TSC ~~ TE
TSC ~~ EE
TSC ~~ DE
TSC ~~ RPA

TE ~~ EE
TE ~~ DE
TE ~~ RPA

EE ~~ DE
EE ~~ RPA

DE ~~ RPA
'
```

The next step is to perform the CFA on the model-building data using the specified CFA model with the following command:

```
CFA <- cfa(model = CFA_model, data =
                     model.building[, var_names],
           estimator = "MLR", std.lv = TRUE,
                  meanstructure = TRUE)
```

Which model should be used is specified by the argument `model`. The arguments `data` and `estimator` are the same as in the code for the EFA. The argument `std.lv` is used to get results similar to the EFA. Since factors are not directly observed, their scale has to be set, which can be done in two different ways: (i) by fixing one of the factor-loadings of each factor to 1 (which will set the scale of the factor equal to the scale of the observed variable which loadings was fixed), or (ii) by setting the variance of the factor equal to 1. Here, the second option was chosen. As already mentioned, the final argument `meanstructure` is necessary if you also want to estimate the intercepts of the observed variables.

After performing the CFA, you can assess how well the model fits to the data. There are two types of fit measures: global and local. You can start with the global fit measures that describe how well the model as a whole fits to the data. Many different global model fit indices exist in the literature. Kline [34] suggests that at least the following four indices should be considered: (1) the Chi-squared significance test, which tests whether the model has a perfect fit to the data, that is, if the model can perfectly recreate the observed relations between variables; (2) the comparative fit index (CFI), which compares the fit of the chosen model to the fit of a model assuming zero correlations between variables; (3) the root mean square error of approximation (RMSEA), which is closely related to the Chi-squared test, but does not test for perfect fit and instead quantifies (approximate) fit between the model and the data in a single number; and (4) the standardized root mean square residual (SRMR) which summarizes the difference between the sample covariance matrix of the variables and the model-implied covariance matrix into one number. Unlike the Chi-squared significance test, the CFI, RMSEA, and SRMR are no tests and assess approximate fit.

Each fit measure is accompanied by rules of thumb to decide whether or not a model fits sufficiently to the data. The Chi-squared significance test should be nonsignificant because the null hypothesis is that the model fits the data perfectly. It is important to note that with increasing sample size, the null hypothesis of perfect fit is easily rejected. Therefore, you should not base your decision on whether the model fits too much on this test.

Regarding the other three measures, the CFI should be larger than 0.9 [35], the RMSEA point estimate and the upper bound of the 95 percent confidence interval should be smaller than 0.05 [36, 37], and the SRMR should be smaller than 0.08 [35]. The fit measures and their interpretation can be obtained with the following command:

```
globalFit(CFA)
```

```
Results------------------------------------------------------------------

Chi-Square (142) = 318.8407 with p-value
          = 1.332268e-15

CFI = 0.9476614
```

```
RMSEA = 0.05332242; lower bound = 0.04589762;
      upper bound = 0.06076663

SRMR = 0.04353874

Interpretations---------------------------------------------------------------

The hypothesis of perfect fit *is* rejected according to the Chi-
          Square test statistics because the p-value is smaller than 0.05

The hypothesis of approximate model fit *is not* rejected according
          to the CFI because the value is larger than 0.9.

The hypothesis of approximate model fit *is* rejected according
         to the RMSEA because the point estimate is larger or equal to
         0.05.

The hypothesis of approximate model fit *is not* rejected according
         to the SRMR because the value is smaller than 0.08.
```

Inspecting the output, you can see that the Chi-squared significance test rejected perfect fit, but that approximate fit holds according to the CFI and the SRMR. Ideally, at least three of the fit measures should indicate appropriate fit, but for the sake of this tutorial's brevity, it was decided to continue with the model without further adjustments. In practice, you may further adjust the model, for example, by including the cross-loading between DE1 and factor 3 again, and re-evaluate fit by fitting the updated model and running the `globalFit()` function again.

All measures above inspect the global fit of the model, so if the model as a whole matches the data well. While informative and necessary, the above measures can miss *local misfit* between the model and the data. If, for example, the 5-factor model describes the relations between all but one of the observed variables well, then the global fit of the model will likely be sufficient even though the estimates of the relations between the ill-fitting observed variable and all others will be completely wrong. Because of this, you should also inspect the local fit of your model; that is, if every part of your model fits the data well. There are many local fit measures that you could use [38], but the most straightforward way of assessing local fit is to look at the absolute difference between the model-implied and the sample covariance matrix. These are the same two matrices that the SRMR is based on, but instead of quantifying the total difference between the two in one number, you obtain the difference between the two matrices for every variance and covariance separately. You can see how much the two matrices deviate for each pair of variables, as well as the maximum difference between the two matrices, by using the following command:

```
localFit(CFA)
```

```
$local_misfit
      TSC1  TSC2  TSC3  TSC5   TE1   TE2   TE3   TE5   EE1   EE2   EE3   EE4
TSC1 0.000
TSC2 0.012 0.000
TSC3 0.007 0.012 0.000
TSC5 0.007 0.002 0.010 0.000
TE1  0.019 0.000 0.009 0.010 0.000
TE2  0.025 0.014 0.031 0.021 0.011 0.000
TE3  0.013 0.010 0.048 0.005 0.003 0.008 0.000
TE5  0.025 0.028 0.032 0.022 0.012 0.026 0.005 0.000
EE1  0.013 0.010 0.004 0.016 0.042 0.044 0.001 0.072 0.000
EE2  0.004 0.009 0.025 0.003 0.029 0.050 0.027 0.043 0.002 0.000
EE3  0.013 0.015 0.039 0.013 0.021 0.042 0.006 0.081 0.012 0.001 0.000
EE4  0.002 0.002 0.000 0.013 0.042 0.021 0.006 0.039 0.017 0.017 0.010 0.000
DE1  0.011 0.019 0.015 0.002 0.010 0.026 0.011 0.036 0.010 0.048 0.042 0.040
DE2  0.014 0.018 0.030 0.011 0.008 0.025 0.032 0.059 0.058 0.031 0.012 0.052
DE3  0.000 0.008 0.041 0.021 0.023 0.006 0.012 0.019 0.048 0.015 0.022 0.012
RPA1 0.008 0.015 0.034 0.011 0.013 0.022 0.001 0.012 0.011 0.018 0.019 0.041
RPA2 0.006 0.008 0.044 0.007 0.021 0.004 0.009 0.008 0.015 0.016 0.002 0.053
RPA3 0.041 0.016 0.012 0.003 0.006 0.010 0.017 0.034 0.035 0.008 0.022 0.009
RPA4 0.020 0.000 0.003 0.031 0.001 0.027 0.031 0.039 0.042 0.035 0.031 0.053
       DE1   DE2   DE3  RPA1  RPA2  RPA3  RPA4
TSC1
TSC2
TSC3
TSC5
TE1
TE2
TE3
TE5
EE1
EE2
EE3
EE4
DE1  0.000
DE2  0.004 0.000
DE3  0.002 0.006 0.000
RPA1 0.008 0.006 0.002 0.000
RPA2 0.010 0.025 0.024 0.024 0.000
RPA3 0.002 0.016 0.021 0.009 0.017 0.000
RPA4 0.006 0.056 0.074 0.046 0.011 0.052 0.000

$max_misfit
[1] 0.08051991
```

Based on the local fit evaluation, you can conclude that no local misfit is present since the biggest difference between the two matrices is only 0.08, which is small compared to the scale of the observed variables. If local misfit is present, for example, if the correlation between two observed variables had been much larger than predicted by the model, further adjustments could be made to your model that specifically address the source of local misfit, like adding an additional covariance between these observed variables. However, these adjustments should always make sense according to theory! Never just add parameters to your model to improve its fit.

This concludes the assessment of fit. The last part of this model-building step is to look at the loadings of the final model with the following command:

```
inspect(object = CFA, what = "std")$lambda
```

```
       TSC    TE    EE    DE   RPA
TSC1 0.657 0.000 0.000 0.000 0.000
TSC2 0.692 0.000 0.000 0.000 0.000
TSC3 0.628 0.000 0.000 0.000 0.000
TSC5 0.726 0.000 0.000 0.000 0.000
TE1  0.000 0.789 0.000 0.000 0.000
TE2  0.000 0.745 0.000 0.000 0.000
TE3  0.000 0.788 0.000 0.000 0.000
TE5  0.000 0.649 0.000 0.000 0.000
EE1  0.000 0.000 0.739 0.000 0.000
EE2  0.000 0.000 0.802 0.000 0.000
EE3  0.000 0.000 0.786 0.000 0.000
EE4  0.000 0.000 0.760 0.000 0.000
DE1  0.000 0.000 0.000 0.665 0.000
DE2  0.000 0.000 0.000 0.640 0.000
DE3  0.000 0.000 0.000 0.738 0.000
RPA1 0.000 0.000 0.000 0.000 0.849
RPA2 0.000 0.000 0.000 0.000 0.854
RPA3 0.000 0.000 0.000 0.000 0.788
RPA4 0.000 0.000 0.000 0.000 0.587
```

The loadings are very comparable to the ones of the EFA, which is not surprising given that only a single cross-loading (with absolute value > .3) was removed. Note that if you would want to extract other model parameters like the factor correlations, you could use the `inspect()` command without the "$lambda‘ at the end.

5.4 Step 3: Assessing Generalizability

The final step is assessing the generalizability of the CFA model from Step 2 by fitting the same model to the holdout sample. If the model fits this alternative dataset, too, you can be more confident that your factor model applies more generally and can capture the underlying structure of your measurement instrument in future studies and samples as well. To assess the generalizability, you use the same code as in Step 2, but now specify your holdout sample under the data argument.

```
CFA_holdout <- cfa(model = CFA_model, data =
                holdout[, var_names],
                      estimator = "MLR", std.lv = TRUE,
                                   meanstructure = TRUE)
```

After fitting your CFA model to the holdout sample, fit measures and their interpretation can again be obtained with the globalFit() command.

```
globalFit(CFA_holdout)
```

```
Results------------------------------------------------------------------------

Chi-Square (142) = 339.7732 with p-value
          = 0

CFI = 0.9429698

RMSEA = 0.05639005; lower bound = 0.04898985;
      upper bound = 0.06383685

SRMR = 0.04161716

Interpretations----------------------------------------------------------------

The hypothesis of perfect fit *is* rejected according to the Chi-
          Square test statistics because the p-value is smaller than 0.05

The hypothesis of approximate model fit *is not* rejected according
          to the CFI because the value is larger than 0.9.

The hypothesis of approximate model fit *is* rejected according
         to the RMSEA because the point estimate is larger or equal to
         0.05.

The hypothesis of approximate model fit *is not* rejected according
         to the SRMR because the value is smaller than 0.08.
```

Inspecting the output, you can see that the fit of the model to the holdout sample is very comparable to its fit to the model-building data. Again, the Chi-squared significance test rejects perfect fit, but approximate fit holds according to the CFI and the SRMR.

Local fit is also tested with the same command as in Step 2 but again applied to your results on the holdout sample.

```
localFit(CFA_holdout)
```

```
$local_misfit
      TSC1  TSC2  TSC3  TSC5   TE1   TE2   TE3   TE5   EE1   EE2   EE3   EE4
TSC1 0.000
TSC2 0.010 0.000
TSC3 0.012 0.015 0.000
TSC5 0.007 0.007 0.005 0.000
TE1  0.023 0.023 0.003 0.014 0.000
TE2  0.027 0.012 0.019 0.008 0.008 0.000
TE3  0.012 0.010 0.024 0.008 0.012 0.002 0.000
TE5  0.019 0.014 0.008 0.002 0.046 0.006 0.013 0.000
EE1  0.037 0.023 0.009 0.005 0.035 0.009 0.002 0.011 0.000
EE2  0.028 0.003 0.003 0.019 0.038 0.016 0.069 0.008 0.033 0.000
EE3  0.032 0.047 0.012 0.017 0.024 0.017 0.004 0.071 0.026 0.019 0.000
EE4  0.006 0.033 0.003 0.002 0.027 0.002 0.002 0.048 0.015 0.020 0.004 0.000
DE1  0.056 0.005 0.007 0.007 0.005 0.020 0.003 0.032 0.037 0.072 0.020 0.036
DE2  0.005 0.029 0.032 0.061 0.012 0.014 0.046 0.006 0.024 0.038 0.034 0.018
DE3  0.012 0.019 0.022 0.002 0.034 0.016 0.014 0.005 0.057 0.032 0.050 0.020
RPA1 0.019 0.009 0.012 0.028 0.018 0.001 0.003 0.004 0.030 0.031 0.037 0.003
RPA2 0.009 0.020 0.023 0.001 0.017 0.016 0.018 0.004 0.003 0.045 0.008 0.051
RPA3 0.000 0.007 0.004 0.009 0.000 0.006 0.000 0.028 0.011 0.021 0.025 0.002
RPA4 0.018 0.015 0.006 0.014 0.021 0.019 0.040 0.036 0.049 0.023 0.046 0.023
       DE1   DE2   DE3  RPA1  RPA2  RPA3  RPA4
TSC1
TSC2
TSC3
TSC5
TE1
TE2
TE3
TE5
EE1
EE2
EE3
EE4
DE1  0.000
DE2  0.020 0.000
DE3  0.019 0.006 0.000
RPA1 0.014 0.029 0.004 0.000
RPA2 0.006 0.010 0.005 0.020 0.000
RPA3 0.030 0.006 0.026 0.016 0.017 0.000
RPA4 0.008 0.028 0.046 0.057 0.005 0.093 0.000

$max_misfit
[1] 0.09310116
```

Results show that the local fit is sufficient for the holdout sample as well (the biggest absolute difference between the two matrices is only .09) and that it is again comparable to the results on the model-building data.

Lastly, you can look at the loadings of the final model when fitted to the holdout sample.

```
inspect(object = CFA_holdout, what = "std")$lambda
```

```
       TSC    TE    EE    DE   RPA
TSC1 0.679 0.000 0.000 0.000 0.000
TSC2 0.689 0.000 0.000 0.000 0.000
TSC3 0.691 0.000 0.000 0.000 0.000
TSC5 0.702 0.000 0.000 0.000 0.000
TE1  0.000 0.694 0.000 0.000 0.000
TE2  0.000 0.772 0.000 0.000 0.000
TE3  0.000 0.819 0.000 0.000 0.000
TE5  0.000 0.677 0.000 0.000 0.000
EE1  0.000 0.000 0.749 0.000 0.000
EE2  0.000 0.000 0.794 0.000 0.000
EE3  0.000 0.000 0.781 0.000 0.000
EE4  0.000 0.000 0.801 0.000 0.000
DE1  0.000 0.000 0.000 0.677 0.000
DE2  0.000 0.000 0.000 0.659 0.000
DE3  0.000 0.000 0.000 0.766 0.000
RPA1 0.000 0.000 0.000 0.000 0.851
RPA2 0.000 0.000 0.000 0.000 0.867
RPA3 0.000 0.000 0.000 0.000 0.700
RPA4 0.000 0.000 0.000 0.000 0.618
```

Again, you can see that results from the model-building data and holdout sample are very comparable, as the factor loadings are similar to before.

Since the model fits the holdout sample sufficiently (for this illustration at least, as mentioned before we would normally want 3 out of 4 fit measures to indicate sufficient fit, which was not the case here) and equally well as the model-building data and parameter estimates are comparable between the two datasets, you can conclude that the model's generalizability is okay. Had the model not fit the holdout sample sufficiently, you would have to conclude that while the CFA model from Step 2 fits the model-building data well, you cannot be certain that it reflects a generally applicable structure of your measure and that the factor structure needs to be further refined. Since you already used your holdout sample in this phase, however, this further refinement would require collecting a new set of data that can be split into a model-building and holdout sample and going through all three steps again.

6 Conclusion

Factor analysis is a great way to study constructs that are not directly observable. Of course, factor analysis has vast applications across several fields that are usually interdisciplinary and has been extended in several ways (e.g., to multi-group factor analysis, which will play an important role in the next chapter, where you will also see a discussion of the important topic of measurement invariance). This chapter serves mainly as a primer to introduce and demonstrate the basics of the method and to get readers interested and confident in applying the method themselves.

7 Further Readings

In this chapter, you have seen an introduction and tutorial on how to apply factor analysis in educational research. To learn more about factor analysis in general, you can consult:

- Kline, R. B. (2015). Principles and Practice of Structural Equation Modeling (4 ed.). Guilford Press.

  To learn more about the difference between EFA and CFA, you can consult:

- Flora, D. B., & Flake, J. K. (2017). The purpose and practice of exploratory and confirmatory factor analysis in psychological research: Decisions for scale development and validation. *Canadian Journal of Behavioural Science, 49*, 78–88.

Finally, to learn more about the history of factor analysis, you can consult:

- Briggs, D. D. (2022). *Historical and Conceptual Foundations of Measurement in the Human Sciences. Credos and Controversies*. Routledge.

References

1. Spearman C (1904) "General intelligence," objectively determined and measured. Am J Psychol 15:201–292. https://doi.org/10.2307/1412107
2. Thurstone LL (2013) The vectors of the mind: multiple factor analysis for the isolation of primary traits. Literary Licensing, Whitefish
3. Jöreskog KG (1969) A general approach to confirmatory maximum likelihood factor analysis. Psychometrika 34:183–202. https://doi.org/10.1007/BF02289343
4. Krijnen E, Steensel R van, Meeuwisse M, Jongerling J, Severiens S (2020) Exploring a refined model of home literacy activities and associations with children's emergent literacy skills. Read Writ 33:207–238. https://doi.org/10.1007/s11145-019-09957-4
5. Hofhuis J, Jongerling J, Van der Zee KI, Jansz J (2020) Validation of the multicultural personality questionnaire short form (MPQ-SF) for use in the context of international education. PLoS One 15:e0244425. https://doi.org/10.1371/journal.pone.0244425
6. Zee KI van der, Oudenhoven JP van (2000) The multicultural personality questionnaire: a multidimensional instrument of multicultural effectiveness. Eur J Person 14:291–309. https://doi.org/10.1002/1099-0984(200007/08)14:4<291::AID-PER377>3.0.CO;2-6
7. Jongerling J, López-Pernas S, Saqr M, Vogelsmeier L (2024) Structural equation modeling with R for education scientists. In: Saqr M, López-Pernas S (eds) Learning analytics methods and tutorials: a practical guide using R. Springer, Berlin

8. Flora DB, Flake JK (2017) The purpose and practice of exploratory and confirmatory factor analysis in psychological research: Decisions for scale development and validation. Can J Behav Sci 49:78–88. https://doi.org/10.1037/cbs0000069
9. Whitelock-Wainwright A, Gašević D, Tejeiro R, Tsai Y-S, Bennett K (2019) The student expectations of learning analytics questionnaire. J Comput Assist Learn 35:633–666. https://doi.org/10.1111/jcal.12366
10. Oster M, Lonn S, Pistilli MD, Brown MG (2016) The learning analytics readiness instrument. In: Proceedings of the sixth international conference on learning analytics & knowledge. Association for Computing Machinery, New York, pp 173–182. https://doi.org/10.1145/2883851.2883925
11. Park Y, Jo I-H (2019) Factors that affect the success of learning analytics dashboards. Educ Technol Res Dev 67:1547–1571. https://doi.org/10.1007/s11423-019-09693-0
12. Kokoç M, Kara M (2021) A multiple study investigation of the evaluation framework for learning analytics: instrument validation and the impact on learner performance. Educ Technol Soc 24:16–28. https://www.jstor.org/stable/26977854
13. Hernández-García Á, Acquila-Natale E, Chaparro-Peláez J, Conde MÁ (2018) Predicting teamwork group assessment using log data-based learning analytics. Comput Human Behav 89:373–384. https://doi.org/10.1016/j.chb.2018.07.016
14. Fincham E, Whitelock-Wainwright A, Kovanović V, Joksimović S, Staalduinen J-P van, Gašević D (2019) Counting clicks is not enough: Validating a theorized model of engagement in learning analytics. In: Proceedings of the 9th international conference on learning analytics & knowledge. Association for Computing Machinery, New York, pp 501–510
15. Baikadi A, Epp CD, Long Y, Schunn C (2016) Redefining" what" in analyses of who does what in MOOCs. In: Proceedings of the 9th international conference on educational data mining. Raleigh, pp 569–570. https://www.educationaldatamining.org/EDM2016/proceedings/paper_128.pdf
16. Wang FH (2021) Interpreting log data through the lens of learning design: Second-order predictors and their relations with learning outcomes in flipped classrooms. Comput Educ 168:104209. https://doi.org/10.1016/j.compedu.2021.104209
17. Lawley DN, Maxwell AE (1962) Factor analysis as a statistical method. J. R. Stat. Soc Series D (The Statistician) 12:209–229. https://doi.org/10.2307/2986915
18. Horn JL (1965) A rationale and test for the number of factors in factor analysis. Psychometrika 30:179–185. https://doi.org/10.1007/BF02289447
19. Schwarz G (1978) Estimating the dimension of a model. Ann Stat 6:461–464. https://doi.org/10.1214/aos/1176344136
20. Prasojo LD, Habibi A, Mohd Yaakob MF, Pratama R, Yusof MR, Mukminin A, Suyanto, Hanum F (2020) Teachers' burnout: a SEM analysis in an asian context. Heliyon 6:e03144. https://doi.org/10.1016/j.heliyon.2019.e03144
21. López-Pernas S, Saqr M, Conde J, Del-Río-Carazo L (2024) A broad collection of datasets for educational research training and application. In: Saqr M, López-Pernas S (eds) Learning analytics methods and tutorials: a practical guide using R. Springer, Berlin
22. Dolan CV (1994) Factor analysis of variables with 2, 3, 5 and 7 response categories: a comparison of categorical variable estimators using simulated data. Br J Math Stat Psychol 47:309–326. https://doi.org/10.1111/j.2044-8317.1994.tb01039.x
23. Rosseel Y (2023) The lavaan tutorial. https://lavaan.ugent.be/tutorial/
24. Patil I, Makowski D, Ben-Shachar MS, Wiernik BM, Bacher E, Lüdecke D (2022) Datawizard: an R package for easy data preparation and statistical transformations. J Open Source Softw 7:4684. https://doi.org/10.21105/joss.04684
25. Bentler PM, Chou C-P (1987) Practical issues in structural modeling. Sociol Methods Res 16:78–117. https://doi.org/10.1177/0049124187016001004
26. Jackson DL (2003) Revisiting sample size and number of parameter estimates: some support for the n:q hypothesis. Struct Equ Modeling Multidiscip J 10:128–141

27. Johnson J (2003) A procedure for conducting factor analysis with large numbers of variables. In: Poster presented at the 18th annual conference of the society for industrial and organizational psychology, orlando, FL
28. Bartlett MS (1950) Tests of significance in factor analysis. Br J Math Stat Psychol 3:77–85. https://doi.org/10.1111/j.2044-8317.1950.tb00285.x
29. Kaiser HF (1970) A second generation little jiffy. Psychometrika 35:401–415. https://doi.org/10.1007/BF02291817
30. López-Pernas S, Misiejuk K, Tikka S, Saqr M, Kopra J, Heinäniemi M (2024) Visualizing and reporting educational data with R. In: Saqr M, López-Pernas S (eds) Learning analytics methods and tutorials: a practical guide using R. Springer, Berlin, in press
31. Hotelling H (1933) Analysis of a complex of statistical variables into principal components. J Educ Psychol 24:417–441. https://doi.org/10.1037/h0071325
32. Rosseel Y (2012) Lavaan: an R package for structural equation modeling. J Stat Softw 48: https://doi.org/10.18637/jss.v048.i02
33. Kiers HAL (1997) Techniques for rotating two or more loading matrices to optimal agreement and simple structure: a comparison and some technical details. Psychometrika 62:545–568. https://doi.org/10.1007/BF02294642
34. Kline RB (2015) Principles and practice of structural equation modeling, 4th edn. Guilford Publications, New York. https://play.google.com/store/books/details?id=3VauCgAAQBAJ
35. Hu L, Bentler PM (1999) Cutoff criteria for fit indexes in covariance structure analysis: conventional criteria versus new alternatives. Struct Equ Model 6:1–55. https://doi.org/10.1080/10705519909540118
36. Browne MW, Cudeck R (1992) Alternative ways of assessing model fit. Sociol Methods Res 21:230–258. https://doi.org/10.1177/0049124192021002005
37. Jöreskog KG, Sörbom D (1993) LISREL 8: structural equation modeling with the SIMPLIS command language. Scientific Software International, Chapel Hill
38. Thoemmes F, Rosseel Y, Textor J (2018) Local fit evaluation of structural equation models using graphical criteria. Psychol Methods 23:27–41. https://doi.org/10.1037/met0000147
39. Jongerling J, López-Pernas S, Saqr M, Vogelsmeier LVDE (2024) Structural equation modeling with R for education scientists. In: Saqr M, López-Pernas S (Eds) Learning analytics methods and tutorials: a practical guide using R. Springer. https://lamethods.github.io/chapters/ch21-sem/ch21-sem.html

Structural Equation Modeling with R for Education Scientists

Joran Jongerling, Sonsoles López-Pernas, Mohammed Saqr, and Leonie V. D. E. Vogelsmeier

1 Introduction

Educational research involves a variety of theoretical constructs that are not directly observable (e.g., motivation, engagement, cognitive development, and self-regulation) and that can only be indirectly studied by looking at participant's responses to observable indicators of these constructs (e.g., participants' responses to questionnaire items). The previous chapter about factor analysis [1] showed how to assess the factor structure of these unobserved, or latent, constructs and, thus, which items are good measurements of these constructs. This is crucial for developing valid instruments to measure (i.e., quantify) the latent constructs encountered in educational research. However, good measurement of latent constructs is typically not the researchers' ultimate goal. Usually, they subsequently want to answer questions about relationships or mean differences between (multiple) of these constructs, such as, "*Is teacher motivation related to student engagement?*" or "*Does student engagement significantly differ for small and large-scale teaching styles?*" If researchers only have (i) *observed variables* (e.g., years of teacher experience) for both predictor and outcome variables and (ii) want to investigate the effect of one or more predictor variables on a *single outcome variable* at a time (e.g., the effect of teacher experience and teacher gender on student GPA), these types of questions can be answered using familiar analysis methods such as multiple regression or ANOVA. However, as soon as questions involve latent variables or testing an entire system of (complex) interrelations between variables (like the ones

J. Jongerling (✉) · L. V. D. E. Vogelsmeier
Tilburg University, Tilburg, Netherlands
e-mail: j.jongerling@tilburguniversity.edu

S. López-Pernas · M. Saqr
School of Computing, University of Eastern Finland, Joensuu, Finland

M. Saqr, S. López-Pernas (eds.), *Learning Analytics Methods and Tutorials*,
https://doi.org/10.1007/978-3-031-54464-4_21

found in most theoretical frameworks), researchers need an analysis technique with more flexibility that allows for modeling a multitude of relationships between (latent and/or observed) variables simultaneously [2]; that is, researchers need SEM.

At its core, SEM is a mix of factor analysis (discussed in the previous chapter) and a method called path analysis, which was invented by Wright [3]. One can think of path analysis as a type of multiple regression analysis because it also allows estimating and testing direct effects between variables. However, while multiple regression merely allows for testing the direct effects of predictor(s) on only a single outcome variable, path analysis can look at both direct and indirect effects between whole sets of predictor(s) on whole sets of other variables simultaneously. This implies that in path analysis, a variable can simultaneously be an outcome and a predictor: a variable might be predicted by one or more variables while also serving as a predictor for other variables. In other words, with path analysis, researchers can pretty much fit any model they can dream of as long as all variables are observed. Fitting a path model to latent variables requires going beyond path analysis. Fortunately, Jöreskog and Van Thillo [4] developed a statistical software called LISREL that allowed the use of latent variables in path analysis and thereby created SEM. Over the years, their work has been integrated into increasingly user-friendly software, contributing to the widespread use of the technique in social sciences today. This widespread use has also led to the definition of a SEM model becoming less concrete as the term is now also used for models with only observed variables, even though that would officially be a path model.

2 Literature Review

Perhaps the most well-known application of SEM in education research is the technology acceptance model (TAM), used to explain teachers' adoption of education technology [5]. The model hypothesizes that a teacher's decision to adopt and use technology is influenced by their perceptions of its ease of use and usefulness. These perceptions shape their attitude toward technology use, which, in turn, affects their intention to use it and their actual usage. Such complex interplay in factors that influence technology acceptance requires using sophisticated methods such as SEM. The model has been extensively applied and validated in several contexts [6], investigating the adoption of multiple technological innovations such as serious games [7], virtual reality [8], or artificial intelligence [9]. The model has also been extended in several ways throughout the years to address the influence of other factors such as personality traits, result demonstrability, and risk, among many others [10].

Many other applications of SEM exist in education research. For example Kusurkar et al. [11] used SEM to investigate how *motivation* affects *academic performance*. They hypothesized that greater levels of self-determined motivation are associated with the adoption of effective study strategies and increased study dedication, resulting in improved academic achievement. Their findings confirmed that the influence of motivation on academic performance is mediated by the use of effective study strategies. The work by Kucuk and Richardson [12] examined the interconnections between the three types of Community of Inquiry presence (*teaching*, *social,* and *cognitive*), four engagement components (*agentic*, *behavioral*, *cognitive,* and *emotional*), and *satisfaction* in online courses. The findings suggest that *teaching presence*, *cognitive presence*, *emotional engagement*, *behavioral engagement*, and *cognitive engagement* were significant predictors of *satisfaction*, explaining 88% of its variance.

The rise of digital learning and the opportunities for unobtrusive data collection have given rise to a new wave of studies that take advantage of the trace log data that students leave behind in online systems [13], instead of relying on self-reported data from questionnaires. For instance, Koç [14] proposed a model that explains the association between *student participation* and *academic achievement*. *Student participation* was measured through attendance to online lectures and discussion forum submissions. His results suggest that "student success in online learning [can] be promoted by increasing student participation in discussion forums and online lectures with more engaging learning activities" [14]. Fincham et al. [15] validated a theorized model of *engagement* in learning analytics using factor analysis and SEM. He found that *affective engagement* (measured by sentiment, sadness, and joy of students' writing) and *cognitive engagement* (measured by syntactic simplicity, word concreteness, and referential cohesion) are not significantly associated with students' final grades but *academic* and *behavioral engagement* (problem submissions, videos watched, and weeks active) are.

The rest of this chapter presents a recap of SEM, an integrated strategy for conducting a SEM analysis that is well suited for Educational Sciences, and an illustration of how to carry out a SEM analysis in R. Like the previous chapter, the presentation in this chapter will be kept applied and focus on how to conduct the analysis in R, instead of diving into technical details (for this, interested readers are referred to the readings listed at the end of this chapter).

3 Recap of SEM

In its most common form, SEM combines confirmatory factor analysis (CFA) with path analysis. Like CFA, SEM relates observed variables to latent variables that are measured by those observed variables, and, like path analysis, SEM allows for a wide range of relations between entire sets of variables (both latent and observed). As mentioned before, SEM is nowadays an umbrella term that also includes path analysis with only observed variables. However, this chapter focuses on SEM with latent variables to show its full potential. For the sake of brevity, it is assumed that the readers of this chapter have already read the factor analysis Chap. 20, where a good description and recap of CFA (and how it differs from exploratory factor analysis) is provided.

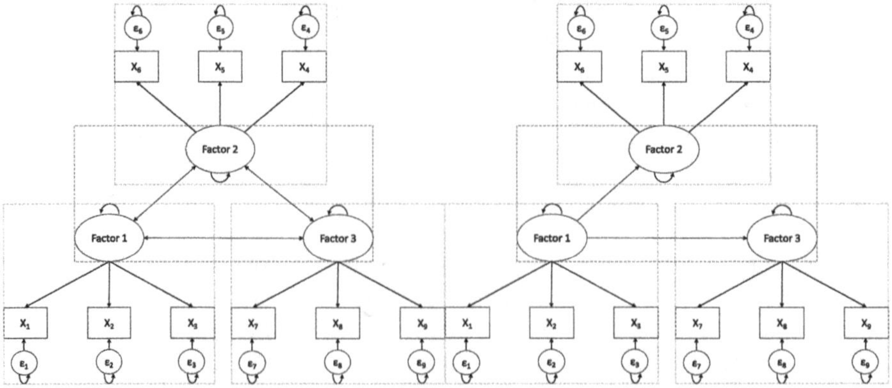

Fig. 1 CFA with covariances between factors vs. SEM with relationships between factors. The straight arrows from the factors to the observed variables indicate factor loadings (the regression weights obtained by regressing the observed variable on the factor) while straight arrows between Factors indicate regression relationships. Double-headed arrows between factors indicate covariance. The ε's represent errors. Curved single-headed arrows indicated variances. Each observed variable can also have an intercept, and each latent variable can also have a mean, but these are not explicitly indicated in the figure The blue dashed boxes indicate the structural part. The gray dashed boxes indicate the measurement part. (**a**) A 3-factor CFA model with correlated factors, where each factor is measured by three observed variables (X_1–X_3, X_4–X_6, X_7–X_9). (**b**) A SEM model with one latent predictor variable and two latent outcome variables. The model is very similar to the CFA model in (**a**). However, now there are regression relationships between Factors 1 and 2, and Factors 2 and 3 (as indicated by the straight arrows between them) while Factors 2 and three are assumed to be unrelated to each other

The similarity between CFA and SEM is illustrated in Fig. 1. It can be seen that both models consist of a part in which the observed and latent variables are related to each other (these relations are referred to as the measurement model) and a part in which the latent factors are related to each other (these relations are referred to as the structural model). The crucial difference between CFA and SEM is that factors in CFA can only *correlate* with each other (see Fig. 1a). In contrast, SEM gives us more flexibility regarding (the set of) relationships between (observed and latent) variables. For example, Fig. 1b shows the same factor model as Fig. 1a. However, in Fig. 1b, Factors 2 and 3 are both predicted by Factor 1a but are unrelated beyond that. Which model researchers specify, that is, what relations between (latent) variables they model, is entirely up to them. However, this flexibility is also a danger of SEM: the fact that researchers can fit any model they can think of does not mean they should. Ideally, the chosen relationships are driven by theory and substantive knowledge of the field.

4 Integrated Strategy for Structural Equation Modeling

In the following, we guide you, the researcher, through the different steps of conducting a SEM analysis. It is important to highlight that the most crucial part of a SEM analysis is checking its assumptions and checking measurement invariance (which was already briefly mentioned in the Chap. 20). Measurement invariance means that the factor model underlying your instrument applies equally to relevant subgroups in your sample. In other words, measurement invariance implies that if you analyzed each of those subgroups with factor analysis separately (using the three steps described in the Chap. 20), you would find the same model in each one. Measurement invariance is crucial: As discussed in the previous chapter, the interpretation of a latent variable depends on the factor model. Therefore, if the factor model of the latent variable(s) in your SEM model differs, for example, across biological sex, you effectively throw apples and oranges together if you analyze both males and females simultaneously in your analysis, which will result in uninterpretable outcomes. At the very least, you would want the same factor model for everyone in terms of the same number of factors and the same pattern of relationships between the factors and the observed variables. This scenario is called configural invariance. However, most research questions require higher levels of invariance, in which not only the number of factors and the pattern of factor loadings are the same across subgroups, but also the actual values of (most of the) parameters such as factor loadings and intercepts are equal across groups. More information about the different levels of invariance and how and when to test for them is presented below. After you have checked the assumptions and measurement invariance, you can safely apply and interpret a SEM analysis. The actual SEM analysis is just the final step of the modeling strategy presented here.

4.1 Step 1: Steps from the Previous Chapter While Assessing Configural Invariance

Since SEM analyses typically look at the relationships of one or more latent variables with one or more other (latent or observed) variables, you first need to ensure that the latent variables represent what you think they represent. To this end, the first step is going through some or all the steps described in the Chap. 20 for every latent variable you include. If you use a new instrument to measure the latent variables, you should go through all the steps from (1) exploring the data structure, (2) building the factor model and assessing fit, and (3) assessing generalizability.[1] As explained in the previous chapter, step 3 should be conducted using a holdout

[1] Note that observed variables (such as age or gender) can just be added to your SEM model. Only multiple-item instruments to assess psychological constructs must be validated. Thus, you can ignore the observed variables in this first SEM step.

sample or a new sample. However, you can use the same (holdout or new) sample for your SEM analysis. More on determining sufficient sample sizes is discussed below in the "SEM in R" section.

Once the factor model of your latent variables has been established and/or verified, you must ensure that this structure applies to all relevant subgroups in your sample. That is, you need to check for measurement invariance. The first level of invariance is configural invariance and, thus, whether the number of underlying factors and the pattern of relations between those factors and the observed variables is the same in each relevant subgroup. To test for configural invariance, you must go through the steps outlined in the previous chapter for each relevant group in your sample separately. Note that this also means you need sufficient observations for model building and generalizability in each subgroup you want to consider. Which subgroups are important to consider separately will always come down to theory.

If you use an existing instrument on a sample from a population for which the instrument had already been validated and configural invariance attained in previous research (i.e., you have strong a priori assumptions about the CFA model), you could skip the three steps from the factor analysis chapter and go straight to assessing the fit of the CFA model to your data. If it fits, you can continue with the SEM analysis. However, even when using existing validated instruments, it is strongly recommended to follow all the aforementioned steps for each relevant subgroup of your sample, as your sample might differ in important ways from the ones on which the instrument was validated, which could potentially introduce biases into your results.

4.2 Step 2: Assessing Higher Levels of Invariance

After establishing configural invariance between relevant groups, it is time to verify that the factor model parameters (specifically the loadings and intercepts) are also invariant across the factor models of each group. To illustrate why these "higher levels of invariance" are important, assume you want to investigate the relationship between children's age and how engaged they are in class. You measure the latent variable "classroom engagement" using a factor model on the three observed variables *engages in group activities*, *asks for assistance when needed*, and *prefers working alone*. Of these three variables, *asks for assistance* might be more indicative of classroom engagement for girls than for boys due to cultural stereotypes. Boys could be way less likely to ask for assistance, not because they are less engaged, but because society taught them that boys should not ask for help and should instead figure things out themselves. In this example, the interpretation of the latent variable would be different between girls and boys: for girls, you truly measure classroom engagement; for boys, you measure an uninterpretable mix of classroom engagement and how much they internalized gender stereotypes. If you ignored this between-group difference in the meaning of the latent variable when including it in a SEM model, you would come to invalid conclusions about the structural relationships between (latent or latent and observed) variables. Fortunately, the type of invariance in this example would show up as a difference in the relation between the factor *classroom engagement* and the variable *asks for assistance* between boys and girls. In other words, it would show up as a difference in factor loadings between groups.

Testing for higher levels of invariance in the factor model between categorical characteristics or "groups" (e.g., gender, nationality) can be done by testing for differences in model parameters between groups using a multi-group approach [16, 17] that compares the fit of a factor model in which all parameters are allowed to differ between groups to one in which one or more parameters are constrained to be equal across groups. If constraining parameters across groups does not cause a substantially worse model fit, invariance holds, and the constrained parameters can safely be considered and modeled as equal across groups. To assess for invariance across continuous variables (e.g., age) one could split the continuous variable into groups (e.g., changing the continuous variable age into the groups "young" and "old"), although one should ideally use more advanced methods like Moderated Non-Linear Factor Analysis [18] since categorizing continuous variables leads to loss of information. This chapter, however, will focus on the multi-group approach because it is the most common and widely available approach.

There are two important points about the invariance of model parameters that make the lives of researchers a little easier. First, not all model parameters have to be invariant. If you investigate relationships between (latent) constructs, as opposed to mean differences, only the factor loadings have to be invariant. For assessing differences in means, the factor loadings and the intercepts must be invariant. Note that the effects of categorical predictors on other variables also pertain to differences in means and that both loading- and intercept invariance are needed to interpret those effects. Note that a factor model comprises not only factor loadings and intercepts but also unique variances. However, unique variances do not have to be invariant across groups when looking at relationships or mean differences across latent variables. Therefore, we will not further consider them in this chapter. Second, the discussion of measurement invariance so far in this section has been about what is called *full invariance,* in which all parameters of a certain type (i.e., all factor loadings and/or all intercepts) are equivalent across groups (or across levels of a continuous variable). While this would represent an ideal situation, full invariance will rarely hold and is also not necessary. Partial invariance in which several, but not all, parameters of a certain type are invariant across groups is typically considered enough (e.g., the factor loadings of 4 out of 6 variables used to measure a factor are invariant across groups). Specifically, as long as at least two loadings are invariant across groups, one can look at relations between the corresponding latent variable and other (observed and/or latent) variables, and as long as at least two factor loadings and at least two intercepts are equal, one can meaningfully look at differences in means too ([19]; but also see [20]).

4.3 Step 3: Building the Structural Equation Model and Assessing Fit

After verifying the required level of measurement invariance for all latent variables that will be used in the SEM analysis, you are ready to investigate structural relations (e.g., regression relationships) between the factors (as well as observed variables if desired). Like with the factor models in the previous chapter, you first have to evaluate if the model fits your data sufficiently. You should only interpret the structural relations if the model fits the data. If the model does not fit, it is not a good description of the data and does not warrant further interpretation.

5 SEM in R

In the following, you will be taken through the essential steps of performing SEM in the open-source software R. To this end, the same dataset [21] will be used as the one to which factor analysis was applied to in the previous chapter [1]. The dataset contains survey data about teacher burnout in Indonesia. In total, 876 respondents have answered questions on five domains: *Teacher Self-Concept* (TSC, 5 questions), *Teacher Efficacy* (TE, 5 questions), *Emotional Exhaustion* (EE, 5 questions), *Depersonalization* (DP, 3 questions), and *Reduced Personal Accomplishment* (RPA, 7 questions). Thus, the total number of variables equals 25. The questions were assessed on a 5-point Likert scale (ranging from 1 = "never" to 5 = "always"). For more information on the dataset, the interested reader is referred to the data chapter of the book [22].

In line with the seven hypotheses about the structural relationships in the original article, the SEM analysis presented below tests whether TE is predicted by TSC and whether EE, DE, and RPA are respectively predicted by TE and TSC. Note that, as SEM builds up on factor analysis, the code presented below also builds up on the factor analysis results and syntax from the previous chapter. Without reading the previous chapter, the steps in this chapter might therefore be harder to follow.

To provide an example of how to assess between-group invariance of the constructs, it will be tested whether the same factor structure holds across gender. More specifically, because the running example is only interested in relationships between constructs and not mean differences, only loading invariance will be tested for. However, code to also test for intercept invariance will also be provided.

5.1 Preparation

To follow all the steps, you have to install the following packages with the function install.packages(). You only have to install them the first time you want to use them; therefore, the commands are commented below. Once you have the packages installed, you must load them with the `library()` function whenever you open the R script.

```
library(lavaan) # install.packages("lavaan")
library(tidyverse) # install.packages("tidyverse")
library(rio) # install.packages("rio")
library(psych) # install.packages("psych")
library(combinat) # install.packages("combinat")
library(devtools) # install.packages("devtools")
library(sleasy) # devtools::install_github("JoranTiU/sleasy")
```

5.1.1 Reading in the Data

The data can be read in, and the variable names can be extracted with the following commands:

```
dataset <- import("https://github.com/lamethods/data/raw/main/4_
                          teachersBurnout/2.%20Response.xlsx")
var_names <- colnames(dataset)
```

In the following, you will also find commands to add a gender variable to the dataset. This variable is only added to demonstrate how to perform invariance tests. Although the original data contains gender, it is unfortunately not part of the publicly shared data.[2] The proportion of men (`gender = 0`) and women (`gender = 1`) in the created variable aligns with the reported values from the article corresponding to the dataset.

```
set.seed(1611)
dataset$gender <- as.factor((sample(c(rep(0, 618), rep(1, 258)))))
```

5.1.2 Are the Data Suited for SEM?

Several data characteristics are necessary for SEM. These largely overlap with the necessary data characteristics for factor analysis, as discussed in the Chap. 20 (i.e., are the variables continuous? Are the correlations between the variables sufficiently large? Is there enough common variance among the variables? Are the data normally

---

[2] We contacted the authors, but they refrained from sharing the variable because they still plan to conduct research using the variable.

distributed?). Therefore, the steps in R to check whether the data are suited will not be presented here, as you can look those up in the previous chapter. To summarize: The data were suited for analysis with factor analysis and SEM, but the distribution of the variables is somewhat left-skewed. Therefore, an estimation method that is robust against non-normality should be used.

The one characteristic that needs additional investigation is the sample size. As mentioned in the previous chapter on factor analysis, Bentler and Chou [23] recommend having 5 observations per estimated parameter, while Jackson [24] recommends having 10, and preferably 20 observations, for each parameter you want to estimate. Therefore, the total number of observations depends on the exact model fitted to the data. If you start your SEM analysis with going through all three recommended model-building steps from the previous chapter (see the "Integrated Strategy for Structural Equation Modeling" above), you need to base your required sample size on the number of parameters estimated by an exploratory factor analysis (e.g., for a two-factor model fitted to 10 variables, you would need about (20 factor loadings + 10 intercepts + 10 error-variances) × 10 = 400 cases) and multiply that by 2 to also have enough observations for a new or holdout sample to check generalizability of your factor-model and for fitting your SEM model. Note that the sample size calculated this way is what you need in each relevant subgroup for which you want to assess measurement invariance. As mentioned above, you could go straight to the SEM analysis if you use an existing instrument on a sample from a population for which the instrument has already been validated and shown to be invariant across relevant groups in previous research. In that case, you would need fewer observations because a SEM model (like a CFA model) will have more constraints than an exploratory factor analysis model and hence fewer parameters to estimate.

You will perform the SEM analysis using the holdout sample from the previous chapter. The sample size for that is 438, which is sufficiently large to estimate 25 intercepts, 25 loadings, 25 unique variances, and 7 relations between factors according to the more lenient rules by Bentler and Chou [23], which, as you will see in Step 2, is the number of parameters of the SEM model that will be fitted to the data.

5.2 *Step 1: Steps from the Previous Chapter*

As explained in the strategy section, it is recommended to go through all the steps mentioned in the Chap. 20 (exploring the data structure, building the factor model and assessing fit, and assessing generalizability), and the R code for these steps can also be found there. For brevity, the steps and code will not be repeated here. Additionally, configural invariance (tested by following the steps from the previous chapter for each relevant subgroup separately) will be assumed to hold so that this section can focus on the code needed for a SEM analysis. We begin with picking up at the code for randomly assigning 438 rows of the data to a holdout dataset:

```
set.seed(19)
ind <- sample(c(rep("model.building", 438), rep(
                    "holdout", 438)))
tmp <- split(dataset, ind)
model.building <- tmp$model.building
holdout <- tmp$holdout
```

For an explanation of the code, please go back to the Chap. 20.

5.3 Step 2: Assessing Higher Levels of Invariance

Before looking into the structural relationships of interest, you have to ensure that the loading (and possibly intercept) invariance holds across groups for which this might be violated according to theory. To this end, you first have to specify the model, which is very similar to the model that was specified in the previous chapter for the CFA model. The model syntax looks as follows:

```
SEM_model <- '
# Regressing items on factors
TSC =~ TSC1 + TSC2 + TSC3 + TSC4 + TSC5
TE =~ TE1 + TE2 + TE3 + TE5
EE =~ EE1 + EE2 + EE3 + EE4
DE =~ DE1 + DE2 + DE3
RPA =~ RPA1 + RPA2 + RPA3 + RPA4

# Relations between factors
TE ~ TSC
EE ~ TE + TSC
DE ~ TE + TSC
RPA ~ TE + TSC
'
```

The first part of this syntax is exactly the same as the syntax that defined the CFA model in the previous chapter, with the syntax elements "=~" indicating that a factor is measured by specific observed variables. The code above defines a model where the factors TSC, TE, EE, DE, and RPA are measured by different sets of variables, which are separated by "+". In addition to the measurement part, this syntax includes the specification of relations between factors using the "~" operator. For example, TE is regressed onto TSC, indicating a directional relationship from TSC to TE. When comparing this syntax to the one from the previous chapter, you will see that the only thing that changes when you move from CFA to SEM is that you specify concrete structural relations (i.e., which latent factors are regressed on each other) instead of solely specifying correlations. Although intercepts are not explicitly mentioned in the model, they can be included by using the `meanstructure = TRUE` argument in the command when estimating the model.

Once the model syntax is specified, you can check whether invariance across groups (here, gender) holds. More specifically, you check this by comparing a model with and without equality constraints on the parameters of interest (here, as motivated before, only the loadings) and check whether adding constraints reduces model fit more than desired. To evaluate the degree of fit reduction, you can look at the changes in the global fit measures described in the previous chapter. These were (1) the Chi-squared significance test, (2) the comparative fit index (CFI), (3) the root mean square error of approximation (RMSEA), and (4) the standardized root mean square residual (SRMR). Recall that, unlike the Chi-squared significance test that assesses perfect fit, the CFI, RMSEA, and SRMR assess approximate fit. Similarly, only the Chi-squared significance test is a formal test for assessing invariance. As for the global fit measures, each fit measure is accompanied by rules of thumb when assessing invariance, allowing you to decide whether or not (approximate) invariance holds. The difference compared to the previous chapter and the rules of thumb presented there for the global fit measures is that the rules of thumb now pertain to differences in the criteria between models with and without constraints. The Chi-significance test should be nonsignificant because otherwise, the model with constraints fits significantly worse. Keep in mind, however, that this test easily rejects the assumption of invariance with increasing sample size. Regarding the other three measures, the rules of thumb are that differences in CFI should be smaller than or equal to 0.01, differences in RMSEA should be smaller than or equal to 0.015, and differences in SRMR should be smaller than or equal to 0.030 [25].

You do not have to perform the invariance assessment by constraining parameters yourself. Instead, you can use the command below. The argument `model` asks you to specify the model syntax, `data` asks for the dataset that you want to use for the analysis, and `group` refers to the variable for which you want to assess invariance. The argument `estimator` indicates which estimation procedure is used. The default is standard maximum likelihood ("ML") estimation. However, for the current data, a robust maximum likelihood ("MLR") estimation is applied to account for small violations of the normality assumption. If the data contain missing values, you can add the argument `missing` and specify it as equal to "fiml", corresponding to a full information maximum likelihood approach. As already discussed in the previous chapter, this is a sensible approach if you have at least missing at random (`MAR`) data (details about missing data mechanisms can be found in the lavaan tutorial [26]). Next, the argument "intercept" indicates whether intercept invariance should be assessed in addition to loading invariance. If put to `TRUE`, the output contains information on both loading and intercept invariance unless two or more of the four fit criteria indicate that loading invariance is violated because then, assessing intercept invariance does not make sense. This information would be provided in the form of a message. Finally, the argument "display" indicates whether output about group-specific parameters and differences are provided (when set to `TRUE`) or not (when set to `FALSE`).

```
invarianceCheck(model = SEM_model, data = holdout,
                group = "gender", estimator = "MLR",
                          intercept = FALSE,
                missing = FALSE, display = FALSE)
```

```
Nested Model Comparison------------------------------------------------------------------

Scaled Chi-Squared Difference Test (method = "satorra.bentler.2001")

lavaan NOTE:
    The "Chisq" column contains standard test statistics, not the
    robust test that should be reported per model. A robust difference
    test is a function of two standard (not robust) statistics.

                     Df   AIC   BIC  Chisq Chisq diff Df diff Pr(>Chisq)
Loadings_Free       320 15711 16283 648.06
Loadings_Invariant  335 15693 16203 660.17     9.8747      15     0.8275

Model Fit Indices -----------------------------------------------------------------

                   chisq.scaled df.scaled pvalue.scaled rmsea.robust cfi.robust
Loadings_Free          632.823†       320          .000         .068       .923
Loadings_Invariant     638.987        335          .000         .065†      .925†
                    srmr
Loadings_Free      .050†
Loadings_Invariant .053

Differences in Fit Indices -------------------------------------------------------

                                        df.scaled rmsea.robust cfi.robust  srmr
Loadings_Invariant - Loadings_Free            15       -0.002      0.002 0.002

Loading Invariance Interpretation ------------------------------------------------

The hypothesis of perfect loading invariance *is not* rejected according to the
      Chi-Square difference test statistics because the p-value is larger or equal
       to 0.05.

The hypothesis of approximate loading invariance *is not* rejected
        according to the CFI because the difference in CFI value is smaller than
        or equal to 0.01.

The hypothesis of approximate loading invariance *is not* rejected
       according to the RMSEA because the difference in RMSEA value is smaller
       than or equal to 0.015.

The hypothesis of approximate loading invariance *is not* rejected
       according to the SRMR because the difference in SRMR value is smaller than
       or equal to 0.030.
```

Inspecting the output, you can see several sections. You may directly go to the last section "Loading Invariance Interpretation" because this provides you with information on whether or not invariance is rejected according to the four fit measures (following the cut-off values provided before). If you are interested in (reporting) details about the Chi-square significance test and the (differences in) model fit criteria, you can also inspect the beginning of the output.[3] It can be seen that invariance is not rejected according to any of the criteria. Similar to assessing model fit in factor analysis, it is advised that no more than one criterion should reject invariance for concluding that invariance holds. If this is initially not the case, you may release one equality constraint at a time (i.e., moving towards partially invariant models) and run the `invarianceCheck()` function again until partial invariance holds. In order to decide which constraints to release, you can consult the information provided when the "display" argument is set to `TRUE`. The output will then show the loadings and intercepts per group as well as the differences in these parameters across groups. You will also get a message indicating which loading (and intercept) differs most between groups. These would be the first targets to estimate freely across groups (while making sure that the minimum requirements for partial invariance mentioned above hold). How to update the model syntax to freely estimate parameters between groups is explained on the lavaan package website [26].

5.4 *Step 3: Building the Structural Equation Model and Assessing Fit*

The next step is to perform the SEM analysis on the holdout data using the specified SEM model with the following command:

```
sem_robust <- sem(model = SEM_model, data = holdout,
                            std.lv = TRUE,
                  estimator = "MLR", meanstructure = TRUE)
```

The arguments are the same as for the `cfa()` function discussed in the previous chapter. The relevant (standardized) output about the structural relations can be extracted from the SEM model with the command below. The first argument asks you to specify the `lavaan` object from which you want to extract the structural relations, `nd` lets you specify the number of decimals you want to display (with default 3).

```
sem_structural_results(sem_robust, nd = 3)
```

[3] Note that you will first see a note from lavaan about 'the "Chisq" column contain[ing] standard test statistics'. You can simply ignore this, as it just gives information on how the difference in fit was tested.

```
  outcome predictor std estimate    se p-value
1     TSC        TE        0.693 0.121       0
2     TSC        EE        0.547  0.17       0
3     TSC        DE          0.4 0.143       0
4     TSC       RPA        0.413 0.111       0
5      TE        EE         0.32 0.114       0
6      TE        DE        0.293 0.088   0.001
7      TE       RPA        0.326 0.084       0
```

Looking at the output from left to right, you can see the outcome variable of the structural relationship, the predictor of the structural relationship, the regression coefficient, the standard error of the regression coefficient, and the p-value of the effect. Here, you see that all structural relations are significantly different from zero and thus that TE, EE, DE, and RPA are all related to TSC and that EE, DE, and RPA are related to TE. As mentioned, the function `sem_structural_results()` displays the most relevant information about the structural relations. For more information and unstandardized estimates, you can use the `summary()` function on the `lavaan` object `sem_robust` (also see the `lavaan` tutorial by Rosseel [27]).

6 Conclusion

SEM is a rich, powerful, and mature statistical framework that has more than a century of evolution and expansion. Several other extensions have been developed to include dynamic structural equation modeling for the analysis of temporal data, latent growth curve modeling for the analysis of systematic change over time, and multilevel SEM for the analysis of hierarchically clustered (or nested) data, to mention just a few. This chapter merely provided a primer that introduces the basics of the methods and an example application. Hopefully, it opens the door for interested readers to explore such a powerful framework that offers an excellent solution to many of the analytical tasks that researchers encounter.

7 Further Readings

In this chapter, you have seen an introduction and tutorial on how to apply SEM in educational research. To learn more about how SEM can be applied to this field, you can consult these resources:

- Teo, T., Ting Tsai, L., & Yang, C. 2013. "Applying Structural Equation Modeling (SEM) in Educational Research: An Introduction". In Application of Structural Equation Modeling in Educational Research and Practice.

- Khine, M. S., ed. 2013. "Application of Structural Equation Modeling in Educational Research and Practice". Contemporary Approaches to Research in Learning Innovations.

To learn more about SEM in general, you can refer to the following:

- Kline, R. B. 2015. "Principles and Practice of Structural Equation Modeling". 4th Edition. Guilford Publications.
- Hoyle, Rick H. 2012. "Handbook of Structural Equation Modeling". Guilford Press.

References

1. Vogelsmeier LVDE, Saqr M, López-Pernas S, Jongerling J (2024) Factor analysis in education research using r. In: Saqr M, López-Pernas S (eds) Learning analytics methods and tutorials: a practical guide using R. Springer, Berlin
2. Teo T, Tsai LT, Yang C-C (2013) Applying structural equation modeling (SEM) in educational research: an introduction. In: Application of structural equation modeling in educational research and practice. Brill, Leiden, pp 1–21
3. Wright S (1921) Correlation and causation. J Agric Res 20:557–585
4. Jöreskog KG, Van Thillo M (1972) LISREL: a general computer program for estimating a linear structural equation system involving multiple indicators of unmeasured. ETS Research Bulletin Series. Educational Testing Service, Princeton, New Jersey, USA
5. Davis FD (1993) User acceptance of information technology: system characteristics, user perceptions and behavioral impacts. Int J Man-Mach Stud 38:475–487
6. Valtonen T, López-Pernas S, Saqr M, Vartiainen H, Sointu ET, Tedre M (2022) The nature and building blocks of educational technology research. Comput Human Behav 128:107123
7. Cardona Valencia D, Betancur Duque FA (2023) Technology acceptance model (TAM): a study of teachers' perception of the use of serious games in the higher education. IEEE Revista Iberoamericana de Tecnologias del Aprendizaje 18:123–129
8. Abd Majid F, Mohd Shamsudin N (2019) Identifying factors affecting acceptance of virtual reality in classrooms based on technology acceptance model (TAM). Asian J Univ Educ 15:51
9. Nja CO, Idiege KJ, Uwe UE, Meremikwu AN, Ekon EE, Erim CM, Ukah JU, Eyo EO, Anari MI, Cornelius-Ukpepi BU (2023) Adoption of artificial intelligence in science teaching: from the vantage point of the african science teachers. Smart Learn Environ 10. https://doi.org/10.1186/s40561-023-00261-x
10. Marangunić N, Granić A (2015) Technology acceptance model: a literature review from 1986 to 2013. Universal Access Inf Soc 14:81–95
11. Kusurkar RA, Ten Cate TJ, Vos CMP, Westers P, Croiset G (2013) How motivation affects academic performance: a structural equation modelling analysis. Adv Health Sci Educ Theory Pract 18:57–69
12. Kucuk S, Richardson JC (2019) A structural equation model of predictors of online learners' engagement and satisfaction. Online Learn 23:196–216
13. Araka E, Maina E, Gitonga R, Oboko R (2020) Research trends in measurement and intervention tools for self-regulated learning for e-learning environments—systematic review (2008–2018). Res Pract Technol Enhanced Learn 15. https://doi.org/10.1186/s41039-020-00129-5
14. Koç M (2017) Learning analytics of student participation and achievement in online distance education: a structural equation modeling. Educ Sci Theory & Pract 17. https://doi.org/10.12738/estp.2017.6.0059

15. Fincham E, Whitelock-Wainwright A, Kovanović V, Joksimović S, Staalduinen J-P van, Gašević D (2019) Counting clicks is not enough: validating a theorized model of engagement in learning analytics. In: Proceedings of the 9th international conference on learning analytics & knowledge. Association for Computing Machinery, New York, pp 501–510
16. Jöreskog KG (1971) Simultaneous factor analysis in several populations. Psychometrika 36:409–426
17. Sörbom D (1974) A general method for studying differences in factor means and factor structure between groups. The Br J Math Stat Psychol 27:229–239
18. Bauer DJ, Hussong AM (2009) Psychometric approaches for developing commensurate measures across independent studies: traditional and new models. Psychol Methods 14:101–125
19. Byrne BM, Shavelson RJ, Muthén B (1989) Testing for the equivalence of factor covariance and mean structures: The issue of partial measurement invariance. Psychol Bull 105:456–466
20. Steenkamp J-BEM, Baumgartner H (1998) Assessing measurement invariance in cross-national consumer research. J Consumer Res 25:78–107
21. Prasojo LD, Habibi A, Mohd Yaakob MF, Pratama R, Yusof MR, Mukminin A, Suyanto, Hanum F (2020) Teachers' burnout: a SEM analysis in an asian context. Heliyon 6:e03144
22. López-Pernas S, Saqr M, Conde J, Del-Río-Carazo L (2024) A broad collection of datasets for educational research training and application. In: Saqr M, López-Pernas S (eds) Learning analytics methods and tutorials: a practical guide using R. Springer, Berlin
23. Bentler PM, Chou C-P (1987) Practical issues in structural modeling. Sociol Methods Res 16:78–117
24. Jackson DL (2003) Revisiting sample size and number of parameter estimates: some support for the n:q hypothesis. Struct Equ Modeling Multidiscip J 10:128–141
25. Chen FF (2007) Sensitivity of goodness of fit indexes to lack of measurement invariance. Struct Equ Modeling Multidiscip J 14:464–504
26. Rosseel Y (2012) Lavaan: an r package for structural equation modeling. J Stat Softw 48. https://doi.org/10.18637/jss.v048.i02
27. Rosseel Y (2023) The lavaan tutorial. https://lavaan.ugent.be/tutorial/

Why Educational Research Needs a Complex System Revolution that Embraces Individual Differences, Heterogeneity, and Uncertainty

Mohammed Saqr, Marieke J. Schreuder, and Sonsoles López-Pernas

1 Introduction

Learning analytics (LA) has emerged to harness the opportunities created by the abundance of data and advanced machine learning methods to improve learning and teaching and offer the much-needed personalized support. The premise was that the availability of massive amounts of data would enable novel insights, improve inferences, and deliver real-life impact [1]. A wide array of learning analytics applications has been developed over the years to realize such aspirations. One of the initial applications of learning analytics focused on predictive modeling: that is, collecting data of online activities, such as clicks, access to educational resources, or forum discussions to create a predictive model that would early flag underachievers. The early identification of an underachieving student in a course paves the way for proactive intervention [2]. Several studies have reported the successful identification of underachievers in individual courses or limited samples. Yet, transferring such models across programs or courses has been a consistent disappointment. All the more so, very few have reported a successful proactive intervention [2].

A recent massive study with data from 250,000 students tried to examine the effectiveness of a large-scale, evidence-based intervention, and reported small benefits. The researchers concluded that interventions are likely more effective when implemented for the right person, at the right moment in time. In fact, such a conclusion is far from new, Gordon Paul's stated in 1967 that the important question is "What treatment, by whom, is most effective for this individual with

M. Saqr (✉) · S. López-Pernas
School of Computing, University of Eastern Finland, Joensuu, Finland
e-mail: mohammed.saqr@uef.fi

M. J. Schreuder
KU Leuven, Leuven, Belgium

M. Saqr, S. López-Pernas (eds.), *Learning Analytics Methods and Tutorials*,
https://doi.org/10.1007/978-3-031-54464-4_22

that specific problem, and under which set of circumstances?"[3]. This requires predictions at a dynamic (i.e., time-varying) and individual level [4]. Another recent large-scale study showed that a low proportion of variance in students' performance was explained by the behavior-based indicators [5], and thus, students should best be identified on internal conditions (e.g., knowledge, self-regulation, and motivation). Recent reviews of intervention using LA methods further emphasize the difficulty of these promises [6]. As Du et al. [7] stated that "every course has different course requirements, it is impossible to identify generalisable thresholds of individual behaviors across courses" (p. 510). This holds even when courses are similar, homogenous, and use the same teaching [8].

The lackluster of predictive LA has led to a wide range of research threads aiming to tap into other methods that help explain and optimize students' learning. Such methods have been used to analyze the relational temporal patterns of students' learning processes. For instance, building on the importance of time, Process Mining (PM) and Sequence Analysis (SM) have gained wide popularity in the analysis of online learning activities to explain the time-ordered patterns of learning activities and to capture patterns of learning strategies [9]. Social network analysis (SNA) has also gained renewed interest and wider application in collaborative learning settings to understand students' roles and interaction patterns and to find SNA measures that help predict performance [10, 11]. Nevertheless, most of such methods—which we covered with examples in the book—require an overarching framework or a theoretical underpinning to better ground the analysis. In this chapter, we discuss the importance and potential of complex systems in understanding learning, learners, and the educational milieu at large.

2 Complex Systems and Education

Most learning theories and frameworks can be conceptualized as systems, that is, composed of multiple components, phases, or elements that interact with each other. Typically, such interactions are non-linear and vary across people, contexts, and time scales resulting in the emergence of a unique learning process [12–14]. For instance, engagement can be considered as a complex system. Engagement is then viewed as the result of interaction between different components, namely behavioral, cognitive, and emotional dimensions [15, 16]. Such interactions vary between tasks, times, and contexts which are often referred to as *interaction-dominant systems* (Fig. 1b) [17, 18]. In interaction-dominant systems, the relationships between components may change intensity and direction across times and situations. For instance, a student may enjoy school in the early days which drives their engagement and achievement and boosts their motivation. These dynamics may change over time, where achievement could be the driving force of future engagement but also results in anxiety rather than enjoyment. In turn, anxiety may negatively affect school enjoyment and engagement. This dynamic view is more realistic than the common b*ox-and-arrow* models where the components of the system are

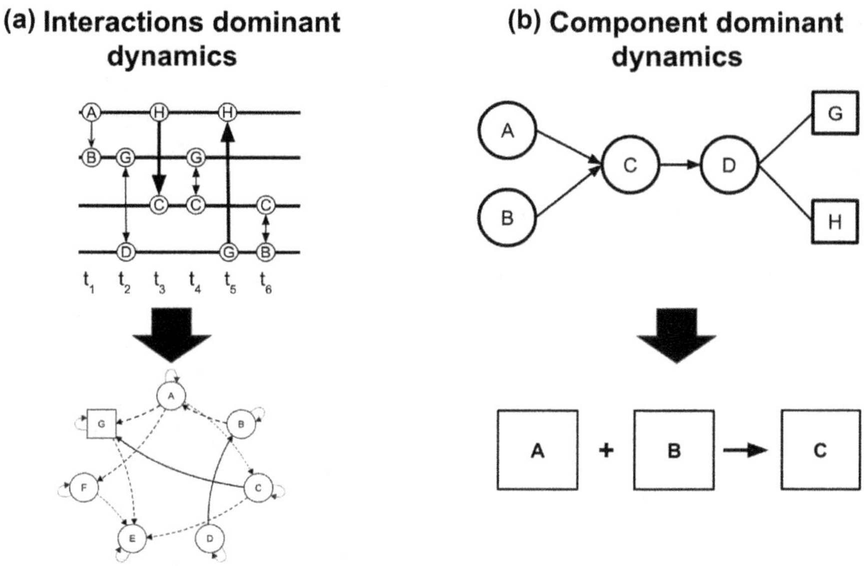

Fig. 1 (**a**) A complex dynamic system where the interactions vary across time, change direction and strength in a soft-assembled manner. (**b**) A box and arrow framework where the interactions are fairly stable and linear

rigidly assembled in a stable manner and the relationships between the components are deemed to be linear (Fig. 1b). Viewing the previous example from a linear perspective would entail that we see that the student will always have a stable relationship: enjoyment always drives engagement and achievement with little changes in the future nor any new patterns emerge. A linear view of engagement is then far from realistic.

Engagement follows Gestalt principles, meaning that engagement is considered more than just the sum of its parts (i.e., engagement ≠ emotions+cognition+behavior) [16], the interactions between these components are often multiplicative rather than simple linear sum and rely on the environment and contextual conditions (family, peers, teachers, school, etc.). Additionally, most engagement theories describe feedback loops (for instance, achievement drives further engagement and vice versa) [19]. Again, such feedback loops fit well with the salient features of a complex dynamic system [14].

Self-regulated learning (SRL), too, follows complex systems principles: the interaction between SRL phases across different learning scenarios and temporal scales results in unique learning strategies which are different mixtures of SRL phases. Such interactions may enhance, impede, or catalyze each other. For example, reflection on performance can lead to improved learning and better goal-setting in a student. In another student, reflection can lead to frustration, and low performance. We can expect vast amounts of variations and complex interactions in the same way. Such conceptualization of SRL as a state of a complex person-environment system is necessary to understand the interplay of the intricate SRL process [13, 17]. Indeed, Boekaerts and Cascallar [20] argue that it is impossible to understand learning and achievement "unless one adopts a systems approach to the study of self-regulated learning". Similarly, other SRL theoreticians share the same conceptualization. For instance, [21] states that SRL is "a complex system of interdependent processes" [21], and so did Winne et al. when they described engagement as complex and dynamically changeable across contexts [22].

In fact, many learning concepts have been already described, operationalized, and framed in complex system terms including motivational [23, 24], achievement motivation theories [25], agency [26], and metacognition theories [27]. Also, the student has been described as a complex system [28], so have small collaborative groups [29], and the classroom as a whole to list a few. The merit of this complex systems view is that it not only accounts for many of the features of learning-related processes (e.g., being interaction-dominant, lacking central control) but also provides a framework for better understanding - and perhaps even predicting - such processes [13]. Nevertheless, endorsing that a system is interaction-dominant means that we also understand that forecasting and prediction of the system's future status may be more uncertain than the simple linear dynamics of component-dominant scenarios [30]. That is, the slightest change in initial conditions may lead to substantial differences in the end state (a phenomenon called "sensitive dependence on initial conditions"). It follows that in order to understand complex systems, one has to look into the dynamics of such systems.

2.1 Dynamics in Complex Systems

The interactions that maintain a complex system tend to give rise to relatively stable configurations, which can be considered attractor states that emerge again and again [31]. Attractors can take many different forms, ranging from chaotic to cyclic to simple point attractors. In the context of learning analytics, for instance, a point attractor could resemble a state of "being relatively engaged". Importantly, the attractors of complex systems may change over time. The aforementioned point attractor could for instance gradually lose its strength, up to a point where the attractor disappears. When reaching such a tipping point, the system switches to an alternate attractor [32]. Such a shift between different attractors is often labeled a transition. Transitions can be harmful—e.g., reflecting a shift from an adaptive state towards a maladaptive state—or beneficial—e.g., reflecting a reverse shift. Relatively well-investigated dynamics are "critical transitions", which entail a shift from one stable regime (i.e., point attractor) towards another regime [31] (i.e., other point attractor). For instance, countries can shift between a state of peace towards a state of war and the climate can shift from a greenhouse to an icehouse state. Similarly, a learning child may shift between a state of engagement towards a state of disengagement. An important premise of complex systems theory is that such transitions - albeit in very different systems - follow the same generic principles.

Among these principles is the idea of critical slowing down [30, 33]. Critical slowing down describes that, prior to a critical transition, it becomes increasingly difficult to recover from perturbations [33, 34]. In the case of engagement, such perturbations can be school problems (e.g., problems with other pupils). When the student is in a stable, engaged state—and thus, unlikely to experience a transition towards a disengaged state—such perturbations only have a brief effect on the student's attention. This means that, upon a perturbation, he/she quickly recovers his/her "baseline" engagement levels [35]. As the resilience of the engaged state declines, however, the student becomes increasingly affected by these perturbations. This means that recovering his/her normative engagement becomes more and more difficult. This in turn translates to altering system dynamics, meaning that the interactions between and within system elements changes. Monitoring such changes may then allow for anticipating otherwise unpredictable transitions in learning processes [30, 33, 34]. Ultimately, this could aid the prevention of harmful transitions or the fostering of beneficial transitions.

An important implication of viewing transitions in learning processes through a complex systems lens is that declining resilience may be detectable *within* systems, which in this case means that inferences are made on the level of the student. This approach contrasts with the common group-level inferences, which may allow for telling *who* is likely to undergo a transition. For instance, group-level approaches may lead to the notion that "individuals with this behavior, this personality, or this socio-economic background are more likely to drop out of school than others". Within-individual approaches, in contrast, may allow for determining *when* a specific individual will drop out. For the purposes of targeted and timely intervention, such insight is invaluable. A related merit of complex systems principles is that they allow for personalization. For instance, it is likely that vulnerability to major changes (e.g., transitions in engagement, school drop-out) manifests in different variables for different individuals. Because declining resilience can be monitored within individuals, such heterogeneity does not pose a challenge. Rather, it can be accommodated by monitoring resilience in those variables that are considered most relevant for this particular student, in this specific context [36]. In conclusion, the possibility to monitor generic indicators of declining resilience may pave the way for deriving person-specific insights in predicting (and potentially, preventing or stimulating) changes in learning-related processes.

2.2 *From Theory to Practice: Measurement and Analyses*

If we agree that the learning phenomena, process or construct can be conceptualized as states in a complex system, then it becomes essential that a complex system lens is used to map the structure and dynamics of the said phenomena [37]. This has profound consequences for both measurement and data analyses. With respect to measurement, a complex systems lens necessitates the collection of time series data. The reason is that systems—and the interactions between elements within those systems—are by definition time-varying, and it is precisely the changes over time that contain information about the system as a whole. Thus, instead of a single, cross-sectional measurement, a complex systems perspective requires collecting repeated measurements for each individual. With advancing technology, collecting such measurements has become increasingly feasible. Broadly speaking, we can distinguish between passively collected data - which includes mobile sensing data (e.g., typing speed, scrolling, app usage, and sometimes also location) and actigraphy data (e.g., movement, heart rate, skin conductance) —and self-reported data—which is gathered through repeatedly prompting students with a questionnaire on their mood, motivation, or other psychological variables. Both modalities have their pros and cons. The main benefit of passively collected data is the amount of data that can be collected without burdening participants. The other side of the coin is that this amount of data often needs aggregation and intensive cleaning, which is far from straightforward, and in that sense the data can be "hard to handle". The main benefit of self-report data is that the content of measurement may be more closely related to the construct of interest. However, self-report data require considerable motivation from participants, and it is not inconceivable that such demanding research designs introduce sampling bias. Put differently, it is possible that the types of individuals who engage in studies involving long-term self-reports are not representative of the general population (e.g., in terms of conscientiousness [38]). At the same time, however, studies that investigated sampling bias in intensive longitudinal studies involving the collection of repeated self-reports did not find evidence for self-selection [39]. As intuitive as it may seem, scientific evidence for the self-selection of participants into intensive longitudinal studies is thus lacking. Besides these relatively practical considerations, the necessity of time series data also comes with more fundamental questions, for instance, related to the timescale of assessments. Ideally, this timescale should be informed by the timescale at which the system's dynamics unfold. This in turn varies between constructs: engagement may shift over minutes, while student's performance may shift over weeks.

Naturally, the focus on time series data has consequences for the analyses that are useful. Not only do we require time series analyses —which can handle the temporal dependency in the data—but we also need methods that can capture nonlinear and person-specific trends. This is because the dynamics of complex systems are typically non-linear, as illustrated by the erratic behavior and sudden shifts that govern complex systems. Examples of such analytical methods include dynamic time-warp analyses, generalized additive models, recurrence quantification analysis, state space grids, and moving window analyses [17]. Despite that most learning theories and processes can be described in complex system terms and the long history of theoretical foundations of complex systems in learning sciences, learners' and learning environments, the uptake of suitable methods and approaches is lagging behind [13, 17]. Furthermore, applications, framing, and operationalization of learning theories as complex systems are rare in educational research [13, 17]. In this book, we therefore provide some theoretical underpinnings of a complex systems perspective on learning and education, and we further included several chapters that deal with methods and analyses that accommodate a complex systems lens e.g., psychological networks, Markovian models, and model-based clustering. In other fields, the adoption of such methods has resulted in the renewal of theories, understanding of human behavior, and the emergence of new solutions to real-life problems [40, 41]. Our aim was to help interested researchers to embrace such methods in their analysis.

2.3 Complex Systems and Individual Differences

Complex systems—as a paradigm—facilitates a better understanding of the heterogeneity and individualized nature of human behavior and psychological phenomena. In fact, many complex system methods, some of which described earlier, have a strong emphasis on person-specific fine-grained dynamics. The next section will offer a more in-depth discussion of the individual mechanisms and how they relate to the general average assumptions.

2.3.1 The Individual

The "individual", or the "self" is a central construct in several learning theories, methodologies, and approaches. For instance, self-regulated learning, self-concept, self-control, and self-directedness to mention a few [42]. Further, the literature is awash with the notion of personalization, student-centeredness, and adaptive learning. Nonetheless, research is commonly conducted using methods that essentially ignore the "individual" process. In that, research is performed using what is known as variable-centered methods where data is collected from a "group of others" to derive generalizable laws. In variable-centered methods, researchers compute standard tendency measures (mean or median) from a sample of individuals (often referred to as group-level analysis) to derive "norms" or "standard recommendations". The average is considered a "norm" where everyone is assumed to fit. What is more, the outcome of such analysis is deemed representative and therefore, generalizable to the population at large. Given that such an average is derived from a sample of others, it rarely represents any single student [43, 44]. An accumulating body of evidence is mounting that humans are heterogeneous with diverse behaviors, attitudes, cognition, and learning approaches. Thereupon, using insights based on group-level analysis has so far resulted in recommendations that don't work, assumptions that fail to hold, and replications that are hard to obtain. Furthermore, intervention programs or procedures based on such samples offered no more than negligible effects, e.g., [4].

The fact that group-level analysis is less representative of the person is far from new and has been recognized for decades. Yet, the methods that are more suited for person-specific analysis may have not progressed fast enough. The last two decades have witnessed a revolution in data collection methods, statistical approaches, and procedures that allow such analysis, collectively known as person-specific analysis.

In many ways, person-specific methods are a paradigm shift in research which according to [45] represent a “brink of a major reorientation” that is “no longer an option, but a necessity”. Endorsing a person-specific approach may change how research is performed and how findings are applied [46, 47]. The person-specific methods—being individualistic—have low potential for generating generalizable recommendations [46]. Therefore, a combination of group-level and person-specific methods may be the best way forward. Such a combination may augment our understanding and provide precise interventions *at the high resolution of the single student* and sharpen our insights of the group level that are generalizable to the wider population [48].

There is an abundance of digital tools and data collection methods that allow the gathering of fine-grained intensive data about students. Such data -where several measurements from the same person are gathered- can allow the analysis of more person-specific insights. In doing so, it can help obtain an accurate view of a student's learning processes and offer more precise personalized support [45–47].

2.3.2 Heterogeneity

As discussed in the previous section, a central assumption of group-level analysis is that “the average individual” represents every individual. Yet, the average individual very often does not exist [49]. To illustrate this problem, let us consider the story of Gilbert Daniels. Daniels was given the task to measure the physical dimensions of more than 4000 pilots who were part of the American Air Force around 1950. The goal was to find the average pilot size, so that cockpits could be re-designed accordingly. However, a remarkable finding of Daniels was that not a single pilot (out of all pilots who were measured) was approximately equal to the average of the 10 most relevant dimensions. Further, for any given combination of three dimensions, only 4% of pilots would match the average. Hence, he concluded that “The tendency to think in terms of the average man is a pitfall into which many people blunder [. . .]. Actually, it is virtually impossible to find an average man”. The consequence of this discovery was that most cockpit material became adjustable so that it would suit everyone [50].

It is not difficult to translate Daniels’ findings to the field of LA. Here, too, students are measured in many dimensions. It is often implicitly assumed that the average of those dimensions will illustrate “a representative student”, but this is not the case. To accommodate this lack of “average students”, we should embrace person-centered methods, similar to how the American Airforce embraced adjustable furniture and clothing. In contrast to group-level analyses, person-centered methods attempt to find patterns where differences are minimal, assumptions are likely to hold and apply to wider groups of people. Recently, the range of available person-centered methods has vastly increased, coupled with improving rigor and potential. Therefore, person-centered methods are increasingly endorsed to model heterogeneity and individual differences across a vast range of empirical designs. In the current book, we have introduced several methods for capturing the heterogeneity of multivariate and longitudinal data, and we encourage researchers to take advantage of such data to capture the diversity and individual differences of learners [51–54].

3 Conclusion

The birth of learning analytics signaled a new wave of educational research that embraced modern computational methods. Whereas the field has matured, several methodological and theoretical issues remain unresolved. In this chapter, we discussed the potentials of complexity theory and individual differences in advancing the field bringing a much-needed theoretical perspective that could help offer answers to some of our pressing issues. In fact, a complex systems view on learning processes can address some of the major barriers that have hampered progress in the field of education and possibly offer a venue for the renewal of knowledge.

References

1. Siemens G (2013) Learning analytics: the emergence of a discipline. Am Behav Sci 57:1380–1400. https://doi.org/10.1177/0002764213498851
2. Ifenthaler D, Yau JYK (2020) Utilising learning analytics to support study success in higher education: a systematic review. Educ Technol Res Dev ETR & D. https://doi.org/10.1007/s11423-020-09788-z
3. Paul GL (1967) Strategy of outcome research in psychotherapy. J Consult Psychol 31:109–118. https://doi.org/10.1037/h0024436
4. Kizilcec RF, Reich J, Yeomans M, Dann C, Brunskill E, Lopez G, Turkay S, Williams JJ, Tingley D (2020) Scaling up behavioral science interventions in online education. Proc Natl Acad Sci 117:14900 LP–14905. https://doi.org/10.1073/pnas.1921417117
5. Jovanović J, Saqr M, Joksimović S, Gašević D (2021) Students matter the most in learning analytics: the effects of internal and instructional conditions in predicting academic success. Comput Educ 172:104251. https://doi.org/10.1016/j.compedu.2021.104251
6. Larrabee Sønderlund A, Hughes E, Smith J (2019) The efficacy of learning analytics interventions in higher education: a systematic review. Br J Educ Technol J Council Educ Technol 50:2594–2618. https://doi.org/10.1111/bjet.12720
7. Xu Du BES Juan Yang, Zhang M (2021) A systematic meta-review and analysis of learning analytics research. Behav Inf Technol 40:49–62. https://doi.org/10.1080/0144929X.2019.1669712
8. Saqr M, Jovanović J, Viberg O, Gašević D (2022) Is there order in the mess? A single paper meta-analysis approach to identification of predictors of success in learning analytics. Stud Higher Educ 1–22. https://doi.org/10.1080/03075079.2022.2061450
9. Saqr M, López-Pernas S, Jovanović J, Gašević D (2023) Intense, turbulent, or wallowing in the mire: a longitudinal study of cross-course online tactics, strategies, and trajectories. Internet Higher Educ 57:100902. https://doi.org/10.1016/j.iheduc.2022.100902
10. Saqr M, Elmoazen R, Tedre M, López-Pernas S, Hirsto L (2022) How well centrality measures capture student achievement in computer-supported collaborative learning? – a systematic review and meta-analysis. Educ Res Rev 35:100437. https://doi.org/10.1016/j.edurev.2022.100437

11. Saqr M, Poquet O, López-Pernas S (2022) Networks in education: a travelogue through five decades. IEEE Access 10:32361–32380. https://doi.org/10.1109/ACCESS.2022.3159674
12. Jörg T, Davis B, Nickmans G (2007) Towards a new, complexity science of learning and education. Educ Res Rev Teach Concerned Progress Educ 2:145–156. https://doi.org/10.1016/j.edurev.2007.09.002
13. Koopmans M (2020) Education is a complex dynamical system: challenges for research. J Exp Educ 88:358–374. https://doi.org/10.1080/00220973.2019.1566199
14. Ladyman J, Lambert J, Wiesner K (2013) What is a complex system? Eur J Philos Sci 3:33–67. https://doi.org/10.1007/s13194-012-0056-8
15. Reschly AL, Christenson SL (2022) Jingle-jangle revisited: history and further evolution of the student engagement construct. In: Reschly AL, Christenson SL (eds) Handbook of research on student engagement. Springer International Publishing, Cham, pp 3–24. https://doi.org/10.1007/978-3-031-07853-8_1
16. Wang M-T, Fredricks JA (2014) The reciprocal links between school engagement, youth problem behaviors, and school dropout during adolescence. Child Dev 85:722–737. https://doi.org/10.1111/cdev.12138
17. Hilpert JC, Marchand GC (2018) Complex systems research in educational psychology: aligning theory and method. Educ Psychol 53:185–202. https://doi.org/10.1080/00461520.2018.1469411
18. Van Orden GC, Holden JG, Turvey MT (2003) Self-organization of cognitive performance. J Exp Psychol General 132:331–350. https://doi.org/10.1037/0096-3445.132.3.331
19. Tinto V (2022) Exploring the character of student persistence in higher education: the impact of perception, motivation, and engagement. In: Reschly AL, Christenson SL (eds) Handbook of research on student engagement. Springer International Publishing, Cham, pp 357–379. https://doi.org/10.1007/978-3-031-07853-8_17
20. Boekaerts M, Cascallar E (2006) How far have we moved toward the integration of theory and practice in self-regulation? Educ Psychol Rev 18:199–210. https://doi.org/10.1007/s10648-006-9013-4
21. Zimmerman BJ, Risemberg R (1997) Becoming a self-regulated writer: a social cognitive perspective. Contemp Educ Psychol 22:73–101. https://doi.org/10.1006/ceps.1997.0919
22. Winne PH, Zhou M, Egan R (2011) Designing assessments of self-regulated learning. Assessment of higher order thinking.
23. Papi M, Hiver P (2020) Language learning motivation as a complex dynamic system: a global perspective of truth, control, and value. Mod Lang J 104:209–232. https://doi.org/10.1111/modl.12624
24. Yuan Y, Zhen H (2021) Teaching and researching motivation. Front Psychol 12:804304. https://doi.org/10.3389/fpsyg.2021.804304
25. Urdan T, Kaplan A (2020) The origins, evolution, and future directions of achievement goal theory. Contemp Educ Psychol 61:101862. https://doi.org/10.1016/j.cedpsych.2020.101862
26. Deakin Crick R, Huang S, Ahmed Shafi A, Goldspink C (2015) Developing resilient agency in learning: the internal structure of learning power. Br J Educ Stud 63:121–160. https://doi.org/10.1080/00071005.2015.1006574
27. Vollmeyer R, Rheinberg F (1999) Motivation and metacognition when learning a complex system. Eur J Psychol Educ 14:541–554. https://doi.org/10.1007/BF03172978
28. Brown JS (1997) On becoming a learning organization. About Campus 1:5–10. https://doi.org/10.1177/108648229700100603
29. Mennin S (2007) Small-group problem-based learning as a complex adaptive system. Teach Teach Educ 23:303–313. https://doi.org/10.1016/j.tate.2006.12.016
30. Wichers M, Schreuder MJ, Goekoop R, Groen RN (2019) Can we predict the direction of sudden shifts in symptoms? Transdiagnostic implications from a complex systems perspective on psychopathology. Psychol Med 49:380–387. https://doi.org/10.1017/S0033291718002064
31. Scheffer M (2009) Critical transitions in nature and society. Princeton University Press, Princeton. https://play.google.com/store/books/details?id=okT_DwAAQBAJ

32. Scheffer M, Carpenter SR, Dakos V, Nes EH van (2015) Generic indicators of ecological resilience: inferring the chance of a critical transition. Ann Rev Ecol Evol Syst 46:145–167. https://doi.org/10.1146/annurev-ecolsys-112414-054242
33. Scheffer M, Bascompte J, Brock WA, Brovkin V, Carpenter SR, Dakos V, Held H, Nes EH van, Rietkerk M, Sugihara G (2009) Early-warning signals for critical transitions. Nature 461:53–59. https://doi.org/10.1038/nature08227
34. Scholz JP, Kelso JAS, Schöner G (1987) Nonequilibrium phase transitions in coordinated biological motion: critical slowing down and switching time. Phys Lett A 123:390–394. https://doi.org/10.1016/0375-9601(87)90038-7
35. Masten AS, Nelson KM, Gillespie S (2022) Resilience and student engagement: promotive and protective processes in schools. In: Reschly AL, Christenson SL (eds) Handbook of research on student engagement. Springer International Publishing, Cham, pp 239–255. https://doi.org/10.1007/978-3-031-07853-8_12
36. Olthof M, Hasselman F, Aas B, Lamoth D, Scholz S, Daniels-Wredenhagen N, Goldbeck F, Weinans E, Strunk G, Schiepek G, Bosman AMT, Lichtwarck-Aschoff A (2023) The best of both worlds? General principles of psychopathology in personalized assessment. J Psychopathol Clin Sci 132:808–819. https://doi.org/10.1037/abn0000858
37. Olthof M, Hasselman F, Oude Maatman F, Bosman AMT, Lichtwarck-Aschoff A (2023) Complexity theory of psychopathology. J Psychopathol. Clin Sci 132:314–323. https://doi.org/10.1037/abn0000740
38. Scollon CN, Kim-Prieto C, Diener E (2003) Experience sampling: Promises and pitfalls, strengths and weaknesses. J Happiness Stud 4:5–34. https://doi.org/10.1023/A:1023605205115
39. Schreuder MJ, Groen RN, Wigman JTW, Wichers M, Hartman CA (2023) Participation and compliance in a 6-month daily diary study among individuals at risk for mental health problems. Psychol Assess 35:115–126. https://doi.org/10.1037/pas0001197
40. Borsboom D, Haslbeck JMB, Robinaugh DJ (2022) Systems-based approaches to mental disorders are the only game in town. World Psychiatry Official J World Psychiatric Association 21:420–422. https://doi.org/10.1002/wps.21004
41. Quintana R (2023) Embracing complexity in social science research. Qual Quant 57:15–38. https://doi.org/10.1007/s11135-022-01349-1
42. Panadero E (2017) A review of self-regulated learning: Six models and four directions for research. Front Psychol 8:422. https://doi.org/10.3389/fpsyg.2017.00422
43. Fisher AJ, Medaglia JD, Jeronimus BF (2018) Lack of group-to-individual generalizability is a threat to human subjects research. Proc Natl Acad Sci USA 115:E6106–E6115. https://doi.org/10.1073/pnas.1711978115
44. Winne P (2017) Leveraging big data to help each learner and accelerate learning science. Teach Coll Rec 119:1–24. https://www.tcrecord.org/books/pdf.asp?ContentID=21769
45. Molenaar PCM, Campbell CG (2009) The new person-specific paradigm in psychology. Curr Dir Psychol Sci 18:112–117. https://doi.org/10.1111/j.1467-8721.2009.01619.x
46. Saqr M (2023) Modelling within-person idiographic variance could help explain and individualize learning. Br J Educ Technol. https://doi.org/10.1111/bjet.13309
47. Saqr M, Lopez-Pernas S (2021) Idiographic learning analytics: a definition and a case study. In: 2021 international conference on advanced learning technologies (ICALT), pp 163–165. https://doi.org/10.1109/icalt52272.2021.00056
48. Epskamp S, Waldorp LJ, Mõttus R, Borsboom D (2018) The gaussian graphical model in cross-sectional and time-series data. Multivariate Behav Res 53:453–480. https://doi.org/10.1080/00273171.2018.1454823
49. Molenaar PCM (2004) A manifesto on psychology as idiographic science: bringing the person back. Meas Interdiscip Res Perspect 2:201–218. https://web-a-ebscohost-com.ep.fjernadgang.kb.dk/ehost/pdfviewer/pdfviewer?vid=1&sid=257bbb8e-f698-4c63-8a11-d87feb3bbf63%40sessionmgr4010
50. Rose T (2016) The end of average: how to succeed in a world that values sameness. Penguin, London. https://play.google.com/store/books/details?id=w7-zCgAAQBAJ

51. Helske J, Helske S, Saqr M, López-Pernas S, Murphy K (2024) A modern approach to transition analysis and process mining with markov models: A tutorial with R. In: Saqr M, López-Pernas S (eds) Learning analytics methods and tutorials: a practical guide using R. Springer, Berlin. https://doi.org/10.1007/978-3-031-54464-4
52. López-Pernas S, Saqr M (2024) Modelling the dynamics of longitudinal processes in education. A tutorial with r for the VaSSTra method. In: Saqr M, López-Pernas S (eds) Learning analytics methods and tutorials: a practical guide using R. Springer, Berlin. https://doi.org/10.1007/978-3-031-54464-4
53. Saqr M (2023) Group-level analysis of engagement poorly reflects individual students' processes: why we need idiographic learning analytics. Comput Human Behav. https://doi.org/10.1016/j.chb.2023.107991
54. Scrucca L, Saqr M, López-Pernas S, Murphy K (2024) An introduction and R tutorial to model-based clustering in education via latent profile analysis. In: Saqr M, López-Pernas S (eds) Learning analytics methods and tutorials: a practical guide using R. Springer, Berlin. https://doi.org/10.1007/978-3-031-54464-4

Index

M. Saqr, S. López-Pernas (eds.), *Learning Analytics Methods and Tutorials*,
https://doi.org/10.1007/978-3-031-54464-4

www.ingramcontent.com/pod-product-compliance
Lightning Source LLC
Chambersburg PA
CBHW062357280826
50469CB00035B/19
* 9 7 8 3 0 3 1 5 4 4 6 3 7 *